I TATTI STUDIES IN
ITALIAN RENAISSANCE HISTORY

Sponsored by Villa I Tatti
Harvard University Center for Italian Renaissance Studies
Florence, Italy

# SUCCESS AND SUPPRESSION

*Arabic Sciences and Philosophy
in the Renaissance*

DAG NIKOLAUS HASSE

Harvard University Press

Cambridge, Massachusetts
London, England
2016

*Library of Congress Cataloging-in-Publication Data is available from the Library of Congress.*
ISBN 978-0-674-97158-5 (hardcover)

*To Mechtild, Henriette, and Ferdinand*

# CONTENTS

# FIGURES AND TABLES

# PREFACE

he Renaissance was a prime time of programmatic statements and professions. The major intellectual trends of the period were all commented upon in prefaces and letters, dedications, orations, pamphlets, invectives, and full-length treatises. As a result, we are well-informed about what some humanist intellectuals thought about the value of Arabic sciences and philosophy: "I will hardly be persuaded that anything good can come from Arabia" (Francesco Petrarca in 1370).[1] "One cannot find anything in the Greeks which is not pure and learned, which is not refined and created with the highest perspicacity, but one will encounter almost nothing in the Arabs which is not rancid and foul" (Leonhart Fuchs in 1535).[2] "I find Averroes' opinion more disgusting than the devil" (Pietro Pomponazzi in 1503–1504).[3]

We know equally well what the defenders of Arabic science thought of the Greek role models of their opponents: "We observe that learned medicine is ruined and that the most dangerous nonsense is brought back from the underworld. Avicenna is condemned, and the lies of Apuleius are embraced" (Lorenz Fries in 1530).[4] "What you can see lucidly and clearly expressed in Avicenna on only one or two pages, Galen in his Asiatic manner hardly manages to comprise in five or six major volumes" (Guillaume Postel in 1539–1540).[5] "I can find nothing more certain and true on the immortality of the soul than what can be taken from Averroes" (Luca Prassicio in 1521).[6]

The humanists emerge from this controversy as winners—if we decide the matter by the degree to which twentieth-century scholarship is colored by the polemical rhetoric of either side. It is not difficult to encounter historians who consider Arabic-influenced thought in the Renaissance to be a "drowning tradition,"[7] classify nonhumanist physicians as "captivated by the Arabic tradition,"[8] and describe "the origin of modern thought" as intricately linked to "the fight against . . . Arabism which controlled the universities all over Europe."[9] One occasionally also meets historians with a pro-Arab bias: they observe a "backsliding" that set in with Petrarca,[10] and censure "the classical reaction" and its "increased worship of Aristotle and Hippocrates, Galen and Dioscorides, at the expense of their Arabic and medieval commentators, continuers, and expanders."[11] But there is no question that the humanists have obtained a win on points. They have persuaded most of us that medical humanism was a historical necessity,[12] that the Greek commentators on Aristotle were justifiedly preferred to Averroes, and that Giovanni Pico della Mirandola was right when he urged his contemporaries to abandon Arabic astrology and read Ptolemy's *Tetrabiblos* instead. But this view of the Renaissance reception of Arabic thought is in need of correction, as the present study will show.

The Renaissance was the crucial period in which the West began to disconnect from its Arabic sources. Today, the famous Arabic scholars who for several centuries were known to every student in Christian Europe have ceased to be part of Western cultural memory. This disappearance was significant enough to prompt a resolution of the Parliamentary Assembly of the Council of Europe in 1991, which was renewed in 2002, that the contribution of Islamic civilization to European culture should be covered more extensively in school textbooks.[13] There are many forgotten heritages in European history, but this issue is particularly precarious because the heritage is owed to a culture that suffers from Western prejudices today. The present study sets out to assess both the success and the suppression of Arabic sciences and philosophy in the Renaissance. It thus seeks to contribute to a more historical attitude toward this important epoch of Western cultural memory.

A sensitive topic such as this calls for a sober historical approach. In recent years, the Arabic influence in Europe has been the subject of biased studies, which minimize or maximize the influence unhistorically. It has been argued in a recent study that the Arabic impact on medieval European sciences was virtually nonexistent, since the Arabic-Latin translations came too late to influence significantly the flourishing mixture of Greek and Christian culture that purportedly is the essence of Europe.[14] This study is motivated by prejudices against Islam and is replete with historical mistakes.[15] But there are also more serious scholars whose work is biased by ideological temptations. On the one hand, there is the temptation to cherish a narrow and comfortably simple ideal of European high culture as formed by Greek, Roman, and Christian traditions, and to marginalize at the same time the contribution of other traditions to that culture—for example, pagan, oriental, Jewish, or Islamic—which make the past history of the West much more complicated. In the eyes of those who cultivate this narrow ideal, scholarship today exaggerates the intellectual achievements of medieval Islam and the Arabic influence in Europe for reasons of political correctness.[16] On the other hand, there is the much less scandalous temptation to overemphasize the blending of cultures and all too loudly advertise Arabic influences that, in the historical measure, turn out to be very thin. Such a tendency colors the attempts to explain the birth of German mysticism and the rise of perspective in Western art from the spirit of Arabic thought.[17] It is important to be aware of these temptations. It is a relief to see that the majority of contemporary studies exhibit the historical and philological rigor that scholarship demands. The present study, which aims to answer questions that continue to be culturally sensitive, is carried by the conviction that the reaction to ideological accounts of the past can only come from the careful historical and philological scrutiny of sources, and from scholarship that does not shy away from judgments based upon such scrutiny.

It is the principal motive of the present book to move beyond the Renaissance and modern polemics and assess the factual influence of Arabic sciences and philosophy in the Renaissance. Its aim is to improve our understanding of the Renaissance as a double-faced phenomenon with respect to the reception of Arabic thought. On the one hand, it is the period in which several Arabic traditions reached the peak of their influence. On the other hand, it is the time when the West began to forget, and even actively suppress, its Arabic heritage. In the first part of this study, I will analyze the attempts of Renaissance scholars to augment the distribution of Arabic sciences by way of editions, biographies, and translations. In the second part, I will contrast the polemical discourse between protagonists of Greek and Arabic traditions with the internal development of the sciences and philosophy. Throughout this study, the emphasis is not on what Renaissance scholars said they intended to do, but on what they actually did. Did they, in the technical detail, use and adopt Arabic scientific theories or not?[18] Did they actively try to augment or diminish the distribution of Arabic sciences? What were the internal, scientific reasons for adopting or discarding Arabic theories, and what were the gains and losses connected with the emergence or disappearance of such theories in the Renaissance? On the basis of these technical studies, it will be possible to describe the role of Arabic learning in Renaissance culture. At the same time, the study is also concerned with the social background of the reception of Arabic science in the Renaissance and, in particular, with the intellectual and social milieu of the various figures involved in this reception.

Chapter 1 introduces the reader to the origin of Arabic traditions in Europe and their continuation in the fifteenth and sixteenth centuries. It describes the considerable printing success of Arabic sources in Latin translation and the firm rooting of Arabic authors in Renaissance university education. Chapter 2 turns to Renaissance biographies of Arabic authors. In this period, the interest in the lives and works of Arabic scholars was greater than ever before in Europe. The chapter investigates the origins of this activity in the historiography of medicine, analyzes the contribution of humanist scholars to Arabic biography, and describes the image of famous Arabic scientists as created by Renaissance biographers. Chapter 3 is an investigation into the achievements of the Arabic–Latin and Hebrew–Latin translators of the time. This translation movement is a remarkable feature of Renaissance culture: it begins in about 1480, more than 150 years after the last major medieval efforts to transport Arabic science into Latin culture. This part of the study analyzes the social background of about a dozen translators, their patrons, and their audience, and contrasts their actual translating techniques with their programmatic statements in prefaces and letters. It will become apparent that humanist scholars were much involved in the translation movement, but that they were not necessarily the best experts on the texts.

Chapters 4, 5, and 6 deal with the three areas of Arabic science most controversial in the Renaissance: pharmacology, Averroes's intellect theory, and astrology. As we shall see, the controversies were long and polemical because the value of the Arabic sciences for Renaissance scholars made it impossible to discard them easily. Chapter 4 traces the attraction of Arabic theories of *materia medica* in the writings of humanist physicians and botanists. Several such authors tried to outdo their Arabic predecessors, but had problems doing so without compromising the truthfulness and reliability of their work. In the end, the humanist attempt of a complete return to Greek medical botany failed. At the same time, it turns out that the prejudiced attempt to suppress Arabic traditions became a catalyst of new developments in pharmacology and botany. The chapter thus complements the picture drawn by the single most important precursor to the present book, Nancy Siraisi's *Avicenna in Renaissance Italy* (1987), which focuses on theoretical medicine. Chapter 5 is devoted to the Arabic philosopher most influential in the Renaissance: Averroes, a philosopher about whom Renaissance intellectuals entertained bitterly opposed opinions. The chapter discusses two aspects of the Renaissance reception of Averroes: his reception as philosopher and as commentator. With respect to his philosophy, the focus is on the Renaissance fate of Averroes's most controversial theory: that there is only one intellect for all human beings. It turns out that, for a limited period, the Averroist position and its protagonists were subject to massive pressure by Church authorities, which was a greater threat to the survival of the doctrine than humanist opposition. In the end, the doctrine survived and lost support only later, when philosophical alternatives were found within the Aristotelian tradition. With respect to Averroes's commentaries on Aristotle, the chapter seeks to assess their intrinsic value for Renaissance readers, in view of the fact that Averroes's commentaries were in competition with the newly retrieved Greek commentaries of late antiquity. Finally, Chapter 6 turns to astrology, a thriving discipline of Renaissance culture. The chapter gives an overview of those parts of astrology that were most disputed among adherents of Ptolemaic and Arabic astrology, and then studies the reception of one particularly influential Arabic theory: the great conjunctions of Saturn and Jupiter, which served as the basis for astrological histories of the world. As will become apparent, Arabic historical astrology, in spite of being hotly disputed, reached the peak of its influence between 1550 and 1650.

In the conclusion, I directly address the consequences of the anti-Arabic polemics in the Renaissance. Did these polemics lead to a decline of Arabic traditions in Western culture? If so, was the suppression of Arabic traditions justified for scientific reasons, that is, if judged from the intrinsic viewpoint of the sciences concerned? At the end of this study, the reader is provided an Appendix, which lists in alphabetical sequence the full range of Arabic authors available in Latin printed editions of the Renaissance. It briefly describes the transmission of each Arabic author in the Latin West and then lists the printed Latin editions of the author's work up to the year 1700.

The Renaissance reception of Arabic sciences and philosophy is a vast field. Let me briefly explain which parts of it fall outside the scope of the present study. I concentrate on Arabic medicine, philosophy, and astrology because these are the three areas that were at the center of the controversies and of the translation and printing activities. In comparison, the influence of Arabic mathematics, astronomy, and the occult sciences was less direct and less obviously tied to famous Arabic names. The impact of Arabic alchemical theory reaches a climax in the sixteenth century and was subject to occasional anti-Arab polemics;[19] this influence still awaits study from an internal, technical point of view. Similarly, much remains to be done to elucidate the circumstances and the achievements of the translation movement, in particular with respect to the Hebrew–Latin translations. I have studied Arabic and Latin texts in printed editions and manuscripts, but Hebrew texts only in print, not in manuscript. For Hebraists, there are many threads to follow up, especially in the recent studies by Giuliano Tamani and Silvia Di Donato, and, of course, in the seminal nineteenth-century works by Moritz Steinschneider.

The present book is about the conscious Western appropriation of Arabic thought, not about influences that were not recognized as Arabic by the scholarly public of the Renaissance. I would like to direct the attention of interested readers to two well-known cases of putative subcutaneous impact of Arabic sciences on the West, which are not treated in this book because they were unearthed only in the twentieth century: the influence of the theory of pulmonary transit of the blood formulated by Ibn an-Nafīs (d. 1288 AD) on Western medicine, and the influence of Arabic astronomers on Nicolaus Copernicus. In both cases, the Arabic and the Western theories betray striking internal similarities that suggest the dependence of the Renaissance authors on their Arabic predecessors. Ibn an-Nafīs had argued in his commentary on the *Canon* that it was impossible for the blood to flow through the septum of the heart, as Galen believed, and that instead it must find its way through the lung to the left side of the heart. It remains a matter of dispute whether the very similar theories of Michel Servet, Juan de Valverde, and Realdo Colombo were influenced by Ibn an-Nafīs (which seems "incontestible")[20] and, if so, in which way. The most probable channel of transmission is Andrea Alpago, who translated other parts of Ibn an-Nafīs's work into Latin (see Chapter 3) and in all probability knew about the new theory of the pulmonary transit of the blood.[21]

The second case, the influence on Copernicus, has received much scholarly attention. The Arabic influence does not concern Copernicus's epoch-making idea to construct a heliocentric system, but mathematical modifications that he applied to Ptolemaic planetary theory, especially with respect to replacing (and thus preserving) the effect of the equant with non-Ptolemaic devices, and with respect to the lunar model and the model for Mercury—modifications that are very similar to the innovations of the Marāġa school, in particular of Naṣīraddīn aṭ-Ṭūsī (d. 1274) and Ibn aš-Šāṭir (d. 1375).[22] In this case, it is difficult to pinpoint a channel of transmission. Possibly, Copernicus got to know Arabic planetary models via Byzantine Greek manuscripts in fifteenth-century Italy that exhibit diagrams of the Arabic models. Recent studies have cast doubt on the hypothesis that Copernicus was influenced by Arabic astronomy, by arguing that Copernicus came to the same results as his Islamic predecessors not because he knew their work, but because he shared their spirit, method, and aims. Just like Ṭūsī and Ibn aš-Šāṭir, Copernicus aimed at a reform of Ptolemaic astronomy, especially with respect to the equant, by developing Ptolemy's geometrical devices into new directions, which is why, it is argued, two Italian astronomers of Copernicus's time came to similar geometrical solutions, independently, it seems, of either Copernicus or Arabic astronomy.[23] It is not yet settled whether the recent arguments for parallel development are strong enough to outweigh the many pieces of evidence that speak in favor of a direct influence of the Marāġa school on Copernicus. In any case, there is no doubt about the important fact that Copernicus's motivation for astronomical reform owes much to Arabic science, in particular to the Arabic critique of Ptolemy's astronomy as deficient from a physical point of view. Copernicus continues the tradition of the Andalusian scholars Alpetragius and Averroes, who were well-known in the Renaissance as critics of Ptolemy's astronomy.[24]

The attitude toward Arabic texts in the Renaissance can be studied from three angles that are not central for the present study: first, with respect to the reception of the Quran; second, with respect to the history of Arabic philology in Europe, which finds expression in the founding of professorships in Arabic at the universities of Paris, Leiden, Cambridge, and Oxford;[25] and third, with respect to the knowledge accumulated by travelers between Europe and the Orient.[26] Research on these topics has advanced much in recent decades through the work of Hartmut Bobzin (especially his book *Der Koran im Zeitalter der Reformation* of 1995), Thomas E. Burman, Alastair Hamilton, and Sonja Brentjes, among others. There are points of contact between these studies and the present investigation: for instance, in the Hebrew–Latin and Arabic–Latin translators and in the works of the intriguing figure Leo Africanus,[27] or in the (albeit rare) religious overtones of humanist anti-Arabic rhetoric. These points of contact, however, are exceptions. It is only at a very late date that the new interest in Arabic philology had a significant effect on the book market: in the late sixteenth and in the seventeenth century, as is apparent from Table 2 in Chapter 3, which lists all Renaissance Latin translations of Arabic scientists and philosophers.[28] The earlier translators, from Elia del Medigo (d. 1493) to Jacob Mantino (d. 1549), had very different motives, which were rarely, if ever, historical, religious, or philological. These translators were physicians or philosophers by profession who translated Arabic or Hebrew works in response to a demand for new translations of Averroes and Avicenna: the doctrinal interest of either the patrons or the translators themselves was the motive for the wave of translations in the decades around 1500. Hence, there are other stories to be told, about religious texts and oriental studies, but the reception of Arabic sciences and philosophy in the Renaissance is a story in its own right.

At the traditional German Gymnasium in Kiel, where I was educated, Anton Raphael Mengs's eighteenth-century copy of Raphael's *School of Athens* adorned a wall of the auditorium, reminding everybody of the true torch-bearers of culture. Later, as a student of Arabic, I realized that my Renaissance heroes of Greek and Latin culture were, on many occasions, bitterly opposed to my new heroes, the great Arabic philosophers and scientists. Was it possible that humanism had deteriorated into ideology and blinded humanists to the intellectual quality of Arabic science? Or was it that, by

1500, the tradition of Arabic science had outlived its historical role and that its supporters, backward-looking as they were, deserved to be overcome? Without a satisfying answer, with sympathies for both sides, and with increasing awareness for the political implications, I studied the phenomenon with the aim of arriving at impartial and historical answers. Nevertheless, the issue never lost its irritating sting to me. *Ira* and *studium* proved difficult to restrain. It is a consoling thought—and perhaps a historical lesson—that ideology, the enemy of true knowledge, grows out of a very humane and productive property: partiality, the incessant motor of cultural change. In this respect, Georg Christoph Lichtenberg is right: "All impartiality is artificial. Man is always partial and is quite right to be."[29] The Renaissance epoch seems to me a prime example of the various grades of partiality, and of the many gains and losses that they entail.

# NOTE ON TERMINOLOGY, ORTHOGRAPHY, AND TRANSLITERATION

brief note on terminology is in order: (1) 'Arabic' puts the emphasis on the language, 'Arab' on the ethnic group, 'Islamic' on the religion and culture.[1] Arabic philosophers and scientists show a variety of ethnic origins and of religions: there are, among others, Arabs, Turks, and Persians, and Muslims, Christians, Jews, Sabians, and even critics of all revealed religions, such as Rhazes. I therefore prefer to speak of 'Arabic,' rather than of 'Islamic' or 'Arab,' sciences and philosophy. (2) 'Science and Philosophy' is taken very broadly and includes the occult sciences and medicine. (3) 'Renaissance' is not used to denote an intellectual affiliation, but to refer to a time span in European history, the two centuries between 1400 and 1600.[2] (4) 'Humanism,' in contrast, refers to the intellectual current, which I understand as tied to the usage of the name *humanista* in the Renaissance. This term was employed to refer to teachers or students of the *studia humanitatis,* that is, grammar, rhetoric, history, poetry, and moral philosophy. 'Humanism' in this sense was, as Paul Oskar Kristeller put it, "essentially a scholarly, educational, and literary movement," with much influence on many disciplines of science and philosophy, but it was not the all-pervading *Zeitgeist* of the Renaissance.[3] (5) Arabic authors are referred to by the Latinized form of their names that was most common in the Renaissance—with the exception of Rabbi Moyses, who is referred to by his modern name 'Maimonides.' In the Appendix, these Latin names are listed alphabetically, which makes it possible to identify the authors. The Arabic form of the name is, of course, used for those authors who were not known in the Latin West.

The Latin is quoted in standardized orthography, with the exception that the spelling is retained in quotations from manuscript, in proper names, and in titles of editions. Punctuation and capitalization are adapted to the expectations of English readers. Bibliographical references in the endnotes quote short titles only; for primary sources, the short title is followed by the year in which the edition appeared. The transliteration of Arabic follows the rules laid down by the Deutsche Morgenländische Gesellschaft, apart from the diphthongs *au* and *ai,* for which I use *aw* and *ay.* English translations are the author's unless otherwise indicated.

# SUCCESS AND SUPPRESSION

# PART I

The Presence of Arabic Traditions

# CHAPTER 1

# INTRODUCTION

*Editions and Curricula*

he Renaissance discussions of Arabic scientific traditions are best understood if viewed against the background of the Arabic influence in the Middle Ages. This influence manifests itself in the many texts translated, read, and quoted by scholastic writers. Texts of Arabic origin were a very tangible reality on European bookshelves, as I will illustrate in an example. In the years 1410 to 1412, the German magister and physician Amplonius Ratinck composed a well-known catalog of his private library, which contained 636 volumes of manuscripts. Since this catalog has survived until today, it is possible to unbind these volumes in the mind's eye and regroup the folios again according to their parentage.[1] We thus obtain circa 40 volumes of Greek, 35 volumes of Pagan-Roman, and 42 volumes of Arabic origin. The rest, about 519 volumes, come from Christian-Latin authors.[2] Amplonius, admittedly, had a distinct affinity for medical literature, which is a genre particularly influenced by Arabic culture. But nevertheless, his bookshelves are typical for the milieu of late medieval scholars. They demonstrate that on the eve of the Renaissance, Arabic culture was a source culture of equal rank with the Greek and the Roman. This is the situation that aroused the protest of humanist scholars.

The presence of Arabic authors on European bookshelves is the consequence of two major translation movements of the Middle Ages: from Greek into Arabic in Baghdad and in other centers of Islamic society from the eighth to the tenth century, and from Arabic into Latin in Spain and Italy, from the late tenth to the early fourteenth century. The Graeco-Arabic translation movement was supported by the financial and political elite of the ʿAbbāsid empire in Baghdad. It covered all those disciplines of Greek learning that were current in the Syriac-speaking Christian communities under Muslim rule in the Fertile Crescent; it transported many volumes of Aristotle's philosophy, Galen's medicine, Ptolemy's astronomy and astrology, and Euclid's geometry—to mention only the more famous names. Some works, astrological ones in particular, were also translated from middle Persian (Pahlavi).[3] This translation movement was the beginning of the impressive development of Arabic scientific and philosophical thought, which, beginning as an elitist form of learning, was gradually adopted into Islamic society and from the twelfth century onward also began to influence the teaching in the Madrasa.[4]

The European translations from Arabic began in late tenth-century Catalonia, where an anonymous scholar rendered several texts on the astrolabe, the astronomical instrument, from Arabic into Latin. In the eleventh century, southern Italy became a center of translation, both from Arabic and from Greek, mostly of texts on medicine and natural philosophy. Best known among these translators is Constantine the African, who died before 1198/1199 as a monk of the Benedictine cloister of Montecassino. He was not the first Arabic–Latin translator in Italy, however; others had already started to translate from Arabic before Constantine arrived from North Africa. The peak of Arabic–Latin translation activity came in twelfth-century Spain, where after the Christian conquest of Muslim cities, such as Toledo in 1085 and Saragossa in 1118, many Arabic manuscripts became accessible to the Latin-speaking world. From about 1120 onward, John of Seville translated more than a dozen astrological and astronomical texts, and other translators followed, focusing on the sciences of the stars, mathematics, religion, and the occult sciences. Philosophical texts followed only later in the century. The city of Toledo, and more specifically, the chapter of Toledo cathedral, became the center of the Arabic–Latin translation movement in the second half of the twelfth century. The famous translators Dominicus Gundisalvi and Gerard of Cremona were both canons of the cathedral. The Toledo translation movement was supported by the archbishop, probably also to promote the languishing Latin culture in Christian Spain, where Arabic and Romance languages dominated.[5] While Dominicus Gundisalvi is important as the principal translator of the philosophical works of Avicenna, Gerard of Cremona is by far the most productive medieval translator from Arabic, with an enormous number of more than seventy translations attributed to him, among them works that were to become central for European science, such as the Neoplatonic *Liber de causis,* Avicenna's *Canon medicinae,* and Ptolemy's *Almagest.* In the Renaissance, Gundisalvi is hardly ever mentioned, but Gerard's fame continued as "a learned and experienced man of medicine" and "most educated in all sciences and languages" *(in omni sci-*

*entia et lingua doctissimus),* as Symphorien Champier described him. For the Renaissance, Gerard of Cremona's fame rested on the translation of Avicenna's *Canon medicinae.*[6]

In the thirteenth century, the translation activity continued in Spain, also fostered by the scientific interests of King Alfonso the Wise of Castile, but the new translating center became southern Italy and Sicily. Frederick II Hohenstaufen attracted many scholars from different Christian, Jewish, and Muslim cultures to his court, among them several translators, and, most notably, Michael Scot from Toledo, who would become his court astrologer and the productive translator of seven (probably even eight) works by Averroes, the commentator on Aristotle.[7] In several Renaissance sources, Frederick II is remembered as a patron of translations. The polyglot and educated ruler is credited with ordering the Arabic–Latin translation of Ptolemy's *Almagest* (this is not correct—the *Almagest* was translated by Gerard of Cremona in twelfth-century Toledo) and with resuscitating the science of the stars by way of translations from the Arabic.[8] With the end of the thirteenth century, after a series of medical translations produced in Montpellier and Barcelona, the medieval Arabic–Latin translation movement came to a close. The influence of Arabic sciences and philosophy was of utmost importance for the development of scientific disciplines in Europe, but in varying degrees: in philosophy, the influence was strongest in natural philosophy, psychology, and metaphysics, and weaker in ethics and logic; in the sciences, there was a strong impact on medicine, astrology, astronomy, trigonometry, algebra, zoology, and the occult sciences, but less influence on other sciences, such as geometry and botany.

This was the medieval background of the Renaissance reception of Arabic sciences and philosophy. In the fifteenth and sixteenth centuries, it was well-known that in the Middle Ages many Arabic works had been translated into Latin—a kind of Latin that was doubly problematic in stylistic terms from a humanist point of view, since it was both scholastic and Arabicized. Renaissance descriptions of the medieval translation movement can differ markedly. They testify to the fact that there was a controversy about the value of Arabic science. In a dedicatory epistle to an edition of Averroes's medical work of 1537, the French physician Jean Bruyerin Champier explains to his readers the routes of cultural transmission, saying that after the demise of the school of Athens and the downfall of the Roman imperium, many Greek volumes of philosophy and medicine were translated into Arabic. Bruyerin praises the Arab people for being most learned in the sciences *(bonarum scientiarum studiosissima)* and continues:

> When Alfonso reigned in Spain, a man with greatest desire for the sciences, especially mathematics, at the time when the Moorish still held Betica *(i.e., the southern province of Spain)*, it easily happened that, partly because of the vicinity, partly because of the frequent commerce between the people, books written in the language of Averroes and the other Moorish were transported into the north-eastern province of Spain, where they were somehow rendered into Latin by some Spaniard or, since the schools of philosophy and medicine were already flourishing in Paris, were brought to Paris, after having been transported from Spain to France.[9]

It is obvious that Bruyerin confuses some facts, since the great translation movement in Spain was in the twelfth, not the thirteenth century when Alfonso the Wise reigned, and most of Averroes's works were translated not in Spain, but in thirteenth-century Italy. But Bruyerin's knowledge of the historical circumstances is impressive enough. In particular, he is aware of the principal transmission routes from Athens to Muslim Spain to Christian Spain to Paris. For Bruyerin, the story ends in France. A translation story with German coloring is told by Jacob Milich, a professor of medicine at Wittenberg University in 1550:

> Even though, after the expulsion of the Greek language, many ancient authors were expelled too, nevertheless some people, thirsty for knowledge, were looking for sources. For this reason, many writings of Hippocrates, Galen and Ptolemy were translated by the Sarracens into the Arabic language, from which not much later, with the help of emperors Lothar and Frederick II, they were translated into Latin.[10]

Here we meet again not only with a compliment for the Arabic culture of learning, but also with Frederick II as promoter of the translation movement, a fame he now shares, fantastically, with emperor Lothar III (d. 1137)—probably because Milich needed to name an emperor of the twelfth century. Neither Bruyerin nor Milich was aware of the fact that an important phase of the Arabic–Latin translation movement was very clerical in character, namely the Toledo translations of the twelfth century. From the viewpoint of these Renaissance scholars, it was Spanish or German kings who supported the translations.

Another example of Renaissance descriptions of the translation movement comes from the other end of the spectrum, that is, from those humanists who see the Arabic and medieval transmission of Greek science as the principal source of intellectual defect. In 1563, Girolamo Donzellini, a physician and philosopher of Verona, describes the transmission from Greek into Arabic into Latin in a paragraph that amounts to a humanist manifesto:

> When the science of medicine was transported from the Greeks to the Arabs, it was shipwrecked *(naufragium fecit)*, and when the Latins received it from the Arabs, they were very unproductively involved in it for a long time. God, finally, having mercy on our fate, brought the sciences back to light, together with the competence in languages, and also illuminated this divine science *(i.e., medicine)*: a number of men were awakened, who taught the science from the clear sources of the Greek.[11]

In this account, the Arabs are not the most learned people, but those who lead Greek medicine to shipwreck, with bad consequences for medieval Latin medicine. At the same time, Donzellini states very clearly what rescued medicine from this situation: the Renaissance study of languages, *linguarum peritia,* and the reading of Greek medical texts in particular. Other humanists, such as Symphorien Champier, entertained similar views about the devastating effects of the medieval translation movements.[12]

## EDITIONS

It is clear that there was an awareness of the Arabic scientific traditions in Europe among Renaissance intellectuals. But were these traditions in any way central to Renaissance culture? This will be a recurrent question throughout the present study. I will make some first steps toward an answer by turning attention to the bookmarket and to printed editions in particular. Table 1 tells more about the presence of Arabic traditions in the Renaissance than many words could do. For each Arabic author the table lists the number of printed Latin editions that appeared before 1700. Altogether, forty-four Arabic authors were available in Latin editions. The table is a distillation of the Appendix at the end of this book, where all editions are listed. The great majority of editions appeared before 1600.

The numbers are impressive, especially for Averroes (114 editions)[13] and Avicenna (78), who were read both as philosophers and physicians, and for the physicians Mesue (72 editions) and Rhazes (67). Note that some of these editions were multivolume enterprises. Astrologers are not printed that often, but the numbers are still considerable, for instance, for Alcabitius, the author of the well-known introduction to astrology (13 editions), and Haly filius Abenragel, whose comprehensive astrological summa was printed eight times. The full significance of the table becomes apparent if

*Table 1.*   Printed Latin editions of Arabic authors before 1700.

| Author | No. of editions |
|---|---|
| Abenguefit (Ibn Wāfid, physician) | 23 |
| al-Abharī (philosopher) | 1 |
| Albategnius (al-Battānī, astronomer) | 2 |
| Albohali (al-Ḥayyāṭ, astrologer) | 4 |
| Albubater (Ibn al-Ḥaṣīb, astrologer) | 3 |
| Albucasis (az-Zahrāwī, physician) | 33 |
| Albumasar (Abū Maʿšar, astrologer) | 8 |
| Alcabitius (al-Qabīṣī, astrologer) | 13 |
| Alfarabi (al-Fārābī, philosopher) | 4 |
| Alfraganus (al-Farġānī, astronomer) | 5 |
| Algazel (al-Ġazālī, theologian) | 2 |
| Alhazen (Ibn al-Hayṭam, scientist) | 2 |
| Alkindi (al-Kindī, philosopher, scientist) | 25 |
| Alpetragius (al-Biṭrūǧī, astronomer) | 1 |
| Avenzoar (Ibn Zuhr, physician) | 15 |
| Averroes (Ibn Rušd, philosopher, physician) | 114 |
| Avicenna (Ibn Sīnā, philosopher, physician) | 78 |
| Costa ben Luca (Qusṭā ibn Lūqā, philosopher, physician) | 25 |
| Ebenbitar (Ibn al-Bayṭār, botanist) | 2 |
| Geber filius Afflah (Ǧābir ibn Aflaḥ, astronomer) | 1 |
| Pseudo-Geber ("Ǧābir ibn Ḥayyān," alchemist) | 16 |
| Haly filius Abbas (al-Maǧūsī, physician) | 5 |
| Haly filius Abenragel (Ibn abī r-Riǧāl, astrologer) | 8 |
| Haly Rodoan (Ibn Riḍwān, physician) | 13 |
| Pseudo-Haly (Aḥmad ibn Yūsuf, astrologer) | 3 |
| Ibn Buṭlān (physician) | 2 |
| Ibn Ǧazla (physician) | 1 |
| Ibn al-Ǧazzār (physician) | 7 |
| Ibn Muʿāḏ (Abhomad) (astronomer) | 5 |
| Pseudo-Ibn Sīrīn (writer on dream interpretation) | 2 |
| Ibn Ṭufayl (philosopher) | 1 |
| Iesus Haly (Ibn ʿĪsā, ophthalmologist) | 3 |
| Iohannitius (Ḥunayn ibn Isḥāq, physician) | 27 |
| Isaac Israeli (Isḥāq al-Isrāʾīlī, physician) | 5 |
| Isḥāq ibn ʿImrān (physician) | 1 |
| al-Isrāʾīlī (Ps.-Almansor, astrologer) | 8 |
| Maimonides (Mose ben Maymon, philosopher, physician) | 15 |
| Messahalah (Māšāʾallāh, astrologer) | 11 |
| Mesue (Ibn Māsawayh, physician), Ps.-Mesue (anonymous) | 72 |
| Omar Tiberiadis (ʿUmar aṭ-Ṭabarī, astrologer) | 7 |
| Rhazes (Ibn Zakarīyāʾ ar-Rāzī, physician) | 67 |
| Serapion (Ibn Sarābiyūn, physician) | 7 |
| Thebit ben Corat (Thābit ibn Qurra, mathematician, astronomer) | 4 |
| Zahel (Sahl ibn Bišr, astrologer) | 9 |

we recall that a good number of twelfth- and thirteenth-century Latin authors who are famous today were transmitted very badly in early prints. Peter Abelard, for example, received only one edition (1616), Roger Bacon only two (1614 and 1618). In contrast, Arabic authors, who entered European culture in the same centuries in which Abelard and Bacon lived, had a much better transmission in printed books.[14]

Printed editions are a good indicator not only of the availability of authors but also of the general demand for these authors and of the distribution of their works. Most incunabula editions of Arabic authors were produced in Italy. In the sixteenth century, however, many editions also appeared in other European countries, such as France and Germany. Venice was the city in which by far the most works of Arabic origin were printed, about 300 in fact. Venice is followed by Lyon (about 100) and Basel (about 50). Other important printing locations were Paris, Nuremberg, Strasbourg, and Pavia. Printing was a commercial business, with decreasing book prizes in the first fifty years after the invention of printing. Authors who did not sell had little chance to be printed again. Table 1 therefore leaves no doubt that the four most often printed Arabic authors—Averroes, Avicenna, Mesue, and Rhazes—were central to fifteenth- and sixteenth-century European culture. Every educated person in the Renaissance will have known their names, even if that person was a humanist of the radical sort who would never dream of reading a text by these authors.

Arabic authors in Latin translation were a Renaissance printing success, so much so that the question arises whether for some Arabic authors and their scientific theories the highpoint of their influence came in the later fifteenth and sixteenth centuries—this will be a recurring theme in the present study. If we read Table 1 with closer scrutiny, and add what can be gleaned from this book's Appendix, we realize that the pace at which medieval translations of Arabic texts reached their readers differed greatly. Some works reached the peak of their transmission in the Renaissance, others lost their significance for the West in this period. Examples of the latter case, that is, for Arabic works much read by the scholastics, but with a very short printing history, come from Avicenna and Thebit ben Corat. The magisterial philosophical works by Avicenna, his *Metaphysics* and *De anima* of the philosophical summa *aš-Šifāʾ* (The Cure), were a great success with the scholastics soon after they had been translated in Toledo in the twelfth century.[15] But they were printed only twice, the *Metaphysics* in 1495 and 1508, the *De anima* around 1485 and 1508. Clearly, these texts have ceased to be central reference points for philosophical discussion. It is true that a good number of Avicennian philosophical doctrines continue to be known and discussed,[16] but Renaissance philosophers only rarely had recourse to Avicenna's philosophical texts themselves. The Renaissance Avicenna is very much the *princeps medicorum*,[17] the famous author of the *Canon medicinae*. The other example is Thebit ben Corat (Thābit ibn Qurra), famous today for his achievements in mathematical astronomy. In the twelfth century, several of his astronomical and mathematical treatises were translated into Latin, some of them by Gerard of Cremona. The printing history, which comprises four editions, is meager, especially if compared to the great manuscript transmission of some of Thebit's texts, for instance, *De quantitate stellarum et planetarum et proportione stellae,* a treatise on celestial and terrestrial measurements, which was widely distributed in manuscript, but not printed at all. In the Middle Ages, Thebit served as a principal authority on Ptolemaic astronomy; this role was taken over by other authors during the Renaissance.[18] At the universities of the fifteenth and sixteenth centuries, one astronomical textbook dominated all others: John of Sacrobosco's Latin *Sphera* of the thirteenth century, which itself draws amply on an Arabic writer: Alfraganus's compendium of Ptolemaic astronomy, the so-called *Rudimenta.* Alfraganus himself is another example of an author widely distributed in manuscript, but rarely printed.

Other authors, in contrast, see the peak of their transmission in the sixteenth century. One such case is Mesue. Under the name of 'Mesue' traveled both the true ninth-century physician Yūḥannā ibn Māsawayh and an anonymous Arabic author of the eleventh or twelfth century, who composed major pharmacological works, which were the basic material for the fourfold *Opera Mesue* compiled in Latin in the thirteenth century. The Latin manuscript transmission of the *Canones universales,* the first part of the *Opera,* was already very significant in the fourteenth and fifteenth centuries (about seventy are known today), but the text was also printed fifty-five times. In addition, the many sixteenth-century commentaries on the text—among the commentators are Angelus Palea and Bartholomaeus Urbevetanus, Johann Dantz, Andrea Marini, and Giovanni Costeo—are a further indication that the sixteenth century saw the highpoint of Mesue's influence.[19] Similar observations can be made with respect to Avicenna's *Canon medicinae* and Rhazes's *Liber nonus ad Almansorem,* two works of medicine, on which the commentary tradition reaches its high point in the Renaissance.

Each of the forty-four authors in Table 1 has his own printing history, which would deserve individual attention, especially with respect to the many persons involved in it: the publishers who printed the book; the editors who composed the volume; the scholars who assisted in publishing, for instance, by retrieving manuscripts; the authors of accompanying texts, such as prefaces, dedications, indices, tables, notes, or even fully fledged commentaries; the revisers or translators of the text; and the dedicatees and the patrons who financed a translation or a publication. In this area much research still needs to be done. Nancy Siraisi's survey of Renaissance editions of Avicenna's *Canon* is the only substantial study so far.[20] In what follows, I will briefly survey the editions of the physician Avicenna and the astrologer Alcabitius and then turn to a fuller account of the printing history of the philosopher Averroes.

Siraisi has shown that there were three main phases in the history of Latin editions of Avicenna's *Canon medicinae (Qānūn fī ṭ-ṭibb)*. In the first phase, the editors presented the text as translated by Gerard of Cremona in twelfth-century Toledo. Often this text was accompanied by the expositions of the principal Latin commentators on the *Canon*. The most amibitious of these publishing enterprises was the five-volume edition of the *Canon* of 1523, issued by the Giunta press in Venice. It printed the entire text together with the fourteenth- and fifteenth-century commentaries by Gentile da Foligno, Jacques Despars, Dino del Garbo, Ugo Benzi, Giovanni Matteo Ferrari, and Taddeo Alderotti. The second phase began in 1527 with the appearance of a *Canon* edition that contained Andrea Alpago's corrections to the entire text, printed in the margins, plus Alpago's dictionary of technical terms. Alpago's corrections had been officially recommended by the University of Padua in 1521.[21] They were based on the evidence of Arabic manuscripts, which Alpago had read in the Near East, as physician of the Venetian embassy in Damascus. Other scholars tried to improve the text of the *Canon* by other means: by translating from the Hebrew or by comparing several previous translations. In Chapter 3 of the present study, I will examine the impressive history of *Canon* translations and revisions in greater detail. In the later sixteenth century, two major editions of the entire *Canon* dominated: that of Benedetto Rinio of Venice, who inserted Alpago's corrections into the text and added references to other medical sources, first published by the Giunta press in 1555, and that of Giovanni Costeo and Giovanni Mongio, first published by Valgrisio in 1564, apparently a competing editorial enterprise. The Costeo–Mongio edition returned to the text of Gerard of Cremona and provided variant readings of Alpago and Mantino in the margins.[22] In a third phase, from 1609 onward, editions and translations appeared that were produced by northern European Arabists, whose interest was increasingly philological and to a lesser extent medical. These works could already draw on the first printing of the *Canon* in Arabic letters by Giovan Battista Raimondi, which appeared in 1593 with the Stamperia Orientale Medicea, the Medici press for Arabic works founded in 1584.[23]

The printing history of astrological texts is, in general, much shorter than that of medical texts, because astrology had a less prominent place in the curriculum of Renaissance universities. Nevertheless, the transmission of Arabic astrological texts in Latin printed editions is considerable. Astrological works of eleven Arabic authors were made available in print: Albohali, Albubater, Albumasar, Alcabitius, Alkindi, Haly filius Abenragel, Haly Rodoan, al-Isrā'īlī (Pseudo-Almansor), Messahalah, Omar Tiberiadis, and Zahel. The most often printed astrological work of Arabic origin was Alcabitius's (al-Qabīṣī's) *Introductorius ad magisterium iudiciorum astrorum (Kitāb al-Mudḫal ilā ṣinā'at aḥkām an-nuğūm)*, which appeared thirteen times, from the first edition in 1473–1474 to the last in 1521 in Venice.[24] Some of these were produced by well-known printers or editors. Erhard Ratdolt of Augsburg, one of the printing pioneers of works of science, who was active in Venice, produced two editions of Alcabitius in 1482 and 1485.[25] The latter is considerably longer because it also contains John Danko of Saxony's fourteenth-century commentary on Alcabitius's work.[26] The medieval commentary tradition on Alcabitius was continued in the Renaissance. Around 1520 an Alcabitius edition appeared in Lyon, which, apart from John of Saxony's commentary, also contained comments, called *additiones,* inserted in Alcabitius's text by Pierre Turrel (d. 1547), principal of the college of Dijon and a well-known astrologer in France. The edition ends with a treatise by Turrel on astrological medicine.[27] Of much greater size, ambition, and influence was a later commentary by Valentin Nabod (1523–1593), a professor of mathematics at Cologne university.[28] This 472-page commentary was published in 1560 by Nabod under the title *Enarratio elementorum astrologiae* as "an exposition of Alcabitius, who published the doctrine of the Arabs in a compendium."[29] In the dedicatory letter, Nabod explains that throughout the work he will compare Alcabitius's doctrines with those of Ptolemy, and that he will refute Alcabitius where his theories are in conflict with *physica doctrina* and hence superstitious.[30] Nabod cites Alcabitius's text in full and comments on almost the entire text, except for the last section of *differentia* 4 and for most of *differentia* 5, that is, the last part of the text. The main reason is that Nabod does not accept the Arabic doctrine of the great number of lots *(partes),* given that Ptolemy recognized only one, the lot of fortune *(pars fortunae).* Turning from the commentaries to the textual tradition of Alcabitius itself, it can be observed that the printing history of Alcabitius also includes an example of textual emendation. Antonio de Fantis, a Scotist philosopher and professor at the University of Padua before its closure in 1509, produced a revision of the text that appeared two times in 1521 with two different printers in Venice, P. Liechtenstein and Melchior Sessa. De Fantis revised the text of the earlier editions by comparing at least one Latin manuscript and perhaps also the commentary of John of Saxony.[31] The most ambitious revision of an astrological text was made in 1551, when Antonius Stupa published a revision in humanist Latin of Abenragel's *De iudiciis astrorum*. This revision was based on Latin witnesses only and was aimed at extinguishing the Gallicisms and Hispanicisms (due to the intermediary Castilian) in the medieval Latin translation, so that the old version "is transcribed into the better diction of the Latin language."[32] It is noteworthy that the printing history of Arabic astrological authors, which was much more modest in total figures than that of medical and philosophical authors, nevertheless shares the same features: the medieval manuscript transmission is continued in print; the medieval commentary tradition does not stop but assumes new formats in the Renaissance; well-known editors care for the publication of the texts; scholars associated with Italian, French, and German schools and universities contribute to the printing history; and several attempts are made to improve the text, also with methods typical of the humanist movement. I am not aware, however, of Renaissance translations of Arabic astrologers into Latin.

My third example from the printing history of Arabic authors is Averroes, the *commentator* par excellence. In fact, the history of Averroes editions is a remarkable chapter of Renaissance culture. Many well-known intellectuals of the time were involved in it.[33] Of the 114 printed editions, 26 concerned Averroes's medical writings, that is, his *Colliget (Kullīyāt)*, the commentary on Avicenna's *Cantica (al-Urǧūza fī ṭ-ṭibb)*, and several short medical treatises. Averroes was printed early. The *Colliget* appeared for the first time in 1482, his Aristotle commentaries in 1472–1474 in Padua. This latter edition was in fact a major enterprise, which set the model for many subsequent editions of Aristotle's works together with Averroes's commentaries. The editor Lorenzo Canozio, also known as an artist, printed seven of Averroes's commentaries on natural philosophy and metaphysics that had been translated in the thirteenth century.[34] The pattern Canozio chose—for each chapter, to print first a Greek–Latin translation of Aristotle, then the Arabic–Latin translation, and finally, in smaller font, Averroes's commentary—would be adopted by the majority of later editors. The next collected edition of Aristotle and Averroes came out in 1483 in Venice under the editorship of Nicoletto Vernia, the Averroist philosopher at the University of Padua, the second successor of Paul of Venice, and teacher of Giovanni Pico, Agostino Nifo, and Pietro Pomponazzi. The edition was produced at the official request of the governor of Padua to carefully clean Averroes's texts of corruptions.[35] Vernia's outspoken Averroism provoked the official reaction of Bishop Barozzi of Padua in 1489 (as will be discussed in

Chapter 5). It is noteworthy that a high-ranking scholar such as Vernia would oversee the Aristotle–Averroes edition of 1483, which is the first to also include the logical and ethical works.

The incunabula editions of Averroes made available the entire corpus of medieval translations of this Arabic author—with the exception of the *Rhetoric* commentary, which was first printed in 1515. In 1488, the first Renaissance Latin translations from Hebrew appeared in print: Elia del Medigo's translations of Averroes's compendium of the *Meteorology* (plus a chapter from the Middle commentary) and of Averroes's preface to book 12 of his Long commentary on the *Metaphysics*. The latter translation had been commissioned by the young nobleman Domenico Grimani, the later cardinal and famous humanist. Grimani was the second patron for whom del Medigo worked; other Averroes translations had been financed by Giovanni Pico della Mirandola. The 1488 del Medigo edition was the starting point for a long series of editions that contained new Averroes translations from the Hebrew: in 1497, again by del Medigo, 1511 by Paolo Ricci, 1521 by Jacob Mantino, 1523 by Abraham de Balmes, 1524 by Giovanni Burana, 1527 by Calo Calonymos, 1539 by Mantino, 1550/1552 by Balmes, Mantino, Burana, and Vitalis Nisso, 1552 by Balmes, 1560 by del Medigo, and 1562 by Mantino. New translations of the medical works *Colliget* and *De spermate* appeared in 1550/1552 and 1560. With very few exceptions, almost all Renaissance Averroes translations were published in print. This shows that the Hebrew–Latin Averroes translation movement was well-linked with the editorial and scholarly circles interested in Averroes. It accords with the fact that the Averroes translations were often supported and even commissioned by aristocratic patrons who were in some way or other connected with the University of Padua, as we will see in Chapter 3.

Another noteworthy feature of the printing history of Averroes is the secondary literature that accompanied the editions. Vernia's pupil Agostino Nifo (d. 1538) published commentaries on Averroes's *Destructio destructionum, De beatitudine animae,* and *De substantia orbis.* Moreover, in his commentaries on Aristotle's *De anima, Physics, De caelo,* and *Metaphysics,* Nifo often comments in detail on Averroes's exposition, so that his texts in many passages assume the format of a super-commentary on Averroes's Long commentaries.[36] The emergence of this genre is particularly noteworthy since super-commentaries on Averroes are widespread in Hebrew,[37] but very rare in Latin. The Averroes text most often commented on in the sixteenth century was his *De substantia orbis,* a short cosmological work that was much read in universities. Other tools for reading Aristotle together with Averroes were written by the philosopher and Averroes editor Marcantonio Zimara, who had been a student of Nifo and Pomponazzi at Padua. In 1508, Zimara published analyses of seemingly contradictory passages in the texts of Aristotle and Averroes: *Contradictiones et solutiones*. The book is designed as a question commentary on the works of natural philosophy and metaphysics, plus *Colliget* and *De substantia orbis*. It appeared in sixteen editions, some of them partial, in the sense that Zimara's comments on a specific work by Aristotle or Averroes accompanied the edition of that book.[38] In 1537 Zimara's second reading tool followed, a massive alphabetical lexicon on Aristotle and Averroes, the *Tabula dilucidationum*. Zimara's two works are important testaments to how Renaissance scholars read Aristotle and Averroes.

The culminating point of the printing history of Averroes came in 1550/1552 with the monumental eleven-volume edition of the works of Aristotle and Averroes, published by Giunta in Venice.[39] The edition is remarkable for its comprehensiveness, since it contains works of Aristotle not commented on by Averroes and works by Averroes that are not commentaries: *Destructio destructionum, De substantia orbis, De animae beatitudine, Epistola de connexione,* and Averroes's three medical works available in Latin. Moreover, it is the first *Opera* edition that brings together the medieval and Renaissance translations of Averroes. This 1550/1552 edition was reissued three times, with some modifications and additions of texts and commentaries: 1560 in Venice by Comin da Trino and in 1562 (reissued 1574/1575) by the Giunta brothers. The significance of this editorial enterprise is apparent in the fact that the 1562 Giunta edition serves, up to the present day, as the reference text for several Latin Averroes translations that have not yet received a critical edition.

A valuable source on the origins of the Giuntine edition is the twenty-folio prefatory fascicle, which forms part of the 1550/1552 edition. The driving force behind the project, as we learn from this fascicle, was Giovanni Battista Bagolino of Verona (not Girolamo, his father, as is sometimes maintained):[40] Bagolino selected the texts, compared the translations, and occasionally corrected them. But he died before the project was finished. The project was then carried on by Marco degli Oddi, a teacher of logic in Padua, with the help of Romulo Fabio of Florence. In the long line of Averroes editors in history up to the present, Bagolino was clearly one of the most important. His father Girolamo Bagolino was a professor of philosophy and medicine at the University of Padua after its reopening in 1517.[41] Among his pupils was the Hebrew–Latin translator Giovanni Burana. Bagolino the father wrote a prefatory letter for an edition of translations by Burana: the Hebrew–Latin version of Aristotle's *Prior Analytics* with Averroes's Short and Middle commentaries, first published in 1524.

Here Bagolino the father advertises the great benefit one can derive not only from the Greek but also from the Arabic scholars in understanding Aristotle's logic.[42] His son, Giovanni Battista Bagolino, was raised in this climate of Paduan interests in Averroes and his logic. We do not know much about Bagolino, but fortunately, in the prefatory fascicle to the Giunta edition, Marco degli Oddi gives a detailed description of how Bagolino had organized the edition. Bagolino, he writes, was selected as editor by the Giunta brothers Tommaso and Giovanni Maria because of his excellent knowledge of Greek and Latin, and because of his profound knowledge of Peripatetic philosophy: "he was admired by everybody for his many highly accurate publications and for his public exposition (of Peripatetic philosophy), greatly esteemed by all, at the University of Padua."[43] From Padua, Bagolino moved to Venice, where he lectured on philosophy in private circles (inter lares domesticos) and became a well-known physician, while at the same time working on the edition project. Bagolino selected what he judged to be the best Greek–Latin translations of Aristotle, corrected some of them, and then, with equal diligence, collected the new Hebrew–Latin translations of Averroes. Where he could not find a Renaissance translation, for example, for the Long commentaries on *Physics*, *De caelo*, and *Metaphysics*, he turned to the medieval translations, which he corrected (in melius reduxit), if they had not been corrected already by Nifo and Zimara. For this purpose, he collected material from other Paduan philosophers, including Pomponazzi and his own father. After Bagolino's untimely death, degli Oddi continued the philological work on Averroes's texts and added further *castigationes* in line with Bagolino's method.

The efforts undertaken by Bagolino and degli Oddi are remarkable. A note at the end of the prefatory fascicle shows that they were in contact not only with the Hebrew–Latin translators in Italy but also with Venetian travelers in Istanbul. In this note, degli Oddi relates that the Venetian ambassador at the Sultan's court in Istanbul in 1550–1552, Bernardo Navagero, an admirer of Averroes, had sent a list with books of Averroes not yet translated, "which can be found in that city among the Jewish and Arab physicians" (apud Iudaeos et Arabas medicos):[44] the Compendia and Middle commentaries on *Physics*, *De anima*, and *Metaphysics*; the Middle commentary on *De caelo*; the Middle commentary on the last nine books *De natura animalium*; the Long commentary on *De plantis*; and the Compendium on Ptolemy's *Almagest*. This information came too late; degli Oddi did not want to postpone the publication of the edition, which had long been in the making. But the list is a testimony to the ambition of the project. The 1562 reprinting of the 1550/1552 edition does not contain degli Oddi's long preface, and all mention of Bagolino has disappeared, as

Charles Burnett has shown.[45] Instead, substantial material is added on Aristotle's *Posterior Analytics* and on Averroes's commentary on it by another Paduan logic professor, Bernardino Tomitano. This again points to the close connections between the Paduan logicians and the production of the Giunta edition of the combined Aristotle–Averroes.[46]

The Giunta edition was reprinted again in 1574?–1575.[47] The last edition of a work of Averroes, the *Epitome of Plato's Republic,* appeared in 1578. This was the abrupt end of the printing history of Averroes. As far as is known today, no seventeenth-century editions of Averroes exist.[48] The Giunta edition, a monument of Renaissance editorship, may have sparked interest in Averroes for a limited period, but, ironically, it also marks the end of the long history of the Western transmission of Averroes.

## UNIVERSITY CURRICULA

The considerable printing histories of Arabic authors would be unexplainable without a great demand for these authors in university circles. As the examples of Avicenna, Alcabitius, and Averroes have shown, there are many links between the universities and the printing of Arabic authors. A good number of university professors were involved in the production of editions. On one occasion, a university institution, the Collegium of philosophers and physicians of the University of Padua in 1521, even officially recommended the usage of a particular textual version, Alpago's correction of the *Canon.*[49] In what follows, I will turn to the university curricula themselves. It is a remarkable feature of the history of European universities that Arabic authors had an important place in university for several centuries, especially in medicine, philosophy, and astrology. The main curriculum authorities were Galen and Avicenna in medicine, Aristotle and Averroes in philosophy, and Ptolemy, Alcabitius, and Albumasar in astrology. The students of universities of the Christian world read very few Christian authors, such as Petrus Hispanus on logic and Sacrobosco on astronomy, in philosophical and medical faculties. The bulk of teaching was on Greek and Arabic authors. The firm rooting of Arabic works in medieval curricula is the backbone of their long-lasting reception in the Latin West.

In the field of medicine, Avicenna's *Canon* is the most influential Arabic text. After the translation by Gerard of Cremona in the later twelfth century, about one hundred years passed until the text assumed a central position in medical education, especially in the dominant universities in the field: Montpellier, Paris, Bologna, and Padua. The Montpellier statutes of 1340, for instance, which were in effect until 1543, prescribe as mandatory reading not only several treatises of Galen, but also Avicenna's *Canon*, book 1 and chapter 6.1.[50] The very detailed Bologna statutes of 1405 divide the medical curriculum into theoretical medicine, based on texts by Galen, Hippocrates, Avicenna, and Averroes, and practical medicine, based on Avicenna only.[51] The chapters of Avicenna's *Canon* most often used in university teaching until the seventeenth century were chapters 1.1 on physiology—a key text for the medieval and Renaissance teaching of theoretical medicine—, 1.4 on principles of therapy, book 3 on diseases from head to toe, and 4.1 on fevers.[52] Another Arabic author who appears in university curricula is Rhazes (Ibn Zakarīyā' ar-Rāzī). In the Latin West, his most successful work was the *Liber ad Almansorem (al-Kitāb al-Mansūrī fī ṭ-ṭibb*, "The Book of Medicine <dedicated to> Mansūr," the ruler of Rayy in Persia), a handbook of medicine, of which book 9 on diseases was the most popular in university education.[53] Book 9 was often commented on by university professors, especially in Montpellier in the fourteenth and in Italy in the fifteenth and sixteenth centuries.[54]

The *Opera* of Pseudo-Mesue, a fourfold work of theoretical and practical pharmacology, had a different role in education. Its first two parts, *Canones universales* (principles of drug preparation) and *De simplicibus* (on simple laxative drugs), seem to be direct translations from an anonymous Arabic source of the eleventh or twelfth century. Much Arabic material from translations that have not yet been identified also appears in the third and fourth parts, *Grabadin* (a pharmacopeia) and *Practica* (a therapeutics). The *Opera Mesue* played a pivotal role in the history of Western pharmacology and were often commented on in the sixteenth century. They are listed in university curricula of the sixteenth century, for instance, in Ingolstadt,[55] but reading Mesue was also institutionalized outside the university: The medieval and Renaissance education of pharmacists did not take place in universities, but had the format of an apprenticeship. The pharmacist was both a craftsman and a scientist, and the pharmacological apprenticeship had a very scientific character—just as in Islamic culture. An indication of this is that in north Italian cities examinations in pharmacy were conducted by the medical faculties of the universities. Saladin of Asculo's fifteenth-century *Compendium aromatariorum* informs us about the most important books that a pharmacist has to know, among them three works of Arabic origin: the *Opera Mesue*, Avicenna's *Canon* book 2 on simple drugs, and Serapion's *De simplicibus*. Mesue and Serapion also appear in the 1430 city statutes for pharmacists of Basel.[56] The first official pharmacopeia to be printed outside Italy, the *Dispensatorium* of Valerius Cordus of 1546, which the Nuremberg council made obligatory for all pharmacies of the city, draws 80 percent of its receipts from Mesue, Avicenna, Rhazes, and the medieval *Antidotarium Nicolai*, which is based on Arabic sources.[57]

As the Appendix to this book shows, the printing of the medical authors Avicenna, Rhazes, and Mesue phases out in the first half of the seventeenth century. This is an indication that these authors gradually lost their influence on the development of medical science. Nevertheless, as Nancy Siraisi has shown, Arabic authors, and Avicenna in particular, remained on the medical curricula in most parts of Europe until the eighteenth century. At Bologna, for example, lectures on Avicenna were abandoned in 1716, at Padua in 1767.[58] This was the case even though humanist scholars in the sixteenth century had tried to eliminate Arabic authors from the curricula. Some of these attempts were successful to a certain degree: statutes were reformed in a humanist manner at Tübingen in 1538, in Heidelberg in 1558,[59] and likewise at other German universities, at the London College of Physicians in 1563, and at the Spanish University of Alcalá in 1565. An example of the humanist reformers' attitude is the following passage from the Tübingen medical statutes of 1538, which is anonymous, but clearly written by the dedicated humanist Leonhart Fuchs:

> Since there is nobody who would not know that the Arabs have copied almost everything from the Greeks, the Arabs will be used only to a very minimal degree for the teaching of this field *(i.e., medicine)*, because it is advisable to scoop the precepts of a science from the sources and not from turbid ditches. Arab and other defective and barbarous authors are therefore to be passed over as far as possible and only Hippocrates and Galen . . . will be taught.[60]

It is a remarkable but little known chapter in the history of Renaissance humanism that the Arabic authors, blocked out by humanist reformers, reappear again in many medical curricula toward the end of the sixteenth century.[61] At Ingolstadt University, for example, where humanists such as Peter Burckard in the 1520s had relegated Arabic medicine to a marginal place, the statutes of 1571 promote the "old medicine *(vetus medicina)*, which rightly can be called the catholic *(catholica)*," and which demonstrates the "falsity of Paracelsian medicine." The statutes reintroduce the reading of Avicenna and Mesue on the same footing with Hippocrates, Galen, and other Greek authors.[62] Freiburg im Breisgau follows the example of Ingolstadt in 1577, after the duke had decreed "that only the old classical authors *(die alten Classici authores)*, such as Hippocrates, Galen, Dioscorides, Avicenna, and Rhazes are to be lectured upon publicly."[63] The confessional and anti-Paracelsian background of the Ingolstadt statutes is interesting, but the more important reason, when one turns to other universities, was that enthusiasm for medical humanism and its narrow curriculum had dwindled. Further examples of the reintroduction of Arabic authors in Germany are the Tübingen statutes of 1601, which prescribe the teaching of Avicenna and Rhazes, with the effect that Leonhart Fuchs's humanist reform was overturned,[64] in Jena 1591,[65] in Wittenberg 1572,[66] and in Würzburg 1610.[67] In seventeenth-century Spain, Avicenna was taught again not only at Salamanca but also at Alcalá, the former humanist stronghold.[68] At Montpellier, lectures on Avicenna and Rhazes disappeared in the sixteenth century, but reappeared in the seventeenth century.[69] The university of Louvain prescribed the teaching of Avicenna's *Canon* again in 1617.[70] In Italy the situation was different, since reforms never led to the complete elimination of the Arabs from the curricula, not even at Ferrara, where anti-Arabic polemics in medicine had originated with the humanists Leoniceno and Manardo, nor at Pisa, which in 1543 received statutes that eliminated the Arabic authors from medical teaching in *theoria,* but not in *practica.*[71]

   This is not to say that the medical curricula of European universities around 1600 returned to their late medieval format. The curricula contained many newly translated treatises of ancient medicine with Galen dominating, had a greater emphasis on anatomy and botany, and occasionally also included modern authors such as Vesalius and Falloppio in anatomy. With respect to the Arabic authors, however, it is important to see that the anti-Arab propaganda, which was most fervent in the first half of the sixteenth century, led to reforms of the curriculum that, for the most part, were later withdrawn. In Chapter 4 of this study, I will show that there is a parallel development in the scientific discussion itself, notably in pharmacology: After the heyday of humanism was over, the physicians began to use and cite Arabic authors again, not out of habit, but for scientific reasons.

   The second field of university education that was influenced by Arabic authors is philosophy, as taught in the arts faculties. Other than in medicine, Arabic authors are hardly ever named in the statutes regulating the philosophical curriculum, even though Averroes's commentaries were part of philosophical teaching all over Europe for several centuries. The educational system in Latin Europe was changed radically in the first half of the thirteenth century: the first universities were founded; an enormous range of Greek and Arabic works became available in Latin, notably the *Corpus Aristotelicum;* and the framework of the seven liberal arts was replaced by a new model that followed the traditional structure of Aristotle's works. In Paris, this process was completed in 1255 when the arts faculty prescribed a long list of Aristotle's works on logic, ethics, natural philosophy, and metaphysics as mandatory reading for all students.[72] Some of these curricular works were not in fact written by Aristotle, but by Arabic authors, notably the following four: first, *Liber de causis,* a Neoplatonic treatise by an anonymous Arabic author of the ninth century, which is based largely on Proclus's *Elements of Theology;* second, *De differentia spiritus et anima* by Costa ben Luca (Qustā ibn Lūqā), which discusses the difference between the material *spiritus* in the body and the immaterial soul; and, third, *De congelatione et conglutinatione lapidum,* an extract from Avicenna's philosophical summa *aš-Šifā',* which was read as the fourth chapter of Aristotle's *Meteorology.* In some statutes we also find a fourth Arabic text: *De substantia orbis,* a cosmological compendium by Averroes, for instance, in the Bologna statutes of 1405.[73] These four works were, in fact, the Arabic treatises most widely diffused in Latin manuscripts. Their appearance in the curricula does not imply that they were necessarily believed to be authentic works by Aristotle. Rather, they were thought to complement the spectrum of Aristotelian disciplines. In the fourteenth and fifteenth centuries, there was an increasing tendency in the arts faculties to concentrate on the *Corpus Aristotelicum* proper, and the philosophical treatises of Arabic origin gradually disappeared from the curricula. They do not appear, for instance, in the Paris statutes of 1366,[74] which influenced other statutes, such as those of Vienna of 1389,[75] which in turn were a model for many middle European university statutes.[76]

The statutes therefore do not reflect what one can witness in the commentary literature produced by masters of arts from the thirteenth to the sixteenth century: that Averroes's commentaries were very much part of everyday teaching in European arts faculties. Averroes appeared in statutes only if the university officials wanted to ban problematic interpretations of Aristotle, as happened at the University of Louvain in 1447. The statutes of the Louvain arts faculty prescribe that Aristotle is not to be interpreted according to Wyclif, Ockham, or their followers, but according to "his commentator" Averroes, "where he does not disagree with faith," or Albert the Great, Thomas Aquinas, or Giles of Rome.[77] Averroes is invoked in the service of orthodoxy against the *via moderna* because he guarantees a reliable interpretation of Aristotle. This is all the more noteworthy in view of the fact that Averroes was known to hold several philosophical theses in conflict with Christian orthodoxy, as we will discuss in Chapter 5. But this did not slow down his reception in the arts faculties, especially not in Italy, where the theological faculties, if they existed at all, were less influential and often functioned merely as promotion boards.[78]

The medieval masters of arts did not adopt the commentary formats from Averroes—the formats had already been developed by logic masters in the twelfth century.[79] But Averroes was of paramount importance for the exposition of Aristotle, especially in the four fields for which there existed Long commentaries by Averroes in Latin translation: *Physics, De caelo, De anima,* and *Metaphysics.* Averroes's influence began around 1230—that is, only very few years after the founding of the first universities and after Averroes had been translated into Latin—and continued all the way up to the late sixteenth century. How and why Averroes's influence as a commentator declined around 1600, that is, about one century before the Aristotelian curriculum was gradually abandoned at European universities in the decades around 1700,[80] will be a matter of analysis in Chapter 5.

Averroes's position within university teaching in the Renaissance was affected by the Latin translation of the Greek commentators of late antiquity into Latin. Theodore Gaza began with Alexander of Aphrodisias in 1452/1453, Ermolao Barbaro continued with Themistius in 1472/1473, and until the mid-sixteenth century, a substantial corpus of late Greek and Byzantine commentaries on Aristotle was available in Latin printed editions. Another factor that influenced the reading of Aristotle and hence the usage of Averroes was the growing interest in Greek language and in reading Aristotle in Greek. Even though the university curricula did not change significantly—in contrast to what we have seen in medicine—the techniques in reading and expounding Aristotle did. Important steps in the rise of the Greek Aristotle were the following: the translating and commenting activity of Byzantine émigré scholars in Italy in the middle of the fifteenth century, such as Johannes Argyropulos and George of Trebizond; Ermolao Barbaro's lectures on Aristotle at Padua in the 1470s, where the Greek text was used to correct medieval Latin translations; Angelo Poliziano's courses on the Greek and Latin texts of Aristotle's *Ethics* and the *Organon* at Florence in the early 1490s; Francesco Cavalli's teaching of the Greek Aristotle at Padua in the same years; Aldo Manuzio's magnificent editio princeps of the Greek Aristotle in Venice 1495–1498; and the establishment of the first chair for lecturing on the Greek text of Aristotle at Padua in 1497, first held by Niccolò Leonico Tomeo.[81]

The translation of the commentators and increasing competence in Greek were real challenges to Averroes's position as the principal commentator on Aristotle in university teaching. Renaissance humanists often pointed out, with full right, that Averroes's commentaries did not rely on the Greek text of Aristotle, but on Arabic translations. It is an open question, however (which will be addressed in Chapter 5), whether the humanists were justified in criticizing Averroes's commentaries for their inferior quality and in calling for his replacement by the Greek commentators. In any event, Averroes survived the challenge. In the first half of the sixteenth century, interest in both the Greek commentators and Averroes was booming. They were read and cited side by side by the university professors. The printing history of Averroes's works, as shown, phased out only in the late sixteenth century, at the same time when the Latin translations of the Greek commentators also ceased to be printed. To be sure, the new interest in Greek therefore had an impact on Averroes's position in the arts faculties: in the sixteenth century, Averroes ceased to be the principal and dominant commentator on Aristotle. But this did not prevent interest in his works and commentaries from reaching new heights in the very same period.

Astrology is the third university discipline in which Arabic authors were read. In addition to grammar and philosophy, the curricula of the arts faculties in the fifteenth and sixteenth centuries also contained a section called *mathematica* or *astrologia,* which had developed out of the early medieval *Quadrivium,* that is, the four liberal arts music, arithmetic, geometry, and astronomy. Typical curricular texts in this field, the *libri ordinarie legendi,* are Euclid's *Elements* in geometry, Sacrobosco's *Sphaera,* and a genre of treatises called *Theorica planetarum* in astronomy, John Pecham's *Perspectiva communis* in optics, Jean de Meurs's *Music,* and arithmetical treatises on *algorismus,* for instance, by Sacrobosco or John of Lignères.[82] The curricula do not change significantly in the Renaissance. When Galileo Galilei became professor of mathematics at Padua in 1592, he still lectured on Euclid, the *Sphaera,* and the *Theorica planetarum.*[83] Arabic authors appear in the curricula both indirectly as the major sources of the curricular texts and also—other than Averroes in philosophy—directly as prescribed reading. In astronomy, optics, and arithmetic, the texts read in the universities were written by medieval Latin authors such as Sacrobosco, but these texts relied heavily on the works of Alfraganus in astronomy, Alhazen in optics, and Alcoarismi (al-Ḥwārizmī) in arithmetic. Also, the thirteenth-century Latin treatise *De compositione astrolabii,* which appears in some statutes under the name of Messahalah,[84] is a compilation from an Arabic treatise on the astrolabe by Ibn aṣ-Ṣaffār.[85] At bottom, what was taught in *mathematica* in the arts faculties was Greek and Arabic science in Latin garb.

In one field, astrology, Arabic authors were even explicitly cited in the curricula. The above-mentioned Bologna statutes of 1405 prescribe a four-year course in the quadrivial sciences under the heading *astrologia,* which was later adopted by the University of Ferrara as well.[86] The third year contains a section on astrology in its narrow sense: "In the third year, first

Alcabitius is read, afterwards Ptolemy's *Centiloquium* with the commentary by Haly."[87] The student therefore starts with Alcabitius's introduction to astrology and continues with Pseudo-Ptolemy's *Centiloquium,* one of the most popular astrological texts in the Latin world, which probably is of late Greek origin. The *commentum Haly* refers to the lemmatic commentary on the *Centiloquium* by Abū Ǧaʿfar Aḥmad ibn Yūsuf of the tenth century (not to Haly Rodoan, that is, ʿAlī ibn Riḍwān,[88] who had written a commentary on the truly Ptolemaic *Tetrabiblos*). The Bologna statutes specify:

> Further, they decreed, ordained, and established that the doctor chosen or to be chosen by the said university at a salary to lecture in astrology, be required to give judgements free to the scholars of the said university within a month after they were asked for, and also in particular to leave a judgement for the year *(iudicium anni)* in writing at the office of the bedells general.[89]

This text indicates a developing practice that will be in full bloom in the late fifteenth and sixteenth centuries: that the universities require one of their professors to produce annual predictions, that is, a horoscope cast for the moment of the entrance of the sun into the first minute of Aries in spring. These predictions often contain astrological weather-forecasting on the coming seasons and prognostication of general events, such as wars, rulers, social uprisings, epidemics, and the like. Since annual revolutions is a branch of general astrology based almost entirely on Arabic sources, in particular, on Albumasar and Messahalah, without any basis in Ptolemy's astrology, the Bologna requirement is another instance of the authoritative position held by Arabic authors at Renaissance universities.

Another example of a detailed astrological curriculum comes from Kraków University, which since about 1448 had a special chair for the teaching of astrology, founded by the scholar Martin Król of Żurawica. The 1476 statutes describe the curriculum:

> &lt;The master of this chair should read&gt; Ptolemy's *Quadripartitum,* Alcabitius, the *Centiloquium* of the words of Ptolemy, Albumasar and other books pertaining to astrology, and shall also present to the university every year a judgement which by the senior &lt;professors&gt; of the same faculty is corrected, revised and approved of.[90]

The reading list here consists of Ptolemy's astrological summa, the *Tetrabiblos,* plus Alcabitius and Pseudo-Ptolemy's *Centiloquium* (the two sources also read in Bologna), and Albumasar, the author of several famous works in astrology. Again, the professor of astrology is advised to produce annual predictions. The importance of this task is underlined by the fact that the senior professors of the faculty have to correct and approve the *iudicium* before it is presented. The Kraków and the Bologna/Ferrara curricula are the two most detailed curricula on astrology extant (as far as I am aware). That these curricula date from the fifteenth century is not a coincidence. In Europe, astrology as a science saw its golden age in the fifteenth and sixteenth centuries, and it is predominantly in this period that astrology was officially taught in universities.

The curricula of Bologna, Ferrara, and Kraków are exceptional. The statutes of most other universities of the Renaissance do not explicitly specify the required reading in astrology, but simply speak of *mathematica*.[91] There is, however, ample evidence from other sources that teaching astrology was a widespread practice; these sources include teaching rolls, student reports, college libraries, lectures, annual predictions, and astrological publications by university professors.[92] At Pavia, a student notebook from the 1480s shows that much more Arabic astrology was read than indicated in the Bologna curriculum—notably, works by Messahalah, Albumasar, Haly Abenragel, and Zahel. Other evidence makes it likely that this broader range of Arabic authors was also read at other Italian universities.[93] At Salamanca, the Colegio viejo, founded in 1410, was particularly well-equipped in astrology,[94] and a chair of astrology was founded in 1561. Chairs of astrology also existed at the universities of Alcalá de Hernares and Valencia.[95] In Paris, the college of Maître Chrétien Gervais, founded around 1370, was equipped by Charles V with two scholarships for the study of astrology and medicine.[96] In fifteenth-century Oxford, permission to teach astrology was given regularly, as surviving university records for the years 1448–1463 show. The Oxford teaching was based on Ptolemy and Alcabitius.[97] We know that astrology was taught at Vienna University from the activity of Georg Tannstetter (d. 1535). There exist student reports of Tannstetter's lectures on introductory astrology and medical astrology.[98] Astrology was a regular part of Renaissance education in Germany, as its inclusion in Gregor Reisch's handbook *Margarita philosophica* attests. Reisch bases his summary on Ptolemy and the Arabic astrologers.[99] Often it was the professors of *mathematica* who did the teaching on astrology. But astrological instruction was also part of medical teaching, since it was a widespread conviction that celestial movements have an influence on the development of diseases and on the effectiveness of medicaments and treatments.[100] Whether in medicine or in *mathematica*, the teaching of astrology was one of those parts of university instruction in which Arabic authors were read regularly.

$\approx$

In order to understand the influence of Arabic thought in the Renaissance, it is sensible to introduce the topic bottom-up, that is, by studying the mere quantitative presence of Arabic authors on the print market and in university curricula. The printing history of Arabic authors is impressive. Some Arabic authors, in particular Averroes, Avicenna, Mesue, and Rhazes, were printed again and again in the great printing locations Venice, Lyon, and Basel. Their works often appeared in expensive multivolume editions, such as the *Opera Mesue,* which alone were printed fifty-five times between 1471 and 1635. The demand for such editions must have been considerable, and it is clear that the distribution of Arabic sciences in Latin Europe reached its high point in the sixteenth century, especially in the three disciplines of medicine, philosophy, and astrology. The great names of Arabic science were to be found everywhere on academic book shelves of Renaissance Europe. This tallies well with the firm rooting of Arabic authors in university curricula. In the faculty of medicine, this is particularly obvious, and it is noteworthy that the humanist reformers were not able to ban Arabic authorities for long from the medical curricula. As to philosophy and astrology, there is ample evidence too of the usage of Arabic authorities in the classroom, especially of Averroes in philosophy, and of authors such as Alcabitius and Albumasar in astrology.

It thus becomes clear that the tradition of Arabic sciences and philosophy in the Renaissance was not restricted to specialist groups of intellectuals. Rather, it was very much part of the general learning of the time. In certain fields, it was even a very vital part: Mesue's pharmacology, Avicenna's medicine, Averroes's philosophy, and Albumasar's astrology were major reference points for the work of Renaissance scholars. The considerable quantitative presence of Arabic authors in early printing and in university curricula is a fact that speaks unambiguously against those historians and pseudo-historians today who, biased against Arabic culture, find the talk about Arabic influence in Europe a mere gesture of political correctness.

It has also become apparent that the transmission and teaching of Arabic authors in the Renaissance differed from the medieval transmission, in spite of all continuities. Some Arabic works lose importance in the Renaissance, if compared with the previous centuries. Such is the case with Avicenna's *Metaphysics* and *De anima,* which much influenced the philosophy and theology of major scholastic thinkers of the Middle Ages. Other examples are Alfarabi's *De scientiis,* which was an influential source for the philosophy of science in thirteenth-century university culture; the *Liber de causis* and Costa ben Luca's *De differentia spiritus et animae,* which are pseudo-Aristotelian texts much read in the arts faculties of the high Middle Ages; and several Arabic texts on arithmetic and astronomy, such as by Alcoarismi, Alfraganus, and Thebit ben Corat. This is not to say that Renaissance mathematicians and astronomers did not build on Arabic algebra and Arabic mathematical astronomy. But they usually did this by improving upon medieval Latin works that had in turn incorporated the accomplishments of Arabic science. Girolamo Cardano praises Alcoarismi as one of the twelve most significant scientists of history, as the "inventor of the Algebraic art,"[101] but Alcoarismi's treatises had long ceased to influence mathematics directly.

On the other hand, many Arabic texts, which had been available in Latin for several centuries, find greater attention in the Renaissance than ever before in the West, especially within university culture. Such is the case, for instance, with the *Opera Mesue,* but also with Arabic astrological texts that now became firmly established in university teaching. It is characteristic of the Renaissance reception of Arabic works that new commentary formats were developed, new translations were produced, and textual criticism was applied to Latin translations of Arabic texts. And, very importantly, the humanist movement influenced the reception in manifold ways. The humanist call for a return to ancient sources provoked opposition from adherents of Arabic scientific traditions. Thus arose a controversy, which left its traces in the medical curricula, at least temporarily, and influenced the usage of Averroes in the classroom. On the other hand, humanists also cared for and promoted Arabic sciences and philosophy, as we will see in Chapters 2 and 3.

# CHAPTER 2

# BIO-BIBLIOGRAPHY

## *A Canon of Learned Men*

orld chronicles of the later Middle Ages are rich sources not only on political and ecclesiastical events but also on famous authors. Many chronicles describe the lives and works of the well-known *magistri,* in particular of the twelfth century: Hugh of St. Victor, Bernard of Clairvaux, Peter Lombard, and others. This biographical tradition continues in the Renaissance. But a good number of Renaissance world chronicles differ from their medieval predecessors in that they also treat Arabic authors. This is remarkable in several respects. It says something about the high respect Arabic authors enjoyed in the Renaissance, about the role they were assigned to in history, and about a specific approach to Arabic science, in which the usage of Arabic sciences is accompanied by biographical and bibliographical inquiries. In what follows, I will inquire into the motives that led to the rise of the biographical study of Arabic authors—whether, for example, historical, philological, medical, religious, anecdotal, or orientalist—and try to situate the Renaissance chroniclers and authors of biographies within the intellectual trends of the time. The primary focus will be on authors who compose whole sets of bio-bibliographies of Arabic scholars inserted in chronicles, treatises on famous men, and histories of scientific disciplines. Other formats of Renaissance biography, such as the vita accompanying an edition, are considered in passing.[1] The present chapter will be concerned with the image that Renaissance biographers created of famous and influential Arabic people, without in fact having access to any significant sources on their lives and works. It will also study the impact of newly translated Arabic biographical sources. For the purpose of illustration, I will repeatedly refer to the Renaissance presentation of three exemplary authors:[2] the physician Avicenna from Buḥārā, the philosopher Averroes from Cordoba, and the astrologer Albumasar, active in Baghdad.[3]

There is a medieval background, albeit meager, to the Renaissance interest in the lives of Arabic authors. Bits of biographical information on Arabic authors appear occasionally in the sources. Guido Bonatti, the thirteenth-century astrologer, relates that Albumasar, the great authority of astrology, had studied in Athens[4]—information that is not trusted today. On Averroes, Giles of Rome reports that his sons were at the court of Emperor Frederick II Hohenstaufen;[5] this story is possible and may help to explain how Averroes's manuscripts came to southern Italy. Less reliable is the claim made by Juan Gil de Zamora in his *De preconiis hispanie* (ca. 1280) that Avicenna, just as Averroes, was of Spanish origin, adding that the true authors of Avicenna's works, according to some sources, were twenty philosophers from Cordoba, who ascribed their work to Avicenna, the son of a king.[6] An unknown Spanish scholar of the fourteenth century, perhaps from Majorca, very convincingly refutes the alleged Spanish origin of Avicenna on the basis of internal information contained in Avicenna's writings, which points to Persia as his home country.[7] The pride of Spanish authors in their home country seems to be one of the driving forces behind the development of biographical material on Arabic philosophers. A fictive story from about 1300 (purportedly translated from Arabic) relates a dispute held in Toledo between philosophical schools; in it appear five scholars from Cordoba: Vergil, Seneca, Avicenna, Averroes, and Algazel.[8] Certain themes of Renaissance biographies are already present in medieval incipits and explicits, such as the kingship of Avicenna *(princeps, rex)*,[9] the royal descent of Mesue,[10] and various places of origin *(Cordubensis, Hispalensis, Aratensis)*. Royal descent was also attributed to other Arabic authors: Geber and Isaac Israeli.[11]

Medieval sources contain bits of biographical material on Arabic authors, but no biographies proper. Arabic authors do not appear in medieval treatises on famous men and in world chronicles—not always two distinct genres—which carry most of the information on the lives of scholars in the Middle Ages, thus taking the place of modern literary histories. The consequence is that Arabic scholars do not enter the biographical tradition in the stricter sense. The watershed in this regard comes with Jacopo Filippo Foresti da Bergamo and his *Supplementum chronicarum* of 1483. The *Supplementum* is the first chronicle, as far as is known today, that gives brief biographies of Arabic authors in various sections on illustrious scholars of the tenth to twelfth centuries. Such biographies are absent in Foresti's main sources, Vincent of Beauvais, Giovanni Boccaccio, Bartolomeo Platina, and Antoninus of Florence.[12] Antoninus's *Chronicon* (ca. 1458) refers to a number of *doctores* in the twelfth and thirteenth centuries—Hugh of St. Victor, Hugh of Folietum, Richard of St. Victor, Bernard of Clairvaux, Helinand of Froidmont, Gratian, Petrus Comestor, Peter Lombard, and Thomas Aquinas—but not to any Arabic scholar.[13] Other late medieval chronicles, such as Vincent of Beauvais's *Speculum historiale* and Ranulph Higden's *Polychronicon,* treat a very similar sequence of Christian scholars. (Some chronicles, of course, do not have sections on famous men at all and concentrate on political history.) When the Arabs enter the tradition of treatises or sections *de viris illustribus,* that is, in Foresti da Bergamo and in the chronicles of Hartmann Schedel and Johannes Naukler who are dependent on him, they are grouped with precisely this circle of scholastic writers. They are not mentioned together with ancient authors, and they are not listed as a special group of recent but pagan writers.

In search of reasons for the late appearance of the Arabs, one may surmise that the traditional biographical genres did not easily lend themselves to incorporating pagan but nonclassical authors. Medieval literary history took its origin with the *libri de viris illustribus* by Jerome and Gennadius of Marseille in late antiquity; most subsequent treatises were confined to Christian authors. In the twelfth century, we meet a number of collections of *accessus* to poets of antiquity, and in the thirteenth century scholars such as Hugo of Trimberg and Konrad of Mure write compendia that mention Christian as well as ancient authors. With Vincent of Beauvais's *Speculum historiale* (which in turn draws on earlier historians) it becomes common practice to include references to authors in chronicles. In the later Middle Ages, many literary histories are exclusively devoted to Christian authors or to authors of a specific Christian order. With the translation of Diogenes Laertius and its reception in Pseudo-Walter Burley's *Liber de vita et moribus philosophorum,*[14] the treatises on famous men take a wider scope and include ancient as well as medieval Christian authors. Examples from the fourteenth century are the biographers Giovanni Colonna and Guglielmo da Pastrengo,[15] who are already motivated by humanist interests. It seems clear from the development of fourteenth- and fifteenth-century literary history that strong Christian or humanist interests—or tendentiousness, if one prefers—proved to be hindrances for the inclusion of Arab authors. For a Christian writer, the Arabs were too pagan, and from the humanist point of view, they were too barbarous and too medieval to deserve attention.

Another background to the emergence of bio-bibliographies of Arabic authors was the developing genre of histories of scientific disciplines, such as medicine or mathematics, whose heyday will come in the sixteenth century. Medieval Latin examples of historical accounts of medicine that include Arabic authors such as Avicenna and Albucasis come from writers on surgery, notably Guy de Chauliac (d. 1368).[16] Another example is the humanist Coluccio Salutati, whose *De nobilitate legum et medicinae* of 1382 contains a brief historical chapter on Arabs. Salutati inserts the Arabs after Galen and before the Italian physicians Taddeo Alderotti and Pietro d'Abano. He mentions Rhazes's *Continens,* praises Avicenna, the "eminent articulator and organizer" of all previous medicine, and briefly refers to Averroes, Mesue, and Maimonides.[17] When some decades later Giovanni Tortelli (d. 1466), humanist and librarian at the Vatican library,[18] wrote what is probably the first Western history of medicine proper, he is influenced by Salutati. Tortelli's *Liber de medicina et medicis* is particularly informative on Greek and Roman authors, but Tortelli—just as Salutati—also inserts, between Galen and Pietro d'Abano, two Arabic authorities in medicine, Rhazes and Avicenna:

> A long time later, Rhazes, a certain Punic,[19] usefully compiled from these authors *(i.e., Galen and Hippocrates)* in the Arabic language a comprehensive work, which meets the needs of the medical art. But after him, Avicenna of Seville produced in the same language a more lucid and learned work; to a very large extent, he followed Galen after having mustered the writings of the other writers; in the first book on cooling he has called Galen 'the prince of the physicians.'[20]

Salutati's and Tortelli's sentences do not come close to either a biography or a bibliography of Arabic authors, but they are significant as early examples of the placing of Arabic scholars in the historiography of the sciences. It is indicative that they do not mention the eleventh- and twelfth-century Salerno school, or any Montpellier physicians, but directly move on from the Arabs to fourteenth-century Italian physicians, among them the Avicenna commentators Dino del Garbo and Gentile da Foligno. Tortelli, in fact, praises Dino for his commentary on Avicenna.[21] The Italian pride in being part of an unbroken Greek–Arabic–Italian tradition in medicine may well have been a motive for including the Arabs in the history of science. Foresti da Bergamo took a different route; he did not write a history of famous physicians, but a world chronicle. Nevertheless, given that Foresti's group of Arabic authors is predominantly medical, it is very likely that he was influenced by histories of the medical discipline. In fact, there is textual evidence that he did use Tortelli as a source, as we shall see soon.

## FORESTI DA BERGAMO

Foresti da Bergamo's *Supplementum chronicarum* is the first printed source to contain a set of bio-bibliographies of Arabic authors. It first appeared in 1483 in Venice and was printed several times with revisions and additions by the author. Jacopo Filippo Foresti, born of a noble family in 1434, entered the order of the Eremites of St. Augustine in 1451 or 1452, and lived in the convent in Bergamo for most of his life; he died in 1520.[22] In the first edition of the *Supplementum* of 1483, the entries on Arabic scholars are shorter than in the later editions from 1486 onward, since Foresti continued to add material. In total, the *Supplementum* contains biographies of eleven Arabic scholars, which appear in the following sequence between the years 982 AD and 1158 AD: Alfarabi, 'Avedadis,' 'Albaterius,' Serapion, Isaac Israeli, Rhazes, Avicenna, Averroes, Avenzoar, and Mesue—plus king 'Evax' of the first century AD.

This list is conspicuous in several respects. Foresti must have had medical interests, although his biography does not point in this direction—he is also known as the author of a book on famous Christian women and of a *Confessionale*—because all Arabic authors mentioned, with one exception, are introduced with the epithet 'physician': *Albaterius natione arabs medicus, Johannes medicus cognomento Serapion,* and so on.[23] Even Averroes, the best known Arabic philosopher in the Renaissance, is introduced as a physician and praised for his 'beautiful book' *(liber pulcher)* on medicine, the *Colliget.* Averroes's activity as a commentator is only mentioned secondarily, but with much praise:

> Also Averroes the physician, entitled "the commentator," a most renowned philosopher *(philosophus clarissimus),* flourished in the region of Cordoba, the Spanish city, in this period *(i.e., around 1149),* as can be inferred from his book *De caelo.* He wrote so excellently on all books of Aristotle that he deserved the title "the commentator." Further, he composed books *De substantia orbis* and *De sectis.* And he collected a beautiful book on medicine *(i.e., the Colliget)* and on theriac *(i.e., De theriaca)* and on floods *(a mistake: De diluviis is a treatise by Avicenna)* and many other books. For he was the rival and utmost enemy of the physician Avicenna.[24]

That Foresti mentions a rivalry between the *medici* Avicenna and Averroes is another indication of his predominantly medical view of Arabic science. Foresti also apparently draws on a medical source when arguing that Averroes lived around 1149. This date was proposed by Pietro d'Abano in the *Conciliator,*[25] on the basis of Averroes's report in the Long commentary on

*De caelo* (chapter 2.111) that he had visited Cadiz before the year 350 after Muhammad.[26] Also noteworthy about Foresti's entry on Averroes is the fact that he portrays Averroes the philosopher as a great authority on Aristotle without any caveat that Averroes was considered an irreligious philosopher by many humanists and theologians.[27] Foresti bequeaths this positive picture to the Renaissance biographical tradition of Averroes. The picture is slightly blurred, however, by Foresti's story that Averroes poisoned Avicenna, as we shall see below.

On Foresti's list of Arabic scholars is one exceptional scholar who is not portrayed as a physician: Alfarabi, the famous Peripatetic philosopher of tenth-century Bagdad. Foresti's very short entry begins: "Alfarabi, by nation an Arab, the philosopher and astronomer of greatest renown is said to have lived and been active in these times." The careful language indicates that Foresti was not abreast of Alfarabi's work; he adds that "some commentaries *(commentaria)* of his are available also to the Latins."[28] The short printing history of Alfarabi's works in the Renaissance is an indicator that Alfarabi was rarely read in the fifteenth and sixteenth centuries—save by Abraham de Balmes, who translated one treatise by Alfarabi from Hebrew into Latin.[29] Foresti's term "commentaries" points to his being known as the author of an introduction to Aristotle's *Rhetoric*. The attribute "astronomer" is strange, since Alfarabi's only astronomical work extant in Arabic, a commentary on Ptolemy's *Almagest*, was certainly not known to Foresti. Perhaps Foresti mistook Alfarabi for the astronomer Alfragani.

There are three strange names on Foresti's list: Avedadis, Evax, and Albaterius. Foresti describes Avedadis as a philosopher and physician who has written many commentaries on Aristotle, "which today are much appraised." This could be a reference to the Jewish philosopher Avendauth, collaborator of Dominicus Gundisalvi in the translation of Avicenna's *De anima* in twelfth-century Toledo. Avendauth, however, is not a commentator on Aristotle. The reference remains obscure and seems not to have been picked up by anybody else.[30]

King Evax had already appeared in Tortelli's above-mentioned history of medicine, and verbal parallels show that Foresti was in fact influenced by Tortelli's account.[31] Foresti describes Evax as "king of the Arabs, famous philosopher, physician and rhetor" of the first century AD, who is said to have sent a medical book on herbs and stones to the emperor Nero.[32] Evax is not an Arabic author, but a legend of ancient origin. Two letters by King Evax to the Roman emperor Tiberius accompany a Latin text of about 500 AD, titled *De lapidibus et eorum virtutibus*, a translation of an originally Greek treatise of a so-called Damigeron.[33] It added to the confusion that a twelfth-century text, Marbod of Rennes's poem *De lapidibus*, was sometimes cited under the name of Evax, as Konrad Gesner realized in the mid-sixteenth century;[34] in fact, the two treatises on stones by 'Evax' and Marbod often traveled together in manuscripts.[35] As we shall see, the spurious Evax appears very regularly in Renaissance bio-bibliographies of Arabic authors. He thus contributed to an orientalizing picture of Arabic science, as an author writing on his country's riches in stones and herbs. This picture is enhanced by another oriental figure often mentioned in chronicles: Queen Saba, the friend of King Salomo and most learned woman in all disciplines, who, just as Evax, lived in an oriental land blessed with *grandi divitiarum copia*.[36]

The third strange name, 'Albaterius,' poses a riddle. Foresti writes:

> Albaterius, an Arab by birth, a most famous physician, flourished in this time *(i.e., 1070 AD)*, wrote and translated many works, and translated among other things all books of the great Galen into the Arabic language, as is testified by Serapion, the physician, chapter 97.[37]

If any Arabic scholar deserves fame as the principal Graeco-Arabic translator of Galen, it would not be "Albaterius," but Ḥunayn ibn Isḥāq, the Latin Iohannitius, who rendered many dozens of Galen's works from Greek into Arabic.[38] The reference to Serapion's chapter 97, as in the above quotation from the 1486 edition of the *Supplementum,* is flawed. In Serapion's *De simplicibus* 97 there is no mention of "Albaterius." The later Renaissance authors, Symphorien Champier and André Tiraqueau,[39] refer to chapters 197 (as does, in fact, Foresti's 1483 edition) and 85,[40] and there in fact we find the name—as in several other chapters. "Albaterius" here is described as a translator of Galen on simple medicines. The person behind the Latin name turns out to be the eighth-century Graeco-Arabic translator al-Biṭrīq,[41] whom Serapion's rough contemporaries Maimonides and Ibn al-Bayṭār also occasionally credit with a translation of Galen's *On Simple Medicines*.[42] A number of subsequent authors pick up the name "Albaterius" from Foresti: Champier, Tiraqueau, Raffaele Maffei (Volaterranus),[43] Simon de Phares,[44] Gian Giacomo Bartolotti,[45] Pedro Mexía,[46] and Bernardino Baldi. The group of names of famous Arabic authors was thus enriched—through Foresti's reading of Serapion—by al-Biṭrīq, the translator, who does not seem to have authored any works himself.

Finally, Foresti's list is conspicuous for the absence of astrologers such as Alcabitius and Albumasar, whose names were well-known in the Renaissance. This omission may be deliberate or simply a result of Foresti's focus on medical authors. It is unlikely that he had a general aversion toward astrology, given that ancient and medieval Latin astrologers such as Guido Bonatti appear in his *Supplementum*.[47] In any case, since Foresti was a source for many subsequent writers, the result was that the Renaissance biographical tradition was much richer in medical than in astrological Arabic authors.

Foresti's entries on Arabic authors are bio-bibliographies in the full sense: each entry contains information at least on the author's profession, floruit, and works. Despite the initial classification of the Arabic authors as physicians, Foresti also lists their astrological, alchemical, and philosophical writings. He has hardly anything to say on the life of some authors, such as Isaac Israeli, and instead concentrates on the works. When possible, however, he adds information on the author's birthplace and home country, his faith, and the proper spelling or meaning of his name, and occasionally relates an episode of his life. The most elaborate entry is on Avicenna. Foresti's biography of Avicenna testifies to the great interest of Renaissance scholars in Avicenna's life and works and to the difficulties they had in placing him correctly in history. Avicenna from Buḫārā on the Silk Road, who died in 1037 AD, is here made an Andalusian contemporary of Averroes, from Cordoba, who died in 1198 AD:

> Avicenna of Seville *(Hispalensis)*, the most famous of all physicians, a man of the greatest prudence, about whose life there is almost nothing certain, was most famous in these times in the entire world. The physicians Mesue and Avenzoar call him 'Aboalim' *(from Arabic Abū ʿAlī ibn Sīnā)*, whereas the ordinary people *(vulgares)* say that he was the king of Bythinia *(in Asia Minor)* and they relate that Avicenna was poisoned by the physician Averroes, but killed Averroes before he died. Since Avicenna was most learned in all disciplines, he wrote a most excellent book *(i.e., aš-Šifāʾ)*, in which he wanted to cover the entire logic and the entire natural philosophy in many volumes; he then most accurately discussed metaphysics. After he had seen the writings of the other physicians in Arabic and of all physicians in general, he discussed—in a more perspicuous and clearer way than all other physicians—the entire medicine in five books *(i.e., the Canon medicinae)*, in which he calls himself the interpreter of Galen. He also wrote *De viribus cordis*, *De theriaca (this is a confusion with Averroes's treatise)*, *De diluviis*, and also produced a book of songs *(i.e., the medical didactic poem Cantica)*. Also, one ascribes to him *On alchemy* dedicated to Assem the philosopher,[48] and a book on colics.[49]

Foresti is an heir to the above-mentioned medieval tradition that makes Avicenna an Andalusian.[50] Foresti's biography of Avicenna is thus mistaken on several points, and the story that Avicenna, the king of Bythinia, was killed by Averroes is fantastic. Nevertheless, it is characteristic of Foresti that he is cautious when it comes to doubtful information. He warns his readers that on Avicenna's life there are no reliable accounts, and he explicitly attributes the Bythinia and murder stories to *vulgares,* that is, to nonscholarly sources, which contrast with the reliable information on Avicenna's name derived from the *medici* Mesue and Avenzoar. As Joël Chandelier has noted recently, the murder story appears in two manuscript sources dating to the decades before Foresti's *Supplementum;* both relate that Averroes killed Avicenna by poisoning the folios of a book, and one adds that Avicenna had Averroes executed by fire when he found out about the fatal poisoning.[51]

The remainder of Foresti's entry is accurate, especially with regard to Avicenna's works. Avicenna's philosophical summa *aš-Šifā'* is well-described as consisting of logic, natural philosophy, and metaphysics—only the section on mathematics is missing, which was not translated into Latin. And Foresti successfully differentiates between authentic works and those only ascribed to Avicenna. In sum, even though Foresti's biographies of Avicenna and other Arabic authors carry material that is legendary, this was certainly against his purpose: he was looking for certain facts about their lives and works. Most of the material in his entries comes from internal evidence and cross-evidence in the writings of Arabic authors, as Foresti indicates when he occasionally refers to his sources. There is no sign of any additional material derived directly from the Arabic or Hebrew traditions: Foresti's biographies are the result of a distinctly Western investigation into and imagination of the lives and works of Arabic scholars.

It cannot be ruled out that Foresti was drawing on an as yet unidentified source for his biographies, for instance, on a history of medicine that is more comprehensive than Tortelli's.[52] That Foresti's biographies are influenced by Tortelli's account of Arabic medicine is clear from verbal parallels. For example, in Foresti's just-cited entry on Avicenna, the sentence "postquam omnium medicorum scripta vidisset" (after he had seen the writings of the other physicians in Arabic) recalls Tortelli's "cum reliquorum scripta viderit." In the entry on Rhazes, Foresti's description "natione Poenus seu Aphricanus" (by nation Punic or African) is an echo of Tortelli's formulation "Rasis quidam Poenus" (Rhazes, a Punic).[53]

Whatever its origins, the group of Arabic authors presented by Foresti acquired the character of a canon in the Renaissance, in the sense of an often-invoked set of illustrious authors: Isaac Israeli, Serapion, Rhazes, Averroes, Avenzoar, Avicenna, Mesue, plus Evax and Albaterius. These were the authors who appeared in learned historical works of the time,[54] in the form of longer bio-bibliographical entries, such as in Staindel, Schedel,

Champier, Gesner, or in the form of a shorter list, such as in Volaterranus (Maffei),[55] Bartolotti,[56] Mexía—and, without doubt, in other works that have escaped my attention.

## HARTMANN SCHEDEL

Among Foresti's readers was Hartmann Schedel (1440–1514), the author of the most famous Renaissance world chronicle, the *Liber chronicarum,* also known as the Nuremberg chronicle, richly illustrated with woodcuts. The Latin original was printed in July 1493 by Anton Koberger. In December 1493 the well-known German translation appeared, under the title *Das buch der Cronicken und geschichten mit figuren und pildnussen.* Schedel's principal source, not only on Arabic matters, was Foresti's *Supplementum,* and we can witness Schedel at work, adopting Foresti's entries for his own purposes. He chooses the very same authors as Foresti, and again introduces most of them as physicians: Evax, Alfarabi, Isaac Israeli, Serapion, Rhazes, Avicenna, Averroes, Avenzoar, and Mesue.[57]

Two authors from Foresti's list are missing: the obscure 'Avedadis' and 'Albaterius' (al-Biṭrīq), probably because Schedel did not know them from his own reading. He takes over the other entries, quotes some of them verbatim, that is, those on Evax, Alfarabi, and Serapion, but changes others. Some of his changes and additions are of bibliographical character: they concern medical books. Schedel adds information on Isaac Israeli's famous dietetic work *De dietis universalibus et particularibus,* on Rhazes's *Continens,* on Mesue's *Canones,* and on the additions to Mesue's *Grabadin* produced by Pietro d'Abano and Franciscus of Piemont.[58] He identifies Averroes's main medical work by the title *Colliget.* These changes testify to Schedel's bibliographical and medical expertise, which are well-attested. He was both a great collector of manuscripts and books, and a physician. In 1466 he returned to Germany from a three-year journey to Padua as doctor of medicine, and worked as a municipal physician in various German towns.[59]

In one case, it is possible to precisely identify the sources for his addition. After the entries on Avicenna, Averroes, and Avenzoar, Schedel adds the following remark:

> 'Aboali Abinceni' are Arabic nouns referring to the father and grandfather. 'Abon' and 'abin' mean 'son' among the Arabs. For, as everybody says, it is common practice among the Arabs that authors write the name of father and grandfather in titles and do not mention their own, so that they may defer the honor to their forefathers.[60]

As verbal parallels show, the sentence on 'abon' and 'abin' is drawn from Hugh of Siena's commentary on Avicenna's *Canon*, whereas the following sentence on the omission of the proper name in titles derives from the *Canon* commentary of Jacopo da Forli.[61] A similarly scholarly addition is made to the entry on Averroes, where it is said that according to Pietro d'Abano's *Conciliator* Averroes flourished around 1150 AD, and that Giles of Rome mentions in his *Quodlibeta* that he had seen the two sons of Averroes at the court of the Emperor Frederick.[62] For this information, Schedel may have relied on intermediate sources; it nevertheless testifies to his scholarly interest.

Schedel has a liking for episodes and fabulous stories, a well-known feature of his world chronicle.[63] This is most apparent in his biography of Avicenna, which differs from the other entries on Arabic authors in various respects: it is much longer, it is much more independent of Foresti da Bergamo, and it discusses a number of legendary traditions on Avicenna's life.[64] Schedel opens the vita with a sentence of praise drawn from Foresti, makes Avicenna a ruler not of Sevilla, but of Cordoba—probably because he thought Avicenna to be a contemporary of Averroes, as we shall see—but soon departs from his source to include more colorful material on Avicenna's life and character:

> He nevertheless invested much labor in the practice of medicine, making it happen that, as I have learned from previous authors *(a prioribus)*, a hospital residence was built in his city, where innumerous sick were laid, which he visited himself. He was of brown complexion and cheerful, since he composed a book of songs, by knocking and singing. He died still very young, as is said, for he did not become fifty.[65]

The hospital story is fictitious, its source unknown. The information about Avicenna's cheerful character was possibly invented by Schedel himself, as a curious extrapolation from the fact that Avicenna is known as the author of the *Cantica (al-Urğūza fī t-tibb)*—a book that Schedel seems to know only by title, since it is a dry didactic poem covering the entire field of medicine in 1,326 double verses. Given Schedel's penchant for lively material, it is remarkable that, with good instinct, he left out Foresti's murder story, and only included a remark on Avicenna's untimely death. In the remainder of the text, Schedel proceeds to discuss at length the dates of Avicenna's life, including a refutation of the story that he exchanged letters with Augustine; in the end, he returns to Foresti's bibliographical remarks on *aš-Šifā'* and the *Canon*.

Schedel's remarks about the alleged correspondence between Avicenna and Augustine again attest to his bibliographical knowledge. For, indeed, a pseudo-Avicennian letter to Augustine of some ten lines is preserved in at least two manuscripts of the late fifteenth and the sixteenth century.[66] Schedel writes, "people say that there exist mutual letters between the two,"[67] and does not show any awareness of the content of the extant letter, but instead attacks the mere possibility of a correspondence on the basis of chronological evidence from the *Canon medicinae* and by arguing that among Augustine's extant letters, none are addressed to princes.[68] If Avicenna was not a contemporary of Augustine, when did he live? Schedel argues:

> Hence it is believed that he lived in the time of Averroes or a bit before him. Algazel and Alfarabi also seem to have been contemporaries of Avicenna; they also lived in the time of Averroes. Avicenna therefore flourished when the Roman empire was declining and when the Arabs were occupying Spain.[69]

Here as before, Hartmann Schedel's treatment of Arabic authors is a curious mixture of scholarly investigation and colorful invention. Schedel, with full justification, mounts arguments against the authenticity of the correspondence with Augustine, and he is very concerned about the correct dating of Avicenna's lifetime—albeit without success. In Schedel's account, Avicenna, Alfarabi, and Algazel, who in fact all lived in different countries and times, become alter egos of Averroes: Andalusians of the mid-twelfth century. The only remedy against this Hispanicizing confusion, which is particularly evident in Schedel's description of Arabic authors, is the advent of new biographical information directly from Arabic sources, which will change the discussion in the early sixteenth century, as we shall see. In spite of Schedel's interest in matters of authenticity and chronology, his entry on Avicenna is more legendary than Foresti's. It is not legendary in an orientalizing sense: Schedel does not stress any typically oriental attributes, nor does he mention the religion of Islam in his biographies of Arabic scholars. The woodcuts that accompany the biographies of Arabic authors in his chronicle all show representations of Western scholars. But Schedel is interested in the anecdotal biographical detail: Avicenna's kingship, his hospital, and his joyful character. Both this fantastic side and the scholarly passages in Schedel's biographies found their imitators.

## SIMON DE PHARES

In comparison with Simon de Phares, Schedel appears a most sober and reliable biographer. Simon de Phares is the author of the *Recueil des plus celebres astrologues,* the first Western history of the discipline of astrology extant. It is in this book that biographies of Arabic astrologers and astronomers appear for the first time. Neither Foresti nor Schedel had mentioned any astrologer or astronomer, save for Foresti's strange qualification of Alfarabi as *astronomus.* The *Recueil des plus celebres astrologues* was written between 1494 and 1498. In the years before, Simon (ca. 1445–ca. 1500), one of the most famous and successful astrologers of his day, had been accused and condemned of divinatory practices, and his large library in Lyon had been confiscated.[70] The *Recueil,* which was not printed in the Renaissance, was meant to justify astrology by surveying the lives and accomplishments of the famous astrologers of history. Foresti da Bergamo's canon of learned Arabic authors has also left its stamp on Simon de Phares's book, even though Simon does not share Foresti's interest in medicine: Simon draws material from Foresti's biographies of Evax, Alfarabi, Isaac Israeli, Albaterius, Serapion, Rhazes, Avicenna, Avenzoar, Averroes, and Mesue—that is, the entire range of Foresti's Arabic authors (with the exception of 'Avedadis').[71] Not one of these scholars was, in fact, an astrologer; Avicenna and Alfarabi were even opponents of astrology. The astrological writings and doctrines attributed to these Arabic authors are Simon's invention. He was willing to create his own history as long as it served his case.

Simon's enormous reading in scientific literature enabled him to add a considerable number of authors to Foresti's group, most of them, of course, astrologers: Abraham ibn Ezra, Albohali, Albubater, Albucasis, Albumasar, Alcabitius, Alcoarismi, Alfraganus, Alkindi, Alpetragius, Almansor, Azarchel, Jafar, Haly filius Abbas, Haly filius Abenragel, Messahalah, Omar Tiberiadis, Thebit ben Corat, and Zahel.[72] To these are added several dozen Arabic names of 'other' authors—some of which simply are different names for the astrologers just mentioned, which are derived from alternative transliterations of the Arabic or from other parts of the Arabic name. The result of this technique is that in Simon de Phares's *Recueil* astrology assumes the appearance of a thoroughly Arabicized art. It is noteworthy that Simon's interest is clearly focused on astrology and not on other sciences: there are no entries, for instance, on the physicians Johannitius or Maimonides, nor on the alchemist Geber, nor even on the astronomers Albategnius and Haly Rodoan, and Simon is very short on the few astronomers he mentions: Alfraganus, Alpetragius, Azarchel, and Thebit ben Corat.[73] It would therefore be misleading to conceive of Simon de Phares as "the first historiographer of astronomy."[74] Simon's *Recueil* is very much a biographical history of the discipline of astrology, and in this it differs clearly from the history of medicine by

Tortelli or the history of astronomy and mathematics by Bernardino Baldi, to which we will turn later.

In Simon's *Recueil*, we encounter for the first time a biography of the famous astrologer Albumasar, whom Simon splits up into different persons.

> Albumasar the Great flourished in Athens in this time, as some say, and with him was Plusardonus, someone called Hermes and Guellius, since in Athens the science of astrology was much practiced. This Albumasar was the teacher of the above-mentioned persons, and he was the one who produced the *Great Introduction* and the *Treatise of the Great Conjunctions,* which our ignorant calumniators take pains to calumniate for something which Albumasar has only quoted from others *(namely, on the animate nature of the planets).*[75]

Simon's source for the legendary story that Albumasar lived in Athens is the thirteenth-century astrologer Guido Bonatti. The motive is not, as one might think, to disguise Albumasar's Arabic origin, since Simon takes any chance to emphasize the Arabic outlook of astrology in other cases, but to invest Albumasar, one of his principal sources, with additional authority. As Jean-Patrice Boudet has noted, the name *Guellius* is likewise drawn from Bonatti, whereas *Plusardonus* and the so-called *Hermes* are completely fictitious.[76] The unwarranted calumniation that Albumasar would teach the animation of heaven is drawn from Pseudo-Albertus Magnus's *Speculum astronomiae.* There is a grain of historical truth embedded in this biography: the attribution of *Introductorium maius* and *De magnis coniunctionibus* to Albumasar, two works of major size. After this passage, Simon remarks that "there are several authors with this name," referring to "Albumazar Abbalachi," "Jaffar," and "Albumazar Tricas," and proceeds to attribute astrological works to these authors. The first two names are transliterations of parts of Albumasar's full name (Abū Maʿšar Ǧaʿfar ibn Muḥammad ibn ʿUmar al-Balḫī), the third comes from Bonatti.[77] Here again, motivated by his desire to make astrology as authoritative as possible, Simon produces a text that combines impressive learning with excessive fantasy.

For all its fictitious material, Simon's *Recueil* is a reminder that the group of authors that Foresti selected and that he bequeathed to the biographical tradition is greatly limited and not fully representative of the current of Arabic learning in the Renaissance. The Foresti tradition does not cover many Arabic names that were well-known among the astronomers, astrologers, and mathematicians of the Renaissance.

## SYMPHORIEN CHAMPIER

The biographies of Arabic authors composed by Foresti and Schedel were written for world chronicles. In his *Chronica,* published posthumously in 1516, Johannes Naukler, chancellor of Tübingen University, follows this tradition and includes entries on Avicenna and Averroes in chapter 'generatio 39,' which extends from 1140 to 1170 AD; Naukler draws directly on Foresti da Bergamo.[78] At the turn of the sixteenth century, there appear the first treatises on famous men that include Arabic authors, by Johann Staindel and Symphorien Champier. Johann Staindel, canon at the cathedral of Passau, writes an addition to Johannes Trithemius's book on ecclesiastical authors, with a focus on philosophers and poets. This *Suppletio virorum illustrium,* composed in 1497 and extant only in manuscript, contains biographies on Evax, Alfarabi, Serapion, Isaac Israeli, Rhazes, Avicenna, Averroes, and Mesue, that is, the entire range of Foresti's authors except Albaterius and Avenzoar.[79] Staindel arranges the authors in rough chronological sequence from Moses to Ulbertino Pusculo, the fifteenth-century neo-Latin poet. Staindel's sole source on Arabic authors is Foresti. His primary interest is in bibliography: he shortens Foresti's biographies, but usually lists all titles mentioned by Foresti—except for Avicenna's philosophical works, which he omits. Very occasionally, Staindel adds bibliographical knowledge of his own, for instance, on Alfarabi. Foresti apparently did not know any book titles of the philosopher Alfarabi, nor did Schedel, but Staindel lists *De ortu scientiarum* and *De divisione scientiarum.*[80] And Staindel also adds some titles by Isaac Israeli and Rhazes. But his overall contribution to the bio-bibliography of Arabic authors is very modest.

Symphorien Champier's treatise, in contrast, is much more original. Champier is an enigmatic and belligerent character, who will appear several times in this study. Around 1506 in Lyon, he publishes a collection of biographies of medical authors, titled *De medicine claris scriptoribus.*[81] Champier will later, in the 1520s and 1530s, become one of the main polemicists against Arabic scientific traditions in the controversy over the value of Arabic medicine,[82] and he will take part in the pharmacological discussion of the time.[83] In this earlier period of his life, he is less biased by anti-Arab attitudes. He has just established himself as a physician in Lyon, after several years of studying in Montpellier, and has gained a wider reputation by publishing two moralizing essays in the tradition of Sebastian Brant, the *Nefs des Princes* of 1502 and *Nef des Dames* of 1503.[84] Champier's biographies of medical authors are not arranged chronologically, as were Johann Staindel's (and, naturally, the biographies inserted in chronicles), but in five curious sections: on kings, philosophers, ecclesiastical men, Italians, and, in the fifth category, Frenchmen, Spaniards, Germans, and Englishmen famous in medicine.[85] The five sections themselves are arranged in roughly chronological order. Champier does not, however, seem to be interested in chronology, in contrast to many other Renaissance writers on Arabic authors. When he adds new biographies to the tradition, or rather to his sources Foresti and Schedel, he does not mention any dates.

From the range of authors described in his treatise, it is apparent that Champier has adopted many entries from Foresti, omitting Alfarabi and adding Sabid, Haly filius Abbas, and Maimonides. Champier mentions as kings: Evax, Sabid, Avicenna, and Mesue—as the alleged nephew of a king of Damascus. Among the philosopher-physicians he groups Rhazes, Haly filius Abbas, Albaterius, Isaac Israeli, Serapion, Avenzoar, Averroes, and Maimonides.[86]

Some of these entries are mere copies from the *Supplementum chronicarum* with minimal changes, such as those on Evax and Mesue. Champier improves upon Foresti's bibliography, sometimes by drawing on Schedel, for instance, when he mentions the title of Averroes's book *Colliget* and adds Franciscus of Piemont as author of the addenda to Mesue's *Grabadin*, sometimes by adding from his own reading, for instance, in the occult tradition, when he refers to the title of the alchemical work *De sublimationibus* (spuriously) attributed to Rhazes.[87]

If in all this Champier differs little from Foresti, Schedel, Naukler, and Staindel, we can observe him twice inserting passages of distinctively rhetorical character: a colorful description of the riches of Arabia when treating King Sabid and a polemical attack on Muhammad on the occasion of Avicenna's mentioning of the Islamic prophet in his *Metaphysics*. Both passages point to Champier's interest, not shared by the previous bio-bibliographers, in the specific cultural surrounding of Arabic science, in the Orient and in Islam.

The first passage forms part of the entry on the obscure *Sabid arabum rex*,[88] which follows the entry on the Arabian king Evax. Champier groups the ancient kings of Arabia together with *noster Avicenna* and inquires into the cultural background of the enormous success of Arabic science:

> Arabia has always produced men excellent in medicine, of many-sided and admirable minds, such as there were king Evax, admirable for his botanical science, and also our Avicenna, the prince of Cordoba and man of the greatest prudence, and Sabid himself, famous physician and king of the Arabs. I would like to say something about this region because it has nations whose power and riches the other nations admire,

being stunned by its plenitude and abundance. I should not omit that, since <that region> is much embellished by sacred things, it does not have a part which does not smell of thyme or myrrh.[89]

Then follows a list of things with which the region is gifted: the mountains have gold, the rivers silver, their benches carry thyme and various fragrant herbs, and their inhabitants possess enormous fortunes and wear clothes of silk or gold. On the one hand, here Champier simply delivers a traditional piece of praise for the land of Queen Saba.[90] On the other hand, his juxtaposition of praise for Arabic physicians and praise for Arabia's material riches is extraordinary. The implication is that Arabia's impressive series of scholars can be explained by the fact that the country's material conditions were excellent[91]—this being a type of explanation for which the locus classicus is Aristotle's *Metaphysics*, chapter 1.1 on the development of the mathematical arts in Egypt.[92] Champier is aware of the impressive scientific output of Arabic countries. His later anti-Arab polemics may well be an attempt to regain cultural superiority for his own culture.

The second conspicuous passage in Champier's biographies appears in the entry on Avicenna. Here Champier inserts a quotation from Avicenna's *Metaphysics*, with significant changes that I have indicated with square brackets:

> He wrote a *Metaphysics* in which he confesses himself to be a Muslim *(machometista)* and seems to have contempt for the law of Muhammad, if it deserves this name, since he says in the ninth treatise at the beginning when speaking about bliss: "You know the delights of the body and the joy, what they are, since our law, which was given by Muhammad, shows the disposition of bliss and of misery with respect to the body." This he says. And later he says in the same chapter: "And there is another promise which is understood by the intellect and by demonstrative argument with which prophecy agrees: this is the bliss of the souls or <their> misery after leaving their bodies. The wise theologians, however, were much more prone to follow the bliss [of the bodies which was given by Muhammad]." This Avicenna says. Note that he affirms that Muhammad was not a prophet and that he had not given intellectual bliss to the Arabs and to his people but only bodily and sensual bliss *(felicitatem corporalem et voluptuosam)*. That most pestilential man <i.e., Muhammad> always was a carnal and stupid dragon, who did not seek the well-being of his home country but his own glory.[93]

The passages Champier refers to come from Avicenna's *Philosophia prima,* chapter 9.7.[94] Champier is correct with his statement that Avicenna describes himself as a Muslim in this passage, and it is also true that Avicenna refers to Muhammad's doctrine of paradise, which promises bodily pleasures. Attacks on the Islamic concept of paradise (Quran 47, 55, and 56) have a long tradition in Christian polemics.[95] Moreover, even though it is a mistake to say that Avicenna "condemned" Islamic law, it is true that he relegated the prophet's laying of religious law to a practical science and a political necessity. The prophet is needed as a legislator for the common people who do not understand the philosophical truth.[96]

Champier is wrong, however, when it comes to the prophethood of Muhammad, which Avicenna would not deny. If in this case Champier simply misunderstands the Avicennian text, he deliberately, it seems, misquotes the second sentence in which Avicenna says that the intellectual concept of bliss is the specific object of the *sapientes theologi (al-ḥukamā' al-ilāhiyūn),* that is, the metaphysicians; Avicenna here refers to his philosophical project of a metaphysics of the rational soul.[97] Champier changes the Avicennian sentence: "to follow the bliss of the souls rather than that of the body," into: "to follow the bliss of the bodies which was given by Muhammad,"[98] and thus completely distorts the sense. He therefore manages to add a castigation of Islamic theologians allegedly based on the high authority of Avicenna.[99] We thus witness Champier's being fascinated and, at the same time, appalled by oriental culture—an ambiguous attitude that perhaps accounts for the vehemence with which, later in life, he attacks the value of Arabic science.

## LEO AFRICANUS

The biographical tradition of Arabic scholars from Foresti to Champier would have looked very different if it had been based on more material from Arabic. How different, is well-illustrated by a text that almost changed the biographical tradition radically, since it was based entirely on Arabic sources: the treatise *De viris quibusdam illustribus apud arabes* by Leo Africanus, written in the 1520s. But Leo's text survived in only one manuscript and seems to have remained unknown until Gerhard Johannes Voss cited the work in 1658 and until it was first printed by Johann Heinrich Hottinger in 1664.[100]

Leo Africanus is a fascinating figure, a wanderer between cultural worlds. His original Arabic name is al-Ḥasan ibn Muḥammad al-Wazzān az-Zayyātī. He was born in Granada, between 1486 and 1488, into a

Muslim family that had to emigrate to Fez after the city's fall (in 1497 at the latest). After extensive diplomatic and commercial travels to Istanbul, Timbuktu, Egypt, Sudan, and Mekka between 1507 and 1518, often in the service of the sultan of Fez, he was captured on the Mediterranean Sea by Christian pirates in the summer of 1518 and given to Pope Leo X as a present. He converted to Catholicism in 1520 in a public ceremony as *Johannes Leo de Medicis* and was released from prison. In Rome, he began to learn Latin and Italian and got in contact with scholars, notably with Cardinal Egidio da Viterbo, one of his godfathers. The literary output of his few years in Italy is considerable. Apart from the treatise on famous men, he also composed, together with the Hebrew–Latin translator Jacob Mantino, an Arabic–Hebrew–Latin vocabulary, a metrical treatise in Latin, and the famous *Descrittione dell'Africa* in Italian. After the sack of Rome in 1527, he returned to Tunis and presumably reconverted. There is no trace of Leo after 1532.[101]

The Florence manuscript of Leo's biographical treatise gives the year 1527 in the colophon (f. 69v): *ex archetypo descriptum anno salutis MDXXVII.* It is not clear whether Leo himself produced this copy, but since on ff. 54r–61v it contains Leo's metrical treatise, which interrupts the biographical work, taking the place of two and a half biographies that are now lost, it seems likely that a scribe rather than the author himself confused the leaves of the archetype (if the confusion was not caused by the binder). The confusion may have occurred because Lisānaddīn ibn al-Ḥaṭīb and Faḥraddīn ar-Rāzī, who was also called ibn al-Ḥaṭīb ("the son of the preacher"), were thought to be the same person, and the copyist conflated the two entries, omitted two authors in between, and put Leo's treatise *De arte metrica* in its place. There is one longer passage that was not included in Hottinger's editio princeps of 1664: a story about how "Mesuah christianus, idest Mesue" healed a peasant suffering from some kind of inflammation of his genitals.[102] The story seems to have been too delicate for Hottinger.

The full title of the work reads: *De viris quibusdam illustribus apud arabes per Io<annem> Leonem Affricanum ex ea lingua in maternam traductis.* This is a twisted title, which points to Leo's difficulties with the newly acquired Latin language: "On some famous men among the Arabs, who are translated by Johannes Leo Affricanus from that language into the mother tongue," that is, from Arabic into Latin. Instead of *ex ea lingua* ("from that language") a more comprehensible expression would have been *ex lingua eorum* ("from their, i.e., the Arabs', language"). The Latin of the treatise is poor—poorer, in fact, than that of most of his medieval predecessors as translators.

The treatise contains in rough chronological order twenty-eight lives of Arabic scholars, plus five biographies of Hebrew scholars at the end. Among the Arabic scholars treated are many who were known in the Latin West through translations of their works: the physicians Mesue, Pseudo-Mesue, Rhazes, Albucasis, Avicenna, and Avenzoar; the philosophers and theologians Alfarabi, Algazel, Avempace, Ibn Ṭufayl, and Averroes.[103] But all other authors must have been completely new to the Latin readers of Leo's treatise. Among them are physicians (Abū l-Ḥasan ibn at-Tilmīḏ,[104] ʿAlī ibn Rabban aṭ-Ṭabarī, Abū Bakr ibn Zuhr, the son of Avenzoar,[105] and Lisānaddīn ibn al-Ḥaṭīb),[106] theologians and philosophers (al-Ašʿarī, al-Bāqillānī, Faḥraddīn ar-Rāzī, Saʿdaddīn at-Taftāzānī, and Ibn Ḥaldūn),[107] astronomers, geographers, and botantists (Abū l-Ḥusayn ʿAbd ar-Raḥmān aṣ-Ṣūfī, aš-Šarīf al-Idrīsī, Ibn al-Bayṭār, Naṣīraddīn aṭ-Ṭūsī), one astrologer (Ibn Haydūr at-Tādilī),[108] and one poet (aṭ-Ṭuġraʾī).[109] From Leo's table of contents, we know the names of the two scholars whose biographies were omitted by the copyist: "Ibnu Banna" and "Ibnu hudeil," who are probably the astronomer and mathematician Ibn al-Bannāʾ al-Marrākušī and the military scientist Ibn Huḏayl al-Andalusī.

It is apparent from these names that Leo's set of authors is distinctively nonoccidental, despite the treatise's very conventional Latin title, *De viris quibusdam illustribus*. The treatise presents only a selection of authors, as the title indicates; we know from his other writings that Leo, as a learned Arab diplomat, knew a much broader range of theologians, historians, and poets.[110] The selection is marked by the predominance of physicians and philosophers—which is remarkable since we do not know of any medical or philosophical writings by Leo—and of North African and Spanish authors—which is not surprising since he comes from that region. Most of his authors lived in the eleventh to fifteenth centuries, that is, in what is often termed the postclassical period of Islamic culture. A number of earlier authors, such as Alkindi, Ḥunayn ibn Isḥāq, and Albumasar do not appear.

What Leo Africanus offered his readers, was a wealth of information on Arabic scholars. But compared to the Arabic tradition Leo was drawing on, his *De viris quibusdam illustribus* is not much more than a miniature version of a biographical work. Biographical dictionaries were a particularly flourishing genre of Arabic culture, which testifies to the enormous cultural activity of the major Islamic cities Baghdad, Damascus, or Cordoba. The biographical dictionary of al-Ḥaṭīb al-Baġdādī (d. 1071), for example, contains about 7,800 entries on the learned men only of Baghdad. The famous biographical dictionary of Ibn Ḥallikān (d. 1282), which is also cited by Leo Africanus, is highly selective, but still contains more than 850 entries.[111] In contrast, the Latin biographical tradition on famous men is much more restricted in scope and ambition. Diogenes Laertius's *Lives and Opinions of Eminent Philosophers* comprises 82 biographies, Jerome's *De viris illustribus,* 135. The difference between the Arabic and the Latin biographical traditions rests not only on rules of genres but also on cultural and social differences: the cities of the Islamic world were of much greater size and hosted a much more numerous educated elite than their counterparts in medieval and Renaissance Europe.

Some of the sources that Leo Africanus has used are named in the text. The most often quoted source is the bio-bibliographer Ibn Ǧulǧul (d. after 994). Since Ibn Ǧulǧul is cited on authors that lived after his death—Avicenna, Ibn Bāǧǧa, and Abū Bakr ibn Zuhr—it seems that Leo here confuses Ibn Ǧulǧul with other bio-bibliographers.[112] I have noted ten other sources quoted in Leo's treatise, among them historians of Andalusia and of Baghdad.[113] Leo's quotations from the Arabic bio-bibliographers are difficult to trace; this can mean either that he knew sources unknown to us or that he was an uninformed and inventive writer.[114] Since Leo Africanus composed this work in Rome, the peculiar usage of sources may reflect the fact that he had very little Arabic material at hand. On the other hand, the material he offers from Arabic bio-bibliographers is so comprehensive that it seems impossible that Leo is recalling everything from memory.

Leo Africanus's *De viris* is not simply an Arabic biographical treatise in Latin garb; it is significantly impregnated by the Latin tradition. On the surface, this is obvious in Leo's usage of standard Latin names for three authors who were very influential in the Renaissance: Avicenna, Averroes, and Mesue. It may therefore be a reaction to the interests of his Latin surrounding that Leo's entries on these three authors are particularly comprehensive. Also, what seems to be Leo's "invention" of a second Mesue, can be explained by the fact that the Mesue whom Leo knew from his Arabic sources, that is, ibn Māsawayh (d. 857 AD), had not written "very useful works on drinkable substances and another book on the composition of medicaments" *(opera utilissima in rebus potabilibus, alium quoque librum de componendis medicinis).* The Arabic Mesue of the ninth century had not written pharmacological works like the *Opera omnia* attributed to Mesue in the Latin world, which had originally been produced by an anonymous Arabic author (or several authors) in the eleventh or twelfth century.[115] Leo recognized the difference between ibn Māsawayh and the Latin *Opera Mesue* and concluded that there had lived a second author of this name. Historically, it is unlikely that the second author was also called ibn Māsawayh, but Leo's distinction between two 'Mesues' is fully justified.

The influence of the Latin tradition is also apparent in the fact that Leo Africanus treats five authors of Foresti da Bergamo's canon: Alfarabi, Rhazes, Avicenna, Avenzoar, Averroes, and Mesue. There are reasons for Leo's silence on the other four: Evax is a legendary author of antiquity, "Albaterius" a hardly known translator, Serapion is a thirteenth-century pseudoepigraph, and the author Isaac Israeli is apparently conflated by Leo with the other Qayrawān authors Isḥāq ibn ʿImrān and Ibn al-Ġazzār (under the entries "Isah filius erram" and "Emran filius Isah"). Hence, there is basic agreement between Foresti and Leo on the choice of the most important Arabic authors, who are all physicians and philosophers. Like Foresti, Leo does not treat astronomers, astrologers, and mathematicians. The most significant element that Leo adds to this canon is the Muslim theologians.

Leo occasionally inserts sentences of his own into the text, introduced with the phrase *dixit interpres* ("the translator has said"), which comment on his activity as a translator or mention poems or books he likes particularly. In the entry on the famous Sunni theologian al-Ašʿarī, for instance, Leo apologizes for not laying out al-Ašʿarī's theology because this would require a longer excursus into the development of theological doctrine. Ašʿarī's works, he adds, are still read by pupils in the Islamic schools. In the entry on Algazel, he praises the poems that he still knows by heart, but refrains from the difficult task of translation.[116] Leo Africanus appears as a learned man, with a gusto for books, theology, and poetry, but not for science.

It is instructive to compare the content of Leo's biographies of Avicenna and Averroes with those of the Foresti tradition. In the entry on Avicenna, Leo first takes issue with a feature of the Western biographical tradition: the meaning of Avicenna's honorary title and the question of whether he was a king of Cordoba. Leo explains that the Arabic term *rahis* means a superior or noble person, a ruler or prince, and that it was a common epithet of Avicenna.[117] And he proceeds:

> And what is said among those who speak the Latin language, that Avicenna was the king of Cordoba, is a lie *(mendacium)*, because it is not found among the biographers of Avicenna in the Latin language. I would not have elaborated upon this topic if I had not seen what had been written by Jacopo da Forli <on the issue>, who did not say anything with many words. Nevertheless, there is a reason in this which was not recognized by him: the fault lies mainly with the person who translated the book by Avicenna into Latin and has interpreted it, because he considered the term *Rahis* to mean nothing other than *Princeps;* and this

poor person, if I may say so, ignorant of Arabic grammar, did not understand or know terms common <in the Arabic language>.[118]

Jacopo da Forli, commentator on Avicenna's *Canon* (d. 1414 in Padua) is
the only Western author mentioned in Leo's entire treatise. Jacopo does
indeed discuss the meaning of Avicenna's name, but he does so without
reference to his kingship.[119] That Avicenna was the king of Cordoba, Leo
could have read in Schedel, Champier, Volaterranus,[120] Ficino,[121] and
others. Note that Leo does not simply refute this legend by giving Avicenna's true birthplace and profession: he argues that the legend is a deliberate lie, since there is no biographical source in Latin that would support
Avicenna's kingship. Given Leo's restricted knowledge of Latin, one wonders whether this statement relies on firsthand reading or on hearsay. Leo
then hits the truth by pointing to an unrecognized reason for the legend:
Gerard of Cremona's misleading translation *princeps* for *ra'īs*.[122]

The remainder of Leo's biography of Avicenna, which covers five pages,
is very different from the Latin biographies, which, as we recall, tend to
make Avicenna an Andalusian involved in legendary struggles with the
rival physician Averroes. Curiously enough, Leo's biography is not reliable
either. He gives the city of *hamadan in persia* as Avicenna's birthplace, but
Hamadān is where Avicenna died, not where he was born. That Avicenna's
father cared much for Avicenna's education is true, but it is wrong that he
sent the very young Avicenna *ad civitatem bagdad*—a city where Avicenna
had never been. Nor did he die in prison, as Leo presumes, following a legendary tradition of Avicenna's life reported in Ibn Ḥallikān's dictionary,[123]
but from falling sick on a military campaign with his master ʿAlā ad-
Dawla. The Arabic biographical tradition on Avicenna, just as present-day
knowledge of his life, is entirely dependent on Avicenna's autobiography
and on the biography of his pupil Ǧūzǧānī[124]—sources that Leo does not
know directly. There is a dim reflection of the autobiography in Leo's description of Avicenna's education in geometry, mathematics, philosophy,
and medicine, but the main bulk of Leo's text is a collection of unreliable
anecdotes from later Arabic sources. Leo gives colorful descriptions of Avicenna's encounters with ordinary people and with the high and mighty, of
his witty replies and astonishing healing successes. Leo Africanus therefore chose badly among his sources. Ironically, the Latin biographies of
Avicenna published by Boneto Locatello in 1505 and Nicolò Massa (before 1544), which both are based on the autobiography–biography complex of sources, will provide the Latin world with material far superior to
Leo's, the only Arab who contributed to the Renaissance biography of
Avicenna.

Leo's biography of Averroes, however, is much more reliable than that of Avicenna: it contains legendary accounts, but also much historical material. Himself a native of Andalusia, Leo had a much better grasp on the sources for the Andalusian Averroes than for Avicenna from Transoxania. It is particularly noteworthy that Leo cites two authors who still count among our principal sources on Averroes's life: Ibn ʿAbd al-Malik al-Anṣārī al-Marrākušī (d. 1303–1304),[125] and Ibn al-Abbār (d. 1260).[126] From Leo's ten-page biography, the Latin West could have learned many authentic details about Averroes's life: about his prestigious family and famous grandfather, the jurist, about his own career as judge of Cordoba, his being accused of heresy and exile to a city with a Jewish population near Cordoba, his return to grace with the ruler and his death in Marrakesh. The intellectual portrait drawn by Leo is not primarily that of the commentator or the physician, though both activities are mentioned, but of "the great and most excellent jurist and theologian" *(tum legista cum theologus magnus ac excellentissimus).*[127] Leo explains, following his source Ibn al-Abbār, that Averroes conceived of himself as a faithful Malikite jurist and Ascharite theologian, but that "according to the true opinion he imitated Aristotle," and that this is why he was accused of heresy by his rivals.[128] Note the difference between the Western image of the *commentator* and the Andalusian image of the judge and theologian, who is secretly a follower of Aristotle. Leo ends the vita with a personal remark:

> Finally Averroes died in the city of Marrakesh and was buried outside the gate of the leather workers in the time of the priest and king Maumed Mansor of Marrakesh *(i.e., the ruler Abū Yūsuf Yaʿqūb al-Manṣūr)* in the year 603 after the Hijra. The translator has said that he has seen his tomb and epitaph.[129]

This piece of personal evidence ties in well with the historical and sober character of Leo's vita of Averroes. Leo does not refrain, however, from enhancing this vita, as well, with fanciful stories. The colorful anecdotes he recites are meant to illustrate Averroes's virtues: his forbearance with his rivals, his generosity, his abstinence from otiosity, and his excellence as a judge. Leo's partly legendary, partly historical account of Averroes's life was often cited by later Western scholars after Hottinger edited the text in 1664. This is the biographical tradition on Averroes against which Salomon Munk and Ernest Renan protested when they first compared several Arabic sources on Averroes's life in the mid-nineteenth century.[130]

Leo's treatise thus emerges as a fusion of Arabic and Latin traditions. It was by far the most detailed source on Arabic scholars in the Renaissance, which by some mischance did not reach the general learned discourse of the time. It is likely that his *De viris quibusdam illustribus* would have found attentive readers; this is suggested by the great reception enjoyed by Leo's *Descrittione dell'Affrica*, and by the attention Leo found among contemporary philologists such as Johann Albrecht von Widmanstetter, Jakob Ziegler, Cardinal Egidio da Viterbo, and the Jewish scholar Jacob Mantino.[131] The kind of information they would have gained may not always have pleased their scholarly minds. Leo Africanus, as we have seen, has a distinct liking for the colorful, hagiographical, and juicy story: about exemplary deeds, witty replies, curious sexual diseases, and financial successes. This aspect of Leo's *Vitae*, however, would have fit neatly into the tradition of Schedel and other Latin biographers who likewise attempted to satisfy the reader's curiosity.

## AVICENNA'S BIOGRAPHY BETWEEN CHAMPIER AND GESNER

In order to illustrate the alternatives among which Western authors were able to choose when writing on the lives of the Arabic scholars, I briefly turn to the fortuna of Avicenna's Latin biography between Symphorien Champier and Konrad Gesner, that is, in the first half of the sixteenth century. Four steps in this story will be mentioned: the Locatello *Canon* edition of 1505, Calphurnius's vita of 1522, Massa's translation of the combined autobiography/biography published in 1544, and Milich's *Oratio de vita Avicennae* of 1550.

In Venice 1505, Boneto Locatello, priest of Bergamo and an experienced printer of the publishing house of Octavianus Scotus and his heirs, published Avicenna's *Canon* together with a one-page life of Avicenna 'which some pupil of this Arab had left among his writings.'[132] It has been suspected that the vita is based on an Arabic source,[133] and, in fact, there can be no doubt that it is based directly or indirectly on the autobiography/biography complex. The vita contains several authentic pieces of information from the autobiography: that Avicenna grew up with one brother in *mageri*, which is a distorted Latin rendering of Buḫārā, that he was a wunderkind pupil, so that he knew the Quran by heart by the age of ten and completed his education in all disciplines including medicine by eighteen. The vita even informs readers of Avicenna's well-known methods of thought enhancement, that is, in Avicenna's words, praying, drinking wine, and dreaming, which in the Latin vita reappear as wine, music, and singing. From Avicenna's childhood the vita jumps right to his death, because, as the reader is informed, everything else would take too long to narrate. Avicenna's death resulting from a colic is then described in extended medical detail, for which the end of Ğūzğānī's biography is the ultimate source.[134] The year of Avicenna's death is correctly given as 428 after the Hijra. Some bits of information get distorted in this Latin version of Avicenna's life: his father, a village governor, is turned into a physician, Avicenna is said to have lived *circa Damascum,* and *Amadem,* that is, Hamaḏān, his place of death, becomes the name of a king. But, nevertheless, for the first time authentic material from Avicenna's autobiography has reached the Latin West—material that contrasts starkly with the Andalusian image of Avicenna drawn in most Western sources.

The translator of this extraordinary vita is unknown. The translation may have been produced in the Orient, just as the contemporary Arabic–Latin translations by Andrea Alpago who worked at the Venetian embassy in Damascus, or in northern Italy with the help of Arabs who came to Venice on diplomatic or commercial journeys. The vita was first reprinted in 1507 in Venice by the Paganini family of printers, who three decades later would publish the first Quran in Arabic letters.[135]

Seventeen years after the Locatello edition, Franciscus Calphurnius of Vendôme published a vita of Avicenna, which appeared in a *Canon* edition produced by Symphorin Champier in Lyon in 1522. Calphurnius was one of the early partisans of the humanist movement in France. Around 1517, he had published a book with epigrams in classical Latin style, the themes of which are both humanist and Christian.[136] The *Canon* edition is a strange publication, since the editor Champier deters readers from Avicenna by inserting before the proper text a seventeen-page refutation of Avicenna's philosophical and medical errors. We can witness Champier gradually developing toward the harsh critic of Arabic medicine that he will later be famous for. Calphurnius's part was to compose a life of Avicenna, which was written in classical Latin style. For this purpose, he consulted both the Schedel–Champier tradition and the Locatello vita. As the following citation shows, Calphurnius thought long about which source is to be preferred and finally declared the most fantastic account to be authentic:

> But because to many people it is not clear how long and when Avicenna lived, I find it appropriate to solve this on the basis of old histories *(antiquis annalibus).* For he died before age, since he had barely completed his forty-eighth year when he was gripped by an unhappy death. Averroes the philosopher and physician was his contemporary, who always fought against Avicenna's and other physicians' fame with hateful missiles of barking language. The well-known physicians Algazel and

Alfarabi were also Avicenna's contemporaries, whom the evil Averroes
(maledicus ille Averroes) also tried to stir up against him . . . Others relate
that he was an Arab and a prince of Abolay, who after the funeral of his
father completely immersed himself in studies . . . But because opinions
about the death of Avicenna among the different authors differ, I therefore do not write anything that I do not dare to assert in every respect.
Nevertheless, after having rejected many opinions, I easily arrive at the
conviction (facile mihi persuadeo) that I do not divert from the truth if I
follow mister Symphorin Champier of Lyon on Avicenna's death . . .
Averroes killed him with a poisonous drink.[137]

The first paragraph cited is based on *annales antiqui,* which turn out to be
Schedel's chronicle, and reports that Avicenna died when he had not yet
reached age fifty. In an earlier passage, Calphurnius quotes Schedel's story
that Avicenna had a hospital built in Cordoba. The fantastic legend about
Averroes stirring up Algazel and Alfarabi against Avicenna—none of the
four were in fact contemporaries—is apparently Calphurnius's own invention. The second paragraph, which starts with "others relate," is based on
the Locatello vita, from which Calphurnius cites the remarkable accomplishments of the ten-year- and eighteen-year-old Avicenna. The "prince of
Abolay," that is, of the alleged town or country "Aboaly," is Calphurnius's
misrendering of the Locatello phrase "the prince Aboaly Avicenna." When
it comes to Avicenna's death, Calphurnius is in a better position than his
predecessors: he has several sources at his disposal. Much evidence militates in favor of the authentic Locatello story: the Islamic context, the
foreign names and the fact that the biography is credited to a pupil of Avicenna. But, apparently, the attraction of the murder story, which tallies
well with the hospital story, was too great for Calphurnius. Proper source
work leads him to the conclusion that Averroes killed Avicenna with a poisonous drink.

The ideological motives behind this distortion surface in Calphurnius's
phrase *maledicus ille Averrois.* That Averroes was an impious philosopher
and an enemy of religion is a common charge of the time, especially in
humanist writings. The story of Avicenna's violent death was attractive
to Calphurnius because it allowed him to depict Averroes as a villain.
Calphurnius's inspiration on this point seems to have been Symphorien
Champier. In his prefatory text to the same *Canon* edition, Champier explains that Avicenna was an Arab Spaniard and a Muslim by faith—which
he deplores—whereas "this impious Averroes was without faith, irreligious,
bare, and pagan. He hallucinated the unicity of the intellect against the
mind of Aristotle."[138] The Christian agenda that is behind Champier's un-

historical depiction of Avicenna and Averroes fits well with Calphurnius's distinct mixture of humanist and Christian interests. The two stand in a longer tradition of a Christian aversion toward Averroes cultivated in the humanist camp, which will receive closer scrutiny in Chapter 5.

A new level of historical precision, even compared with the Locatello vita, was reached when the Venetian physician Nicolò Massa produced a Latin version of the combined autobiography–biography sometime before 1544, when it was first published as part of a *Canon* edition.[139] In the introductory passage to his life of Avicenna, Massa relates that Paolo Alpago, nephew of the well-known Arabic–Latin translator Andrea Alpago (d. 1522), had found among the papers of his deceased uncle "the life of Avicenna by the Arab Sorsanus *(i.e., Ǧūzǧānī)*" in Arabic and had complained to him that he could not find a translator. Massa then asked an interpreter of the Venetian merchants in Damascus, named Marco Fadella, to render Ǧūzǧānī's words in Italian, and on the basis of this material produced his vita. The result is not a literal translation, but a renarration of the autobiography–biography complex, which he attributes in its entirety to Ǧūzǧānī. Massa's text follows the Arabic original, but adds occasional remarks, such as that Avicenna was Persian and Muslim, and skips pieces of information, such as Avicenna's acquaintance with the Quran and his study of Ptolemy's *Almagest* together with his teacher. Nevertheless, the amount of historically authentic material transported was considerable. Its effect was increased by Andrea Alpago himself, who had prefaced his dictionary of transliterated terms in Avicenna's *Canon,* first printed in 1527, with a note protesting against the legend of Avicenna's Spanish origin, insisting that Arabic sources know him as a Persian born in Buchara *(fuit Persicus ex civitate Bochara).*[140]

An early reader of Massa's biography was Jacob Milich (1501–1559), the first occupant of an anatomy chair of the faculty of medicine in Wittenberg.[141] In 1550, Milich publishes an *Oratio de Avicennae vita,* which is meant to incite the young to study medicine and to turn to Avicenna in particular. The impact of authentic Arabic material on Avicenna's life is very apparent in the first part of Milich's vita, which gives a description of Avicenna's youth based on Massa's text. But apparently Milich was not content with his historical source and therefore invented a more fashionable account of Avicenna's life after the age of eighteen. The father, says Milich, sent Avicenna to study medicine with the luminary Rhazes in Alexandria (the historical Rhazes lived a century before Avicenna in Persia and Baghdad), where the remnants of the ancient medical schools were still to be found. After four years, Avicenna realized that Rhazes was too often diverting from the ancient sources and therefore left for Cordoba to study under Averroes, who had earned his enormous fame by detracting the authority of other scholars. Avicenna returned to Alexandria, studied the true Hippocratic and Galenic sources of medicine in Arabic translation, saw the whole of Egypt, and finally returned to his home country, where he wrote his masterpiece, the *Canon,* a summa of Galenic medicine—the book that with full justification is taught in all universities *(in omnibus publicis academiis)* for four hundred years. The vita ends with praise for Avicenna's advances over Greek medicine, especially in pharmacology.[142] Milich certainly knew from Massa's vita that Avicenna never traveled farther West than Persia, but, from an educational viewpoint, it seemed preferable to make Avicenna meet other Arabic authorities whom the students should know and to directly link Avicenna to the Greek schools of medicine, or whatever was allegedly left of them in Alexandria. Boneto Locatello, Leo Africanus, Andrea Alpago, and Nicolò Massa had all furnished material against the Andalusian life of Avicenna. But for the humanist Milich, who wanted to defend Avicenna against prejudices among his humanist colleagues, the Middle Eastern scholar Avicenna could not serve his case as well as a fictitious Hellenized student of Alexandrian medicine. Jacob Milich is an exception in the biographical tradition on Avicenna: other biographies were tendentious, but not unblushing fabrications such as Milich's.

## KONRAD GESNER

With Konrad Gesner (1516-1565), the Swiss scientist and philologist, we return to the Foresti-Schedel-Champier tradition. Gesner considerably enlarges the tradition by treating a great number of Arabic scientists of all disciplines. One reason for Gesner's wider interest is that his *Bibliotheca universalis* of 1545 is designed to be an all-inclusive guide to and survey of books, another reason that he was one of the leading botanists and zoologists of the time—apart from being a teacher of Greek in Lausanne and chief physician in Zurich in different phases of his life. His interest in the natural sciences clearly surpassed that of Foresti and Schedel.[143] In addition to the traditional canon of Arabic scholars, Konrad Gesner also mentions the physicians Haly Rodoan, Albucasis, Johannitius, Ibn Buṭlān, and Jesus filius Haly, the astrologers and astronomers Abraham ibn Ezra, Albategnius, Albubater, Albumasar, Alcabitius, Alfraganus, Almansor, Alpetragius, Alfarabi, Abenragel, Omar Tiberiadis, Messahalah, Thebit, and Zahel, the optical scientist Alhazen, the philosophers Algazel, Alkindi, Avempace, Maimonides, and the alchemist Geber.[144] Gesner's book thus mirrors the considerable presence of Arabic authors on the international book market, which I have described in Chapter 1.

Gesner's main source on the traditional set of Arabic scholars is Symphorien Champier's *De claris scriptoribus medicinae,* as he himself acknowledges. Occasionally, Gesner shortens Champier's text, most notably in the case of Avicenna, where he quotes Champier on Avicenna's having confessed himself a Muslim in the *Metaphysics,* but omits the scathing remarks on Muhammad and Islamic theologians, which we have discussed above.[145] For all his learnedness, Gesner does not bother to consider any alternative accounts of Avicenna's life and works. As a result, he prolongs the outdated legend of Avicenna's Andalusian origin and of his alleged death by the hands of Averroes. The entry on Averroes himself, however, has a positive tone, as it had in the early work of Champier. At the end of the entry, Gesner adds a less sympathetic remark of his own:

> In his Aristotle commentaries, Averroes has, in great measure, imitated the Greek commentators, such as Alexander of Aphrodisias and Themistius. Some people call Averroes an offense to physicians *(stimulus medicorum),* since he disagreed with Galen on a good number of issues.[146]

These remarks show that the intellectual atmosphere has changed gradually since the beginning of the sixteenth century. Averroes is now read alongside the Greek commentators in the universities, and Gesner here mildly echoes the humanist charges that Averroes has "stolen" from his Greek predecessors (see Chapter 5). Galen, in turn, is now available in Greek and Latin editions, and has reached the peak of his reputation. Gesner himself contributed to this current by editing the complete Latin Galen and writing a programmatic bio-bibliography of him.[147] Averroes's Aristotelianizing medicine and especially his critique of Galenic physiology was not to Gesner's taste.

If Gesner's entries on the Arabic authors of the Foresti tradition are conventional or even outdated, he is more impressive on the authors outside this tradition. He gains information on them mainly by drawing on the available Latin editions, in particular, by studying prefaces of the translators, editors, or authors themselves.[148] On the basis of a very broad reading, he attempts to solve the riddles concerning the identity of Mesue and Serapion. Today we know that the true Mesue, Yūḥannā ibn Māsawayh (d. 857 AD), was not identical with the anonymous Arabic author (or authors) of the fourfold *Opera Mesue* on pharmacology, who worked in the eleventh or twelfth century. Likewise, the true Serapion, Yūḥannā ibn Sarābiyūn (late ninth century AD), the author of a medical compendium called *Practica,* is not identical with "Serapion," the author of the *Liber aggregatus,* who in fact is the Andalusian scholar Ibn Wāfid (d. ca. 1074 AD).[149] For Gesner there is only one Mesue and one Serapion. The confusion he is tackling arises from the fact that both Arabic authors were called *Joannis* (Yūḥannā) and hence could be identical with the Christian theologian John of Damascus. In his entry on Mesue, Gesner refutes the identification of Mesue and John of Damascus by arguing that Mesue was not a contemporary of John of Damascus since he was born and educated in Damascus several centuries after John, namely, in the time of Frederick Barbarossa.[150] In the entry on Serapion, he notes that Serapion too is conflated with John of Damascus. Gesner perceptively argues that Serapion is the youngest of the Arab physicians, since he quotes many of them in his writings, among them "Ioannes Mesuaei filius" who lived around 1158 AD; it is likely that the two were contemporaries because they quote each other. John of Damascus, in contrast, is a Christian theologian who lived around 400 AD (historically, this is three hundred years too early) and is not attested as a physician; whereas Serapion is a Muslim physician.[151]

Gesner's discussion is best understood, it seems, as a reaction to the 1543 edition of Alban Thorer's *Iani Damasceni Decapolitani summae inter Arabes auctoritatis medici therapeutici methodi . . . libri vii*—which, in fact, is an edition of Serapion's *Practica* under a new title. Gesner erroneously thought that this edition was Mesue's *Opera*, but he correctly criticized the invention of the name *Ianus Damascenus*: "It seems that someone deliberately has changed the name <of the author>, so that a new title should appear."[152] Even though Thorer's preface does not openly suggest the identification of this author from Damascus with Mesue or with John of Damascus, Gesner obviously thought the edition a source of considerable confusion. The confusion of the three Johns—John of Damascus, Mesue, and Serapion—is a phenomenon already of the medieval manuscript tradition. Konrad Gesner rightly insisted on keeping them apart; his relative chronology is not entirely off the mark (his absolute chronology surely is), and he makes the best of his material when demonstrating the different intellectual profiles of Serapion and John of Damascus.

We can thus witness Gesner going into considerable detail on the identity of Arabic physicians and botanists, in accordance with his own scientific interests. He has much less respect for the medieval translators of Arabic authors. Gesner censures Omar Tiberiadi's *De nativitatibus* for its *rudior stilus*, which is the fault of the translator (*est interpretis vitio*), and the Ibn Butlān translation for its *sermo . . . rudis et barbarus*, while the translation of Johannitius's *De oculis* earns the superlative *barbarissima*.[153] And he castigates the translator John of Seville directly for his Latin style: "John of Seville has turned Alcabitius's Introduction to the doctrine of the stars into a Latin of barbarous style."[154] This is the voice of the humanist Gesner, the teacher of Greek and author of books on the ancient languages. In contrast to Symphorien Champier, whose polemics targeted Muhammad and Islamic theologians, Konrad Gesner takes offense at the style of the Latin translations. It is noteworthy that Gesner, only a few years after the attacks on Arabic learning had reached their culmination in the 1530s, clearly distinguishes between the original works of the Arabs, on which he does not pass judgment, and their medieval translations, which he criticizes severely.

Gesner's accomplishments and prejudices are well apparent in his entry on the astrologer Albumasar. Before Gesner, only Simon de Phares had written a biography of Albumasar, who, as we recall, was split up by Simon into different personalities and was made to live not in Baghdad, but in Athens. Gesner, who does seem to know Simon's text, begins soberly by stating that Albumasar is the same person as Iaphar, but later inserts polemical comments:

> The astrologer Albumasar, with another name, Iaphar, wrote <a work of> eight books on the great conjunctions, annual revolutions and their profections. <This work is printed in> 30 folios, quarto, in Augsburg 1489. Elsewhere *(namely, in the edition of 1515)* the title is as follows: 'This is the book of the individual higher <planets>, in general, on the indications on events which occur in the world, on their appearance with respect to conjunctionalist beginnings and other <beginnings>.' Even though it is written in obscure and barbarous language *(obscuro et barbaro sermone scripta)*, we have added the table of contents of the different chapters, in order that one may judge from it the style of the entire work. I was not able to trace the Latin translator. *(Then follows half a page with a table of contents.)* And this until one gets sick of it, nevertheless we have quoted it. Often inserted are depictions of the zodiac and the planets. <Also there is> the volume "Selections of Astrology" *(Flores astrologiae)* by Albumasar, printed in Augsburg 1488, quarto, in five folios, with figures of the zodiac and the planets.[155]

Gesner's bibliographical approach to the history of science had helped him in the cases of Mesue and Serapion, and here again he is able to give at least some information on the astrologer Albumasar and his work by consulting the editions available to him. In spite of his apparent disgust with the translator's language, Gesner obviously thought the book *De magnis coniunctionibus* important enough to quote extensively from it. It is interesting to see that the two works by Albumasar that are known to Gesner belong to general or world astrology. Not mentioned is Albumasar's *Great Introduction,* a handbook of all parts of astrology, which was particularly successful in the Middle Ages and also cherished by Simon de Phares. Gesner's presentation of Albumasar as an author on general rather than individual astrology tallies well with the boom of Arabic general astrology in sixteenth-century Europe.

Gesner's *Bibliotheca* is also an important testimony to the limited biographical interest of Renaissance scholars in Arabic astronomy. Gesner has much information on astrologers, such as Albumasar, but little on astronomers. The Arabic astronomers he is able to say the most about are Alfraganus and Thebit ben Corat, because he has seen their works in print. But he knows about others, such as Alpetragius, Albategnius, and Alhazen, only from hearsay. He conflates the astronomer Geber filius Afflah with the alchemist Geber, and what he says about his works mainly concerns alchemy. The relative silence of the biographical tradition on Arabic astronomers is broken toward the end of the sixteenth century by Bernardino Baldi.

## BERNARDINO BALDI

Histories of scientific disciplines are an important part of the Renaissance biographical tradition on Arabs. As we have seen, Foresti's canon was partly inspired by histories of medicine, and Simon de Phares's collection of astrologers and Champier's book on famous physicians transport much biographical material on Arabic authors. This is not the rule, however. Often histories of disciplines have rich information on ancient or contemporary authors, but are brief on their Arabic successors. Such is the case with Luca Gaurico's *Speech on the Inventors of Astronomy* of 1507, a succinct history of astrology that merely enumerates the many Arabic astrologers,[156] and with Otto Brunfels's *Catalogue of Famous Physicians* of 1530, which has only three lines each on Avicenna, Averroes, Mesue, Rhazes, and Albaterius based on the Foresti tradition.[157] And remarkably enough, Renaissance histories of astronomy and mathematics hardly ever include Arabs, in spite of the fact that Arabic astronomers were readily available in manuscripts and printed editions. Arabic authors appear in Erasmus Reinhold's half-page table of famous "philosophers and artists,"[158] but they do not figure in Petrus Ramus's, Federico Commandino's, Christophorus Clavius's, and Tycho Brahe's historical outlines of mathematics.[159] I am aware of only two exceptions to this trend: Johannes Regiomontanus and Bernardino Baldi.

In the spring of 1464, the famous astronomer Regiomontanus (d. 1476) gave a lecture course on the Arabic scientist Alfraganus at the University of

Padua, of which the opening lecture is still extant: *Introductory Speech to Reading Alfraganus,* or as the title page of the 1537 edition puts it: *Introductory Speech to All Mathematical Sciences.* The lecture gives a detailed outline of the history of the exact sciences, with an emphasis on astronomy, from the Greeks to the Arabs to the Latins. Regiomontanus is full of praise for the Arabic astronomers, of whom he explicitly names Albategnius, Geber Hispalensis (i.e., Ǧābir ibn Aflaḥ) and Alfraganus, before he turns to the Italians. But his remarks on their scientific merits are too brief to contain any biographical or bibliographical information proper.[160]

This changes only with Bernardino Baldi (1553–1617), born in Urbino, pupil of Commandino, abbot of Guastalla since 1586, and most notable for a commentary on the pseudo-Aristotelian *Mechanica problemata.* Baldi wrote 202 lives of mathematicians in the years 1587–1596, among them fourteen Arabic authors.[161] *Le vite de' matematici* was not printed until the nineteenth century[162]—except for short extracts under the title *Cronica de matematici,* which appeared in 1707.[163] In 1576 and again in 1595–1596, Baldi studied Arabic with Giovan Battista in Rome.[164] In view of this, it has been argued that it was Baldi's knowledge of Arabic that enabled him to give more detailed lives of Arabic mathematicians than any European writer before him.[165] In fact there are no traces of any usage of sources in Arabic in Baldi's *Vite.* His main sources are, apart from the Latin translations of Arabic works, other Renaissance scholars and, in particular, astronomers, as is apparent from the references that Baldi gives himself: he often uses the astronomers and astrologers Francesco Giuntini, Johannes Stöffler, and Erasmus Reinhold, and the historian Gilberto Genebrardo.[166]

Baldi's entries are very uniformly structured: they first give the birthplace and home country of the author and then add information on the different parts of his name. Here Baldi was clearly helped by his knowledge of the structure of names in the Arabic language.[167] Then follows the central part of each biography, an assessment of the author's contribution to the history of mathematics—which in most cases means astronomy. At the end of the entry, Baldi always proposes dates for the time of flourishing of the person in question, often considering the author's dates of astronomical observations. His biographies are less colorful than those of the other authors in this survey. This is partly because there was much less material available on the lives of mathematicians than on those of physicians, but partly also because of a lack of interest on Baldi's side. He is mainly interested in correct names, correct dates, and, as we shall see, Arabic astronomical theories.

Baldi discusses an impressively comprehensive list of Arabic astronomers and astrologers, and, in fact, knows much more about them than

Gesner, whose knowledge did not cover astronomical theory proper. The authors discussed by Baldi are the following, in the sequence in which they appear in the manuscript:[168] Messahalah, Albategnius, Haly filius Abenragel, Azarchel, Haly Rodoan—who is discussed twice under different names—Almansor, Alhazen, Alkindi, Alpetragius, Geber—here Baldi conflates, just as Gesner, the astronomer Geber filius Afflah and the alchmist Geber—Alfraganus, Thebit ben Corat, and Albumasar.[169]

The tradition inaugurated by Foresti is not present in Baldi's *Vite*, with one exception. Baldi's and Foresti's texts both contain a biography of 'Albaterius,' who in fact turned out to be the Graeco-Arabic translator al-Biṭrīq. Baldi wrongly identifies 'Albaterius' with Albategnius: "That he *(i.e., Albategnius)* was a physician and an excellent one, has been said by Fra Filippo <Foresti> in his *Supplementum*, even though he mutilated and distorted the name into *Albaterius*."[170] Baldi supports his thesis that Albategnius was a physician with a reference to Pedro Mexía and his assertion that 'Albetenio' had translated the works of Galen from Greek into Arabic. Pedro Mexía, however, is entirely dependent upon Foresti's entry on 'Albaterius,'[171] and thus Baldi's attempt to solve Foresti's riddle of 'Albaterius' only created further confusion.

On three Arabic authors, Baldi gives much more information than on others, and dwells at length on their contribution to the history of astronomy: Albategnius, Alpetragius, and Thebit ben Corat.[172] Note that these three authors had been neglected by Simon de Phares; Baldi is particularly interested in them because they are astronomers—if he discusses astrologers, it is only with respect to their contributions to astronomy. Albategnius (al-Battānī, d. 929 AD) is praised for the exactness of his observations of the planetary movements and for having found many emendations to Ptolemy. Baldi had not seen any of Albategnius's works in print, and it seems that he thought it important to collect everything known about Albategnius to rescue his contribution.[173] Alpetragius (al-Biṭrūǧī, late twelfth century) is said to have been influenced by his friend Aventasele (Ibn Ṭufayl) and perhaps also by Averroes to postulate the simplicity and concentricity of the orbits and to dismiss the Ptolemaic eccenters and epicycles, which do not agree with physical reality. In Baldi's view, both Alpetragius's theory of the planets and his theory of movement have successfully been refuted.[174] Tebitte (Ṭābit ibn Qurra, d. 901) receives appreciation for the succinctness of his description of the principles of astronomy, of the Azimut and of the Almicantarat, for his value of the length of the year, and for his theory of the movement of the eighth sphere, which the moderns call movement of trepidation.[175] Historically, this very last contribution rests on the common misattribution of the treatise *De motu oc-*

*tavae sphaerae* to Thebit ben Corat, which dates around 1080 AD, that is, about 180 years after Thebit's death.

Baldi's comments on the contribution of Arabic astronomy show, on the one hand, that he valued the Arabs for the exactness of astronomical data and for inventing theories for the movements of the planets and the fixed stars, and, on the other hand, that he had a keen sense for a progressive history of astronomy leading from the Greeks to the Arabs to the Latins. What he describes is the cross-cultural Ptolemaic tradition.[176] The language of historical progress is much more prominent in Baldi than in previous biographers. Champier and Gesner are well aware of the Arabic–Latin translation movement in the eleventh to thirteenth centuries, as is obvious from their entries on Constantinus Africanus[177] and Gerard of Cremona,[178] but the gradual transmission of the sciences is not their focus.

To illustrate Baldi's biographical method, I turn again to the biography of Albumasar. Baldi begins his entry with a statement about the various spellings of the name and remarks that Albumasar was mainly concerned with astrology. He then discusses Albumasar's opinion on the order of the planets and the position of the sun in particular, as exhibited in his *Great Introduction,* and proceeds to mention several other works of his. For information on the *Great Conjunctions* and the *Flowers of Astrology,* Baldi relies silently on Gesner. He then turns to biographical details proper, beginning with the surprising information that Albumasar had a son "Abalachio." This legend seems to originate from misreading Albumasar's nisba "al-Balḫī," that is, his patronymic name "from the city of Balḫ" in Persia:

> Albumasar had a son with the name 'Abalachio,' who also was a well-known mathematician, as Francesco Giuntini relates *(sc., in his Speculum astrologiae).*[179] Albumasar was a Spaniard by birth and is counted among the famous astrologers of Spain by Damianus de Goes. He flourished, according to Reinhold's historical table, 844 years after Christ. Hence Giuntini is mistaken in believing that Albumasar lived around 540. But if Albategnius, who has epitomized the *Almagest* enlarged by Albumasar, flourished around 880, as we have proved in his entry, it is clear that Albumasar flourished either in his time or a bit after him. Giuntini therefore errs.[180]

The sources of this passage have been unearthed by Moritz Steinschneider and Elio Nenci. Iberic pride is behind the statement that Albumasar was a Spaniard (the historical Albumasar was born in Balḫ on the Silk Road, now in Northern Afghanistan).[181] This information comes from the Portuguese humanist Damiano de Goes, who also declared Avicenna, Rhazes,

Almansor, and Messahalah to be Spaniards.[182] Baldi simply accumulates material on Albumasar's life, family, and origin without further inquiry. His sole concern is that the sources do not disagree with each other.

There was a scholarly conflict on Albumasar's floruit, and here Baldi begins to argue with his sources. His conclusion is that Albumasar lived around 880 AD or slightly later. This is correct as far as the century is concerned: Albumasar lived 787–886 AD. But Baldi's argumentation is flawed. Erasmus Reinhold had stated in a table of "famous philosophers and artists" that Albumasar lived around 844 AD and Albategnius (i.e., al-Battānī) around 880 AD, which is not a bad estimate, since al-Battānī in fact died in 929 AD.[183] It is not true, however, as Baldi argues, that Albumasar had written an amplification of the epitome of the *Almagest* produced by Albategnius, and hence must have lived after Albategnius. Here Baldi misunderstands a note by Rudolph of Bruges, the twelfth-century translator of Ptolemy, who had mentioned that Albategnius epitomized the *Almagest,* whereas Albumasar wrote a lengthy exposition *(ampliavit)* of Ptolemy's *Tetrabiblos*—and not of Albategnius's work, as Baldi believes.[184] Albumasar therefore lived a generation before Albategnius, not after him. Baldi's argumentation is nevertheless noteworthy, not only because Francesco Giuntini's dating of 540 AD is successfully refuted,[185] but also because Baldi tries to establish a relative chronology of the Arabic astronomers.

With his chronological interest, Baldi is a child of his epoch, the second half of the sixteenth century, when Joseph Scaliger makes chronology a thriving discipline.[186] The bio-bibliographical tradition both suffers and profits from this development. On the one hand, a number of world histories, for instance, by Johann Thomas Freigius and Elias Reusner, now focus on the chronology of events and rulers and neglect biographies proper. On the other hand, a scholarly discussion develops about the exact dates of the lives of Arabic mathematicians and astronomers, in which Luca Gaurico, Bernardino Baldi, Giuseppe Biancani, and Gerhard Voss, among others, participate.[187]

The Foresti tradition of bio-bibliographies, however, declines noticeably in the second half of the sixteenth century. The writing of world chronicles undergoes dramatic changes,[188] and Arabic scholars disappear again from world history: they are not mentioned in the world histories of Sebastian Franck, Johann Carion, Christian Egenolff, Johannes Sleidanus, Johann Thomas Freigius, Michael Neander, and Elias Reusner.[189] One reason is to be found in the general tendency of sixteenth-century world historians to keep history and biography separate: Franck, Carion, Egenolff, and Sleidanus, who in many respects prolong medieval traditions, rarely mention *viri illustres* or *viri disciplinis excellentes,* in contrast to their immediate predecessors Antoninus of Florence, Foresti, and Schedel. It seems that the Protestant and anti-Scholastic interests of these four writers allowed for biographies of ancient scholars such as Plato and Aristotle, but not for the traditional section on famous men of the tenth to fourteenth centuries.

᠅

In the final analysis, it was remarkably difficult for Renaissance scholars to write about the lives and works of Arabic authors. Reliable material from Arabic sources was extremely rare. This changed only in 1624, when Jacob Golius, the Dutch scholar of Arabic, acquired an Arabic manuscript of Ibn Ḥallikān's biographical dictionary in Morocco, and in 1650, when the Oxford scholar Edward Pocock Sr. published Barhebraeus's history of the Arabs, adding much bio-bibliographical material from Arabic sources in his commentary.[190] And even when, in the sixteenth century, new material from Arabic sources became available, the situation remained difficult, as the example of Leo Africanus showed: Leo's biography of Averroes came close to the historical truth, but his biography of Avicenna was far from it. The reason is that the Arabic bio-bibliographical sources themselves are problematic and can only be trusted after careful source criticism. In addition, the Renaissance difficulty with Arabic biography was increased by various forms of tendentiousness. Leo's and Schedel's liking for the anecdotal led them astray. Calphurnius and Champier cultivated a Christian aversion toward the allegedly impious Averroes. Simon de Phares and Jacob Milich Hellenized their Arabic heroes and made them live in Athens and Alexandria. Champier, biased against Islam, turned Avicenna into a Muslim critic of Muhammad and Islamic theology. Hence, authentic material on Arabic authors, such as that provided by Alpago and Massa, was not only scarce, but also had a slow reception, since it was hampered by tendentious attitudes.

In view of these difficulties, the development of a considerable set of biographies of Arabic scholars in the Renaissance is a remarkable accomplishment. Renaissance biographers are at their best when they set out to refute errors: as when Schedel argues against Avicenna's correspondence with Augustine, Gesner insists on keeping the three 'Johns' apart, and Baldi mounts arguments against dating Albumasar's flourishing to the sixth century on the basis of relative chronology. In the final analysis, it proved a fruitful technique to base the new knowledge about Arabic authors on the evidence of their works, especially since printing allowed for a much better overview of the field. The potential of bibliography for biography was best exploited by Konrad Gesner, who wrote numerous entries on astronomers, mathematicians, and astrologers that had received hardly any biographical attention before.

The social background of the Renaissance scholars involved in Arabic biography is diverse—which is also indicative of the diverse grounding of Arabic sciences and philosophy in general in Renaissance society. One can discern several groups: first, learned scholars well-versed in the academic disciplines, such as Foresti and Schedel; second, practicioners of a specific discipline, such as Simon de Phares and Champier; third, scholars of the Arabic language, such as Leo, Alpago, and Massa; and fourth, humanists such as Calphurnius and Gesner. The largest group is the first: learned scholars (such as Foresti, Schedel, Staindel, and Champier, followed by Naukler, Volaterranus, Tiraqueau, Mexía, and others), who are well-acquainted with the authors read at the universities, but nevertheless share some ideals of the new humanist movement. The crucial figure in this group is Foresti, the author of the standard world chronicle of the Renaissance. In an unprecise sense only, Foresti is a representative of the historiography of Italian humanism.[191] It is true that Foresti successfully replaced the medieval world chronicles by adopting some humanist standards of historiography, such as the usage of ancient sources and models, the inclusion of *viri illustres,* and, in particular, a mildly classicizing Latin style. And it is important to note that the bio-bibliography of Arabic scholars in the West did not originate in a strictly scholastic surrounding. However, the *Supplementum chronicarum* clearly continues the tradition of medieval world chronicles, and the Arabic scholars appear precisely in the group of famous scholastic authors who have had their place in medieval histories since Vincent of Beauvais and Ranulph Higden. Foresti was not a humanist in the narrow sense: he was not a man of letters or a Latin scholar, but a friar of the order of the Eremites of St. Augustine and author of a *Confessionale.* His world chronicle does not care for the humanists' ideals of a narrative history, the distinction of sources, and the differentiation of legend from truth.[192]

As to the other groups of authors writing on Arabs, it is noteworthy that humanists also belong to them: Calphurnius, Milich, and Gesner. In contrast to their radically humanist colleagues, these humanists are interested in the content of the Arabic scientific tradition. Calphurnius's humanist profile, for example, is very visible in the exquisite Latin style of his vita of Avicenna; also apparent is his Christian aversion to Averroes, a typical attitude among humanists. Gesner, as a professor of Greek, is a humanist by profession. His leanings are very apparent in the repeated comments on the bad Latin style of medieval Arabic–Latin translations.

And Milich makes Avicenna a direct disciple of the great authorities of ancient medicine. Calphurnius and Milich compromise with historical truth because of their humanist convictions, whereas Gesner is an example of a humanist who, in spite of stylistic reservations, considerably contributes to the study of Arabic scientific traditions in Europe.

The overall image of Arabic authors created by the Renaissance is Hispanicizing: in many Renaissance sources, the Arabic scientists were all declared Andalusians—partly as a result of Spanish pride, and partly because Arabic authors were thought to come from the country where they had been translated from Arabic. What is the more specific image created of the three authors Avicenna, Averroes, and Albumasar? Avicenna is the most famous of all Arabs, the paramount authority in medicine. His philosophical work is mentioned but little known in the Renaissance. The fictive Andalusian image of Avicenna's life conflicts with the Persian image transmitted by the newly translated Locatello, Alpago, and Massa sources. Thanks to these translating activities, the biography of Avicenna is the only Renaissance biography of an Arabic scholar that is developing toward a historically accurate account.

Averroes, physician and most renowned philosopher (*philosophus clarissimus,* as Foresti puts it), receives much praise for his commentaries on Aristotle and for his *Colliget.* Hence, the biographical tradition transports a positive image, which, however, is marred by one very negative element: Averroes is often described as filled with jealousy about other physicians' success, which, in some versions, lead him to murder his rival Avicenna. In the background of this legend is the Christian depiction of Averroes as an impious and pagan philosopher. This image is transported into the seventeenth century, when his name was still used to refer to irreligious philosophy and to the denial of providence in particular, even though his works were hardly ever read.[193] The information that Leo Africanus offered would have lifted the biography of Averroes to a completely new level; unfortunately, it did not find readers.

The image of the astrologer Albumasar remains faint throughout the Renaissance, since biographical information was gained only from his writings. Speculation made him a Spaniard or a denizen of Athens. But Renaissance scholars were successful in collecting material about his important astrological writings. The image presented by Gesner, for example, is of an important astrologer, who sadly is only accessible in terrible medieval translation. Giovanni Pico della Mirandola describes Albumasar as a "professor of the science of grammar" and hence as a mere dilettante in astrology.[194] In the course of the sixteenth and seventeenth centuries, 'Albumasar' becomes a popular label for the unreliable, deceiving astrologer, the central character in the plays *Astrologo* by Giambattista Porta

(1570s) and *Albumazar* by Thomas Tomkins (1615).[195] In Giuseppe Maria Buini's Opera *Albumazar* of 1727, the name is used for a Turkish tyrant, and the astrological overtones have disappeared. The negative image of Albumasar has one source in Gregor Reisch's handbook *Margarita philosophica,* completed around 1496 and often printed, which describes Albumasar as an irreligious philosopher: "Albumasar, heaping impiety upon impiety, says: Who prays to God in the hour when Moon with the Head of the Dragon is joined with Jupiter, obtains whatever he asks."[196] This sentence about the "divellish" Albumazar is translated in John Melton's anti-astrological treatise *Astrologasters* of 1620.[197] The popular image of Albumasar as a villainous or even irreligious charlatan contrasts starkly with the undisputed reputation he enjoys with the scholars Simon de Phares, Gesner, and Baldi.

In the Renaissance, the interest in the lives and works of Arabic scholars was greater than ever before in Europe. As shown above, the historiography of medicine was at the origin of this biographical tradition, when it first emerged in the chronicle of Foresti. Conceptually, the condition of this development was the idea, already present in Salutati and Tortelli, that the Arabic physicians were the bridge between Galen and the Italian physicians around 1300. It is a sense for the three-step development and transmission of medicine from antiquity to the Arabs and from the Arabs to Italy that caused the Arabic authors to appear in historiography. This sense was not shared by all historiographers. Some Renaissance accounts, such as, in fact, most histories of astronomy and mathematics, deal with ancient authors only or move directly from ancient authors to contemporary authors. The historians of such works have a different attitude toward history. They believe that the cultural and scientific peak reached by antiquity was never reached again and that modern progress can only be achieved by reforms made in the spirit of ancient science.

It was decisive for the rise of Arabic biography that Foresti and Schedel did not yet share the ideals of the influential movement of medical humanism, as we call it today, which developed fully only after 1500. It is unconceivable that radical humanist physicians such as Leoniceno (d. 1524), Ruel (d. 1537), or Fuchs (d. 1566) would have included Arabic authors among the *viri illustres* of their discipline, since it was an inherent feature of the biographical genre that the famous authors were praised as models for imitation. The scholars in the Foresti tradition, however, did not propose to supplant Avicenna with Galen, or Mesue with Dioscorides. They valued Arabic physicians for reasons of academic tradition, but also because they believed in the gradual development of medicine.

# CHAPTER 3

# PHILOLOGY

*Translators' Programs and Techniques*

he biographies of Arabic authors were the subject of much investigation by Renaissance scholars, but this scholarly activity was greatly surpassed by the many Arabic–Latin and Hebrew–Latin translation efforts, which began in the 1480s.[1] The rise of a new wave of translations in the Renaissance is a remarkable phenomenon, in particular because it cannot be explained as a mere continuation of medieval activities. The medieval Arabic–Latin translation movements, impressive as they were, break off around 1300, after the completion of a number of medical translations in Montpellier and Barcelona. Around 1480, a new wave of translations begins, many of them via the Hebrew in Italy, and some of them directly from Arabic in the Near East. The movement lasts about seventy years, until the death of the translator Jacob Mantino in 1549—save for some *Canon* excerpts translated by Jean Cinqarbres printed in 1570–1586. It is characterized by two major projects: first, the Latin rendering of Averroes's commentaries, which were not yet available in Latin, and second, the replacing of the medieval version of Avicenna's *Canon medicinae* by more modern and reliable translations. The results of these efforts are impressive: nineteen commentaries of Averroes were translated for the first time, in addition to the sixteen commentaries translated in the Middle Ages, and six new versions of Avicenna's *Canon* or parts of it were produced. And apart from these major projects, there were other important translation enterprises, such as translations of Alpetragius's *De motibus caelorum*, Alhazen's *De mundo et coelo,* and Avicenna's short psychological tracts.

Many of these Latin translations, and the Averroes translations in particular, were produced from Hebrew versions. The Renaissance Hebrew–Latin translators thus relied on the previous work done by the Arabic–Hebrew translators of the Middle Ages. This translation movement started in 1210 when Samuel ibn Tibbon translated Aristotle's *Meteorology* from Arabic into Hebrew. About two decades later, Jacob Anatoli and other scholars began to translate Averroes into Hebrew. Medieval Hebrew philosophy is much influenced by Averroes, and this is why it was possible for Renaissance Jews in Italy to find Hebrew manuscripts of Averroes, because Averroes was still part of their own intellectual culture.[2] In contrast, two Renaissance translators and revisers of Avicenna's *Canon* worked directly from the Arabic: Girolamo Ramusio and Andrea Alpago. They did this in Damascus, where they worked as physicians to the Venetian embassy, and they made journeys in the Near East in search of manuscripts.

This first wave of translations of Arabic works of science was followed by a second. It begins with Jakob Christmann's translation of the astronomer Alfraganus in 1590 and continues until the Arabic–Latin edition of Ibn Ṭufayl in 1671, which was produced by the Oxford Arabist Edward Pocock Sr. together with his son. These later translations were produced already in the context of Arabic studies at European universities. They were still motivated partly by interest in the scientific content, but historical and philological concerns dominate. Moreover, Latin as the language of academic translations from Arabic was gradually supplemented, and eventually replaced, by other European languages in the seventeenth to nineteenth centuries. In what follows, I will be concerned only with the first wave of translations.

The Renaissance translation movement from 1480 to 1550 was impressive in its output, but its quality, influence, and integration into Renaissance society have not yet become clear. Past research has established a reliable picture of the Arabic–Latin translations of the Middle Ages. These translations successfully transmit a great amount of Arabic philosophy and science to Latin Europe. Their influence is particularly strong in the thirteenth century and in some fields continues for several hundred years, as we have seen in Chapter 1. This much is clear, the Renaissance translation movement did not influence European thought on a similar scale, since in the decades around 1600 the interest in Averroes's philosophy and Avicenna's medicine declined. What, then, did the Renaissance translations achieve? It will be a matter of future studies to illuminate the full impact of the translations. In the present chapter, I will approach the question from a different angle and study the translations' philological quality and social setting. What did the numerous Renaissance translations achieve in terms of quality of transmission, if compared with each other and with their medieval predecessors? Did the Hebrew intermediary have any tangible effect? For this purpose, I will compare the translators' programmatic statements with what they produce in actual effect. The translation movement is also a social phenomenon, in which several different milieus meet: academic professors, Jewish translators, aristocratic patrons, humanist scholars, and travelers in the Near East. To what extent was the translation movement integrated into Renaissance society? I shall give answers to this question by studying the translators' biographies and social context. Special attention will also be paid to the involvement of humanist scholars and humanist ideals in the translation activity.

To this end, five translation projects are singled out for closer inspection: Averroes's preface to *Metaphysics* 12, Averroes's Middle commentary on the *Topics*, Avicenna's *Canon*, Averroes's medical opus *Colliget (Kitāb al-Kullīyāt)*, and Pseudo-Mesue's *Opera omnia*. This choice is the result of a combination of systematic and contingent reasons. I concentrate on multiple translations, that is, on an Arabic text that was translated both in the Middle Ages and in the Renaissance or was translated at least twice in the Renaissance. All five projects are of this kind. This will allow us to compare the translators at work and thus to arrive at a clearer image of what is specific for them. Contingent factors are involved insofar as I have chosen texts that are still extant in the Arabic original (Pseudo-Mesue is the only exception) and of which the Hebrew intermediary, if there was one, is accessible in a printed edition. In the case of Averroes, this is a serious limitation since many of his commentaries are lost in one of the three languages of transmission, Arabic, Hebrew, or Latin. For instance, an attractive object for a comparative study of working methods would be Averroes's Long commentary on the *Posterior Analytics,* which was translated three times in the Renaissance, by Burana, Balmes, and Mantino; the Hebrew intermediary, however, exists only in medieval Hebrew manuscripts, and I therefore leave this interesting task to the Hebraists.[3] The focus on multiple translations has the advantage that it will lead us to the Arabic texts that were considered particularly important for Renaissance culture—Averroes's commentaries and his *Colliget,* Avicenna's *Canon,* Mesue's *Opera*—and it will introduce us to a broad range of translators, ten altogether. The full range of the Arabic–Hebrew–Latin translation activity is visible in Table 2, which lists all Latin translations of scientific and philosophical works between 1450 and 1700 that I am aware of. For full bibliographical references to the Renaissance editions of these translations, the reader should consult the author entries in the Appendix to this study. Note that the table does not contain anonymous, theological, grammatical, or literary works; it thus leaves out, for example, the translators of the Neo-Platonic *Theologia* of Pseudo-Aristotle,[4] and the translators of the Quran.[5]

As is a recurrent method of this book, I will confront what the translators say they do with what they in fact do in the philological detail. In this context, it is unavoidable to address a persistent problem of studies in the history of philology: the problem of qualitative judgments. Histories of

*Table 2.*    Renaissance Latin translations of Arabic sciences and philosophy
(1450–1700).

| | |
|---|---|
| Girolamo Ramusio (d. 1486), active in Damascus, trans. from Arabic | *Avicenna:* Canon 1 (Ms. Paris BN arabe 2897) [interlinear translation] |
| Elia del Medigo (d. 1493), Venice, Padua, Florence, trans. from Hebrew | *Averroes:* Comp. Meteor. + Comm. med. Meteor. (fragm.), 1488 Comm. mag. Metaph. Prooem 12, 1488 [twice] Quaest. in An. pr., 1497 Comm. med. Animal. (Ms. Vat. lat. 4549) Epitome of Plato's Republic, 1992 (ed. A. Coviello) Tractatus de intellectu speculativo (Ms. Vat. lat. 4549) Comp. An. (fragm.) (Ms. Vat. lat. 4549)[1] De substantia orbis (lemmata, with his own commentary) (Ms. Vat. lat. 4553) Quaestio de spermate, 1560 |
| Anonymous **H** | *Averroes:* Comm. med. An. (Ms. Vat. lat. 4551) |
| Anonymous **H** | *Algazel:* Liber intentionum philosophorum (Maqāṣid al-falāsifa) (with commentary by Moses Narboni) (Ms. Vat. lat. 4554) |
| Anonymous Hebrew scholar attached to Pico della Mirandola (before 1493) **H** | *Ibn Ṭufayl:* Ḥayy ibn Yaqẓān (Ms. Genua Bibl. Univ. A.IX.29) |
| Andrea Alpago (d. 1522), Damascus **A** | *Avicenna:* Canon 1–5, Cantica, De virtutibus cordis, 1527 [corrections] Compendium de anima . . . , 1546 De removendis nocumentis, De syrupo acetoso, 1547 *Ibn an-Nafis and Quṭb ad-Dīn aš-Šīrāzī:* commentaries on the Canon, 1547 [selections] *Ebenbitar (Ibn al-Bayṭār):* De limonibus, 1583 *Serapion:* Practica, 1550 [corrections] |
| Giovanni Burana (d. before 1523), Padua **H** | *Averroes:* Comp. An. pr., 1524 Comm. med. An. pr., 1524 |

*Table 2.* (continued)

| | |
|---|---|
| | Comm. med. An. post., 1550/1552<br>Comm. mag. An. post., 1550/1552 |
| Abraham de Balmes<br>(d. 1523), Venice and<br>Padua **H** | *Avempace:*<br>Epistola expeditionis (Ms. Vat. lat. 3897)[2]<br>*Alfarabi:*<br>De intellectu (Ms. Vat. lat. 12055)<br>*Alhazen:*<br>Liber de mundo et celo (Ms. Vat. lat. 4566)<br>*Averroes:*<br>Comp. Org., 1523<br>Quaesita logica (Int., An. pr., An. post.), 1523<br>Comm. mag. An. post., 1523<br>Comm. med. Top., 1523<br>Comm. med. Soph. El., 1523<br>Comm. med. Rhet., 1523<br>Comm. med. Poet., 1523<br>Comp. Gen., 1552<br>Comp. An., 1552<br>Comp. Parv. nat., 1552<br>Comm. med. Phys. (Ms. Vat. lat. 4548)<br>Quaesita naturalia, De substantia orbis 6–7,<br>1550/1552 (and Ms. Vat. ottob. lat. 2060)[3]<br>Liber modorum rationis de opinionibus legis [Kašf<br>ʿan manāhiǧ] (Ms. Vat. ottob. lat. 2060, Ms. Milan<br>Ambros. G. 290)<br>*Arabic logicians:*<br>Diversorum arabum quaesita ac epistolae [short<br>treatises by four Arabic authors not yet identified][4] |
| Calo Calonymos ben<br>David (d. after 1526),[5]<br>Venice **H** | *Alpetragius:*<br>Theorica planetarum, 1531<br>*Averroes:*<br>Destructio destructionum, 1527<br>Epistola de connexione intellectus abstracti cum<br>homine, 1527<br>Libellus de viis probationum in opinionibus legis,<br>1527 [Kašf ʿan manāhiǧ, quotations in Calo's De<br>mundi creatione]<br>Libellus de differentia inter legem et scientiam in<br>copulatione [Faṣl al-maqāl, quotations in Calo's De<br>mundi creatione][6]<br>De qualitate esse mundi [one of the Physical<br>Questions] (Ms. Vat. ottob. lat. 2016)[7] |

(continued)

*Table 2.*    (continued)

| | |
|---|---|
| Vitalis Nisso (d. ?) **H** | *Averroes:*<br>Comp. Gen., 1550/1552 |
| Paolo Ricci (d. 1541),<br>Padua and Pavia **H** | *Albucasis:*<br>Liber theoricae, 1519 (first two books of the Kitāb<br>at-Taṣrīf li-man ʿağiza ʿan at-taʾlīf ) [ed. by Ricci,<br>translator uncertain]<br>*Averroes:*<br>Comm. med. Cael., 1511<br>Comm. mag. Metaph. Prooem. 12, 1511 |
| Jacob Mantino (d. 1549),<br>Bologna, Venice, and<br>Rome **H** | *Averroes:*<br>Comm. med. Animal., 1521 (also Vat. lat. 4547)<br>Comp. Metaph., 1521<br>Comm. med. Isag., 1550/1552<br>Comm. med. Cat., 1550/1552<br>Comm. med. Int., 1550/1552<br>Comm. med. Top. 1–4, 1550/1552<br>Comm. med. Poet., 1550/1552<br>Comm. med. Phys., 1550/1552<br>Comm. mag. Phys. Prooem., 1550/1552<br>Comm. mag. An. 3.5+36, 1550/1552<br>[retroversion, not from an Arabic–Hebrew, but from<br>a Latin–Hebrew version][8]<br>Comm. mag. An. post. (fragm.), 1562<br>Epitome of Plato's Republic, 1539<br>Colliget 3.57–59, 1550/1552<br>*Avicenna:*<br>Canon 1.1, 1.1.3.29, 1.4, 1530, 1538, ca. 1540 |
| Jean Cinqarbres<br>(d. 1587), Paris **H** | *Avicenna:*<br>Canon 3.1.4, 3.1.5, 3.2, 1570, 1572, 1586 |
| Jacob Christmann<br>(d. 1613), Heidelberg **H** | *Alfraganus:*<br>Chronologica et astronomica elementa, 1590 |
| Jean Faucher (d. before<br>1630), Beaucaire **A** | *Avicenna:*<br>Cantica, 1630 |
| Tommaso Obicini of<br>Novara (d. 1632), Rome **A** | *Al-Abharī:*<br>Isagoge . . . in scientiam logices, 1625 |
| Peter Kirsten (d. 1640),<br>Breslau **A** | *Avicenna:*<br>Canon 2, 1609 or 1610 |
| Johann Buxtorf Jr.<br>(d. 1664), Basel **H** | *Maimonides:*<br>Liber mōre nevūkīm, 1629 |

*Table 2.* (continued)

| | |
|---|---|
| Antonius Deusing<br>(d. 1666), Groningen **A** | *Avicenna:*<br>Cantica, 1649<br>*Mesue:*<br>Aphorismi, 1649 |
| Edward Pocock Sr.<br>(d. 1691), Oxford **A** | *Algazel:*<br>*Interpretatio confessionis fidei orthodoxorum,* 1650 |
| Jacob Golius (d. 1667),<br>Leiden **A** | *Alfraganus:*<br>Elementa astronomica, 1669 |
| Pierre Vattier (d. 1667),<br>Orléans and Paris **A** | *Avicenna:*<br>Canon 3.1, 1659 |
| Vopiscus Plemp<br>(d. 1671), Louvain **A** | *Avicenna:*<br>Canon 1.2, 4.1, 1658 |
| Georg Hieronymus<br>Welsch (d. 1677),<br>Augsburg **A** | *Avicenna:*<br>Canon 4.3.21–22, 1674 |
| Edward Pocock Sr.<br>(d. 1691) and Edward<br>Pocock Jr., Oxford **A** | *Ibn Ṭufayl:*<br>Epistola . . . de Hai Ebn Yokdhan, 1671 |

1. Puig Montada, 'Eliahu del Medigo, traductor,' 713–729.

2. Di Donato, 'La traduzione latina della Risāla al-wadāʿ,' 301–314.

3. The titles and opening paragraphs of the altogether eleven chapters, which are preserved in Balmes's autograph, are edited in Tamani, 'Una traduzione,' 91–101.

4. Aristotle and Averroes, *Omnia . . . opera* (1562), 1:2b, f. 120v–128r. The names and titles are: (1) Abbulkasis Benadaris, *Quaesitum de notificatione generis et speciei,* and *Quaesitum de nominis definitione;* (2) Alhagiag bin Thalmus (Ibn Ṭumlūs?), *Quaesitum de mistione propositionis de inesse et necessariae;* (3) Abuhalkasim Mahmath ben Kasam, *Quaesitum de modo discernendi demonstrationes propter quid et demonstrationes quia,* and *Castigatio quartae speciei secundi generis congerierum demonstrationum ipsius Alpharabii;* and (4) Abuhabad Adhadrahman ben Iohar, *Epistola de negativa de necessario et negativa possibili,* and *Epistola De termino medio quando fuerit causa maioris an imaginabile sit secundum aliquem locum non esse causam minoris.*

5. On this translator see DBI s.v. 'Calonimo, Calo' (by Hill Cotton); Di Donato, 'Traduttori di Averroè,' 38–40.

6. Di Donato, 'Il *Kašf . . .* confronto,' 241–248.

7. Di Donato, 'Traduttori di Averroè,' 39n41.

8. See Zonta, 'Osservazioni,' 15–28, who argues that Mantino was translating from the Latin–Hebrew version of Baruch ibn Yaʿish (fifteenth century).

philology, more often than not, abound in marks given to colleagues of the past.[6] Philology is neither a subjective nor an objective science: on the one hand, its rules and goals are not fixed, but subject to change, while on the other hand the problems posed and the solutions chosen are not all arbitrary, but differ in quality. It would be hermeneutically unproductive to renounce all qualitative judgements, while it would be historically unsound to insist on a fixed ideal of "pure philology." It seems sensible, therefore, to steer a middle course and judge the translators by their own measures or, at least, by those of their colleagues or opponents.

In the history of philology, the Latin translators and revisers of Arabic scientific and philosophical texts have not yet found a place—either in the history of Latin or of Arabic philology. In historiographies of Arabic studies, this has been justified by arguing that the work of these scholars did not lead to a systematic study of the Arabic language or Islamic culture, since their primary interest was in the scientific content of the texts.[7] In the field of Latin studies, the term *philology* is usually reserved for the study of ancient texts—if it is not more narrowly defined as the "invention" and "resuscitation" of what is called textual criticism.[8] As a consequence, it still seems an oxymoron to speak of "medieval" or "scholastic philology," in spite of the fact that modern scholarship has shed much light on the working methods and achievements of medieval translators, commentators, and revisers of texts, such as on Theodulf of Orléans's ninth-century editions of the Bible. To be sure, Renaissance humanists, and Poliziano in particular, developed new forms of textual criticism that were not practiced in the Middle Ages, especially with respect to the evaluation of textual witnesses and to source criticism.[9] But there is no question that the work of the Toledan translators,[10] or William of Moerbeke in the thirteenth century,[11] or Andrea Alpago in the Renaissance is philological in the sense that these scholars sought to improve the transmitted text of another author by solving technical textual problems. It is this wider sense of philological activity that is the topic of the present chapter.[12]

## AVERROES: PREFACE TO *METAPHYSICS* 12

Among the pieces of Arabic learning that were translated several times into Latin in the Renaissance is Averroes's preface to book 12 of his Long commentary on the *Metaphysics:*[13] it was translated twice in the 1480s by Elia del Medigo, sometime before 1511 by Paolo Ricci and between about 1520 and 1549 by Jacob Mantino. These translations filled a gap that had been left by Michael Scot, the medieval translator of Averroes's commentary.[14]

It is with Elia del Medigo that the story of Arabic–Latin and Hebrew–
Latin translations in the Renaissance opens. From 1480 to about 1490, del
Medigo, a Jew native from Candia (Crete), lived in Italy, notably in Venice,
Padua, and Florence.[15] After his arrival in Italy, del Medigo produced several
treatises in Latin on topics of cosmology and psychology, which largely
adopt positions by Averroes.[16] But he also worked as a translator. By 1486, he
had already produced a number of translations of works by Averroes, most
of them for Giovanni Pico della Mirandola: the compendium of Aristotle's
*Meteorology,*[17] the Middle commentary on *De animalibus,* a *Tractatus de intellectu speculativo,*[18] questions concerning the *Analytica Priora,*[19] and the epitome
of Plato's *Republic.*[20] Del Medigo's *Commentatio media* on the *Metaphysics,*
books 1–6, in turn, is not a straightforward translation, but apparently a
commentary written by del Medigo himself based on Hebrew versions of
Averroes's Middle and Long commentaries on the *Metaphysics.*[21] When Pico
fled to France and was arrested there, del Medigo worked for another young
nobleman: Domenico Grimani of Venice. In 1487, at the age of twenty-six,
Grimani had received his doctorate at the University of Padua; later, when
he was cardinal and patriarch of Aquila, he would become famous as a humanist cleric and dedicated bibliophile, who acquired Pico's library after his
death.[22] It is fortunate that a letter from del Medigo to Grimani survives, in
which he explains his motives for translating the preface to *Metaphysics* 12.
Del Medigo recalls that Grimani, when philosophizing, as he liked to do,
had often spoken to him about the sequence of the books of Aristotle's *Metaphysics* and about the universals not existing outside the soul.[23] Both topics
are addressed in Averroes's preface to *Metaphysics* 12, and that is why del Medigo decided to translate the preface for Grimani. He had already translated
the preface previously for Pico, says del Medigo, but does not have it at
hand.[24] He concludes the letter with a remark about his method of translation: "Because I know that you are able to know many good and difficult
things by yourself, I have in this <case> preserved not only its meaning, but
also the words in the best way possible. Farewell."[25] In another preface, del
Medigo mentions a recent discussion of his with Grimani and Pico about
the relation of physics and metaphysics.[26] It is interesting to see that Grimani was a Paduan acquaintance of del Medigo.[27] Despite the humanist
interests of his later years, Domenico Grimani retained his liking for the
Aristotelian tradition and for Averroes, a liking that in the 1480s in Padua
he shared with many other intellectuals. Later in his life Grimani commissioned translations of Averroes from Abraham de Balmes, his personal physician, as we will see below. Grimani's interest in Averroes is also reflected in
the fact that in the Hebrew section of his famous library, Averroes was the
most important author. Of the 193 Hebrew manuscripts that have survived,

34 are Averroes manuscripts. Grimani had inherited a substantial part of these manuscripts from Pico, but Pico's collection, as far as can be reconstructed today, contained only a few Averroes manuscripts.[28] From Pico we know that he had discussed problems of natural philosophy with Grimani.[29] In view of all this, there is much evidence that the University of Padua, its teaching of Averroes and its aristocratic students, is the climate in which the Hebrew–Latin translation movement for Averroes's works began.

The second translation of the *Metaphysics* preface was produced without awareness of del Medigo's previous rendering. The translator, Paolo Ricci, who died as a Habsburg court physician in 1541, was born into a Jewish family of Northern Italy and baptized as a young man.[30] Many of his printed works are concerned with cabbalistic themes—in fact, he is a famous promoter of *cabbala christiana*—but in 1511 a volume also appeared with several translations of works by Averroes. The translator and editor of the volume calls himself "Paulus Israelita"; he is clearly identical with Paolo Ricci.[31] Paolo dedicates this work to Étienne Poncher, archbishop of Paris from 1503 to 1519,[32] whom Paolo addresses as "Maecenas" *(praesidem Mecoenatemque)*. It had long been his intention, says Ricci, to offer a present to the archbishop, which he is now going to fulfill by presenting to him Averroes's *De coelo et mundo*—which turns out to be Ricci's Hebrew–Latin translation of Averroes's Middle commentary on Aristotle's *De coelo,* which, before Ricci, had not been accessible in Latin. To this he adds two small treatises of his own, plus Averroes's prooemium to the Long commentary on the *Physics* in the thirteenth-century translation of Theodore of Antioch.[33] Unfortunately for our purposes, Ricci does not say anything about the final piece of his volume, the translation of Averroes's preface to *Metaphysics* XII, which is announced on the title page as: *Averois in duodecimo Metaphisice prooemium quoque de hebraico decerptum exemplari.* In order to understand Ricci's motives and methodical principles, we have therefore to extrapolate from what he says about Averroes's commentary on *De caelo:* that the text is a present worthy of the archbishop because of the celebrity of its author, its thematic nobility, the clarity of its meaning, its novelty in the library of the Latins, and its great accordance with the theologians.[34] A further motive emerges from Ricci's prefatory words to the *De caelo* commentary: "Not only these three prefaces, but the entire Latin version of Averroes, as I have once explained to some of our companion philosophers *(conphilosophis),* abounds in many corruptions and errors *(crebris corruptelis erroribusque).*"[35] He insists on this lest he be accused as a bad translator—by people who do not bother to consult the many experts in Hebrew or the few in Arabic.

That Ricci makes this last philological point—which is missing in del Medigo—testifies, first, to his humanist, or at least antischolastic attitude, and second, to the fact that he was expecting criticism of his translation because his *Averroes latinus* differed in Latin style from all previous renderings. We will have to investigate later whether the humanist tone in Ricci's prefaces is reflected in the technical details of his translation. The two translators also differ in that del Medigo responds to the philosophical interests of his patron Grimani, whereas Ricci translates Averroes's *De coelo* because he finds the treatise a noble philosophical text suitable to be dedicated to the archbishop.

The translators, however, coincide in one important respect: the intellectual context of their activity. Paolo Ricci's volume of 1511 was dedicated to the Paris archbishop, but was intended to be read by the *conphilosophi* mentioned in the preface; with this Ricci refers to the acquaintances of his student years in Padua. It was in Padua that Ricci had studied philosophy under Pietro Pomponazzi, until about 1506, when the latter had not yet turned against Averroes. Again, as one generation earlier with del Medigo, the motives for translating Averroes are connected to the intellectual climate of Padua. In 1511, at the time of the publication of the volume, Ricci was teaching medicine and philosophy in Pavia, the University of Padua was closed,[36] and Pomponazzi had moved to Bologna. The ties between Ricci and his teacher remained close, however, as appears from the fact that in 1514 Pomponazzi wrote a preface of recommendation to Ricci's main philosophical work, *In apostolorum simbolum dialogus*.[37] Also, we know that Pomponazzi in a lecture in 1514 quoted Ricci's edition of 1511 on the authenticity of Averroes's prooemia to the *Physics*.[38]

Another indication that Ricci was aiming at a Paduan audience is that the Paduan Aristotelians, such as Nicoletto Vernia and Agostino Nifo,[39] had long been very interested in the *De caelo* tradition. Pomponazzi himself conducted a lecture course on *De caelo* in Bologna in 1515–1516, in which he first signaled disagreement with Averroes's theory of the unicity of the intellect. With respect to Averroes's preface to *Metaphysics* 12, we can surmise that Ricci's translation could expect to attract readers in the same philosophical context. The text fits well into the 1511 volume because Ricci himself discusses the function of prefaces in a small treatise. The *Metaphysics* preface forms a counterpart to the *Physics* preface also included in the volume.

The third translator of the *Metaphysics* preface is the most prolific and most acclaimed among all Renaissance translators of Averroes: Jacob Mantino.[40] His version was printed only after his death in 1549: it was included in the 1550/1552 Giunta edition of the combined Aristotle and Averroes (on this editorial enterprise see Chapter 1). Mantino's motives for translating Averroes can be gathered from three dedications that accompany his editions of the translations: to Pope Leo X in 1521 (paraphrase of

*De animalibus*), to Ercole Gonzaga, bishop of Mantua, in 1521 (epitome of the *Metaphysics*), and to Pope Paul III in 1539 (paraphrase of Plato's *Res publica*).[41] Mantino served as personal physician to Pope Paul III, probably already before Paul's election in 1534. In these prefaces, Mantino points out, just as Ricci had before him, that he is translating works of Averroes that are unknown to the Latins,[42] that his translations are of great use to natural philosophers, and that teaching Aristotle is impossible without recourse to Averroes.[43] He adds the critical remark that it is characteristic of all older, that is, medieval translations that "much appears unkempt and mutilated" *(inculta et mutilata)*: Averroes's works were rendered *foede, barbare, obscure*. In typically humanist manner, Mantino here castigates the uncultivated Latin style of Averroes's medieval translators. This is the reason, says Mantino, that many in these days condemn Averroes's doctrine *(doctrinam Averroys damnent)*. It is clearly one of Mantino's motives to rescue Averroes for the humanist movement. At the same time, he admits that his own translations are not characterized by *latina eloquentia*, because "I have not attained it" *(fateor enim me eam non esse assecutum)*. He promises, however, to avoid the old style and to translate clearly and reliably.[44]

In view of these humanist tones, one might suspect that Mantino was aiming at an exlusively humanist audience and thus different from that of del Medigo and Ricci. But this is not the case, because the most important patron of his translation was the second of the three dedicatees mentioned, Bishop Ercole Gonzaga of Mantua, and Gonzaga, in turn, was an admirer of Pietro Pomponazzi. Apart from the 1521 dedication to Gonzaga, we also have from Mantino a reference to Gonzaga in a dedication to Guido Rangoni, at that time commander of the Pope's army and a Hebrew pupil of Mantino. Mantino had translated Maimonides's *Eight Chapters* on ethics for Rangoni and offered to produce more translations for him:

> And even though I have translated Averroes' philosophy for my most noble patron, the revered Ercole Gonzaga, doubtlessly the most generous prince and most decorated with all virtues, I shall, nonetheless, not stop—hindered by these tasks—to fulfill your wishes.[45]

This text was printed in 1526, that is, five years after the printing of the epitome of the *Metaphysics* that Mantino had translated for Ercole Gonzaga. We can gather from the later text of 1526 that Mantino is still, or again, engaged with translating "Averroes' philosophy" into Latin for his patron Gonzaga and that he conceives of this enterprise as a longer-lasting project that will consume much of his time in the future ("hindered by these tasks"). We can therefore assume that a considerable number of

Mantino's Latin translations of Averroes were produced in the 1520s and promoted by Ercole Gonzaga. Not all of them, however, originated in this time and context: we know from the editors of the 1550/1552 Giunta edition that Mantino's death in 1549 prevented him from finishing four translations: the epitomai of the *Organon* with Levi ben Gerson's super-commentary, the Middle commentaries on *Topics* and *Physics,* and the Long commentary on the *Posterior Analytics.*[46] These then seem to be enterprises Mantino started later in his career, in the 1530s and 1540s, when he was physician to Pope Paul III. The editors of the Giunta edition, Bagolino and degli Oddi, may have invited or asked Mantino to produce these new translations or revisions—they do not mention this expressly, however. In any case, it remains very plausible that Ercole Gonzaga was the single most important patron of the translators of Averroes in the Renaissance—equaled, if by any-body, by Domenico Grimani, to whom we shall return below when dealing with Abraham de Balmes. It adds to Gonzaga's influence on the Averroes translation movement that another translator, Calo Calonymos ben David, dedicated a volume of Averroes translations to Gonzaga: the 1526 edition with Calo's Hebrew–Latin translations of *Destructio destructionum* and the *Epistola de connexione intellectus abstracti cum homine.*[47]

Ercole Gonzaga, member of a family that ruled Mantua for many centuries, was born in 1505, that is, he was only sixteen years old when Mantino dedicated Averroes's epitome of the *Metaphysics* to him; later in life Gonzaga would become a cardinal and preside over the Council of Trent.[48] His interest in Averroes goes back to his student years in Bologna, when Pietro Pomponazzi was his teacher. The most infamous and famous philosopher of his time is known to have escorted the young nobleman every day that he lectured from his house to the university.[49] The presence of the young student once prompted Pomponazzi to insert a eulogy on the family of Gonzaga in his lecture.[50] When Pomponazzi died in 1524, he was buried in Mantua, his hometown, and Ercole Gonzaga, now bishop of Mantua, erected a statue in bronze in memory of his teacher in the church San Francesco.[51] Mantino himself mentions in his dedication of 1521 that Gonzaga "drinks from the clear water of the great spring of the most famous philosopher Pietro Pomponazzi, the patrician from Mantua."[52] Ercole Gonzaga was studying philosophy under Pomponazzi when the latter had already turned against Averroes's unicity thesis and had triggered the most spectacular philosophical debate of the Renaissance by advancing his mortality thesis in 1516.

It is likely that Jacob Mantino himself also knew Pomponazzi directly. In 1530, he writes to the Venetian doge Andrea Gritti, that he had "always since his youth been addicted to your most flourishing studio of Padua"[53]—which seems to imply that Mantino had studied at the studio before its closure in 1509 (it was reopened in 1517). Also, as a young physician Mantino seems to have lived in Bologna in the early years of the 1520s,[54] where Pomponazzi was teaching philosophy.

We may conclude that the intellectual context of the new wave of Latin translations of Averroes is thoroughly colored by the Aristotelian and Averroist currents of Padua. This is true in spite of the fact that only del Medigo lived in Padua for a longer time, but Grimani in Venice, Ricci in Pavia, Pomponazzi in Bologna, Mantino in Bologna or Venice, and Ercole Gonzaga in Bologna or Mantua—all of them, in one way or another, had close ties to the philosophical climate of the University of Padua. This is remarkable because it is characteristic of the Paduan milieu after 1500 that the leading teachers of the university, Nifo and Pomponazzi, slowly turned their back on Averroes's key doctrine of the unicity of the intellect. However, they continued to use and lecture on his commentaries, and as the evidence of patronage for the translation movement clearly shows, there existed a broad academic milieu connected to Padua that was very sympathetic toward Averroes—and remained so after leading Paduan Aristotelians had attacked doctrines of Averroes publicly. At the same time, the patrons Pico and Grimani are known for their typically humanist interests and style: Pico as the author of the *Oratio de dignitate hominis* and of elegant letters to his humanist friends,[55] and Grimani as a collector, among other books, of Greek literary works.[56] The patronage by Pico and Grimani is a first indication that scholars with humanist interests also participated in the Arabic-Hebrew-Latin translation movement.

Let us now turn to the translation of the preface to *Metaphysics* 12 proper and see whether the varying motives and programmatic statements of the translators—del Medigo's wish to feed the philosophical interests of his Paduan clients, Ricci's antimedievalism, Mantino's humanist interests, but modest aspirations in Latin style—are reflected in the actual details of their work as translators; and let us see what these details say about the quality of the translations.

Del Medigo's translation was not printed in the Renaissance, but was known to Jacob Mantino, as will appear below. Del Medigo's Latinity has been judged to leave "much to be desired."[57] Translated into historical language, this can only mean that del Medigo's Latin differs markedly from the Latin of the humanists of his day, and this, in fact, is true. The prooemium is translated from Hebrew. That del Medigo did not know Arabic, is apparent from a note he inserted into the text concerning the letters of the alphabet attached to the books of Aristotle's *Metaphysics:* according to the Latin alphabet, there is no book missing, but, says del Medigo, he cannot tell with respect to the Arabic alphabet, which he does not know.[58]

Averroes begins his preface to book Lambda by saying that no ancient commentaries on the *Metaphysics* have survived, except a partial commentary on book Lambda by Alexander. And, he adds: "We have found on it *(i.e., book Lambda)* by Themistius a complete exposition according to the sense."[59] Del Medigo translates: *Et invenimus in ipso a Themistio declarationem completam secundum rem seu sententiam.*[60] As in this sentence, del Medigo translates very literally,[61] and preserves the Hebrew word order. If a word is ambiguous or if he is unsure about its meaning, del Medigo adds an alternative Latin term with *seu: secundum rem seu sententiam,* or as other examples go: *signum seu ratio, significatio seu intellectio,* or *fundamentum seu subiectum.* In the present case, the ambiguity goes back to the Hebrew version of Averroes's text: The Hebrew term עניין *('inyan)* may mean "sense," but also "thing, affair, issue," a connotation excluded by the Arabic term *ma'nā* in this context, where it clearly means "sense."[62] Del Medigo's translation is overcorrect: It preserves the original sense better than the later translations, which omit the qualification "according to the sense" and simply write *paraphrasis* (Ricius) or *expositio* (Mantino); but del Medigo adds a lexical ambiguity.

The extreme literalness of del Medigo's translation has left its stamp on the Latinity. Del Medigo imitates the usage of prepositions in Semitic languages, for example, when writing *considerare in principiis* instead of *considerare principia,*[63] or *incipere in solutione dubitationum* instead of *incipere solutione dubitationum.*[64] When del Medigo employs an unclassical *cum* in the phrase *non est possibilis cum ipso ratiocinatio* ("reasoning is impossible for him"), Mantino supplants this with a sentence in classical style: *nulla debet ab eo acceptari ratio.*[65] Particularly conspicuous is del Medigo's imitation of Arabic and Hebrew syntax when using a complete sentence as predicate, as in this example: *Nam duo tractatus qui sunt post hunc, non declarat in eis aliquid primaria intentione,* which translates as: "For, the two treatises which follow upon this one, he does not explain in them anything pertaining to his primary aim."[66] This is a nominal sentence, where the subject and predicate are not bound together by a copula, but by a pronominal suffix in the predicate part of the sentence. In this case, the personal pronoun *eis* refers back to the nominal phrase *duo tractatus.* Latin lacks such syntactic flexibility. Del Medigo's literal renderings therefore produce anacolutha, as is obvious also in the following case: *Et ideo negans hoc principium, non est possibilis cum ipso ratiocinatio,* which means: "And therefore who negates this principle, reasoning is impossible for him."[67] Del Medigo's anacolutha, therefore, are not the result of modeling the Latin upon the usage of the spoken vernacular, as has been claimed,[68] but of his imitation of Semitic syntax.

Del Medigo increases the syntactic disarrangement of his Latin, when opening his nominal sentence with a noun in the accusative: *Nam reliquos tractatus quos posuit in isto sermone, quidam eorum continent dubitationes,* meaning: "For, the remaining treatises, which he wrote in this work, some of them contain doubts."[69] In this sentence, the original genitive of the antecedent noun *reliquorum tractatuum* is attracted into an accusative by the relative pronoun *quos*. This kind of inverted attraction of the relative is extremely rare in classical Latin.[70] In such cases, when del Medigo's sentences breach syntactic rules of Latin, the translation does not serve its purpose: it hinders understanding. There was a reason, therefore, for Jacob Mantino to substantially revise the translation. In the case of this particular passage, Mantino follows Averroes's word order, but remains within the limits of Latin grammar; this he achieves by opening the sentence with a prepositional phrase instead of a bare noun: *Nam inter alia, quae dicta sunt in hac scientia, quaedam habent nonnullas dubitationes.*[71]

Del Medigo's literal translation technique has its predecessors in the Middle Ages. The mentioned characteristics of his style are found, occasionally, also in the medieval Arabic-Latin translations produced in Spain and Sicily: John of Seville, Dominicus Gundisalvi, Gerard of Cremona, and Michael Scot. Del Medigo also resembles these translators in his usage of medieval Latin: he likes to construe the perfect participle passive with *fuit* instead of *est*, as in *locutus fuit, apprehensum fuit, dictae fuerunt,* and so forth,[72] and may leave out the subjunctive in indirect questions, for example: *quaerere quae sunt*[73] ("inquire which are"), corrected by Mantino into *investigare quaenam sint.*[74] But del Medigo's text differs from that of his medieval predecessors in that the mentioned stylistic peculiarities appear very regularly. As a consequence, his Latin is more Arabicized and scholastic than that of the Toledo translators. As such, it must have been particularly unwelcome to humanist readers. But del Medigo's patrons Pico and Grimani, remarkably enough, were not choosy. They must have known what kind of text del Medigo would deliver, and apparently were simply glad to get access to Averroes's text. Del Medigo's principal aim was not to lose any bit of information or any bit of sense of the text he was translating. In view of the strained Latinity that was the result of his endeavor, one cannot say that his approach was fully successful.

When we now turn to Paolo Ricci, we come to the only translator of Averroes who writes an ostentatiously humanist Latin. I have not found any indication that Paolo Ricci knew del Medigo's translation,[75] when he produced his own version sometime before 1511, the year in which it was printed. Ricci translates from a Hebrew manuscript that in a number of significant cases diverges from the readings of the transmitted Arabic text and the Hebrew text translated by del Medigo and Mantino. It is apparent from Bouyges's critical apparatus to the Arabic edition that Ricci's translation often shares readings with the Hebrew manuscript *a* (Paris, Bibl. Nat. hébr. 886).[76] The Hebrew manuscript, however, can explain only very few of the many divergences between Ricci's version and those of del Medigo and Mantino. The most obvious difference in technique is Ricci's usage of an elaborate classical style. The opening sentence, which contains the reference to Alexander and Themistius, strikes a tone entirely unheard of in the history of *Averroes latinus:*

> Quoniam nec Alexandri nec posteriorum in aliquam huius scientiae dictionem glosa expositiove reperta est nisi in hanc dictionem, cuius quasi duae tertiae partes glosae Alexandri et completa Themistii paraphrasis mihi ablata est, ideo in singulis articulis Alexandri sententiam recensere ratum fore censui, et quod Themistius et nos apposuimus aut dubitavimus adicere.[77]

Ricci's Latin translates as:

> Because there has not been found any commentary or exposition by Alexander or any later author on any part of this science except on this part, of which about two-thirds of the commentary by Alexander and the entire paraphrase of Themistius have come down to me, therefore I thought it would make sense to muster Alexander's text in its single parts and to mention what Themistius and myself have added or cast doubt upon.

The most obvious difference between Ricci and del Medigo is that Ricci has turned into one long period what del Medigo, in imitation of the original syntax, had expressed in five sentences and many more words. Del Medigo had followed the parataxis of Arabic and Hebrew in his translation: *Dico . . . Nam nos invenimus . . . Et invenimus . . . Et iam visum est mihi . . . Dicemus etiam.* Ricci, in contrast, makes full use of the possibilities of hypotaxis offered by the Latin language. The drawback of Ricci's method is that he introduces logical relations between the sentences that are not in Averroes's text. In this case, the connection *quoniam . . . ideo* is his own invention. There are more examples of Ricci's introducing, for the sake of hypotaxis, precise logical relations between sentences:[78] causal relations, concessive relations, and so on. Similarly, in the twelfth century, when

Hermann of Carinthia tried to translate Arabic astrology into a classicizing Latin acceptable to the flourishing centers of Latin learning at the
French cathedral schools, he also added disjunctive, adversative, causal,
and other particles—without an explicit correspondence in the Arabic text.
This is an unavoidable feature of all classicizing Latin translations from
Semitic languages, as I have tried to show elsewhere.[79] At the same time,
it is a tactic of doubtful value for translating scientific or philosophical
texts.

In general, Ricci has a tendency to shorten the text. In this passage, he
also omits the above-mentioned qualification "according to the sense"
and Averroes's statement that he is going to summarize Alexander "as
clearly and briefly as possible." His abbreviation technique, however, is
gentle if compared with the more drastic techniques employed by some
medieval translators such as Michael Scot.

Paolo Ricci's Latin is clearly inspired by humanist classicizing ideals.
He has a predilection for intricate Latin expressions, as in the above-
quoted sentence: *sententiam recensere ratum fore censui,* meaning: "I thought
it would make sense to muster Alexander's text." Another example comes
from a sentence of the preface that is concerned with book Alpha of the
*Metaphysics:* "And since the ancient authors *(i.e., before Aristotle)* have held
false opinions on the principles of beings, it is necessary for him to refute
them."[80] Jacob Mantino, the third translator, writes in classical, but
simple style: *antiqui de principiis entis falsas protulerunt sententias, quas refellere
oportebat.* Ricci's formulation is more elaborate. He writes *prisci* instead of
the more regular *antiqui, primordia* instead of the more technical *principia,
obruere* instead of the more common *refellere.* This is Ricci's sentence: *At
quia prisci de rerum primordiis falsa quaedam enuntiant, condecens quoque erat
haec ipsa illorum dicta obruere.*[81] The most conspicuous element of the sentence is *condecens,* a rare late antique term.[82] It testifies to Ricci's liking for
the exquisite, but also shows that he is occasionally carried away with his
Latin: in the view of Averroes, it was not "apt" *(condecens)* for Aristotle
to criticize his predecessors in book A of the *Metaphysics,* but necessary
*(waǧaba ʿalayhi)* (ראוי עליו), as Mantino correctly put it.

In sum, therefore, Paolo Ricci produces an impressive example of
humanist learning applied to Averroes's text, but it is doubtful that he
achieves all his goals. As was explained above, Ricci mentions various motives for his publishing of Latin translations of Averroes, among them that
they present material hitherto unknown to the Latins and that all previous Latin translations of Averroes abound "in corruptions and errors."[83]
He is clearly successful in providing new material to his readers—given
that he did not know del Medigo's translation—and to write a Latin that

appeals to the gusto of humanists. But his classicizing translation is marked by imprecision and introduces rather than avoids errors.

The third translation of the preface comes from Jacob Mantino and first appeared in print in the 1550/1552 Giunta edition of the combined Aristotle and Averroes, that is, after Mantino's death in 1549.[84] Mantino knew del Medigo's translation and read it closely. We have seen above that Mantino does not translate as literally as del Medigo and that he occasionally classicizes del Medigo's Latin style, without however shying away from adopting expressions untypical of classical authors.[85] Mantino's Latin is mildly classicizing, but does not aspire to Ricci's exalted stylistic ideals. Ricci's translation, in fact, seems to have been unknown to Mantino. Some of del Medigo's sentences are hardly altered at all. In others, Mantino introduces minor changes, such as when writing *notificatio substantiae* instead of *declaratio substantiae* (for Arabic *bayān,* "elucidation, explanation," Hebrew ביאור).[86] For the most part, Mantino does not stop at the alteration of one term in del Medigo's sentences. It is his technique to work with del Medigo's vocabulary, but to introduce many changes in syntax and semantics. Del Medigo's nominal sentences, for example, are all transferred from Semitic syntax into regular Latin syntax. Mantino's changes are so substantial that the translation can properly be called his own.

A typical example is the following sentence, in which Averroes announces that he finds it appropriate to give an exposition of Alexander's summary of the remaining books of the *Metaphysics* other than book 12:

> Arabic: "What he (*sc., Alexander*) said, however, with regard to understanding the contents of the rest of the books written on this science, is ambiguous; perhaps this is the passage that had best be explicated."[87]

> del Medigo: "Et id, quod dixit <Alexander> in significatione seu intellectione eorum quae continentur in reliquis tractatibus positis in hac scientia, habet aliam declarationem seu alia declaratio pertinet ei, et iste locus est dignior aliis in declaratione huius."

> Mantino: "Id vero, quod dixit pro intelligentia eorum quae in aliis libris positis in hac scientia continentur, potest aliter exponi, et forte huic loco magis debetur expositio quam alibi."[88]

Mantino has extinguished the alternative translations connected with *seu* typical of del Medigo, and he has transformed the Arabic–Hebrew word order into a syntax more acceptable to Latin readers: he has moved the verb *continentur* to the end of the clause and has found alternative expressions for del Medigo's clumsy phrases *habet declarationem* and *locus est dignior in*. At first sight, Mantino's technique of revision resembles the *novae versiones* of Averroes's *Colliget,* Mesue, and Abenragel produced by the humanist authors Jean Bruyerin Champier, Jacobus Sylvius, and Antonius Stupa without any knowledge of Arabic and Hebrew. In contrast to these revisers, Mantino compared del Medigo's translation with the Hebrew. In this sentence, this is obvious from the insertion of the word *forte* (from Hebrew אולי, Arabic *la'alla*),[89] which hardly could have been guessed at. Mantino worked with both a previous Latin translation and a Hebrew version not only in this case but also when translating other Averroes commentaries: the *Topics* commentary, as we shall soon see, and the *Categories* and *De interpretatione* commentaries, as Roland Hissette has shown.[90]

The quoted passage is a reminder that the Hebrew intermediary occasionally may have an effect on the sense of the text. Instead of the phrase "what he said . . . is ambiguous," the Latin translators write *habet aliam declarationem* (del Medigo) and *potest aliter exponi* (Mantino).[91] Del Medigo and Mantino follow the Hebrew ביאור אחר,[92] which means "a different exposition," whereas the Arabic reads *fi-hi iḥtimāl* ("there is an ambiguity in it").[93]

In the final analysis, all three translators aimed at a successful transmission of Averroes's work. But it is Jacob Mantino who was most effective in attaining his goals. Del Medigo succeeded, of course, in making an unknown Arabic text available to his patron, and Ricci produced a Latin version of Averroes in an entirely new garment, a lavishly classicizing style. But we have witnessed both falling short of "preserving meaning and words" of Averroes's text (del Medigo), and of avoiding and extinguishing "errors and corruptions" (Ricci). Ricci's "humanist Averroes" possibly attracted new circles of readers, but, for a philosophical reader, his vocabulary is too idiosyncratically stylish and, as a result, too imprecise. Mantino had declared his intention to preserve the meaning and render it intelligibly. In view of this, his technique of translation is practically and sensibly chosen: he did not produce an entirely new translation, but worked with the textual material offered by del Medigo, classicized it mildly, and corrected it against the Hebrew. While Ricci's translations are the most impressive attempt to transport Averroes into the age of Renaissance humanism, Mantino's version achieves the most: reliability, philosophical clarity, and readability for a Renaissance Latin audience.

## AVERROES: MIDDLE COMMENTARY ON THE *TOPICS*

The three most productive Hebrew–Latin translators of the Renaissance were del Medigo, Mantino, and a person we have not yet discussed: Abraham de Balmes. In order to put Balmes into the context of the translation movement, we turn to a text that was translated by Balmes and also by Mantino, and of which the Arabic and a chapter in Hebrew have been published: Averroes's Middle commentary on Aristotle's *Topics*.

Abraham de Balmes was born in Lecce, Puglia, into a Jewish family; his grandfather was a physician at the court in Naples.[94] When defending his Latin style, Balmes explicitly praises a fellow countryman of his who was born in Rudiae, the ancient Lecce—Ennius (d. 169 BC)—because it was typical of Ennius, says Balmes, to concentrate on the meaning rather than on the splendor of words.[95] By permission of the pope, Balmes was allowed to acquire a doctorate in 1492. He died in 1523, and it is reported by an eyewitness to the funeral that Balmes had taught in Padua—apparently not at the university, but in the context of the Hebrew community—and that many of his students followed his bier.[96] We do not know in which year Balmes left Naples for the Veneto; perhaps a consequence of the expulsion of the Jews from Naples in 1510. The main results of his work as a translator are collected in a volume that appeared in Venice in 1523, the last year of his life. In the manuscripts, three logical texts are dedicated to Alberto Pio, prince of Carpi, a humanist Aristotelian and student of Pomponazzi,[97] but the printed volume bears a dedication to Cardinal Domenico Grimani—who thirty-five years earlier had been the patron of del Medigo. The connection between Grimani and Balmes was particularly close: Balmes was Grimani's personal physician; and Balmes writes to Grimani about the Averroes texts: "we have produced them upon your commission *(te iubente)*, we have dedicated them to your name, and we have published them with your support *(te favente)*."[98]

Hence, Abraham de Balmes was linked in two ways to Padua: he was teaching in Padua, and his patron Grimani had studied there in the 1480s. The intellectual bond, however, that united Balmes and Domenico Grimani, was a predilection for logic: this, rather than the natural philosophy of Pomponazzi and Pico della Mirandola, seems to be the intellectual context of Balmes's translations.

The Middle commentary on the *Topics* did not appear in print until the combined Aristotle and Averroes edition of 1550/1552.[99] Nevertheless,

Balmes's motives for translating the commentary are clearly apparent in the 1523 edition of his translations and its accompanying dedication. Balmes explains that he is translating for the benefit of the students of the *bonae artes,* and that he has produced corrected versions of older translations of Averroes, new translations of the same, and a number of works of his own. For financial reasons, he cannot publish everything, and he has therefore decided to publish Averroes's logical treatises first, since logic precedes all other parts of philosophy.[100] Nobody can be an Averroist without knowing this logic. In logic, where there is no religious problem to be feared, Averroes deserves full trust. He occasionally even surpasses Aristotle.[101]

Abraham de Balmes argues here against two groups of intellectuals: the mainstream Averroists, who read only the books on natural philosophy and metaphysics by their master, but bypass logic, and against those "blown up with snootiness rather than philosophy who polemicize against our Averroes in order to boast about speaking Greek" *(ut graecizare se ostentent).*[102] In contrast to Mantino, who at the same time, in the early 1520s publishes his first translations, Balmes takes an openly antihumanist position, also with respect to language:

> Nobody shall believe that we have written this for the embellishment of eloquence *(eloquentia)* and the pretense of good expression; rather we have translated it by use of our commonly used words and the forms which appear everywhere, since I prefer to be criticized for the wrong usage of eloquence rather than for changing the meaning or the order <chosen by> the author; first, because from my very first years I was immersed in my Hebrew letters in my talmudic schools, in which truth was given priority to eloquence, which was despised, but second and more importantly because the difference between languages usually gives rise to frequent breaches of the norms of speech. For the Romans thought that witty eloquence *(dicacitas)* always deserves to be honored first, the Hebrews thought the same about truth *(veritas).*[103]

Latin style obviously was a permanent and important topic of the Renaissance translators of Averroes. We have seen that Ricci fully embraces humanist ideals and presents Averroes in a new Latin garment and that Mantino aims at writing humanist Latin, but admits of his own linguistic shortcomings. Balmes rejects the new concepts of good Latin style as, on the one hand, superficial and detrimental to the reliability of the translation, and, on the other hand, as the expression of a particular cultural attitude, which he labels Roman.

Jacob Mantino's motives for translating were discussed above. With respect to the Middle commentary on the *Topics* there does not exist any particular information on the circumstances of the project. The commentary is unfinished and covers only the first four books. The Giunta editors mention that they took care of Mantino's unfinished enterprise and decided to publish it, also because Mantino's translation was much to be preferred to Balmes's because of its splendor and truthfulness. In their view, Balmes's translation is "full of obscurity and errors" *(obscuritate et erratis referta).*[104]

When we turn to the actual text, we find that Balmes, in a manner similar to del Medigo, follows the Arabic–Hebrew text word for word. A sentence from Averroes's first chapter, which echoes Aristotle's opening sentence of the *Topics* on the purpose of the science (100a18), may serve to illustrate Balmes's technique of translation. Averroes's Arabic sentence translates into English as:

> This science, in general, is the science in virtue of which we are able, when we are the questioners, to form a syllogism out of acceptable *(mašhūr)* premises, in order to nullify every proposition which the respondent undertakes to sustain, and in order to sustain every universal proposition which the questioner wishes to nullify, when we are the respondents.[105]

Balmes's translation is:

> Et ut universalius [inquam] dicam, haec ars est illa ars qua possumus, quando quaerimus, instituere syllogismum ex promulgatis praemissis, ad destruendum positum quod respondens admittit in sua cautione et ad cavendum omne positum universale quod quaerens procurat destruere, quando nos responderemus.[106]

In view of this queer Latin sentence, Balmes had much reason to defend his style against potential critics. A humanist reader would have protested, to begin with, against Balmes's wavering between construing *quando* first with indicative *(quaerimus)* and second with imperfect subjunctive *(responderemus),* and against the strange usage of *cautio* and *cavere* in the sense of "defending" (for Arabic *ḥafiẓa,* "to defend, sustain," Hebrew שמר). Moreover, *procurat,* which usually means "to look after, attend to," is an infelicitous choice for a verb meaning "to want, try" (Arabic *rāma*) or "to strive" (Hebrew שדל). In these cases, Balmes's problem is not his literal method of translation, but his restricted command of Latin.

For most of this passage, Balmes translates *verbum de verbo*. The exception is the introductory phrase *ut universalius dicam,* which renders one term in Arabic and Hebrew (*bi-l-ǧumla,* בכלל).[107] The passage points to a difficulty of Semitic syntax that we have met before and that also poses problems for Balmes: the construction of relative clauses. Balmes's translation for "proposition which the respondent undertakes to sustain" is *positum quod respondens admittit in sua cautione,* which translates literally as: "a proposition which the respondent sustains in its/his defense," that is, Balmes all too literally translates the pronominal suffix with the misleading *sua.* Jacob Mantino's translation is far more convincing: *problema . . . quod ipse respondens assumit sustentare.*[108] The pronominal terms have disappeared, and the choice of the term *sustentare* for holding a thesis is clearly more fitting than Balmes's terms *cautio* and *cavendum.* Also, Mantino replaces *procurare* with the more adequate *conari,* "to try." This is Mantino's sentence in full:

> Generatim autem haec ars est qua, cum nos argumentamur, possumus syllogismum efficere ex propositionibus probabilibus ad destruendum omne problema universale propositum, quod ipse respondens assumit sustentare, et ad sustentandum quodlibet problema universale propositum, quod quidem conatur destruere ipse argumentator, quando nos fuerimus respondentes.[109]

Mantino opens the sentence with the classical term *generatim* ("in general"). From a humanist point of view, this is preferable to Balmes's unclassical expression *ut universalius dicam.*[110] There is much evidence elsewhere in the text that Mantino had Balmes's translation at hand when he produced his own version of the Middle commentary on the *Topics.* In this sentence this is not particularly obvious, but still tangible; we can witness Mantino retaining some combinations of terms: *haec ars est . . . qua . . . possumus, ad destruendum, quod . . . respondens, quando nos.* He changes Balmes's translation when it comes to logical key terms: he writes *propositionibus probabilibus* instead of *promulgatis praemissis* and *problema* instead of *positum.* In other passages, Mantino corrects Balmes's translation of the Hebrew עניין (*'inyan*) for Arabic *ma'nā,* "sense," "meaning," a term that had posed problems for del Medigo too: Balmes often uses the untechnical translation *res,* where the context requires *significatum.* An example is Balmes's sentence *hoc autem dicitur de duabus diversis rebus,* meaning "this is said about two different things," which Mantino corrects to: *quod quidem habet duo significata distincta,* "which has two different meanings."[111]

Mantino is very consistent in employing a technical vocabulary different from Balmes's in other chapters of the commentary as well. It is, in fact, in logical vocabulary that the two translations differ most characteristically. In the sentence quoted above, Mantino's translation *propositionibus probabilibus* seems, at first sight, less faithful to the Arabic and Hebrew, which speak of "widely known" or "acceptable" premises (*mashūra,* מפורסם). But, in fact, the translation *probabilis* is in tune with both the medieval translator Michael Scot and Boethius. Michael Scot uses *probabilis* in the sense of "acceptable" as a translation for *mashūr,*[112] and Boethius translates Aristotle's definition of the dialectical syllogism as a reasoning "from reputable opinions" (ἐξ ἐνδόξων) with the phrase *ex probabilibus* (100a30).[113]

Throughout the commentary, Mantino writes *propositio* where Balmes translates literally with *praemissa* (for *muqaddima,* הקדמה). The differences in terminology are very apparent in chapters 1,10 and 1,11 (1,8 and 1,9 in the Giunta edition), where Aristotle defines the three main terms of the technique of dialectic: "dialectial proposition" (πρότασις διαλεκτική), "problem" (πρόβλημα), and "thesis" (θέσις). Compare the Latin terms of Boethius's translation with the anonymous translation of the twelfth century (which exerted hardly any influence) and with Balmes and Mantino:[114]

| Boethius | propositio dialectica | problema dialecticum | positio |
| Anonymous | propositio dialectica | quaestio dialectica | thesis |
| Balmes | praemissa topica | Quaestio/-situm topica/-um | positio |
| Mantino | propositio dialectica | problema dialectica | positio |

Mantino clearly uses Boethian vocabulary. It is important to note that the topical tradition in the early sixteenth century still worked with the Boethian terms. We find the terms *propositio, problema, positio,* for instance, in the logical section of Gregor Reisch's widely read encyclopedia *Margarita philosophica.*[115]

The correspondence between Boethius and Mantino is close enough to suggest that Mantino's choice of vocabulary was a deliberate move toward Boethius. He chooses Boethius's terminology for the two modes of argument proper to dialectic in chapter 1,12: *inductio* and *syllogismus,* where Balmes writes *inquisitio* and *syllogismus.*[116] Particularly telling is the description of the four means of finding syllogisms of chapter 1,13 (1,11):[117]

| Boethius and Anonymous | propositiones sumere | quotiens unumquodque dicitur posse dividere | differentias invenire | similitudinis consideratio |
|---|---|---|---|---|
| Balmes | sumere praemissas | posse discernere singula nomina et quot modis dicantur | differentiarum eductio | ex verisimili scrutinium |
| Mantino | inveniendi propositiones | elicere iudicium quo unumquodque nomen dignoscatur quot modis dicatur | adinvenire ipsas differentias | similitudinem considerare |

In this last example the correspondences in wording indicate that Mantino was using a copy of Boethius's version of the *Topics* when translating Averroes's Middle commentary. Mantino adopts the following phrases from Boethius: *unumquodque, (ad)invenire differentias,* and *similitudin(em) considera(re).* One of Mantino's motives, it seems, for producing a new translation of the commentary was to coin it in a vocabulary that would make it easily adaptable to the logical tradition of his time.

With respect to Latin syntax and nontechnical vocabulary, Mantino's translation moves some steps toward humanist ideals—in the sense that, for instance, he searches for the more classical expression and he turns nominal phrases into fluid sentences, such as when he replaces the queer expression *orationes inter duos* ("speeches between two") with *disputationes quae inter duos homines contingunt.*[118]

Hence, the working methods of Balmes and Mantino correspond to their programmatic aims: Balmes successfully "avoids" humanist Latin by writing the standard scholastic Latin of his day, and he faithfully copies the word order chosen by the author, while Mantino, as in the case of the *Metaphysics* prooemium, moderately classicizes Balmes's Latinity and produces a very clear and fluent text. The philological quality of Balmes's and Mantino's versions differs markedly. In many passages, Balmes's rigid *verbum de verbo* technique obscures the syntactic structure of the text, and his limited command of Latin results in misleading vocabulary. These findings agree with Silvia Di Donato's in-depth analysis of Balmes's Latin translation of Averroes's *Kitāb al-Kašf (Liber modorum rationis de opinionibus legis):* here again Balmes imitates the Semitic structure of relative clauses, writes *res* where the context requires *significatum,* uses imperfect subjunctive instead of indicative, imitates the Semitic usage of prepositions such as in *significare super,* and makes infelicitous choices of vocabulary. Again, as in the translation of the *Topics* commentary, Balmes's translation is seriously hampered by his extremely literal method of translation and by his lack of knowledge of Latin.[119] Di Donato's textual specimens also contain several passages where Balmes's source, the Arabic–Hebrew translation of the Middle Ages, is mistaken on the syntax of the passage, as when the translator does not recognize the beginning of an apodosis introduced by the particle *fa-* ("then") and instead translates with ו ("and"), or when the Arabic construction *ammā . . . fa* ("as to . . .") is transformed into "however . . . And."[120]

As we have seen, a major drawback of Balmes's version of the *Topics* commentary is that it is not coined in the standard logical vocabulary of the medieval Latin tradition—which contrasts with his profuse praise of logical expertise in the preface. That Balmes was not entirely abreast of the Latin logical tradition has also been observed with respect to his translation of the Long commentary on the *Posterior Analytics*.[121] It may be a reflection of the fact that Balmes was teaching philosophy in Padua not within a Latin, but in a Hebrew context. It is exactly this drawback of Balmes's translation that was convincingly remedied by Jacob Mantino, as were many of its linguistic shortcomings.

To conclude: The picture that emerges from the analysis of the two Averroes translations—the *Metaphysics* preface and the *Topics* commentary—is disillusioning and impressive at the same time. Del Medigo's and Balmes's translations are seriously restricted in quality. In fact, they do not reach the level of the medieval Arabic–Latin Averroes translations by Michael Scot, William of Luna, and Hermann of Carinthia. Their problem is not the Hebrew intermediary. It is true that any intermediary version increases the number of corruptions, and we have met with some of them caused by the Hebrew intermediary. But Mantino's example shows that a Hebrew–Latin version of Averroes can almost reach the level of precision that the medieval Arabic–Latin versions had achieved. Del Medigo's and Balmes's main problem is their poor command of Latin. It is all the more impressive to see what Ricci and Mantino accomplish. While Ricci's translations are stylish to a degree that they tend to be unreliable, Mantino's texts are major philological achievements: Mantino's methods result in a reliable Hebrew–Latin transmission of about a dozen texts by Averroes. Renaissance readers could use these texts with confidence, and the high praise of the Giunta 1550/1552 and 1562 editors for the "extremely lucid" and "golden" translations of this "most learned" person show that the quality of Mantino's work was in fact recognized.[122]

## AVICENNA: *CANON*

The other great translation enterprise, the attempt to improve upon the twelfth-century translation of Avicenna's *Canon* by Gerard of Cremona, originated in a different context. The first two new versions of the *Canon* were produced from the Arabic, and not in Europe, but in Damascus: by Girolamo Ramusio (d. 1486) and, one generation later, by Andrea Alpago (d. 1522). In the course of the sixteenth century, important chapters were translated from the Hebrew by Jacob Mantino (d. 1549) and Jean Cinqar-bres (d. 1587), and there was one other translation that purported to be from the original Arabic by Miguel Jerónimo Ledesma (d. 1547), as well as a revision of the Latin with the help of a Jewish scholar by Andrea Gratiolo (fl. ca. 1540–1580).[123]

The new wave of Avicenna translations started with Girolamo Ra-musio.[124] Ramusio, a descendant of a well-established family in Rimini and a composer of poems in Italian and Latin, studied philosophy and medicine in Padua in the 1470s, worked for some years as a ship's doctor in merchant trade, and in 1483 settled in Damascus as physician to the Venetian embassy. In the three years before his death in Beirut in 1486, Ramusio produced an interlinear translation of large parts of the *Canon medicinae*. His manuscript has survived (Paris, Bibl. Nat. Arabe 2897): Ramusio copied the Arabic text of the *Canon* from another manuscript in what is clearly a beginner's handwriting, and noted a Latin equivalent over each Arabic word—with the curious effect that the Latin text runs from right to left. Many of the Latin renderings are his own, others are taken from Gerard of Cremona's old version. There is some information on Ramusio's motives and methods in a letter to a friend "Flavius" at the end of the manuscript.[125] Ramusio explains that in copying the Arabic he has become aware of the omissions and errors of the Latin translation. He has copied and translated those sections of the book that are "more useful," leaving the rest for later. His translation has not, for the most part, received "a careful revision," writes Ramusio, adding: "I hope to return to everything in the next year, examine it with utmost care and turn it into its best form." Death prevented Ra-musio from fulfilling this promise; a century later, the manuscript was used by Andrea Gratiolo, as we shall see below. Ramusio's tactic of translating useful things first is also echoed in a remark in the manuscript at the beginning of the section *De virtutibus:* "I wanted to see first what is publicly lectured upon at the university of Padua."[126] It is obvious, then, that Ramusio's project, unfinished as it remained, was aimed at the Italian market and its medical faculties, where Ramusio had received his education. Can the same be said of the later translators?

In about 1487, Andrea Alpago, descendant of a noble family of Belluno north of Venice, became Ramusio's successor in Damascus,[127] and served for about thirty years as physician to the Venetian embassy. In Damascus, Alpago was well-connected with Arabic physicians. As his surviving political letters show, Alpago sympathized with the new Safavid dynasty of Persia, which rivaled the Ottomans for control of Syria.[128] In late 1520 or early 1521, Alpago returned to Padua, where he briefly occupied a professorship in medicine beginning in September 1521, until his early death on January 22, 1522. Andrea's nephew Paolo Alpago published his uncle's corrections to the *Canon* in 1527 in Venice, together with a dictionary of technical terms for which Andrea Alpago had used Arabic commentators of the *Canon.* The date 1527 marks a turning point in the history of the Western reception of the *Canon:* Andrea Alpago's substantial corrections, which were printed in the margins of Gerard of Cremona's medieval version, much influenced the Western reading of the *Canon* in the following decades, for instance the commentary on the *Canon* by the Paduan physician Oddo Oddi.[129] Alpago's motives and aims as a translator can only be surmised from the preface that his nephew wrote for the edition: The older Latin text of the *Canon,* due to its many faults *(errata),* does not express exactly what Avicenna means and prescribes, says Paolo Alpago, and this is particularly worrying in a medical context, where the life and health of human beings are concerned. Hence came his uncle's decision to revise the *Canon.* He consulted several "very old" manuscripts *(plurimis collatis antiquissimis exemplaribus).* The previous Latin Avicenna offered only a false and empty image of himself. But now the physicians have much occasion to be proud of their Avicenna.[130] These sentences of the preface seem to be colored with antimedieval, humanist tones. We shall see below, however, that Alpago's actual textual work does not follow humanist ideals.

Who were the prospective readers of Alpago's new version? From 1479 onward, Alpago had studied philosophy and medicine in Padua, and he kept in contact with this center of learning in the Veneto when he lived in the Near East. In January 1521, one year before his death, the Collegium of philosophers and physicians of Padua decided to officially recommend Alpago's corrections of the *Canon,* which they had examined and found *utilia et necessaria.*[131] The university's offer of an extraordinary professorship of practical medicine followed in September 1521.[132] Obviously, then, Alpago had sent or brought his emendations to Padua in order that they be officially endorsed by the Collegium. His philological work was not the lonely enterprise of a private scholar. Alpago had a specific audience in mind, and this audience taught and studied medicine at the universities of northern Italy.

The third *Canon* translator, and the first to translate from Hebrew, was Jacob Mantino. In the later 1520s, when Mantino had already produced and published a number of Averroes translations, he rendered *Canon* 1.4 (*fen quarta primi libri*) into Latin. It appeared in print in 1530 in Venice with a dedication to doge Andrea Gritti of Venice, who had studied at Padua. Mantino begins this dedication by praising Avicenna as the principal author in medicine, especially in a teaching context, and continues:

> But because Avicenna in writing had used a pagan and idiosyncratic Arabic language, which is not that easily acquired by Latin people, the translation of his works abounds in many and greatest errors, which Andrea of Belluno, the famous physician of our time and equally learned in the Arabic and Latin languages, has laudably emended for the most part. He was not, however, able to purify the field entirely from foreign seeds; rather, many remain, which cloud the truth of reading (*veritatem lectionis*) as with some fog.[133]

It is for this reason, says Mantino, that he decided, after having produced many Hebrew–Latin translations in various disciplines, to turn now to Avicenna's texts, purge them of all faults, as far as possible. Since three sections of the *Canon* are most often lectured upon in the universities (*in gymnasiis publice legantur*), namely, sections 1.1, 1.4, and 4.1, he has chosen these for translation and started with section 1.4 (on forms of treatment) "because this part seems to be of greater usefulness for the general art of healing than the other parts."[134] The fact that the translation of *Canon* 1.4 was printed several times in different countries (Ettlingen 1531, Paris 1532, Haguenau 1532,[135] and Paris 1555), shows that Mantino's assumption that there is a demand for the chapter was correct. Mantino does not seem to have translated *Canon* 4.1, as announced, but his version of *Canon* 1.1 appeared around 1540 in Venice, and once again in 1547 in Padua.

This later translation of *Canon* 1.1 by Mantino is accompanied by a dedication to Marcantonio Contarini, member of the Venetian senate and humanist orator at the court of Pope Paul III, to whom Mantino served as physician.[136] Mantino here names an implicit premise of his translation enterprise: that it is possible and sensible to translate from the Hebrew, rather than from the original Arabic: "I have translated into Latin the first Fen, as they call it, of the first book of Avicenna from Hebrew, of which even the less educated knows that it does not at all differ (*nihil penitus*) from the Arabic sense itself."[137] There is indeed some justification for this claim,[138] because of the similarities between the two Semitic languages Arabic and Hebrew and because of the extremely literal method of translation employed by many medieval Arabic–Hebrew translators.[139] There are, however, clear limits to the Hebrew–Latin approach, as we have seen above when discussing the Hebrew–Latin Averroes translations. We have to return later to the question of whether Mantino was able to improve upon Alpago's version of the *Canon*.

In 1570, 1572, and 1586, Jean Cinqarbres published Hebrew–Latin translations, which will be passed over in the present study, since they are three short excerpts only. Two other "translators" of the *Canon* did not, in all probability, translate, but revised the text on the basis of Latin sources only: Miguel Jerónimo Ledesma and Andrea Gratiolo. Ledesma, professor of medicine at Valencia, was a major proponent of medical humanism in Spain; he received his education at a center of the humanist movement, the University of Alcalá.[140] In the dedication to Tomas de Villanueva, archbishop of Valencia, Ledesma praises Avicenna, in truly humanist attitude, for "always behaving as a translator of Galen" *(Galeni se usquequaque faciens interpretem)*, and he utterly deplores the fact that Avicenna had "a barbarous translator and even more barbarous commentators" *(nactus est barbarum interpretem barbarioresque multo enarratores)*. Ledesma therefore promises to restore Avicenna to his "Arabic truth" *(ad arabicam veritatem . . . emendare)*. For this purpose, he has used "an extremely old manuscript," which differs significantly from the commonly used version *(vetustissimus codex Avicennicus manu scriptus longe a vulgato dissidens)*.[141] To speak of a vulgate text makes sense only if a textual version exists that is well-known among scholars and physicians—and this can only be a Latin text. Also, it would be pointless to qualify the *codex* as *manu scriptus* if it was an Arabic text—because the printing history of Arabic had hardly begun.[142] Ledesma proceeds to mention that he has made "efforts to search for the peculiarities of the languages"—which may mean that he has tried to learn Arabic—and that he has had the help of a person with expertise in both the Arabic language and the art of medicine. Andrea Alpago's version is criticized by Ledesma for following the medieval commentators on the *Canon*.[143] I believe it is a mistake to conclude from this preface that Ledesma translated the *Canon* "directly from Arabic into Latin," his starting point being an "'ancient manuscript' in Arabic . . . whose content was rather different from the published version."[144] We will have to investigate below whether the textual details of Ledesma's translation show any knowledge of the Arabic tradition. When Ledesma died in 1547, he had only completed the translation of book 1, Fen 1 of the *Canon*.

Not much is known of Andrea Gratiolo di Salo apart from what can be gleaned from his works and translations.[145] He must have lived in the decades between 1530 and 1580. Born in Toscolano in the province of

Brescia, he studied philosophy and medicine at the University of Padua and later was active as a physician in the 1550s and 1560s in Desenzano, Mantua, and Montagnana.[146] We are thus led back to the Paduan intellectual climate of the earliest Renaissance translations of the *Canon* by Ramusio, Alpago, and Mantino. Apart from his translation of Avicenna's *Canon*, book 1, which was printed in 1580, Gratiolo is also known as the Latin translator of the late Greek commentaries on the *Posterior Analytics* by Alexander of Aphrodisias, Eustratius Nicaenus, and John Philoponus, and as the author of a *Discorso di Peste* printed in Vinegia in 1576. Gratiolo had been persuaded to participate in the *Canon* project by Oddo Oddi, his teacher in medicine, who had already sat in the meeting of the Paduan faculty of philosophers and physicians in 1521 that decided to examine Alpago's emendations to the text of the *Canon*. Oddo Oddi has adopted the new translations by Alpago and Mantino into his own commentary on the *Canon*,[147] and seems to have been a promoter of new *Canon* translations throughout his life.[148] Gratiolo must have begun the translation before Oddi's death in 1558.

Gratiolo's main motive for investing philological care in the text of the *Canon* is the many defects of Gerard of Cremona's medieval translation: Gerard "has obscured many originally clear sentences, has not understood many passages, has turned everything barbarous and ugly, has, against the sense, inserted a number of words in the course of the text, which do not appear at all in the author's text itself."[149] Gratiolo is very clear about his own approach, which consists in consulting all previous translations and emendations, especially those of Andrea Alpago, and in seeking the assistance of the experienced physician Girolamo Donzellini,[150] and of the Jewish physician Jacob Anselmi, who possessed a Hebrew copy of the *Canon*. In the last days of his project, Gratiolo received the Arabic manuscript of the *Canon* with Ramusio's interlinear Latin translation, the most important renderings of which Gratiolo prints in the margins of his own text.[151] Gratiolo claims to have "investigated the true and real sense (if I am not wrong) of the author *(verum ac germanum sensum scriptoris)*, which I have expressed in Latin and clear words, as far as I could achieve this linguistically."[152] In contrast to Ledesma, Gratiolo does not pretend to be translating.

All five scholars surveyed attempted a revision or new translation of the first Fen—from Arabic *fann,* "field, discipline"—of the first book of Avicenna's *Canon*. The first Fen is concerned with theoretical medicine, in particular, with the definition of medicine (section/*doctrina* 1), the theory of elements (*doctrina* 2), and with the physiology of the Galenic tradition (*doctrinae* 3–6). This part of the *Canon* played a major role in the medical teaching of

Renaissance universities, and this seems to have been the primary reason for the philological attempts by Renaissance scholars to improve its textual basis. For the purpose of confronting the programmatic aims of the translators with the result of their work, I have selected chapter 5 of Book 1, Fen 1, *doctrina* 6, *capitulum* 5, on the animal faculties of perception, that is, the external and internal senses.[153] With respect to content, this chapter is important because in it Avicenna comments on conflicting views between the medical and philosophical traditions (as he does occasionally in the previous *doctrinae* 3–5 on physiology). Moreover, it is important within the broader context of Avicenna's oeuvre, because he addresses topics that he also discusses in philosophical works. And finally, the chapter deserves attention because its psychology was of no little influence on medieval scholasticism.[154]

It was mentioned above that the first Renaissance translator, Girolamo Ramusio in Damascus, began to work on the section *De virtutibus* because he wanted first to concentrate on what was lectured on in Padua. Danielle Jacquart has pointed to several characteristic features of Ramusio's technique of translation, especially with respect to vocabulary: that he deliberately chooses vocabulary different from Gerard of Cremona's, that he prefers Latin vocabulary that is closer to the root meaning of an Arabic word, that he shuns Greek expressions, and that he preserves most of Gerard's Latin transliterations of Arabic words. With respect to the technical vocabulary of medicine, Jacquart observes that Ramusio was ill-prepared since he did not use any Arabic or Latin commentaries or glossaries to the *Canon*.[155]

It is difficult to imagine a more literal translation than Ramusio's. Since his translation is written on top of the Arabic terms, his Latin sentences follow the Arabic direction from right to left, thus visualizing the one-to-one correspondence between Arabic and Latin terms (see Figure 1).[156]

There are limits to the extreme literalness of Ramusio's translation. Ramusio's Latin words are inflected, which means that he tries to construe proper Latin syntax, as far as possible. Occasionally, he leaves Arabic terms untranslated if they do not add anything to the meaning of the sentence, but only serve syntactic purposes in the Arabic language. An example is the relative clause *quam memorabimus [eam] postea*, "which we will mention [it] again later":[157] Ramusio usually translates pronominal suffixes, but in this case, he leaves out *eam* because it is superfluous in Latin syntax. Note that on this specific point Ramusio's technique of translation is less literal than that of Elia del Medigo, the translator of Averroes, obviously because Ramusio was more at home in Latin than del Medigo.[158]

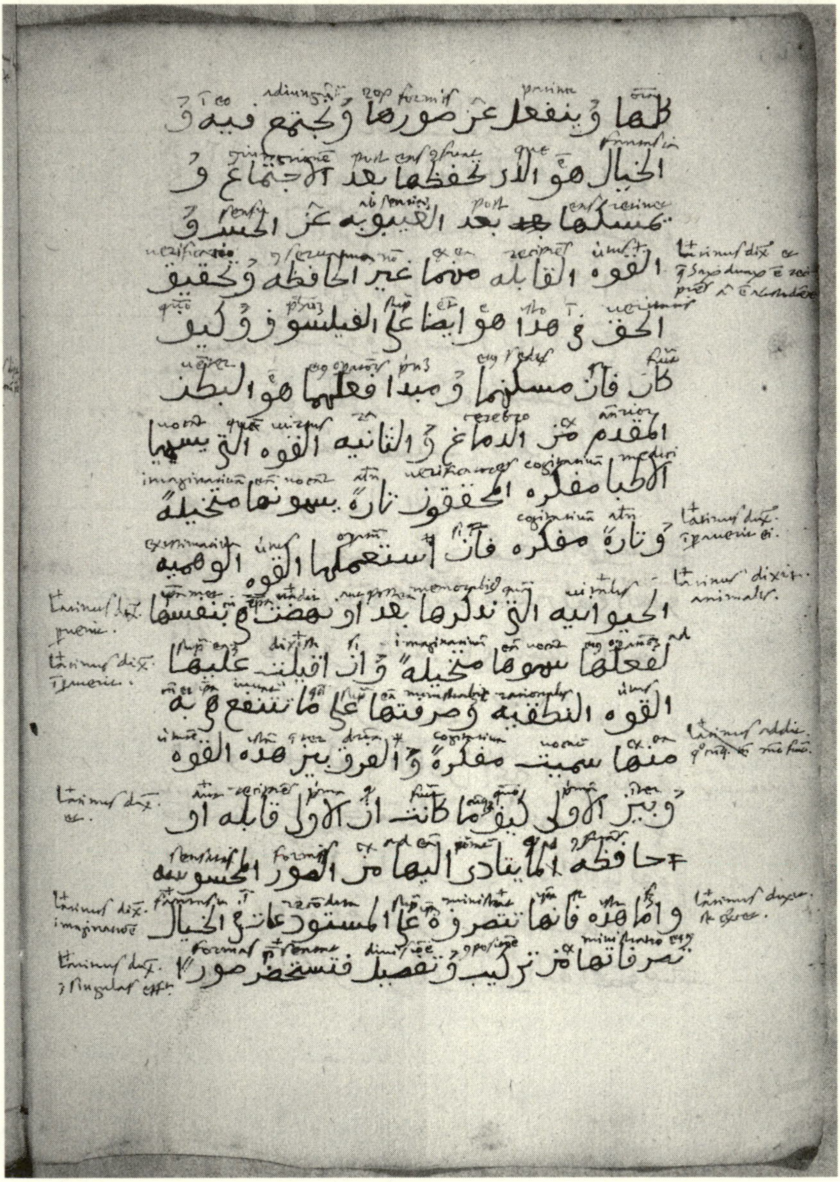

FIGURE 1.  Avicenna, *Qānūn*, with interlinear Latin translation by Girolamo Ramusio, f. 169v. Paris, Bibliothèque nationale, MS arabe 2897.

The main advantage of Ramusio's rigid technique is a sensitiveness to inconsistencies in the medieval Latin translation of the *Canon*. In the chapter we are surveying, *Canon* 1.1.6.5 on the senses, one example of such sensitiveness concerns the nomenclature of the so-called internal senses.[159] According to Avicenna, the philosophers count five internal faculties: common sense and imagination in the first ventricle of the brain, the imaginative/cogitative faculty and estimation in the second ventricle, and memory in the third. The physicians, in contrast, assign only one faculty to each of the three ventricles, because their focus is on the possible areas of injury only.[160] The Arabic terms used for imagination, the second faculty of the philosophers, which stores sense data, and for the imaginative/cogitative faculty, the third faculty, which combines and separates sense data, stem from the same Arabic root: *ḫayāl* (imagination) and *mutaḫayyila* (imaginative <faculty>).[161] The similarity of these expressions proved a source of confusion for many readers of Avicenna's *De anima* and *Canon*.[162] Gerard of Cremona's translation of the *Canon* is unambiguous for the most part: he chooses *fantasia* for *ḫayāl,* and *imaginativa* for *mutaḫayyila.* But in one sentence he puzzles his readers by introducing the ambiguous term *imaginatio:* "Et ista est illa quae se exercet in eis quae in imaginatione recondita sunt exercitio componendi et dividendi," that is, "and <the third faculty> operates upon what is deposited in imagination in the way of composing and dividing."[163] Ramusio, in contrast, writes: *in fantasia* and notes in the margin: *latinus dixit: imaginatione,* "the Latin translator (i.e., Gerard) said: in imagination."[164] Ramusio's correction is justified: Gerard's *imaginatio* translates *ḫayāl,* which for the sake of consistency and clarity he should have translated with *fantasia.*

It is true that in very few cases Ramusio's translation results in a better text than that of the other translators. But he deserves credit for being the first to propose a significant number of corrections, which are usually thought to be the achievement of Andrea Alpago. In our chapter, Ramusio, for example, corrects the misleading phrase *sicut hominem qui in monte volat smaragdino,* which has the strange meaning: "like a man who is flying in a mountain of emerald,"[165] by translating more appropriately *sicut homo volat et mons ex smaragdo,* "like a man who is flying and <like> a mountain of emerald."[166] That is, he distinguishes, just as the Arabic, between two different unreal images, which Gerard's translation (or its medieval manuscript tradition) had fused into one. This correction was also proposed by Alpago and Mantino.[167] And there are more examples of valuable improvements of the text in this chapter.[168]

But all this cannot hide the fact that Ramusio's translation fails to produce intelligible sentences in cases of more complex Arabic syntax. This is tangible in the just cited inconspicuous example *praesentat formas . . . sicut homo volat et mons ex smaragdo,* because the phrase *sicut homo volat* renders what contains in Arabic an indefinite relative clause: "such as a person *who* is flying." In Arabic no relative pronoun is needed for this kind of relative clause;[169] in Latin it is needed. Ramusio, however, is not willing to add the pronoun *qui* to his Latin translation, for which there is no Arabic equivalent. As a consequence, indefinite relative clauses are not discernible in Ramusio's translation.

A more drastic example of an unfortunate translation may serve to illustrate Ramusio's problems in translating Arabic syntax. The sentence comes from a paragraph in which Avicenna demarcates philosophical and medical knowledge:

فإن كانت المضرة تلحق فعل قوة بسبب مضرة لحقت فعل قوة قبلها وكانت تلك المضرة تتبع سوء مزاج أو فساد تركيب في عضوٍ ما فيكفيه أن يعرف لحوق ذلك الضرر بسبب سوء مزاج ذلك العضو او فساده

> If a damage has afflicted the operation of a faculty because of a damage which has afflicted the operation of a<nother> faculty in front of <this faculty>, and <if> this damage has resulted from a bad mixture or a corruption of the composition in some organ, then it is sufficient for <the physician> to know that this damage has afflicted because of the bad mixture of this organ or its corruption.[170]

That is, the physician does not need to know about the exact relations between the faculties, as long as he knows which organ is afflicted. Ramusio renders this as follows:

> Si fuerit nocumentum adiungitur operationi virtutis propter causam nocumenti adiungitur operationi virtutis ante eam et fuerit illa nocumentum sequi(?)tur malitiam complexionis huius membri aut corruptionem compositionis in membro aliquo, sufficit ei quod sciat adiunctu(?)m huius nocumenti propter causam malitiae complexionis huius membri aut corruptionis eius.[171]

This Latin sentence is incomprehensible, as Ramusio himself surely knew. For this reason, the term "translation" does not appropriately describe the product of Ramusio's efforts. There is no doubt that he ultimately wanted to write a translation for the Italian market, but the stage of his work that is documented in the Paris manuscript has the character of rough notes, which were eventually to be transformed into a translation proper.

One wonders, however, whether these notes would have served their purpose. In the sentence quoted, the notes could be a helpful tool for improving the vocabulary of a future translation, but they obscure the meaning of the sentence in at least three respects. One feature ignored by Ramusio again is an indefinite relative sentence—just as with "a mountain who is flying." A relative sentence follows *nocumenti*, but Ramusio omits the pronoun: *propter causam nocumenti [quod] adiungitur operationi virtutis,* meaning: "because of the damage [that] afflicted the operation of the faculty." The second problem concerns Ramusio's verbal phrases *fuerit... adiungitur* and *fuerit... sequitur,* which are impossible constructions in Latin. They are literal renderings of the Arabic *kāna* ("to be") plus imperfect, that is, the combination of an auxiliary verb with a full verb. This combination in the protasis of an Arabic conditional clause is usually expressed with the present tense in English and Latin.[172] There is no point in rendering *kāna* with *fuerit,* and it may well be that Ramusio did not understand the construction. There are, in fact, more indications that Arabic syntax was not Ramusio's strength, for instance, the repeated rendering of *innamā* ("only") with *quod,* where *tantum* or *non... nisi* would have been appropriate.[173]

Third and finally, Ramusio chooses the term *adiunctum* as the translation of the verbal noun *luḥūq* ("afflicting"), that is, he tries to imitate in Latin the nominalization of a complete sentence, which in Arabic can be achieved with a verbal noun *(maṣdar).*[174] In this case, the sentence "this damage has been afflicted because of the bad mixture..." is expressed with the Arabic nominal phrase: "the afflicting *(luḥūq)* of this damage because of the bad mixture...." Ramusio's translation *adiunctum huius nocumenti propter causam malitiae complexionis...* is more or less understandable, but grammatically problematic, because the Latin participle *adiunctum* has only the nominal force of an adjective and is not capable of nominalizing a complete sentence. Ramusio's phrase *adiunctum huius nocumenti* has the odd meaning "<what is> added of this damage." The other translators very reasonably choose the other verbal noun offered by the Latin language, the infinitive: *impedimentum... accidisse* (Gerard and Alpago) and *laesionem provenire* (Mantino).[175] Ramusio preferred the participle to the infinitive probably because he wanted to retain the following genitive: *adiunctum huius nocumenti*—at the expense of obscuring the meaning of the sentence.

Let us turn to Gerard of Cremona and Andrea Alpago to see how they cope with this difficult sentence. Gerard starts his translation with the words: *Quod si laesio acciderit operationi...,* that is, in contrast to Ramusio, Gerard does not render the auxiliary term *kāna* in Latin. The indefinite relative clause omitted by Ramusio appears as well; it is introduced by the pronoun *quod.* In view of the fact that Gerard also found an elegant solution for translating the Arabic verbal noun, one can conclude that the syntax of the sentence did not prove an obstacle for Gerard—as we would expect from the most experienced Arabic–Latin translator of the Middle Ages. Nevertheless, some aspects of his translation leave room for improvement, as Ramusio and Alpago realized. This is Gerard's translation in full, with the emendation in brackets:

> Quod si laesio acciderit operationi alicuius virtutis causa impedimenti [*Ramusio:* nocumenti], quod accidit prius operationi alterius virtutis quam ei [*Alpago:* virtutis quae est ante ipsam], fueritque impedimentum [*Ramusio:* nocumentum] illud secutum malitiam complexionis alicuius membri aut corruptionem compositionis, sufficit medico ut sciat impedimentum [*Ramusio:* nocumentum] illud propter malitiam complexionis membri aut suae compositionis corruptionem accidisse.[176]

Thoughout this chapter, Gerard uses the term *impedimentum* for the Arabic *maḍarra,* which has the more specific meaning: "harm, damage, loss." Ramusio rightly prefers *nocumentum,* and he noted this difference to Gerard in the margin of the Paris manuscipt: "The Latin translator always said 'obstructions,' this person *(i.e., he himself)* says 'damages' *(latinus semper dixit: impedimenta, vir iste dicit nocumenta).*[177]

Andrea Alpago's annotations to the text of the *Canon* appeared in 1527. What did he achieve with the twelve corrections that he proposed to our chapter?[178] Alpago is not interested in correcting stylistic features of Gerard's translation, and hence does not change the phrase *fuerit . . . secutum,* which is a nonclassical expression that a humanist would certainly have changed to *secutum sit.* Nor does Alpago emend the term *impedimentum.* In the quoted passage, Alpago proposes one correction that concerns the phrase *quam ei* ("another faculty than it"). Alpago's alternative is: *quae est ante ipsam* ("another faculty which is in front of it"). This is a sensible emendation. The Arabic here clearly refers to a faculty that not only is different from the first faculty but also is located in its vicinity, namely, in front of it: ". . . a damage which afflicted the operation of a‹nother› faculty in front of ‹this faculty›" *(qabla-hā).*[179]

The other emendations to this chapter are of similar character. Alpago does not interfere with Gerard's psychological vocabulary—with the one exception that he once corrects the term *existimationem* to *existimativam.* Alpago makes this correction for the sake of consistency, because Gerard himself in all other passages writes *existimativa* for the faculty of *wahm* ("estimation"). The term *existimatio*—a mixture of the terms *estimatio* and *existimativa*—may well have resulted from a corrupt Latin manuscript transmission. We cannot tell as long as we do not have a critical edition of the *Canon*. Girolamo Ramusio, as we have seen, had also corrected another inconsistency in Gerard's text, the one misleading occurrence of *imaginatio* instead of *fantasia*, but this was not spotted by Alpago. The rest of Alpago's emendations to this chapter concern the correct rendering of Arabic phrases and syntax into Latin. It is noteworthy that Alpago's motives in annotating this chapter were neither medical nor philosophical, but philological in a nonhumanist manner—in the sense that he aimed at a more reliable translation, but was not concerned about Gerard's Latinity and did not fill his annotations with learned material. In this regard, his approach resembles Ramusio's—which explains the remarkable fact that five of the twelve corrections were spotted by both of them. But since Alpago also had no difficulties improving on Gerard's syntax, he achieved much more than Ramusio, to the benefit of the sixteenth-century readers of the *Canon*.[180]

Jacob Mantino was one of them, but he was convinced that Alpago's corrections did not go far enough; problems remain that cloud "the truth of reading." Mantino's translations of *Canon* 1.1 and *Canon* 1.4 are done from the Hebrew, just as his renderings of Averroes. In many parts, the latter are close to revisions of earlier Averroes translations, as we have seen, but his translations of the *Canon* do not at all resemble the older versions by Gerard and Alpago: they differ in vocabulary and style. For illustration we return to the complex sentence about damages in ventricles of the brain. I cite Mantino's translation from the 1547 edition, which is marred by printing errors. The text first appeared around 1540 in Venice; both editions are rare today:

> Laeditur tamen eius operatio propter laesionem accidentem actioni virtutis illi praecedentis, quae laesio insequitur mala\<m\> temperiem aut corruptam constitutionem formationis membri, nam satis est illi si norit illam laesionem provenire propter malam temperaturam illi\<u\>s membri vel corruptionem constitutionis eius.[181]

Remember that the Arabic original contains a conditional clause: "If a damage has afflicted . . . , then it is sufficient for \<the physician\>." This conditional clause has disappeared in the Hebrew–Latin transmission and has been replaced by an adversative *tamen* in the former protasis and a causal *nam* in the former apodosis.[182] The occasional corruption of Arabic syntax in Hebrew transmission is a phenomenon we have met before when

discussing the Hebrew transmission of Averroes. In spite of the great si-
miliarities between the Arabic and Hebrew languages, it is clear that the
Hebrew intermediary was an additional source of corruption. In this regard, Mantino's Hebrew–Latin *Canon* project had its obvious limitations.

It is all the more noteworthy that the three syntactical problems that
bothered Ramusio did not prove problematic for Mantino. The indefinite
relative clause after "damage" is elegantly rendered with a participle construction: *propter laesionem accidentem actioni virtutis.* The auxiliary verb
*kāna* plus imperfect is rendered with present tense, as is appropriate *(lae-
ditur, sequitur),* and the verbal noun *luḥūq* ("afflicting") is well-translated
with an expression involving the infinitive, as mentioned above: *laesionem
provenire.* When we turn to Gerard's translation itself and its problematic
passages, we again find Mantino offering felicitous solutions. Remember
Alpago's emendation *virtutis quae est ante ipsam* for the phrase "another faculty in front of this faculty." Mantino aptly translates with a participle:
*virtutis illi praecedentis,* which shows that Mantino, influenced by humanist
ideals of Latin style, chooses among a greater variety of stylistic means of
the Latin language than his predecessors. With respect to vocabulary,
Mantino prefers *laesio* to Gerard's misleading term *impedimentum,* and,
more significantly, introduces changes in medical vocabulary: he writes
*temperies* and *temperatura* instead of *complexio,* and *constitutio* instead of *com-
positio.* Here we witness Mantino partaking in an important development
of Renaissance medicine: a gradual transformation of medical nomenclature in the sixteenth century, as part of humanist endeavors to depart
from medieval Latinity.[183] Remember that Mantino's translation of Averroes's *Topics* commentary also aimed at up-to-date vocabulary, in that case
in the field of logic.

Mantino also works on the philosophical and psychological vocabulary
of the text, which is not surprising given that he has much expertise with
translating the philosophical works of Averroes. He does not use Gerard's
term *fantasia* for *ḫayāl,* but instead writes *imaginatio,* which is the more
common name in philosophical literature. He distinguishes consistently
between *imaginatio,* Avicenna's second internal sense, and *virtus imaginativa,* the third. Unfortunately, he blurs the clarity of his translation by
using the term *imaginaria* for both faculties (*ḫayāl* and *mutaḫayyila*); the
reader has to guess from the context which faculty of the two is meant.[184]
The internal senses themselves are referred to as *virtutes interius apprehen-
dentes,* and not *comprehensivae occulte,* as Gerard put it. Again, this is a translation in tune with current terminology; the most common expression is
*sensus interiores,*[185] or, in the tradition of Avicenna's *De anima, apprehensivae
deintus.*[186] We thus encounter a translation technique for the *Canon* similar

to that for the *Topics* commentary, as far as vocabulary is concerned: a principal concern of Mantino's is to use up-to-date technical terminology.

Because of the Hebrew intermediary, this approach cannot be successful in all instances. Avicenna differentiates the objects of the faculty of estimation, *maʿānī*, "meanings" or "connotational attributes," from the objects of sense-perception: *ṣuwar*, "forms." All previous translators had rendered *maʿānī* as *intentiones*. Mantino offers three different translations: *res, aliquod,* and *formae*.[187] He obviously had not realized that in this context the Hebrew word עניין (*ʿinyan*) was translating an Arabic technical term.

Mantino's translation of *Canon* 1.1 was first printed around 1540 and again in 1547. In the same decade in Spain, Miguel Jerónimo Ledesma produced his version of *Canon* 1.1, apparently without any knowledge of Mantino's efforts. Ledesma's text is the most readable of all translations of the *Canon* composed in the Renaissance. He has achieved this by writing a fluent humanist Latin and by shortening Avicenna's text. The following passage from chapter 1.1.6.5 on the imaginative faculty may serve to illustrate this. The Arabic translates as:

> And the <other faculty> operates upon what is deposited in imagination operations of composition and division. It makes forms present according to what is transmitted by sense-perception, and forms different from these, such as a person who is flying and a mountain <made> of emerald.[188]

Ledesma's translation fuses the two sentences into one:

> Haec autem tum formas ex sensu ad se pervenientes dividit atque componit, tum alias effingit, veluti hominem in monte volantem smaragdino.[189]

> ("This <faculty>, at one moment, divides and composes the forms that come to it from sense-perception, and at the other moment creates other forms, such as a man flying in a mountain of emerald.")

Ledesma omits Avicenna's remark that the imaginative faculty works on forms that are "deposited in imagination." But it is more surprising that Ledesma, in contrast to all other Renaissance translators, does not correct Gerard's corrupt phrase *hominem qui in monte volat smaragdino*. He does not adopt Alpago's accurate emendation *hominem volare et montem smaragdinum* ("such as a flying man and a mountain of emerald"), even though he claims in the preface that he knows Alpago's emendations—which he criticizes as being part of the medieval tradition.[190] Even if Ledesma, which is unlikely enough, worked with an Arabic exemplar that already contained the corruption, he should have followed Alpago's emendation. It seems the emendation escaped Ledesma's attention.

Conspicuously, Ledesma also ignores an emendation of Alpago's when he comes to translate the complex sentence with the three syntactical problems quoted above. Remember that Alpago's correction of Gerard of Cremona concerned the fact that the damage was caused not only by a "different faculty," but by a "faculty in front of this faculty" *(virtutis quae est ante ipsam)*. The Arabic runs:

> If a damage has afflicted the operation of a faculty because of a damage which afflicted the operation of a<nother> faculty in front of <this faculty>, and <if> this damage has resulted from a bad mixture or a corruption of the composition in some organ, then . . . [191]

Ledesma translates:

> Quod si facultati<s> alicuius functioni nocumentum ex alterius facultatis prius laesae noxa, eaque aut ex membri distemperie, aut partium <in>commoderatione advenerit, . . . [192]
>
> ("If the operation of some faculty receives a damage from the damage of another faculty that has previously been injured, and this either from the bad mixture of an organ or from the bad composition of parts, . . .")

It is possible that Ledesma omitted Alpago's emendation for the sake of brevity. Two sentences before this passage, the same Arabic phrase *qabla-hā,* "in front of it," rendered by Gerard with *quae sunt antequam ipsa,* was also skipped by Ledesma. Note, however, that in the present sentence he keeps Gerard's term *prius,* thus adopting the temporal interpretation of the phrase that was not followed by either Ramusio, Alpago, or Mantino. This may be an indication that Ledesma did not pay as much attention to the Arabic as he claimed. Here he seems to be more interested in introducing modern vocabulary: *distemperies* and *commoderatio* replace *malitia complexionis* and *corruptio compositionis*—in this he resembles Jacob Mantino.

Another instance of a significant abbreviation of Avicenna's text by Ledesma concerns the intellective faculty. In the last paragraph of chapter 1.1.6.5, Avicenna explains that it is not necessary for the physician to know whether memory and recollection are one faculty or not:

Here is the occasion for a philosophical investigation into whether the memorizing faculty and the recollecting faculty, which retrieves what among the stored things of estimation is concealed from memory, are one or two faculties.[193]

Ledesma writes:

Offert sese in hoc loco perscrutari sitne idem memoria et reminiscentia.[194]

("It is apt in this context to inquire whether memory and recollection are the same or not.")

In Ledesma's sentence, the reference to the philosophical character of the problem has dropped out, and Avicenna's intricate definition of the faculty of recollection has disappeared. The sole advantage of Ledesma's translation is that the two faculties are clearly named. His translation thus advances over the transmitted text of Gerard's translation, which adds a confusing *et* after *memorialis: virtus conservativa et memorialis [et] virtus quae reducit,* and so on,[195] thus turning two faculties into three. The Arabic text continues:

But this is not necessary for the physician, since the damages occurring to these two are of the same kind, being damages which occur to the posterior ventricle of the brain either in respect to mixture or in respect to composition.

Caeterum id medicus non moratur, quoniam nocumenta utriusque eiusdem sunt rationis, vel ex distemperie vel ex incommoderatione.

("Besides, the physician does not care for this, since the damages of the two are of the same kind: they originate either from bad mixture or bad composition.")

Ledesma omits Avicenna's information that the damages happen to the last ventricle of the brain. This omission is possible since it was mentioned before by Avicenna that memory is located in the last ventricle, and the fact is not essential to the argument. The passage finishes with a brief reference to the faculty of the intellect:

The remaining faculty among the perceiving faculties of the soul is the human intellective faculty. Because the physicians's investigation omits estimation for reasons we have explained, it also omits this faculty; rather, their investigation is only directed at the operations of the three faculties and not of any other.

> quod dictum est in extimativa, quoniam trium tantum in universum facultatum habenda est ratio et non plurium, sic et in memoriae apprehensiva dicendum est.

> ("What has been said with respect to estimation, namely that only three faculties in all are to be considered and not more, likewise has to said about the perceiving faculty of memory.")

Here we can witness Ledesma substantially shortening and also changing Avicenna's text: the faculty of the intellect has disappeared in the translation. Instead, Ledesma adopts the second sentence on the intellect to a different context, namely, to a further explanation of why the distinction between memory and recollection is not the physician's task. It is obvious that Ledesma's modifications not only concern the linguistic surface of the text, but extend to the content: he attempts to cut out information that he finds either not essential for the argument or repetitive. This is intelligently done, at least in the chapter surveyed here, but the drawbacks of this technique are serious: Ledesma's translation may be practical as a medical source, but it is not a reliable guide to what Avicenna meant. And what is worse, there is no sign at all of the usage of Arabic sources, which Ledesma himself had boasted about.[196]

The last sixteenth-century attempt to improve the Latin version of Avicenna's *Canon* was made by Andrea Gratiolo. His "translation" of book 1 appeared in print in 1580. As was mentioned above, Gratiolo does not pretend to know Arabic. He produced his version by comparing the translations of Gerard of Cremona, Ramusio, Alpago, Mantino, and by consulting the Jewish physician Jacob Anselmo who had a Hebrew copy of the *Canon* at his disposal. Hence, Gratiolo's version cannot be called a translation proper. Nevertheless, it deserves attention because it carefully combines the achievements of earlier scholars.

The principal source of Gratiolo's text in chapter 1.1.6.5 is Jacob Mantino's version: many words, phrases, and half-sentences are taken over. Gratiolo does not, however, follow Mantino on the structure of the sentence with the three syntactical problems. Remember that here Mantino had inserted an adversative *tamen* in the former protasis of the conditional clause and a causal *nam* in the former apodosis, apparently because of a defective Hebrew exemplar. Gratiolo returns to Gerard's original sentence structure: "If a damage has afflicted . . . then it is sufficient." Let me quote the fifth and last Renaissance version of this sentence. Note that Gratiolo, in good humanist manner, retains Mantino's modernized vocabulary such as *temperatura* and *temperies,* but enhances his Latin style, for instance, by supplanting the relative clause *quae laesio insequitur . . . aut* ("which damages result from either . . .") into an elegant construction with *sive ex . . . sive,* or by replacing *propter* with the more exquisite *ob:*

> Quod si laedatur actio alicuius facultatis ob laesionem aliquam, quae prius actioni alterius praecedentis facultatis acciderit, sive ex membri intemperie sive depravata compositione, satis est illi si norit eam noxam oriri ob vitium temperaturae membri illius aut compaginis depravationem.[197]

For the problematic phrase "before another faculty" Gratiolo uses both *prius* and *praecedentis,* and hence misses Avicenna's original sense of "in front of"—unhappily, since Alpago had hit on the correct solution.

In the passages on the internal senses, Gratiolo continues to follow Mantino's text, but does not adopt his technical vocabulary. Mantino had called the second internal faculty *imaginatio* instead of *fantasia;* Gratiolo retains *fantasia.* Unfortunately, Gratiolo is twice misled by previous translators into choosing a different name for the same faculty of *fantasia:* once, he replaces *fantasia* with *imaginatio,* prompted by Gerard of Cremona, who had correctly been criticized by Ramusio on this point,[198] and once, he writes *imaginativa* instead of *fantasia,* this time in imitation of an unhappy choice of Mantino.[199] Gratiolo also once errs with Mantino—and, possibly, with the Hebrew intermediary—when choosing *in thesauro cogitativae* instead of the correct *aestimativae* for Arabic *al-wahm.*[200]

It was remarked above that Girolamo Ramusio, the first Renaissance translator, was the most consistent of all on the terminology for the internal senses. In view of this, it is regrettable that Gratiolo did not make much of this source—as he said in the preface, he received Ramusio's text only at a very late stage of his project. In chapter 1.1.6.5, Gratiolo notes one translation by Ramusio in the margin: *\*Ramusius quinta aut quarta.* In this passage Avicenna introduces the faculty of memory with the words: "The third faculty, which the physicians mention, this being the fifth or fourth faculty among the philosophers . . . etc." Mantino, in contrast to all other translators, had chosen "and" instead of "or": *quae est quinta et quarta . . . ,* and this obviously prompted Gratiolo to check the corresponding passage in Ramusio's interlinear translation, which, against Mantino and with the Arabic, reads *aut* and not *et.* Gratiolo himself obviously felt unsure about the correct solution, and finally and deploringly decided to leave out the phrase *aut quinta* altogether; he writes: *quae quarta a philosophis putatur.*[201]

Passages like this show that in the end it was impossible to improve upon previous translations without recourse to the Arabic original. Both Ledesma's and Gratiolo's versions of chapter 1.1.6.5 apparently are attempts to offer a better text by primarily drawing on earlier translations; I have not encountered any indication in the textual details that either of the two was seriously using an Arabic or Hebrew source. Apart from this, Ledesma and Gratiolo do not have much in common: Gratiolo's version by far surpasses Ledesma's in reliability, first, because his text is not shortened, and second, because Gratiolo has diligently compared Gerard's and Mantino's version and has payed attention to Alpago's emendations. Ledesma, in constrast, did not do much more than to rework Gerard's translation, and thus missed several sensible corrections by Alpago. He is far from reaching his self-proclaimed goal: to restore Avicenna to his "Arabic truth": *Avicennam . . . ad arabicam veritatem . . . emendare.*[202] In this case the most humanist and rhetorical scholar of the five translators betrays the greatest difference between his programmatic claims and the actual result of his work.

In view of the above evidence, what did the three translators from Arabic and Hebrew achieve: Ramusio, Alpago, and Mantino? In retrospect, it emerges that the medieval translation by Gerard of Cremona is an achievement of considerable quality. Gerard, with all his experience, easily overcomes problematic syntactic hurdles. His text nevertheless leaves room for improvement, also because the three centuries of Latin manuscript transmission resulted in distortions. Girolamo Ramusio's interlinear Arabic–Latin "translation" is not much more than a rough draft for future translation, but as such it achieves a considerable improvement in vocabulary and consistency, albeit being of very little help with the meaning of whole sentences. Andrea Alpago's posthumously published annotations to the *Canon* count among the most impressive philological achievements of the Renaissance translation movement. His emendations for the chapter surveyed are not concerned with style or technical vocabulary, but only with philological realiability, both in syntax and semantics. When Benedetto Rinio integrated Alpago's emendations into his edition of the *Canon* in 1555, the Latin transmission of the *Canon* was definitely lifted to a higher level of precision. Jacob Mantino, finally, is impressive in his attempt to compose an entirely new translation of the *Canon,* based on the Hebrew and on his expertise in medicine and philosophy. The result is a text in moderately humanist Latin, which sensibly adopts contemporary vocabulary. However, the direct comparison with Ramusio's and Alpago's versions shows that, on the basis of the Hebrew and without access to Arabic, it was impossible to increase the *veritas lectionis,* the "truth of reading," as had been Mantino's self-proclaimed aim.

## AVERROES: *COLLIGET*

Averroes's main medical work, the *Kitāb al-Kullīyāt fī ṭ-ṭibb* (Book on the Generalities of Medicine), called *Colliget* in the Latin West, is an interesting object for investigation for several reasons: it allows for the comparison of translators' techniques, since it was translated once and in its entirety in the Middle Ages, in 1285 in Padua by the Italian Jew Bonacosa,[203] and twice, but only partially, in the Renaissance, by Jean Bruyerin Champier and by Jacob Mantino. The case of the *Colliget* introduces us to a translator who has not yet been mentioned in the present survey: the French physician Bruyerin Champier. Just as Ledesma, he worked in a different milieu than the great majority of the Renaissance translators who were Italians. It is an open question whether Bruyerin simply reworked a Latin translation,[204] or translated from a Hebrew manuscript.[205]

This problem calls for a close analysis of Bruyerin's text, but it can provisionally be settled by interpreting his preface to the 1537 edition, which is reprinted in the Giunta edition of the combined Aristotle and Averroes.[206] Bruyerin is clearly aware of the usefulness of consulting Hebrew manuscripts to improve the understanding of Arabic–Latin medicine, as can be gathered from a story he relates about Denys Coronné, his teacher in Greek, who turned to his Hebrew manuscript whenever a passage in Avicenna's *Canon medicinae* was controversial.[207] With respect to his own manuscript sources for the translation, Bruyerin writes:

> When I reflected upon the fact that almost all works of our prince Galen were translated into Latin, together with the works of later Greek authors, due to the steady effort of learned men, it came to my mind that I too can help medicine if I invest the most careful effort possible into the matter which I am concerned with, which was either not tried by others or greatly corrupted and distorted. In this matter I have spent enormous labor because the manuscripts which have been printed *(codices quos typis excusos habemus)* are corrupted to a high degree, so that I could not derive any or hardly any help from these. But some months ago, a very old manuscript *(codex vetustissimus)* fell into my hands, which comprised three or four sections of these *Collectanea,* and which recalls that time in which the philosophy and the medicine of the Arabs and Moors entered France.[208]

The reference to the manuscripts of the *Colliget* that are accessible in print is important. It shows that Bruyerin was searching for better Latin—and not Hebrew—manuscripts. We know of seven editions of the Latin *Colliget* that precede Bruyerin's translation (Ferrara 1482, Venice 1490/1491, 1496, 1497, 1514, 1530, and Lyon 1531), but of no edition in Hebrew.[209] With the exemplars of these editions, Bruyerin contrasts the *codex vetustissimus*—a suspicious term, as we know from Ledesma[210]—which, he implies, is much less corrupted. The reference is obviously to a Latin manuscript. Later in the preface, he repeats his assertion that his task was laborious because he had no assistance in "rendering and restoring" the text except from the old age of his manuscript, "which more reliably renders the text than the books which are printed in letters."[211]

What Bruyerin says about his aims in translating the text accords well with his search for Latin manuscripts only: since students of medicine are deterred from Averroes, says Bruyerin, because of the terrible Latin of the text, he decided to render it "more polished" *(exultiorem)*. He claims to be the first to undertake "such a translation *(interpretatio)* of barbarous <authors>," a translation that in the end hopefully will be judged "more suitable for promulgation and more applicable and useful in healing" *(ad eloquendum aptior et ad medicandum commodior utiliorque)*.[212] Bruyerin's accusations of uncultivated language are not aimed at Averroes, but at his translators. His principal concern is to remedy this situation by improving the Latin style of Averroes's text.

Bruyerin wrote this preface in the house of his uncle,[213] Symphorien Champier,[214] editor of Avicenna and, in the later decades of his life, fierce opponent of Arabic medicine and Islamic religion; we have met Champier before as the author of a treatise on famous men in medicine. As the title page of his Averroes translation indicates, Bruyerin Champier himself was the physician of Cardinal François de Tournon (1489–1562), at times bishop of Lyon, cardinal since 1530, influential adviser of the French king, and patron of humanist studies in general and of Symphorien Champier in particular.[215] Bruyerin's intellectual context is very much that of medical humanism, as is also obvious from the preface, in which he presents himself as a successor to the humanist ambitions of Lefèvre d'Étaples and Nicolò Leoniceno.[216] Later in his life, Bruyerin publishes the main result of his learned studies, *De re cibaria libri XXII*, printed in 1560 in Lyon, an encyclopedia of contemporary food and drink in exquisite humanist Latin. Hence, with Jean Bruyerin Champier we meet with another example, after Paolo Ricci and Miguel Ledesma, of a full-blooded humanist engaged in the Arabic–Latin translation movement.

What does Bruyerin's text itself tell us about his technique of translation? One chapter of the Arabic *Kitāb al-Kullīyāt* has received a full critical edition, with commentary and German translation by Johann Christoph Bürgel in 1967: chapter 2.19 on the organs of breathing *(fī ālāt at-tanaffus)*. It will therefore serve as the textual basis for the following analysis; unfortunately, there is no edition of either of the two Hebrew translations. The Arabic original exists in two redactions, and Bürgel has drawn attention to the fact that Bonacosa (whom Bürgel confuses with Mantino) and Bruyerin both translate the longer redaction, since they render paragraphs 10 and 16–18 of chapter 2.19, which do not appear in the short redaction extant in the Arabic manuscript Granada, Colección del Sacro Monte, nr. I.[217]

There is no evidence in Bruyerin's version of chapter 2.19 that he had access to Arabic or Hebrew manuscripts of the *Colliget*. Rather, there is evidence that Bruyerin worked only with Bonacosa's Latin translation. This translation diverges from the Arabic text in several respects: it omits, adds, and transposes phrases, sentences, and whole sections. These differences between Bonacosa and the Arabic may be the result of the Latin manuscript transmission, or go back to Bonacosa himself, or (least probable) to an Arabic exemplar unknown to us. Whatever the origin of the differences between Bonacosa and the Arabic, the important point is that Bruyerin's text always sides with Bonacosa. For example, Bruyerin omits a paragraph on Galen's method of elimination, he omits references to the nutritive faculty and the epiglottis,[218] he adds a logical excursus on Galen's deficiencies in logic, and he adds a sentence on the art of anatomy in Aristotle's time,[219] just as Bonacosa does.[220] One could argue that on one occasion Bruyerin corrects a mistake in Bonacosa's text by eliminating a negation that is not in the Arabic, but this mistake could be spotted by any alert reader of the Latin only.[221] Most indicative are, in fact, the omissions shared by Bonacosa and Bruyerin, because some of these omissions obscure the text. Any reader of an Arabic or Hebrew manuscript would have been happy to restore the sense of Latin passages that are obscured by missing references to the nutritive faculty and to the epiglottis—references that appear in all Arabic manuscripts used by editors so far, including the shorter Granada redaction.[222] It is thus safe to conclude that Bruyerin worked with Latin texts only, at least in this chapter.

It is characteristic of Bruyerin Champier's technique of revision that his vocabulary and syntax differ considerably from Bonacosa's—to an extent that it may have been one of his aims to conceal that Bonacosa was his only source. Where possible, Bruyerin chooses a different anatomical vocabulary, for instance, as in the opening sentence. Bonacosa lists the instruments of breathing as follows:

> Instrumenta anhelitus sunt diaphragma, pulmo, canna, trachea arteria, epiglottis et uvula.

Bruyerin changes this into:

> Respirationi autem partes deservientes septum transversum, pulmones, aspera arteria, epiglottis, columella.[223]

Many changes on Bruyerin's part, as the one just quoted, are obvious attempts to turn Bonacosa's language into up-to-date humanist Latin. In some cases, his humanist interests lead Bruyerin to depart considerably from his medieval Latin exemplar. When Averroes explains that the epiglottis exists in animals not because it is necessary, but "for the sake of a better <condition>," Bonacosa translates *propter bene esse,* whereas Bruyerin transforms this into the flowery sentence: *ut melius ac felicius vitae curriculum transigeremus,* "so that we may spend a better and happier course of life."[224]

Bruyerin does not refrain from adding sentences of his own: he gives the additional information that "the diaphragm separates the nobler from the less noble parts" of the body; that the windpipe is also called *spiritalis fistula;* that *uvula* (epiglottis) is called *columella* by others.[225] His overall tendency, however, is to shorten and straighten the text. In this, he much resembles Miguel Ledesma, who had tried the same with Avicenna's *Canon.* Just as Ledesma, he omits phrases or half-sentences that seem unnecessary for the general argument of the text. For instance, Bruyerin omits the information that the movement of the arteries comes from the heart, that some have argued that breathing belongs to the same faculty as pulse, and that the eyebrows are an example of nonvoluntary movement.[226]

One might be inclined to think that all humanist revisers more or less work like Bruyerin and Ledesma. But, in fact, there was a great variety of techniques, as we have seen when analyzing Andrea Gratiolo's strategy of revising the text of the *Canon:* Gratiolo arrives at his text by comparing two previous translations (Gerard and Mantino) with two previous attempts at emendation (Ramusio and Alpago) and usually follows one of the four. He refrains from any omissions or additions. Bruyerin, in contrast, changes technical vocabulary, inserts explanatory sentences, adds phrases and half-sentences for the sake of style, omits information allegedly not necessary for the line of thought, and attempts to write syntax clearly different from

Bonacosa's. In the chapter investigated, there is no indication that these changes were motivated by medical reasons; rather, they were introduced to make the text "more polished" *(excultior)* and thus to rescue the *Colliget* for Bruyerin's fellow humanist physicians and their students. The Giunta editors of 1552 understood his purpose. They praise Bruyerin as "a learned man and very polished in his expression,"[227] and announce his version as follows: "Three parts of the *Collectanea,* which correspond to three books of the *Colliget,* . . . most elegantly *(elegantissime)* rendered into Latin by Jean Bruyerin Champier, are added after the old translation for the convenience of students *(ob studiosorum commodum).*"[228] It was Bruyerin's professed aim to improve upon the printed versions of the *Colliget,* which he found most corrupt. Instead of doing so, he produced a version of his own liking, which is a good read, but not a good guide to Averroes's text.

Jacob Mantino, the second translator of the *Colliget* in the Renaissance, demonstrated that it was indeed possible to improve upon Bonacosa's translation. The relative chronology of the two Renaissance translations is unclear. Apparently, they were produced independently of each other. Bruyerin's text appeared in 1537, Mantino's was not printed before 1552 when it was included in the Giuntine edition. It is indicative, however, that Mantino already mentions the project in the dedication to Ercole Gonzaga that accompanies his Averroes translations of 1521. He hopes to translate more of the commentaries of Averroes, Mantino says, "when I have finished (as I have begun) translating the first book of Averroes' *Colliget;* for this book as a whole is translated in a distorted and corrupted manner, as can easily be judged if our translation is compared with the earlier one."[229] The Latin versions of the *Colliget* that Mantino produced in the end, were not of the first book, but of the fifth: the last three chapters 57, 58, and 59 (on the composition on medicaments). This is unfortunate for our investigation since the techniques of Bruyerin and Mantino cannot be compared directly because they chose different parts of the *Colliget* for translation. Chapters 3.57–59 contain Averroes's critique of Alkindi's theory of degrees for compound medicines, and since this topic had assumed a certain popularity among late medieval and Renaissance physicians,[230] Mantino seems to have selected it for translation.

Mantino's translation of the *Colliget* is clearly a revision of Bonacosa's medieval translation: Mantino borrows many phrases and syntactical structures from Bonacosa. We therefore have to answer the question whether in this case Mantino departed from his usual technique of translating from the Hebrew and produced a simple revision on the basis of Latin texts only. The following analysis of a sample passage, a section on the effects of compound medicines of chapter 5.57, will show that he also worked with a Hebrew exemplar.

In this chapter Averroes mentions that an advantage of compound medicaments over simple ones is that some illnesses can be cured only by a medicament that has several effects different or even contrary to each other. This does not mean, however, that these effects happen at the same time in the same place, as occurs in sense-perception, when we perceive white and black together. And Bonacosa proceeds: "and when we perceive warm and cold together and once, due to a mixture of hot and cold water, as Galen says" *(et sicut sentimus calidum et frigidum simul et semel ex commixtione aquae ferventis et frigidae, sicut dicit Galenus)*. Mantino offers a different text, which is more detailed on how the same body can perceive hot and cold at the same time, namely, in different parts of the body: "and when we perceive heat and cold in one moment in our entire body, if it happens that small parts of our body are washed partly with warm water, partly with cold water, as Galen says" *(caloremque et frigidatem <scil. percipimus> in instanti per universum corpus nostrum, si contingat ut particulae nostri corporis partim frigida, partim calida aqua abluantur, ut dicit Galenus)*.[231] Mantino's translation accords literally with the Arabic original, both in the shorter and the longer redactions. It is clear that Mantino followed Bonacosa's translation only as long as it did not divert from the reading of his Hebrew manuscript.

Averroes then proceeds to argue that contrary effects are possible in sense-perception because they are not material. In the translation of Mantino:

> Even though this happens to the senses, it in fact only happens because they are not material. The reason for this has already been related in a different discipline.

Bonacosa offers a longer text, which again diverts from the Arabic. In the following quotation, the additions are put in brackets:

> hoc non accidit virtutibus sensibilibus cum sensatis nisi propterea quia non sunt materiales [permutabiles, alterabiles, passibiles in seipsis a sensatis passione manifesta, eo quod illae non sunt coniunctae cum materia forti coniunctione]. Et causa huius redditur in alia scientia [nobiliori ista, hoc est in tertio De anima.][232]

The different readings in Bonacosa's text are clearly additions: they supply further explanatory information and a reference. Similar additions appear again in the final sentences of the passage, where Averroes explains that the effects of compound medicines are material and hence do not occur at the same time and in the same place. Mantino's translation again closely follows the Arabic text as we know it today, whereas Bonacosa inserts three passages with additional information. As was remarked above, we do not yet know whether these additions originate from Bonacosa, a Latin scribe, an Arabic scribe, or Averroes himself.[233] Whatever the origin of Bonacosa's different readings, the effect of Mantino's revision was that he restored the text to a more original format, which was not yet augmented by additions. It is very likely, therefore, that Mantino was working from a Hebrew manuscript. His text is far superior to Bruyerin Champier's, and it also makes a tangible advance over Bonacosa's. Remember that Mantino could not achieve such progress when working on Avicenna's *Canon*. The reason for this discrepancy is that Gerard of Cremona's translation of the *Canon* is much more literal and reliable than Bonacosa's translation of the *Colliget,* or at least than the printed versions of Bonacosa's text, which Mantino had access to. In the case of the *Colliget,* the textual differences between the Arabic and the Latin were so great that Mantino could achieve much by correcting the Latin against the Hebrew.[234]

Before we turn to our final translator, Jacobus Sylvius, a concluding word on the *doctissimus* translator Jacob Mantino is in order. His translation technique is impressively pragmatic and flexible. When working on Averroes's preface to *Metaphysics* 12 and on the Middle commentary on the *Topics*, Mantino adopts the previous Renaissance Hebrew–Latin translations by del Medigo and Balmes and corrects them against the Hebrew. In the case of Avicenna's *Canon,* Mantino produces an entirely new translation from the Hebrew, for which he also consulted Gerard's translation and Alpago's emendations. Finally, for his *Colliget* translation, Mantino produces a thorough and highly accomplished revision of Bonacosa's medieval Latin translation against the Hebrew. The Hebrew intermediary, which was Mantino's main basis, did not allow him to reach the same reliability as Renaissance translators working from Arabic. He was nevertheless able to contribute greatly to the Latin transmission of Arabic works. All translations analyzed are written in a moderately humanist Latin and are characteristically sophisticated in their medical and philosophical vocabulary. For all his philological quality, which much surpasses that of his fellow translators from Hebrew surveyed here, Jacob Mantino is a worthy successor to the great medieval translators Dominicus Gundisalvi, Gerard of Cremona, and Michael Scot.

## MESUE: *OPERA*

The year 1542 marks a turning point in the transmission of the works of Mesue. It is in this year that Jacques Dubois—better known under his latinized name Jacobus Sylvius—published a new version of the entire *Opera* in Paris under the title: *Joannis Mesuae Damasceni De re medica libri tres Jacobo Sylvio medico interprete.* Sylvius's new version was enormously successful: it was reprinted twenty-one times, while the older, medieval texts, which had a long printing history before Sylvius, were printed only four times after 1542, and always together with Sylvius's version.[235]

Sylvius, born in Amiens in 1478, teacher of medicine in Paris beginning in 1531, was well-known in his time as a prominent humanist and medical writer, especially on anatomy and *materia medica.* The most famous of his students was Andreas Vesalius. When the latter published his own anatomical magnum opus, the *Fabrica* of 1543, Sylvius, as an ardent admirer of the classics, was deeply irritated by Vesalius's criticism of Galen and reacted with harsh attacks in print. Most influential among Sylvius's pharmacological works was his *De medicamentorum simplicium delectu, praeparationibus, mistionis modo* of 1542. In this book Sylvius takes Mesue as a starting point for developing methods of selecting and correcting all kinds of *simplicia,* and of mixing medicaments.[236] Sylvius died in 1555.

In view of the fact that pharmacology was at the center of the long controversy over the value of Arabic science, it is noteworthy that it was a convinced partisan of medical humanism who produced the new version of Mesue's *Opera.* In his dedicatory letter to the bishop of Bayonne, Sylvius defends his project by arguing that Mesue, in spite of occasional mistakes, closely followed Hippocratic and Galenic principles in the description of simple and compound drugs, and hence that the critique launched against Mesue by the "Greek family" *(familia graeca)*[237] of his day could not but target the ancient authorities as well.[238] It is impossible, says Sylvius, to ignore the Arabs, and Mesue in particular who has invented so many medicaments and who has described mild laxatives much less dangerous than those used by the Greek physicians.[239] The dedicatory letter is primarily meant to defend and praise Mesue, as Sylvius himself says,[240] and that is probably why he does not say much about his aims and methods in preparing his new version. It is his aim, explains Sylvius, "to apply or rather restore the brilliance" *(splendor)* of a text that has long been immersed in darkness and dirt.[241] In the future, he will perhaps also "refine" *(expolire)* the fourth and last part of Mesue's *Opera,* the *Practica.*[242] On the title page and in the dedication, Sylvius refers to his own work with the terms *interpretatio* and *interpretari,* but in contrast to Ledesma and Bruyerin Champier, he does not pretend to be translating from the Arabic or Hebrew. It is clear from the dedication that *interpretari* in this case means purifying, polishing, and refining the Latin text.

The *versio Sylvii,* therefore, is not a translation proper, but a revision. This revision is worth closer inspection especially since it is the work of a fervent admirer of Hippocrates and Galen. The analysis of Sylvius's techniques will therefore round out the picture we have attained of the involvement of humanists in the Latin transmission of Arabic works in the Renaissance. Sylvius's text is interesting for two other reasons as well: first, because Mesue's widely distributed *Opera* are central for the reception of Arabic learning in the Renaissance, and second, because the quality of Sylvius's work is disputed in modern scholarship: on the one hand, it was argued that Sylvius has radically altered the medieval text by turning it into humanist Latin, leaving out whatever appeared unclear to him and changing the sequence of the recipes without good reason;[243] on the other hand, it was replied that such omissions were very rare and that the extraordinary value of Sylvius's version was apparent to everyone who attempted to understand the medieval Latin texts without Sylvius's intelligent interpretation.[244]

To understand Sylvius's work, it is necessary to explain briefly the structure and provenance of the *Opera Mesue.* The entire text, as it was printed in the Renaissance, consists of four parts: (1) *Canones universales,* a set of rules pertaining to the preparation of laxative drugs, (2) *De simplicibus,* fifty-four monographs on laxative drugs, (3) *Grabadin,* a pharmacopeia with twelve chapters on different kinds of composite drugs, and (4) *Practica,* a therapeutics with medicaments *a capite ad calcem.* According to the present state of research,[245] these texts are clearly the product of one or several Latin compilers of the thirteenth century. They contain much material that was obviously translated from the Arabic, but they also have parallels in wording with the twelfth-century translation of Avicenna's *Canon.* The structure of the *Opera Mesue,* especially of the *Grabadin,* shows similarities to that of the *Canon,* which indicates that the compilers have taken Avicenna's book as a model. The first three parts of the *Opera* are connected to each other by cross-references; moreover, they cite very similar sets of Greek and Arabic authorities; the fourth part is less well-connected to the rest. Since several of the authorities cited lived after the historical Joannes Mesue, that is, Yūḥannā ibn Māsawayh, who died in 857 AD, and since we know of no other Arabic author with this name, the attribution to "Mesue" is pseudoepigraphic. Strictly speaking, the author of the *Opera* should be referred to as "Pseudo-Mesue." Even if a few passages have possibly been inserted by the compilers, it is clear for linguistic reasons that the *Opera* in general are of Arabic origin. But the Arabic treatises from which these translations are made have not yet been identified.

It is impossible, therefore, to check Sylvius's work against the Arabic original: we can only judge it by comparing the two Latin versions. Sylvius's revision of the thirteenth-century Latin text covers the first three books: *Canones, De simplicibus,* and *Grabadin.* It is a thorough revision, in the sense that Sylvius has not left any sentence untouched. It was obviously his aim to transform the medieval and arabicizing sentence structure into proper humanist syntax and to modernize the vocabulary. An example is Sylvius's nomenclature for the four humors of the Galenic tradition: he retains *sanguis* ("blood"), but writes *pituita* instead of *phlegma* ("phlegm"), *bilis* instead of *cholera* ("yellow bile"), and *bilis atra* instead of *cholera nigra* ("black bile").[246] In what follows, I will compare a passage from the *versio antiqua* and the *versio Sylvii,* namely, the monograph on the plant senna in the second book, Mesue's *De simplicibus,* of which I cite the full text with an English translation in Table 3. The shrub senna and the medicinal properties of its leaves and pods will reappear in our next chapter, which is concerned with the conflict of Greek and Arabic traditions in Renaissance pharmacology.

To understand the changes that Sylvius introduces, it is important to know the standard structure of a monograph on a drug in Mesue's *De simplicibus.* Mesue discusses fifty-four plants that can be used as laxative drugs. Each monograph has six parts: it begins with a brief botanical identification of the plant, and is then followed by the parts "selection" (electio), "temperament and property" *(complexio et proprietas),* "correction" *(correctio),* "effect" *(posse),* and "dose" *(dosis).* Accordingly, in the case of senna, Mesue first identifies the drug, then explains which parts of the plant to select, namely, the pods rather than the leaves *(electio),* describes the temperant of senna as warm and dry *(complexio et proprietas),* explains how to administer senna and how to strengthen its laxative effect *(correctio),* describes its medicinal effect on the body *(posse),* and closes with a brief statement on the dose to administer *(dosis).*

Sylvius's most apparent alteration consists in a regrouping of these sections. He combines two of them into one: temperament/property *(complexio et proprietas)* and effect *(posse).* This alteration has been criticized as unfounded,[247] but it is not arbitrary: Sylvius applies it to all monographs of *De simplicibus.* It is justified to a certain degree, insofar as the two categories *proprietas* and *posse* are closely related: the first states in a general manner the property of the drug that depends on its temperament, that is, on the mixture of the elemental qualities hot, cold, wet, and dry, while the second lists the specific medicinal effects of the drug. It is true, though, that the combination would not have found Mesue's approval because he

Table 3. The monograph on Senna in Pseudo-Mesue's *De simplicibus*.

| Mesue, *Opera* (1549), f. 64ra (**versio antiqua**) | Translation of the old version | Mesue, *Opera* (1549), f. 63vb–64ra (**versio Sylvii**) |
|---|---|---|
| Sene est folliculus plantae quam vocant Persae abalzemer, et eius ortus est secundum semitam orobi, et invenitur ex eo domesticum et sylvestre. | Senna is the pod of a plant that the Persians call "abalzemer." It grows in the way of vetch. There exists a domestic and a wild senna. | Senna est folliculus plantae Persis dictae abalzemer, orobi modo nascentis. Sativa est et agrestis. |
| Electio. Melior pars plantae eius est folliculus, deinde folia, sed tamen in eis est virtus debilis valde. Et melior folliculus est cuius color accedit ad viriditatem, et subnigredinem quandam, et in quo est de amaritudine res modica cum stipticitate, et qui est magis completus, et in quo sunt semina ampla compressa. Subalbidus autem non est bonus et completus[1] similiter. Et meliora folia sunt viridia; subalbida vero et tenuia non bona. Et antiquum ex eo est sine spiritu, et stipites eius sunt inutiles. | Selection. The better part of the plant is the pod, than the leaves, but the leaves' power is much weaker. The better pod is that whose color comes close to green and to a certain blackness, and in which there is a limited amount of bitterness together with astringency, and which is filled better and in which there are big, compressed seeds. The whitish pod, however, is not good, and also it is not filled. The better leaves are the green ones; the whitish and thin ones are not good. An old <pod> is without power, and its twigs are useless. | Folliculo quam foliis est efficacior, praesertim si is ex viridi nigricat, modice amarus, subadstringit absolutus, recens in quo semen amplum compressum; vetustate enim exanimatur, subalbidus et imperfectus improbantur. Folia autem viridia sunt praestantiora subalbidis et tenuibus. Surculis est inutilibus. Calfacit initio secundi ordinis, siccat primo. Folia etiam primo calfaciunt. Terget, expurgat, digerit, |

(continued)

*Table 3.* (continued)

| Mesue, *Opera* (1549), f. 64ra (**versio antiqua**) | Translation of the old version | Mesue, *Opera* (1549), f. 63vb–64ra (**versio Sylvii**) |
| --- | --- | --- |
| Complexio et proprietas. Calidum est in principio secundi, siccum in primo, et folia sunt in primo calida. Et est abstersivum, mundificativum convenienter, et resolutivum. | Temperament and property. It is warm in the beginning of the second <degree>, dry in the first, and the leaves are warm in the first. It expells and purges moderately, and digests. | purgat clementer melancholiam et bilem ustam a cerebro, sensoriis, pulmone, corde, hepate, liene. Proinde morbis dictarum partium ab humore huiusmodi proficiscentibus succurrit, ut melancholicis febribus et antiquis, gaudium gignit ablato humore sine causa externa tristante; et floridum corpus efficit, obstructiones viscerum aperit. Foliorum huius et chamaemeli decoctum lotione cerebrum, nervos roborat. Eadem quovis modo usurpata, visum et auditum firmat. |
| Rectificatio. Est solutionis debilis et tardae et debilitat stomachum. Verum confortatur eius operatio in permiscendo cum eo aliquid ex rebus acutis sicut zingiber proprie, et sal gemma, et sal indus; et medicinae cordiales cum eo et stomatichae felicitant opus eius. Et dixit Galenus: decoquatur cum iuribus gallorum aut gallinarum aut carnium, et solvit sine molestia, sed oportet ut eius quantitas sit multa. Et si infundatur in aqua casei cum spica, deinde coquatur ebullitione aliqua, est medicamen bonum. Et si propinetur pulvis eius in lacte dulci, fiet medicamen bonum. | Correction. It has a weak and slow solubility, and weakens the stomach. But its effect is strengthened by mixing it with something acrid such as ginger proper, rock salt and Indian salt. Medicaments for the heart and for the stomach mixed <with it> make its effect more successful. Galen said: It should be boiled with broth of cock or hen or meat, and it has a laxative effect without any trouble, but a great quantity is needed. If it is poured in cheese-water (*i.e., whey*), together with spikenard, then cooked in some boiling, it is a good medicament. And if its powder is poured in sweet milk, it becomes a good medicament. | Eius purgationem tardam et imbecillam celerant et intendunt mista acria, ut zingiber, sal gemmeus, sal Indus. Ne vero stomachum laedat, cardiaca et stomachica sunt miscenda. Ob id Galeni decreto coquenda largo pondere ex iure galli, gallinae, vel aliarum carnium, ut citra molestiam purget; vel sero lactis infusa cum spica, deinde parum diu fervefacta; vel pulvis eius ex lacte dulci sumendus est. |

Vino musteo albo miscebat quidam vim magnam sennae, post tres menses potandum dabat, sicque purgabat cerebrum et sensoria, et gaudium augebat. Sunt qui decocto eius cum prunis et spica faciliter[2] purgant. Mediocrem autem coctionem sustinet.

Infunditur ab aureis tribus ad unciam unam.

Similarly, someone advises that a quantity of it should be immersed in must, it being white must, and that after three months it should be administered, where necessary, because wine has a laxative effect, purges the brain and the senses and generates happiness. There are people who administer it when it is boiled together with plums and spikenard, and it becomes a good laxative. It bears boiling in a moderate way.

Effect. In soluble form, it easily expells black bile and burned bile. It purifies brain, heart, liver, spleen, the organs of the senses, the lung. It is of help in their diseases. It loosens constipation in the bowels. It strengthens those who are young when using it. It produces joy and abolishes the reason for sadness. Its leaves, <bathed> in water, are applied to the head, in particular with camomile, and <this> strengthens the brain and the nerves. In all applications, it assists the sense of sight, and it strengthens hearing. It is a good medicament against melancholical and old fevers.

Dose. The cup of it, which is poured, ranges from three florins to one ounce.

Similiter praecepit quidam ut quantitas eius submergatur in musto, et sit mustum album, et post tres menses propinetur ubi oporter; est enim vinum solutivum mundificans cerebrum et sensus et generans laetitiam. Et sunt qui propinant decoctionem eius et prunorum et spicae cum eis, et fit solutivum bonum. De decoctione autem sustinet mediocriter.

Posse. Solutione educit cum facilitate melancholiam et choleram adustam. Et mundificat cerebrum, cor et hepar et splenem et membra sensuum et pulmonem, et confert aegritudinibus eorum. Et aperit oppilationes viscerum. Et confortat utentem eo in iuventute; et generat gaudium, causam autem tristitiae amputat. Et ponuntur folia eius in lavacris ad caput, et proprie cum chamaemilla, et confortat cerebrum, et nervos. Et secundum omnem administrationem eius addit in visum et corroborat auditum. Et est medicina bona febrium melancholicarum et antiquarum.

Dosis. Eius infusi potio est ab aureis iii usque ad unciam i.

1. Cf. Mesue, *Opera* (1570), f. 86ra: *incompletus.*
2. Cf. Mesue, *Opera* (1548), 132: *foeliciter.*

distinguishes in the *Canones* between *virtus communis* and *virtus propria* of laxative simple drugs: the former depends on the temperament, the latter, which is responsible for the specific effects, exceeds the power of the temperament and is due to celestial influence.[248]

It has been shown that Sylvius, in good humanist manner, usually avoids Arabic nomenclature and attempts to replace it with Latin or Greek terms.[249] He does not do this, however, in the present chapter on senna: the pharmacological vocabulary of the Arabic–Latin tradition is left untouched for the most part. Sylvius retains the terms *abalzemer, orobus, chamaemelum, zingiber, sal gemmeus, sal Indus, vinum mustum, pruna,* and *spica,* and only changes *aqua casei* into *serum lactis.* But he tries to transform some medieval expressions into more humanist Latin. A typical example of a change in Latin style comes from the botanical description at the beginning: Sylvius replaces the medieval expression *domesticum et sylvestre* ("domestic and wild") with *sativa et agrestis.* These terms are regularly used in antiquity for cultivated and wild plants or animals; the term *sativum* appears particularly often in Pliny's *Naturalis historia.*[250]

Sylvius's main interpretative technique in this monograph does not concern the vocabulary, but the arrangement of the material: he changes the sequence not only of the main sections but also of single sentences and longer phrases. For example, in the section on *electio,* that is, on the parts of the plant to select for medicinal purposes, Sylvius regroups the material so that the paragraph is more clearly divided into three parts on pods, leaves and twigs. Another example of such rearrangement can be found in the section on the medicinal effects *(posse)* of senna on various organs of the body. Here Sylvius combines two sentences that originally stood apart in the *versio antiqua* by adding phrases, which I have put in brackets in the following quotation: senna "is of help in diseases of the aforementioned organs, <diseases which originate from this kind of humor>, <such as> through melancholical and old fevers" ( . . . *morbis dictarum partium <ab humore huiusmodi proficiscentibus> succurrit, <ut> melancholicis febribus et antiquis).* Sylvius thus fuses two statements that are unrelated in Mesue's text: first that senna, as Mesue says, "easily expells black bile" *(melancholiam)* and is helpful with the many diseases, the second that senna is a good medicament against melancholical fevers—hence Sylvius's conjecture that the diseases referred to by Mesue are those caused by *melancholia* and that melancholical fever is one of them. This is a possible interpretation, but it remains a conjecture.

In a similar manner, Sylvius regroups the material in the section on the drug's correction *(correctio).* Again, he establishes a medical connection between two sentences, which in Mesue's text do not follow upon each other:

the first being that senna weakens the stomach, the second, that medicaments for the heart and the stomach "make its effect more beneficial" *(felicitant opus eius)*. In Sylvius's version, the combined sentence reads as follows: "lest it *(i.e., senna)* damages the stomach, one should mix it with medicaments for the heart and for the stomach" *(Ne vero stomachum laedat, cardiaca et stomachica sunt miscenda)*. This is a clever interpretation, since it is well-supported by the parallel structure of the original text: Mesue mentions two disadvantageous properties of the drug, first its weak and slow effect, and second its weakening of the stomach, and then proceeds to describe two techniques of combating these problems, first, a method to strengthen its effect, and second, mixing it with medicaments for heart and stomach "to make its effect more beneficial." Since the first method clearly remedies the first disadvantage, its weak effect, Sylvius's guess is well-founded that the second method is meant to remedy the problems that senna causes in the stomach. It is well-founded, but, of course, it implies an interpretative step that we do not find with the translators proper.

Our analysis shows that Sylvius's technique of revision differs noticeably from that of his humanist colleagues Ledesma, Gratiolo, and Bruyerin Champier. We have seen that Gratiolo refrains from any omissions and additions, but tries to present a text very similar to a true translation. Ledesma and Bruyerin Champier, in contrast, are primarily concerned with humanist Latin style; they shorten and straighten the text to a considerable extent. Sylvius shares both the translator's and the humanist's motives, but the principal advantage of his new version lies in the medical competence of the reviser: Sylvius's version aims at a better understanding of the medical content by extinguishing ambiguities and by reordering Mesue's textual material. Since this is intelligently done, at least in the chapter analyzed, it is no wonder that the *versio Sylvii* became a great printing success in the sixteenth century.

⁊⁊

The Arabic–Hebrew–Latin translation movement in the Renaissance did not evolve in marginal corners of Italian and French society. It grew in and was supported by influential milieus. In northern Italy, several aristocratic patrons financed the new Hebrew–Latin translations of Averroes's commentaries: Giovanni Pico della Mirandola, Cardinal Domenico Grimani, and Bishop Ercole Gonzaga. All three of them sympathized with the intellectual climate of the University of Padua, and later of Bologna, where Gonzaga listened to the lectures of Pomponazzi. At the same time, Pico and Grimani were well-respected among humanist scholars, Pico for his elegant Latin and classical learning, Grimani for his collection of Greek manuscripts. Andrea Gritti, doge of Venice, is chosen as dedicatee of Mantino's 1530 *Canon* translation because of his ties to the University of Padua,[251] where he had studied. Other patrons of Hebrew–Latin translators were Guido Rangoni, the papal military commander, and Alberto Pio, prince of Carpi and humanist Aristotelian.[252] In Damascus, the two translators Ramusio and Alpago themselves, the physicians of the embassy, were descendants of Northern Italian noble families. They too had studied in Padua, and Alpago returned to the city at the end of his life to take up a professorship in medicine. The learned Jewish translators of Averroes were physicians by profession—with the possible exception of del Medigo, who seems to have lived on teaching philosophy—and as such they formed an integral part of Italian society. Jacob Mantino was one of the most prominent Jewish scholars of the time, a friend and physician of aristocratic patricians, ambassadors, and popes. In France, the milieu in which the revisions of Averroes and Mesue grew was that of medical humanism. Bruyerin Champier worked as physician to a cardinal and at the French court; Jacobus Sylvius was professor of medicine at Paris; both were enthusiastic about classical sources.

It has become clear that many humanist scholars were involved in the translation movement: Paolo Ricci, who translated Averroes's commentaries into stylish humanist Latin; Miguel Ledesma, the author of a humanist revision of the *Canon,* disguised as a translation; Andrea Gratiolo, the reviser of the *Canon* who was also a translator from the Greek; and the two French humanists Bruyerin Champier and Jacobus Sylvius. Humanist ideals also influenced the translation techniques of other translators and revisers: Jacob Mantino modestly said of himself that he was lacking in *eloquentia,* but his Latin nevertheless appealed to humanist readers. Also, many translators worked on a better transmission: they decided to correct a translated text or to translate it anew. For this purpose, they turned to or even searched for Arabic, Hebrew, or Latin manuscripts that could be the basis of a better text. The concern for extinguishing corruptions and for retrieving better source material was much more widespread among Renaissance translators than among their medieval predecessors. Here the Renaissance translation movement is clearly influenced by the new humanist attitudes toward the text. Hence, in spite of the humanist polemics against Arabic sciences in the Renaissance, it would be wrong to conceive of humanist and Arabist traditions as two antagonist movements. In the translation movement, they were intertwined.

This does not mean, however, that the most humanist of the translators were the greatest experts on the texts. Ricci's, Ledesma's, Bruyerin's, and Sylvius's translations cannot aspire to render the "Arabic truth," as Ledesma had claimed: in varying degrees, all four of them introduce new features into the text and suppress others, which makes these versions interpretations rather than translations. In these cases, the gulf between programmatic rhetoric and actual results can be drastic. As interpretations, which do not pretend to be compared to the Arabic, these works are significant achievements, especially so, if the reviser is equipped with technical expertise, as was Jacobus Sylvius.

The Hebrew intermediary, which was important for so many translations of the Renaissance, was an opportunity and a problem at the same time. It was an opportunity because it gave access to a great range of new texts, especially by Averroes. It was a problem because it did have a tangible effect on the transmission. In spite of the close relationship between Arabic and Hebrew, the corruptions in syntax and the ambiguities in vocabulary were considerable enough to influence the Latin versions. The effect varied from text to text. The Hebrew exemplar was of greatest help when it was used to correct a problematic or defective medieval Latin version of an Arabic treatise, as was the case when Jacob Mantino corrected Bonacosa's medieval rendering of Averroes's *Colliget*.

The efforts undertaken by the translators and the sheer number of translations produced are impressive. The *latina editio* of Averroes was considerably enlarged, many mistakes in the transmission of Avicenna's *Canon* removed, the reliability of Averroes' *Colliget* much increased, and Pseudo-Mesue's *materia medica* rendered more intelligible. On the other hand, the products of the translators differ greatly in quality. A repeated claim of the Renaissance translators is that they restore the true meaning of the Arabic author, that they translate omissions and extinguish additions made by the previous translators, and in general remedy their errors. In the actual result, the philological work of the Renaissance translators did not always reach these aims. Stylistic revisions of earlier translations, without consulting the Arabic or Hebrew, tend to increase corruptions rather than to extinguish them, and the Hebrew intermediary sometimes proved to be problematic. Another problem was a lack of competence in Latin. The translations by Elia del Medigo and Abraham de Balmes successfully transport unknown material into Latin, but the quality of their translations suffers from their restricted command of Latin. Often, del Medigo's and Balmes's Latin syntax is difficult to understand, and the vocabulary is infelicitously chosen. The translators who achieved most in terms of quality are Andrea Alpago and Jacob Mantino, the worthy successors of the great translators of the Middle Ages. Alpago and Mantino stand out for many reasons: one reason is that they reach their aims by choosing pragmatic philological techniques that fit well with the source material;

another reason is that they have full command of the two languages involved: Arabic and Latin, or Hebrew and Latin. It is a particular merit of Mantino's translations that he developed a language of translation that was literal and at the same time moderately classicizing, and thus met the expectations of his contemporary readers. These comments on the qualitative merits of Alpago's and Mantino's translation should not obscure the fact that del Medigo's and Balmes's texts found their readers too, not only among their patrons but also among Italian professors of philosophy.

The influence of the Renaissance Arabic-(Hebrew)-Latin translation movement is not on the scale of that of the Middle Ages. But it would be false to assume that the Renaissance translations were produced into a void and came too late to exert any serious impact. The few examples of their influence that have already been unearthed indicate that there is much to be expected from further research, especially from studies on the works of sixteenth-century university professors in Italy. One area of influence is zoology. As has been shown recently, Pietro Pomponazzi in his 1521–1524 Bologna lectures on Aristotle's *De partibus animalium* makes ample use of Jacob Mantino's translation of Averroes's *Paraphrasis* of the work, which had just been printed in Rome in 1521.[253] Another originally Paduan philosopher, Agostino Nifo, also uses Averroes's *Paraphrasis* for his own comprehensive commentary on Aristotle's books on animals, composed in the 1530s. Nifo in fact often adopts Averroes's position silently in order to criticize positions of Galen.[254]

In the field of logic, the influence of Averroes's newly translated treatises must have been even greater, given the numerous Hebrew–Latin translations of logical treatises. Some cases of reception are already known. The 1562 Giunta edition of the combined Aristotle and Averroes prints three works by Benardino Tomitano (d. 1576), professor of logic at the University of Padua. His texts all deal with the *Posterior Analytics* and explicitly address issues raised in Averroes's newly translated Long commentary on the *Posterior Analytics* and his *Logical Questions*. The third text, in fact, is even a proper commentary on Averroes's *Logical Questions*.[255] Another professor of the University of Padua, Giovanni Giacomo Pavese (d. 1566), published in 1552 a lengthy commentary on Averroes's introduction to the Long commentary on the *Posterior Analytics,* citing Balmes's and Burana's translations as the *textus* of his own comments.[256] Other Italian professors too commented on parts of Averroes's commentary: Giovanni Bernardino Longo of Naples (d. 1599) and Annibale Balsamo of Milan (late sixteenth century).[257] Given that the discussion of scientific method was particularly vivid in the sixteenth century, also in the context of the emergence of modern science,[258] it is likely that these cases of reception of Averroes's logic of demonstration are only the tip of the iceberg.

The new versions of Avicenna's *Canon,* and especially that of Andrea Al-pago, quickly found their readers. An example of this reception is the Paduan professor of theoretical medicine Oddo Oddi (d. 1558), who wrote a commentary on the *Canon* based on Alpago's corrected text. Oddi often compares Alpago's version with the versions of Gerard of Cremona and Jacob Mantino.[259] The *Canon* commentator Santorio Santorio, who held the same chair of theoretical medicine in Padua between 1611 and 1624, also pays attention to the differences between the three versions, and expresses his respect for Alpago as the one "whom we follow mostly" on Avicenna's biography and on his text.[260] With regard to medical content, the most important contribution of the translation movement was probably Alpago's *Interpretatio Arabicorum nominum,* that is, his lexicon of Arabic proper names in Avicenna's *Canon,* which was first printed in 1527. Alpago also composed an index to Arabic names of simple medicines in Serapion, that appeared in 1550. There are indications that Alpago's lexica were used and estimated by sixteenth-century physicians and botanists.[261] *Materia medica*—the field to which we shall turn in the next chapter—was a thriving and hotly debated field in the sixteenth century, and it is very likely that Alpago's influence was much greater than is visible to us at the present stage of research.

# PART II

Greeks versus Arabs

# CHAPTER 4

# MATERIA MEDICA

## Humanists on Laxatives

he physician is a profession chosen by a large number of humanist authors: it is rivaled only by the secretary or chancellor and the teacher of Latin and rhetoric.[1] In accordance with and perhaps also in consequence of this social phenomenon, it is the science of medicine that more than any other science was subject to the most substantial and dramatic attempts of humanist reform or refurbishment. The polemic that accompanied this process was colored by anti-Arabic tones, which reached their climax in an international debate in the 1520s and 1530s.

The main stages of this development are well-known. The founding book of the learned challenge of Arabic medicine was Nicolò Leoniceno's *De Plinii et plurium aliorum medicorum in medicina erroribus,* published in installments from 1492 to 1507.[2] Giovanni Manardo, pupil of Leoniceno in Ferrara, continued the attack in his *Epistolae medicinales,* which appeared in ever more comprehensive editions beginning in 1521.[3] In the 1520s, the French physician Symphorien Champier made a public turn against the Arabs and embraced medical humanism;[4] in 1530, the first German anti-Arabic and pro-humanist pamphlet appeared: Leonhart Fuchs's *Errata recentiorum medicorum.*[5]

A single quotation drawn from this ample literature may serve to illustrate the acerbity of the polemics; the quotation comes from Fuchs's *Paradoxorum medicinae libri tres* of 1535:

> In the past, I had never imagined that studying Arabic authorities could be so pernicious as I realize now. I understand more clearly every day how much they are able to jeopardize, and I confess freely that I have treated them much too mildly in the past. One has to treat them much

stricter, at least for the sake of posterity to prevent them from entering blindly the Arabic brothels *(ne arabica illa lupanaria incidant improvidi)*. I declare publicly in front of everybody that I shall be the toughest enemy of the Arabs, that I will invest greater labor than before and besiege their strongholds and do not cease to do so until I have completely taken them, if God blesses me with a long life. For who could endure witnessing this pest raging any longer among the people? Nobody, to be sure— except those who desire the extinction of men, and of the Christians in particular. Instead let us proceed to the sources *(ad fontes potius pergamus)* and scoop the pure water, which is free of any dirt.[6]

In this passage, Leonhart Fuchs combines the humanist appeal to return to antiquity *(ad fontes)* with anti-Islamic polemics: he conjures up the ruin of Christendom—this religious tone, in fact, is rather untypical of the medical controversy in general and also of Fuchs in particular. What is characteristic of Fuchs, is the belligerent, furious language, the metaphorical usage of the vocabulary of war, prostitution, and dirt.

There exist several useful surveys of the history of the controversy over the value of Arabic science, and of Arabic medicine in particular.[7] Instead of retelling this story, I will briefly point to the types of polemical arguments employed by the anti-Arabic party. It is the general line of humanist criticism that the medical sources of Arabic provenance abound in confusion and errors. Such errors are particularly worrying in medicine because they are dangerous to health.[8] An important source of these errors, they say, is the confusion of terminology in the Arabic–Latin tradition and the corruption of vocabulary in the course of transmission. The Renaissance physicians use Greek plant names for the wrong plants, complains Leoniceno, which is mainly due to "the Arabic authors, who among themselves do not agree either."[9] Often it was claimed, as we have seen in the preceding chapter, that the medieval translations were not reliable and that they were written in a barbarous Latin style.[10] Similarly, it was argued that the Arabic translations of the Greek were not reliable, since in these translations, as Fuchs puts it, "many things were added by the translators against the intention of the Greeks, and many things were omitted." It was also complained that the Arabic scholars did not, in general, know Greek and Latin. The latter criticism was often directed against Averroes (as will emerge in Chapter 5).[11] A further charge against the Arabs was plagiarism: that they have stolen everything from the Greeks and, like parasites, live on the work of others. Manardo complains that Avicenna has copied his entire medicine from Greeks and *barbari,* that is, other Arabs. And Fuchs charges: "The Arabs, many as they were, have not aimed at anything else than to consume, just as drones, the supplies of someone else's labor and to wear foreign and even stolen plumes."[12]

These arguments are part of the external history of the debate. The present chapter is an attempt to investigate the controversy internally by concentrating on the changing approaches to a specific technical problem. For the purposes of selective illustration, it is sensible to focus on a problem central to the Renaissance disputes. The most hotly debated field in medicine was *materia medica*,[13] a field in which medical humanists were particularly active, as is apparent from the many humanist commentaries on the greatest pharmacognostic text of antiquity, Dioscorides's *De materia medica* (Περὶ ὕλης ἰατρικῆς) of the first century CE. Danielle Jacquart and Dietlinde Goltz have much advanced our understanding of the field by focusing on groups of pharmacological terms traveling through different linguistic cultures and from Arabic to Latin in particular.[14] I will concentrate on one specific drug and its botanical and medicinal properties: the plant and laxative drug senna.

Senna is a suitable object for such a study since from a humanist standpoint it has the serious defect that it does not seem to be mentioned in any antique source; it is described by Arabic authorities only—and by medieval sources who draw on them. Senna appears regularly in Arabic pharmacognostic treatises,[15] some of which were accessible in Latin translation: Pseudo-Mesue's *De simplicibus,* Pseudo-Serapion's *Liber aggregatus in medicinis simplicibus,* and Averroes's *Colliget.*[16] Senna was a common good on the Renaissance medical market, which made it a difficult target for the humanist project of cleansing *materia medica* from medieval authorities. Together with half a dozen other drugs, senna held a prominent place in the controversies, as is illustrated in the following quotation from Guillaume Postel (of ca. 1539/1540):

> How many of the happiest successes of our time do we owe alone to the Arabs, and not to Galen?! I do not want to name the temperamental quality of all those medicines: sugar, rhubarb . . . turbit, senna and manna, which are most excellent medicines, and all those compound drugs . . . which are so much in use and of such extraordinary convenience that they can only be dismissed with the greatest disadvantages.[17]

To give a firm grounding to the technical discussion, it is sensible first to see how the plant and its medicinal properties are described today. Botanically, senna is a shrub growing wild in tropical regions in Africa and Asia. Until recently, the different senna plants were thought to belong to the several hundred species of the genus *Cassia,* but lately the botanical discussion has elevated senna from a subgenus to a genus. Modern pharmacognosy recognizes two different senna plants: *Cassia acutifolia* (Delile), "Alexandrian senna" or "Khartum-senna," a plant that grows wild in the Sudanese Nile River area, and *Cassia angustifolia* (Vahl), from the Arabian peninsula, the eastern coast of Africa and India.[18] Around 1800, *Cassia angustifolia* also began to be cultivated in Tinnevelley, India, and is therefore often called "Tinnevelley senna."[19] The plant is a shrub of 1 to 1.5 meters in height with low branches, leaflets in four to six pairs and yellow flowers in racemes. The Arabian *Cassia angustifolia* grows a little higher than Sudanese senna. Senna leaves and pods are widely used as laxatives. As a drug it is applied against acute constipation and in all cases in which smooth defecation is indicated. The usual dose is 0.5–2 grams.[20] The senna that reached Europe in the Middle Ages seems to have been *Cassia acutifolia* from the Nile, "Alexandria senna," insofar as late sixteenth-century botanists begin to refer to senna using the term *Alexandrina sena*. The import of *Cassia angustifolia* from the Arabian peninsula or Africa, however, cannot be ruled out. The relationship between *Cassia* and senna was not realized until about 1700, as we will see at the end of this chapter.

In the fifteenth century, senna appeared very regularly in Latin pharmacological handbooks and in herbals such as the *Hortus sanitatis*.[21] It is treated not only separately as a simple drug, for example, in Saladin of Asculo's *Compendium aromatariorum* (dating ca. 1450), but also as the most important ingredient of the compound drug *Diasene,* for instance, in Nicolaus Prevost's *Dispensarium* and in Manlius de Bosco's *Luminare maius*. The standard ingredients of *Diasene,* already in the *Antidotarium Nicolai* of the late twelfth century, are senna itself, hazelnuts, minium, lazulite, and sugar.[22] The compound drug *Diasene* is not of much interest to us since it did not become a topic in the debate we are investigating, but it points to the ways in which senna was used in the Renaissance.

As a simple medicine, senna was known primarily for its laxative powers: "It expels and purges moderately, and digests" *(est abstersivum et mundificativum convenienter et resolutivum)*, as Saladin of Asculo says,[23] quoting literally from the most influential Arabic authority on the drug, Pseudo-Mesue's *De simplicibus*—a text we have met before when studying its revision by Jacobus Sylvius.[24]

The handbooks by Saladin, Nicolaus Prevost, and Manlius de Bosco were often consulted by pharmacists of the fifteenth and sixteenth centuries, and there can be no doubt that *Sene* and *Diasene* were part of everyday medical culture. In the notebooks of German physicians and pharmacists we can find *Sene* and *Diasene* in several lists of drugs offered on the Venetian and Genoese markets.[25] Throughout the late Middle Ages the drug

was a very common good of Levant trade. Francesco Pegolotti, the Florentine merchant and author of a comprehensive manual of commerce, the
*Libro di divisamenti di paesi e di misure di mercatantie* (dating ca. 1340), is a
good witness to the presence of senna in late medieval trade. Pegolotti,
who is also well-informed of Levant trade because of a longer sojourn
on Cyprus from 1324 to 1329, remarks that senna can be bought in Constantinople, Alessandria, Famagusta on Cyprus, Messina, Majorca, and
Venice.[26] It counted among the cheap articles if compared with spices.[27]
While in the fourteenth century a number of trading nations competed
for supremacy in Levantine trade, by the end of the fifteenth century, one
nation, Venice, had eclipsed all rivals.[28] And hence, wherever senna appeared in Europe around 1500, it is likely to have been imported from the
East by Venetian merchants.

## THE ITALIAN HUMANISTS: BARBARO, MARCELLO VIRGILIO, AND MANARDO

Such was the situation when Leoniceno and other humanists set out to
give a new foundation to pharmacology by applying their philological
skills to ancient pharmacological sources. Senna is not among the drugs
that achieve attention from the very beginning, such as aloe and rhubarb.
It does not appear, to the best of my knowledge, in Nicolò Leoniceno's
*De Plinii . . . erroribus* (dating 1492–1507), the pioneer among humanist
pharmacological books. Humanist scholars such as Leoniceno and Barbaro concentrate on drug names known to Pliny, Dioscorides, and Galen;
when they criticize medieval sources, they usually do this in regard to drugs
known since antiquity. Because the Latin term *senna* is of Arabic origin—the
plant name is *sanā*—senna had a late and indirect entry into the debate.
Nevertheless, it became a matter of dispute.[29]

Ermolao Barbaro (d. 1493), the Venetian humanist, scathing critic of
scholastic philosophy and opponent of the more philosophically minded
Pico della Mirandola, is among the first to discuss a possible identification
of senna with an ancient plant, the herb *peplis* mentioned by Dioscorides. In
his *Supplement (Corollarium)* to Dioscorides,[30] which appeared posthumously
in 1517, but was composed in the 1480s,[31] Barbaro refers to an obscure
Greek author named "Isacius,"[32] who favors this identification. Barbaro
does not agree: "I think they are of different kinds, since senna has pods
with seeds in them, whereas peplis does not" (*Ego diversum puto genus, cum
senna siliquas et in iis semen ferat, peplis non ferat*). The fact that Barbaro implicitly acknowledges the existence of an herb for which he does not know
an ancient synonym, is in silent disagreement with his conviction that the

ancient authors had expert knowledge on everything, as he puts it in one of the prefaces: to the ancients "nothing was untried, nothing unknown from experience" *(quibus intentatum nihil, nihil inexpertum fuerit).*[33]

Barbaro's discussion of the *peplis* identification is the prelude to a series of humanist attempts to identify senna with a plant known to the ancient Greek botanists. Three proposals were particularly prominent in the discussion: *delphinion* by Marcello Virgilio, *empetron* by Giovanni Manardo, and *colutea* by Jean Ruel.

Barbaro's supplement to Dioscorides appeared in 1517, the year in which Jean Ruel published a new Latin version of Dioscorides. In October 1518 Marcello Virgilio published yet another translation together with an elaborate commentary. These works signal the opening of an impressive series of commentaries on Dioscorides, amounting to almost forty in the sixteenth century.[34] In contrast, the twentieth century saw only one commentary by Julius Berendes. There are many motives and causes of this extraordinary development, some of them apparent in the work of Marcello Virgilio Adriani (d. 1521).[35] The chancellor of the Florentine Republic explains that he aims at transmitting and confirming Greek antiquity by using Greek arguments only;[36] that the discipline of the plants and of *materia medica* is difficult and intricate because of the changes of time and places and also because of the barbarous additions of the "new," that is, Arabic and medieval, medicine;[37] that he is not concerned with the reality of plants, which is known anyway, but with the ancient names and the history of *materia medica (pro medicae materiae antiquis nominibus et historia).*[38] As a consequence of this approach, Marcello Virgilio mentions post-classical plants only if they serve to illuminate Dioscorides's text. In contrast to Leoniceno, but in line with his rival commentator Barbaro, Marcello Virgilio avoids mentioning Arabic and medieval authorities and their nomenclature. Later in the century, his translation of Dioscorides is praised as *Romana* and *elegans,*[39] a praise that can easily be extended to his commentary.

Marcello Virgilio discusses the plant *peplis* without any reference to senna, thus distancing himself from Isacius and Barbaro. But he proposes, on the authority of *aliqui* whom he does not name, to identify *delphinion* with senna.

> <This plant> which our time calls senna, which is known to all pharmacies and among the people, was called *delphinion* in antiquity, as some have thought. The meaning and origin of the younger name *(i.e., senna),* the one which is used today in medicine, is uncertain to us. The older name *(i.e., delphinion),* very well known to sailors, has fishes at its origin.[40]

He proceeds to explain the name of the plant by the dolphin-like shape of its pods. After this etymological explanation, which is typical of Marcello Virgilio's commentary, he admits that *delphinion* is a strange and dubious plant:[41] it is not mentioned by Plinius and Paulus Aegineta; the oldest manuscript of Dioscorides's text does not have an entry on *delphinion*;[42] and, finally, the identification with senna is open to criticism:

> Someone may quite rightly be astonished about the fact that Dioscorides does not say anything sub voce *delphinium* on the various properties of senna, nothing that pertains to the production of the medicine for people except in the case of a scorpion's sting. On the basis of the shape of senna pods, which correspond to the full shape of dolphins, and employing some later authorities we have declared that senna was called *delphinium* and *sosandrum* by the ancients. We admit freely, however, that we do not have anything else on senna which is certain.[43]

We meet here with the full dilemma of Marcello Virgilio's humanist project: modern pharmacological knowledge, for instance, of "the various properties of senna" *(in multiplici senae proprietate)*, is indispensable for understanding ancient *materia medica,* but the most informative younger sources, barbarous as they are, cannot be quoted. Marcello Virgilio escapes this dilemma by simply presupposing that everyone knows about the properties of senna and about the shape of its pods. The passage also shows the gulf between the pharmaceutical market—goods known "to all pharmacies and among the people" *(officinis omnibus et vulgo)*—and the humanist discourse about *materia medica.*

Giovanni Manardo (d. 1536), the famous Ferrarese physician, is among the first critics of Marcello Virgilio's translation and commentary. He argues against specific renderings of Greek terms, identification of plants, and the restricted usage of Galen.[44] His critique appeared as letters 8.1–3 of the *Epistolae medicinales.* Of greater influence on the doctrine of senna were Manardo's annotations on Pseudo-Mesue's *Simplicia medicamenta,* written in Hungary between 1513 and 1519, and first published around 1521.[45] To find Manardo commenting on an Arab authority may come as a surprise: together with his teacher Nicolò Leoniceno, Manardo counts among the early heralds of anti-Arab propaganda, and not without reason. He accuses Avicenna of plagiarizing Greek authors and recommends turning to Greek sources whenever one is in doubt about a medical problem.[46] His ideal is the true medicine of antiquity, *antiqua et vera medicina,* cleaned of all barbarous additions; his enemies are the many acclaimed physicians who prefer to err with the Arabs rather than "to say the truth with those who draw their medical knowledge from Greek sources."[47] On closer inspection, however, Manardo resists straightforward classification as a humanist hard-liner. Occasionally his critique is also aimed at Greek authors such as Galen and Aristotle, while his medical teaching makes ample use of Arabic sources.[48]

Manardo's running commentary on Pseudo-Mesue's entry on senna opens with general information about further sources on the drug. The passage attests to Manardo's wide reading and to his up-to-date knowledge of contemporary sources; we meet again with Marcello Virgilio's *delphinion* and Isacius's *peplis*:

> On Senna. Averroes groups also this <drug> with the new medicaments. Avicenna does not write anything on the drug directly, but refers to it at the end of the chapter on *fumus terrae* and in the chapter on five-day-fever. Serapion writes about it, but does not cite the opinions of Dioscorides and Galen, as he usually does with other medicines. Some <call it> Delphinion, others Pelecinon, Isacius <calls it> Peplion. I sometimes wondered whether it was Empetron, sometimes whether Analypon. It does not matter much, however, to know what it was called by the ancients, as long as we know its properties, which are without doubt noble and effective.[49]

This is an impressive opening—if contrasted with what had been written before on senna, not only by humanists but also by scholastic authors such as Manlius de Bosco.[50] Manardo conquers the field by giving reliable references to essential literature on the drug: to Averroes's *Colliget* 5.56, Avicenna's *Canon* 2.2.282, and to Pseudo-Serapion's *Liber aggregatus in medicinis simplicibus* sub voce.[51] These few sentences demonstrate impressively that Arabic authorities were the most informative sources available on *materia medica,* with respect to the number of medicines described and to the botanical and medicinal information given—and this was true not only with respect to oriental plants. Manardo's brief remark about Serapion not citing Dioscorides and Galen, as he usually does, is a very attentive observation; it is Manardo's first hint to the reader that this plant was perhaps not known to the ancients. The attempts at identification that follow, those of his colleagues and his own, present the latest state of contemporary knowledge on senna. Manardo signals openly that the method of identification of ancient plant names is perhaps not the best way to discuss senna. In contrast to Marcello Virgilio, Manardo is ready to turn from ancient to medieval and Arabic sources if they offer information on the properties of plants.

Since Manardo's commentary is a precious example of the meeting of Arabic science and humanist scholarship, and since it was very influential, I quote it in full (the complete text of Pseudo-Mesue's monograph on senna,[52] together with an English translation, can be found in Table 3 in Chapter 3):

Pseudo-Mesue: *Melior pars plantae est folliculus*

Manardo: Experience itself shows that the leaves are more powerful laxatives <than the pods>. Senna grows in Apulia, but the one imported from Egypt is better.[53]

. . . *Calida est in principio secundi, sicca in primo*

Averroes says 'dry in the second <degree>'; I believe Serapion, who says 'of little heat and dryness.'[54]

. . . *Et debilitat stomachum*

Experience shows the opposite, and the taste is slightly bitter and astringent; but this is not said by Serapion.[55]

. . . *Et dixit Galenus*

If only you <i.e., Mesue> had said where and under which name; I suspect, however, that you are fantasizing, here as often elsewhere.[56]

. . . *Decoquatur cum iuribus gallorum*

Dioscorides says about Empetron that it should be boiled in broth or in sweet water, and we know that a broth in which this common <drug> has been boiled or soaked overnight works easily as a laxative. Serapion says that the drug is more effective when boiled than when pulverized.[57]

. . . *Solutione educit cum facilitate melancholiam, et choleram adustam*

Serapion simply says "<red> bile" without adding "burned," and <he also says> that it is good against melancholy, cracks, nervous disorders, loss of hair, chronic headaches, scabies and epilepsy. Dioscorides says about Empetron that it purges phlegm, <red> bile and watery substances. Averroes denies that Senna is able to purge phlegm. I have learned by experience, however, that it is able <to do this> and that it is very good against the disease called "French," <i.e., syphilis>.[58]

. . . *Dosis eius infusi potio usque ad unciam unam*

Serapion gives one drachm of the pulverized, five of the boiled <drug>. I know that both <measures of the> boiled substance can safely be exceded, and that one ounce of the boiled substance purges <only> in a mediocre way.[59]

It is the strategy of Manardo's commentary to juxtapose Pseudo-Mesue's doctrine with a number of other authorities—Pseudo-Serapion, Averroes, Dioscorides—and with his own experience. It is not a philological commentary in a strict sense; rather it attempts to assess the value of Pseudo-Mesue's information. The result of this approach is a series of important arguments on senna that give a new basis to the discussion. Mesue is wrong, Manardo finds, on the powers of the pods since the leaves are better laxatives; Egyptian senna is better than the senna from Apulia; it is Serapion rather than Mesue who is to be followed on the temperamental properties of senna; senna does not weaken the stomach, as Mesue maintained; Mesue's reference to Galen cannot be trusted; both Dioscorides (on *empetron*) and Serapion confirm that senna works better when boiled in broth; in addition to the effects listed by Mesue, senna purges phlegm and is good against syphilis; the doses recommended by Mesue and Serapion are too small.

Of these arguments, two were particularly influential in the sixteenth century, both directed against the authority of Pseudo-Mesue on the basis of Manardo's own experience: that the leaves are better laxatives than the pods, and that senna does not weaken the stomach. Is it possible to decide the matter of disagreement positivistically? Supposing that both authors, Pseudo-Mesue and Manardo, wrote about Alexandria senna, it was Manardo who took the correct view on the respective powers of pods and leaves. Senna pods contain slightly less aloe-emodin glycoside than senna leaves; the laxative effect of the pods is milder, which makes them a more appropriate drug for children.[60] The other issue is less clear-cut: if the dose is too high, senna may indeed cause stomachache;[61] this could be behind Pseudo-Mesue's phrase *debilitat stomachum*. Perhaps Manardo found that this is only an accidental property of the drug. In any case, it is possible that Manardo's arguments rest on experience and everyday practice. One difficulty in assessing the value of Manardo's and Pseudo-Mesue's empirical information lies in the fact that the Egyptian merchants are likely to have mixed the good Alexandria senna with less effective leaves of other kinds of senna or of *Cassia* (of which there are hundreds) and of other plants.[62] Adulteration of senna was still common practice around 1800.[63] Moreover, Manardo does not say whether he used Alexandria senna or, alternatively, the leaves and pods of senna cultivated in Apulia, an inferior kind of senna, as he points out. We will come back to Italian senna when returning to Brasavola's critique of Manardo's censure of Pseudo-Mesue.

Empirical arguments are only one of several intellectual strategies in Manardo's commentary. It is obvious that he implicitly suggests an ancient equivalent to senna: *empetron*. One could suspect that this identification is as random as Marcello Virgilio's *delphinion*, especially in view of the fact that today *empetron* is considered an unidentifiable plant.[64] But there are subtle differences between the two writers. Marcello Virgilio's sole reason for identifying senna with *delphinion* is that the shape of senna pods corresponds to the shape of dolphins—which, one could argue, is true of many pods. Manardo, in contrast, juxtaposes statements on boiling—Pseudo-Mesue: *decoquatur cum iuribus gallorum,* Dioscorides: *decoquendum in iure vel aqua dulci*—and on purging phlegm—Averroes: *negat senam posse purgare pituitam,*[65] Dioscorides: *purgat pituitam.* He does no more than point to these textual correspondences, and leaves it to the reader to draw the conclusion. Manardo's philological arguments are stronger, his strategy of identification more modest than Marcello Virgilio's. In spite of this, Marcello Virgilio's *delphinion* theory fared much better in the sixteenth century.

## THE EARLY 1530S: CORDUS, CHAMPIER, AND BRASAVOLA

In the late 1520s and early 1530s, the discussion passed beyond the boundaries of Italy. In this case, it was not Leonhart Fuchs, the Tübingen professor of medicine and author of *De historia stirpium* (1542), who opened the discussion in the North. His early *Errata recentiorum medicorum* of 1530 do not mention senna. Instead, others pick up the thread of the Italian pharmacologists: Symphorien Champier and Jean Ruel in France, Euricius Cordus in Germany.

The passage on senna in Cordus's *Botanologicon* of 1534 rests primarily and perhaps also exclusively on Marcello Virgilio's commentary on Dioscorides. Cordus's book itself is not a commentary but a loosely structured dialogue between the author and three friends on the occasion of a botanical excursion in Hessia. If the format of the book would have allowed for an approach to the topic that bypasses ancient authorities, the utterly humanist motives of the author did not. Cordus, a professor of medicine in Marburg, had won fame as a humanist poet and witty epigrammatist before he published his medical works. He had received his doctorate in medicine from Leoniceno in Ferrara (in 1521),[66] and his writings testify to the conviction that the apothecaries of his time were very badly informed about the simples they were selling,[67] and that the standard pharmacological handbooks were of very low standard.[68] It was Cordus's determination to remedy this situation by means of improved knowledge of ancient sources,[69] of Dioscorides in particular,[70] and through the experience he had won in botanological excursions.[71]

Cordus's view on senna is that Marcello Virgilio's identification with *delphinion* is not supported by Dioscorides's description, and even if it was, he does not believe that anything was written about senna by the ancients—for *delphinion* is missing in Galen, Paul of Aegina, Pliny, and the oldest manuscripts of Dioscorides. Senna is a very powerful and healthy drug, which is why it is called *sosandros,* that is, "rescuing men."[72] And he adds in the epilogue to the book: "Senna: probably unknown to the ancients" *(Sene forte veteribus ignota).*[73] Almost every bit of information in this passage is drawn from Marcello Virgilio. It does not appear consistent of Cordus to criticize his source for the identification with *delphinion* and, at the same time, to use *sosandros,* an alternative plant name in Dioscorides for *delphinion,* as a name for senna. Nevertheless, Cordus makes an important point that passes beyond the tight humanist boundaries of Marcello Virgilio's commentary: his experience with German plants had taught Cordus that "many of these herbs, especially vulnerary herbs and those of the genus Solidagines, were discovered after antiquity" *(multae enim huiusmodi, vulnerariae praesertim et ex solidaginum genere, herbae post illa tempora repertae sunt).*[74]

From here it is only a small step to acknowledging the contribution of the Arabs to pharmacology. Cordus takes this step. The occasion on which he does this suggests that among Arabic drugs, laxatives and digestives were particularly well-known. At some point on their botanological expedition, one of Cordus's friends, the Frenchman Gulielmus Bigotius, complains of stomachache and flatulence: he had eaten carrots. His companions make learned jokes about the audibility of his problem and give him galingale. One of the friends remarks that he had not been able to find this simple drug in Dioscorides. This is not surprising, rejoins Cordus:

> There are many other medicines of the Arabs which neither Dioscorides nor anybody else among the ancient Greeks mentioned, to the best of my knowledge. Many of these names I have written down on a paper which I have with me in my bag, if I am not wrong. Megobacchus: Do not hesitate to repeat them to us. Cordus: I cannot find the list; instead I shall enumerate some of them from memory: ambra, anacardion or anacardus, behen, bellirici, bel, camphora, casia fistula (as it is also called today), coton (gossipium in Latin), culcul, dadi, dend, doronici, emblici, fagre, feleng, faufel, galanga (which is the one you are wondering about), kadi, kanabel, lingua avis, moschus, manna, macis (if it is

not machir), mirobalani citrinae, and chabulae Indae (if not someone counts them among the palmulae of Disocorides), michad, mummia, musa (or maum), nersin, nux muscata, nux metel, nux mechil, and perhaps nux hende, nux indica (if this is not the one which Galen calls nux regia and which, he says, contains seven ounces), nux vomica (if this is not the fruit of thitymalus caryites, as some claim), phel, piperella (which, however, is also called the seed of vitex), ramech, ribes, sadervam, sandalum, sel, sene, tamarindus, turbith, uiret, zarnabum, zeduaria, zelim, zurumbeth.[75]

This, obviously, is a carefully compiled list in alphabetical order, valuable for its comprehensiveness, impressive for the complex problems lurking behind each term. Megobacchus's reaction is understandable: "With these monster terms one could draw the moon down from heaven, I believe" *(His vocabulorum monstris lunam e coelo deduci posse crediderim)*. In the following passages of Cordus's text, some of these items, for example, senna, are discussed, most of them are not. The list demonstrates that there were humanists who saw the necessity to take or rather leave on board simples that were unknown to ancient authorities. The hard-liners among Cordus's humanist colleagues, such as Jean Ruel, did not share this position, as we will see soon.

One of the very first readers of Manardo's commentary on Pseudo-Mesue was Symphorien Champier, the Lyonnais physician, whom we have met repeatedly in the present study.[76] Manardo and Champier exchanged letters,[77] and were on amicable terms.[78] Manardo's *Epistolae medicinales* and Champier's *Symphonia Galeni ad Hippocratem* were grouped together as the most prominent anti-Arabic tracts of the 1520s by Lorenz Fries, when in 1530 he wrote his *Defensio medicorum principis Avicennae* and thus triggered an international dispute.[79] In spite of his open anti-Arab polemics, Champier takes a much more flexible approach than other humanists to Arabic pharmacology. In his *Castigationes* of 1532, in a chapter on senna not yet influenced by Manardo, Champier—in good humanist vein—censures Mesue for false citations of Galen on senna, since senna is a plant not mentioned by any Greek author. Champier admits that senna is a noble medicine and cites Serapion and Mesue on the drug. The true victim of his polemics in this chapter are *nostri pharmacopolae,* the pharmacists of the time, who use the leaves of some tree in France instead of the true senna leaves, with deadly consequences for the patients, Champier claims.[80]

In a letter to Manardo of the following year, Champier explains his doctrine of "praiseworthy medicines" *(medicinae benedictae),* as he puts it: "In the time of Galen, those medicines which we today call praiseworthy, were unknown, for example *manna roris, cassia fistula,* rhubarb, senna, and perhaps the good *agaricum.*"[81] The concept of the praiseworthy medicines reappears in Champier's *Cribratio medicamentorum* of 1534, now developed into a full-fledged theory of purgative medicaments: they are either poisonous *(pharmaca venenosa)* or praiseworthy *(benedicta);* only the former were known to the ancients. Champier accuses Arabic authors and their Western followers and commentators of misunderstanding a number of poisonous medicines such as scammonium and turbith, but he cites Arabic authors approvingly on rhubarb, senna, and other drugs.[82] The chapter on senna is a collage of citations, mainly from Pseudo-Serapion and Manardo's annotations on Pseudo-Mesue.[83] Of Manardo's arguments, Champier does not quote the objections against Pseudo-Mesue but rather the additions to him: he mentions that senna purges phlegm and is good against syphilis and that one may safely use higher doses than those recommended by Pseudo-Mesue and Pseudo-Serapion. As to the possible identifications of senna—Marcello Virgilio's *delphinion* and Manardo's *empetron*—Champier does not favor either of them.[84] In sum, it is remarkable to see that Champier takes a much more pragmatic approach to the Arabic drug senna than his fierce anti-Arabic polemics would suggest. In actual pharmacological practice, he was not a humanist hard-liner.

Manardo's theses did not remain unchallenged. In 1536, one of his Ferrarese pupils, Antonio Musa Brasavola (d. 1555), published a book on simples used in the pharmacies of Ferrara, titled *Examen omnium simplicium medicamentorum quorum in officinis usus est.* The book, composed around 1534,[85] takes a format very similar to that of Cordus's *Botanologicon:*[86] an herbalist, an old apothecary, and Brasavola himself discuss various drugs on the occasion of an excursion to the mountains. When the old apothecary asks Brasavola about senna, he is told that senna was certainly unknown to the ancients: *Senam apud antiquiores incognitam esse certum habeo.*[87] This statement prompts serious objections among Brasavola's companions. There follows a passage remarkable for its attempt to explain the lacuna in Dioscorides:

> Old man: Is it not strange that an herb so much worthy of praise and referred to with so many names was not known to the Greeks? Brasavola: In any case, it was not, it seems. Old man: Did it not exist in those times, or was it generated later by the earth? Brasavola: It did exist also back then, but perhaps it was not tried or seen by the ancients. Old man: How

could it not have been seen by Dioscorides, who traveled almost through the entire world? Brasavola: It is true that Dioscorides served in the military, but it is not certain that he wandered through the entire world; at least, he did not go to the area where <senna> grows, or it was brought later to the provinces nearer by, as clearly happened with many other things which the ancients did not know or which the works extant omit. It is certain that not a hundredth part of the herbs existing in the entire world was described by Dioscorides, nor were the plants <described in their entirety> by Theophrastus and Pliny; rather, it is from day to day that we improve our knowledge and that the medical art grows *(in dies addiscimus et crescit ars medica)*.[88]

Passages such as this demonstrate why pharmacology was a hotly disputed topic in the sixteenth century. The obvious tension that existed between everyday pharmaceutical experience and the humanist textual tradition of *materia medica* posed a challenge to the basic principles of many humanist scholars: to the priority of philology, to the infallibility of ancient authorities, to the uselessness of barbarous, that is, medieval sources, and to the concept of decline and rebirth. Despite the humanist format and language of his book, Brasavola's standpoint conflicts with many of these principles: senna is traded on the market but not described in ancient sources, says Brasavola; Dioscorides is far from being perfect; to fill the lacunae in ancient texts one has to turn to later sources, especially to Arab authorities; medical science progresses daily and has left the ancients far behind.

It is consistent with Brasavola's approach that he refutes all attempts "to discover senna in Dioscorides."[89] He argues against three identifications of senna that had been listed by Manardo: *empetron* (Manardo), *delphinion* (Marcello Virgilio), and *peplion* (refuted by Barabaro). Remember that Manardo had proposed the identification with *empetron* on the basis of textual parallels that concern the boiling of the drug and its power to purge phlegm. "This conjecture is so weak that it does not even hang on a string," replies Brasavola, "for likewise I could prove that a cow is a cock because both are boiled before they are consumed."[90] Brasavola here exposes the rather thin analogical structure of Manardo's arguments. And he goes on by comparing the rest of Dioscorides's description of *empetron* with his knowledge of senna: the leaves of *empetron* have the shape of lentils, senna leaves do not; *empetron* tastes salty *(salsa)*, while senna tastes slightly bitter *(subamara)*.

Turning to Marcello Virgilio, Brasavola proceeds: "Even worse is the reasoning of those who maintained that senna is *delphinion*, for senna leaves are not similar to dolphins; nor is there any need to take refuge in the <similarity of the> pods, for Dioscorides speaks of leaves."[91] Brasavola's arguments are very well-phrased, and to a certain degree, they are convincing. The information Brasavola draws from Dioscorides is correctly cited: *empetron* does taste salty, it has the shape of lentils, the name *delphinion* is given to the leaves and not to the pods of the plant. This last point is indeed a valuable one; Marcello Virgilio had addressed the problem himself, suggesting rather desperately that the manuscript tradition of Dioscorides must be faulty and that some phrase about pods had disappeared.[92] In this respect, Brasavola's argument against Marcello Virgilio is as devastating as his exposition of Manardo's problematic analogies.

Brasavola's own information on the botanical features of senna, however, remains weak: instead of giving information about what senna leaves look like, he only says that they do not resemble lentils or dolphins. The only positive information the reader gets about senna is that it tastes slightly bitter *(subamara)*—which, curiously enough, is a phrase drawn from Manardo and turned against him. Brasavola gives a hint to the sources of his information when he adds that senna from Apulia and Liguria tastes more bitter than other kinds of senna because it grows closer to the sea. Manardo had already mentioned that senna was cultivated in Apulia; both authors agree that the kind of senna imported from Egypt is of higher quality.[93] The way Brasavola refers to Italian senna suggests strongly that he knows it firsthand. The fact, however, that he silently quotes Manardo on the slightly bitter taste of senna and that he avoids any positive description of the leaves, shows that he does not feel comfortable with his knowledge—perhaps because it was restricted to senna cultivated in Italy, perhaps because he did not have a paradigm for discussing a plant unknown to the Greeks and growing wild in a country he had not seen. The only serious model available was Manardo's commentary on Pseudo-Mesue, the very source he was criticizing.

Brasavola feels more secure when finally turning to the medicinal properties of the plant. It is in this context that he formulates his attitude toward Arabic sources. When Brasavola says that senna is a drug with very many and very good effects, a drug that should never be omitted, one of his companions, the old apothecary, replies:

> How do you know? Brasavola: Primarily on the authority of the Arabs, an authority which should not be dismissed easily, as some contemporary authors do, especially where <the Arabs> do not disagree

with the Greeks; secondly, through daily experience, the teacher of everything.[94]

And, with apologies to his revered teacher *(tam celeberrimi viri pace dixerim et praeceptoris colendissimi),* he counters Manardo's two main objections against Pseudo-Mesue: that senna does not weaken the stomach and that the leaves are stronger laxatives than the pods. On the basis of his own experience, Brasavola recommends always mixing senna with other substances in order to avoid stomachache. With respect to the problem of pods versus leaves, Brasavola again claims experience as witness: senna pods are of greater effectiveness, as is also clear for the theoretical reasons that the pods are of older age than the leaves, that the substance of pods is thicker and that the seeds are powerful themselves.

It is difficult to assess the value of Brasavola's arguments because they rest on a mixture of personal experience and general medical principles. An additional difficulty is that Brasavola and Manardo do not indicate when they speak about Italian senna and when about Egyptian senna. It seems that it was not until 1586 that the two kinds of senna were depicted separately (by Joachim Camerarius the Younger).[95] The pictorial tradition before Camerarius was also influenced by ideology, as we will see below.[96] All this makes it hard to follow the suggestion by Friedrich Flückiger that Italian senna was perhaps identical with mediocre senna from the Nile area, a thesis recently revived in scholarship on Fuchs's *Historia stirpium*.[97] We will never be able to have firm knowledge about Italian senna, the cultivation of which had long stopped in the nineteenth century.[98] We only know that it was generally held to be inferior to oriental senna.

This much can be said of Brasavola: his discussion of senna was a serious attempt by an experienced botanist and humanist scholar—for that he was in language and conviction[99]—to reestablish the views of Pseudo-Mesue on the powers of the drug. The strongest part of Brasavola's section on senna is the refutation of possible identifications of senna with plants in Dioscorides: he demonstrates the deficiencies of Manardo's and Marcello Virgilio's arguments. Euricius Cordus, Symphorien Champier, and Antonio Musa Brasavola form a group of scholars within the humanist tradition who acknowledge that Arabic authors made valuable and indispensable additions to Greek pharmacology. But among them only Brasavola continued the approach taken by Manardo—although he came to different conclusions—to directly evaluate the doctrines of the Arabic–Latin tradition philologically, botanically, and pharmacologically.

## THE *COLUTEA* THEORY:
### RUEL, BOCK, FUCHS, AND MATTIOLI

But the scene was not won for this approach. More conservatively humanist authors continued to limit the range of plants to those known in
antiquity, and the commentary on Dioscorides remained a very popular
genre with much influence on the illustrated herbals that appeared in
Germany in the 1530s and 1540s. A new chapter in the history of senna
was opened in 1536 when Jean Ruel (d. 1537), physician to King Francis I,[100]
proposed to identify senna with *colutea,* a tree described by Theophrastus only. For this purpose, Ruel has very cleverly weaved together a
series of sources. The crucial passage on the identification of *colutea* is the
following:

> The Arabs use their vernacular name 'senna' for it *(i.e., for colutea)* and
> make two kinds of it, one <growing> spontaneously and one <culti-
> vated> in the garden *(this Ruel takes from Pseudo-Mesue on senna: "invenitur
> ex eo domesticum et sylvestre"),*[101] for it grows also from <planted> seeds,
> especially in sheep droppings *(from Theophrastus, who says about colutea:
> "nascitur semine et fimo praecipuo ovillo").*[102] They add that oblong pods, in
> which the seeds, arranged in both directions, are enclosed, hang on thin
> hangers, and that for this reason they are easily thrown down by the
> wind *(from Pseudo-Serapion on senna: "habet vaginas oblongas et obtortas,"
> etc.),*[103] while those pods that hang more firmly are knocked down with
> sticks *(inspired by Theophrastus on colutea: "baculos decidunt");* the shepherds
> collect the pods which have fallen down *(Pseudo-Serapion)* because they
> are very useful for fattening sheep *(Theophrastus: "oves mirum in modum
> pinguefacit").*[104]

This passage cites the Arabs on three features of senna that cannot be
found in any Arabic authority: that senna grows particularly well in
sheep droppings, that the firmly hanging pods are knocked down with
sticks, and that senna pods are very useful as fodder for fattening sheep.
These pieces of information are drawn from or inspired by Theophrastus's description of *colutea,* which Ruel had used some paragraphs earlier
to describe *colutea.*[105] Through this textual manipulation, Ruel suggests to
his readers that Theophrastus and the Arabs are speaking about the same
plant. The Latin of the passages is stylishly classical, with a gusto for rare
words: instead of *pastores* (shepherds), Ruel uses the rare *opiliones,* instead
of *baculum* (stick) he writes the exquisite *pertica.*[106] The smooth, classicizing
surface of the text is meant to deceive the readers, so that they do not re-

alize that Ruel has adulterated the texts of the Arabic authorities to make it possible to identify senna with *colutea.*

The next paragraph of Ruel's text is concerned with the medicinal properties of *colutea.* He now omits any mention of Arabic authorities and instead refers to the Byzantine author Actuarius (d. ca. 1330), whom Ruel himself had translated into Latin.[107] Ruel purports to speak about *colutea,* "this tree which is celebrated by Theophrastus,"[108] but in fact draws on material from Actuarius not on *colutea,* but on senna—a fact that is not apparent in Ruel's text:

> We shall briefly explain what the later Greek *(i.e., Byzantine)* physicians have left on this matter. Actuarius reports that a pulse *(fructus siliquosus)* is mentioned by the Barbarians which purges phlegm and <red> bile without any trouble, if drunken at the weight of one denar *(up to this point, the quotation from Actuarius on senna is quite correct; but the following sentences in Actuarius do not refer anymore to senna, but to two other drugs, Cassia nigra and Manna: "Quae his deinceps succedunt," etc.—in English: "The following drugs purge in a very mild way, i.e., Cassia nigra and Manna."[109] Ruel, however, proceeds:)* After these, the <pulse> purges the remaining humors in a very mild way and removes the burned and black bile and their solutions in chicken broth *(this last feature comes from Pseudo-Mesue, who had proposed taking the drug not in water, but in chicken broth).*[110] It helps with chronic headache, scabies, epilepsy and mange. It is better administered in broth of something warmed up rather than in powder of something ground *(from Pseudo-Serapion).*[111] It loosens constipation in the bowels *(again from Pseudo-Mesue).*[112]

Jean Ruel's strategies of textual manipulation are very apparent in this passage. Theophrastus, as with other plants in his book, does not mention any medicinal properties of *colutea.* This gap is filled with the laxative properties of senna. As a late Greek authority on the matter, Ruel cites Actuarius, who, unfortunately, does not say more than one sentence on senna; moreover, the sentence has the disadvantage that it does not contain anything about the purging of black bile, which is the main property of senna according to traditional Arabic theory. That is why Ruel borrows some phrases about black bile from the following passage in Actuarius about *manna* and *cassia nigra,* and then elegantly proceeds to silent quotations from Pseudo-Serapion and Pseudo-Mesue on senna. The Latin retains the classicizing tone and thus covers the Arabic origin of the doctrine. There can be no doubt that Ruel is cheating his readers. He also deceived a modern scholar who believed that "the course which Ruel took *(on colutea)* is that which any man of scholarly caution would have taken."[113]

Note that Ruel's sinister intellectual strategies are without parallel among the authors discussed in this chapter. One could argue that Marcello Virgilio identified *delphinion* with senna mainly because he could not come up with any other plant more or less similar to *delphinion;* but Marcello Virgilio did not try to deceive his readers, he even pointed to flaws in his argument, for example, to the fact that Dioscorides's text has "leaves" where he himself would prefer to read "pods." Jean Ruel's manipulation of the textual tradition, in contrast, is a silent one: it works only if not realized by the reader.

Ruel remains faithful to his methods when he refutes the *delphinion* thesis later in the book: "Those who thought that *delphinium* was called senna by the Arabs were hallucinating heavily."[114] There follows a comparison of Dioscorides' *delphinion* with "senna"—which, in truth, is a comparison with *colutea,* for Ruel describes senna in terms he had drawn from Theophrastus to describe *colutea:* round leaves, parchment-like pods that carry dark and hard seeds of lentil-like shape, etc.[115] Ruel concludes with the statement: "What senna was for the ancients, we have shown <in the section> on trees."[116] As a result of Ruel's textual manipulation, a hardly known Greek tree has received medicinal properties, an annoyingly unavoidable Arabic plant is Hellenized, at the price of a fictitious botany.

The reaction to Ruel's identification of *colutea* with senna was immediate: *falsissimum esse constat,*[117] wrote Luigi Mondella, physician of Brescia, in a letter to a friend in May 1537:

> There are two points in particular which make me doubt Ruel's opinion: first, that Theophrastus says that *colutea* is a tree, and a large one; <second>, that he does not attribute to *colutea* (as he does in other cases) any power of loosening the stomach and purging black humor, this being proper to and typical of senna, as the Arabic authors say.[118]

It is clear to the experts, explains Mondella, that senna should be called an herb rather than a tree or a bush *(herba magis appellanda est quam arbor aut frutex).*[119]

These are powerful arguments, and Mondella is not the only one to advance them. Similarly, Amatus Lusitanus, the well-known Portuguese physician, writes in 1554 that Ruel is wrong because *colutea arbor est, sena vero planta.*[120] *Planta* is the term used by Pseudo-Mesue, which is almost all the information, meager as it is, on the size of senna to be found in the sources.[121] Ruel seems to have thought that Pseudo-Serapion's phrase "the shepherds collect the windfalls"[122] *(cadunt et pastores colligunt eas)* pointed to the treelike size of senna.[123] Many Renaissance authors refer to senna as *herba,* among them Saladin de Asculo, Manardo, and Brasavola.[124] Hence, the *colutea* theory had the major disadvantage of swimming against the current concept of senna. For the rest, it is structurally similar to the *delphinion* theory: the one feature that connects senna with *delphinion* and *colutea* is that all of them carry pods with seeds inside. Both theories are open to the same criticism: Marcello Virgilio himself had admitted that one would like to read something about the laxative powers of *delphinion*— just as Mondella argues when criticizing Ruel's identification with *colutea*. While *delphinion* had the flaw that it appeared only in late manuscripts of Dioscorides, Marcello Virgilio's theory had the one advantage over Ruel's that it did not make senna a tree.

But there were also authors who found Ruel's identification appealing. The very first is a famous name in the historiography of botany: the schoolteacher Hieronymus Bock (d. 1554), whose *Neu Kreüterbuch* of 1539 treats herbs that grow in Germany.[125] That the question of identification with an ancient plant was not dear to Bock's heart, however, is apparent from the very first sentence of the chapter on senna:

> Even though with respect to senna *(der Senet)* scholars still dispute in which chapter it belongs in Dioscorides, senna has become well-known in Germany to a degree that I did not want to ignore it.[126]

A description of senna follows that differs from previous ones in motivation: it does not follow the Arabic–Latin tradition of describing senna in the terms of Pseudo-Mesue and Pseudo-Serapion, nor does it resemble humanist attempts at description such as Brasavola's, which are inspired by a comparison with antique plants. Bock does not actually compare senna to *colutea;* he simply states that he follows Ruel and understands *colutea* to be senna "because of the characteristic description and effect of <senna>" ("von wegen der artlichen beschreibung unnd wuerckung des selben").[127] He does not even bother to mention that *colutea* is a tree.

Since the features of senna that Bock describes are so different from those of the textual traditions, it is more than probable that his information derives from personal experience. This accords well with his claim that he has grown small specimens of senna from the seeds ("von welchem *(i.e., the seed)* ich hab etwan junge stoecklein gezielet"). He tells his readers that senna does not survive frost, that the leaves are similar to pointed clover or to fenugreek but larger, that the stems have white flowers at the top, which eventually turn into black, curved pods, that the seeds are broad and gray. He contrasts this kind of senna, which grows wild in Italy and France, with the best kind of senna that is imported from Alexandria: it is small and dainty, and the flowers are pale yellow.[128]

Bock offers an argument in favor of *colutea* theory that he himself did not exploit in any way (and that had not been used by Ruel): his description of senna leaves as similar to but larger than fenugreek agrees very well with Theophrastus's statement: <*colutea*> *habet folium non absimile foenograeco*. Ruel's *colutea* theory may well have prompted Bock to mention fenugreek in the chapter on senna. To sixteenth-century discourse on senna, Bock's description adds a fresh ingredient that is motivated by botanical experience and not by any kind of humanist interest in classical authors or texts. The other side of the coin is that Bock is not fully abreast of the contemporary discussion: when turning to the medicinal properties of the drug, he follows Pseudo-Mesue and Pseudo-Serapion[129]—for instance, on the preference of pods over leaves. There is no trace of the critical discussion begun by Manardo.

Such was the situation in 1542 when Leonhart Fuchs (d. 1566) published his *De historia stirpium*.[130] As was mentioned in Chapter 1 above, Fuchs was instrumental in expelling Arabic authorities from the medical statutes of Tübingen University in 1538. As the arch-propagandist against Arabic learning, he could not give senna a separate entry in his botanical *magnum opus,* for that would have forced him to describe senna in terms of Arabic and medieval sources. After a full decade of disputes with old-fashioned scholars Fuchs had acquired a sound knowledge of contemporary sources, some of which are listed in the preface: Barbaro, Ruel, Marcello Virgilio, Brunfels, Cordus, and Bock. Hence, he could choose among a variety of approaches to senna. His choice fell on Ruel's *colutea*.

Fuchs had reasons for it. In spite of his hatred of everything Arabic and medieval and in spite of his emphasis on language skills as the basis of medicine, Fuchs did not continue the commentary tradition on Dioscorides. This is obvious from many passages of the preface, for instance, when Fuchs criticizes (without much justification) Hieronymus Bock for imposing Dioscoridean terms on every plant, as if Dioscorides had described the plants of all regions;[131] or when he accuses Marcello Virgilio, as a nonphysician, of a deficient knowledge of herbs;[132] or in the following passage:

> Hence, whenever we could use a Greek term which corresponds totally to the German name, we have given this term to the plants rather than a wrong one from Dioscorides.[133]

In the case of several plants, Fuchs inverts the traditional approach: instead of employing contemporary botanical knowledge to explain Dioscorides, Fuchs uses the Greek conceptual framework to describe plants common in his time and country unknown to the ancients. Moreover, he is willing to include foreign *(peregrinis)* plants and apologizes for using barbarous, that is, Arabic, terminology *(barbaris nomenclaturis):*

> Also, since in this work we have inserted monographs concerning many plants which neither Dioscorides nor any other ancient author knew—for they are mostly vulnerary medicines, being so much in daily use, especially among surgeons, that we found it impossible to omit them—we were forced to employ the common and barbarous nomenclature, because we were lacking in Latin terms.[134]

Remember that the vulnerary herbs *(vulnerariae)* had already appeared in Euricius Cordus's *Botanologicon* as a group of plants unknown to the ancients. It is possible that the long lists of Arabic plants in Cordus, Champier, and Brasavola had influenced Fuchs. The Tübingen professor says it less openly than his humanist predecessors, and he is less clear about it than in the case of plants growing wild in Germany, but nevertheless he acknowledges that the Arabic contribution to the knowledge of plants cannot entirely be ignored. The reader-oriented approach he takes—which aimed at giving a reliable tool to the scholars of herbal medicine *(herbariae medicinae studiosi)* by the combined usage of ancient sources, his own experience and skillful depictions of plants—forced Fuchs to include non-European plants.[135]

This was enough of a concession, however. Fuchs did not miss the occasion to insert some anti-Arab propaganda in his preface when crediting Ruel for the explosion of Arab errors *(proscriptis et explosis tandem Avicennae et aliorum Arabum insulsis erroribusque),*[136] and in the entire work he never referred to any Arabic author by name. In the history of humanist teaching on senna, one can form two groups of authors: those who are willing to cite Arabic authorities by name—Manardo, Cordus, Champier, and Brasavola—and those who only speak of *Mauritani* or *Arabes,* if they mention Arabs at all—Barbaro, Marcello Virgilio, Ruel, and Fuchs. Barbaro had treated senna only in passing, Marcello Virgilio had simply assumed that his readers would know about senna, Ruel had cited *Mauritani* and Actuarius on the topic. How would Fuchs write about senna?

The answer is: by rewriting Ruel. It is possible that Fuchs was aware of the critical replies to Ruel's making senna a Greek tree. He evaded this criticism by creating two kinds of *colutea: colutea* proper, which reaches the

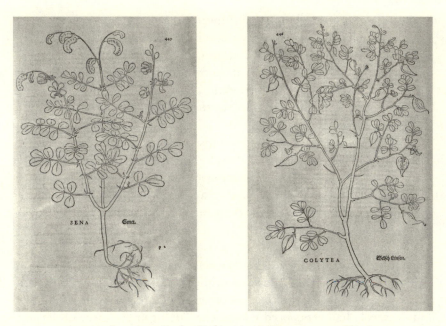

FIGURE 2.  Leonhart Fuchs, *De historia stirpium commentarii insignes*
(Basel, 1542, repr. Stanford, 1999), 446, 447.

size of a tree in four years, and senna, which is smaller. Other differences
between *colutea* and senna concern the shape of the pods—*colutea* pods are
swollen, senna pods are moon-shaped—and the shape of the seeds—*colutea*
seeds are round, of lentil-like shape, senna seeds are oblong and pointed
just as a human heart. For the features of *colutea,* Fuchs refers to Theophrastus, but the wording is that of Ruel:[137] *quadriennio se in arborem
efferat . . . turgido spiritu distenduntur . . . siliquas praetumidas.* These phrases
derive from a sentence that Ruel had coined himself as a description of a
descendant of the ancient *colutea,* which, as he maintained, was growing in
France. The sentences sound like Theophrastus Latinus, but they are
Ruel.[138] For the features of senna, in contrast, Fuchs's text is without prece-
dent; it is possible that these additions are based on firsthand knowledge
of senna pods and seeds, since the Alexandria senna that we know today
indeed has moonlike pods and heart-shaped seeds.[139]

With this we come to the problematic question of the woodblock
figure of senna in Fuchs's herbal (see Figure 2).[140] It has been argued that
the egg-shaped leaves of the figure indicate that Fuchs was depicting
senna cultivated in Italy in the sixteenth century.[141] This is not impossible, especially in view of the fact that later in the century the botanical
works of Camerarius and Jakob Theodorus print two pictures of senna,

one very similar to Fuchs's, called common senna (*<Sena> vulgaris*) or Italian Senna *(Sena italica),* with egg-shaped leaves, the other called Alexandrian *(Sena Alexandrina)* or oriental senna *(Sena orientalis),* with sharply pointed leaves.[142]

We should be careful, however, not to give Fuchs too much credit. While he shows much interest in pods and seeds, he does not say anything about the leaves of either *colutea* or senna—the exception being a very general phrase referring to the entire genus *colutea* (i.e., both to *colutea* proper and to senna): the plant is equipped with *folio foenigraeci,* "with the leaves of fenugreek," a phrase that Fuchs draws either from Ruel, who in turn had taken it from Theophrastus's description of *colutea,* or, less probably, from Hieronymus Bock.[143] In full accordance with the text, Fuchs's pictures show senna and *colutea* with very similar, egg-shaped leaves. In fact, the leaves are the one feature that make the observer believe that the two plants belong to the same genus. It would have been impossible for Fuchs to depict Alexandria senna, because its sharply pointed leaves looked too different from *colutea.* And for us, this means that the ideology of Fuchs's text does not allow us to say anything certain about the botanical material he had in his hands when designing the article on *colutea.*[144]

Fuchs's monograph comprises the sections *forma, locus, tempus, temperamentum, vires,* and *appendix.* In all of them, he treats *colutea* proper and senna as one plant: the section *forma* contains Ruel's description of French *colutea* (this is the passage with the reference to fenugreek); the sections *locus* and *tempus* make *colutea* a garden plant flowering in May and June; with regard to *temperamentum,* Fuchs quotes a sentence from Pseudo-Mesue but uses the phrase *Arabes . . . statuunt* to keep his text clean from Arabic proper names; the section *vires* purports to derive from Actuarius—who, as a Greek author, is quotable—but in fact is drawn literally from Ruel along with his textual manipulations, that is, the silent quotations from Actuarius on the laxative powers of *Cassia nigra* and from Pseudo-Mesue on head diseases, chicken broth, and constipation in the bowels; the final section *appendix* states again the basic assumption of Fuchs's monograph, the similarity between the two subkinds of *colutea:*

> Since these two types of *colutea* are similar to each other not only in appearance, but in taste, it is easy to gather from this that they differ very little in properties. Therefore, both draw out *(educit)* black and parched bile *(atram bilem retorridamque)* with ease, and purge the head, brain, and the sense organs of noxious humors. Why go on? They are effective against all diseases arising from black bile.[145]

While the phrase "black and parched bile" derives from Ruel's description of *colutea,* taken in turn from Actuarius's description of *cassia,*[146] the main thrust of the theory is adopted from Pseudo-Mesue who had written on senna: "In soluble form, it easily expels black bile and burned bile. It purifies brain, heart, liver, spleen, the organs of the senses, the lung."[147]

The final section again shows that Pseudo-Mesue, even if Fuchs avoids any open reference, is one of the two sources of his article—the other being Jean Ruel, who is not credited either. The text does not attest to any first-hand knowledge of Theophrastus and Actuarius, the two authors invoked by Fuchs; both are cited in Ruel's corrupted version.[148] Many times Fuchs had censured his colleagues for philological carelessness;[149] here he has fallen victim to the very same vice. Had he read the sources himself, instead of relying on Ruel, he could have avoided continuing a fabricated botany.

On the other hand, one can easily see the strengths of his approach. It allows Fuchs to include both the obscure Greek plant *colutea,* on which there is hardly any information in Greek sources, in particular with respect to its application, and the popular exotic plant senna, which was well-known for its laxative powers but needed to be Hellenized to be acceptable for Fuchs. A number of features make Fuchs's article on *colutea* much stronger than Ruel's: the idea of dividing *colutea* into two subkinds, the reader-oriented presentation of the information, and the very suggestive pictures. Manardo, Cordus, and Champier had realized that senna was too important a drug to be simply identified with this or that plant, as Marcello Virgilio had done. Instead, Manardo reluctantly entered on barbarous terrain by writing a commentary on Pseudo-Mesue, Cordus criticized Marcello Virgilio but did not seem to know how to proceed further, and Champier offered quotations of Arabic authorities hedged around with anti-Arab polemics. In contrast, Fuchs found a way to write directly on senna without compromising his humanist conviction. Making senna a subkind of an antique plant allows Fuchs to acknowledge differences between the two plants. He integrates the plant not into a commentary on Dioscorides or a dialogue between humanist friends but into a pharmacognostic compendium organized after Greek models. In the views of Manardo, Cordus, and Champier, senna had remained an exotic and barbarous but useful drug; in Fuchs's interpretation, it is transformed into the vulgar version of an antique plant. Fuchs underlines this Hellenizing interpretation by suppressing any credit to Arabic authorities by name. In the history of humanist attitudes toward senna, Fuchs's article in *De historia stirpium* marks a climax: it is a great mixture of ideological motivation, philological unscrupulousness, personal observation, and conceptual ingeniousness. If Fuchs had been successful, we would today speak of "Colutea vulgaris" and not of senna.

One year later, in 1543, Fuchs published a reworking of his *De historia stirpium* in German, the *New Kreuterbuch*. The arrangement of the plants is the same, but since all plant names are in German instead of Greek, the original alphabetical order is lost. To write a book in a vulgar language for a broader audience, was not a project dear to Fuchs's elitist and humanist heart; it seems to have been the idea of the printer.[150] It may be that he was not particularly happy about the adaptations he had to make. In the case of *colutea* and senna, much of the original Hellenized guise disappeared. The original entry had opened "De Colytea. Cap. CLXVIII. Nomina. Kolytea Graecis, Colytea et Colutea Latinis dicitur . . ." It now runs:

> About "Senet." Chapter 169. Names: "Senet" we call here the herbs which are called "Coluteae" by Theophrast, because they resemble each other in particular with respect to the leaves. Kinds: These herbs have two kinds. One is describe in particular by Theophrast and is called "Colutea." It is not known in the pharmacies and is called by us "welsch Linsen." The other is called "Sena" by Actuarius.[151]

By heading the chapter "Von Senet," Fuchs has given senna a much more prominent place than in *De historia stirpium*. He compensates this loss in antique outlook by extinguishing all references to the Arabs: *Mauritano ore* is replaced by *von dem Actuario*, references to *Arabes* and *barbari* have disappeared altogether. Note that Fuchs now speaks of the similarity of the leaves at the very beginning of the article; in *De historia stirpium*, this was implied but not formulated openly. The reason may be that the description of the differences between *colutea* and senna is more detailed in the German book so that Fuchs found it necessary to counterbalance this by stating the similarity of the leaves at the opening of the article. Again, it is not impossible that Fuchs had gained additional information on the appearance of senna from his own observations, for he now includes a description of the flowers of senna: yellow with five petals.[152] The senna plants that we know today bear clusters of numerous yellow flowers.[153] Again, we are not in a position to say anything about the material Fuchs had in his hands. The appearance of senna in the *New Kreuterbuch* is an indication that the distorting forces of radical humanist ideology were at play primarily in the learned discourse of humanist Latin, and that their influence on the vernacular was not on the same scale.

Fuchs's version of the *colutea* theory—the third version after Ruel and Bock—influenced other botanists such as Adam Lonitzer, who treated

Theophrastus's *colutea* and vulgar senna as subkinds of senna.[154] Two of the theory's opponents were mentioned above: Luigi Mondella and Amatus Lusitanus. It was decisive for the theory, however, that one of the botanical heavyweights of the sixteenth century turned against it: Pietro Mattioli (d. 1577).[155] His commentary on Dioscorides was an extraordinary printing success and dominated the field until Caspar Bauhin's *Pinax* of 1623. It appeared in Italian in 1544, with subsequent editions in 1547, 1548, 1549, 1552, before the first Latin reworking of the commentary went to press in 1554. Here we will concentrate on the 1565 edition in Latin, which stands out for its comprehensiveness and high-quality illustrations.[156]

Senna and *colutea,* as plants not known to Dioscorides, are treated in the chapter on *delphinion;* in this, Mattioli's commentary betrays the influence of Marcello Virgilio. The structure of the remarkably long entry is:

1. The entry on *delphinion* in Dioscorides is spurious. *Delphinion* is not identical with larkspur ("Rittersporn, id est equitis calcar"). (16 lines)

2. *Delphinion* is not identical with senna. (5 lines)

3. Description of senna. (14 lines)

4. Senna is not identical with *colutea.* (12 lines)

5. Description of *colutea.* (15 lines)

6. On Mesue's claim that senna pods are more effective as laxatives than senna leaves. There are two kinds of senna pods. (15 lines)

7. On Mesue's claim that senna weakens the stomach. (17 lines)

8. The preparation of senna. (14 lines)

9. The medicinal properties of senna. (9 lines)

10. The synonyms of *delphinion* and senna. (3 lines)

It is very obvious that Mattioli's commentary, despite the chapter's title, is not concerned with *delphinion,* but with senna and *colutea.* Mattioli believes that the entry on *delphinion* in later manuscripts of Dioscorides is fictitious (*fabulosa*) or at least not authentic.[157] He directly replies to earlier theories about senna by Manardo, Ruel, and Fuchs, using empirical arguments as his main scholarly weapon. He claims to have personal experience of senna and also of *colutea,* which he has seen growing wild in Trent.[158] If planted earlier than May, senna is easily destroyed by frost, as he has experienced himself several times.[159] Mattioli finds that Ruel is wrong in making senna a tree and explains the error as caused by Ruel's ignorance of Tuscany where the true senna is planted, especially in the Florentine area.[160] Against Fuchs's claim that *colutea* and senna have similar medicinal properties, Mattioli claims that he knows for certain that the seeds of *colutea* do not provoke purging, but vomiting no less than broom—apparently, he had firsthand experience.[161] He admits that senna imported from Alexandria or Syria is more effective, but insists on the very good laxative properties of a syrup that he produces from senna growing in Italy.[162] Replying to Pseudo-Mesue's claim that senna weakens the stomach, he assures his readers that he does not remember ever having met anybody complaining of stomachache after drinking a solution of senna; if pain in the bowels occurs, it is not due to senna but due to the expulsion of fat excrements.[163]

There is much information on Mattioli's personal experience with senna in the section on Pseudo-Mesue's theory that senna pods are better laxatives than senna leaves. Mattioli explains that two kinds of senna pods exist, those that fall down by themselves and those that are picked before they are fully ripe: the latter are fat and heavy but rarely sold.

> This latter kind, as I have experienced a thousand times, has laxative powers no less than the leaves. The first kind, however, which can be found almost everywhere in Venetian pharmacies, does not only have less power but almost no power. From this one can easily conclude that in this matter Mesue is not to be damned without qualification since he is referring to pods which are picked and not to windfalls, which in turn are the only ones used by those who falsely reproach Mesue. I once belonged to this group, but when later I had planted an entire field of senna, so that I could test the picked pods, which are green and full of juice, as well as the dried pods, I easily realized in testing that things are different.[164]

After many decades of discussion, this is a remarkable return to the position of the Arabic authority of Mesue, justified with arguments based on experiment. Since many colleagues contributed to Mattioli's work by communicating their discoveries and opinions, especially in the later editions of the commentary, it is difficult to determine how much of Mattioli's rhetoric actually reflects personal experience.[165] If it is only rhetoric, it is very well done and sets Mattioli's commentary apart from the older philological approach to botany.[166] The language of experience and experiment serves Mattioli to give answers to several major issues of the current discussion of senna: Ruel's identification of senna with *colutea,* the question of the relative value of Italian and Alexandrian senna, the criticism that Manardo, "this most learned man from Ferrara" *(Ferrariensis vir doctissimus),*[167] had advanced against Pseudo-Mesue on the weakening of the stomach and the preference for pods over leaves. Mattioli's commentary is truly a summa of Renaissance theories on senna.

It becomes obvious that the discussion has changed noticeably since the time of Marcello Virgilio, when we turn to the latter's identification of *delphinion* and senna.[168] Mattioli refutes it as *gravis error* by comparing the thickness of the leaves and the color of the fruits. Remember that Brasavola had tried a very similar refutation, but did not offer enough botanical information on senna to make his arguments entirely convincing; Brasavola was strongest when arguing philologically or logically. Mattioli, in contrast, gives a detailed botanical description of senna, probably the most detailed so far. While Dioscorides describes the leaves of *delphinion* as thin, long, and parted and the flowers as purple, senna leaves are rather thick and slightly fat, says Mattioli, and the flowers are golden.[169] In a similar vein, Mattioli argues against Ruel's *colutea* theory: the pods and seeds differ much in their appearance, and senna, apart from not being a tree, does not live longer than a couple of months.[170]

In the hands of Mattioli, the trend toward better descriptions of senna—with its modest beginnings in Bock's *Neu Kreüterbuch* of 1539 and Fuchs's *New Kreuterbuch* of 1543—is turned against the humanist project of identifying senna with plant names of antiquity. Ironically, Mattioli's contribution to the theory of senna is hidden away in an entry on *delphinion,* which for Mattioli is a hollow phrase without any botanical or pharmacological significance. The unhappy format of a Dioscorides commentary is remedied by very good indexes, which make Mattioli's exuberant information on *materia medica* accessible to readers.

As a preliminary conclusion one can note that the decline of the humanist reductionist approach did not lead to a return to pre-humanist descriptions of senna, that is, to the scarce information on the appearance of senna leaves, pods, seeds, and flowers in Arabic sources. The long controversies in the humanist camp about the correct identification—which covered not only *delphinion* and *colutea* but also Manardo's *empetron* and Isacius's *peplis*—had led to an ever more refined discussion of the features of the plant. Eventually, this discussion ceased to be entirely bookish. A late and somewhat hybrid example of a philological approach to plant description is Leonhart Fuchs's quotation of Ruel's description of *colutea.* Fuchs's own additions to Ruel, however, especially in the *New Kreuterbuch* of 1543, show that he was ready to go beyond the boundaries of philology not only in the case of German plants but also when treating Mediterranean and Oriental plants, which constitute the bulk of the Graeco-Arabic tradition. The future would belong to a science based on ever more accurate differentiations of the outer properties of plants. Mattioli still finishes his description of senna with a quotation from Pseudo-Serapion on the shape of the pods and on the shepherds collecting them.[171] But it was only a matter of time until the precision of Western botanical diagnoses could dispense with the assistance of Arabic sources.

These sources remained indispensable, however, on the medicinal properties of senna. With the enormous success of Mattioli's commentary it was clear that the Arabic authorities on *materia medica* had survived half a century of anti-Arab propaganda and would continue to influence pharmacology in the following centuries. The example of Ruel and Fuchs had shown that it was impossible to write on the properties of the drug without drawing at least silently on Pseudo-Mesue's *De simplicibus*. Pseudo-Mesue was the only source available with ample information on the preparation of the drug, on its improvement through mixing with other drugs and on its internal and external medicinal properties. In other words, he offered a proper theory of senna as a medicine. Even Leonhart Fuchs, in a late work of his, would give in and cite Pseudo-Mesue on the purgative effects and many other properties of senna, "which everybody can read in <Mesue's> book" *(quae apud illum legere quivis potest)*.[172]

## FROM MATTIOLI TO LINNAEUS

In the second half of the sixteenth century, scholarship on *materia medica* lost much of the anti-Arabic acerbity that had colored the discussion for several decades. It was commonly acknowledged that senna was not known to the ancients, that it was a *res nova*.[173] Some, such as Mathias de l'Obel and Pierre Pena, gave credit to the Arabs:

> To the great disadvantage of the sick, senna became known to the Greeks only very late *(i.e., in the Byzantine era)*; we owe senna together with its name to the Arabs.[174]

As a consequence, senna was now given a separate chapter, usually after or before *colutea* (thus in l'Obel, Camerarius the Younger, Daléchamps, and Bauhin).[175] With the decline of the old model of Dioscorides commentaries, scholars were forced to develop their own methods of plant classification, culminating in Carl Linnaeus's eighteenth-century classificatory system that is based on the botanical features of the flower. In what follows, first I

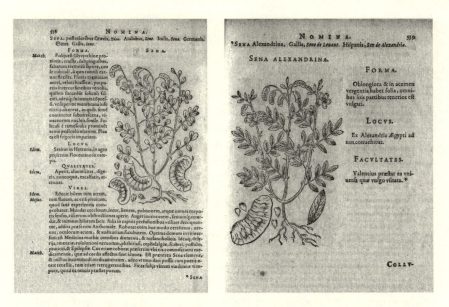

FIGURE 3.  Joachim Camerarius (the Younger), *De plantis epitome utilissima Petri Andreae Matthioli* (Frankfurt am Main, 1586), 538, 539.

review the botanical discussion of senna, since it is instructive to see how botany disconnected from Arabic sources, and then I turn to the Arabic influence on pharmacology, which continues into the eighteenth century.

A first important step in the botanical understanding of senna was taken by Joachim Camerarius the Younger in his Epitome of Mattioli's *Commentarii,* published 1586. It seems that Camerarius was the first to print two pictures of senna on facing pages (see Figure 3), one with egg-shaped leaves headed *Sena* and one with sharply pointed leaves headed *Sena Alexandrina*. The text accompanying the picture of *Sena* consists of quotations from Mattioli and Pseudo-Mesue (the latter is cited on properties). On *Sena Alexandrina,* Camerarius writes:

> Shape: It has rather oblong and almost pointed leaves; it is in all its parts finer than the common <senna>. Place: It is transported to us from Alexandria in Egypt. Properties: It does everything effected by common senna, but in a stronger way.[176]

Ever since Manardo's commentary on Pseudo-Mesue, scholars had pointed to the differences in medical value between senna from Italy and from Alexandria. Camerarius takes these differences so seriously that he refrains from treating them as one plant. Another remarkable feature of his pictures is the separate depiction of the pods and the seeds of both kinds of senna. In the footsteps of Camerarius, Jakob Theodorus writes in 1613 that "Senet ist zweyerley / eine so Sena orientalis genennt wird / die ander Sena Italica" and prints two accompanying woodcuts.[177] With Caspar Bauhin's magnum opus *Pinax* of 1623, the distinction between *Senna Alexandrina sive foliis acutis* and *Senna Italica sive foliis obtusis* becomes canonical for two centuries.[178] The botanical discussion of senna was also advanced by several attempts to situate senna in classificatory systems. In 1583, Andrea Cesalpino treats senna among trees that carry their seeds in pods.[179] In 1680, Robert Morison groups senna with leguminous plants equipped with hanging pods.[180] In 1687, Paul Hermann lists several Egyptian, American, and Brazilian plants as subkinds of senna that all have flowers of four or six leaves[181]—a list picked up a few years later by John Ray in a section on plants with pods and five petals.[182]

But the key to a classification accounting for the relationship of senna to other plants was found, as far as we can see today, by Joseph Pitton de Tournefort (d. 1708). In his *Elemens de Botanique* of 1694 Tournefort makes senna one of three genera of the section "trees and shrubs with roselike flowers, whose pistil turns into a pod," the other genera being *Poinciana* and *Cassia*.[183] It should prove influential that Tournefort had assigned senna and cassia to the same group on the basis of detailed descriptions of both pods and flowers.[184] In the following decades, scholars realized that the different kinds of cassia and senna had much in common.[185] In Carl Linnaeus's seminal *Species plantarum* of 1753, senna has become a species of the genus *Cassia*—which remained the standard view for more than two hundred years. Only recently have botantists returned to the older idea of Tournefort that *Cassia* and senna are both genera of their own that are closely related. In Linnaeus, the genus *Cassia* is defined as having pods and flowers, a flower cup with five leaves, five petals, three upper sterile anthers, and three lower, beak-shaped anthers.[186] Cassia senna is one of twenty-six species of *Cassia;* it is described as "Cassia with nearly egg-shaped leaflets in three–four pairs . . . It grows in Egypt."[187] Linnaeus acknowledges a variation of this species by referring to Bauhin's "Senna italica sive foliis obtusis." Among the other species of *Cassia* one can find several of Hermann's subkinds of senna, which Linnaeus now calls, for instance, "Cassia Tora" or "Cassia hirsuta."

In the early nineteenth century, it was realized that one could further differentiate between senna with sharply pointed leaves and senna with narrow leaves. This led to the differentiation of the species "Cassia acutifolia" and "Cassia angustifolia"; the latter replaced Linnaeus's "Cassia senna."[188] These are the pharmacologically most important kinds of senna that are still in use today. It was remarked above that in the nineteenth century the cultivation of senna in Italy had long stopped. The name "Cassia italica," however, continued to be used for a species of cassia growing in Africa and India, whose leaves are less effective laxatives than senna leaves. Whether this plant has anything to do with the Italian senna, which was so important for the discussion in the period between Manardo and Bauhin, is a question that is unlikely ever to be solved.[189]

The contribution of Arabic sources to post-Renaissance botanical discussions of senna was very limited. Occasionally, Pseudo-Mesue's distinction between a domestic and a wild growing kind of senna is cited, for example, by l'Obel (1576), Bauhin (1623), and Morison (1680).[190] The above survey of the botanical discussion of senna between Mattioli and Linnaeus has shown that the crucial idea, which transformed the discussion, was to base a classificatory system on precise descriptions primarily of the appearance of the flowers, as in Tournefort, with Linnaeus perfecting this approach by focusing on petals and anthers. The few features of senna pods mentioned by Pseudo-Mesue and Pseudo-Serapion were of no help in this regard. Among the last authors to quote their descriptions extensively are Andrea Cesalpino (1583)[191] and Jacques Daléchamps (1586/1587).[192] Note, however, that the contribution of humanist sources to the botany of senna was very limited too. None of the attempts to identify senna with a plant in Dioscorides and Theophrastus seriously influenced the post-Renaissance botany of senna.

While the botanical discussion of senna was disconnected from Arabic sources, the pharmacological one was not. It was mentioned above that even Leonhart Fuchs later in his career referred his readers to Pseudo-Mesue and that, in the eyes of Joachim Camerarius, Pseudo-Mesue was the authority on the medicinal properties of the drug, Mattioli the authority on all other features. That Pseudo-Mesue continued to be an important source for pharmacology, is reflected in the appearance of a new commentary on Mesue's *Opera* in 1568, composed by Giovanni Costeo (d. 1603), which is particularly rich on the simples.[193] The author addresses all the main issues of the sixteenth-century controversies about senna, as is apparent from the structure of his commentary:

1. Senna is not mentioned by the Greeks, except by Actuarius.

2. Senna is not identical with *colutea*.

3. It is doubtful whether our common senna is identical with that of Actuarius who calls it "fructus."

4. Mesue seems to use the name "sene" only for the pods.

5. Mesue judges the medicinal value of pods higher than that of leaves, because he speaks of pods that are picked and not of windfalls (an argument by Mattioli).[194]

6. Mesue rightly recommends the mixing of senna with certain salts in order to avoid stomachache.

7. Whether senna is able to purge phlegm (a question raised by Manardo).

8. Why Mesue groups senna, which is a mild laxative, together with strong laxatives.

These topics—pods versus leaves, stomachache, the purging of phlegm—are discussed in other works of the late sixteenth century in an explicit and comprehensive manner, which contrasts significantly with the cautious approach to Arabic authorities typical of the first half of the century. Mathias de l'Obel, for instance, frankly castigates the opponents of Pseudo-Mesue:

> The person, therefore, who claims against Mesue that <senna> should not be corrected since it accommodates the stomach with its dryness and with some astriction, does not know what he is commenting upon.[195]

In Jacques Daléchamps's *Historia,* Manardo and Mattioli are given credit for their criticism of Pseudo-Mesue, but in the end the author favors compromise positions that do not detract from Pseudo-Mesue's and Pseudo-Serapion's authority (clearly under the influence of l'Obel).[196] Jacob Theodorus's *Kreuterbuch* of 1613 mentions Pseudo-Mesue only on the external application of senna leaves, such as against headaches, but much of his teaching on the medicinal properties is silently drawn from Arabic authorities.[197]

That this is still the case at the end of the seventeenth century is apparent from the works of Robert Morison (of 1680) and John Ray (of 1688). When Morison writes that senna purges black and yellow bile and phlegm *(educit bilem tum atram tum flavam ac etiam pituitam)* and that it purges various organs including the senses and opens the bowels *(mundat cerebrum, iecur, lienem, pulmonem, atque omnes corporis sensus; viscerum obstructiones aperit),* we meet not only with Pseudo-Mesue's doctrines but also with his vocabulary.[198] The Arabs have much praise for the pods, but the Europeans recommend the leaves, Morison says, adding Mattioli's explanation of the difference.[199] John Ray is less indebted to Arabic sources than

Morison, but still cites Pseudo-Mesue as an authority on the preparation of the drug: *Debet autem decoctio esse mediocris, ut Mesue praecipit.*[200] It is only with the gradual decline of the Galenic system of humoral pathology that Arabic doctrines on the purging of black and yellow bile through senna became less significant.[201] Two university dissertations on senna may serve to illustrate the development. In 1733, a medical dissertation at the University of Altdorf cites Pseudo-Mesue as the main authority on the specific medicinal properties of senna: "it mildly purges black bile from brain, the organs of the senses, lung, heart, liver, spleen" *(clementer purgat melancholiam a cerebro, sensoriis, pulmone, liene).*[202] In 1856, a dissertation from Dorpat refers only to contemporary chemical literature on senna.[203] Apparently, a new step in the development of a theory of senna as medicine was not made before the nineteenth century, when senna leaves began to be analyzed chemically.

❧

The example of senna clearly demonstrates the deficiencies of the radical forms of medical humanism. Marcello Virgilio aims at confirming the authority of ancient Greek texts, but is forced to use medieval pharmacological sources when confronted with obscure passages in Dioscorides. Jean Ruel does not refrain from drastic textual manipulations that allow him to use material from Arabic sources in order to resuscitate a tree described by Theophrastus. Just as Marcello Virgilio, Ruel tries to eliminate medieval pharmacology by identifying "barbarous" drugs with antique ones. Leonhart Fuchs's approach was perhaps the most dangerous to the survival of Arabic pharmacological teaching: his aim was not simply to restore antique *materia medica,* but to Hellenize modern *materia medica.* This meant that texts of the Arabico-Latin tradition had to be rewritten and assigned to Greek and Roman authorities. The prejudiced convictions underlying this approach, however, were not shared by a sufficient number of humanists. When scholars from within the humanist camp such as Manardo, Brasavola, and Mattioli returned to reading and evaluating Pseudo-Serapion and Pseudo-Mesue, the humanist hard-liners' campaign was lost.

Arabic *materia medica* proved indispensable from an internal, scientific point of view. When the tactic of identifying all contemporary plants with antique predecessors was exposed as unsound in philological, logical, and empirical respects, scholars realized that there was only one significant textual tradition to which they could relate their findings—the comprehensive and detailed accounts of the features and properties of plants in sources of Arabic provenance. Arabic pharmacognosists had added significantly to ancient knowledge of *materia medica*. Dioscorides, in his influential work *De materia medica*, had described about 1,000 plants, animals, and minerals. In comparison, the greatest Arabic work of *materia medica*, *The Comprehensive Book of Simple Medicines and Food* of Ibn al-Bayṭār (d. 1248), contains 2,324 entries and is based on more than 260 different sources, incorporating botanical and pharmaceutical knowledge from many cultures and regions, much of it unknown to Greek culture. If we do not count the many synonyms with Greek names in Ibn al-Bayṭār's work, there still remain about 200 plants and 200 animals and minerals unknown to the Greeks.[204] Ibn al-Bayṭār's text was not available in the Latin Renaissance, except for an excerpt on lemons translated by Andrea Alpago, but a good number of earlier texts of this tradition were, notably by Avicenna, Pseudo-Serapion, and Pseudo-Mesue. Medical botany is an area of medicine in which the Arabic contribution is particularly impressive.

The humanist failure to bring about a complete return to Dioscorides echoes a development several centuries earlier in Arabic science: After the translation of Dioscorides into Arabic in the ninth century, faithfulness toward the Greek author became a matter of dispute among Arabic scientists. The dispute was sparked by a treatise on non-Dioscoridean drugs written by Ibn Ǧulǧul in the tenth century. In order to preserve the authority of Dioscorides, it was argued, for instance, that Dioscorides had omitted the description of certain drugs because they were well-known anyway in his time.[205] In the end, Arabic scholars added hundreds of drugs to those known from Greek science. Of course it was sensible to continue in Dioscorides's footsteps and to try to identify synonyms of the drugs described. But a purely reductionist approach was doomed to fail. The West still had to learn this lesson. The difficulty of the lesson was increased by a characteristically humanist addition that was missing in the Arabic discussion: humanist scholars not only strove to identify every plant with an ancient equivalent, they also cherished an ideal of linguistic purification. The ideological conviction that a proper plant name can only be a name of classical culture was a serious hindrance to the development of Renaissance *materia medica*.

Arabic authorities on *materia medica* survived several decades of polemical attacks by radical humanists and continued to be cited, as we have seen, until the eighteenth century. There was no "downfall of Arabism,"[206] as is sometimes claimed (for instance by George Sarton, Heinrich Schipperges, and Peter Dilg), except in humanist propaganda. The findings of this chapter correspond to a parallel development in the university curricula of medicine that I have described in Chapter 1: that Arabic authorities reappear in many university statutes of the late sixteenth and early seventeenth centuries, after they had been extinguished several decades earlier by humanist reformers. In some universities, the Arabs had never disappeared from the curriculum. Another background to the return of Arabic authorities in *materia medica* is the apothecaries. It was not in the economic interest of sixteenth-century apothecaries to return to ancient *materia medica,* which lacks many important drugs of the orient trade and which, in its Hippocratic traditions, concentrates on simple medicines rather than on recipes. Arabic medicine, in contrast, had developed ample literature on composite medicines. It is indicative that the dispensatories of the Renaissance, such as the *Ricettario fiorentino* of 1498 and the famous Nuremberg *Dispensatorium* by Valerius Cordus of 1546, do not follow humanist trends, but continue to list Arabic recipes. Valerius Cordus's work relies up to 80 percent on Arabic or Arabicized sources. In view of this, it is not surprising that, as we have seen, humanists such as Symphorien Champier and Euricius Cordus complain of their contemporary pharmacists, who do not follow their ideals—ideals that were bookish and did not account for the realities of drug commerce.[207] The humanist contempt for apothecaries is well-expressed in a dictum of 1543: "Most of today's pharmacists are idiots; they only care for that from which they gain greater profit. There are no books in their workshops."[208]

Hence, the humanist reductionist program was a failure. The active suppression of Arabic material by some humanists was not a "historical necessity,"[209] but resulted in serious damages to medical botany as a scientific discipline. Fortunately, much material on senna that had been suppressed by Marcello Virgilio, Ruel, and Fuchs, was recovered by other humanist scientists. Thus, the losses in the knowledge about senna were not dramatic, and they lasted only a limited period of time, as long as Jean Ruel's fabricated drug description enjoyed a certain popularity. But given the tenacity of Ruel's and Fuchs's prejudiced research, there is much reason to suspect that tangible and even unrecoverable losses occurred in the case of other Arabic additions to *materia medica.*

In spite of all this, the productivity of anti-Arab attitudes is beyond question. As was pointed out above, the discipline changed much in the disputes of the 1510s to 1540s: humanist scholars not only recovered Greek and Latin sources of medical botany but also developed a linguistic technique of ever refined descriptions of plants and their properties as the only convincing way to argue against or in favor of an identification of a plant. In the case of senna, it was not so much the illustrations that gave the crucial impetus to the slowly developing science of botany,[210] but improved textual techniques that were sometimes related to experience and sometimes not. Indirectly, the humanist defenders of ancient texts gave an empirical impulse to the field of medical botany because it was not possible to substantially improve the precision of botanical and pharmaceutical description without precise observation and actual testing—this impulse is obvious in the texts of Giovanni Manardo, Hieronymus Bock, Antonio Musa Brasavola, and Pietro Mattioli. Here we meet an example of the productivity of prejudiced research. The Dioscoridean challenge to Arabic and medieval *materia medica* did not lead to a return to Dioscorides, as the radical humanists had hoped; but it was a great catalyst of new developments. Hence, the history of *materia medica* in the Renaissance was marked by both gains and losses, as with many dramatic changes in the history of the sciences.[211]

The humanists were right on a crucial point: They were also justified from an internal, scientific point of view in insisting on the importance of linguistic and philological skills in medical botany. In its history, *materia medica,* this very old and international textual discipline, had bridged several linguistic borders and had covered very diverse botanical regions. It is important to note that medieval pharmacologists were aware of the philological problems too. In an attempt to cope with the Graeco-Arabic–Latin tradition of *materia medica,* they composed texts of the *Synonyma* genre, such as the *Synonyma Avicennae, Synonyma Rasis* and the *Synonyma Serapionis,* which list the Latin transliterations of Arabic drug names and give brief explanations of them. On the basis of these texts, Simon de Gênes in the late thirteenth century united all information on pharmaceutical terminology in his widely read *Clavis sanationis.*[212] In the fifteenth century, Gentile da Foligno and Jacques Despars, the commentators on Avicenna, successfully improved the knowledge of Arabic transliterated drug names by analyzing Latin sources. Toward the end of the fifteenth century, the influential pharmaceutical handbooks by Saladin of Asculo, Nicolaus Prévost, and Manlius de Bosco listed synonyms that had become a widely accepted standard. At this point, the humanist reform program set in, and with much justification. To be sure, the humanist rhetoric was exaggerating the terminological chaos that allegedly reigned in the discipline, given that medieval pharmacology had in fact reached a high degree of clarification and acculteration of foreign drug names.[213] But the medieval *Synonyma* tradition rested on Latin sources only, and as such, it had reached its limits, as the humanists clearly saw. The most important textual source of the entire field, whose vocabulary and doctrines had been transported into several cultures, was neither a Latin nor an Arabic text, but a Greek one: Dioscorides's *De materia medica* (Περὶ ὕλης ἰατρικῆς). The humanists demonstrated that firsthand knowledge of this text was essential for any progress in medical botany. This was even true in the case of senna. Without a proper philological understanding of Dioscorides it was impossible to tell whether any of the plants named by Dioscorides was a synonym of the oriental plant senna. Dioscorides, as a much-traveled military physician, could have known senna. It was only on the basis of the Greek text that humanist scholars could show, to their own disappointment, that senna was a *res nova*, that it was not mentioned in ancient sources. Hence, while the reductionist program of medical humanism was a failure, its philological turn was very fruitful for medical botany.

Most members of the humanist movement, however, were blind on another eye: they did not recognize the usefulness of learning Arabic for the progress of *materia medica*. It was the initiative of a few dedicated individuals that the potential of the Arabic language and of traveling to the Orient was exploited for Western pharmaceutical knowledge. Pierre Belon (d. 1564) and Leonard Rauwolf (d. 1596) are examples of scholars whose travel accounts of the Middle East are replete with botanical and medical information.[214] More influential was the translating and commenting activity of Andrea Alpago (d. 1522), the longtime physician of the Venetian embassy in Damascus. His *Interpretatio Arabicorum nominum,* a lexicon of transliterated Latin names for drugs, plants, minerals, animals, and so forth in Avicenna's *Canon,* which was first printed in 1527, gives explanations of 2,050 Arabic terms of varying length.[215] Among them mere *id est* identifications with Latin terms, but also long discussions, in which Alpago cites Arabic authorities such as Ibn al-Bayṭār (Ebenbitar), the botanist, and aš-Šīrāzī, the commentator on the *Canon,* and reports his own knowledge of apothecaries, physicians, and plants in Arab countries. Alpago very wisely did not replace Gerard of Cremona's transcriptions of Arabic terms, but explained them, and thus continued the tradition of medieval *Synonyma,* albeit on a completely new level of precision and knowledgeability.[216] From Alpago's pen there also exists an index to transliterated Arabic drug names in Serapion's *Practica,* with Latin equivalents, which was printed in 1550.[217]

An indication of the influence of these lexical works is a passage in the *Historiae Aegypti Naturalis* by Prospero Alpini (d. 1617), the traveler in Egypt and well-known botanist. Alpini remarks that in Cairo he had met Jacopo Manni, a physician in Venetian service, who had translated an Arabic medical dictionary into Latin, "to excite the spirit of the students for the more important" and in order to correct the false spelling of Arabic names in Andrea Alpago's indices to Avicenna and Serapion.[218] This episode must have taken place during Alpini's sojourn in Egypt from 1581 to 1584. It shows that Alpago's lexica were apparently much in use,[219] and that attempts were even made to improve upon them.

It is also noteworthy that early modern Arabist scholars often adver-
tised the study of Arabic by pointing to the great benefits for medical
botany. Guillaume Postel, in his much-read *Grammatica arabica* of about
1539/1540, set the tone when alerting his readers to the many simple and
composite medicines that the West owes not to Antiquity, but to the
Arabs.[220] Other Arabists followed: Jacob Christmann in 1582, Giovan Battista Raimondi at about the same time,[221] William Bedwell in 1612,[222] and
Thomas Erpenius in 1620.[223] Christmann puts it succinctly:

> The Arabs have discovered many simple medicines of very great
> usefulness, and they have produced perfect mixtures of drugs which
> the Greeks used to administer to the sick not without cruelty and
> aversion . . . How many simple and composite medicines bear Arabic
> names? But only a few of us understand their properties. Indeed, I should
> hope that through the knowledge of the language of Arabic we obtain
> descriptions of those <drugs> whose <Arabic> names continue to be used
> in medicine.[224]

A further argument comes from Thomas Erpenius, who adds that many
more "most useful simple remedies hitherto unknown to you and the
Greeks" can be derived from new Arabic sources, such as by Ibn al-Baytār
and others, and that he himself could add more than three hundred such
medicines.[225] In the end, however, academic Arabists themselves did not
contribute much to the Western knowledge of *materia medica,* but medical
and botanist scholars with knowledge of Arabic did, such as Alpago and
Alpini.

This chapter has shown that the antagonism between the humanist
and Arabist parties should not be overstressed—an observation we have
made before when studying biographies and translations. The humanist
scholars Champier and Cordus, in spite of their anti-Arabic rhetoric, openly
adopt Arabic theories on senna. Manardo composes a critical commentary
on Pseudo-Mesue's *De simplicibus* and Brasavola in fact reintroduces Pseudo-
Mesue's position in the debate about senna. This is all the more noteworthy
because Manardo and Brasavola come from the famous Ferrara school of
medical humanism. And we have met with other humanist scholars before
who care for the Arabic tradition in *materia medica,* such as Jacobus Sylvius,
the humanist reviser of Pseudo-Mesue's *Opera.* While some humanist
scholars work on the destruction of the Arabic inheritance in Europe, others
actively work on retrieving its contribution. Here as before, the polemical literature is not a good guide to the true development of Renaissance science.

Much evidence points to the fact that Arabic *materia medica* reached the high tide of its influence in Europe in the sixteenth century. Medieval traditions continued, in particular in the genre of pharmacological handbooks and manuals for apothecaries, which are based on Arabic and medieval sources,[226] such as the dispensatories of Valerius Cordus of 1546 and of Johannes Pacotomus of 1560. Translators like Andrea Alpago improved the Western understanding of Arabic *materia medica*. Travelers contributed firsthand knowledge. Arabist scholars stressed the importance of understanding Arabic sources of medical botany. Even humanist scholars tried to integrate Arabic traditions into their works. Many scholars wrote commentaries on pharmacological works of Arabic provenance, such as Manardo, Marini, and Costeo on Pseudo-Mesue, or other scholars on Avicenna's *Canon*, book 2, on simple medicines.[227] Still others, such as Walther Ryff in 1543 and Pietro Mattioli in 1544,[228] wrote commentaries on Dioscorides that are replete with the discussion of Arabic authorities. The philological, medical, and botanical efforts to understand Arabic *materia medica* were never greater in the West than in the epoch between 1450 and 1600.

## CHAPTER 5

# PHILOSOPHY

### *Averroes's Partisans and Enemies*

he Renaissance reception of Averroes is a meeting of extremes. Around 1470, Benozzo Gozzoli, on a panel painting for Pisa Cathedral, depicts the defeated heretic Averroes lying below the feet of Thomas Aquinas. But other intellectuals of the time view Averroes very differently. The philosopher Nicoletto Vernia praises him as Aristotle's *fidelissimus interpres*. Valla castigates the barbarian Averroes for his ignorance of Greek, Marsilio Ficino for his undermining of religion. Pico della Mirandola and Domenico Grimani commission the Jewish scholar del Medigo to translate more of Averroes's commentaries into Latin. The bishop and the inquisitor of Padua forbids by threat of excommunication the public teaching of Averroes's theory of the unicity of the intellect. The philosopher Luca Prassicio censures his colleague Agostino Nifo for apostatizing from the true doctrine of Averroes. As these examples illustrate, the Renaissance educated world was deeply divided over the Arabic philosopher. Averroes, therefore, is much more than a widely read philosopher of the Renaissance. He is a symbol of Christian Europe's openness to philosophical thinking and of the great tensions that this openness created in a heterogeneous society. As historians we have grown accustomed to the fact that the students at universities all over Europe read Aristotle with the help of this Arabic commentator, who was so uncompromisingly philosophical. Indeed, this was common academic practice. But it remains a remarkable phenomenon, and one that posed great challenges to the learned world of the Renaissance.

Averroes was the Arabic author most influential in the Renaissance. He was printed and translated most often, and his works were not only read and cited but also commented upon. In the earlier chapters of the present study, the Renaissance reception of Averroes has already been investigated from different angles. I have studied the long printing history, the image drawn of Averroes in the biographical tradition, and the new translations and revisions of his medical and philosophical works in the Renaissance. In what follows, the focus will now be on his reception as a philosopher and as a commentator on Aristotle.

The principal aim of the present chapter is to assess the justification of the Renaissance polemics against Averroes's philosophy and his qualities as a commentator. I shall proceed in five steps. After three introductory sections, which are concerned with the popular image of Averroes, with the historiographical label "Averroism," and with controversies among the partisans of Averroes, I will analyze two examples of the Renaissance reception of Averroes, which illustrate his role first as a philosopher and second as a commentator. The philosophical example is the most controversial thesis of Arabic philosophy in the Renaissance—Averroes's theory of the unicity of the intellect, that is, the theory that there is only one intellect for all human beings. The analysis will focus on the question of whether the reception of the unicity thesis was impeded or promoted by nonphilosophical motives: did the polemics surrounding the figure of Averroes influence the philosophical discussion? Which factors led to the decline of the unicity thesis in the second half of the sixteenth century? The second example concerns the quality of Averroes's Long commentary on the *Metaphysics*. Did Juan Luis Vives, the famous Spanish humanist, demonstrate the ignorance of Averroes as a commentator when attacking his exposition of Aristotle's *Metaphysics,* chapter Alpha 5? Renaissance humanists called for a replacing of Averroes through the Greek commentators of late antiquity. The example of the *Metaphysics* commentary may serve as a test case to what extent this call was justified. In a first step, I now turn to the programmatic utterances by partisans and enemies of Averroes and to the Janus-faced image of Averroes that resulted from the polemics.

## THE PUBLIC IMAGE

The great esteem enjoyed by Averroes in academic circles is evident in the editorial and translating efforts invested in his work, in which leading Aristotelian philosophers played a prominent role. Particularly noteworthy among the editions are the second collected edition of Aristotle and Averroes, which was produced by the philosopher Nicoletto Vernia in 1483/1484 at the official request of the governor of Padua, and the monumental eleven-volume edition published by the Giunta press in 1550/1552. In the prefaces to many of these editions, Averroes is praised as the best commentator on Aristotle,[1] and as a philosopher in his own right,[2] and the useful-

ness of his texts for students and natural philosophers is emphasized.[3]
The polemics against Averroes are historically and philosophically unjustified and dishonest, as Benedetto Rinio pointed out in 1556:

> He awakened the studies of the Latins and provided the occasion to
> investigate nature in a serious manner for many <scholars> who are now
> being celebrated for this very reason. For Averroes' doctrine was so full
> of learning that he has found partisans who even in our time still most
> seriously defend his opinions.[4]

We have seen in Chapter 3 of this study that the time's great interest in
Averroes is also reflected in the many new Latin translations of Averroes's
works. The Giunta edition of 1550/1552 incorporates many of these translations and thus testifies to what has been called the second revelation
of Averroes.[5] A noteworthy feature of the translation movement was that
it was initiated and supported by a series of aristocratic patrons who
had intellectual links to the climate of Paduan Aristotelianism. The high
esteem of Averroes passed beyond the restricted milieu of the classroom.

If these editions and translations were in themselves a clear and publicly visible sign of the time's great interest in Averroes, there were also authors, many of them teaching in Padua, who took Averroes's side openly by
writing commentaries or treatises on him or even in his defense. These
authors praise Averroes as "Aristotle's most faithful interpreter" (*fidelissimus ille Aristotelis interpres*), a phrase used by Nicoletto Vernia,[6] or as "the
most elegant philosopher," "Aristotle transposed" (*Aristoteles transpositus*),
and "Aristotle's priest" (*sacerdos Aristotelis*), as Agostino Nifo referred to
him.[7] Nifo once calls him "the Arab Aristotle" by virtue of his total and
laudable concordance with Aristotle's views.[8] In 1508 and 1537, Marcantonio Zimara published his often-printed two reading tools for the combined Aristotle and Averroes, the question commentary *Solutiones contradictionum* and the alphabetical lexicon *Tabula dilucidationum,* which also
served as a defense of the Averroist interpretation of Aristotle.[9] A very
clear attempt to defend Averroes against his critics was made in 1521 by
the Averroist Luca Prassicio, who announced his standpoint in the title of
his work: "Question on the immortality of the intellective soul according
to the opinion of Aristotle, who was never more truly interpreted than by
Averroes."[10] The Platonist Francesco Patrizi valued Averroes as *Aristoteli-
cissimus,* the most Aristotelian, of all commentators on Aristotle.[11] The
Aristotelian philosophers Pietro Pomponazzi and Jacopo Zabarella were
ready to excuse Averroes for errors as a commentator because, they argued, he did not know Greek and was forced to rely on a problematic

translation.[12] On the occasion of a logical problem, Zabarella even extended this praise:

> Averroes has to be excused because he was an Arab and ignorant of the Greek language, and in particular because, even if he erred somehow with respect to language, he did not err on the matter. Rather, he performed so well that nobody has understood Aristotle's intention in this matter more profoundly and better than he, because he has also detected and refuted the errors of the Greeks and has elucidated the truth of the matter most beautifully.[13]

In Zabarella's eyes, the quality of Averroes's understanding of Aristotle was such that Averroes occasionally surpassed the Greek commentators even though he was relying only on Arabic translations.

But the Renaissance image of Averroes was much determined also by the many enemies he had found among humanists, theologians, Church officials, and even Aristotelian philosophers. The polemical attacks of these groups were colored by religious tones because Averroes's name was associated with two important theses in conflict with Christian faith: the unicity thesis, which allowed only for a collective, but not a personal immortality,[14] and the Aristotelian thesis of the eternity of the world.[15] The humanist disdain for Averroes reaches back to Francesco Petrarca, who repeatedly attacked Averroes in his letters and invectives for being in conflict with Christian faith.[16] In a late letter he advised the addressee: "Be the enemy of Averroes, who is the enemy of Christ!" *(Christi autem inimico esto hostis Averroi!).*[17] In 1370, Petrarca send a letter to a young Italian scholar in Paris, Luigi Marsili, asking him to write a treatise "against that frantic dog Averroes who is prompted by an undescribable fury to bark at his lord and master Jesus Christ and the catholic faith."[18] Even Petrarca's defamation of Arabic poetry as ugly and immoral is connected to Averroes: it very probably rests on a misleading Latin translation of Averroes's Middle commentary on Aristotle's *Poetics.*[19]

The two anti-Averroist themes of irreligion and bad style remained standard among humanist authors; another motive was Averroes's ignorance of Greek. Coluccio Salutati declared Averroes to be *irreligiossimus,* in view of his terrible views about God and the eternity of the soul.[20] Lorenzo Valla rejected both Avicenna and Averroes as barbarians for their ignorance of Greek and Latin. The linguistic incompetence undermines their authority in philosophical questions, argues Valla, since they cannot be consulted on the meaning of words.[21] As a result of this attitude, for humanists such as Erasmus and Bembo, the term "Averroista" was tantamount to a philosopher with barbarous language and confused thinking.[22]

Ermolao Barbaro, like Valla and Poliziano a representative of the philological approach to the classics, was a major pioneer in humanist *materia medica,* as we saw in Chapter 4; but he also contributed to the growing rivalry between the Greek and Arab commentators in Renaissance intellectual culture. In 1481 he published his translation of Themistius's *De anima* paraphrase, which influenced the subsequent psychological discussion (as did the commentaries by Simplicius and Alexander of Aphrodisias). The extent to which Barbaro valued Averroes's qualities as a lowly commentator is obvious from a letter of 1485, where he accuses Averroes of stealing from the Greek commentators: "if you compare his writings with those of the Greeks, you will find that, word for word, they are stolen from Alexander, Themistius, and Simplicius."[23] The same attitude is expressed by Girolamo Donato in the preface to his translation of Alexander of Aphrodisias's *De anima,* book 1, of 1495: Upon word-for-word inspection of Averroes's texts one will find, says Donato, that his doctrines are taken from the excellent authors Alexander, Themistius, and Simplicius.[24] In 1517, the humanist Ambrogio Leone echoed these allegations in his lengthy *Castigationes adversus Averroem,* which describe Averroes as a plagiarist and thief of Greek texts, immoral philosopher, and irreligious thinker.[25] A few decades later, in 1543, the translator of Simplicius's commentary on *De anima,* Giovanni Faseolo, furnished his edition with a dedication and three letters, in which he finds drastic words for Averroes's bad style *(obscurus, ieiunus, incultus, horridus atque omnino barbarus),* adding that whatever is convincing in Averroes, especially in his psychological works, is drawn from Simplicius.[26] Faseolo passionately appealed to his fellow scholars and students: "Make the decision to only read Simplicius, day and night . . . and take <Averroes> out of your hands!" *(de manibus deponite).*[27]

The same motives appear in those branches of philosophical literature that are influenced by humanist ideals. Marsilio Ficino, in a letter of 1484 and in the preface to his translation of Plotinus in 1492, launched a well-known attack against the two sects of the Peripatetics, which, he says, occupy almost the entire world: the Alexandrians and the Averroists. The psychology of both sects undermine religion totally, in particular because they deny divine providence; in addition, they desert their Aristotle.[28] In his earlier magnum opus, the *Platonica theologia,* Ficino had written a long chapter against Averroes's doctrine of the unicity of the intellect, arguing that Averroes, since he was ignorant of Greek, fell victim to the perversities of the Arabic translations of Aristotle.[29] Similarly, in the Christianizing and moralizing Aristotelian philosophy of Jacques Lefèvre d'Étaples, there was no place for the *impius Arabs.*[30] Lefèvre did not refute but ignored Averroes: he found him, just as Alexander of Aphrodisias, guilty of stirring up the intellectual world with his insanities.[31] The humanist philosopher Juan Luis Vives puts special emphasis on Averroes's ignorance of Greek when devoting a separate chapter, which we will analyze below, of *De causis corruptarum artium* of 1531 to an invective against Averroes.[32] There is no way to get around the conclusion that anti-Averroist attitudes were widespread among Renaissance humanists.[33]

Polemical criticism of Averroes was not confined to the humanist tradition. It was also advanced by scholars teaching and writing within a traditional Aristotelian context. The Paduan philosophers Nicoletto Vernia, Agostino Nifo, and Pietro Pomponazzi in their later years, while still using Averroes amply, wrote refutations of principal tenets of Averroes's philosophy and doubted his reliability as a commentator. The later Nifo was particularly harsh with Averroes: his Aristotle interpretation is faulty because of the defects of the Graeco-Arabic translation, and hence "Averroes interpreted hardly any word correctly."[34] The disdain for Averroes common among traditional scholastic philosophers is well apparent in the popular philosophical handbook *Margarita philosophica* of Gregor Reisch, which was written in the 1490s. Reisch justifies his polemics against Averroes as follows:

> Pupil: Why have you called Averroes damned and hallucinating *(maledictum et somniantem)?* Teacher: Because this man, mocking catholic truthfulness, has turned to the dog-like dishonesty of Muhammad and, even though in many cases has explained Aristotle's works in a not unlearned way, nevertheless has not only obscured the passage which treats the perpetual existence of the rational soul, but has also tried to corrupt this passage with reasons and arguments hardly stronger than the web of a spider, so much that he maintained that all human beings have one and the same intellect numerically.[35]

Criticism also comes from Thomist and Scotist theologians,[36] who of course could amply draw on Thomas Aquinas's refutation of Averroes's unicity thesis in *De unitate intellectus.* The most polemical of these was a treatise by Antonio Trombetta (d. 1517),[37] a professor of metaphysics in *via Scoti* at the Paduan faculty of theology, titled *Tractatus singularis contra Averroistas . . . ad catholice fidei obsequium.* Dedicated to Bishop Barozzi and printed in 1498, the treatise is written as an attempt to rescue those who insanely follow the path of Averroes by a clear demonstration of Averroes's erroneous and brainless position.[38] The paratexts that accompany Trombetta's publication cite Church authorities, Pietro Barozzi, Julian Cardinal of Ostia, and the inquisitor John Antonio of Padua.

The reaction of the Church officials to Averroes is more direct than in thirteenth-century Paris, when theses of Averroes were condemned, but the author was not named. On May 4, 1489, Bishop Pietro Barozzi of Padua and the inquisitor of Padua, Marco da Lendinara, issued an edict, the central passage of which is:

> We decree that none of you by threat of excommunication, which you undergo as soon as you act against this very sentence, dares or presumes to publicly discuss in any kind of question style the unicity of the intellect; even if this <thesis> had been <derived> from the opinion of Aristotle—according to Averroes, a clearly learned, but criminal man; by whose force also Avicenna was drawn into an argument—Avicenna from Seville, the most eminent of the physicians and, as many think, king of Bythinia, who was murdered by Averroes with poison, but before he died killed Averroes, as is reported.[39]

The decree's most prominent addressee was withouth doubt Vernia, who recanted in the following years; apart from writing an anti-Averroist treatise, Vernia also declared in his testament that he had once taught that Averroes's thesis of the unicity of the intellect was in agreement with Aristotle, but that he found later, after reading very learned Greek and Arabic authors, that this thesis deviates not only from catholic faith and the truth, but also from Aristotle: "I assert by God that I have never believed in this opinion."[40] In the same year 1499, Barozzi writes a letter to Vernia congratulating him on composing his anti-Averroist treatise, recalling that Vernia in his younger years "had led almost the entire Italy into error" with the unicity thesis (ut totam paene Italiam errare feceris).[41] In 1498, the inquisitor of Padua renews Barozzi's edict by declaring some of Averroes's theses heretical and, by threat of excommunication, forbidding anyone to maintain that they are true. One is allowed to maintain only that the theses belong to Averroes.[42]

The high point of the Church's reaction came in 1513, when the Fifth Lateran Council issued the bull Apostolici regiminis, on the basis of a draft produced by a commission of which Antonio Trombetta, enemy of the Averroists, was a member. Among the cardinals in charge was Domenico Grimani, the supporter of Averroes translations. The general of the Dominican order, Cajetan de Vio, protested against the idea that the philosophers should be forced to publicly teach the truth of faith and hence voted against the bull.[43] The diverging interests resulted in compromise formulations such as *pro posse* and *pro viribus*.[44] The bull condemned all those who asserted that the soul is mortal or that it is one in all human beings, and it decreed that such persons are to be punished as detestable heretics. In addition, it prescribed that all philosophers teaching such issues in schools and universities had to teach plainly and as persuasively as possible *(pro posse)* the truth of Christian faith, and to refute the opposing arguments of the philosophers to the best of their ability *(pro viribus)*.[45] The decree not only declared partisans of Averroes's theory of the intellect heretical, it also attempted to direct the philosophical teaching in the schools toward the teaching of the Church.

The image of Averroes, therefore, is marked by extremes. It is noteworthy that the biographical tradition, as surveyed in Chapter 2, portrayed Averroes for the most part positively, while the pictorial tradition is dominated by depictions of Averroes as a heretical philosopher. The positive picture transported in the biographies of the great commentator, praiseworthy physician, and *philosophus clarissimus*—as Jacopo Filipo Foresti put it—is blemished only by reports of his extreme jealousy of Avicenna. When Bishop Barozzi in the quotation above calls Averroes a criminal man and relates the legendary story of his murdering the physician and king Avicenna, he in all probability is quoting from Foresti's entry on Avicenna in the chronicle *Supplementum chronicarum* printed in 1483 and 1486: "It is commonly reported that Avicenna was the king of Bythinia and that he was poisoned by the physician Averroes, but that he killed Averroes before he died."[46] The motif of envy, in fact, reappears in many accounts of Averroes's life in the Renaissance, even if the murder is not related.

As to the pictorial tradition, a few images exist of Averroes as a turbaned philosopher; most notable is a very fine depiction in the Pierpont Morgan Library exemplar of the 1483 edition of Aristotle's *Opera,* which shows a splendidly gowned and turbaned Averroes in discussion with Aristotle. Averroes is sitting at Aristotle's feet, a book between his legs. A more fully attested tradition of representing Averroes is traceable in pictures of Averroes's defeat by Thomas Aquinas.[47] Averroes is shown lying below or huddling near the feet of Thomas Aquinas on a panel painting by Maestro del Biadaiolo (second quarter, fourteenth century), a big panel painting in Santa Caterina, Pisa (of ca. 1365), a fresco in the chapterhouse of the Dominican cloister Santa Maria Novella, Florence (of 1366–1368), a panel painting in the Dominican cloister San Marco, Florence (mid-fifteenth century) (Figure 4), a small panel painting by Giovanni di Paolo now in the

Saint Louis Art Museum (of ca. 1445–1450), Benozzo Gozzoli's panel painting in Pisa Cathedral, which is now in the Louvre (of ca. 1470),[48] Filippino Lippi's mural painting in the Carafa chapel of Santa Maria sopra Minerva, Rome (of 1488/1490—here Averroes appears in a lunette that depicts him as a nonbeliever in the miracle scene of Thomas Aquinas),[49] an anonymous altarpiece of the Dominican cloister of Syracuse (of ca. 1500), Mario di Laurito's altarpiece of Santa Rita, Palermo (first quarter of the sixteenth century), and on several woodcuts in editions of the early sixteenth century, where Averroes is shown crawling in front of Thomas Aquinas's teacher's desk (see Figure 5).[50]

I have listed these paintings of Thomas's triumph because they present a stable tradition of depicting Averroes as a heretic, which lasted about two hundred years. The tradition is obviously inspired by Thomas Aquinas's treatise *De unitate intellectus*. The context of these paintings is clerical and to a significant extent Dominican. On some of these paintings, Averroes is depicted together with well-known heretics, whom Thomas Aquinas has attacked in his writings. On the Santa Maria Novella painting, Averroes is flanked by the early Christian heretics Arius and Sabellius. On the panel painting in San Marco, Averroes sits between Sabellius and William of Saint-Amour, the thirteenth-century theologian and polemicist against the mendicant orders, who was accused of heresy (Figure 4). While these Christian theologians had long lost support in contemporary European culture, Averroes was a thinker with many supporters. The paintings, therefore, not only describe the triumph, they also appeal to the observer to follow Thomas's example and fight against contemporary heresy. There is an enormous tension between this appeal and that of Renaissance university professors who urge their students to trust Averroes, the *fidelissimus interpres* of Aristotle.

The above survey, then, has shown that Averroes was a subject of polemical dispute in the Renaissance,[51] a polemics that continued even in the seventeenth century.[52] There was fierce opposition against him by humanists, philosophers, and Church officials, and there was outspoken partisanship in his favor. The polemical arguments surveyed are sweeping and complex, but also historically indicative. On the one hand, they obscure our view of the technical philosophical discourse in the Renaissance, and on the other hand, they are the starting point for the historian's questions to the past. Later I will turn to the issue of whether Averroes as a commentator was dependent on or even "stealing" from his Greek predecessors and whether he was hampered by the shortcomings of the Greek–Arabic translations. I begin with his *fortuna* as a philosopher, inquiring into the

FIGURE 4. Triumph of Thomas Aquinas: panel painting in the Dominican cloister San Marco, Florence (mid-fifteenth century).

FIGURE 5. Triumph of Thomas Aquinas: Thomas Aquinas, *In decem libros Ethicorum Aristotelis profundissima commentaria* (Venice, 1526 or 1531), f. 1r.

reception of his major theses. For this purpose, a brief note on the historiographical label "Averroism" is in order.

## WHAT IS AVERROISM?

"Averroism" is a term used by modern historians to describe philosophical movements in Europe.[53] In contrast to many other historiographical labels such as "Augustinisme avicennisant," which are useful, but a modern invention, "Averroism" has a medieval Latin predecessor: the term *Averroista* (Averroist). Its first occurrence, as far as we can tell today, is in Thomas Aquinas's *De unitate intellectus* of about 1270 AD, a treatise that in the later manuscript tradition received the addition *contra Averroistas*.[54] Except for exceedingly rare cases, *Averroista* refers to a follower of Averroes, not to Averroes himself.[55] Some Renaissance scholars, for instance Nifo and Ficino, also speak of *Averroica familia*, *secta*, or *schola*, and thus give a historical grounding to the modern term "Averroism." In general, a historian may define his or her historiographical terms freely, as long as they are fruitful for analysis, but when the term has been used before in the epoch studied, a free definition is bound to create confusion. I am convinced, therefore, that our usage of the terms "Averroist" and "Averroism" should be tied to medieval and Renaissance usage—just as the term "humanism," if used sensibly, is tied to the Renaissance usage of *humanista* and *studia humanitatis*.[56] There does not yet exist an inventory of the occurrences of *Averroista* from the thirteenth to the sixteenth centuries, but two recent articles are helpful steps in this direction.[57] I will not go into the details of these surveys, but draw my conclusions from them.

*Averroista* is hardly ever used as a self-description—an exception is Pietro Pomponazzi who once speaks of "many Averroists, among whom I count too."[58] Normally, *Averroista* refers to a member of a group maintaining positions that the author does not share. Hence, inevitably, the term tends to have a neutral or negative connotation: the Averroist is a partisan of controversial theories by Averroes. For some authors, such as Raimundus Lullus, *Averroista* can be tantamount to a heretic philosopher. In the fifteenth and sixteenth centuries, the medieval usage of *Averroista* for partisans of contentious theses by Averroes continues to be dominant, but the term may also assume the positive meaning of "expert on Averroes."[59]

A whole set of theories is attributed to Averroists, predominantly in psychology, natural philosophy, and metaphysics, such as the theory of the unicity of the intellect, the thesis that matter does not enter the essence of things, the (Aristotelian) thesis of the eternity of the world, the denial of God's infinite power, the denial of God's knowledge of particulars, the theory that first matter is characterized by an indeterminate dimensionality coeval with first matter, and the theory of happiness as reached through knowledge of the separate substances. This set of theories is still too small and will be enlarged by future discoveries of passages containing the term *Averroista*. Also, it is not without exceptions: Raimundus Lullus ascribes many more doctrines to "Averroists," all of them somehow related to the Parisian condemnations of 1270 and 1277; only some of them can be attributed to Averroes.[60] And further theories on matter and elementary mixtures are ascribed to Averroists in the sixteenth century. In general, however, from the thirteenth to the fifteenth centuries, the theories of the Averroists usually belong to the mentioned smaller set of contentious or even heretical theses.

Medieval and Renaissance authors do not use *Averroista* for an author who simply relies on Averroes when interpreting Aristotle. They would not, for example, call Thomas Aquinas, Duns Scotus or John Buridan "Averroists" for the reason that their *Metaphysics* commentaries are based on Averroes as their principal source. To be sure, the historical notion of *Averroista* is not precisely limited, but it is nevertheless clear that the medieval and Renaissance *Averroista* is more than a reader of Averroes: he is a partisan of or an expert on Averroes. It is therefore misleading when historians use "Averroism" broadly for the reception of Averroes's thought.[61] Instead, I see two historically legitimate usages of "Averroism"— designating a movement of partisans or referring to expertism on Averroes. The first usage is based on a regular meaning of *Averroista* since the thirteenth century, and the second draws on the occasional positive connotation of the term in the sixteenth century. In the present study, I use the first definition, "partisanship of Averroes," which has the longer historical pedigree. "Averroism" thus differs from other historiographical labels such as "Avicennism" or "Augustinism," which are used broadly for the influence of Avicenna and Augustine. There are Avicennisms in Thomas Aquinas, but no Averroisms—at least, they have not yet been detected. The influence of Averroes in the Latin West is an ocean, Averroism is a small river.

It is problematic that research on the reception of Averroes has been much more substantial on psychology than on natural philosophy, metaphysics, and logic. This has created the misleading impression that Averroism is mainly concerned with the thesis of the unicity of the intellect.[62] On the other hand, the unicity thesis is the position most often attributed to Averroists, as far as we can see today, and it clearly is the most famous and most infamous thesis in the sense that it greatly influenced the popular picture of Averroes as heretic and was the main reason for the polemics against Averroes and his followers. Thomas Aquinas's *De unitate intellectus* was very influential in this regard, as the paintings of his defeat of Averroes show.

Now, did Averroist movements exist? Or were they only an invention of Averroes's enemies, as Fernand Van Steenberghen has argued with respect to the thirteenth century and Paul Oskar Kristeller with respect to the Renaissance?[63] Van Steenberghen and Kristeller do not see sufficient evidence for philosophers embracing a distinctive set of doctrines from Averroes and hence prefer to speak of "secular Aristotelianism" or of radical followers of Aristotle instead. In answering this question, it is sensible to differentiate between mere traditions of thought and philosophical movements. For a movement, more is required than that the protagonists share a doctrine or a philosophical practice. Movements should be characterized by some sort of group coherence, that is, the group members' activity in the same time and region, as well as personal relations between the members. At least, there should be an awareness of one's immediate predecessors and cognates in mind. For the field of psychology, it can be shown that these criteria are fulfilled by masters of arts in fourteenth-century Bologna: Taddeo da Parma, Angelo d'Arezzo, Matteo da Gubbio, and Giacomo da Piacenza. All of them hold the unicity thesis, which they hedge around with declarations of faith, and betray an awareness of and dependency on preceding Averroists. Likewise, these criteria are fulfilled for a veritable line of philosophy teachers in fifteenth- and sixteenth-century Italy, notably at Padua: Paul of Venice, Niccolò Tignosi, Nicoletto Vernia, Alessandro Achillini, Agostino Nifo, Luca Prassicio, Marcantonio Genua, Francesco Vimercato, and Antonio Bernardi. All of them hold the unicity thesis in at least one of their writings, even though some declare that it is in conflict with faith or they distance themselves from it in other writings.

Also, many of them are connected to each other by teacher–pupil relations. Tignosi was a student of Paul of Venice at Padua. Vernia was the student of Paul's pupil Gaetano da Thiene, and in turn was the teacher of both Nifo and Pomponazzi at Padua. Vimercato and Bernardi studied at Padua, and Genua was teaching there for over four decades. In light of this evidence, one can safely conclude that Averroist movements existed in late medieval and Renaissance Italy. This finding agrees well with the testimony of contemporary critics who refer to the movement's protagonists using the term *Averroistae*.[64] Note that this is a conclusion based on the evidence of psychological works only. It is possible that future research will detect groups with similar coherence holding other Averroist theses in natural philosophy or metaphysics.

## FOUR CONTROVERSIES ABOUT AVERROES'S DOCTRINE

Averroism as a movement—not the influence of Averroes as such, which is too complex to allow for judgment at the present stage of scholarship[65]—reaches its culmination in the Renaissance in the decades around 1500.[66] Statistically, this is difficult to prove since sources for the intellectual history of the Renaissance are much more exuberant than for the Middle Ages. But several specific features of the Averroist movement around 1500 indicate that it is stronger than ever: the popularity of the term *Averroista;* the extraordinary efforts in editing and translating Averroes; the composition of commentaries on Averroes's works; and several heated discussions among the members of the movement about its proper direction. I have discussed the first three indications before in this book,[67] and I now turn to the last, the rise of controversies between partisans of Averroes. Four such controversies will briefly be described: between Trombetta and Nifo (1497/1498), between Nifo and Zimara (1497/1508), between Pomponazzi and Nifo (1516/1518), and between Nifo and Prassicio (1518/1521). Here I summarize the findings of a previous study,[68] which continues research done by Edward Mahoney.

The debates have a focal point in John of Jandun, the Parisian master of arts of the early fourteenth century.[69] His commentaries are a model for many generations of Italian philosophers in the fourteenth and fifteenth centuries, who adopted not only the format of the commentary from John of Jandun but also many interpretations of Aristotle, and with them, also of Averroes. But John of Jandun's interpretations are challenged by other experts of Averroes at the end of the fifteenth century. Nicoletto Vernia lauds John of Jandun as Averroes's "best defender,"[70] but also criticizes

him for positions that are not *ad mentem Averoys*.[71] In the generation of
Vernia's pupils, John of Jandun becomes the target of heavy attacks, for
instance, by Giovanni Pico della Mirandola, who charges John with cor-
rupting the doctrine of Averroes on almost all philosophical matters.[72]
Agostino Nifo is the most fervent critic of John of Jandun: "This man's
deficient knowledge of the works of Averroes makes him commit errors."[73]
Nifo's principal objection is that John of Jandun does not understand the
*littera,* the wording, of Averroes's texts,[74] and that the fame he enjoys is unwarranted: "John of Jandun has been the cause for many errors, because he
is so famous, to such a degree that nobody was thought to be an Averroist
if he was not a *Gandavensis*."[75]

The first of the four controversies arose because Nifo challenged John
of Jandun's theory of intelligible species for not being in accord with Averroes.[76] Averroes's true position, Nifo claims in his early commentary on
Averroes's *Destructio,* is that the quiddity of a thing is known without the
mediation of intelligible species.[77] This reading of Averroes, however, is attacked by the Scotist Antonio Trombetta, who defends the theory of intelligible species and attributes it to Averroes: "The possible intellect has an
intelligible concept only by receiving it from the phantasm, as is held
unanimously by all those who understand correctly the opinion of the
Commentator . . . Those *(others)* are perverting the opinion of the Commentator as well as that of the Philosopher, and, what is most pernicious,
they are stirring up the most pestilential errors against faith and truth."[78]
It is interesting to see that Trombetta, for all his anti-Averroism, does not
blame Averroes himself, but his badly informed interpreters for promul-
gating heretic doctrines.

The second controversy was a follow-up to the first.[79] Sometime before
1508, a third Paduan intellectual entered the controversy with a *Quaestio* on
the topic: Marcantonio Zimara, who later became well-known for his above-
mentioned handbooks on both Aristotle's and Averroes's doctrines. In the
line of Trombetta, Zimara attacks the pseudo-experts on Averroes who deny
that Averroes holds the theory of intelligible species. "All ancient Averroists,
such as Albertus Magnus" agree on this, he claims.[80] It is very likely that
Zimara's target is Nifo, whom he also criticizes on other issues of the interpretation of Averroes, especially on intellect theory.[81]

The third and fourth controversies concern Averroes's unicity thesis, to
which I will return later. Here I want to draw attention to the fact that the
controversy about the unicity thesis is concerned not only about whether
Averroes was right or wrong but also about how to interpret Averroes correctly. The opponents in this controversy about the correct reading of
Averroes refer to each other by name: Pietro Pomponazzi, Agostino Nifo,

and Luca Prassicio. In 1516, Pietro Pomponazzi published his provocative and much-read treatise *De immortalitate animae,* which, as part of the argument for the mortality of the soul, also contains a rejection of Averroes's unicity thesis. In 1518, Nifo reacted with his own *De immortalitate libellus,* in which he counters, among other things, Pomponazzi's arguments against Averroes.[82] Pomponazzi misunderstands Averroes, Nifo says, as holding that human beings consist of a double soul: an immortal intellective soul, which is one for all human beings, and a mortal *anima sensitiva.* This, he continues, is not Averroes's opinion, but that of John of Jandun.[83] Nifo proceeds to demonstrate that John of Jandun's interpretation is very controversial by pointing to the many diverging positions of *Averrois sectatores* and *interpretes* on the issue, and he surveys the different interpretations of Siger of Brabant, John Baconthorpe, and Thomas Wylton.[84] The correct interpretation of Averroes is of the greatest concern to Nifo, even though he had long stopped being an adherent of the unicity thesis itself. Nifo's main target is John of Jandun's claim that for Averroes the intellective soul is not the *forma informans* of the body, as would be correct, Nifo claims, but a form assisting the body.[85]

The fourth controversy arose because Luca Prassicio, an outspoken Averroist, reacted to Nifo's text in 1521. Prassicio, to whom I shall return below, does not miss any occasion to criticize Nifo for misunderstanding Averroes: When Nifo argues that there are two kinds of intellection according to Averroes, Prassicio replies that Nifo reads Averroes with the eyes of Thomas Aquinas.[86] When Nifo claims that the proper activity of the human soul is to know God, Prassicio censures this as a diversion from Averroes's theory of conjunction with the active intellect.[87] When Nifo maintains that he is the first to interpret Averroes correctly on the unicity of all celestial spheres in one primary sphere, Prassicio criticizes this as a brainless mistake.[88] The harshest attack on Nifo comes on Averroes's theory of the union of soul and body:

> It is astonishing that Agostino <Nifo> holds that the intellective soul is the lowest of the intelligences and that he maintains that there is an additional union in <the intellective soul>; this means nothing else but breaking heretically with the doctrine of Averroes *(non est aliud nisi apostetare in doctrina Averrois)* . . . <Averroes's true> position in fact was admirably swallowed and tasted by John of Jandun. What a madness, what a melancholic humor or spirit would have gripped Averroes, had he designed such an obvious nonsense: that the intellect or the intellective soul, as the lowest of the intelligences, which is essentially and in reality separate and incorruptible, would be united as a form <to the body>, as is

claimed by Agostino <Nifo> ... From Averroes's words very clearly emerges that according to him the soul in no way is united with us in a formal and univocal way, as Agostino <Nifo> maintains, who together with Siger of Brabant and Roger <Bacon>[89] has to be condemned on this matter.[90]

Prassicio continues to censure Thomas Wylton's and John Baconthorpe's positions, as cited by Nifo. Prassicio presents Thomas Wylton as holding that the intellect is a *natura communis* related to all individuals, which is a position he rejects as Platonizing.[91] And he argues that Baconthorpe's idea of a double conjunction *(copulatio bifaria)*[92] between intellect and body is compatible with Christian religion, but not with Averroes. Prassicio instead adopts John of Jandun's interpretation that the soul is united to the body only *per assistentiam ad phantasmata.*[93]

These are only four of doubtlessly many more controversies among Renaissance philosophers about how to interpret Averroes correctly. It would be wrong to believe that such controversies developed only in the Renaissance. In fact, Nifo and his colleagues continued a tradition that began in the early fourteenth century—if not earlier—when John of Jandun censured anonymous interpreters of Averroes for misunderstanding Averroes on the relation between intellect and body.[94] On the same issue, Antonius of Parma attacked Thomas Aquinas and Giles of Rome for misunderstanding Averroes,[95] and John Baconthorpe openly criticized Thomas Wylton.[96] The scale of the Renaissance discussion, however, is entirely different. Prassicio, for instance, does not simply give references to earlier interpretations; he enters into a heated debate with his contemporary Nifo. Prassicio's critique of Nifo pervades his entire treatise, and it concerns many details of Averroes's intellect theory, not only the problem of substantial form. In a similar manner, Nifo's critique of Pomponazzi's understanding of Averroes covers many folios of his treatise on immortality.

In what sense do these controversies witness to the liveliness of Averroism as a movement? It is true that some of the protagonists were not Averroists with respect to intellect theory: Trombetta, Zimara, the older Pomponazzi and the older Nifo did not adopt the unicity thesis. It is clear, however, that the debate was also so fierce because the young 'Sigerian' Nifo and the young Pomponazzi had been followers of Averroes on the unicity of the intellect. It was the young Nifo who prompted the response by Trombetta and Zimara, and it was Pomponazzi's desertion of Averroes that also motivated Nifo to write his reply. Most indicative are Nifo's protest that "nobody was thought to be an Averroist if he was not a follower of John of Jandun *(Gandavensis)*!" and Prassicio's exclamation: "this means nothing else but breaking heretically *(apostetare)* with the doctrine of

Averroes!" Prassicio's condemnation of Nifo as a heretic from the Averroist cause is a truly Foucault-like procedure of exclusion. Against this kind of attitude Pietro Pomponazzi protests: He does not criticize Averroes or anybody else for the sake of contradicting, but because "whoever wants to find the truth, needs to be a heretic in philosophy"—meaning, deviate from the mainstream position of the Averroists.[97] These testimonies show that Renaissance Averroists were fighting not only about the true *positio* or *doctrina* or *mens Averrois,* but also about who deserved to be called an "Averroist" and who was the true leader of the movement. In addition to all the other evidence, this makes Renaissance Averroism a very lively movement. The significance of Averroism should therefore not be diminished to a mere continuation of a decrepit scholastic tradition,[98] or to Averroes's role as a secondary source on Aristotle in university teaching.[99]

## UNDER PRESSURE: THE UNICITY THESIS

This chapter is concerned with the most controversial thesis of Arabic philosophy in the Renaissance: Averroes's theory of the unicity of the intellect, or, more precisely, of the material intellect. Many major Renaissance thinkers supported the unicity thesis, from Paul of Venice in the early fifteenth century until Antonio Bernardi in 1562. There are many causes for the success of Averroes's thesis. One of them is an ambiguity in Aristotle. Aristotle's theory of the rational soul, as developed in his *Peri psyches (De anima)* and his zoological and metaphysical writings, has a double tendency: in some passages, Aristotle emphasizes the dependency of the rational soul on the body, for instance, in the well-known definition of the soul as the form of a natural body that has life in potentiality (412a19–20), or in the explicit statement that the soul is not separable from the body (413a3), or in the epistemological theory that the soul never thinks without an image (ἄνευ φαντάσματος, 431a16–17). In other passages, however, Aristotle stresses the separability of the rational soul from the body, as in the theory that the intellect alone is divine and enters from outside (*Generation of animals,* 736b27–28, cf. 408b18–19), or in the remark that the soul may be related to the body as a sailor to the ship (413a3–9), or in Aristotle's categorical statement in *De anima* Γ.4—which was particularly important for Averroes—that the thinking part of the soul has to be unmixed (ἀμιγής), pure capacity, and without blending with the body (429a18–25). In modern accounts of Aristotle's hylomorphism, these latter passages, which stress separability, tend to be downplayed, but in the Greek and Arabic commentary tradition, they were very influential.[100]

Averroes belongs to a separabilist tradition of understanding Aristotle's theory of the human intellect. Just as many philosophers before him, Averroes adopts Aristotle's distinction, developed in *De anima* Γ.5, between an active and a potential/material intellect, and identifies the active intellect with the lowest of the celestial intelligences.[101] The material intellect therefore is the human intellect. In his Long commentary on *De anima,* which was translated by Michael Scot, Averroes argues in the famous comment 3.5, sometimes called the *digressio magna,* that the material intellect also has to be one and eternal. Averroes here comments on Aristotle's doctrine in *De anima* Γ.4 that the intellect is unmixed with the body. It is important to note, however, that comment 3.5 is only one of many utterances by Averroes on intellect theory, and that many of his psychological writings were not available in Latin.[102] The Hebrew reception of Averroes's psychology, for instance, was to a large extent determined by the Middle commentary on *De anima,* in which Averroes does not contend that the material intellect is incorporeal and unique.

In the *digressio magna* Averroes develops his own position by criticizing earlier commentators, in particular the ancient Greek commentators Themistius and Alexander of Aphrodisias. Themistius, a proponent of a separabilist interpretation, is criticized by Averroes for holding that both the material intellect and the grasped intelligibles are eternal. Alexander of Aphrodisias, a naturalist interpreter of Aristotle's psychology, is rejected for maintaining that the human intellect is generated and corruptible.[103] Averroes's own position starts with the assumption, shared by Themistius, that for Aristotle the material intellect is pure potentiality to receive intelligible forms, and therefore must be incorporeal and eternal.[104] The material intellect is the ontological place and receiver of the intelligible forms, but not the medium through which the human being is joined to the intelligible. This role is taken by the actualized imaginative forms (the phantasmata): we grasp the intelligibles via the faculty of imagination.[105] Hence, in contrast to Themistius, Averroes insists that the intelligibles are grasped by each single individual insofar as they have their epistemological basis *(subiectum)* in imagination. They are eternal only with respect to their ontological basis *(subiectum),* the eternal and unique material intellect, which is their incorporeal receiver.[106] The *duo-subiecta* theory ensures that Averroes avoids the absurd consequence that Themistius was facing: that all men know if one man knows.[107] For Themistius had assumed that the actualized intelligible is the result of the actualization of the material intellect through the active intellect, that is, without involvement of the phantasmata[108]—and not, as Averroes assumes, the result of the actualization of the phantasmata in imagination.

In two previous publications,[109] I have studied the philosophical attraction that this theory exerted on six Renaissance authors: Niccolò Tignosi, Nicoletto Vernia, Alessandro Achillini, Agostino Nifo, Luca Prassicio, and Francesco Vimercato. My aim was to show that the unicity thesis was successful in the Renaissance for philosophical reasons. The thesis was not "the most flagrant nonsense," *maxima fatuitas,* as Pomponazzi put it,[110] but could convince philosophically in several respects. The unicity thesis takes seriously Aristotle's claim that the human intellect is unmixed with the body and explains how intellectual knowledge can be truly universal, even though it is grasped individually. If the comprehended intelligible did not reside in a universal intellect, the partisans of Averroes argue, it would be impossible that a pupil learns from a teacher.[111] Also, if the intellect was many, it would follow that the intellect, due to its individuation, was a faculty in the body and could not grasp universal intelligibles—true intellectual knowledge would be impossible.[112] The strength of this last argument was admitted by Thomas Aquinas, who remarked "that in this argument Averroes seems to have laid the principal force *(praecipuam vim)* <of his theory>."[113]

In what follows, the philosophical reception of the unicity thesis will not be surveyed again—it has been studied in detail by Bruno Nardi, Eckhard Kessler, and others.[114] Rather, in line with the principal tenor of the present book, I will be concerned with the relation between polemics and technical discourse. Do the polemics pro and contra Averroes influence the philosophical discussion? Or, in other words, do nonphilosophical motives influence philosophical solutions? One can easily surmise that at least three nonphilosophical factors—religion, humanism, and scholasticism—had an impact on the reception of the unicity thesis: its compatibility with Christian faith, its rootedness in Greek sources, and its verification as truly Aristotelian. All three factors are nonphilosophical, or even ideological, in the sense that they are irrelevant for the quality of a philosophical argument or a philosophical idea.

A second issue of the following analysis is the influence of the Greek commentators on intellect theory. Themistius's paraphrase on *De anima* was translated into Latin by Ermolao Barbaro in 1472/1473; his pupil Girolamo Donato translated Alexander of Aphrodisias's commentary, book 1, which was printed 1495; and in 1543 Giovanni Faseolo published his translation of Simplicius's *De anima* commentary. In the same decades, philosophers became increasingly interested in and capable of reading Aristotle in Greek (see Chapter 1). We have seen above that this development was accompanied by polemics against Averroes and by the demand to return to Greek Aristotelianism. It is unclear, however, whether the accessibility of the Greek Aristotle and of the Greek commentators significantly influenced the philosophical discussion of Averroes's unicity thesis: some modern scholars say it made Aristotelians turn against Averroes,[115] others warn that Renaissance thinkers should not be trusted too easily when boasting about their knowledge of Aristotle in Greek and their reading of the Greek commentators.[116]

### Paul of Venice, Gaetanlo da Thiene, Niccolò Tignosi

Our investigation begins with a founding figure of Renaissance Aristotelianism: Paul of Venice (d. 1429), professor of arts at Padua, who had studied in Paris and Oxford.[117] There is no doubt that Paul of Venice, as a promoter of the unicity thesis, can be called an Averroist.[118] In his *Summa naturalium* of 1408, he holds, among other arguments, that the unicity thesis is the only Aristotelian way to account for Aristotle's statement that "the intellect comes from outside" *(intellectus venit de foris),*[119] and that the human intellect, if material, would not be comprehending universals.[120] Paul does not accept Averroes's theory uncritically, but insists contrary to Averroes that the intellect is the substantial form of the body.[121] Paul of Venice in later years composed a commentary in question format on Aristotle's *De anima* that, in one question, repeats the same arguments in favor of the unicity thesis.[122] But in the next question Paul declares and supports with arguments that "under the terms of the standpoint of faith *(secundum opinionem fidei)* the intellect is multiplied according to the multiplicity of the individuals of the human species."[123] Paul of Venice here borrows a practice from fourteenth-century Averroists, such as Taddeo da Parma, Angelo d'Arezzo, and Giacomo da Piacenza, who insert a conventional declaration of faith after concluding that the intellect is one in all human beings, for example: "even though this is against faith" *(licet hoc sit contra fidem).*[124] Note that these declarations do not have an effect on the philosophical solutions of either the Bologna masters of arts or Paul of Venice. The solution remains philosophical, while the declaration of faith serves to signal distance as a Christian believer.

Paul of Venice's student and successor, Gaetano da Thiene (d. 1465) is an example not of an Averroist—since he explicitly disagrees with the unicity thesis[125]—but of the presence of the unicity thesis in the classroom. In his commentary on *De anima* (dating from 1443),[126] Gaetano gives a lucid summary of the two long digressions of Averroes's comments 3.5 and 3.36, which apparently played a major role in teaching. Gaetano explicitly aims at a presentation adequate for the introduction of beginners *(ad iuniorum introductionem).*[127]

Niccolò Tignosi (1402–1474) received his philosophical and medical education in Bologna and Perugia, but called himself, just as Gaetano, a

pupil of Paul of Venice.[128] For many years, Tignosi taught natural philosophy at the Studio di Pisa, where Marsilio Ficino was among his students.
He was aquainted with the humanist currents of his time and chose the
new translation of *De anima* by John Argyropulos as the textual basis for
his commentary, but his interests and works are inspired by the tradition
of scholastic Aristotelianism. Tignosi shows great respect for "the most
famous" Averroes,[129] whom he praises as *subtilissimus commentator,*[130] and
also for Albertus Magnus—a combination of authorities common among
Renaissance Aristotelians.[131] The *De anima* commentary is his last work,
finished some months before his death in 1474.[132] In this work, the conventional religious caveat takes on a new role. When addressing the question
of whether there is only one intellect for all human beings, Tignosi answers that "according to the standpoint of faith" *(secundum opinionem fidei)*
the intellect is many.[133] Here he claims to follow Paul of Venice, whom he
quotes stating the following methological position (from Paul's late commentary on *De anima*):

> All of these <arguments against Averroes> may give rise to the most
> serious objections, and it would perhaps not be difficult to sustain the
> opposite of these arguments, but the faith of the Christian insists on a
> different attitude, which is most true according to conscience. For
> those who argue on the basis of natural philosophy *(loquentes natural-
> iter),* the plurification of the intellective soul can only be held with great
> difficulties.[134]

Tignosi admits that the philosophical arguments against Averroes and in
favor of the *plurificatio* thesis are weak; "speaking as a natural philosopher"
*(naturaliter loquens)* one can only hold Averroes's thesis of the unicity of the
intellect. But in contrast to Paul of Venice, Tignosi does not decide the
matter philosophically in the solution. The solution he states is *secundum
opinionem fidei,* in spite of the fact that Tignosi is writing a philosophical
treatise.

   Tignosi proceeds similarly when coming to the related question of
whether the soul is *forma substantialis* of the body. He discusses at length
Averroes's position (read through the lens of John of Jandun) that the intellect is joined to the body not as form but through the phantasmata that
are intelligible in actuality,[135] mentions a series of counterarguments, and
finally settles the question with Thomas Aquinas's theory of substantial
form.[136] This solution, however, is seriously undermined by Tignosi's remark that *naturaliter loquens* the arguments against Averroes can easily be
refuted and that Averroes's position can thus be sustained.[137] Hence,

Tignosi's reliance on Thomas Aquinas seems influenced by nonphilosophical motives: Aquinas is invoked as a Christian authority in order to safeguard the individuation of the human intellect as the substantial form of the body and thus to safeguard personal immortality. There are further indications in Tignosi's commentary that he was a convicted partisan of Averroes's intellect theory, for instance, when he assumes that the active and the possible intellects are ungenerated and incorruptible, that is, eternal.[138] This evidence indicates that Niccolò Tignosi was a closet Averroist, who streamlined the solutions of his *De anima* commentary in order to make them compatible with Christian faith.

### Nicoletto Vernia

Nicoletto Vernia (1420–1499) was a student of Gaetano and in 1468 became his successor in Padua.[139] Unlike his teacher, Vernia was an outspoken Averroist,[140] who defended the unicity thesis not only in the classroom but also in writing, in particular in a *Questio* that dates to around 1480 and is titled: *Utrum anima intellectiva . . . eterna atque unica sit in omnibus hominibus.*[141] As was mentioned above, Bishop Pietro Barozzi of Padua issued a decree in 1489 "that no one under pain of excommunication . . . dares or presumes to publicly discuss in any kind of question style the unicity of the intellect." Vernia was without doubt the principal target of this decree. One year before, Barozzi had postponed a Church benefice for Vernia, apparently in an attempt to hurt Vernia financially.[142] Vernia recanted and in 1492, already an old man of seventy-two years, composed a treatise "against the perverse opinion of Averroes about the unicity of the intellect" *(Contra perversam Averroys opinionem de unitate intellectus),* which he prepared for print in 1499 (the year of his death) but appeared only in 1504. The printing was accompanied by testimonials of orthodoxy by three theologians, which Vernia himself had requested, and by a letter of approval from Bishop Barozzi.

In the Averroist *Questio* of 1480, Vernia is not ignorant of the position of Christian faith. The third and last section of his *Questio,* Vernia says at the beginning, will deal with the Latin commentators—presumably Thomas Aquinas, Albertus Magnus, and others—and the true doctrine of Christian faith. But this section is missing in the manuscript, and perhaps was never written. Of the exant text, the first, very short section is on Plato's position, and the second on Aristotle according to the best Greek and Arabic commentators. In the latter section, Vernia defends the unicity thesis of *acutissimus Averroes* against a series of counterarguments. It is remarkable that Vernia signals from the very beginning of his treatise that he is going to explain the *fidei Christianae veritas* on the issue. He thus continues the conventional Christian declaration of faith that had accompanied Averroist positions since the Bologna masters of arts in the fourteenth century. But apparently, this was not enough for Bishop Barozzi. The bishop decrees that it is forbidden to discuss the unicity thesis in public disputations, even if it should turn out to be Aristotle's opinion.[143] This clearly is a new form of pressure on the masters of arts at the University of Padua.[144]

In Vernia's treatise of recantation *Contra perversam Averroys opinionem* (ca. 1492), the conclusion has the style of a confession:[145]

> I say in accordance with the holy Roman church and with the truth that the intellectual soul is the substantial form of the human body, giving it existence formally and intrinsically, that it is created by God and infused into the human body, and that it is multiplied in the bodies according to their number ... And I do not only believe all these statements on the basis of faith alone, but I also say that all this can be proved in natural philosophy *(physice)* and that one can state these things in accordance with Aristotle.[146]

For a late medieval and Renaissance master of arts, this is a very unusual statement, both in style and content, since it claims an astonishing identity of religious doctrine, philosophical truth, and Aristotelian exegesis—an identity on issues that have been subjects of complex disputes since the early thirteenth century. These doctrines derive from Thomas Aquinas's theory of the soul and are formulated in his language (as in the expression *forma dans esse*), but they cannot as such be found in Aristotle's works. Vernia himself admits at the opening of the treatise that Aristotle had not explicitly settled these issues in his *De anima*. Vernia therefore turns to the principles of Aristotle's philosophy and tries to show that they are not in conflict with the doctrines stated. The unusual conclusion, therefore, clearly reflects the pressure exerted by the bishop.

It is a noteworthy feature of Vernia's recantation treatise that, other than in the earlier *Questio,* several Greek commentators are cited. The "famous Peripatetics" Theophrastus, Alexander of Aphrodisias, Alfarabi, Themistius, and Simplicius all hold, claims Vernia, that the soul is the substantial form of the body, giving it existence, that the human souls are created by God and that they are individuated. Hence all Greeks and the most eminent Arabs Alfarabi, Avicenna, and Algazel disagree with Averroes. Averroes's only support, Vernia says, are "Albubacher et Avempache"—that is, two Arabs mentioned in Averroes's *digressio magna,* who are actually the same person: Abū Bakr ibn Bāǧǧa.[147] In his testament of 1499, Vernia again mentions that it was "the most learned Greek and Arab scholars" who convinced him that the unicity thesis was not, in fact, Aristotle's teaching.[148]

Much has been made of this appearance of the Greek commentators in Vernia's treatise: a "major turning point" (as Edward Mahoney put it), indicating a truly Renaissance shift toward the Greek commentators as the true guide to Aristotle.[149] I do not think this is true. In fact, it would have been a sad turning point. The doctrines attributed to the Greek and Arabic commentators are not their historical positions, which in fact differ considerably from each other, but those of Thomas Aquinas. The principal function of the mentioning of the Greek commentators is to give additional weight to the orthodox position. Vernia's main inspiration and source is not the new humanist translations by Barbaro and Donato, but Thomas Aquinas's *De unitate intellectus*. Vernia himself mentions that he is quoting an argument against the unicity thesis by Themistius "which is put forward by the holy doctor against the Averroists" *(quam adducit sanctus doctor contra averroistas)*.[150] Thomas Aquinas, in the second chapter of *De unitate*, discusses the psychological positions by Themistius, Theophrastus, Alexander of Aphrodisias, Avicenna, and Algazel, and concludes that "also the Greeks and Arabs were holding that the intellect is a part or faculty or power of the soul, which is the form of the body."[151] Note that Thomas claims less than Vernia: he does not, for instance, attribute the doctrine of God's creation of the soul to the Peripatetics. Thomas, who had his own problems with the Parisian bishop, was under considerably less pressure than Vernia. Hence, to mount the Greek commentators against Averroes is a truly Thomistic tactic and not yet a symbol of a new humanist era.

Vernia, remarkably enough, finds a way to introduce Averroes through the back door. At the end of his treatise, Vernia cites Albertus Magnus's *De intellectu et intelligibili* when answering the question of how the intelligible forms remain universal in an individual intellect:

> The intellect is individual insofar as it is something belonging to the nature of the soul. But insofar as it sends out acts of intellection, it is in a universal power and thus receives universals, which are not in the intellect like an accident in a subject or a form in matter, and therefore <are> not made individual through the subject.[152]

Vernia holds with Albertus that the intellect is individual only in one sense, and universal in another. This smacks suspiciously of Averroes's *duo-subiecta* theory. If Barozzi—or the three theologians who examined Vernia's recantation treatise and approved of the complete extinction of

Averroes, the enemy of Christ[153]—had opened Albertus's *De intellectu,* they would have been pleased to see that the passages form part of a refutation of Averroes's unicity thesis.[154] But if they had also followed Albertus's reference to his earlier *De anima* commentary, they would have been surprised to find out that the passage quoted by Vernia recalls an "Averroist" position by Albertus,[155] as Albertus himself remarks in his *De anima:* "In this matter, Averroes, in his commentary on *De anima,* agrees with us."[156] Albertus holds that the *intellectus speculativus,* that is, the actually grasped intelligible, is individual and corruptible with respect to the particular form, whereas with respect to the possible intellect it is incorruptible and universal. "And hence comes the solution to the first question of the three about how the active intellect and likewise the possible intellect can be a permanent substance and how the speculative intellect can be generated and changeable."[157] This indeed is Averroes's solution:[158] The grasped intelligible (the speculative intellect) has two subjects, one particular, namely, the phantasma in the faculty of imagination, and one universal, namely, the material intellect, which is unique and eternal. In these passages, Albertus Magnus clearly endorses an Averroism avant la lettre. Albertus's *De anima* was written in 1254–1257, his *De intellectu* around 1258, that is, about a dozen years before the term *Averroista* appears for the first time in Thomas Aquinas's *De unitate intellectus.* Here, as in other cases, Albertus was a source for ideas that later became known as "Averroist."[159] Agostino Nifo, Vernia's pupil, was therefore right when he complained that Vernia's *Contra perversam* smacks of the idea that "somehow there could be a unicity of the intellectual soul for different bodies."[160]

Vernia's quotation of Albertus's Averroist theory of universal intellection does not, of course, turn his entire treatise *Contra perversam Averroys opinionem* into a defense of the unicity thesis. But the passage points to the secret success of the unicity thesis after its condemnation through Barozzi. The thesis did not need to be dropped once it was camouflaged as good Albertist doctrine. In view of the confession-like and philosophically simplistic solution of the treatise and in view of the clandestine Averroist coloring of its epistemology, it is very probable that Vernia was writing under great pressure. Pressure, and not the Greek commentators—which would not have led him to Thomas Aquinas's doctrines anyway—made him turn against Averroes. It is understandable that historians have wrongly stylized Vernia as a pioneer of Greek Aristotelianism in order to rescue his reputation as a sincere philosopher who is driven only by philosophical motives and not opportunistically complying with the Church. But this has closed our eyes to the most likely interpretation that Vernia was neither an opportunist,[161] nor a freethinker, nor a humanist pioneer, but an ordinary philosopher and a victim. The violent intervention of the bishop and the inquisitor broke with a long tradition in which the Italian masters of arts could pronounce the unicity thesis as long as they declared that it conflicted with the truth of Christian faith. There are indications that Vernia's sad case impressed the next generation of Aristotelian philosophers.

### *Alessandro Achillini*

Alessandro Achillini, born forty-three years after Vernia in 1463,[162] was a young professor of philosophy at the university of Bologna, when Barozzi banned the public teaching of Averroes's unicity thesis in Padua. In 1494, he publishes his main psychological work, the *Quolibeta de intelligentiis* in Bologna. Later scholars, such as Pomponazzi, Zimara, and Vimercato, refer to Achillini as a partisan of Averroes and his unicity thesis.[163] When one turns to the content of *Quolibeta,* however, one realizes that Achillini distances himself from Averroes's thesis.[164] His conclusion on the issue is that according to Aristotle, every human being has the same intellect, but that this is not true *(non est vera).*[165] And Achillini explicitly says in the following discussion that he does not share Averroes's fundamental thesis, the unicity of the intellect.[166]

One reason that Achillini was nevertheless regarded as an Averroist is that he argues forcefully in favor of Averroes's interpretation of Aristotle, holding that the unicity thesis can be attributed to Aristotle himself. The other reason is that Achillini's arguments against the unicity thesis remain brief and hardly convincing, while his arguments pro Averroes are formulated with much diligence and persuasive power. Achillini's focus is on epistemological arguments, and he leaves no doubt that Averroes's thesis offers a strong solution to the problem of how individual people acquire truly universal knowledge and how a teacher and a pupil may share this knowledge.[167] If the intellect were many, it would not be *cognoscitivus universalium* ("able to know universals"), argues Achillini with a phrase he adopts from Paul of Venice.[168] It is noteworthy that Achillini proceeds in a similar manner four years later when discussing the problem of the eternity of the world in his treatise *De orbibus.* He states that Aristotle's eternity thesis is not true, but lays out the arguments in favor of Aristotle's position at great length.[169]

Achillini does not address the compability or incompatibility of the unicity thesis with Christian faith directly in the *Quolibeta,* which remain a purely philosophical work, but in a later treatise, the *De elementis* of 1505:

> You may ask: How does Aristotle's opinion accord with the faith? For according to natural reason, either the intellect is one, as Averroes

understood Aristotle, or the intellect is many <and> its existence has a beginning, as Alexander of Aphrodisias meant. But none of these <opinions> is in accordance with the faith. <My> answer is: this is the reason why I have said that it is necessary to relinquish Aristotle, while choosing among these two false opinions the one which is more probable: the opinion of Averroes.[170]

Achillini stays with his previous conviction that the unicity thesis is the most probable interpretation of Aristotle. But now, instead of arguing philosophically against it, he explicitly says that it is in conflict with faith. There is much reason to suspect, therefore, that the true motive for Achillini's opposition against the unicity thesis in the *Quolibeta* was not its philosophical weakness, but its falsity from the vantage point of Christian faith. This would explain his lukewarm philosophical refutation of the unicity thesis. From an inquisitor's point of view, there was nothing to incriminate in Achillini's work, except that his interpretation of Aristotle is still too pagan. But for the philosophical readers and students of Achillini, there was much Averroism to be found lurking below the surface of the *Quolibeta*. That the unicity thesis was hidden, rather than defended openly, may well have been a repercussion of the Vernia affair.

### Agostino Nifo

For Agostino Nifo (1469/1470–1538), a student of Vernia in Padua,[171] Averroes was without doubt the most important thinker in his long and productive life as a philosopher. From his early commentary on Averroes's *Destructio destructionum,* printed in 1497,[172] until his late commentary on Aristotle's zoological writings, finished in 1534, for which he uses Mantino's recent translation of Averroes's Middle commentary, Nifo remained a diligent reader and user of Averroes's works. He was involved in publishing a two-volume reprint edition of Aristotle's *Opera* together with Averroes's commentaries, which appeared in 1495–1496.[173] He wrote commentaries on Averroes's treatises *Destructio destructionum, De substantia orbis* and *De animae beatitudine,*[174] but his most impressive contribution to the study of Averroes is his exegesis of Averroes's Long commentaries: Nifo's commentaries on Aristotle's *De anima, Physics, De caelo,* and *Metaphysics* in many passages amount to super-commentaries on Averroes.[175] Agostino Nifo was doubtless one of the greatest Averroes scholars of all times, even though he could not read Averroes in Arabic *(in lingua sua)*—a deficiency he acknowledged himself.[176]

In spite of all this, Nifo's attitude toward Averroes is exceedingly complex. While the early commentary on *Destructio destructionum* contains the most high-flown praises of Averroes,[177] later works defy Averroes as a "great drunk" *(magnus temulentus)* who does not know Greek and hence mistakes Aristotle on almost everything.[178] Nevertheless, Nifo continues to defend Averroes against misguided interpretations of his commentaries by other philosophers, such as Pomponazzi and Zimara, as we have seen before.

Nifo's attitude toward Averroes's unicity thesis is also complicated and marked by heavy revision. Nifo's main psychological work is the well-known *De intellectu,* which was first printed in 1503, but completed in August 1492, when Nifo was twenty-two or twenty-three years old and was appointed extraordinary professor of philosophy at Padua. The 1503 edition contains an important document, the dedication to the respected Venetian statesman and humanist Sebastian Badoèr. The dedication is undated, but must have been written before Badoèr's death in 1498. Here Nifo writes:

> Lest the labors of my youth perish, I would have cared to publish the question on the intellect if there had not been adversaries *(aemuli)* of mine who had accused me of heresy, and I preferred to wait until today rather than to incur the charge of such a misdeed *(huiuscemodi criminis culpam subire).* Now the accusations have stopped. The mischief of my adversaries, as my faith demands, is visible to everybody. Those, therefore, who have prosecuted me with their snares, can impute it to their benefit if they have gained some good, and they may have learned by now that they have to read more carefully what they seek to be incriminated, so that they appear to proceed more cautiously. But enough about these! It is sufficient for me that I have had among my judges and defenders Pietro Barozzi, the bishop of Padua, the glory and splendor of the Christians of our age, you, to whom I pay respect in faith as well as in philosophy, and many other theologians and philosophers, who will always be witnesses of my innocence. I had treated this most noble matter, which is drawn from the sources of all ancient philosophers, in the style of our times, recommended, it seems, by almost everybody, except for some among which Hieronymus Malelavellus was the most distinguished . . . He encouraged me to divide the whole text into chapters, arguing that an ancient topic should be treated in an ancient style. Convinced by this one reason of his . . . I have changed the text. I found it sensible to omit some things, to change others, to add many. I have deleted nothing which is against orthodox faith, since what does not exist cannot be destroyed.[179]

This passage is central for understanding Nifo's intellectual biography, but strangely absent in a number of modern studies of Nifo.[180] The *Questio de intellectu* to which Nifo refers is the early treatise not extant today (save for its 1503 revision), which was finished 1492, three years after Barozzi's decree. Adversaries *(aemuli)* of Nifo accused him of heresy, but, says Nifo, they did not read carefully enough what he had produced. Pietro Barozzi, as chancellor of the university and as the bishop in charge, had to judge in this affair, as had other theologians and philosophers, and they decided that Nifo was innocent. The incriminated text obviously was *De intellectu.* The accusation of heresy may have concerned several doctrines, but the unicity thesis certainly figured prominently. This is clear not only in view of the 1489 decree but also in many passages on the unicity thesis in Nifo's subsequent works that show signs of reworking and self-censorship. In the dedication to Badoèr, Nifo tries to give the impression that the difference between the early *Questio de intellectu* and the *Liber de intellectu* concerned mainly the structure of the work, whose question format is replaced by classical chapters. But the redressing—*quedam tollere, mutare alia, addere plurima*—was clearly designed to make the work safe from the charge of heresy.

The reverberations of the accusations against him are tangible in his later works. In the commentary on Averroes's *Destructio,* on which he worked from 1494 to 1497, Nifo argues that the intellect, according to Averroes, is joined to the body only through its activity *(operando),* that is, when the human person by virtue of its cogitative power apprehends a universal form.

> You should know that this is the opinion of Aristotle and Averroes and that this is pure error according to us Christians. One has to decide this question differently, as I have explained in the book *De intellectu,* where I have argued against these philosophers with the strongest arguments.[181]

In his commentary on *De anima,* which he completed in 1498, Nifo again asserts, in the footsteps of Siger of Brabant,[182] that both Aristotle and Averroes hold that the intellective soul is "one by number in all human beings,"[183] but signals several times that a refutation is to come: "But you should know that this opinion of Averroes is false according to us Christians; I will argue against it in the third part, so God will. For now I have quoted it to explain Aristotle."[184] The third part, however, does not contain any such refutation, but an explicit reference to *De intellectu.* Note that Nifo now says that the unicity thesis is in conflict not only with "us Christians" but also with philosophy:

You should know that everything I have said here was written by me only as an interpreter *(ut expositor)*. That is why after this commentary I have written another book with the title *De intellectu*, in which I take back everything which Averroes here says *(omnia retracto quae hic Averroys dicit)*, and in which I show in which way Averroes's utterances do not accord either with philosophy or with any kind of truth. Hence, read that book.[185]

Nifo is moving slowly in the direction in which Vernia had been pressed by Barozzi: to argue against Averroes philosophically. He takes first steps in the early commentaries on *Destructio* and *De anima* by declaring the unicity thesis to be "an error according to us Christians" and to be refutable philosophically, but he does not yet mount philosophical arguments against it. This he will do in *De intellectu* of 1503, which is why he appeals to his readers to turn to this book. The *De anima* commentary was published in the same year as *De intellectu*, but Nifo later, in the reprint edition of 1522, complains that the early publication of the commentary was against his wishes: "I had not destined the text to be published had it not been kept back for nine years,"[186] quoting the precept of Horace's *Ars poetica* (v. 388) that a manuscript should remain hidden and tested for nine years before publication. Nifo explains that in the early *De anima* commentary his exposition of Aristotle was not yet mature and that he often missed the true sense of Aristotle's words because he had followed in almost everything the sometimes faulty interpretations of Averroes.[187]

In the *Liber de intellectu* of 1503 Nifo has come full circle. He now states that Averroes's unicity thesis is against faith, against Aristotle, against philosophy, against the consensus of the wise men of history, and against religious experience,[188] and, for examples of the latter, Nifo refers his readers to a work by Bishop Barozzi titled *De ratione bene moriendi*.[189] With this conclusion, Nifo follows the line that Vernia had taken after his recantation: to show that Averroes on the intellect is refutable philosophically, while Aristotle is not. Like Vernia, Nifo invokes the authority of Thomas Aquinas's *De unitate intellectus* for this new position.[190] Nifo's *De intellectu* is divided into six sections *(libri):* on the definition of the soul, on the separateness of the soul, on the unicity of the intellect, on possible and active intellects, on the speculative intellect, and on intellectual bliss.[191] The principal target of the first section is Alexander of Aphrodisias and his theory that the soul is mortal, but the rest of the book is essentially a very critical discussion of the main points of Averroes's intellect theory. Several times Nifo remarks that he has changed his mind on Averroes: just as many other followers of the Peripatetics and Averroes, he had long thought, erroneously, that the potential intellect is a separate intellect[192] and that Averroes's theory of the union of soul and body is true and cannot be refuted with arguments of natural philosophy.[193] Here Nifo clearly admits that in earlier times he had been an Averroist in the sense of having been a partisan of the unicity thesis. He also remarks that he "had long reflected whether there is some rescue for Averroes" *(evasio pro Averroe),* that is, whether Averroes too might have taught that the rational soul is the specific form of the body—but he had reflected in vain. Nifo says that he does not care about this kind of *evasio* anymore. He obviously does care, however, about the correct interpretation, the *vera mens* of Averroes, which John of Jandun and others have missed—the new theme of Nifo's lifelong work on Averroes. With his commentaries on Aristotle and Averroes of the next decades, Nifo will mold himself into a champion of Averroes's interpretation. He has turned from a defender of Averroes's philosophy into a defender of the correct interpretation of Averroes's philosophy.

At the beginning of this development stood the accusation of heresy by his adversaries at the University of Padua. Nifo managed not to suffer the same fate as Vernia, his teacher, who had to recant publicly. He rescued himself by changing his position on Averroes sometime before 1498, the latest possible date for his dedication to Badoèr, and by redressing his texts: by substantially revising the early *Questio de intellectu,* transforming it into the *Liber de intellectu,* a multipage refutation of the various facets of Averroes's intellect theory, and by inserting references to the *Liber de intellectu* into his commentaries on *Destructio* and *De anima,* when they touch on potentially heretical issues. Without Barozzi's edict, without the Vernia affair, and without the charges of heresy against him, Nifo's later intellectual development is inexplicable. Bishop Barozzi complained in 1504— Nifo had left Padua for southern Italy around 1500—that at the University of Padua "the errors on the eternity of the world, on the unicity of the intellect, on nothing coming from nothing, and so forth" abound among the philosophers and that without the Scotist theology professor "nothing in this university is taught which is not taught in a university of pagans" *(studio de' pagani).*[194] Barozzi was still not content with what he had achieved. But the effects of his repression are visible enough among the partisans of the unicity thesis. His intervention of 1489 had a direct impact on Vernia, it apparently impressed Achillini, and it changed Nifo's intellectual biography.

I now turn to the content of Nifo's *Liber de intellectu,* with an eye to the question of whether other factors were also involved in Nifo's shift in attitude toward Averroes, in particular, his reading of the Greek commentators and of Aristotle in Greek. I am also interested in investigating whether the unicity thesis still enjoys a hidden success with Nifo. Nifo's conclusions on the main questions of intellect theory are clear, at least on the surface: The rational souls are as many as there are human beings, and the active and possible intellects are powers of the rational soul. The intellected form is different from the intelligible species through which we acquire knowledge and which are a real accident of the soul.[195] In addition to these conclusions, which are in obvious conflict with Averroes, Nifo often claims to have refuted successfully the arguments of Averroes and his partisans.[196] On the other hand, the philosophical attraction by Averroes can be felt on many pages. Nifo admits that for him Averroes is the paramount authority on the rational soul, albeit an authority in error:

> Since among those who have written about questions of the rational soul there is nobody who has invested as much labor as Averroes did, it will be an honor for me to fight with him among others. I wish for you to know that I will mention everything which can be mentioned to strengthen this one man, and that I will not pass over a single word... For when I have shown that his utmost strength is weak and repulsive, everybody will have to be content with believing that this position is wrong and against reason.[197]

This program is, in fact, carried out. Nifo searches for the *argumenta fortissima* that can be advanced in favor of Averroes's intellect theory.[198] Also, in the course of *De intellectu*, Nifo time and again refutes traditional arguments against Averroes that he does not consider convincing, and then proceeds to forward his own, demonstrative arguments against Averroes. This also happens in the context of the discussion specifically devoted to the unicity thesis. Nifo mentions three common counterarguments, which he himself had employed previously, but does not find convincing anymore: If the intellect was one and not many, the rational soul would be an individuum. If the intellect was one, a person would know something known by a different person. If the intellect was one, the rational soul could not be the first perfection of the body, since only one human being would exist.[199] Of these three arguments against Averroes, the second, which concerns the common knowledge of two persons, is of particular interest, since it is a prominent argument in the discussion of the unicity thesis. In the eyes of Nifo, this argument is not convincing, for Averroes would reply that the intelligible form of, for example, a stone that is grasped by two persons, on the one hand is the same because it is known through the same intellect, but on the other hand, the two persons know it individually because the intelligible form of the stone coincides with and is connected to forms of the imagination. That is why my knowledge is not identical with yours, argues Nifo, and that is why the argument against Averroes is not convincing.[200]

There are many such passages in Nifo's book that swim against the current of his conclusions. In fact, the unicity thesis has a life of its own in Nifo's work, as it had in Vernia's: in quotations from Albertus Magnus.[201] In his *solutio* to the second book on the soul as the form of the body (2.22), Nifo quotes, without attribution, the very same proto-Averroist passage from Albertus's *De intellectu et intelligibili* that Vernia had quoted too. Both authors, in fact, combine the passage with a well-known formulation by Albertus,[202] which ultimately derives from the Neoplatonic *Liber de causis,* that the human soul is located "on the horizon of eternity and temporality" *(in orizonte aeternitatis et temporis).*[203] After the quotation from Albertus Nifo concludes: "The rational soul therefore is individual insofar as it is a form of a human being, but as a power of the spiritual light it is indeed universal."[204] These reverberations of Averroes's *duo-subiecta* theory do not turn Nifo into an Averroist, but they indicate that Nifo continues to be impregnated by Averroes's thought.[205]

It is conspicuous that the final arguments that Nifo offers (in chapter 3.28) to settle the question of the unicity of the intellect and to refute Averroes, do not come from the *De anima* tradition, but from "moral philosophy" *(philosophia morali).* Nifo adopts this idea from Thomas Aquinas, who had also argued in *De unitate intellectus* that "the principles of moral philosophy are destroyed" in consequence of Averroes's unicity thesis.[206] But Thomas and Nifo differ entirely on the content of these principles. Thomas argues that the human being, with only one intellect and hence one will *(voluntas)* shared by many, could not decide about his own actions and hence be remunerated for them. Nifo will forward similar arguments in a later chapter (3.30), but here he refers to the following "moral principles": God has to be honored; the souls derive from God; the human being is a divine miracle; the divine law derives from God; human beings cannot live together without God.[207] In support of these principles Nifo quotes Plato, Porphyrius, Avicenna, Pseudo-Aristotle's *De mundo,* Aristotle's *Politics,* and Algazel, among others. It may seem strange that Nifo mounts these statements against an epistemological and psychological thesis. The missing link is immortality, as is apparent from Nifo's conclusion at the end of the chapter: If Averroes's position was true, Nifo says, "no animal would be more miserable *(infelicius)* than me or any other human being, and in this opinion I have found our Ficino to be in full accordance with me."[208] This echoes Marsilio Ficino's opening sentence of the *The Platonic Theology on the Immortality of the Souls,*[209] that the human being, if bereft of its difference with other animals, would be more miserable than any other animal. Since Ficino finds immortality threatened by Averroes's unicity thesis, he devotes the fifteenth book of his work to the refutation of it. Nifo's description of the four moral principles, in turn, not only amply draws on several works by Ficino, as has recently been shown,[210] but also presupposes the basic assumption of Ficino's critique: that Averroes's unicity thesis removes the individual immortality of the soul and thus threatens the religious orientation of mankind.

In the next chapter of his conclusion (3.29), Nifo argues that in addition to the violation of moral principles, the unicity thesis also runs against two principles of natural philosophy *(praecepta naturalis scientiae):* First, a single mover uses exactly one appropriate object in motion. Second, no mover produces different effects of the same kind in the same time.[211] These principles are in fact drawn from Albertus Magnus and Thomas Aquinas.[212] While these latter arguments engage with Averroes on a philosophical level, Nifo's "principles of moral philosophy" come in a philosophical garment, but are motivated religiously. In the footsteps of Ficino, Nifo argues against Averroes's unicity thesis stating that it violates religion, and that religion is postulated by philosophy. For the latter thesis he cites authorities, but does not give arguments. This again is an indication that Nifo found the unicity thesis difficult to refute philosophically and, moreover, that his refutation of the unicity thesis is motivated by religious concern—a concern that influences his philosophical argumentation. In comparison, Thomas Aquinas's *De unitate intellectus* is a more strictly philosophical treatise.

Nifo declares in 1508, in his commentaries on the *Physics,*[213] and on *De beatitudine animae,*[214] that in his youth he had defended Averroes, but later found his position to be nonsense *(deliramentum)* when reading and examining Aristotle in Greek. In the *Physics* commentary, after having censured Averroes for his lack of Greek and for a resulting misinterpretation of Aristotle and Themistius, Nifo proceeds to formulate a humanist credo, or rather, a piece of humanist ideology: "I prefer to err with the Greeks in expounding Greek authors rather than to be right with the Barbarians, who do not know the languages except in dreams."[215] On the basis of these statements and in view of Nifo's quotations of Greek commentators, it has been argued that in *De intellectu* Nifo made a dramatic shift from Averroes to the Greek commentators as the true guides to Aristotle, and that this shift was epoch-making in the history of Renaissance Aristotelianism.[216] It is true that Averroes's position as a commentator was seriously challenged in the Renaissance by the accessibility of the Greek commentators and of the Greek Aristotle, but Nifo's *De intellectu* is not yet the text where this challenge is visible. Nifo does not cite the Greek text of *De anima* once, as he will do in later works, for instance, in his *De immortalitate animae* of 1518. Nor do the Greek commentators Alexander, Themistius, and Simplicius, whom Nifo cites much more often than Vernia, influence his solution on the plurality of the human intellects.[217] Instead, Nifo's solution is inspired by Marsilio Ficino, Thomas Aquinas, and Albertus Magnus—and by other Arabs, as Nifo himself remarks (chapter 3.31): "Hence, what we adopt is the middle way which Avicenna, Algazel and Costa ben Luca and other Arabs propounded, namely that the rational souls are many and that they are the enliving and intrinsic forms of the body, whose intellects are separate"[218]—a truly "Barbarian" solution to the unicity problem. Nifo's Greek turn against Averroes in *De intellectu* is a fiction created by Nifo himself in 1508, when he fashioned himself as a torchbearer of Greek Aristotelianism. In view of his earlier remark that he postponed the publication of *De intellectu* because of charges of heresy, it is very likely that Nifo's story of the Greek turn is meant to cover up the most important factor in his early intellectual biography: the pressure of orthodoxy. In the last part of this chapter, I will study how the older Nifo of 1508 and later employs the Greek text of Aristotle's *Metaphysics* to demonstrate the deficiencies of Averroes's interpretation. The younger Nifo had to cope with a very different problem: his survival as an orthodox philosopher.[219]

### Pietro Pomponazzi

It is difficult to discern whether Barozzi's intervention influenced philosophers other than Vernia, Achillini, and Nifo. Barozzi died in 1507. In 1513, the Fifth Lateran Council declared the unicity thesis and the thesis of the mortality of the soul to be heretical and decreed that the philosophy professors of the universities argue against them. Since the later adherents of the unicity thesis, such as Prassicio, Genua, and Vimercato, return to the traditional practice of keeping philosophical argumentation and Christian faith apart, as do Cajetan and Pomponazzi on the immortality of the soul, it seems that the pressure of the Church was not able to influence the philosophical solutions anymore, as it had done with Vernia and Nifo.[220] On the other hand, it may also be that, without Church pressure, the distribution of the unicity thesis would have been much greater. It is possible that the intervention of the Church thwarted the distribution of the thesis exactly in the decades when it was about to reach the peak of its influence.

Pietro Pomponazzi's (1462–1525) attitude toward Averroes is less enigmatic than Nifo's.[221] His treatise *De immortalitate animae* of 1516, which provoked the most famous debate of Italian Renaissance philosophy, contains an outright refutation of Averroes. In the eyes of Pomponazzi, Thomas Aquinas has sufficiently demonstrated that Averroes's intellect theory is false. Pomponazzi leaves it at this and concentrates on showing that the theory is in conflict with Aristotle.[222] The refutation of Averroes's reading of Aristotle is an important step in Pomponazzi's argumentation, which culminates in the thesis that according to Aristotle and to natural reason, the intellective soul is essentially dependent on the phantasmata and hence dependent on bodily functions, with the consequence that the soul is essentially mortal.[223] In the last chapter he qualifies his position by reserving true certainty in this matter to the Christian faith. Whether the disclaimer is sincere or not, in the following debates with his critics,[224] Pomponazzi successfully insists that the immortality of the soul cannot be proved philosophically and that it can be held only on the basis of divine revelation.[225] It is impressive to see that Pomponazzi, after having witnessed the pressure exerted on his teacher Vernia and his fellow student Nifo, uncompromisingly advocates the autonomy of philosophy.

The 1516 refutation of Averroes is the standpoint of the elder Pomponazzi. The younger Pomponazzi, however—as we know from manuscripts preserving reports of his Paduan lectures on the soul of 1503–1504[226]— maintained that Averroes's position, albeit wrong, was nevertheless identical with Aristotle's. Averroes and Aristotle are described as holding that the unity between soul and body is of a "middle" or loose kind: not as close as between form and matter, but not as independent as between captain and ship.[227] Pomponazzi was not happy with this conclusion, as the phrasing of the passage in a student's report shows:

> With regard to Averroes's opinion, I shall say the following: I believe that this opinion is beyond imagination, fatuous and fictitious. Oh, you have said that it is the opinion of Aristotle! I say that this is true, but I also say that even Aristotle was a human being[228] and was able to make mistakes.[229]

Or, in the wording of a different student listening to the same lectures:

> With regard to Averroes's opinion, it seems to me that it was that of Aristotle. However, I cannot by any means adhere to it, and it seems to me the most flagrant nonsense (*maxima fatuitas*). People may say what they like, but I find Averroes's opinion more disgusting than the devil.[230]

Pomponazzi was stuck in a dilemma: the true and holy position of the theologians is based on the light of faith. "Without the light of faith, I am very uncertain *(valde perplexus)* about this matter."[231] The position of Averroes and Aristotle is wrong or even nonsense; and the position of Alexander of Aphrodisias,—that is, the soul's complete dependency on the body—is too difficult to defend. The reason that Pomponazzi was not yet ready in 1504 to accept Alexander's theory, lay in the strength of Averroes's refutation of Alexander, as Pomponazzi himself admits: "the argument about the universal weighs heavily against Alexander" *(contra Alexandrum multum valet argumentum illud de universali).*[232] With this he refers to Averroes's argument that the intellective soul would not be able to know universal intelligibles if it was immersed in matter.[233] For Pomponazzi, Averroes's standpoint had the advantage of offering a theory of universal intellection, and this proved an obstacle to Pomponazzi's way toward adopting Alexander's philosophy.

We can witness Pomponazzi turning against Averroes in lectures on *De caelo,* which he held in Bologna in 1515–1516 (again preserved in manuscript). Based on the principle that what is created is corruptible and what is not created is incorruptible, Pomponazzi concludes that Aristotle could not have argued for the soul's being generated but incorruptible—as Thomas Aquinas had done. And he proceeds to signal accordance with Duns Scotus's interpretation of Aristotle as holding that the soul is generated and corruptible—which agrees well with Alexander's thesis of the soul's complete dependency on the body.[234]

When Pomponazzi in 1516 finally sided with Alexander's position, he was forced to present a theory of universal intellection in harmony with the soul's dependency on the body. His solution is expressed in the sentence: "The universal is comprehended in the particular" *(universale in singulari speculatur).*[235] That is, Pomponazzi sacrificed the idea that human beings would be able to know universal intelligibles *simpliciter,* as eternal.[236] Rather, it is only through phantasmata that the human being grasps universals. This theory differs from Averroes's contention that we comprehend the intelligibles only via imagination, since Averroes insisted that the material intellect had to be entirely incorporeal in order for it to receive universals as such; imaginative forms and material intellect then form a union in the moment of intellection. Pomponazzi has given up the idea that an incorporeal intellect is a necessary condition for grasping universal intelligibles. He has finally turned against Averroes, but it took him a long time and cost him a certain philosophical price.

In comparison with Vernia and Nifo, Pomponazzi's philosophical solutions are much less influenced by nonphilosophical motives. He does not hesitate because of religious reasons to formulate a theory of the mortality of the soul, but he hesitates because he has not yet solved the epistemological problems of Alexander's position. Nor does Pomponazzi appear to be driven by a humanist bias toward the Greek when embracing Alexander's reading of Aristotle instead of Averroes's. The solutions of Vernia's and Nifo's anti-Averroist treatises draw amply on Thomas Aquinas and Albertus Magnus, whereas Pomponazzi makes full use of the newly translated *De anima I* of Alexander of Aphrodisias in the translation of Donato, printed in 1495.[237] From then on, Averroes's interpretation of Aristotle's *De anima* had a serious rival.

It is noteworthy, however, that Averroes remains a principal point of reference for Pomponazzi, in the immortality treatise of 1516 as well as in his other philosophical works. He even uses new translations of Averroes, as in his 1521–1524 Bologna lectures on Aristotle's *De partibus animalium*. This work draws much on humanist translations, such as Theodore Gaza's Latin rendering of Aristotle's text, but often also on Jacob Mantino's translation of Averroes's commentary, which had just been printed in Rome in 1521.[238] Pomponazzi's attitude toward Averroes always remains critical in the sense that he believes "that the Commentator has made mistakes and is not a God."[239] And in the *Metaphysics* commentary of 1517 Pomponazzi complains: "If someone charges <me>: 'You then do not defend Averroes?' it will soon happen that I do not have any friend in Padua and Ferrara."[240] These testimonies show on the one hand that the position of Averroes in Northern Italy was stronger than ever and not shaken by religious or humanist attacks, and on the other hand that among some partisans of Averroes, the belief in the truth of his position deteriorated into ideology. There is evidence that one such partisan was Luca Prassicio.

### Luca Prassicio

In the year 1521, the Neapolitan philosopher Luca Prassicio (d. 1533) enters into a full-blooded controversy with Agostino Nifo about the correct interpretation of Averroes's intellect theory.[241] Prassicio's *Questio de immortalitate anime intellective* is an interesting testimony not only to the liveliness of the Averroist movement in the field of intellect theory, but also to the fact that the bull *Apostolici regiminis* of 1513 was of limited influence. For example, Prassicio outspokenly asserts in the preface: "As far as I can see, I can find nothing more certain and true on the immortality of the soul than what can be taken from Averroes"—even though Aristotle's and Averroes's positions on immortality are false from the infallible standpoint of *catholica nostra religio*.[242] The overall conclusion of Prassicio's treatise is that not only the active intellect is separable and eternal, but also the material intellect is, and that hence Aristotle and Averroes teach the immortality of the soul.[243] With this conclusion, Prassicio counters Pomponazzi's rejection of Averroes and at the same time corrects Nifo's "hallucinating" defense of Averroes against Pomponazzi. Prassicio's interpretation of Averroes is strictly separabilist:

> Since <according to Averroes> the soul has one nature in number in a metaphysical constitution *(i.e., separated in reason and essence)*, as we have often said, the soul has its own intellection which is separate from the phantasma, continuous and eternal and which is the substance and not the power of the <person> who acquires knowledge. When it happens that <the intellection> is united with a human being—but not in a univocal way and through formal inherence, as Nifo falsely hallucinates, rather through effective assistance to the phantasmata, by making the phantasm actually intelligible—then a universal form comes about in the material intellect.[244]

Prassicio rejects the idea that Averroes could hold both the unicity of the intellect and the thesis that the soul is the substantial form of the body. The soul is the form of the body only in an equivocal sense, that is, by attending the phantasmata of the human being *(per assistentiam ad phantasmata)*.[245] Hence, Prassicio castigates Nifo's interpretation of Averroes as "apostatizing from the true doctrine of Averroes" *(apostetare in doctrina Averrois)*.[246]

The rigidity of Prassicio's reaction against Nifo is an indication that motives foreign to philosophy are also at play: the attempt to claim leadership within the intellectual movement of Averroism, and a prejudiced conviction that Averroes is an orthodox philosopher. Prassicio claims, against a very old tradition, that Averroes is the best defender of the immortality of the soul. It is true that Averroes does not hold the soul to be mortal and that he explicitly argues that the species human being is eternal.[247] But Averroes does not recognize personal immortality in the Long commentary on *De anima*. Prassicio circumvents all discussion of personal immortality, apparently in an attempt to rescue Averroes as a champion of immortality theory. The belief that Averroes is an orthodox philosopher colors Prassicio's one-sided presentation of Averroes's philosophy. With respect to Greek sources, it is remarkable that Prassicio is an exponent of Averroism who does not care at all about the ancient commentators. His principal concern is the *mens Averrois,* to a degree that even Aristotle is relegated to the background. His main points of reference are Nifo, Pomponazzi, Thomas, and the medieval interpreters of Averroes. The future of natural philosophy in Italian universities, however, will belong to a different set of sources, in which Averroes and the Greek commentators are read alongside.

## Marcantonio Genua

In the decades after Pomponazzi, the most influential philosophy teacher at the University of Padua was Marcantonio Passero, called "il Genua," the Genoese (d. 1563). He taught philosophy at Padua for over forty years, from 1517 onward, and had among his students Zabarella, Tomitano, and probably also Vimercato. In 1542–1544 Genua held lectures on *De anima,* which were printed only after his death, but often copied in manuscript.[248] In these lectures, Genua repeatedly praises Simplicius's *De anima* commentary as a *divina expositio.*[249] Simplicius's text had been printed in Greek in 1527 and in a Latin translation by Genua's student Giovanni Faseolo in 1543.[250] As was remarked above, the translator Faseolo attacks Averroes in an introductory letter as barbaric and exclaims: "Make the decision to only read Simplicius, day and night . . . and take <Averroes> out of your hands!" *(de manibus deponite).*[251] This sentiment was not shared by his teacher Genua, who praises Averroes as the *princeps* of all Peripatetics and as the "second Aristotle" *(alter Aristotelis).*[252] Genua, in marked contrast to Agostino Nifo, also thinks highly of John of Jandun, as the greatest Aristotelian and the greatest Averroist *(summus Averroicus),*[253] and in fact continues Jandun's interpretation of Averroes as holding that the intellect is joined with the body *per assistentiam* and not as a substantial form.[254] Genua's exposition of Aristotle follows Averroes so closely that the text almost takes the format of a super-commentary. In his long exposition of the crucial comment 3.5, which covers forty-five columns in the 1576 edition, Genua counters in detail Thomas's and Albertus's arguments against Averroes's unicity thesis and concludes: "We say, arguing in Peripatetic manner with Aristotle, that there is one intellect in all human beings."[255] And he proceeds to show that Themistius and Simplicius support this position. Genua acknowledges, however, that Averroes disagrees with Themistius on the speculative intellect.[256] Then Genua asks whether this position of Aristotle is "true in itself." It is not, says Genua:

> The true conclusion, both in philosophy and in theology, is that the created intellectual souls are many in number . . . For that philosophy is more to be trusted as true and truthful which is in greater accordance with the theological truths, in which the highest truth is to be found. Only the philosophy professed by Plato is closest and most

similar <to the truth>, which is why Plato was called "the Athenian Moses."[257]

One may recall that Nifo had invoked Ficino's presentation of Plato to formulate his philosophical conclusion, with the effect that Nifo offered a philosophical proof of religion. In contrast, Genua continues the traditional distinction between philosophical and religious truth, as did Pomponazzi and Prassicio before him. Plato is quoted by Genua only to give a philosophical coloring to what is essentially a brief disclaimer of orthodoxy. Genua, the philosophy professor at Padua university for decades, is the clearest indication that Barozzi's edict and the bull of 1513 had lost their impact, for Genua is without doubt an Averroist in the sense that he is a defender of the unicity thesis. The evidence for other external influences on his philosophical standpoint is weak as well. He does not share Faseolo's humanist disdain for Averroes, in spite of his marked interest in the Greek commentators, or Prassicio's pro-Averroist bias. Genua thus returns to the sober traditions of Paduan natural philosophy. Note that he has a wide reading of the Greek commentators in Latin translation, but that he does not use the Greek text of Aristotle for his exposition.

Bruno Nardi and Eckhard Kessler have shown that in the later chapters 3.16–18 of his commentary Genua connects Averroes's theory of the active intellect with the Simplician notions of *intellectus manens* and *intellectus progressus*.[258] But, with the enthusiasm of pioneers in the field, Nardi and Kessler exaggerate the influence of Simplicius when describing Genua as the foremost representative of a "Simplician" and hence Neoplatonic current of Aristotelianism and when arguing that Genua accepted no other than Simplicius's interpretation of Aristotle. As we have seen, Genua describes himself as an heir to Averroes and to John of Jandun in the first place, and not to Simplicius. And, in fact, Genua's commentary in general remains much closer to Averroes's text than to Simplicius's, not only on the unicity thesis. One should also note that Averroes's intellect theory has many Neoplatonic traits in itself, especially in the theories of the gradually increasing intellection of the separate intelligences and of the gradual conjunction between speculative and active intellect (as in chapter 3.36, which comments on Γ.7, 413b16–19). It is true that the Neoplatonist reading of Aristotle received a new impulse in the Renaissance, but this impulse only added to a long Arabic and scholastic tradition. We have met an example of this tradition in the works of Vernia and Nifo, who adopt from Albertus Magnus the Neoplatonic idea of the rational soul being situated on the horizon between eternity and temporality, when they quote proto-Averroist passages from Albertus. Hence, it would be anachronistic to detect a "breakthrough of Neoplatonism" in Genua's theory of the intellect, since Neoplatonism had never ceased to be a vital part of the commentary tradition on Aristotle.[259]

What is more characteristic of Genua is his mastery of the commentary tradition. Genua had many sources of the *De anima* tradition at his disposal. His reading in the Greek, Arabic, and Latin commentators on *De anima* is most impressive and surpasses that of his predecessors. In view of the many choices he had among the commentators, it is all the more remarkable that, in the eyes of Genua, Averroes remains the *princeps* of them all. The importance that Averroes enjoyed at the University of Padua in Genua's time is the humus in which the great editorial and translating enterprises between 1520 and 1560 could grow. The pivotal example of this effort is the monumental Giunta edition of 1550/1552, which was prepared in the 1540s by Giovanni Battista Bagolino, the son of a Paduan philosophy professor, and Marco degli Oddi, logic professor at Padua. The 1562 reprint edition adds much material on Averroes's logic by one of Genua's students, Bernardino Tomitano, likewise a professor of logic. Hence, Genua's attitude toward Averroes, just like that of Prassicio in Naples, is a clear indication that Averroes was not replaced by the newly available Greek commentators.

### Simone Porzio and Francesco Vimercato

This is not to say that the Greek commentators did not pose a challenge to the Averroist reading of Aristotle. Several Italian philosophers, in the footsteps of Pomponazzi, took their inspiration from Alexander of Aphrodisias's commentaries, such as Giulio Castellani (d. 1586) and Simone Porzio (d. 1554).[260] Porzio's attitude toward Averroes deserves some attention. In 1551, Porzio, who was professor at Pisa and Naples, publishes *De humana mente disputatio,* in which he musters the positions of Philoponus, Themistius, Simplicius, Averroes, and Alexander. He finds points of critcism with all of them, but in the end develops a position close to Alexander's. "The Averroists, Simplicianists, and Themistianists," he charges, deviate too much from Aristotle's text.[261] In Porzio, we can observe a tendency to simplify Averroes's position by fusing it with that of Simplicius and Themistius. Porzio does not say much about Simplicius, whom he finds difficult to understand. But as to Themistius, he argues that Averroes, even though criticizing Themistius vehemently, "thought almost the same," namely, that the active, possible, and speculative intellects are all one.[262] It is true that Averroes's position is, to a certain degree, similar to Themistius's, but the crucial difference is that Averroes holds the speculative intellect to be one only with respect to all human beings; with respect to the phantasmata,

the speculative intellect is individual and not eternal. Averroes thus eliminates the most problematic part of Themistius's theory. Porzio's conflation of Themistius and Averroes makes Averroes's intellect theory an easy
target for criticism.

Porzio uses the Greek text of Aristotle's *De anima* in his argumentation,
but these first steps toward a philologist interpretation remain rare occasions in his text. A much more humanist approach to Aristotle's *De anima*
was attempted by one of the last authors who adopts the unicity thesis,[263]
Francesco Vimercato of Milan (d. ca. 1571), who had studied at several
Italian universities. Vimercato spent the greater part of his academic
career in Paris, from about 1540 to 1562 as a teacher of logic and as the
first teacher of Greek and Latin philosophy at the Collège Royal.[264] In his
commentary on book 3 of *De anima,* which was printed together with the
Greek text in 1543, Vimercato presents for each lemma the expositions of
Alexander of Aphrodisias, Themistius, Simplicius, Philoponus, and Averroes. Vimercato was among the first to systematically compare Averroes's
commentary to the Greek text of Aristotle's *De anima.* He is able to identify
several passages in which Averroes's exposition is seriously hampered by a
faulty translation or transmission. An example is Aristotle's chapter Γ.5,
430a14–17, where several corruptions cause Averroes to believe that Aristotle speaks of three and not of two intellects: "he has a text which is much
removed from the Greek truth" *(habuit textum a veritate Graeca . . . valde diversum),* as Vimercato rightly remarks.[265] Vimercato's criticism of Averroes's
exposition can be biting, but he does not employ it to challenge Averroes's
intellect theory directly. Instead, when he comes to doctrinal questions of
intellect theory, Vimercato refers his readers to another treatise of his, *De
anima rationali peripatetica disceptatio,* which was printed together with his *De
anima* commentary in Paris in 1543.

In the *Disceptatio,* Vimercato occasionally quotes the Greek text, but the
argumentation in general is not philological, but philosophical. Aristotle's
theory of the nature of the human soul, says Vimercato, is particularly obscure and has thus caused great controversy among the commentators.[266]
By pondering the arguments of the Greek, Arabic, and Latin commentators, Vimercato arrives at the following conclusions: the soul is substance
and form of the body; the intellect is immortal; the active intellect is part of
the soul, but one for all human beings; and the possible or human intellect
is one for all human beings.[267] This is Vimercato's concluding statement on
the unicity question:

> Whoever considers in detail the arguments that prove the unicity—to
> say frankly what I mean—will find that they are much more cogent

*(multo efficaciores)* than the others and in fuller accordance with Aristotelian principles.[268]

Seven principal reasons are given in support of the unicity thesis, three by Themistius and four by Averroes. Vimercato demonstrates their cogency by refuting Thomas Aquinas's and Nifo's counterarguments. In the eyes of Vimercato, two arguments among these seven are of particular value. The first is attributed to Themistius (but was also developed by Averroes):[269] if the intellect was not one, human beings would not share the same knowledge, they would not understand each other, and a pupil would not learn from his teacher. The second argument comes from Averroes and concerns the infinity of souls or intellects. This is the best of all arguments offered by Averroes, says Vimercato.[270] The infinity argument appears in scholastic sources since the thirteenth century,[271] but Vimercato remarks that it was already used in Averroes's *Destructio destructionum*.[272] If it is maintained with Aristotle that the world is eternal and that the intellect is immortal, then in actuality an infinite number of intellects exists—since to the present an infinite number of human beings had lived and died. But an actual infinite regress is not allowed with Aristotle.[273] It is indicative that Vimercato is particularly taken by these two arguments, which do not presuppose a specific psychological theory. He is a partisan of the unicity thesis without accepting or even discussing the technical details of Averroes's theory, such as the distinction between the three intellects—speculative, material, and active. Again, as in Porzio's treatise, the presentation of Averroes tends to be simplifying. The differences between Averroes and the other commentators, which for Averroes had been of central importance, are glossed over. Vimercato's philosophical discussion, as a whole, does not reach the level of that of Nifo, Pomponazzi, or Genua. While Vimercato's philological criticism begins to undermine Averroes's competence as a commentator on Aristotle, Vimercato's simplistic presentation of the unicity thesis is bound to undermine Averroes's authority philosophically.

This undermining, however, does not yet have an effect on Vimercato's philosophical position. It is remarkable that the unicity thesis also continues to attract philosophers in a truly humanist surrounding. Vimercato's reception of Averroes was not hampered by humanist prejudices, nor was he influenced by religious qualms. He is well aware of the problem of heterodoxy and closes his treatise with the remark that Aristotle's doctrine of the rational soul does not accord with Christian faith and that the unicity thesis is absurd, but that there is no reason why he should refrain from freely presenting Aristotle's opinion.[274] If we wanted to leave the Peripatetic gymnasium *(lyceum)* and not adhere to Peripatetic doctrines anymore, he says, we could easily show that the theory is in conflict with *theologia christiana.*[275] Vimercato suspects that Aristotle was convinced of the unicity thesis, but refrained from stating it openly—probably because the thesis would not have found acceptance among the common people of his time and because he feared that it would undermine moral and civil life.[276] The Aristotle of Vimercato shares, anachronistically, Thomas Aquinas's concerns about the moral consequences of the unicity thesis. Vimercato addresses the heterodoxy problem, but at the same time, it is clear that he does not want to exit the Peripatetic gymnasium and concern himself with the religious problems of the common people. He was proud to be a Peripatetic philosopher and a Greek scholar who could say what he wanted on philosophical issues.

## The Decline of the Unicity Thesis

Vimercato's *Disceptatio* appears in 1543, and Antonio Bernardi's *Eversionis singularis certaminis liber* in 1562; Bernardi is among the very last to advance the unicity thesis in print: he finds it philosophically cogent, but false from a Christian point of view.[277] In the second half of the sixteenth century, the thesis meets a noticeable decline. Since it is a principal thesis of the Averroist movement, its decline is an important test case for the decline of Averroism as a whole. We have seen that the pressure of the Church was a serious threat to the distribution of the unicity thesis, but that Averroism survived the attacks and sailed into calmer waters again after about 1520. Hence, the presumed heterodoxy of the thesis was not the direct cause of its decline. One of the previously advanced explanations for the end of Averroism is the success of the intellect theories of the Greek commentators. But this can explain only part of the story, since, as the above survey has shown, Themistius and Simplicius were quoted in support of the unicity thesis. Alexander was the sole Greek rival to Averroes in intellect theory, albeit a serious rival: there is a veritable line of Alexandrists in intellect theory leading from Pomponazzi over Porzio and Zarabella to Cremonini. Another, but hardly satisfactory explanation advanced in scholarship is that Averroism received the 'first shock' when in 1497 a chair for teaching the Greek Aristotle was established in Padua (Antonino Poppi, Maurice-Ruben Hayoun, Alain de Libera).[278] But the Greek text of Aristotle is remarkably marginal in the development of Renaissance intellect theory. Nor do more general explanations convince, such as that Averroism succumbed to the new modes of thought of the scientific revolution (Ernest Renan),[279] or that it drowned together with Aristotelianism as a result of a new philosophy of nature (Bruno Nardi).[280] These explanations presuppose that Averroism disappeared in the course of the seventeenth century, but, as far as the unicity thesis is concerned, the decline took place in the second half of the sixteenth century. Other explanations date it too early, to the beginning of the sixteenth century—when Averroism was as vigorous as ever.[281]

I should like to suggest two other explanations for the decline. The first concerns the classroom usage of commentaries. The unicity thesis, even if easy to formulate, ceases to be convincing if not spelled out in technical detail. For this we have seen indications in the works of Porzio and Vimercato, who present simplified versions of Averroes's theory. A technical exposition, however, can only be built on a word-for-word exegesis of Aristotle's text under the guidance of Averroes's Long commentary. Averroes the philosopher disappeared when Averroes the commentator ceased to be important for everyday teaching in Italian universities. Marcantonio Genua still taught *De anima* with Averroes's commentary as his principal guide, which enabled him to present Averroes's intellect theory in its technical details. But later philosophers cannot compete with Genua in this competence. This development finds expression in the fact that in 1574/1575 and 1578, the last two Averroes editions appeared. Averroes's printing history found a much more abrupt end than, for instance, that of Avicenna and Mesue (see Chapter 1).

The decline of Averroes as a commentator in university teaching is a complex process that involves many factors. One factor was the accessibility of the Greek commentaries, which, however, was not a decisive factor: the Greek commentators came out of fashion in the classroom roughly parallel to Averroes, in the later sixteenth century. An indication of this is that the printing history of the Greek commentators, both in Greek and in Latin translation, phased out in the same decades as the printing history of Averroes. A more important factor for changes in university teaching was the rise of the Second Scholastic, that is, of the Aristotelianism of Counter-Reformation Catholicism, in which the Thomistic commentary tradition was preferred over both the Greek and the Arabic, as the Jesuit *Ratio studiorum* of 1599 clearly stated.[282] Moreover, the demise of Averroes's commentaries was also the result of general changes in Aristotelian commentary culture: the rise of the purely philological commentary and the rise of the Aristotelian textbook in teaching, especially in Northern Europe.[283] The details of these developments have not yet been satisfactorily explored in scholarship.[284] It is nevertheless clear that these changes of commentary culture had an impact on the fate of Averroes the philosopher.

The second important cause for the decline of the unicity thesis is that it was rivaled by philosophical alternatives within the current of Aristotelianism. These, rather than external alternatives such as Humanism,

Platonism,[285] or experimental science, made the unicity thesis disappear. In order to show this, I will describe the situation in methodical idealization as if the authors of the late sixteenth century had been choosing freely between varying theories of the intellect. How was universal intellection explained if not, with Averroes, as resulting from a contact between the phantasmata in the individual person and the intelligible forms in the separate material intellect? In what follows, I will briefly present the different answers of three well-known philosophers: Philipp Melanchthon (d. 1560), Jacopo Zabarella (d. 1589), and Francisco Suarez (d. 1617). These authors represent philosophical alternatives to Averroes insofar as they are important partisans, or even founders of philosophical currents that were very influential in seventeenth-century philosophy.

Philipp Melanchthon's *Liber de anima* appeared in its revised format in 1552 in Wittenberg. In this book, Melanchthon accepts the Aristotelian thesis that the human intellect proceeds from the senses to higher forms of knowledge.[286] But in contrast to many other philosophers,[287] Melanchthon openly dismisses the principle that "nothing is in the intellect which had not been in the senses before." For neither universal notions *(universales notitiae)* nor judgments have their origin in the senses, whose sole function is to "move and excite" the intellect.[288] With *universales notitiae* Melanchthon refers in this context to eternal, inborn ideas that are received from God; examples are: the numbers, the rules of syllogistic logic, and the principles of the theoretical and practical sciences. This is a Platonic, un-Aristotelian position, as Melanchthon himself notes.[289] These inborn concepts guarantee, together with universal experience and the understanding of the rules of logic, that we have certain knowledge of universals that are not inborn.[290] By combining Aristotelian and Platonic elements in one theory, Melanchthon is able to explain the intellection of universal concepts without recourse to Averroes's unicity thesis. Melanchthons *Liber de anima* was an extremely successful and influential book. Together with Juan Luis Vives's *De anima et vita,* it inaugurated a new trend in philosophical psychology that shunned the technical vocabulary of the Peripatetic tradition and put new emphasis on physiology and the organic soul.[291] Melanchthon and Vives share a contempt for the "fruitless" disputes of the medieval commentary tradition,[292] and for Averroes in particular.[293] Melanchthon's alternative to Averroes found hardly any supporters south of the Alps. Let us therefore turn to two Aristotelian positions that were also of influence in seventeenth-century Italy.

The Paduan philosopher Jacopo Zabarella (d. 1589)[294] differs from Melanchthon in that he continues the Peripatetic tradition. One could even say: the Peripatetic tradition culminates in him, for Zabarella's knowledge of the entire Graeco-Arabic-Latin Peripatetic tradition is hardly paralleled by that of anyone in history, not even of his teacher Marcantonio Genua. He is probably the last author to offer a comprehensive discussion of Averroes's intellect theory—in order to arrive at an entirely different position in the end. With respect to universal intellection, the most informative of Zabarella's works is his *De mente agente* (first published posthumously in 1590). In this treatise, he explains knowledge of universal concepts in a traditional Peripatetic manner, as the result of an abstraction that is achieved through the cooperation of active and possible intellect (Zabarella prefers the term *intellectus patibilis* for the latter). Averroes would have agreed with this general thesis, but not with the details of Zabarella's theory of abstraction. The active intellect, argues Zabarella, makes objects of intellection distinguishable from each other, so that they are rendered intelligible; the active intellect illuminates the essences *(quidditates)* and natures that inhere in the phantasmata. The possible intellect chooses one among these clearly distinguishable objects, in an act of abstraction.[295] As to the active intellect, Zabarella argues that it cannot be a human power, for then the human intellect would illuminate itself.[296] By invoking the authority of Alexander of Aphrodisias, Zabarella argues that the active intellect is identical with God, since its function, namely, to render objects distinguishable and their essences intelligible, can only be achieved by the highest heavenly intelligence.[297] Whether we are able to have abstract knowledge or not, does not depend on the active intellect, whose light is always present, but on the disposition of the faculty of imagination.[298] Hence, Zabarella guarantees the intelligibility of universals by assigning the most important step in abstraction, namely, the discovering of essential structures in the phantasmata, to the eternal competence of a divine entity. In this, as in many other details of his intellect theory, he will be followed by his successor Cesare Cremonini (d. 1631).[299]

Let us turn to our third author, Francisco Suarez (d. 1617),[300] the most prominent representative of the Second Scholastic. It is characteristic of this current that Averroes played a stereotyped role as 'impious philosopher.' His unicity thesis is refuted by many authors of the Second Scholastic, for example, by Francsicus Toletus and the Coimbra commentators, but the exposition of his doctrine and the arguments employed against it take on a standardized and brief format.[301] This is also true for Suarez.[302] What is the theory of universal intellection that he offers as an alternative? Suarez's intellect theory differs radically from Zabarella's in that it presupposes that our souls are able to acquire intellectual knowledge on their own account and that they therefore have to be equipped with everything necessary for intellection. Without the active intellect, without this power that produces intelligible forms, there is no intellectual knowledge; this is why the active intellect has to be a power of the human intellect.[303] The soul is not a perfect image of the world, and hence it is dependent on forms that resemble the external objects: intelligible forms, *species intelligibiles*.[304] These forms are produced when the human active intellect illuminates the phantasmata:[305] the soul first knows an object *phantasiando* and then, on the basis of this exemplar, represents the same object with the help of the active intellect in an intellectual way,[306] that is, by abstracting from particular conditions.[307] Hence, for Suarez, universal intellection is possible for individuals by virtue of a special power of the intellect, its power of representation. This standpoint rests on two presuppositions, which Suarez himself names: first, that a correspondence *(adaequatio)* exists between the object and its representation in the intellect, so that every intelligible can be known, and second, as noted above, that the intellect is equipped with everything necessary in order to know the intelligible object.

As this brief survey of later Aristotelian intellect theories shows, Averroes's intellect theory was replaced not by radically different modes of thought, but by related theories that operate with the same vocabulary and work on the same problems, in other words, by new forms of Aristotelianism, or by new "Aristotelianisms." Averroes's intellect theory, therefore, was not an outlived, decrepit, medieval, and barbarous form of Aristotelianism that succumbed with full right, but much too late, to the new modes of thought of the modern world.[308] As the above examples have amply shown, the unicity thesis was thought to be strong philosophically—so strong that it was not only defended publicly but also enjoyed a hidden success with Vernia and Nifo, that Pomponazzi worked long and hard on avoiding it, and that Vimercato was sure that Aristotle was fully convinced of it, but did not dare to say so publicly.

If a theory that is supported by strong arguments is replaced, it is always to be expected that the philosophical gains of this process are accompanied by philosophical losses. This is also true for the three philosophical alternatives to the unicity thesis mustered above: all three of them, just as Averroes's doctrine itself, show argumentative strengths and weaknesses. Melanchthon, Zabarella, and Suarez are not able to guarantee the possibility of universal intellection in as equally unlimited a fashion as Averroes did, since they forgo a universal human intellect. The solutions offered instead have the great advantage that they ensure the individuality of intellection and the individuality of the soul. This obvious advantage, however, is bought at a certain price—the price of assuming the existence of inborn truths (Melanchthon), of postulating a suprahuman entity of abstraction (Zabarella), and of presupposing a correspondence between abstracted forms and reality (Suarez).

We have seen, therefore, that the philosophical discussion of intellect theory in the Renaissance was not autonomous: it was much strained and subject to substantial external influences. The unicity thesis in particular was under much pressure. The most serious pressures came from the Church, but humanist prejudices and pro-Averroist biases also influenced and distorted the discussion. Remarkably enough, the philosophical discussion itself always regained its strength and assertiveness, and in the end, the unicity thesis was not suppressed, but disappeared gradually, to the benefit of other Aristotelian explanations of universal intellection. This then is a first conclusion to be drawn about the actual reception of Averroes's thought, when contrasted with the rhetoric and polemics surrounding his figure. Before I turn to a proper conclusion, in a final step, I will inquire into the relevance of the humanist attack on Averroes's qualities not as a philosopher but as a commentator.

## HOW JUAN LUIS VIVES DEMONSTRATES AVERROES'S TOTAL IGNORANCE

It is an often repeated charge by Renaissance humanists that Averroes did not know Greek and that his commentaries are therefore inferior to those of the Greek commentators of late anquity. Averroes was even accused of plagiarizing the Greek commentators, for instance, by Ermolao Barbaro, Girolamo Donato, and Giovanni Faseolo, the Latin translators of Alexander, Themistius, and Simplicius, respectively. This drastic charge is unwarranted: Averroes often credits his Greek predecessors, and in many cases, there were no predecessors accessible to him, as for almost the entire *Metaphysics*. But it is true, of course, that Averroes depended entirely on Graeco-Arabic translations, and that the value of his commentaries for medieval and Renaissance philosophers in turn depended on the Latin translations done in the thirteenth century.

An example of such criticism comes from Agostino Nifo. When commenting on the *Metaphysics,* Nifo remarks that Averroes's interpretation is faulty because of the defects of the Graeco-Arabic translation *(fallitur defectu translationum)*.[309] In view of this, Averroes's fame as a commentator is a mystery to Nifo:

> It is very astonishing why this man was trusted so much by the Latins <that is, the Scholastics> on the interpretation of Aristotle's words, since Averroes interpreted hardly any word correctly; <this is

particularly astonishing> given the fact that Averroes himself, in the third book on *De caelo* complains about the translations, saying that the translations he is using were done by Alcindi *(in fact Yaḥyā ibn al-Biṭrīq)* and are <all> false, that the more correct ones were by Isaac *(in fact Isḥāq ibn Ḥunayn),* which he did not have.[310]

With Nifo's attitude contrast those Renaissance authors who find Averroes the most reliable and most Aristotelian of all commentators. He may err with respect to language, Zabarella claims, but is nevertheless right about the matter.[311]

How convincing is the humanist critique of Averroes's commentaries? In spite of the historical significance of Averroes as a widely read commentator from the thirteenth to the sixteenth centuries, should we not pity Averroes's Latin readers for their lack of access to better commentaries, for instance, to commentators who could read Aristotle in the original Greek? One way to avoid the question of the quality of Averroes's commentaries would be to say that all interpretations are prejudiced to a certain degree. The Greek commentaries, for example, are tainted by the Neoplatonic and Peripatetic currents of late antiquity. But this does not settle the question. All commentaries may be alike in that they are prejudiced, but they nevertheless differ greatly in quality. I am convinced that the question of the quality of Averroes's commentaries can only be answered by looking at textual details. All commentators make mistakes, either by their own faults, or because of a faulty transmission. It is unavoidable that the number of mistakes increases through translation and manuscript transmission. The central question is whether in Averroes's case the many small mistakes sum up to a degree that the quality of his commentary is seriously affected and the corruption cannot be compensated by his experience and philosophical acumen.

For a detailed analysis, I have selected chapter Alpha 5 of Averroes's Long commentary on the *Metaphysics* (which is the beginning of chapter Alpha 6 in the Bekker edition). The reason for this choice is that from the humanist scholar Juan Luis Vives there exists a detailed critique of Averroes's interpretation of this chapter, as part of his work *De causis corruptarum artium* of 1531.[312] Vives, the most famous Spanish humanist of the Renaissance, who in his later years lived mainly in Bruges, wants to demonstrate Averroes's ignorance and incompetence as a commentator: Averroes had no knowledge of Greek culture, he argues, nor of the Greek schools of thought; he has never read a text by Plato; he did not read Aristotle in Greek; and the few Greek commentators he knew were accessible to him only in extremely faulty translations. Vives concludes that Averroes stultified the minds of his readers *(hominum mentes dementare)*.[313] In addition to Vives's critique, there also exists another Renaissance comment on the chapter. In about the same years, in the 1530s, Agostino Nifo commented on Alpha 5 and on Averroes's commentary. The comment is part of Nifo's *Expositiones in Aristotelis libros Metaphysices,* a running commentary, which in fact is also a super-commentary on Averroes's commentary. This text was written in the late phase of Nifo's life, in Salerno and Naples, and was published only posthumously. Nifo had, of course, long dropped the unicity thesis, but he still values Averroes highly—so highly that he writes a super-commentary. At the same time, his attitude toward Averroes is very critical.

Alpha 5 is the chapter in which Aristotle begins his discussion of Plato's theory of ideas. In the opening chapters of the *Metaphysics,* Aristotle defines wisdom, the highest form of philosophy, as knowledge of the first causes *(aitiai)* and principles *(archai)*. He then turns to a discussion of the four causes, refers his readers to his *Physics,* and begins a discussion of his predecessors and their theories of causes. Aristotle's discussion of the previous philosophers is long; it covers the remainder of book Alpha. He begins with those who acknowledge the material and efficient causes, and then turns to the Pythagoreans—or the Italians, as he also calls them (987a10 and a31)—who recognize a formal cause in a superficial manner. At this point our Alpha 5 passage begins.

Vives prints the Greek text of Alpha 5 and also provides a Latin translation, which in fact is Vives's own, as a comparison with previous translations shows. He has not copied William of Moerbeke's thirteenth-century translation nor Cardinal Bessarion's translation of 1447–1450. Vives knew Greek, which is worth mentioning since many humanists did not know the language. Let us first turn to Aristotle's Greek text and see how it translates into English:

μετὰ δὲ τὰς εἰρημένας φιλοσοφίας ἡ Πλάτωνος ἐπεγένετο [Vives: ἐγγύσατο] πραγματεία, τὰ μὲν πολλὰ τούτοις [Vives adds: τοῖς Πυθαγορείοις] ἀκολουθοῦσα, τὰ δὲ καὶ ἴδια παρὰ τὴν τῶν Ἰταλικῶν ἔχουσα φιλοσοφίαν. ἐκ νέου τε γὰρ συγγενόμενος [Ross with Aᵇ Alex.: συνήθης γενόμενος] πρῶτον Κρατύλῳ καὶ ταῖς Ἡρακλειτείοις δόξαις, ὡς ἁπάντων τῶν αἰσθητῶν ἀεὶ ῥεόντων καὶ ἐπιστήμης περὶ αὐτῶν οὐκ οὔσης, ταῦτα μὲν καὶ ὕστερον οὕτως ὑπέλαβεν· Σωκράτους δὲ περὶ μὲν τὰ ἠθικὰ πραγματευομένου περὶ δὲ τῆς ὅλης φύσεως οὐθέν.[314]

After the aforementioned philosophies came the doctrine of Plato, which in most respects followed these [τούτοις—here Vives adds τοῖς

Πυθαγορείοις: these Pythagoreans], but also had peculiarities beyond (παρὰ) the philosophy of the Italians. For, in his youth (ἐκ νέου), he first was associated with Cratylus and the Heracliteian doctrines—that all sensible things are always in a state of flux and that there is no scientific knowledge about them—and he held these <views> also later (ὕστερον). Socrates however occupied himself with ethical matters, but not with the whole of nature.

Vives then cites Michael Scot's Latin translation of the Arabic version of Aristotle's text. The Arabic–Latin version of Alpha 5 translates as follows:

> Et post hoc quod dictum fuit de modis philosophiae inventa fuit philosophia Platonis, et sequebatur illos in multitudine, in unitatibus autem erat secundum opinionem Italorum. Et primus qui contigit post Democritum fuit opinio Herculeorum, secundum quod omnia entia sunt semper in fluxu et quod nulla est in eis scientia. Istas igitur opiniones secundum hoc accepimus in postremo. Socrates autem locutus fuit in moralibus et nihil dixit de natura universali.

> After these aforementioned kinds of philosophy Plato's philosophy was devised. And it followed those in many respects, but in some respects it was in accord with (secundum) the belief of the Italians. And the first which came up after (post) Democritus was the belief of the Herculeans (Herculei), according to which all beings are always in flux and there is no knowledge of them. Now, we have received (accepimus) these beliefs in this manner in the end (in postremo). As to Socrates, he spoke on moral matters and did not say anything about nature as a whole.[315]

In Vives's eyes, the differences between the Greek and the Arabic–Latin versions are appalling. Cratylus is replaced by Democritus; the followers of Heraclitus have become *Herculei*; Plato's disagreements with the Italians have become agreements; the reference to Plato's youth has disappeared altogether; and Plato's adherence to the doctrine of Heraclitus in later years has been turned into our reception of the doctrine. And Vives exclaims: "If Aristotle was resuscitated, would he understand this <text>?"[316]

Vives proceeds to demonstrate Averroes's ignorance in detail by quoting Averroes's commentary. "Listen to this famous man, the second Aristotle," he begins.[317] One of the quotes concerns Plato's disagreement with the Italians. Averroes writes in his commentary (in Michael Scot's translation):

Et dixit [Vives: dicit] quod consequebatur [Vives: sequebatur] illam [Vives: istam] in multitudine, id est, quod Plato consequebatur [Vives: consectabatur] in pluribus suis opinionibus opinionem Pythagoricorum et in paucioribus opinionem Italorum. Et ibi [om. Vives] fuerunt primi naturales in Italia, scilicet Anaxagoras et Empedocles et Democritus.

And Aristotle said that Plato followed it in many respects, that is, that Plato followed in more of his beliefs the Pythagoreans and in fewer the Italians. And there, i.e. in Italy, lived the first natural philosophers, namely Anaxagoras, Empedocles and Democritus.[318]

Remember that Aristotle's Greek text says that Plato followed the Pythagoreans in many respects, but on particular issues disagreed with the Italians—that is, the Pythagoreans! Vives comments: "Averroes distinguishes the Pythagoreans from the Italians, as if the Italians were different from the Pythagoreans! Not even our pupils would make that mistake (*quod nec pueri nostri ignorant*)." And he further complains: "Why do you, Averroes, transfer Anaxagoras from Ionia to Italy?"[319]

Vives is even more repelled by the corruption of the names Cratylus and Heraclitus. Where the Greek text reads: "in his youth Plato first was associated with Cratylus and the Heracliteian doctrines," the Arabic–Latin text has, "And the first which came up after Democritus was the belief of the Herculeans (*Herculeorum*)." Vives finds it superfluous to seriously discuss this sentence, since it is completely alien to Aristotle. "These words are only there to be driven off the stage by hissing and clapping (*ad exibilationem atque explosionem*). Where does the text mention Democritus? Who, the devil, are these *Herculei*? Because Hercules is *hêraklês* in Greek, is this the reason why the *heraclitici* become *herculei*?"[320] Vives certainly has a point here: The corruption of the supporters of Heraclitus into the supporters of Hercules is fantastic.

A final point of critique concerns the statement of the Greek text that Plato also held these views (i.e., those of Cratylus) in later years. Averroes comments: "These aforementioned beliefs then are what has come down to us of those who have applied themselves to philosophy up to now."[321] Averroes speaks of our reception, Aristotle of Plato's adherence. Vives exclaims: "Yes, this is Averroes whom some people in their madness have estimated as much as Aristotle, and higher as the divine Thomas! . . . Nothing is more horrible, more lacking in culture, indecent and childish."[322] Why has Averroes been so successful? It is not his fault, it is our fault, Vives says. It was too convenient to remain in the dark, in ignorance and pretension.

And some people have welcomed Averroes because of his impiousness, says Vives:

> Because the doctrine of Averroes, the metaphysics of Avicenna, and in fact everything Arabic seems to me to smack of the hallucinations of the Quran (deliramenta Alcorani) and of the insane blasphemies of Muhammad.[323]

This is exceptional. Anti-Islamic polemics in Renaissance writings about Arabic sciences and philosophy are rare—Symphorien Champier being a notable exception (see Chapter 2). Presumably, there is a specifically Spanish background to Vives's words. Since the last decade of the fifteenth century, Spanish Muslims and Jews were forced to convert to Christianity. Spanish Islam and Spanish Judaism were driven underground. Vives comes from a Jewish family that had converted to Christianity in the early fifteenth century and that continued to suffer enormously from the inquisition—his father was executed by the Spanish Inquisition in 1524. Nevertheless, Vives shares the anti-Islamic and anti-Jewish prejudices of the Spanish Christians, as his late work De veritate fidei christianae shows (published posthumously in 1543), which contains harsh refutations of Judaism and Islam.[324]

With this polemical climax, Vives finishes his demonstration of the ignorance of Averroes. In a second step of my analysis, I proceed to examine the justification of Vives's critique. Were the incriminated mistakes in fact mistakes? If so, who is to be blamed for them? There exist several possible causes: a corruption of the Greek manuscript transmission; or a corruption of the Arabic or Latin manuscript transmission; or a mistake by Nazīf ar-Rūmī (late tenth century), the translator of Alpha from Greek into Arabic,[325] or by Michael Scot, the translator from Arabic into Latin;[326] or a mistake by Averroes himself in his comments.

In what follows, I analyze the origin of errors and misunderstandings generically: from Greek into Arabic into Latin. The difficulties in this case do not begin with the Greek manuscript transmission. There is no reason to suppose that the Greek versions that Nazīf and Vives were using differed dramatically. In general, there exist two major recensions in the Greek transmission of Aristotle's Metaphysics, one represented by the Florence manuscript A$^b$, the other by the Paris manuscript E. In chapter Alpha 5, there is only one major difference between the two recensions: the Florence branch reads συνήθης γενόμενος ("become familiar"—i.e., Plato has become familiar with Cratylus) instead of συγγενόμενος of the Paris branch, which means about the same. Nazīf and Vives seem to have read the Paris version.[327]

The translator Naẓīf, however, did in fact make a mistake, which later misled Averroes. Remember that Aristotle wrote that Plato held peculiar doctrines "beyond (παρὰ) the philosophy of the Italians." Naẓīf translates: bi-ḥasab, "according to" the opinion of the Italians. The Greek preposition παρὰ with accusative is an easy source of error. It may mean either "along, next to," as Naẓīf understood it, or "beyond," as the sense of Aristotle's sentence requires. The result for Averroes was dramatic: he understood the sentence as saying that Plato was following the Pythagoreans in most part, and was "in accordance" with the Italians in some other parts. This made Averroes believe that the Italians must be philosophers different from the Pythagoreans and caused him to guess that "the Italians" were the natural philosophers whom Aristotle had discussed earlier in book Alpha: Anaxagoras, Empedocles, and Democritus.[328]

Another mistake that also probably originated with Naẓīf concerns the reference to Plato's youth. Aristotle writes: "In his youth (ἐκ νέου), Plato first (πρῶτον) was associated with Cratylus and the Heracliteian doctrines." Naẓīf writes: "The first thing which happened after Cratylus was the opinions of the Heracleans" (wa-kāna awwal mā ḥadaṯa baʻda Qrāṭilus ārāʼ al-hiraqliyīna).[329] There is no reference to Plato's youth in the Arabic sentence. It is possible that the phrase had disappeared in the Greek or Arabic manuscript transmission, but it is more probable that Naẓīf took ἐκ νέου to mean "from the beginning," that is, to mean the same as πρῶτον ("first"). Again, the consequences for Averroes were significant. Averroes could not understand that this sentence was about Plato's youthful adherence to the doctrines of Heraclitus, of which Aristotle in the next sentence says, "Plato held these <views> also later (ὕστερον)."

This is then the condition in which Alpha 5 reached Arabic culture. The passages about the Italians and about Plato's youth were already corrupted. Now, did the Arabic manuscript transmission add any mistakes? Yes. The confusion of Cratylus with Democritus, which Vives found so appalling, very probably originated in the Arabic transmission. Naẓīf will have transliterated Cratylus's name as Qrāṭilus. Arabic scribes corrupted this very unfamiliar name into Dīmuqrāṭīs, a well-known name that shares many letters with Qrāṭilus, especially in unpointed Arabic script. The manuscript that Averroes was using and that he quotes in his lemmata apparently contained the corrupt "Democritus." Vives was even more enraged about the fantastic corruption of the followers of Heraclitus into the followers of Heracles. Again, this seems to be a matter of scribal corruption. Naẓīf will have transliterated Ἡρακλειτείοις as Hiraqlitiyīna. The crucial letter "ta," however, which distinguishes Heraclitus from Heracles, has disappeared in all Arabic and Hebrew witnesses, which instead write:

*Hiraqliyīna.* The disappearance of the "ta" is not surprising given that it looks like a "ya" in unpointed Arabic script.

The textual condition of Alpha 5 was therefore far from ideal when Averroes began to comment on it. But unfortunately he increased, rather than limited the confusion. He misread the phrase "he (i.e., Plato) held these views" as meaning "we received these views," that is, he read Arabic *na'ḫuḏu* (we receive) or *'aḫaḏnā* (we received) instead of *aḫaḏa* (he received). In unpointed Arabic, the two forms look very similar, and it may well be that Averroes had to guess. Averroes's guess was the wrong one, since Aristotle's point was not that Heraclitus's doctrine was transmitted to us, but that Plato also in his later years remained a follower of Heraclitus. Averroes, however, seems to have smelled the corruption of the term *Hiraqliyīna.* He does not mention the name in his commentary. When we now turn to the Latin translator Michael Scot, we see that he writes *Herculeorum.* With this phrase, the ambiguity of the Arabic term *Hiraqliyīna* is turned into a full-fledged mistake: "Heracliteans" have finally become "Herculeans." Michael Scot was probably at a loss when reading the corrupt *Hiraqliyīna.* Since Heraclitus's doctrine is discussed in book Gamma, he could have detected that *Hiraqliyīna* must be a reference to the followers of Heraclitus, and not of Herakles. Michael Scot's unfortunate translation *Herculei* therefore would have been avoidable. However, for the rest of Alpha 5 Michael Scot's Latin translation faithfully reproduces all sound and corrupt passages without adding any significant errors.

In retrospect, we can only agree with Vives that the corruption of the text is dramatic. The majority of mistakes were made before Averroes—by the translator Nazīf and by Arabic scribes. It is telling to see that the mistakes concern small things: the mistranslation of a preposition and of an adverbial phrase ("since his youth"), the confusion of single letters in unpointed Arabic script. The Arabic translators and scribes were not incompetent. But the difficulties were such that distortions of this kind were unavoidable. What Vives and many other humanists claimed is true: that Averroes relied on translations of Aristotle's works that contained many mistakes. The humanists had a point when demanding that one should read Aristotle in Greek.[330]

It is instructive to compare Averroes's reading of Alpha 5 with that of Alexander of Aphrodisias, the ancient Greek commentator. Was Alexander any closer to Aristotle's original text? He writes:

> Next Aristotle sets forth the doctrine of Plato, who, he says, followed the Pythagoreans in many respects but also took certain positions that are peculiar to him; one of these distinctive features was that

concerning the Ideas. And he tells the source which led Plato to posit Ideas, and reports that from Cratylus, a Heraclitean, whose associate he was, he took <the belief> that all sensible things are in flux and never stand still, and that Plato continued to maintain this opinion as true. Socrates, however, occupied himself with ethical questions and in seeking the universal.[331]

Alexander, of course, gets the Greek names right, but he also correctly understands the preposition παρὰ and the reference to Plato's continuing to hold Heraclitean doctrines. The contrast with Averroes is stark. The mistakes of the Arabic–Latin tradition have disppeared altogether as if extinguished with a reset button. When the humanists first read the Greek commentators, among them Alexander and his *Metaphysics* commentary, it must have been an impressive experience. Alexander's *Metaphysics* commentary was not printed in Greek in the Renaissance, but only in a Latin version by Juan Sepulveda, which first appeared in 1527 in Rome.[332] Among its first readers was Agostino Nifo.

But does this mean that Averroes, the commentator, ought to be debunked—at least on Alpha 5? The answer is no. Averroes's commentary is severely hampered by a faulty Aristotelian text, but the commentary is nevertheless an informative reading. Vives ignores this because he omits important passages of Averroes's commentary. The strengths of Averroes's commentary are more apparent in Agostino Nifo's above-mentioned *Metaphysics* commentary. Nifo is more gracious to Averroes. With respect to Averroes's misreading of "he accepted" as "we accept," Nifo notes: *Errat errore translatoris,* he errs because of the error of the translator, who does not refer this phrase to Plato.[333] Nifo's excusal of Averroes is infelicitous, since the faulty reading "we accept" most probably originated with Averroes and not with Naẓīf, as we have seen, but it remains noteworthy that Nifo at least tries to distinguish between the faults of the translator and those of Averroes. And Nifo has an eye for Averroes's overall interpretation of Aristotle's passage. I will first quote the initial half of Averroes's commentary on Alpha 5, and then turn to Nifo's comments. The translation is from Arabic; passages not translated into Latin are in square brackets:

> He (i.e., Aristotle) says: *And after these kinds of philosophy there was the philosophy of Plato,* i.e. [after the philosophy of the adherents of numbers, who are], the Pythagoreans and generally those who posited mathematical things to be the principles of beings, and after the philosophy of the natural philosophers, who are Anaxagoras, Empedocles, and Democritus. As to his saying [about Plato's philosophy] that it followed them

in many respects, he means that Plato followed in more of his beliefs [about beings the doctrine of the adherents of numbers, i.e.] the Pythagoreans [and those close to them], and in fewer the Italians, [who are known today <to live> in the land of the Franks]. And there, [God knows it best,] lived the first natural philosophers Anaxagoras, Empedocles, and Democritus [and the followers of each of them]. And he said that Plato followed in most of his philosophy those who posited mathematical things to be the causes of the perceptible things [or of the individual perceptible things], because Plato claimed the existence of forms and believed that the nature of forms and the nature of number were the same, as will be explained in the treatises on substance [of this science], and believed that the four elements were composed of planes of equal sides and angles, these being the five bodies mentioned in Euclid's last book, and followed the natural philosophers by claiming the existence of first matter and of the four elements, [i.e., that all perceptible compounds are composed of them].[334]

In this passage, we can again observe the consequences of the faulty translation and of Averroes's lacking knowledge of the history of Greek philosophy: the Pythagoreans and the Italians have become two different groups. And he adds further information on Plato's sources: Plato adopts much from the Pythagoreans, because his theory of forms, which are also numbers, is influenced by the Pythagorean idea that numbers are the principles of perceptible things; and Plato also adopts a few doctrines from the natural philosophers: about first matter and the four elements.

Nifo explains that Averroes makes three points in this commentary, which he criticizes in sequence. First, he scolds Averroes for claiming that Anaxagoras, Empedocles, and Democritus adhere to the doctrine of numbers. Second, he criticizes that Averroes makes these thinkers the first natural philosophers in Italy. With respect to the first point, Nifo has misunderstood Averroes's commentary: Averroes in fact does not attribute the doctrine of numbers to the natural philosophers, but to the Pythagoreans. But Nifo rightly complains that "the Italians" in Aristotle's text are not to be identified with the natural philosophers. Third, as to the passage about Plato's doctrine of forms and its sources, Nifo comments: *non est ad propositum textus,*[335] this does not pertain to the topic of the text, as Nifo claims by contrasting Averroes's exposition with the *expositio Alexandri*. Nifo has read Sepulveda's translation of Alexander's commentary and uses it amply in his own commentary, also silently.[336]

Nifo is not fully justified in his criticism that Averroes's commentary leads one astray. Averroes realizes that Alpha 5 is a passage about Plato's sources. He gives his readers reliable information about Plato's sources by naming doctrines Plato has adopted from the Pythagoreans on the one hand and from the natural philosophers on the other hand. Averroes does what he very often does: he makes good Aristotelian sense out of faulty passages. The sense is not what Aristotle said in this passage, but something that Aristotle *could* have said, in this or other contexts. Averroes is not a perfect guide to Aristotle's texts, but he is a very good guide to Aristotelian philosophy. His commentaries always are also expert manuals of Aristotelian thought. This quality of Averroes's commentaries was probably the cornerstone for his great success in medieval and Renaissance Europe.

To conclude: The analysis has shown that the basic argument of Vives's and Nifo's critiques of Averroes's commentary on Alpha 5 is justified to a large degree: Averroes is not a good guide to what Aristotle wanted to say in this passage. As a commentator on the passage, Alexander is clearly to be preferred to Averroes. Because Vives knew Greek, he was able to show this. Since he did not know Arabic, however, he was not in a position to pass judgment on the origin of the mistakes in Averroes's text. Vives did not have a sense for the considerable problems of translation and transmission, and hence made Averroes's intellectual qualities his central target—which he was not able to evaluate. As a result, much of his polemic is merely an expression of prejudices.

Alpha 5 is a difficult test case for the quality of Averroes's commentary. The understanding of the passage requires historical knowledge rather than philosophical expertise. It contains transliterated Greek names, and with Cratylus even a rare name. Averroes, with all his experience, was not able to remedy the mistakes made by translators and scribes of Alpha 5. And hence the question arises whether the result concerning Alpha 5 is in any way representative of the qualities of Averroes's commentaries as a whole. From my experience as an Averroes editor I can say that, apart from the transliterated names, the textual corruptions are very typical of the *Metaphysics* transmission in general. Many chapters are translated and transmitted very faithfully, but many chapters are also distorted in the way Alpha 5 is. The problem is not usually Aristotle's philosophical vocabulary. An experienced reader of Aristotle's text like Averroes is often able to counterbalance a translator's infelicitous choice of vocabulary. The greatest problems for understanding Aristotle's text come from the distortion of small words that have an effect on the syntax of the text. The following list of such small corruptions in book Gamma may illustrate the point. In Gamma 9, the Arabic text contains an additional conjunction "if"; in Gamma 10 the Greek text contains a negation that is missing in the

Arabic; in Gamma 15 and 27, Greek "either . . . or" constructions are rendered with "and" in Arabic; in Gamma 18, the Arabic *fa-idan,* "hence," has become *fa-id,* "since" by way of scribal corruption; similarly, in Gamma 27, *lā abyad,* "not-white" has been corrupted into *al-abyad,* "the white," and in Gamma 29, the Greek "impossible" corresponds to Arabic "possible."[337] Such corruptions are difficult to come by, and, mostly, Averroes's interpretation falls victim to them. Only sometimes is he able to circumvent them, as in Gamma 24, where the omission of the name Cratylus in Aristotle's text does not allow him to differentiate between Heraclitus and Cratylus; nevertheless, his overall interpretation of Heraclitus's doctrine remains unaffected, clear, and pertinent.[338] It is true that small corruptions that distort the meaning of the text can happen in all phases of transmission, and also within the Greek transmission itself. But such mistakes accumulate with every additional step in the transmission. And hence, the Greek commentators have an advantage over Averroes when it comes to matters of textual interpretation, as the humanists rightly pointed out. This explains why the Renaissance revival of the ancient commentators was a real challenge to Averroes's dominating position as a guide to Aristotle in university education. It was with good reasons that the humanists complained about their university colleagues who did not want to learn Greek.

This being said, it has also become clear that Averroes often evades the problems of textual reliability by offering doctrines that Aristotle *could* have expressed on the topic of the chapter. In this respect, Averroes offered more than the Greek commentators. He was an all-around Aristotelian philosopher, from logic to zoology and ethics, to a degree unparalleled in antiquity. And hence, the Middle Ages and the Renaissance were very fortunate in having him as a guide to the Aristotelian textbooks of the universities. This quality is the principal reason why, in the Renaissance, Averroes continued to be read when the Greek commentators became accessible.

The above analysis is not meant to exaggerate the problems of translation and transmission from Greek into Arabic into Latin. Cross-cultural and cross-linguistic transmission has always been a great challenge to translators, scribes, and commentators in any epoch, and the Graeco-Arabic translation movement is among the most remarkable achievements in history: the translators successfully adopted foreign sciences into their own culture by developing a scientific and philosophical Arabic language that did not exist before.[339] Because of this, the Arabic versions of Aristotle's texts are assessed today as valuable witnesses for the philological reconstruction of the Greek text, as has recently been exploited for the *Poetics* and the *Nicomachean Ethics.*[340] Hence, with respect to the wording of Aristotle's text, the Graeco-Arabic transmission is two very different things at the same time: a source of both illumination and confusion.

✣

The Renaissance dissension about Averroes was deep. The Arabic commentator had dedicated partisans, especially in the philosophical faculties of Italian universities, and he had bitter enemies among humanists and theologians in many parts of Europe. The humanist protest against Averroes, the commentator who lacks Greek, stretches over two centuries. Equally tenacious, but more vehement is the religious polemic against *maledictus* Averroes, which appears in many different sources: theological treatises, official Church documents, biographical literature, and humanist treatises. The long-lasting pictorial tradition of Thomas Aquinas triumphing over Averroes is a particularly vivid expression of the fact that for a certain section of Renaissance society Averroes was a heretic in the first place—and a heretic who even had his followers among contemporary philosophers. The disgust about this situation is well captured in Bishop Barozzi's remark that the University of Padua appeared to him in no way different from "a university of pagans" *(studio de' pagani)*—the university whose chancellor he was.[341]

The Renaissance movement of Averroism was a reality and not an invention of nineteenth-century French historians. From the thirteenth century onward, the term *Averroista* was used to designate followers of Averroes who adhere to contentious doctrines. About a dozen doctrines were attributed to Averroists, among them, and very prominently, the unicity thesis. In the Renaissance, the self-confidence of Averroists was such that the term *Averroista* occasionally took on a positive connotation meaning "expert on Averroes." A reasonable definition of the historiographical label "Averroism," I have argued, should be tied to the medieval and Renaissance usage of the term. Then two historical uses of the term "Averroism" are possible: meaning either a controversial philosophical movement or expertise on Averroes. The first sense has a longer tradition in the Middle Ages and in modern scholarship and is therefore used in the present study. In view of this, it is inappropriate when modern historians use the term "Averroism" broadly for any kind of Averroes influence in Europe.

Averroism was not only a tradition of controversial thought, but a proper movement. Renaissance Averroists formed a group: they referred to each other, they were related to each other as teacher and student, and they debated about the proper direction of the movement. On the evidence of the unicity thesis, we can identify an Averroist movement at Italian universities reaching from at least the early fifteenth century to the middle of the sixteenth century. Much evidence shows that this movement was particularly strong in the decades around 1500. The reception of other Averroist theses, especially in physics and metaphysics, still awaits proper research, as does the overall influence of Averroes, which is enormous and much broader than the Averroist movement itself.

The philosophical discussion of Averroes's unicity thesis in the Renaissance was subject to considerable religious pressure. This pressure was much more dangerous for the survival of Averroism than the humanist charges against Averroes's incompetence. The very tangible impact of the 1489 condemnation is without much precedence in the Middle Ages. Past research on the famous condemnations of 1270 and 1277 has shown that the condemnations had some impact on the theological faculty of Paris, but that the masters of arts remained largely unimpressed.[342] It is all the more remarkable that several Renaissance philosophers reacted very sensibly to the pressure exerted by Bishop Barozzi, by the inquisitors of Padua, and by theologians such as Antonio Trombetta. Before about 1480 and after about 1520, the philosophers usually declared the unicity thesis to be untrue from a Christian point of view, but did not compromise their philosophical solutions. In the decades in between, however, the climate was different. Niccolò Tignosi was among the first to modify his philosophical conclusion on the unicity thesis in accordance with theological considerations. The orthodox pressure on the old and respected philosophy professor Nicoletto Vernia was substantial. Vernia's anti-Averroist treatise of 1492, his anti-Averroist testament, and his appeal to three theologians for testimonies of orthodoxy are desparate attempts to evade the pressure. The confession-like "philosophical" conclusion of his 1492 treatise is a negative high point in the history of Western philosophy faculties. In Bologna, Alessandro Achillini apparently was impressed by Barozzi's intervention and hid his Averroist position under an orthodox surface. The young Agostino Nifo, in turn, was accused of heresy by unknown adversaries, an episode that changed his intellectual biography. Nifo did not have to recant publicly, as did Vernia, but he substantially redressed and censored his writings. His later utterances on the unicity thesis are clearly written with orthodox scissors in mind. Barozzi and his allies created a climate of denunciation, intimidation, pressure, recantation—and secret persistence. Vernia and Nifo both retained parts of their earlier philosophical conviction by quoting Albertus Magnus's implicit adoption of the unicity thesis. This does not turn Vernia and Nifo into freethinkers.[343] They remained ordinary philosophy professors who tried to retain a certain amount of philosophical consistency. In view of all this, it is remarkable that the unicity thesis survived the attempt at repression. Prassicio, Genua, and Vimercato returned to the traditional practice of keeping religious and philosophical argumentation apart.

It has also emerged in the investigation that the pressure of orthodoxy was not exerted by "the Church" as such, but was due to some individuals among the clerics, in particular Pietro Barozzi, the Paduan bishop, and Antonio Trombetta, the Paduan theologian, who made anti-Averroism their trademark of orthodoxy. Other leading Church officials had a different attitude toward Averroes: Cardinal Domenico Grimani, the collector of Hebrew Averroes manuscripts, was an important patron of translations of Averroes, as was Bishop Ercole Gonzaga of Mantua in the 1520s. One may add also that the bull of the Lateran Council of 1513, which demanded the orthodoxy of philosophical teaching, was voted against by Cajetan de Vio, general of the Dominican order, and Nicolas Lippomani, bishop of Bergamo,[344] and that Pietro Pomponazzi, when facing charges of heresy, found support with Cardinal Pietro Bembo.[345] There was significant clerical resistance against the idea that the Church should interfere with the teaching of the faculties of philosophy. The attempts to suppress Averroes, therefore, which clearly existed and hit Vernia badly, were not carried out by a large group of clerics, but by dedicated individuals, who were able to gather the weight of the institution behind their cause for a limited period.

It is time to abandon a myth about Averroes's influence in the Renaissance that has found some distribution in the past decades—that the Arabic commentary tradition and Averroes in particular were supplanted by the Greek commentators in the Renaissance. In a series of studies, Charles Lohr, Edward Mahoney, and Eckhard Kessler have unearthed the influence of the Greek commentators, building on the earlier research by Bruno Nardi, Paul O. Kristeller, and others.[346] In their enthusiasm about the rebirth of the Greek commentators they overemphasized their influence in one respect: it did not lead to the replacement of either Averroes the philosopher or Averroes the commentator. It is true that Vernia and Nifo claimed to have abandoned Averroes because they had read the Greek commentators or Aristotle in Greek. But, as we have seen, the Greek turn advocated by Vernia and Nifo was also a means to cover up another cause of their shift, the pressure of orthodoxy. The new position adopted by the two philosophers owes little to the Greek commentators and very much to Thomas Aquinas's *De unitate intellectus*.

With respect to intellect theory, the only real alternative to Averroes among the Greek commentators was Alexander of Aphrodisias, an alternative chosen by several later philosophy professors from Pomponazzi onward. But other colleagues continued to prefer Averroes's intellect theory, among them a central figure in sixteenth-century Paduan philosophy: Marcantonio Genua. Averroes the commentator, in turn, was not expelled from the classroom in exchange for his Greek forerunners. There is ample evidence that Averroes was read alongside the Greek commentators. He remained the principal point of orientation, even if much criticized, for Nifo in the 1530s and for Genua in the 1540s. To be sure, Barbaro, Donato, and Faseolo called for the replacement of Averroes through the Greek commentators in the prefaces to their translations, and they did this in an aggressive manner. These programmatic statements may have influenced the judgment of modern scholars, but they do not reflect the reality of Italian university teaching—their primary function seems to have been the advertisement of humanist translations. As to the Greek text of Aristotle's *De anima,* its impact on the reception of the unicity thesis and on the *De anima* commentaries surveyed here was remarkably limited. The only serious discussion of the Greek text encountered above was undertaken by Vimercato in 1543. This result agrees with the findings of other scholars who have pointed to the delayed influence of Greek texts in the Renaissance—to "Hellenism postponed," as Vivian Nutton put it.[347] Scholars clearly liked to pride themselves on their knowledge of Greek, but we should be careful not to trust them easily. Here the physician Lorenz Fries may have hit the mark when complaining in 1530: "In fact, there are few who know Greek, but many who as soon as they are able to sketch a iota, unhesitatingly boast about being ardent Greek scholars."[348]

The abandoning of the myth of Averroes's replacement by Greek authors closes an explanatory gap. If it were true that Vernia and Nifo marked the major turning point toward Greek Aristotelianism, it could not be explained why the history of Averroes editions and Averroes translations reached a climax with the Giunta edition of 1550/1552 and the translations of Jacob Mantino, whose translations date from the 1520s to his death in 1548. Instead of being replaced, Averroes was everywhere in Italian philosophy faculties of the first half of the sixteenth century, so much so that Pomponazzi complained in 1517 that he will not have "any friend in Padua and Ferrara" if he does not defend Averroes.[349]

The controversies between partisans and enemies of Averroes have left visible traces in the technical philosophical and exegetical discourse. What were the merits and demerits of these interferences? In view of the evidence laid out above, the answer will be different for the controversies about the unicity thesis and about Averroes's quality as a commentator. The attacks on the unicity thesis by Church officials and humanists are difficult to justify from an internal, that is, a philosophical point of view.

Averroes's theory, whether heretical or not, had important philosophical and explanatory advantages. It not only offered an explanation of intellectual knowledge but also safeguarded the individuality of thinking by linking it inextricably with the actualization of the phantasmata in imagination. Averroes's theory was also attractive because his *duo-subiecta* theory resonated with a tenet popular among Renaissance philosophers: that the human being partakes in both the inferior world of the animals and the higher world of the intelligences and God. This is how Vernia and Nifo understood Averroes's intellect theory, via the interpretation of Albertus Magnus.

It contributed to the strength of the unicity thesis that it could not easily be rejected as disagreeing with Aristotle's text. Averroes offers a separabilist interpretation of Aristotle's intellect theory, which emphasizes passages in Aristotle that have traveled without much distortion from Greek into Arabic into Latin—that the thinking part of the soul is unmixed and pure potentiality. A refutation of Averroes's unicity thesis, therefore, was bound to be technical and philosophical, rather than philological. It was part of the fascination exerted by this theory that it supported a simply formulated thesis with a highly technical apparatus of concepts and arguments: the differentiation between several different intellects, the *duo-subiecta* theory, the distinction between first and final actuality, the rejection of previous intellect theories by Theophrastus, Alexander of Aphrodisias, Themistius, and Avempace (Ibn Bāǧǧa). In this sense, the unicity thesis was elitist, a theory for real experts in Aristotelian philosophy. There are indications that the elitist attitude of Averroists occasionally deteriorated into ideology, as in Prassicio's prejudiced belief in Averroes's orthodoxy and his attack on deviators from the true Averroist path. An indirect testimony to the ideological petrification of Averroism comes from Pomponazzi who insists that he has the right and freedom to disagree with Averroes and hence to be "a heretic in philosophy." It is possible that the prejudiced defenses of Averroes are reactions to the prejudiced attacks on him, and perhaps the elitist attitude of Averroists contributed to or was even necessary for the survival of Averroism. In the end, the unicity thesis survived all nonphilosophical pressures. It disappeared not because it was suppressed, but because new Aristotelian solutions were preferred and because the commentary culture, on which Averroism was growing, changed.

One factor in the decline of the unicity thesis as a technical theory was the decline of Averroes the commentator. The humanists were right, at least in their general line, when criticizing the reliability of Averroes's commentaries. The quality of his exposition suffers severely from the accumulated errors of transmission and translation. Averroes is able to balance some of the textual corruptions with his enormous knowledge of the

Aristotelian corpus. But his commentaries cannot compete with the level of textual precision offered by the Greek commentaries on Aristotle's Greek text. The medieval Latin Averroes translations again added corruptions. The humanists were the first who could pass judgment on the differences between the Greek Aristotelian sources on the one hand and the Arabic–Latin tradition on the other, and their protest against the pivotal position of Averroes as the dominating commentator on Aristotle was justified. The humanist enemies of Averroes were not justified, however, in calling for a complete replacement of Averroes. Averroes, as an expert on virtually all parts of Aristotle's philosophy, has qualities as a commentator that are without parallel in the ancient world.

Medieval and Renaissance Aristotelians did not take full advantage of this. Nifo in one text admits the obvious, namely, that he cannot read Averroes in the Arabic.[350] The humanist credo *ad fontes* was heavily biased toward antiquity. In view of the quality of Averroes's commentaries, the humanists should have advocated the learning not only of Greek but also of Arabic in order to be able to read Averroes in the original. Some scholars, in fact, did advise the learning of Arabic for this reason, for instance, Thomas Erpenius, the Leiden professor of Arabic, in 1620:

> Philosophers, consider how important it would be to be able to hear
> that other Aristotle, ابن راشد *(sic)* Ibno-Rasjidum, teaching most elo-
> quently in his own language. You name him ineptly Averroes and,
> because of the way he stammers in Latin, you rightly complain that you
> can scarcely understand him and that you need an interpreter for the
> interpreter himself.[351]

In a similar vein, Jacob Christmann, the Heidelberg professor, argued in 1582 that it is worth reading the Arabic philosophers in Arabic because they often expound Aristotle much better than the Greeks themselves.[352]

But these calls were not heard, or they proved impossible to follow. Averroes was much read in Hebrew and Latin in the Renaissance, but a serious study of the Arabic Averroes in Europe began as late as the mid-nineteenth century with the work of Salomon Munk.[353] As we have seen in Chapter 3, when studying the Hebrew–Latin translation movement of Averroes, there was a great demand for better Latin versions of Averroes's commentaries in the Renaissance. Hebrew–Latin translations significantly increased the number of commentaries available in Latin, but it was hardly possible to improve existing Latin versions on the basis of the Hebrew only. Unfortunately, Averroes's commentaries did not have a reviser like Andrea Alpago, who was able to lift the Western understanding of Avicenna's *Canon* through recourse to the Arabic. The Collegium of philosophers and physicians of Padua in 1521 officially recommended Alpago's "translations and corrections of passages" in Avicenna's writings, and it is very likely that an Arabist *correctio locorum* for Averroes would have been very welcome too. Alpago, as a former student of Averroist teachers in Padua,[354] may well have pondered the idea of improving not only the Latin Avicenna but also the Latin Averroes. However, he would have had many difficulties finding, in Damascus or Cairo, an Arabic manuscript of Averroes's commentaries, which are extremely rare up to this day,[355] or meeting an Arab scholar who knew a commentary by Averroes. The very meager reception of Averroes's commentaries in Arabic, and the dearth of manuscripts resulting from it,[356] may well have been the reason for the lack of Arabist studies of Averroes in the Renaissance. Hence, the European reception of Averroes, unlike that of Avicenna, did not benefit from the growing competence of Renaissance scholars in the Arabic language.

# CHAPTER 6

# ASTROLOGY

*Ptolemy against the Arabs*

Astrology, as a practical and theoretical science, saw a remarkable success in fifteenth- to seventeenth-century Europe, a success hardly paralleled in history. Astrological advice was valued by kings and popes, citizens and peasants, the educated and the uneducated, and it was offered by a great range of astrologers, from university teachers to city physicians and mere fortune-tellers. It was common astrological practice to cast horoscopes for the moment of birth of a specific person, to inquire about the best moment for a specific action, or to formulate a prediction for the general events of the coming year.[1] The influence of astrology was much widened by the discovery of printing. Because of the mass production of astrological predictions in pamphlets, astrology mattered to a much wider public than before, especially since many predictions were written in the vernacular. A well-known example of such wide attention was the prediction of a catastrophic flood in 1524, which caused great anxiety among the German and Italian public about a second deluge.[2] In the course of the sixteenth century, astrologers several times aroused the public, especially in Protestant societies, by prognosticating the end of the Roman Empire, the second coming of Christ and the end of the world, for instance for the year 1583, as indicated by a Saturn–Jupiter conjunction.[3] The production of apocalyptic predictions declined only toward the very end of the sixteenth century.[4]

In addition to the great and visible influence of astrology as a practical art, astrology also thrived as an intellectual discipline. Many thinkers of the period tried to improve astrological doctrine, and thus continued Greek and Arabic traditions, among them capital figures such as Philipp Melanchthon, Girolamo Cardano, and Johannes Kepler. The development of the discipline profited from the fact that at many universities astrology formed part of the regular curriculum in *mathematica* or medicine, as we have seen in Chapter 1 of this study. In Bologna, Ferrara, and Kraków, the statutes even explicitly designate the mandatory astrological reading for students: texts by Ptolemy, Alcabitius, and Albumasar. Astrology was thus given a firmer institutional basis than it had in previous centuries, when it was taught only irregularly in the context of the quadrivial sciences. That astrology was a booming discipline is also visible in the many editorial efforts concerning astrological sources by Greek and Arabic authors, such as Ptolemy's *Tetrabiblos* and Alcabitius's *Introductorius*. Several attempts were made to improve the transmitted texts, and new forms of commentaries were developed.

Astrology was not left unchallenged, however. Throughout its history, it has always been accompanied by criticism. The arguments came from different angles, as a few names may illustrate: in antiquity, pagan and Christian thinkers like Cicero and Augustine attacked astrology for advocating determinism and for negating free will. The Arabic philosopher Avicenna (d. 1037) did not doubt the influence of the stars, but argued that it was impossible for human beings to acquire certain knowledge about their influence—an argument adopted later by Pico della Mirandola.[5] In the Latin world, Thomas Aquinas (d. 1274) was convinced that astrological prediction was possible and sensible, but not for events that happen by chance or are dependent on the free will of human beings. Nicole Oresme (d. 1382) tried to refute astrology by demonstrating that basic astrological tenets were impossible astronomically and mathematically. In the Renaissance, the controversy over the value of astrology continued. Famous critics of astrology were Pico della Mirandola and Bernardino Telesio.[6]

The Catholic Church showed an ambivalent attitude toward astrology. While many popes between Sixtus IV (1471–1484) and Paul III (1534–1549) favored the science and employed professional astrologers,[7] the index of prohibited books of 1557 banned all books on "judicial astrology" *(astrologia iudiciaria)*. Subsequent indices of 1559 and 1564 repeat the earlier formulation, but explicitly exempt astrological publications aimed to benefit navigation, agriculture, and medicine. Pope Sixtus V, in a bull of 1586 against judicial astrology, adopted the formulation of the index, again exempting predictions in the above-named fields. Sixtus's ban was renewed by Pope Urban VIII in 1631.[8] It is noteworthy that the prohibitions do not condemn astrology in its entirety. They do not cover the great part of general astrology that is concerned with weather, seasons, illnesses, and so forth. The general tone of the prohibitions is such that all judgments seem to be allowed that do not concern contingent or chance events or freely willed actions. Moreover, the 1564 index limits the prohibition to those judgments claiming that something will happen "with certainty" *(certo)*. As a result, the exact range of the prohibitions remained controversial even within the Roman institutions. As to the impact of the prohibitions on the book market and on the development of the discipline, it is not possible to say much at the present stage of research. Naturally, they did not have an effect on the Protestant world, where astrology had been a thriving discipline since the beginning of the Reformation. Martin Luther grew increasingly skeptical of astrology over time and in his later years disliked the entire science. But, very influentially, his friend Philipp Melanchthon, the academic mastermind of the Reformation, actively fostered astrological work in Wittenberg and elsewhere. It is a well-known chapter of Reformation astrology that a dispute arose about the correct nativity and the correct birth date of Martin Luther.[9]

In spite of the significant opposition against astrology, the fifteenth to seventeenth centuries were without doubt the golden age of this science in Europe. In the eyes of most intellectuals who adhered to it, astrology needed hardly any justification. It was obvious to them that the lower planets, the sun and moon, were influencing the earth, its seasons, weather, and tides, and it seemed unlikely that the other planets should not exert any influence, even if it was difficult to detect.[10] Long-distance effects, such as those that later centuries explained by gravitation or magnetic fields, have always been among the most difficult puzzles for scientists, and they still are. Renaissance scholars did not yet have satisfactory explanations for many such effects.[11] A refutation of astrology was a difficult undertaking, and astrological doctrines appealed to the best minds of the time, especially since they were based on the traditional teaching of Aristotle that the supralunar world has physical effects in the sublunar world. When studying Renaissance astrology, therefore, we ought to free ourselves from modern attitudes toward astrology. To understand the development of astrology in the Renaissance, and to value its Greek and Arabic components, we ought to accept, for the sake of method, the basic framework of an astrologer of the time and to treat astrology as a serious science.

## THE GREEK–ARABIC ANTAGONISM

When humanist intellectuals turned their attention to astrology, they realized that the discipline, as it had developed in the Middle Ages, was drawing not only on Greek sources but also, and even more heavily, on Arabic ones. Greek and Arabic were, in fact, the two main source cultures of medieval and Renaissance astrology. To a lesser degree, Hebrew was also a source language, since several works by the Hebrew astrologer Abraham ibn Ezra (d. ca. 1161) were available in medieval Latin versions by Pietro d'Abano.[12] The historical roots of astrology lay in ancient Babylonian culture, but the development of the art into a scientific discipline was an achievement of Greek astrologers who were active in the two centuries before and after the beginning of the common era. Among these astrologers were Dorotheus of Sidon and Ptolemy. The major Arabic source texts, which in turn were influenced by Greek, Indian, and Persian sources, were written in the eigth to tenth centuries CE, most of them in Baghdad, the city of the caliph. In twelfth-century Spain, a comprehensive corpus of Arabic astrological treatises was translated from Arabic into Latin, prompting the rise of astrology as a scientific discipline in the Latin Middle Ages. Previously indigenous Roman traditions of astrology and some tenth-century translations from Arabic had also existed, but the major texts were translated in the twelfth century, mainly by John of Seville: Albumasar's *Great Introduction to Astrology,* Alcabitius's introduction to judicial astrology, and several technical works in specific branches of astrology by Albumasar, Messahalah, Zahel, and other Arabic authors. In the mid-thirteenth century, Abenragel's astrological summa was translated, again in Spain. Together with Ptolemy and Abraham ibn Ezra, Arabic astrologers formed the basis of late medieval and Renaissance astrology. The successful printing history of Arabic astrological authors in the fifteenth and sixteenth centuries is but one of many indicators that Arabic traditions in this field were in full flourish in the Renaissance.

This is the situation that led to the confrontation between Greek and Arabic astrological traditions in Renaissance culture. This antagonism had a certain irony to it, as we will see, since it was not known in the Renaissance that Ptolemy's *Tetrabiblos* gives a select picture of Greek astrology and that many "Arabic" additions condemned by humanist authors were in fact perfectly Greek in origin. As in the field of *materia medica,* there is a founding book to the controversy about the value of Greek and Arabic astrology: Giovanni Pico della Mirandola's *Disputationes adversus astrologiam divinatricem,* published two years after his death in 1496, apparently with heavy editing by Pico's nephew Gianfrancesco and by his assistant, the young Giovanni Manardo, who will later become a prominent humanist physician and fervent critic of Arabic medicine.[13] Pico's book is a full-blown attack on the entire science of astrology. It served as a storehouse of arguments for later critics of astrology. But Pico's text is multilayered, as recent research has shown.[14] Some passages criticize the Arabs for deviating from Ptolemy's astrology, other passages criticize the Latins for not knowing their Arabic and Hebrew sources, still others criticize Ptolemaic principles of astrology, but not Ptolemy himself, who is treated respect-fully throughout. Moreover, Pico is well-versed in the details of astrolog-ical theory, so much so that the attraction exerted by this science can be felt in the technical passages of the treatise. Hence, the argumentative aims of the *Disputationes* do not seem to be stable. While much reads like a rejection of astrology as a whole, other parts seem to advocate a humanist reform that purges Ptolemaic astrology of medieval and Arabic additions. The reformist passages in Pico's work often concern astrological doctrines that Pico called "Arabic." He attacked Arabic astrology for deviating from the methods and doctrines of the Greek astrologers, in particular in the field of general astrology. When criticizing the doctrine of the great conjunctions of Saturn and Jupiter, he denounces Albumasar as "the source or inventor of this error."[15] On another doctrine of general astrology, the *magnus orbis* cycle of 360 years, he comments: "Nothing like this was ever thought of by Ptolemy or by any other ancient author; rather, it is a fiction by the more recent Arabs."[16] Pico becomes particularly agitated on the sub-ject of the nativities of Mary and Christ: "Where did Ptolemy or any other ancient author mention these nativities? They are the purest follies and fic-tions of the Arabs" *(meracissimae nugae sunt Arabumque figmenta).*[17] It is dif-ficult to tell today whether the reform of astrology was a true aim of Pico's, or whether the differentiation between Greek and Arabic astrology only served him to discredit astrology as a whole. Whatever his aims, the *Dispu-tationes* were read by many as a reformist treatise.

Pico's call for a return to Ptolemy and for an abandoning of the as-trology of the Arabs found support among a good number of sixteenth-century astrologers. Lynn Thorndike and, more recently, Paola Zambelli have unearthed the rhetoric of controversy between partisans of Greek and Arabic astrology, especially among the authors participating in the heated arguments about the flood prediction for 1524. The Paduan phi-losopher Agostino Nifo adopts Pico's line when stating in a 1505 treatise *De nostrarum calamitatum causis:* "On this topic *(i.e., general astrology)* more recent authors, Albumasar and many others, write much that we consider to be senseless, superstitious and in conflict with philosophy and Ptole-my's theory of the stars."[18] And he calls Albumasar "the prince of these fabulists" *(princeps horum fabulantium).*[19] The anti-Arab tone intensifies with Albert Pigghe, a Flemish critic of the 1524 flood prediction and of the contemporary practice of annual predictions in general: "If there are some who claim to be astrologers, they are in fact ignorant of the astrological sciences and are only filled up with Punic *(i.e., Arabic)* fictions and supersti-tions, against which we have begun this war in the first place."[20] He calls on the students of astrology "to give up the fables of Albumasar and the *(other)* Punics and to read with great diligence our Ptolemy."[21] Other followers of Pico sang the same tune throughout the sixteenth century and beyond: Cardano attacked the "follies of the Arabs" *(nugas Arabum)*,[22] and Campanella proclaimed in the very title that his manual of astrology is cleaned of any "superstition of the Arabs and Jews" *(superstitione Arabum et Judaeorum eliminata)*.[23] The Greek–Arabic antagonism is well expressed by Philipp Melanchthon in a preface of 1553:

> Antiquity possessed a doctrine about the movements of the stars which, without doubt, was most scholarly erected upon most ancient observations, and it possessed a most accurate description of the succession of time periods and of ancient emperors. But when in Egypt the barbarism of the Sarracens *(Sarracenica barbaries)* destroyed the academy in Alexandria and the arts *(studia)*, the old documents were ruined. And the entire doctrine about the movement <of the stars> would have perished, by being dispersed into many different books, if Ptolemy, briefly before the Sarracens' turbulence, had not summed up the entire art in one volume.[24]

This preface accompanies the Latin translation of Ptolemy's *Tetrabiblos,* which was produced in part by Joachim Camerarius and in part by Melanchthon himself. Melanchthon's denigration of the Saracens must be understood as an advertisement of his edition of Ptolemy in the first place: Ptolemy's *unum volumen* is a manual of ancient astrology in its entirety, and hence to be preferred to all the other books on the market, especially to the "many different books" of the Arabs, Ptolemy's epigones. We have seen in Chapter 1 that Renaissance authors were divided over the question of whether the Greek–Arabic–Latin transmission was responsible for the shipwreck or for the rescue of ancient sciences. Melanchthon's Wittenberg colleague Jacob Milich describes the Arabs in the most positive terms, but Melanchthon makes them ruin ancient astrology. Whether or not Melanchthon believed in his unhistorical story, it certainly served him to direct his readers toward his own edition of Ptolemy.

But these attacks on Arabic astrology, of which one could cite many more,[25] were not left unanswered. Giambattista Abioso, among others,[26] replied to Pico and Nifo that there was progress in astronomy and astrology after Ptolemy and that "the Arabs and Saracens perfected what the Greeks were not able to perfect in the sciences."[27] The result was that "from the doctrine of Ptolemy we can draw much that was written by Messahalah and Albumasar, who are not in conflict with the wisdom of Ptolemy; rather, they by way of consummation and experiment make manifest what

Ptolemy has said in truncated and brief manner in the *Tetrabiblos*."[28] Many authors, in fact, specifically defended the doctrine of the great conjunctions, and argued that the Arabic authors are unduly attacked since this doctrine was in fact held by Ptolemy, either in his extant works or, possibly, in works not transmitted to us.[29] After Ptolemy, the doctrine was much developed by observation and experience, such as on the conjunctions of Saturn and Jupiter, as Sebastian Constantinus of Taormina argued.[30]

The rhetoric of controversy should not impress us too much. We cannot judge from these programmatic statements whether Arabic astrological traditions were in their rise or decline, and we cannot judge the relevance of the arguments employed. It is sensible, therefore, to turn our attention to the actual development of theoretical astrology in the Renaissance. This is a difficult undertaking, given that the astrological literature of the fifteenth to seventeenth centuries is extremely rich, and research on astrology of this epoch still has a long way to go. The following analysis therefore combines a survey with a more detailed study. The survey is concerned with the reverberations of six disputes between supporters of Greek and Arabic traditions in astrological handbooks. The study, in turn, is devoted to the most prominent object of anti-Arabic polemics: the theory of great conjunctions.

## SIX MATTERS OF DISPUTE

The dispute between supporters of Ptolemaic and Arabic traditions was focused on a number of issues, almost all of which had been discussed by Pico in his *Disputationes*: interrogations, elections, anniversary horoscopes, calculations of the length of life, lots, and revolutions of the years of the world. Other issues were disputed too, such as the fardars (*firdariae*) and house division,[31] but the above-named six are certainly among the most prominent and hotly debated. In order to understand the impact of the Greek–Arabic antagonism on the general development of astrology at the time, the following survey will be based on astrological handbooks, that is, works with an encyclopedic or comprehensive character, in which the above six topics are likely to appear. From the large group of astrological handbooks, which includes introductions to ephemerides and almanacs,[32] I have selected the following works for inspection, partly because of their fame and influence, and partly because they are representative of a genre: Pedro Ciruelo's *Apotelesmata* of 1521, a comprehensive handbook of all parts of astrology; Pietro Pitati's *Almanach novum* of 1544, which is representative of the genre of introductions to ephemerides and almanacs;

Johannes Schöner's famous summa of natal astrology *De iudiciis libri tres* of 1545; Girolamo Cardano's commentary on Ptolemy's *Tetrabiblos* of 1554, as the best known example of a commentary on Ptolemy; Claude Dariot's *Ad astrorum iudicia facilis introductio* of 1557, an introduction to astrology in humanist style; Francesco Giuntini's *Speculum astrologiae* of 1573, a very influential summa on nativities; and Tommaso Campanella's *Astrologicorum libri VII* of 1614,[33] a comprehensive handbook in the tradition of Ptolemaic reform.

These handbooks continue a long tradition, which is much influenced by the organization of Ptolemy's *Tetrabiblos*. The first book of the *Tetrabiblos* treats introductory and general topics. The second is concerned with general astrology, that is, effects concerning the people and the weather, predicted mainly from comets and eclipses. The third and fourth books treat matters of individual astrology, and nativities in particular. Medieval and Renaissance handbooks of astrology often follow this Ptolemaic division, but they add, under Arabic influence, other branches of general or individual astrology.

A good witness to this medieval development is Leopold of Austria's *Compilatio de astrorum scientia,* a very popular handbook of astrology of the late thirteenth century, which is organized in ten chapters *(tractatus)*. Leopold first explains the basic principles of astronomy and astrology (chapters 1 to 4) and then turns to general astrology, for which he uses a term inherited from the Arabs: *annorum revolutiones,* revolutions of the years of the world, with a brief section on planetary conjunctions (chapter 5). A separate section is devoted to astrometeorology (chapter 6). Then follows the much longer part on individual astrology; here Leopold treats not only nativities *(nativitates,* chapter 7) but also two branches adopted from Arabic sources: interrogations, that is, the answering of questions on the basis of the chart drawn for the moment of asking (chapter 8), and elections, that is, the determining of the best time to begin an action (chapter 9). The chapter on elections also contains a brief section on astrological talismans *(imagines)*. The book ends with a short chapter on how to know the intentions of a questioner.

The other great handbook of medieval astrology, Guido Bonatti's *Tractatus astronomie,* likewise of the later thirteenth century, treats a very similar series of astrological topics: principles of astrology (chapters 1 to 3), conjunctions of the planets (chapter 4), judgments in general (chapter 5), interrogations (chapter 6), elections (chapter 7), revolutions of the years of the world and lots (both chapter 8), nativities (chapter 9), and astrometeorology (chapter 10).[34] This overall division of astrology was still popular around 1500, as shown by Gregor Reisch's university textbook *Margarita philosophica,* which was written in the 1490s. There are five parts to astrology, he says: the first an introductory part, the second on annual revolutions of the world, the third on nativities, the fourth on interrogations, and the fifth on elections, and some, Reisch explains, add a sixth part on talismans.[35] The handbooks by Leopold, Bonatti, and Reisch all follow, with minor variations, the standard Arabic arrangement, which was known in the Latin West since the twelfth century: introductory principles, revolutions, nativities, interrogations, and elections. This division is also followed by the much read *Speculum astronomiae,* a thirteenth-century text often attributed to Albertus Magnus.[36]

It was immediately apparent to anyone who compared medieval handbooks with the *Tetrabiblos* that entire branches of medieval astrology had not been treated by Ptolemy. This is particularly true of interrogations, elections, lots, and revolutions of the years of the world. Given the very Arabic appearance of medieval Latin handbooks, we can expect that the Renaissance handbook tradition changes visibly when influenced by Ptolemaic reformers. Humanist critics argued that the branches not treated by Ptolemy are inappropriate Arabic additions to Greek astrology. While they rejected some doctrines in their entirety, they called for the replacement of other theories by their Ptolemaic counterparts. It is a particular interest of the following survey to trace the technical and nontechnical arguments that played a role in these disputes. I will also pay attention to the historical origin of the doctrines at stake. This should help us to improve the understanding of the rise or decline of "Arabic" and "Ptolemaic" theories in the Renaissance.

### Interrogations

The first area of conflict between Greek and Arabic traditions is the part of astrology that is called *masā'il* ("questions") in Arabic, *interrogationes* or *quaestiones* in Latin, and horary astrology today. As the research of David Pingree has shown, interrogational astrology was an Indian contribution to astrology, which was formed in the second century AD on the basis of Greek sources and Indian indigenous traditions. Greek astrologers practiced catarchic astrology, in which the astrologer determines the best time for the beginning of an action such as when to depart for a journey, when to buy a slave, or when to marry. Catarchic astrology reached India via an anonymous Greek astrological treatise written in Egypt but not extant today, which was translated into Sanskrit in the early second century AD.[37] In India, catarchic astrology was practiced but also transformed into interrogational astrology, in which the astrologer does not determine the time to begin an action, but makes predictions about the client's actions in the future, for example, about his journeys or purchase of a slave. The horoscope is of the moment in which the question is posed, not of a beneficial moment in the future. Interrogational astrology was further developed in Sassanian Persia. From there, the technique reached Arabic culture, where it became a great success. The transformation of Greek catarchic astrology into interrogations is also visible in the two Arabic translations, one of them fragmentary, of a Persian version of Dorotheus's Greek treatise.[38] In the Latin West, two Arabic sources on interrogations were particularly influential: Messahalah's *De receptione* and Zahel's *De interrogationibus*. Interrogations are also treated in Haly Abenragel's astrological summa *De iudiciis astrorum,* books 2–3. Since almost all Persian sources are lost today, it is impossible to determine the exact contribution of Arabic astrologers to interrogational astrology, for instance, whether it was a Greek, Persian, or Arabic idea to arrange the subjects of the questions according to the twelve houses, as in Zahel's treatise. It is clear, however, that Arabic astrologers added material that pertained to their own culture.

The medieval Latin handbooks of astrology by Bonatti and Leopold contain substantial sections on interrogational astrology, but this tradition is interrupted in the Renaissance. Pico directly attacks interrogations. He is not convinced of the idea that a horoscope of the moment in which a client consults an astrologer is in any way indicative. How can the heavenly constellation of this moment be a cause or a sign of events in the future about which the client is asking the astrologer? For interrogational astrology, the intention of the client at the moment of his consultation is an important factor. But Pico protests that there is no connection between our desire and the thing we desire to know. He also argues that there might be a thousand accidental reasons why a client arrives earlier or later at his astrologer, with potentially enormous effects on the horoscope.[39] A few decades after Pico, in 1521, Pedro Ciruelo continues the attack on interrogations and claims that there is hardly "a greater and more dangerous error in the Christian religion," both from the theological and philosophical standpoints. Ciruelo argues with Thomas Aquinas that the celestial bodies "do not signify naturally anything which they cannot cause." The image of an absent thing—that is, in the mind of the client—cannot be caused by the stars. And hence, the entire art of interrogations is false and dangerous.[40]

It is difficult to see the extent to which these arguments in fact influenced other astrologers. But it is clear that many sixteenth-century handbooks on individual astrology do not deal with interrogations: they do not

appear in the reference works by Schöner, Pitati, and Giuntini. One critic
of interrogational astrology is Girolamo Cardano. Cardano takes issue
with the eleventh-century Arabic astrologer Haly Rodoan ('Alī ibn Riḍwān)
over the question of why Ptolemy had not treated interrogations and elections in his *Tetrabiblos*. Haly argues that this would have been superfluous
since the rules for prediction from a chart drawn for a contemporary or
future event, as in interrogations and elections, are the same as for prediction from a nativity chart; what differs is only the matter to which the
chart is related. And Haly concludes: "Since this is the case, interrogations
and elections are included in the words of Ptolemy."[41] Cardano has a very
different opinion on the matter. He does not find it surprising that Ptolemy
ignores interrogations and elections, since interrogations are entirely orac-
ular *(sortilegae)* and inappropriate for Christian and decent people, and
since elections are useless and only serve the astrologer's greed for money
*(avaritia)*. And Cardano continues to argue with much historical justification that the presence of interrogations and elections in the (Pseudo-) Ptolemaic *Centiloquium* proves that this text is not by Ptolemy in reality.[42] On
the other hand, Cardano's example shows that the attraction of interrogational astrology was not entirely lost to Renaissance astrologers. For despite his critique, Cardano composed a *libellus* on interrogations, which he
introduces as follows:

> Since, in other places, I have often condemned the branch of interro-
> gations, because it is fortuitous, in conflict with our religion and an oc-
> casion for many evils, but since I judge the matter itself to be necessary,
> because many people desire to know a question, but not all, therefore I
> want to add these few things without diverging from what was said
> before.[43]

Then follows a treatise with twenty chapters on traditional questions concerning journeys, harvests, marriage, and so on. Cardano's text is a good
testimony to the fact that interrogational astrology continued to be practiced in the sixteenth century and that a gap was developing between theoretical astrology as influenced by Pico and traditional practice. The great
success of interrogational astrology in Arabic and Latin culture can largely
be explained by the fact that medieval clients usually did not know the day
of their birth—not to mention the exact time. The advantage of interrogations was that the astrologer could make predictions for individuals
without a nativity chart.[44] It is not a surprise, therefore, that interrogations continued to flourish in the sixteenth century, when even Martin Luther was unsure about the year of his birth.

## Elections

Elections is the second area of dispute among Renaissance astrologers. This is the above-mentioned Greek technique of καταρχαί ("beginnings"), the determination of the most beneficial moment for an action. Catarchic astrology was not discussed by Ptolemy in the *Tetrabiblos,* deliberately, it seems, since other Greek sources give ample evidence of its existence, most notably the astrological treatise by Dorotheus of Sidon (fl. 25 to 75 AD). This branch of astrology reached Arabic culture via Persian intermediaries, especially through translations of the fifth book of Dorotheus's treatise. The Arabic term for it is *iḥtiyārāt,* the Latin term *electiones.* In the Latin West, the main Arabic sources on the doctrine were Zahel's *De electionibus* and Haly Abenragel's *De iudiciis astrorum,* book 7. The thirteenth-century handbooks of astrology by Bonatti and Leopold have long chapters on elections. In the Renaissance, this technique, just as interrogations, comes under attack, but it is noteworthy that the Renaissance reception of interrogations and elections differs markedly: elections continue to be covered in several Renaissance handbooks, while interrogations largely disappear from this genre. Among the critics of elections were Giovanni Pico, Pedro Ciruelo, and Girolamo Cardano. Pico admits that elections are often thought to be the predictions most useful for daily life,[45] but he counters that the natal horoscope makes elections useless: either the nativity is unfavorable toward the chosen action, and then the election of an hour will not help, or it is favorable, but then one's own choice of time will be fully sufficient.[46] Cardano, as we have seen, censures elections as non-Ptolemaic, useless, and purely commercial.[47] Ciruelo distinguishes carefully between licit and illicit elections. He rejects all elections that involve voluntary human actions, since, as spiritual actions, they are not dependent on the stars and since, as corporeal actions, they are influenced by the nativity and not by a later constellation, as Pico had argued before.[48] Nevertheless, Ciruelo advocates the truth of all elections for purely natural and corporeal actions, such as when choosing fortuitous times in agriculture and medicine, as well as for settling in a city and in a society. These elections Ciruelo then discusses in detail.[49]

Other Renaissance handbooks likewise continue to discuss elections. Pitati's *Almanach novum* contains a detailed treatise on elections, as do Dariot's *Introductio* and Campanella's *Astrologicorum libri* (book 6). Campanella's range of topics is fairly traditional: he discusses favorable times for doing theoretical philosophy, for medical treatment, conception, political and military actions, journeys, construction, and agriculture. Even Johannes Schöner, whose interest is mainly in nativities, includes two and a half pages on elections on how "to choose favorable times for actions or enterprises."[50] Only Giuntini's *Speculum,* with its exclusive focus on nativities and natal revolutions, omits elections entirely. Pico's main argument that elections are useless because they interfere with nativities is in fact countered by Dariot. It is true, Dariot says without naming Pico, that a good election cannot avert something bad predicted in the nativity. But elections are nevertheless useful, because a good election can increase a good prediction of the nativity and can at least diminish a bad prediction.[51] The major authors of astrological handbooks, therefore, did not share Cardano's praise of Ptolemy for not treating elections, and they were not impressed by his protest that elections are only invented for making money. One factor for the survival of elections clearly was their usage in agriculture and medicine, areas in which there is no potential conflict with the contingency of human action, as Ciruelo emphasized. As was mentioned above, even the index of prohibited books and pope Sixtus V's bull against astrology explicitly allowed astrological prediction on navigation, agriculture, and medicine, thus permitting the practice of a major part of elections.

### Anniversary Horoscopes

The third area of conflict is a subbranch of nativities: anniversary horoscopes *(revolutiones nativitatum, taḥāwīl sinī l-mawālīd),* also called natal revolutions, the drawing up of charts not for the moment of birth, as in classical nativities, but for each anniversary of the person's birth. Today this technique is referred to as "solar returns" or "solar revolutions." The astrologer draws a chart for the sun's return to the birth point of the ecliptic, and then compares this chart to the original nativity (at least, in a very common version of this technique). The information gained concerns the fate of the individual in the coming year.[52]

Anniversary horoscopes are not mentioned in Ptolemy's *Tetrabiblos.* Ptolemy instead briefly discusses three other forms of lifelong prediction, when treating the length of life and the phases of life (chapters 3.10 and 4.10): "prorogations" (ἀφέσεις), "annual chronocrators," and "planetary transits." In prorogation technique, the most relevant of the three in our context, an imaginary point is moving along the zodiac and thus symbolizes the passing years of the client. Ptolemy here adopts an astrological doctrine that can be found already in Dorotheus, apparently for the first time. But Dorotheus's *Carmen astrologicum,* book 4, also contains the technique of anniversary horoscopes, which is missing in Ptolemy.[53] From Dorotheus it reached Indian, Persian, and finally Arabic culture. In the works of the first-generation Arabic astrologer Messahalah, especially in his *Book of Aristotle,* we find a most refined exposition of the various methods of lifelong prediction, including prorogations and anniversary horoscopes. Unfortunately, Hugo of Santalla's Latin translation of Messahalah's treatise found very few readers, so that the Latin astrologers drew mainly on other, more general Arabic sources by Albohali, Omar Tiberiadis, and Alcabitius.[54] As a result, Bonatti's and Leopold's medieval handbooks mention natal revolutions only in passing—when they discuss techniques of prorogation or of revolutions of the years of the world, a subbranch of general astrology.[55] For a comprehensive treatment of natal revolutions, later medieval authors could turn to Haly Abenragel's *De iudiciis,* book 6. Renaissance reformers of astrology tried to promote the technique of prorogations, which was also used in the Middle Ages, as truly Ptolemaic.[56] We will see, however, that this development did not lead to a decline of Arabic methods of revolution.

Pico does not single out anniversary horoscopes for special attack, but criticizes the Arabic additions to Ptolemy's theory of prorogation for the specific topic of predicting the length of life—as we shall see below. Cardano's attitude again is ambivalent, as it was on the topic of interrogations. On the one hand, Cardano is the author of a treatise on natal revolutions.[57] On the other hand, in his commentary on *Tetrabiblos* 4.10, Cardano argues that Ptolemy does not treat revolutions, and rightly so, because revolution is characterized by inequality and because prorogation may properly start only from five places—that is, Ptolemy's five starting points of prorogation, which are horoscope, lot of fortune, moon, sun, and mid-heaven.[58] Campanella will later reply to these arguments that the revolution does not have the same power as the prorogation, but that its power is nevertheless considerable.[59]

In spite of the humanist support for Ptolemaic prorogation, the popularity of anniversary horoscopes increases tangibly in the sixteenth century, as far as can be judged from the genre of handbooks. An indication of this is, on the one hand, the appearance of succinct chapters on revolutions in the works of Ciruelo and Pitati, and, on the other hand, the long discussion of anniversary horoscopes in the very popular works of Johannes Schöner and Francesco Giuntini—comprehensive discussions without parallel in medieval Latin astrology. Schöner justifies the treatment of anniversary horoscopes in the third and final book of his *De iudiciis nativitatum libri tres* of 1545. If the exact time of a future event is to be determined, one turns to a branch of astrology that some people call "revolutions," Schöner says. Ptolemy does not pay attention to anniversary horoscopes, observes Schöner, and adds: "I do not easily move away from Ptolemy" *(Ego non facile a Ptolemaeo discedo).* That is why he gives the reader a very brief summary of what Ptolemy had taught about prorogations and transits of planets, and then turns to "the transmitted doctrines of the Arabs and others"*(Arabum et aliorum traditiones)*. These doctrines are not to be condemned, but the preeminence is always given to Ptolemy, says Schöner. This statement of Schöner's is meant to justify the integration of an entire branch of Arabic astrology into a mainly Ptolemaic framework.[60] In the remainder of the book, Schöner quotes extensively from Haly Abenragel's *De iudiciis astrorum,* book 6 on natal revolutions, without naming his source.[61]

Francesco Giuntini's *Speculum astrologiae* of 1573 opens with the statement that almost the entire aim of astrology is directed at two things, which will constitute the two parts of his work: nativities and natal revolutions.[62] Giuntini's second part covers 217 folios. It is the longest sixteenth-century treatment of anniversary horoscopes I am aware of. In contrast to Schöner, Giuntini often names his sources, two of which stand out: Haly Abenragel's *De iudiciis,* and "Hermes," that is, Albumasar's *De revolutionibus nativitatum.* This latter text, called *Kitāb Taḥāwīl sinī l-mawālīd* in Arabic (On the Revolutions of the Years of the Nativities), had made a cross-cultural journey: it had reached the West not directly from Arabic, but through a Byzantine translation, which served as the source for Stephen of Messina's Latin translation of the later thirteenth century. Giuntini probably used the only printed edition of the work, which came out in Basel in 1559.

Hence, anniversary horoscopes were a matter of dispute and a thriving discipline at the same time in the sixteenth century.[63] The reasons for its rise are surely manifold, but two seem evident in the first place: first, from a technical point of view, anniversary horoscopes run parallel to revolutions of the years of the world, which is an extremely popular branch of general astrology that even the Church did not object to. The idea of relating the revolution chart to the nativity chart is simple and elegant at the same time, and had the great advantage that it was less complicated than many forms of prorogation. And second, anniversary horoscopes apparently profited from the fact that in the Renaissance, the number of clients who knew their exact birth dates was increasing continually. This was, of course, a development of importance for the entire branch of nativities, but it also made the genre of anniversary horoscopes much more attractive for the practicing astrologer than it had been in the centuries before.

### The Length of Life

The fourth area of dispute concerns an issue of lifelong prediction that has already been mentioned: the calculation of the length of life. Many Renaissance authors contrast Ptolemy's and the Arabs' methods on the issue,

and Pico makes it a matter of polemics. Ptolemy treats the quantity of life
in chapter 3.10 of the *Tetrabiblos*. He calculates the life span by determining, for the birth chart, two points on the ecliptic, a starting point or
"releaser" (ἀφετικὸς τόπος, or: ἀφέτης, Arabic *hīlāğ*, Latin *hyleg*) and a destructive point (ἀναιρετικὸς τόπος), and by measuring, in two alternative
methods of prorogation, the degrees that separate the two points, one degree signifying one year. In contrast, many Arabic and Hebrew authors,
among them Alcabitius (4.4–5),[64] Omar Tiberiadis, and Haly Abenragel
(4.3–4), use a method that also involves determining the Alcocoden (Arabic: *al-kadḫudāh*, originally a Persian term), which is the planet ruling over
the Hyleg. The Latin calque Alcocoden and similar forms of the term became widely known in the Latin West. In the spelling *colcodea* or *colcodrea*,
the term even enters philosophical literature as an equivalent for Avicenna's concept of the "giver of forms" *(dator formarum)*.[65] For determining the
Alcocoden, the Hyleg—which for some authors is a point, for others a
planet—needs to be found first. If the Alcocoden, the planet ruling over the
Hyleg, has a powerful position in the chart, the full amount of its planetary years will deliver the maximum life span of the individual; if it is less
powerfully placed, it will grant fewer years. Other properties of the Alcocoden may further modify the years granted. This is the basic version of
the Hyleg and Alcocoden theory, but on the details there is no consensus
in the Arabic–Latin astrological tradition. To Pico and other critics of the
Arabic tradition it appeared that the Alcocoden method was not a Greek
invention, but an Arabic one. This is not true, as we know today. The idea
of using a ruling planet to determine the quantity of life is older than
Ptolemy and appears in several Greek sources, among them Ps.-Nechepsos-
Petosiris, Dorotheus, Vettius Valens, and Firmicus Maternus.[66] The basic
difference between the Ptolemaic and the "Arabic" method is that the
former derives information about the length of life by way of prorogation,
the latter by analyzing the absolute and relative properties of a specific
planet on the birth chart. The Arabic and Latin texts on the issue offer
many combinations of the two techniques, with the result that the astrological tradition on the length of life is enormously complicated, not to
say muddled. The long treatment of Hyleg and Alcocoden in Bonatti's and
Leopold's handbooks is only one example of this.[67]

Renaissance criticism of the Alcocoden technique is prominently
placed in the works of Pico, Cardano, and Campanella. Pico censures Haly
Rodoan, the Arabic commentator of the *Tetrabiblos,* for ascribing an Alcocoden theory to Ptolemy: "What do you hallucinate, Barbarian? What do
you dream about? Where have you read Alcocoden in Ptolemy?" The Alcocoden, Pico argues, is an addition of the *barbari,* the Arabs, and of the

*nostrates,* our own astrologers.[68] The ascription of a specific Alcocoden technique to Ptolemy can, in fact, be found in several Arabic and Latin sources, for instance in Omar Tiberiadis, Haly (Aḥmad ibn Yūsuf), Ciruelo, Schöner, and Giuntini. They apparently rely on Pseudo-Ptolemaica, such as the remark in the *Centiloquium* (no. 66) that one should not use prorogation only, but also the ruling of the *significatores*—that is, Hyleg and Alcocoden, as Haly explains.[69] This shows again that the textual transmission of Ptolemy's astrological works proved a major difficulty for any puristic reform of astrology along Ptolemaic lines. Girolamo Cardano's critique of the Alcocoden method is more technical than Pico's. The Arabs wrongly believe, he says, that the "releaser" alone (*apheta* or *hyleg*) does not suffice for calculating the length of life and that a *significator vitae* has to be added, the Alcocoden. What deceived the Arabs was that a powerful ruler of the *apheta* makes the *apheta* and hence life more robust and longer in general, as Ptolemy would concede too, but it does not prolong life by a precise number of years.[70] Or, as Campanella puts it in the footsteps of Cardano: the Alcocoden corroborates life, but does not measure it: *roborat vitam, non autem metitur.*[71]

Rejecting the Alcocoden method, however, was only one of several Renaissance reactions to the problem, and in general one can observe that the Alcocoden theory survived the humanists' attacks. Dariot gives a brief description of Ptolemy's theory, but advises his readers to turn to Haly Abenragel and Bonatti for further details—that is, on Alcocoden theory. Pitati, at the other end of the spectrum, simply continues the Arabic tradition, as does Ciruelo, who compares several Alcocoden theories, but warns that with the Alcocoden techniques one cannot know the span of life *nisi grosso modo.*[72] Most interesting is the discussion of the problem in Schöner and Giuntini, who present both theories in sequence, the Ptolemaic and the Arabic. In Johannes Schöner's presentation, even the Ptolemaic method involves the Alcocoden, albeit in a simplistic fashion. On both Hyleg and Alcocoden, he compares several alternative theories, and also gives an example for an Alcocoden technique by predicting the length of life of Maximilian I. At the end of the chapter, Schöner discloses his preference for Ptolemy's method, which he praises as more precise.[73] Francesco Giuntini, who draws on Schöner, likewise recommends Ptolemy over the Arabs, but when he comes to discuss an exemplary horoscope at the end of book 1, he calculates the life span twice: according to Arabic and according to Ptolemaic practice.[74] Schöner's and Giuntini's reaction to the problem shows that the antagonism of Greek and Arabic astrology, even if motivated by partisanship, could be very productive. On the present topic, it led to a discussion that was to the benefit of astrological theory. Schöner's and

Giuntini's treatment of the problem, and in some sense also Cardano's, can be regarded an advance from a technical point of view: it helped to distinguish several Greek and Arabic models in an area of astrology that had long suffered from the confusion of models.

## Lots

The fifth area of dispute to which I should like to turn concerns the theory of the lot or part (κλῆρος, Arabic *sahm,* Latin *sors, pars,* or *cehem*). The lots are points on the ecliptic circle that are determined by adding the number of degrees between two planets to the ascendant, that is, the degree where the ecliptic cuts the horizon. The lot of fortune *(pars fortunae)*, for instance, is counted from the sun to the moon—or, for a person born at night, from moon to sun—and then is added to the ascendant, or, in some exceptional cases, to the position of a planet. The theory of lots much increases the range of issues that can be predicted from a nativity chart, but its drawback is that it presupposes accurate knowledge of the ascendant degree. Nevertheless, probably because of the richness of material, lots were a popular astrological doctrine in Arabic culture. Albumasar's *Abbreviation of Astrology,* for instance, discusses over fifty different kinds of lots, Alcabitius about seventy-five lots: the lot of marriage, the lot the father, the lot of the sea journey, and so on.[75] Ptolemy, in contrast, had recognized only one lot, the lot of fortune in his *Tetrabiblos* (3.10: κλῆρος τύχης), which is a much older doctrine that appears already in Ps.-Nechepsos-Petosiris.[76] But as on so many other points, Ptolemy is not representative of Greek astrology. Many other Greek astrologers discuss a greater range of lots, among them Dorotheus, Vettius Valens, and Paul of Alexandria. Dorotheus, for instance, mentions about fifteen different lots. Arabic astrologers drew on these Hellenistic sources via Sanskrit and Persian intermediaries, and themselves considerably enlarged the list of lots. This tradition was continued in the Latin West. The handbooks of Bonatti and Leopold contain substantial sections on a large range of lots, which are based on Albumasar and Alcabitius for the most part.[77]

In the Renaissance, however, the doctrine sees a dramatic decline, at least in handbook literature. As far as I can see, there are no lists of lots in the reference works of Ciruelo, Pitati, Schöner, Cardano, and Campanella—in great contrast to the Arabic and medieval Latin tradition. The omission of the doctrine does not seem to be commented on in these texts. The lots disappear without much noise from handbook literature. Other texts, however, attack the Arabic doctrine of lots directly, most notably Pico's *Disputationes.* Pico's criticism of the doctrine of lots is very impressive, because in this case Pico has realized that Ptolemy's position is

not shared by other ancient astrologers. In antiquity, Pico argues, the original doctrine of the parts was simple: *pars* did not mean more than the
part of reality that a planet signifies, such as when Mars signifies brothers.
Because the ascendant degree, that is, the horoscope, corresponds to the
sun, some ancient astrologers also wanted to construct a horoscope for the
moon *(horoscopus lunae)* and argued that its degree would be found by way
of calculating the distance between the sun and moon—they called this
degree the lot of fortune, which Ptolemy also recognizes, says Pico. Ptolemy
with full right disapproved of all other lots, but he should have disapproved of the lot of fortune as well. Unfortunately, complains Pico, this
was not the end of fabrication, because Paul of Alexandria and other ancient astrologers took this as a model and invented further lots, a tendency
continued by the Arabs and Latins who added innumerable lots even of
silly things: "For you find lots of chickpea and beans and lentil and barley
and onions and of a thousand such things in all the Arab and Latin <as-
trologers>."[78] Pico did not recognize that it was impossible for Ptolemy to
leave out the lot of fortune as well, since this lot was necessary as one of the
five possible starting points for calculating the length of life and for prorogation. But even if Pico did not fully understand the origin of the doctrine
and its historical development, he did see that the Greek astrologers differed much on the issue and that the list of lots was growing continuously
when transported through the cultures. Pico's strongest argument was
physical: if, as the doctrine demands, an arc between two planets is trans-
ferred to and counted from the ascendant, one assumes a power in the
heaven "where there is no star, no light, nor any other natural quality."[79]
How, then, should this point be influential?

These arguments were powerful, but the handbook literature of the
sixteenth century, as surveyed, does not offer much evidence on the
measure in which Pico contributed to the decline of the doctrine. One
piece of evidence comes from Valentin Nabod's comprehensive commentary on Alcabitius's *Introductorius* of 1560, which I had mentioned before
when discussing Alcabitius's textual transmission in the Renaissance in
chapter 1. Nabod cites and comments on almost all parts of Alcabitius's
treatise, but omits most of the fifth, last section, which Alcabitius devotes entirely to the lots. Nabod explains, in the footsteps of Pico, that
Ptolemy held the lot of fortune to be effective because it is, as it were, the
horoscope of the moon. Just as the sun illuminates the ascendant, the
moon illuminates the lot of fortune. The other planets, however, are neither luminaries nor do they reflect the light of the sun. "Hence it did not
seem in any way necessary or useful to me to spend more time on examining the lots."[80]

It is important to see, however, that the disappearance of the lots from astrological handbooks, although dramatic, was not complete. The lots appear, without any introduction or comment, in the influential handbooks of Schöner and Giuntini in the sections devoted to specific aspects of human life. They also appear in Giuntini's illustrative horoscope for a prince born in 1561, whose name is not given: the lot of sickness, the lot of life, the lot of death, the lot of law, the lot of fortune, the lot of marriage, and the lot of children.[81] It is obvious that Schöner and Giuntini did not want to do without the richness of prediction that the lots allow for. At the end of his treatise, Giuntini adds a one-page index that leads his readers to the definitions of twenty-eight lots that he has given in his treatise. At the end of the index he adds further definitions of eleven lots that he had not defined before.[82] Giuntini therefore prolongs the tradition of a proper theory of lots in astrological handbooks, and he even demonstrates how to use them for nativities. Two quotes from Giuntini's work may serve to illustrate the attitude that is behind his adoption of the doctrine. They show that Giuntini is aware of the antagonism that other scholars see between Ptolemy and the Arabs, and that he has a practical attitude toward the problem. On the lot of fortune, Giuntini observes that Ptolemy only recognizes the direction from sun to moon, and not from moon to sun. In his youth, Giuntini says, he had found this issue irritating, but now, on the basis of long observations, he has abandoned Ptolemy's position: *magis amica veritas,* he remarks.[83] And when predicting the actions of the prince with regard to his achievements in the military, political, and religious realm, Giuntini writes: "With respect to actions, I ignore the opinion of Ptolemy, and also of Firmicus Maternus, and follow the Arabs, who in this matter were better observers *(meliores observatores)* than Ptolemy and the Latin astrologers."[84] It is remarkable that even the doctrine of lots, despite its weak theoretical basis and its obvious absence from the *Tetrabiblos,* survived in the practical parts of two influential handbooks on nativities. This is not simply a sign of a continuing tradition, but points to a dilemma of the Ptolemaic reformers: the *Tetrabiblos,* because of its puristic and theoretical character, could not by itself serve as the basis of astrological praxis. In one way or another, it had to be supplemented by the doctrines of other astrologers.

## Revolutions of the Years of the World

The sixth and final area of dispute concerns a branch of general astrology: revolutions of the years of the world *(revolutiones annorum mundi* in Latin, *taḥāwīl sinī al-ʿālam* in Arabic), also called *annual revolutions.* Just as anniversary horoscopes for individual persons, this technique is based

on a revolution chart. It offers predictions for the coming year on the basis
of a horoscope cast for the entrance of the sun into the first minute of Aries,
that is, for Aries ingresses, as the term is in modern astrology. The prediction
concerns general matters, such as weather, agriculture, wars, diseases,
earthquakes, kingship, and social uprisings. The doctrine of annual revolutions is of Sassanian origin. It was developed in Persia between the third
and eighth centuries AD. Dorotheus, the Greek astrologer, influenced the
Persian doctrine to a certain extent in that he discussed the casting of anniversary horoscopes of a person's birth date, as we have seen, but the Sas-
sanians extended this method to general astrology and cast horoscopes of
vernal equinoxes.[85] The two principal sources on annual revolutions in
Latin astrology were Albumasar's *Flores* and Messahalah's *Liber revolutionum,* plus a one-page summary of the doctrine in Messahalah's *Epistola
in rebus eclipsis lunae.* Annual revolutions were treated in substantial chapters in Bonatti's and Leopold's astrological handbooks,[86] which is also remarkable in view of the fact that the theory of great conjunctions, as will
be apparent below, did not receive much attention in these handbooks.
The most important technical device of annual predictions is the finding
of the lord of the year *(dominus anni),* the planet with the most dignities on
the chart. Bonatti holds, in accordance with Albumasar, that if there is a
planet in the ascendant house with enough dignities, it will be the *dominus
anni,* if not, the search proceeds, according to Albumasar, with the ruler of
the tenth house, or rather, according to Bonatti, with some other candi-
dates first, such as a planet that has its exaltation in the ascendant house.[87]
The events of the year are then predicted from the nature, position, and
aspect of the lord of the year, as well as from many other properties of the
revolution chart, especially from conjunctions of planets.

Pico attacked the doctrine in chapter 5.5 of his *Disputationes.* He rejects
the Arabic idea that conjunctions not only of sun and moon but also of
other planets such as Saturn and Jupiter are indicative of general events in
the future. His main argument is that general effects can only be attributed to causes of general and maximal efficiency, and hence only to the
sun and the moon, that reflects the sun's light.[88] Pico thus concedes that
annual predictions are possible, but only on the basis of a method mentioned by Ptolemy in *Tetrabiblos* 2.10:[89] predictions for the coming seasonal
quarter of the year can be made by interpreting conjunctions of sun and
moon—that is, full moon or new moon—which immediately precede the
sun's ingresses into Aries, Cancer, Libra, and Capricorn. Full eclipses of
sun and moon, Pico argues, allow prediction of events in the remote future
as well.[90] With this criticism, Pico successfully founded an astrological
tradition that advocated the abandoning of revolution charts in general

astrology. This trend may be illustrated by reference to three of its supporters: Agostino Nifo, Albert Pigghe, and Valentin Nabod.

In his treatise *De nostrarum calamitatum causis* of 1505, the Paduan philosopher Agostino Nifo argues, in the footsteps of Pico, that eclipses of sun and moon have greater impact than comets, which in turn have greater impact than the conjunctions of the higher planets. But he criticizes Pico for annihilating the influence of the higher planets, since the true Ptolemaic position, Nifo argues, assigns secondary powers *(vires subministras)* to them.[91] Nifo finishes his treatise with a programmatic statement:

> It is no wonder that so many erroneous and idiotic things are written by those who make annual predictions, since they are not versed in Ptolemy's rules, because of the difficulty of Ptolemy, and because of the deficiency of the translations of Ptolemy, and in particular because of the easiness *(facilitatem)* of Albumasar's doctrine.[92]

The Flemish astrologer Albert Pigghe draws both on Pico and Nifo in his *Astrologie defensio* of 1519,[93] when declaring a "war" *(bellum)* against the astrologers who follow the fables of the Arabs *(Punicis fabulis)* instead of Ptolemy.[94] Pigghe argues that there is no single moment in the year on which the fate of the entire year would depend; rather there are four starting points, the beginnings of the four seasonal quarters. And even if there was such a single moment, it would be the full moon or new moon immediately preceding the vernal equinox.[95] While he follows Pico on this point, Pigghe adopts Nifo's modification concerning the higher planets.[96] Another argument advanced by critics against the revolutionary chart concerns the exactness of measuring: the time of the sun's Aries ingress "can hardly be calculated with sufficient exactness that it could be employed for the making or construction of the celestial chart without enormous error," as Valentin Nabod puts it.[97]

Not all sixteenth-century handbooks consulted for this chapter are representative in the case of this doctrine, since they do not cover general astrology. Such is the case with the works by Schöner, Dariot, and Giuntini. Indicative, however, is the treatment of the topic by Ciruelo, Pitati, Cardano, and Campanella. Pedro Ciruelo concedes that the Aries ingress has superiority over the other equinoxes and solstices, for systematic reasons, but also because the God of the Old Testament has made the month of Passover the first month of the year (Exodus 12).[98] Ciruelo, however, proceeds to argue that it is "safer" *(tutius)* to predict on the basis of charts drawn for the latest full or new moon, rather than for the sun's Aries ingress.[99] He nevertheless also recommends drawing a chart of the Aries ingress, since this is the "general method" *(modus communis),* and making separate predictions from both charts. In case of conflicting predictions, the Ptolemaic chart is to be preferred, since it is of "greater power" *(maioris virtutis)* and because the Ptolemaic method is "more probable."[100] He then proceeds to explain the Ptolemaic and the common methods separately and at length.[101] Ciruelo's discussion is telling in several respects: it testifies to the fact that revolution charts, though un-Ptolemaic, were common practice, and it shows that Ptolemy's original doctrine is transformed. For even though Ciruelo advises, in Ptolemaic fashion, to concentrate on sun–moon conjunctions and to draw a chart for that moment, he then maintains that the *dominus figurae* and planetary conjunctions are the most indicative parameters on the chart. Arabic theories are thus introduced through the back door. Another reflection of the fact that revolution charts are common practice is Pitati's *Almanach novum,* which contains an entire chapter on annual and natal revolutions, with tables for computing the moment of the sun's entrance into Aries between the years 1500 and 1600. Pitati gives instructions for finding the lord of the year "according to Bonatti" *(secundum Bonatum),* thus transporting Albumasar's theory of the *dominus anni.*[102]

With Cardano and Campanella, we come to two authors who prefer Ptolemaic techniques on the issue. Girolamo Cardano is very skeptical of revolutions, his main reasoning being that the times for full moon and new moon can be known with much greater precision than that of the sun's Aries ingress. Full eclipses have the most powerful effect, followed by the sun's ingress and, third, full moon and new moon. Note that Cardano concedes that the sun's ingress has a greater effect than full and new moon. The problem with the sun's ingress is epistemological: it cannot be known exactly.[103] Campanella in *Astrologicorum libri VII* accepts three causes in general astrology: Ptolemy's eclipses and comets, plus the Arabic doctrine of great conjunctions.[104] Campanella rejects annual revolutions because he does not see a reason to single out one of the two equinoxes and the two solstices as the starting point of the year.[105] Campanella then proceeds to use the Ptolemaic new and full moon method for discussing predictions of the four seasons.

The discussion of annual predictions in these Renaissance handbooks is conspicuously thin if compared with Bonatti's and Leopold's full-length treatment of annual predictions. The influence of Pico's arguments is very tangible. This picture changes, however, when we consider further sources: revolutions of the years of the world are treated in several introductions to ephemerides of the fifteenth to the seventeenth centuries, for instance, in Giovanni Magini's ephemerides printed in 1610, where attempts are made to determine the moment of the sun's Aries ingress more exactly.[106] In this literature, the topic is often discussed in purely astronomical terms, but the interest is clearly astrological. Hence, while there is tangible retreat from the topic in handbooks due to Ptolemaic reformers, the Arabic doctrine of annual predictions nevertheless survived.

The survival of the doctrine is not surprising in view of the fact that it was extremely popular in sixteenth-century astrological praxis. This is very apparent from the long lists of sixteenth-century predictions extant in manuscript or print that have been collected by Gustav Hellmann.[107] Often universities required their astrology or mathematics professor to write predictions for every coming year; cities asked their town physician, and courts their astrologer in residence. Many of these annual predictions are astrometeorological in character, but often they also touch on matters of general astrology in the Persian–Arabic tradition: epidemics, wars, rulers and their subjects, and so forth. Some of these predictions claim in their titles to follow the doctrine of Ptolemy, such as Johannes Schöner's *Iudicium* for 1537, whereas other predictions are clearly based on Arabic doctrine. As Hellmann and Thorndike have pointed out, annual predictions continued to be a very lively practice throughout the sixteenth century, especially in German-speaking countries, a practice that only declined in the course of the seventeenth century.[108] It remains to be studied how much Ptolemaic and how much Arabic astrology are transported in this prognostic genre, but it can already be observed that basic Arabic ingredients such as the *dominus anni* and the Saturn–Jupiter conjunctions accompany the genre until its end.[109]

In terms of astrological technique, the revolution chart worked like a nativity chart for the coming year. On the one hand, this was a theoretical disadvantage because the world does not have a date of birth that everybody could agree on, which is why a dispute developed in the Renaissance over the excellence of the spring equinox. Also, the difficulty of calculating the exact vernal equinox resulted in prognostica for the same year with different rulers.[110] On the other hand, the revolution chart was an attractive idea since it allowed the astrologer to consider all celestial properties of the moment of the sun's Aries ingress, and not only the properties of a sun and moon conjunction or of an eclipse. Hence originated the tendency in Ciruelo and other Renaissance astrologers to interpret Ptolemy as holding that the latest sun and moon conjunction serves as the moment for the year's "birthday" chart—and that the chart itself contains the essential information. There are passages in Ptolemy's *Tetrabiblos* 2.6–8 that can be mobilized for such an interpretation, but the idea of a "nativity" of the year is clearly a Renaissance adoption from the Persian–Arabic tradition. Moreover, the theory of revolution charts allows for the prediction of political, religious, and military events. These topics are central concerns of Renaissance clients, but play a marginal role in the *Tetrabiblos,* whose general astrology is focused on issues of meteorology and agriculture.

## GREAT CONJUNCTIONS

Conjunction theory is the most prominent topic in the Renaissance controversy about the value of Arabic astrology. It is the main target of Pico's attack on the Arabs, and Pico's critique is echoed and enlarged upon by subsequent authors such as Albert Pigghe and Valentin Nabod. At the same time, conjunctionism was an "Arabic" theory with extraordinary success in the sixteenth century. It was remarked above that conjunctions appear in the many annual predictions that flooded the Italian and German book markets. Even the most widely known astrological event in the sixteenth century was a great conjunction: the prognostications of the 1524 flood, which caused great anxiety in society, concern a conjunction of Saturn and Jupiter, as well as of other planets, in the sign Pisces in February 1524. Great conjunctions were also discussed in scholarly works. A good number of astrological histories were written in this period, from Pierre d'Ailly in the early fifteenth century to Giambattista Riccioli in the seventeenth century.

The topic of Saturn–Jupiter conjunctions is related to that of the revolutions of the years. Both theories concern general events and not individuals, both originate from Sassanian Persia, and both were adopted into Arabic around 800 AD. Greek astrology also knew the prediction of general events, but did not develop theories about Saturn–Jupiter conjunctions or about the stars influencing the fate of religions and dynasties. These are features introduced by Sassanian astrologers, as far as can be reconstructed today from Arabic sources.[111] Technically, conjunction theory and revolution theory have a common ground, at least in the version of Messahalah and Albumasar. For in order to predict the effects of a particular Saturn–Jupiter conjunction, a horoscope is drawn of the sun's Aries ingress of that year in which the conjunction occurs, just as it is practiced for annual predictions.[112]

The Arabic textual tradition on conjunctionist astrology is much richer than the Latin. In Arabic manuscripts, many longer and smaller texts on the topic are attributed to Albumasar, who is the dominant author of the genre, but only the *The Book of Religions and Dynasties (Kitāb al-Milal wa-d-duwal)* was translated into Latin, known under the title *De magnis coniunctionibus.*[113] Albategnius,[114] Abraham ibn Ezra, Omar Tiberiadis, and Messahalah wrote treatises on conjunctions,[115] of which only Messahalah's short *Letter on the subjects of lunar eclipses and planetary conjunctions (Epistola in rebus eclipsis lunae et in coniunctionibus planetarum)*,[116] Abraham ibn Ezra's *Book of the World (De mundo vel seculo)*,[117] and the brief tenth chapter of Abraham ibn Ezra's *Beginnings of Wisdom (Principium sapientiae)* were translated into Latin.[118] Messahalah's original doctrine has been reconstructed and is, in fact, much more substantial than the Latin treatise suggests.[119]

The medieval Latin world was fortunate that with Albumasar's *Book of Religions and Dynasties* at least one comprehensive Arabic conjunctionist treatise was available in Latin. Latin astrologers, however, did not take full advantage of this opportunity. While Albumasar's general astrological handbook, *The Great Introduction*, much influenced Latin handbooks of similar size in the Latin West, Albumasar's *Book of Religions and Dynasties*, as far as we know today, was not imitated by Latin authors. An exception may be seen in John of Ashenden's *Summa iudicialis de accidentibus mundi* of 1347/1348, which is exclusively devoted to general astrology, treating annual revolutions and conjunctions. But, due to its patchwork character, Ashenden's compilation cannot aspire to the theoretical level of Albumasar's *Book of Religions and Dynasties*.[120] Only summaries of Albumasar's doctrine appear in medieval astrological handbooks. Guido Bonatti devotes a brief chapter to conjunctions, and Leopold's manual has half a page on the topic.[121] This was probably due to the fact that the Arabic handbooks available in Latin treat the topic either very briefly, such as Alcabitius's *Introduction*,[122] or not at all, such as Albumasar's *Great Introduction*. Nevertheless, a good number of judgments on great conjunctions of specific years were written, and several astrologers composed theoretical texts on great conjunctions, among them Roger Bacon, Bartholomew of Parma, Pietro d'Abano, Simon of Couvin, Jean de Bruges, and, as mentioned, John of Ashenden.[123] Cardinal Pierre d'Ailly (d. 1420) stands out as the author of several astrological histories that fuse Arabic conjunctionist theory with the Christian history of salvation.[124] Apart from Pierre d'Ailly and Nicolaus Cusanus,[125] however, authors of astrological histories are few in the Middle Ages. The great time for Arabic astrological history was still to come in the late Renaissance.

As an introduction to the content of the doctrine, consider a Renaissance source, Pietro Pitati's *Almanach novum* of 1544. Pitati begins by saying that the great conjunctions incline human beings and their character to various things in general, and proceeds with a standard list of six conjunctions:

Of these the greatest is the conjunction of Saturn and Jupiter in the beginning of Aries, which happens every 960 years. The second is the conjunction of these two in the beginning of every triplicity, that is every 240 years. The third is the conjunction of Saturn and Mars in the beginning of Cancer, which happens every 30 years or around 30 years. The fourth is the conjunction of Jupiter and Saturn in each sign of the same triplicity. The fifth is the entry of the Sun into the vernal equinox, that is, which happens every year in the beginning of Aries. Which is why the astrologers take this as the beginning of the year. The sixth is the conjunction of the luminaries *(i.e., of the Sun and the Moon)* in the monthly lunar conjunctions.[126]

As the wording shows, Pitati's mediate or immediate source is Alcabitius's *Introduction to Astrology,* chapter 4. Alcabitius's list in turn is very close to Albumasar's sixfold division of conjunctions in *The Book of Religions and Dynasties,* chapter 1.1. As Charles Burnett and Keiji Yamamoto have noted, Albumasar's division integrates previously distinct traditions: the Saturn–Jupiter conjunctions of the Persian tradition, the Saturn–Mars conjunction in Cancer that is typical of the astrology of Alkindi and his circle, and, last on the list, the conjunctions of sun and moon that are the cornerstone of Ptolemy's general astrology in the *Tetrabiblos.*[127] The fifth conjunction listed, the sun's Aries ingress, does not seem to be a conjunction proper at first sight, since it simply marks the point on the ecliptic essential for annual predictions. The original passage in Albumasar's *Book of Religions and Dynasties,* however, mentions the latest conjunction or opposition of sun and moon before the Aries ingress as the moment for which the horoscope is cast; this is the moment that is central for predictions in the tradition of Ptolemy (and as such was favored over the Aries ingress itself by some Arabic astrologers, such as Albategnius).[128]

Arabic conjunctionist astrology is much richer than Pitati's short abstract suggests. Albumasar's voluminous book discusses many more celestial features that influence general history, such as: the possible conjunctions of all the planets with each other; the properties of the planets and the places in themselves with regard to general events; the transits of planets over each other (transits depend on how high planets are in their own orbits when they meet in a conjunction); the *dawr,* a cycle of 360 years; and lots that pertain to general matters like rulership. Among these doctrines, however, the Saturn–Jupiter conjunctions are given prominence by Albumasar, who treats them first in his *Book of Religions and Dynasties.*[129]

The theory of Saturn–Jupiter conjunctions is based on the astronomical fact that the two highest planets pass each other on their way through the ecliptic about every twenty years. These conjunctions always occur in the

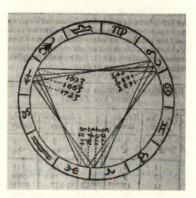

FIGURE 6.  Giambattista Riccioli, *Almagestum novum astronomiam
veterem novamque complectens* (Bologna, 1651), 7.5, 672.

same three signs of the zodiac, which are 120 degrees apart, thus forming a
triangle, or "triplicity," such as Aries, Sagittarius, and Leo (see Figure 6).

The triplicity beginning with Aries is called the "fiery triplicity," while the
other three triplicities correspond to the elements earth, air, and water.
The conjunctions remain in the same triplicity only for twelve successive
conjunctions—more or less—depending on how close to the beginning of
a sign the first conjunction was. After that, that is, after 240 years—if we
calculate by 12 times an average interval of 20 years—the following twelve
conjunctions move on to another triplicity, such as Virgo, Taurus, and Capricorn. After 960 years, the conjunctions return to the original triplicity. The
first conjunction in the original sign is often called "greatest conjunction"
(*maxima coniunctio*). Since the astrologer's year begins with Aries, the *maxima
coniunctio* is the first conjunction to happen again in Aries. "Middle conjunctions" mark a shift of one triplicity to another every 240 years, "great conjunctions" are the ordinary conjunctions that happen every 20 years.[130] The
terminology, however, is not stable. Cardano, for instance, speaks of minor,
middle, and great conjunctions.[131]

This standard version of the doctrine, straightforward as it seems, is
problematic in several respects. The interval of 20 years between two
successive conjunctions is not an exact approximation. The average is
19.86 years—with the effect that the complete cycle of four triplicities does
not take 960 years, but less than 800 years. Moreover, the exact boundary
between two triplicities is difficult to determine. After the first or second
conjunction in a new triplicity, an intermediate conjunction in the earlier

triplicity often follows. The conjunctions may even move back and forth twice between two triplicities before settling definitely with the new triplicity. It can also happen that a conjunction in fact consists of three successive conjunctions that are some days or weeks or even months apart; this happens when Jupiter is in apparent retrograde movement for a limited amount of time. A final problem concerns the fact that the Arabic authorities on the doctrine were calculating the time of the conjunctions with values for mean motions of the planets, that is, supposing that the planets move at uniform speed in their own orbits. In the Renaissance, scholars such as Ciruelo—as we shall see soon—will also calculate *true* conjunctions. They did this by using the Alfonsine tables, which are an expert product of their time, but do not yet enable the astronomers to determine the exact day of a conjunction. A good example for the imprecision that can arise from these problems is the shift from the airy to the watery triplicity in the fourteenth century AD, which Stephan Heilen has drawn attention to. If calculating with mean motions, the first conjunction in Scorpio occurred in 1385. The true shift, according to Renaissance calculations on the basis of the Alfonsine tables, was in 1365. But in fact, the first conjunction in Scorpio had already occurred in 1305; the next two conjunctions in 1325 and 1345 returned to the airy triplicity; only from 1385 onward did the conjunctions stay with the watery triplicity.[132]

Albumasar and Messahalah were aware of the fact that the figure of 20 years for the mean interval between subsequent conjunctions is an approximation. Both authors give more precise figures. For successive conjunctions, Albumasar mentions the value of 19y(ears) 314d(ays) 14h(ours) 23′ 37″ 18‴ 6⁗ 48⁗‴, for years of exactly 365 days. This value was accessible in the Latin translation.[133] Messahalah, according to the testimony of Ibn Hibintā, holds that the duration between two conjunctions was 19y 10m 11d—a figure that was not known in the Latin world.[134] The Latins for a long time focused on the number 20 and its multiples 240 and 960, as we shall see, even though they had access to Albumasar's better values.

In what sense was the Arabic theory of the great conjunctions successful in the Renaissance?[135] John North has remarked that "Pico's attack marks the beginning of the end of the doctrine of great conjunctions," since, in spite of its being used in the sixteenth century more than ever before, Pico's critique must have impressed intelligent individuals.[136] It is true that Pico's arguments much influenced the discussion about the validity of the doctrine. But it is important to emphasize that the years 1500 to 1650 saw the climax of the doctrine in Western history, in at least three respects: the doctrine itself was the object of much controversial discussion, due to Pico; it was amply used in annual predictions and apocalyptic prognostications, which I have discussed above; and it was employed to construct elaborate astrological histories.

Pico's critique of the doctrine of great conjunctions is much more substantial than that of any other theory of Arabic origin. It covers the entire book five of his *Disputationes*.[137] Experience militates against the doctrine, he says, especially in the case of the prognostications by Arnald of Villanova, Abraham ibn Ezra, and Pierre d'Ailly, which have in the meantime proved wrong (chapter 5.1). Pico criticizes astrological theories working with cycles of 360 and 300 years (5.2–3), but his main target is the Saturn–Jupiter conjunctions. It is not clear, Pico argues, why conjoining planets should have a greater effect than separate planets. The ancient authorities, and Ptolemy in particular, do not treat conjunctions of the higher planets. Ptolemy, in fact, assigns general effects only to the conjunctions of sun and moon. Pico proceeds to explain the truly Ptolemaic doctrine of conjunction by interpreting aphorisms 50, 58, and 65 of Pseudo-Ptolemy's *Centiloquium*, which he believes to be authentic. These aphorisms, which are clearly post-Ptolemaic, mention small, middle, and great conjunctions, as well as Aries ingresses—theories that Pico tries to reinterpret in purely Ptolemaic lines. The term "conjunctions" itself, he says, refers to nothing but conjunctions of sun and moon. Small conjunctions allow predictions for one month only. Middle conjunctions are those conjunctions of sun and moon that precede the sun's ingresses into Aries, Cancer, Libra, and Capricorn, that is, they concern the four seasons. Great conjunctions are full eclipses that indicate important matters of a year (5.5).[138] Pico passes in silence over the fact that aphorism 63 of the *Centiloquium* clearly speaks of conjunctions of Saturn and Jupiter, both in the Greek and in Pico's favorite Latin translation by Giovanni Pontano.[139] He thus interprets the text rather oddly along Ptolemaic lines.

Pico then advances several arguments against the theory of Saturn–Jupiter conjunctions. Most authorities, he says, calculate these conjunctions with abstract mean rather than true motions of the planets. As a result, they arrive at the faulty number of 960 years for the greatest conjunction and miss the true dates of triplicity shifts in history.[140] Furthermore, there is no justification for why the effect of a conjunction should be retarded by some or even numerous years, as conjunctionist authors often maintain—a retardation that some authors extend over a period of several other conjunctions whose effect is then ignored (5.6).[141] As to the conjunction used to explain the occurrence of the biblical Flood, it is a mystery why the world is not destroyed on the return of the same conjunction (5.11). The nativity of Christ, which Arabic astrologers offer, is pure fiction and absent from ancient sources (5.14). Pico is particularly dismissive of

Roger Bacon's elaboration on conjunction theory, according to which six religions are equated with six conjunctions of Jupiter with other planets. This merely shows, protests Pico, that such Latin astrologers are almost completely ignorant of the ancient Arabic and Hebrew books on the topic (5.17).[142] On this respectful tone toward Arabic and Hebrew astrology, Pico ends his critique of conjunctionism. It is apparent again that the *Disputationes* are multilayered. On one level, Pico is critical of astrology as a whole, on another he advertises Ptolemaic astrology and deprecates Arabic traditions and Albumasar in particular, the *princeps* of conjunction theory,[143] and on still another level he has respect for Arabic astrology and censures Latin astrologers who do not know their sources.

From Lucia Bellanti's *Defense of Astrology against Giovanni Pico della Mirandola* of 1502 onward, Pico's arguments found many defenders and opponents, in particular among authors discussing the catastrophic flood prediction of 1524—Abioso, Pigghe, Nifo, Scepper, Tannstetter, Giannotti, Stöffler, and others. But the discussion continued much beyond 1524 well into the seventeenth century to authors such as Johannes Kepler and Giambattista Riccioli. If Pico's attack on the doctrine is remarkable for the force of its arguments, Riccioli's account in his *Almagestum novum* of 1651 is an impressively lucid exposition and critical examination of conjunctionism, with a magisterial grasp of the sources in the 150 years since Pico.[144] It is worthwhile, therefore, to stay with Riccioli for a moment, to see how the discussion had developed since Pico, and then to proceed to an analysis of the astrological histories written between Pierre d'Ailly and Riccioli.

Riccioli weighs the arguments and authorities for and against the doctrine. He does not take sides on the issue, but clearly signals his general aversion to judicial astrology.[145] Riccioli mounts nine arguments against conjunction theory, which in fact summarize many centuries of argumentation: first, the doctrine is in conflict with Ptolemy; second, the astronomical tables do not offer data precise enough to draw an exact chart for the time of a conjunction; third, the astrologers tend to use mean instead of true motions of the planets; fourth, the classification of the triplicities as fiery, earthy, airy, and watery is a human invention and not a physical fact, as Kepler has pointed out; fifth, Albumasar and the other Arabic authorities offer rules for detailed prediction that are as ridiculous as the cabbala or the Quran; sixth, the chronology of historical events employed by these authors is often questionable, if not false; seventh, the historical falsity of many predictions about the rise and decline of religions and the end of the world is ample evidence of the falsity of the doctrine—here Riccioli continues with a devastating list of false or purely speculative predictions, ranging from Albumasar to Pierre d'Ailly, Cusanus, the 1524 prediction, and Cardano, and on to the seventeenth century; eighth, there is no reason for the considerable delay of effect that many conjunctionist authors admit between the conjunction and the historical event; and ninth, many utterly capital events in history have happened without a corresponding conjunction.[146]

Some of the arguments are old—the mean motion argument had already been formulated by Abraham ibn Ezra in the twelfth century[147]—but others testify to more recent developments: to Johannes Kepler's and Tommaso Campanella's critique of "unnatural" astrology; to the rise of chronology as a scientific discipline, especially since Joseph Scaliger's *De emendatione temporum* of 1583;[148] and to the disillusion provoked by the numerous apocalyptic prognostications of the sixteenth and early seventeenth centuries that were proved wrong by history.

It is impressive to see that Riccioli, against all the arguments, is nevertheless ready to pinpoint the strength of the doctrine of great conjunctions. He refers to the many prestigious astrologers, who have advocated the doctrine since Albumasar and Alcabitius, among them Pierre d'Ailly, Leovitius, Spina, Origanus, Argoli, and others, and gives particular attention to Kepler's ambivalent attitude toward the doctrine, which Kepler embraces in a number of passages. Then Riccioli concludes:

> The most important and extraordinary argument (if we leave out those that are generally advanced for the efficacy of every aspect and alignment of the planets) is the connection of important events and changes in the world with these great conjunctions.[149]

The strongest support for the doctrine of great conjunctions is empirical. It is the evidence of history that great events coincide with great conjunctions, especially with the *maximae coniunctiones* at the beginning of Aries. This argument may seem unplausible at first, but it gains plausibility in view of the development of historical astrology from 1400 to 1650. As we will see, astrology worked with increasingly refined astronomical figures, which happened to be roughly contemporaneous with two important epochs of Christian world history: Christ and Charles the Great.

Table 4 shows nine different astrological histories based on great conjunctions from the early fifteenth to the mid-seventeenth centuries: by Pierre d'Ailly,[150] Pedro Ciruelo,[151] Girolamo Cardano,[152] Cyprian Leovitius,[153] Johannes Kepler,[154] David Origanus,[155] Giovanni Francesco Spina,[156] Andrea Argoli,[157] and Giambattista Riccioli.[158] The table is preliminary; it is likely that other such histories will be found.[159] An inspiration for such

Table 4. Astrological histories from Pierre d'Ailly to Giambattista Riccioli.

| Pierre d'Ailly (*Concordantia*, 1414) | Pedro Ciruelo (*Apotelesmata*, 1521) | Cardano (*Segmenta*, 1541) | Leovitius (*De coniunctionibus*, 1564) |
|---|---|---|---|
| 5023 BC (320 anno mundi) Murder of Abel (in 100 AM). The evils introduced by Cain | | | |
| 4063 BC (1280 AM) A little after the birth of Methuselah. | | | |
| 3103 BC (2440 AM) Two years before the Flood. | | | |
| 2143 BC (3200 AM) After the discovery of the magic arts by Zoroaster. After the birth of Abraham. After changes in the kingdoms of Assyria, Egypt and Greece. | | | |
| 1183 BC (4160 AM) 100 years after the fall of Troy. Many changes in kingdoms both before and after | | 880y 98d BC | |
| 223 BC (5120 AM) = "ca. 225 BC" Translation of the Septuagint. First Punic War. | 5 BC The 36th year of the reign of Augustus. | 5y 249d BC | 6 BC Incarnation of Christ. |
| 735 AD (6080 AM) | 789 AD (or 809 true conj.) Charles the Great. The empire transferred. Flourishing of the sciences and of monasticism. Retreat of the infidels. | 788y 330d AD | 789 AD Beginning of the reign of Charles the Great. Title of the Roman emperor. |
| 1695 AD (7040 AM) "1693" Great changes in law and sects, probably the advent of Antichrist. | ca. 1600 | 1583y 179d | 1583 End of the Roman Empire. The second coming of Christ. |

| Kepler (*De stella nova*, 1606) | Origanus (*Ephemerides*, 1609) | Spina (*De maximis*, 1621) | Argoli (*Pandosion*, 1644) | Riccioli (*Almagestum*, 1651) |
|---|---|---|---|---|
| 4000 Adam. Creation of the world. | 3957 Creation of the world. | 3954 Creation of the world. | 3957 Beginning of the world. | 3980 Adam. Beginning of the world or fall of Adam. |
| 3200 Enoch. Robberies, cities, arts, tyranny. | 3163 | 3131 | 3163 | 3185 Enoch. Robberies, hence destruction of cities. The invention of arts. |
| 2400 Noah. The Flood. | 2368 The Flood. | 2337 The Flood. Noah. | 2638 (*sic*) The Flood. | 2390 Noah. The Flood. Restoration of the world. |
| 1600 Moses. Exodus from Egypt. The Law. | 1574 Moses. Exodus from Egypt. | 1543 Moses. Exodus from Egypt. | 1574 Moses. The Law. | 1595 Egyptian Plagues. Exodus. The Law. |
| 800 BC Esaias. The Era of the Greeks, Babylonians, Romans. | 779 BC Olympics. Romulus and Remus. Foundation of Rome. Assyrians destroy Israel. | 749 BC End of Assyrian and beginning of Medan reign. Olympics. Foundation of Rome. | 779 BC Romulus. Foundation of Rome. Olympics. Assyrians destroy Israel. | 801 BC Romulus. Olympics. Foundation of Rome. The new reign of Nabonassar. |
| 0 BC Christ. The Roman monarchy. Renovation of the world. | 15 AD Briefly before a great change in politics and religion. | 15 AD Great changes in politics and religion. | 6 BC Augustus rules. | 6 BC Augustus. Augustus's monarchy. Census of the world. Birth of Christ. |
| 800 AD Charles the Great. The empires of the West and of the Saracens. | 809 AD Charles the Great reconstitutes the Roman Empire. | 809 AD Charles the Great reconstitutes the Roman Empire. | 800 AD Charles the Great reunifies the empire. | 789 AD Charles the Great. The Roman Empire being transferred to the Franks. |
| 1600 Rudolph II. Our life, fate, and aims. | 1601 | 1603 Disputes. Wars. The destruction of the Turkish and Muslim Empire. | 1600/1700/1780 The coming of a new, universal monarchy. | 1583 or 1603 Calendar reform. The Japanese kings' embassy to the pope. Three new stars. |

astrological histories was a passage in Albumasar's *Book of Religions and Dynasties,* in which he presented a brief history of the world covering the creation of Adam, the Flood, and the occurrence of Islam (which, however, was not based on a 960-year, but a 360-year cycle).[160]

The table is indicative in several respects. It shows that astrological histories were particularly popular in the first half of the seventeenth century and that Johannes Kepler's *De stella nova* much influenced the historical information given in subsequent astrological histories. It is also apparent that Girolamo Cardano popularized a sequence of great conjunctions that comprised the years 800 BC, 5 BC, 789 AD, and 1583 AD. The interpretation of the past was much less contentious than that of the present and the future. The greatest disagreement is about how to interpret the triplicity shifts in 1583 and 1603. While Leovitius prognosticates the second coming of Christ for 1583, other authors link these dates with the end of the Roman or Muslim empires. Finally, Cardano, like Ciruelo before him, abandons Albumasar's figure of 960 years for the interval between great conjunctions, a figure still used by Pierre d'Ailly in his *Concordantia* of 1414.

It is one of the merits of Renaissance astrologers that they realize that the figure 960 is in need of correction. The figure was based on the assumption that Saturn–Jupiter conjunctions appear every twenty years, which, as mentioned before, is a rough approximation based on mean motions of the planets. It had been known since antiquity that the planets do not move with uniform speed, but calculating with mean motions was a very useful idealization. Renaissance authors demonstrated two things: that the true motion of Jupiter and Saturn differed from their mean motion so significantly that the validity of the conjunctionist doctrine was undermined, and second, that even in terms of mean motion, Albumasar's figures of 20, 240, and 960 years were in need of correction.

As to the first criticism, Pedro Ciruelo formulated a particularly convincing argument in his *Apotelesmata* of 1521.[161] Ciruelo criticizes Albumasar for using mean motions "which are purely imaginative and do not have any effect in the world."[162] The use of mean motions by Albumasar is evident, Ciruelo says, from the fact that he calculates with fixed intervals of twenty years between great conjunctions, whereas in fact the interval between true conjunctions varies. To support his claim, Ciruelo lists the intervals between the four great conjunctions of 1464, 1484, 1504, and 1524, which are: 20y 7m 10d, 19y 6m 20d, and 19y 7m 22d, respectively. A modern calculation of these intervals yields similar results: 20y 7m 10d, 19y 6m 7d, and 19y 8m 6d.[163] Where many authors, including Pico, simply deplore the use of mean motions, Ciruelo proceeds to demonstrate its weakness. The true motion of Saturn and Jupiter is such that the interval between successive conjunctions may be several months longer or several months shorter than twenty years.

This is one of several arguments Ciruelo mounts against the conjunctionist doctrine and against the many "senseless, absurd and false" doctrines in Albumasar's *De magnis coniunctionibus*. The Parisian theologians were right, says Ciruelo, referring to the thirteenth-century theologians William of Auvergne and Thomas Aquinas,[164] when they condemned Albumasar, the "archprophesier" *(archidivinator)*, together with his work.[165] But Ciruelo nevertheless embraces a reformed version of conjunctionism, since Albumasar's book also contains "much that is true." The practical art of astrology is the same whether you follow Ptolemy's preference for sun and moon or Albumasar's preference for the higher planets, argues Ciruelo. Because of the slowness of Saturn and Jupiter it is impossible to draw a chart for the exact hour of conjunction; rather, one should draw a chart for the immediately preceding conjunction or opposition of sun and moon.[166] Ciruelo thus finds a compromise between the Greek and Arabic traditions of mundane astrology. In fact, his solution echoes methods of 'rectification' used in natal astrology, if the exact date of birth of a client was unknown.[167] For illustrating the doctrine of great conjunctions, Ciruelo gives historical examples, using the approximate figure of 800 years for the interval between great conjunctions: 6 BC, briefly before the birth of Christ, 809 AD, a time when the imperial honor was transferred from the Greeks to the Latins and Charles the Great, and finally around 1600 AD, of which he does not give an interpretation.[168] Ciruelo is thus a witness to the refined criticism of and unbroken adherence to the theory of the great conjunctions.

In the history of sixteenth- and seventeenth-century conjunctionism, the most influential argument was formulated not by Ciruelo, but by Girolamo Cardano.[169] Cardano's attitude toward Arabic astrology in general is ambivalent, as we have seen on several occasions. This is also true with respect to conjunctionism. In his youthful *Pronostico generale* of 1534, he unhesitatingly uses the scheme of small, middle, and great conjunctions and discusses the conjunctions of 1524, 1544, and 1564. The *coniunctione maxima* of 1564 signifies "the renewal of all laws, the Christian as well as the Muslim," as Cardano claimed.[170] And in his *Libelli duo* of 1538, Cardano uses the great conjunctions again to predict the success of both Charles V and Islam.[171] A different attitude appears in the prooemium to Ptolemy's *Tetrabiblos* of 1554, where Cardano attacks the "innumerable crowd of idiots" *(innumberabilis turba nebulonum)* who have ruined astrology. The list of "idiots" added by Cardano is remarkable for its distinct Arabic character: "So many Albumasars, Abenragels, Alcabitiuses,

Abubaters, Zahels, Messahalahs, Bethens, Firmicuses, Bonattis—o good gods, what has remained, what is left of so many imposters? Of so many follies?"[172] Astrology can only be saved by a divine gift: the one and only Ptolemy. In his commentary on the Ptolemaic text, Cardano explains that it is impossible to observe "the exact day, not to speak of the hour" of a Saturn–Jupiter conjunction and thus to draw a chart.[173] "For this reason everybody will clearly see the small importance of the science of these conjunctions, which up to the present day has been praised so much. Nevertheless, some general things can be gained from the great conjunctions."[174] And Cardano proceeds to explain general features of the doctrine, such as, the linkage of the triplicities' return to Aries with the rise and fall of empires: in about 800 BC the rise of the empire of the Medes, at the time of Christ the Roman Empire under Augustus, in 790 AD Charles the Great, and from 1583 to 1782 a new empire.[175] It is obvious that Cardano's rejection of conjunctionism was only halfhearted.

The historical scheme of his Ptolemy commentary, however, was not Cardano's most influential contribution to the doctrine of the great conjunctions. In his *Aphorismorum astronomicorum segmenta septem* of 1541, an astrological compendium that was useful for astrological practice because of its technical character,[176] Cardano devotes a paragraph to conjunctionism, which transformed the discussion for many years to come. Here Cardano presents his astronomical interpretation of the doctrine, as based on the mean motions of Saturn and Jupiter. He first gives a figure for the distance between two successive conjunctions and then explains how these conjunctions "travel" through the signs of the zodiac. On this basis he then presents mean figures also for middle and great conjunctions, which considerably differ from the traditional figures of 240 and 960 years:

> The small conjunction of Saturn and Jupiter happens every 19 years, 315 days and 19 hours. From there it proceeds through 242 degrees, 59 minutes and 9 seconds, that is, through 8 signs of zodiac, 2 degrees, 59 minutes and 9 seconds, and thus the nature of the signs proceeds in the same triplicity, such as from Aries to Sagittarius, by almost three degrees. Hence we realize that the nature of the signs moves from the fiery to the earthy, hence to the airy and then to the watery nature, after 10 revolutions have been completed under the same nature and quality of the signs. Thus they remain in the same triplicity of the signs for 198 common years and 236 days . . . Therefore, there began a great conjunction 800 years and 98 days before Christ, which returns after 794 years and 214 days.[177]

Cardano is using a good value for the gain in degrees per mean conjunction: 242°, 59′, 9″, which is close to the figure of 242°, 41′, 33″, which we calculate today. On the basis of his new value Cardano judges that, on average, it does not take twelve conjunctions until a triplicity shift occurs, as the Persian–Arabic–Latin tradition held, but ten. Moreover, Cardano does not calculate with the traditional value of twenty years between two successive conjunctions, but with the value of 19y 314d 19h—a value close to 19y 313d 12h (19.85838 years), which we calculate today. By multiplying this value ten times, Cardano arrives at the value of 198y 236d for middle conjunctions. By multiplying it forty times, he arrives at 794y 214d for great conjunctions.[178] One may recall that Albumasar himself had worked on more precise figures. His value for the gain in degrees between two successive conjunctions was 242° 25′ 17″, his value for the time difference between two conjunctions 19y 314d 14h.[179] Albumasar and his Latin followers, however, did not use these figures for the calculation of middle and great triplicity shifts and hence continued to apply the average values of 240 and 960 to world history.

The full strength of Cardano's approach becomes obvious when he turns to the astrological explanation of history. On the page following the above citation, he inserts "a table with all great and middle conjunctions according to mean motions" from 800 BC to 3967 AD.[180] The four great conjunctions are placed in 800y 98d BC, 5y 249d BC, 788y 330d AD, and 1583y 179d AD. They are not linked explicitly to historical events, but this can be done very easily, as the reception history of Cardano's chronology shows. The intervals between Cardano's four great conjunctions on the table are 794y 214d, just as the calculated value demands. The striking advantage of Cardano's conjunctionist theory is that the values for the mean interval between small, middle, and great conjunctions are calculated in a way that Cardano's historical dates for mean conjunctions come very close to the dates of true conjunctions. True conjunctions had been the subject of discussion by previous authors, such as by Ciruelo, as we have seen. Ciruelo had, for instance, discussed a true conjunction on March 3, 789 AD. This date is close to Cardano's mean conjunction at the end of 788 AD. Previous figures for the period of *maximae coniunctiones*, such as Albumasar's and Pierre d'Ailly's period of 960 years, or Ciruelo's rough figure of 800 years, were too imprecise to allow astrologers to place true conjunctions in an astrological history. Cardano's figure of 794 years and 214 days makes it possible.

It was remarked above that the borders between triplicities are not clear-cut, in the sense that, for a limited period, conjunctions tend to jump back into the previous triplicity. The conjunction of the year 789 AD, for example, was, in fact, the second to last conjunction in the watery triplicity and not the first conjunction in the fiery triplicity, which had already taken place in 769. In spite of these problems, Cardano's figure was very helpful. It guaranteed that the astrologer could work with dates that were always close to true conjunctions and relatively close to true shifts of triplicity.

Later authors, such as Nabod,[181] Leovitius, Kepler, Argoli, and Riccioli are clearly influenced by Cardano's figures, whether directly or indirectly, as one can see in the Table 4. The years 800 BC, 5 BC, 788 AD, and 1583 AD became points of reference, even if they were corrected to more precise dates. We are now in a position to understand why Riccioli could claim that the strongest argument in support of conjunction theory was the connection of important events with Saturn–Jupiter conjunctions. He was clearly thinking of the dates 5 BC or 6 BC and 789 AD, that apparently served as empirical proof of the validity of the doctrine: the dates appeared to indicate important religious and political events. This connection, which was made possible by Cardano's new astronomical figures, helps to explain the success of the doctrine between Cardano and Riccioli. The doctrine, in fact, never found more adherents in the Latin world than in this period.

It was remarked above that Johannes Kepler was an important step in the reception of Albumasar's and Cardano's theories. He extended the astrological interpretation of history back to the creation of the world, on the basis of Cardano's new figure of 794 years, just as Pierre d'Ailly had done with the old figure of 960 years. It cannot be ruled out that Kepler was drawing on an as yet unidentified source, but it remains clear that it was his astrological interpretation of history that was cited by later authors. This is true even though Kepler's attitude toward conjunctionism was as ambivalent as Cardano's. In *De trigono igneo* of 1606, Kepler directly refers to Pico's censure of conjunctionism saying that he is not fully convinced by Pico's critique.[182] Kepler summarizes his own standpoint as follows:

> I do not say this because I want to defend the astrologers' conclusions up to every detailed prediction, but in order to claim that at the time of the great conjunctions there occurred changes of the natures and of their natural effects in the human beings of such magnitude that the astrologers were deceived by them and believed that these things, which happened on acount of these changes, derived from this celestial principle *(i.e., from the great conjunctions)*.[183]

This means that Kepler conceded the existence of a parallelism between great conjunctions and major changes in the world, but he did not accept a causal connection. In the same text, he moves further toward conjunctionism by claiming that "the very great power of planetary conjunctions in provoking the powers of sublunar things has been observed by the ancients and is today observed."[184] Kepler carefully uses the terms *exciere* (provoke) or *excitari* (excite) instead of *cogi* (be forced): the great conjunctions provoke the sublunar nature, but they do not force it. This relation of "provoking," however, is real, and Kepler approves of the empirical, that is, historical foundation of conjunctionism.[185] What Kepler mainly protests against is the extension of this general rule to particular predictions. In a late astrological treatise on the conjunction of 1623, Kepler again says much on the topic. He approves of the doctrine as long as it does not result in a determinism of the human will. It is based in nature and in God, who arranges great conjunctions, Kepler argues, such as the conjunction at the time of the birth of Christ, which the magi saw.[186] The Arabic authorities nevertheless err on the topic, Kepler says, because they link the doctrine with the division of the zodiac into twelve signs, which is helpful for memorizing, but baseless and superstitious; also, they assign elements to the four triplicities, which again is gratuitous.[187]

Kepler's most influential contribution to conjunctionism was his astrological histories, which appear in at least two places: in the aforementioned *De trigono igneo* of 1606 and in the *Epitome Astronomiae Copernicanae* of 1618. In the later text, the *Epitome,* Kepler presents a brief conjunctionist history of the world, which he bases on Cardano's figure of 794y, with succinct information on the historical events. The historical scheme can be reconstructed as follows:[188] creation at 3975 BC, Enoch 3182 BC, the Flood 2388 BC, Exodus from Egypt 1514 BC, Babylonian exile of Israel, Olympics, foundation of Rome 794 BC, Christ's coming 6 BC, Charles the Great 789 AD, and stella nova 1583 AD.[189]

More influential was the astrological history presented as a handy table in the first text, *De trigono igneo*. Here, Kepler uses the figure of 800 years between *coniunctiones maximae*. Since he was convinced of this periodization of history, but not of more detailed astrological theories, he is content with these "round and hardly precise numbers."[190] Astrological history, in Kepler's view, was most useful for memorizing. The result is a simple scheme, which extends, in periods of 800 years, from Adam in 4000 BC to Rudolph II in 1600 AD. As can be seen in Table 4, Kepler's historical scheme covers creation, flood, exodus, the Roman empire, Christ, the Frankish Empire, and the Saracens; it influenced several subsequent authors, among them Andrea Argoli and Giambattista Riccioli.

When in 1653 Riccioli wrote his masterful summary of the doctrine of great conjunctions, its content, its defenders, and its critics, Kepler was given an important place. Riccioli offers quotations from Kepler's *De trigono igneo* to demonstrate that major authorities have supported the doctrine. Riccioli's own table of historical events is computed with the four-triplicity period of 794y 133d,[191] taken from Erasmus Reinhold, who had tried to improve upon Cardano's figure of 794y 214d; Reinhold's figure was also adopted by David Origanus and others.[192]

It is remarkable to see that the doctrine of the great conjunctions is in full flourish around 1600—about a thousand years after its invention in Sassanian Persia. It has survived the attacks of Pico and other humanist reformers of astrology and sees the highpoint of its influence in early modern Europe. This influence is noteworthy not only for its extent, both in popular and theoretical astrology, but also for its quality. Astrologers such as Pico, Ciruelo, Cardano, Kepler, and Riccioli enhanced the understanding of the doctrine and transformed it to a considerable degree. On the one hand, they exposed the doctrine to fundamental criticism that made its weaknesses obvious: that the existing tables did not allow computing the precise moment of a true conjunction; and, in particular, that there are dramatic time gaps between true conjunctions and those calculated with the average figure of twenty years. Also, they argued historically that major turns of world history happened without accompanying great conjunctions. On the other hand, they considerably improved the astronomical understanding of the phenomenon and provided much better figures for the mean interval between conjunctions. This, in turn, sparked the flourishing of the genre of astrological world histories between 1550 and 1650, which was much supported by the "empirical" argument that the 794-year cycle between great conjunctions coincided with important events of history as viewed from a Christian standpoint.

It is not surprising that conjunctionism resonated particularly with Christian authors of the early modern period—and, as far as we know, even more so than with Zoroastrian, Muslim, and Jewish authors of earlier times. For the chronological construction of world history was a major concern of Western scholars of the time and a driving force of important intellectual trends.[193] Even more important, Christianity, among the four religions mentioned, is the most historical: it involves not only creation, prophets, and the end of the world, but also the incarnation of God at a historical moment. Christian presuppositions, therefore, clearly aided the reception of the conjunctionist doctrine. Humanist reserverations, on the other hand, though uttered by Pico, Riccioli and others, were not strong enough to prevent the success of the doctrine, which, remarkably enough, was not supported by any ancient authority, save for a spurious passage in Pseudo-Ptolemy's *Centiloquium*. The impact of humanism on the discussion, in general, is less obvious than in other parts of astrology, where there exist ancient counterparts to Arabic doctrines. In this case, the reformist pressure, which led, for instance, to the disappearance of the figure 960, is inspired by other motives that were technically astronomical or astrological in the first place.

Hence, the Persian and Arabic conjunctionist doctrine, which had been available to Western scholars since the twelfth century, came to full fruition as late as the sixteenth and seventeenth centuries. It is true that the doctrine was transformed in these centuries, but its main assumptions remained intact. Some authors such as Kepler divest the doctrine of its more speculative parts, for example, the classification of the triplicities as fiery, earthy, airy, and watery, but the basic idea of the Persian–Arabic tradition continued to appeal—that Saturn–Jupiter conjunctions indicate great political and religious changes in world history. The continuation between oriental and Renaissance astrology is apparent not only in the construction of astrological histories but also in the attempts to reform the doctrine astronomically. When Ciruelo and Cardano tried to arrive at better astronomical figures, they follow the example of Albumasar, Messahalah, and others, who had also worked on the doctrine's astronomy.

The Western decline of the doctrine still awaits proper study; it seems that it did not lose its attraction before it declined together with the entire science of astrology in the Enlightenment era.[194]

⁂

It has emerged from the above survey of disputes that the Greek–Arabic antagonism was a catalyzer of many changes in Renaissance astrological theory. The criticism voiced by Pico, Ciruelo, Cardano, and others against Arabic astrological traditions resulted, most visibly, in the gradual disappearance of interrogations and lots from astrological handbooks. It also provoked a very skeptical attitude toward all those elections that do not concern agricultural and medical matters. Moreover, it sharpened the awareness of the fundamental difference between Ptolemy's method of determining the length of life by means of prorogation and Arabic methods that proceed from the absolute and relative properties of a specific planet. In general astrology, anti-Arabic attitudes led to the disappearance of annual revolutions from several handbooks, because of a preference for Ptolemy's doctrine of sun and moon conjunctions. It is true that Arabic and medieval Latin astrologers had pointed to differences between astrological authorities before, but in Renaissance astrology, this tendency was much intensified.

On the other hand, it is also obvious that the idea of a Ptolemaic reform of astrology was only of limited success. Even Girolamo Cardano, an outspoken partisan of Ptolemy, composes treaties on interrogations and anniversary horoscopes, that is, on techniques that he rejects in his commentary on the *Tetrabiblos*. And Cardano contributes an important doctrinal precision to conjunctionist astrology, but rejects the doctrine in other places. In a similar fashion, Johannes Schöner and Francesco Giuntini continue to employ the theory of lots in many passages of their handbooks, even though they avoid treating them prominently as part of their astrological theory. In general astrology, reformers move away from the doctrine of annual revolutions, but the production of such predictions reaches an unprecedented climax in the sixteenth century. Apparently, in some of these areas, a gap is developing between astrological theory proper and astrological praxis. Moreover, on the level of theory, many Arabic traditions continue to thrive in spite of the criticism voiced against them by advocates of a return to Ptolemy: elections remain a standard chapter in astrological handbooks, as do anniversary horoscopes and the Alcocoden theory. Anniversary horoscopes, in fact, far from being suppressed, gain a popularity in sixteenth-century astrological literature that is unparalleled in the centuries before, in spite of the attempts to revive Ptolemaic methods of prorogation. Nor did humanist attacks prevent the conjunctionist theory of history, which is lacking any Greek credentials, from becoming extraordinarily successful in the sixteenth and seventeenth centuries.

From a technical point of view, it can be observed that the doctrines of interrogations and lots, which suffered most from the polemics, are comparatively simple and uncomplex. Their theoretical weaknesses are readily exploited by Renaissance critics. With respect to interrogations, it is difficult to justify the celestial importance of the moment a question is asked. With respect to the doctrine of lots, it is unclear why one should assign celestial importance to a degree on the ecliptic found by adding the distance between two planets to the ascendant. At the same time, however, both doctrines answer to specific practical needs, which explains why they do not disappear completely from Renaissance astrology. As to the other matters of dispute, there are indications that the survival and popularity of Arabic doctrines was not simply a concession to tradition, but a development motivated by internal and technical reasons. First, using anniversary horoscopes, or revolutions, is a method of lifelong prediction allowing predictions that are exact and detailed for a specific year of the client's life.

In Ptolemy's rival technique of prorogation, predictions can be of greater detail and specificity, but at the cost of a complicated calculation of trajectories on the ecliptic starting at five different points. Hence, in comparison to prorogations, revolutions, while being sufficiently rich in prediction, have the advantage that they are based on an elegant and simple idea: a double nativity chart. Second, the Alcocoden theory, as we have seen, is technically a real alternative to Ptolemy's calculation of releaser and destructive points. Third, the election of beneficial moments for actions is a doctrine absent from Ptolemy but with close relations to other sciences and arts such as medicine, agriculture, politics, and the military. Fourth, the revolution chart used in Arabic astrology for predicting general matters contains many more parameters for interpretation—such as the lord of the year and conjunctions of the higher planets—than Ptolemy's horoscopes for sun and moon conjunctions; and Ptolemy's method is more limited in content, since it is very brief on political and military issues. And fifth, the theory of great conjunctions was flourishing in an unprecedented way, not only because it was supported by trends in chronology and Christian views of history but also because its astronomical basis was transformed successfully by Renaissance astrologers, so that it became much more easily applicable to history. In sum, there is much reason to believe that the Arabic traditions in Renaissance astrology developed as they did not only for reasons of context but also for reasons internal to the science.

This allows us to draw some conclusions about the influence of nonastrological factors on the development of astrology. As was pointed out above, the humanist preference for Ptolemy triggered productive transformations of the science of astrology. But on some issues, the same preference blinded the followers of Ptolemy to the strengths of Arabic doctrines: on anniversary horoscopes, Alcocoden, and elections. In these cases, the advocation of a return to Ptolemy is mainly motivated by a biased belief in the superiority of Greek science and of Ptolemy in particular.[195] Johannes Schöner's attempt to describe anniversary horoscopes as "Ptolemaic" in their principles is an example of such ideological motivation. As has become clear, there is a historical irony in calling Ptolemy's astrology "Greek," in view of the fact that many Arabic doctrines attacked—elections, anniversary horoscopes, a ruling planet on the length of life, and many lots—are perfectly Greek in origin, as has been fully visible only since the twentieth century. It is ironic that Pico's idea of a return to Greek astrology was focused on the *Tetrabiblos,* a text that, as the black sheep of Greek astrology, is far from being representative. One disadvantage of the astrology of the *Tetrabiblos* has emerged several times in the above survey: it is a puristic version of astrology, which is original and theoretically challenging in it-self, but hardly practical. The astrology of the *Tetrabiblos* cannot be trans-ferred into astrological practice without adding or inventing detailed tech-niques, which would then probably be similar to what could be read in Dorotheus or Vettius Valens or their Indian, Persian, and Arabic successors.

On the other hand, the scientific productivity of the humanist idealiza-tion of Ptolemy is very apparent: Ptolemy's theoretical foundation of as-trology made Renaissance astrologers reflect again about the theoretical validity of the astrological doctrines current in their time. It was difficult to defend interrogations and lots without seeking refuge in arguments of tradition and Arabic partisanship: that they should be practiced because they have always been practiced, and by famous Arabic astrologers at that. We have met with some indications that the practicing astrologers con-tinued to use these techniques even though Renaissance handbook litera-ture ignored them, for instance in Cardano's remark that people continue "to ask questions." In general astrology, revolutions of the years of the world remained a popular practice in spite of the warning, often voiced by schol-arly astrologers, that the calculation of the vernal equinox cannot be precise enough to allow for detailed prediction. Apparently, the expectations of Renaissance clients also played a conservative role in the development of astrological theory. It is likely that studies in the sociology of Renaissance astrology will be able to buttress these findings.

In the final analysis the question may arise whether the label "Arabic" can be sensibly employed in the field of astrology, given that so much of the allegedly Arabic doctrines ultimately are of Greek, Indian, or Persian or-igin. This is true to some extent. But in another sense the label "Arabic" is perfectly appropriate. Hardly any other science in medieval and Renais-sance culture was based as massively on Arabic sources. It is the Arabic version of interrogations, elections, anniversary horoscopes, and so forth that the reformers had to confront. In addition, Arabic astrologers clearly added to and transformed astrological theory, also in the centuries rele-vant for the Latin West, that is, in the period between Messahalah in the eighth century and Haly Abenragel in the eleventh century. But modern scholarship on this development is still in its infancy. It is to be expected from the coming decades of research that the contributions of Greek, Indian, Persian, and Arabic scholars to astrological theory will be much clarified and will thus also offer a better basis for understanding the devel-opment of Renaissance astrology.

# CHAPTER 7

# CONCLUSION

I t was a precondition for the development of Renaissance scientific culture that knowledge had been migrating on a large scale in the Mediterranean and Near Eastern region for more than two thousand years. Translation movements and personal contacts guaranteed the transfer of sciences and philosophy even in times of political conflict. As a result, Renaissance thought did not rest merely on a fusion of Greek logos and Christianity, as an oversimplified account has it; rather, it inherited the knowledge traditions of Mesopotamians, Egyptians, Greeks, Romans, Jews, Christians, Muslims, and Germanic and Slavonic pagans—to name only a few obvious ancestors. A historically adequate account of the genesis of Renaissance culture is therefore bound to be complex and heterogeneous.

But if one takes such an approach to the Renaissance, it may seem self-contradictory, or at least unwise, to speak of "the Arabic influence" on "the Renaissance." Such vocabulary can run the risk of implying the existence of two antagonistic worlds, one Oriental, one Occidental. Would it not be advisable to drop the concept of cultures influencing each other altogether, in order to avoid an "imaginative geography,"[1] a simplistic essentialism of cultural blocks? Despite the risk, the answer is no, for at least two reasons. First of all, the concept of cultures, in the sense of groups of human beings united by language or religion or other practices, remains a hermeneutically fruitful tool for the historian, if handled carefully. Among the intellectual ancestors of Renaissance philosophy and science are numerous individuals and groups, but their contributions differ in quantity and quality. As historians, we are trying to delineate and demarcate these differences. Thus we distinguish, among the founders of Renaissance culture, a substantial Arabic-speaking group of individuals from various religious and ethnic backgrounds, who were active in Islamic societies. In demarcating this group and its role as a source culture for the Renaissance, we do not and need not create an essentialist notion of the science of the Orient. Second, the result of such an endeavor, if it is successful, is historical fairness. It is a principal task of the historian to determine who was responsible for what in the past. We thus credit individuals and groups with what they, and not others, achieved intellectually. Such fairness is not always a central aim of historical research, but it becomes pivotal whenever we see that the historical contribution of individuals and cultures has been largely lost from memory, or even suppressed, for an extended period of time.

How then did Arabic sciences and philosophy influence the Renaissance? The main thesis of the present study, namely, that Arabic traditions experienced both success and suppression in the Renaissance, may seem vulnerable to attack on both counts. On the one hand, one may doubt that there was success, and argue that the Arabic influence on Renaissance sciences and philosophy has been exaggerated by modern scholars, partly for reasons of political correctness.[2] In this vein, it has been argued that Arabic traditions, once significant in the thirteenth century, declined after 1300; by the time of the Renaissance, they were outmoded and clearly deserved to be set aside.[3] On the other hand, one may doubt that there was suppression, and argue that polemics, being superficial, do not say much about the true attitude of humanists and clerics toward Arabic traditions, and that this attitude was, in fact, generally very positive and supportive.[4] In what follows, I will argue against both these contentions. In doing so, I will draw on the results of the preceding chapters to present an overall image of the Arabic influence in the Renaissance.

## SUCCESS

It is widely known that Arabic sources had a great influence on medieval sciences and philosophy, especially in the thirteenth century, a few decades after these texts had been translated in Spain and Italy and had then reached the centers of European learning. It is much less widely known that a good number of these texts developed their full impact as late as the sixteenth century.

Chapter 1 of the present study contains a table that, as I have remarked before, says more about the Arabic influence in the Renaissance than many words could do. Table 1 lists the number of printed Latin editions of Arabic authors up to the year 1700, among them, most prominently, 114 editions of Averroes, 78 of Avicenna, 72 of Mesue, and 67 of Rhazes. All in all, 44 Arabic authors were made available in Latin prints. The scholarly preparation, the editorial work, and the financial means invested in these editions were considerable. Much philological care went into textual improvements, into humanist revisions of the Latin, and even into new translations of texts from Arabic or Hebrew. The textual transmission of several important Arabic works reached its peak in the sixteenth century. Examples of such transmission peaks are Avicenna's *Canon,* several commentaries by Averroes, Mesue's pharmacological *Opera,* and Rhazes's *Ad Almansorem,* book 9. That these texts were central sources of Renaissance academic culture is also indicated by the substantial number of commentaries written on them. Lexical tools were composed that helped the reader to use these sources, such as Marcantonio Zimara's *Tabula dilucidationum,* a comprehensive lexicon on the vocabulary of Aristotle and Averroes, or Andrea Alpago's two dictionaries of technical terms in Avicenna and Serapion. In astrology, Arabic authors dominated the field along with Ptolemy. The market for books on astrology was not as big as that for medicine and philosophy, but the printing history of an Arabic astrologer such as Alcabitius, who was printed thirteen times, is still considerable. Arabic astrological texts, just like their medical and philosophical counterparts, were commented on and improved philologically in the Renaissance.

The main reason for the great interest in Arabic authors was their firm rooting in Renaissance university culture. Averroes, the commentator on Aristotle, was a regular part of Renaissance education, especially in Italy, in spite of the polemics against Averroes as an allegedly irreligious philosopher. He continued to be read intensively even after the Greek commentators on Aristotle had become accessible in Latin translation. Averroes—and the Greek commentators—were read in the universities without being officially mentioned in the curricula. Arabic medical authors, in contrast, appear as prescribed reading in many university statutes of the Renaissance. In a remarkable episode of the history of universities, humanist reformers successfully expelled Arabic authorities from many medical curricula in the first half of the sixteenth century, especially in German, Spanish, and French universities, but were not able to ban these authors for long. In the decades around 1600, Arabic authorities were reintroduced into the medical curricula of many universities. Astrology, in turn, was never taught more regularly in universities than in the period between 1450 and 1650, as part of the teaching in the faculties of *mathematica* and medicine. Wherever we have evidence about the Renaissance curriculum in astrology, we can see that Arabic texts, together with Ptolemy, formed part of it.

As a consequence, the names of academic authorities such as Averroes, Avicenna, Mesue, and Rhazes were known to every educated person in the Renaissance. This is also reflected in the rising interest, unprecedented in the Middle Ages, in the biographies of Arabic scholars. World chronicles, treatises on famous men, and histories of scientific disciplines often describe the lives and works of the most famous Arabic scholars, such as Avicenna, the towering authority of medicine, or Averroes, the *philosophus clarissimus* and well-known physician. Reliable biographical material on Arabic authors was exceedingly sparse, and a good number of Renaissance biographies are imaginative rather than accurate. But the interest was so great that for some authors, and for Avicenna in particular, Renaissance scholars tried to get access to new and solid biographical information from the Arabic. One motivation for this interest was an Italianizing conception of the history of medicine: Renaissance physicians in Italy saw themselves as heirs to a history of medicine that led directly from the Greeks to the Arabs to Italian medical scholars such as Pietro d'Abano and his successors.

With respect to content, the great influence of Arabic theories in the Renaissance has been demonstrated in this study for three areas: epistemology, *materia medica,* and astrology. In all three, the discussion of Arabic theories was marked not only by a great number of Renaissance contributions but also by high levels of sophistication and scientific quality. In philosophy, the Renaissance discussion of the unicity thesis is an example of such quality. Averroes's theory of the one intellect for all human beings was, if stated as such, both understandable to everybody and counterintuitive, and hence the philosophers participating in the discussion could only be thoroughgoing experts who were able to base the theory on a complex argumentative and exegetical structure. There is ample evidence for the great attraction exerted by Averroes's theory on philosophers and commentators in the fifteenth and sixteenth centuries, even after the introduction of rival theories such as that of Alexander of Aphrodisias. Agostino Nifo, the commentator on Averroes's long commentaries, or Marcantonio Genua, the Averroist philosophy professor at Padua for more than forty years from 1517 onward, who acquired an unprecedented mastery of the Greek, Arabic, and Latin commentary tradition, are but two examples of the very high level of Averroes's reception in the Renaissance. Intensive editorial and translating activities further stimulated this process. Far from being outmoded or set aside, Averroes was everywhere in Italian philosophy faculties of the first half of the sixteenth century.

In *materia medica,* the scientific discussion in the 1510s to 1540s had the appearance of a crisis in which several trends conflicted with each other: the tradition of Arabic pharmacology, the new humanist call for a return to ancient pharmacology, and the empirical trend toward testing and botanical depiction. The heated controversy about the value of Arabic *materia medica* and the refined Renaissance commentary tradition on Arabic pharmacologists greatly helped to master this complexity: Renaissance botanists and pharmacologists, in particular the humanist scholars Giovanni Manardo, Antonio Musa Brasavola, and Pietro Mattioli, successfully connected the information in Arabic sources with Greek and contemporary knowledge. Thus, Arabic–Latin *materia medica* not only remained part of the Western tradition, but received unprecedented philological, medical, and botanical attention in the sixteenth century. Academic and humanist physicians such as Giovanni Manardo, Andrea Marini, and Giovanni Costeo composed commentaries on the pharmacological works of Mesue, Avicenna, and Serapion. Humanist scholars revised medieval translations of Arabic works of *materia medica,* such as Jacobus Sylvius's reworking of Mesue's *Opera* and Niccolò Mutoni's revision of Serapion's *Liber aggregatus.* The well-known sixteenth-century dispensatories for apothecaries relied greatly on Arabic sources. Medical students and apprentices in pharmacy read Arabic authors on *materia medica* in school. Travelers to the Orient, such as Leonhard Rauwolf or Prospero Alpini, reported about Near Eastern botany in their travel accounts. Andrea Alpago, who lived in Damascus for most of his life, contributed philological corrections to the *Canon* that were officially endorsed by the University of Padua in 1521, and composed a dictionary of 2,050 transliterated Arabic terms in Avicenna as well as a list of simple medicines in Pseudo-Serapion. As a result of all these efforts, the Western understanding of Arabic *materia medica,* and of its synonyms in Greek and Latin, reached an entirely new level in the second half of the sixteenth century.

In astrology, Arabic sources, together with Ptolemy, formed the basis of the Renaissance boom in astrology as well as the controversial and high-level discussion that involved many of the foremost thinkers of the time. The great range of Arabic sources translated in the Middle Ages and their theoretical and practical potential now came to bear full fruit. In general astrology, Arabic doctrines—or Persian doctrines developed by their Arabic successors—played a fundamental role, both in the genre of annual revolutions and in astrological history. Annual revolutions—that is, predictions of the coming year on the basis of a chart for the vernal equinox—were an extremely popular genre of Renaissance astrology. Here, Arabic doctrines such as those concerning the *dominus anni,* planetary conjunctions, and weather forecasting appeared in predictions all over Europe. Astrological history, in turn, as based on the theory of the great conjunctions of Saturn and Jupiter, was never more popular in Europe than in the sixteenth and seventeenth centuries. In individual astrology, many specifically Arabic doctrines were used in the classical genre of birth charts or nativities, such as the Alcocoden technique for calculating the length of life. Other branches of individual astrology based on Arabic sources were thriving too: elections, which aim at finding fortuitous times for specific actions, and anniversary horoscopes, that is, charts drawn and interpreted not for the date of birth but for its anniversaries. It is important to note that these branches of astrology were not only practiced widely but also discussed on a theoretical level that has very few, if any, parallels in history. Sixteenth-century authors such as Johannes Schöner and Girolamo Cardano had a superb understanding of both Greek and Arabic techniques of astrology, which enabled them to compare several rival theories on the same issue and to advance their own technical solutions. By way of these much-read authors, a great deal of Arabic astrological theory was transported into the coming century, even in humanist text genres such as commentaries on Ptolemy.

In these three areas, then, epistemology, *materia medica*, and astrology, there is much evidence for an intense Renaissance reception of Arabic thought. But the full picture is broader than that, as we know from other evidence and other studies. On the one hand, there is the silent influence of originally Arabic theories that had long been adopted into Western thought, such as in algebra, or had exerted a new but silent influence on Renaissance scientists, such as Ibn an-Nafīs's theory of pulmonary circulation, which was known to several sixteenth-century physicians, or the critique of Ptolemy's astronomy by Arabic scholars, which influenced Copernicus.[5] On the other hand, there are also many more examples of obvious reception of Arabic thought in the Renaissance. In the discipline of logic, the new Hebrew–Latin translations of Averroes's logical works and, in particular, of his Long commentary on the *Posterior Analytics* led to a discussion of Averroes's logic that contrasts greatly with the meager reception of his logic in the preceding centuries. In zoology, too, the first full translation of Averroes's commentary on Aristotle's *De partibus animalium* influenced the commentary tradition, especially in Padua. Arabic theories of miracles—in particular Alkindi's explanation of long-distance effects through the extramission of rays and Avicenna's theory of the nonmaterial causation of material effects through powerful souls—found many supporters and critics in the Renaissance, among them Marsilio Ficino, Andrea Catani, Pietro Bairo, Pietro Pomponazzi, and Thomas Erastus. While these theories resonated with authors on magic, they also contributed to naturalistic explanations of miracles and prophecy in the Renaissance and to the academic critique of religious superstition.[6] In theoretical medicine, there is the enormous evidence of Arabic influence collected by Nancy Siraisi on the basis of more than sixty Latin commentaries and lectures on Avicenna's *Canon* by sixteenth- and seventeenth-century authors. Siraisi analyzes six of these commentaries in detail with respect to *Canon* 1.1. Her analysis sheds light on many Renaissance debates on subjects bordering on medicine and natural philosophy that were influenced by Arabic authors: these include debates on the elements, on celestial influences, on the human soul and its faculties, as well as on many issues of theoretical medicine, such as the temperaments and the humors.[7] The sheer wealth of philosophical and medical discussion in this small section of the Renaissance commentary tradition on Avicenna is astounding. Other commentary traditions on Arabic authors remain to be explored, such as the commentaries on *Canon,* book 2 on pharmacology, on Rhazes's *Liber ad Almansorem* 9 on pathology, on Pseudo-Mesue's *De simplicibus,* and on Averroes's *De substantia orbis.*[8] Whatever the eventual outcome of such research, it has already become clear that the Renaissance reception of Arabic authors was not a decrepit leftover from the Middle Ages, but a very active and intellectually challenging endeavor of its own kind.

The humanist movement was much involved in the reception of Arabic sciences and philosophy in the Renaissance. On this issue, modern scholars have often been too impressed by the humanist polemics against Arabic authors and by their dramatic call for a return to the Greeks. As a consequence, it was readily believed that Arabic theories deserved to be rejected and that humanists in general had a hostile attitude toward the Arabic traditions in Europe. This characterization, however, is not historically accurate. In fact, humanist scholars contributed a great deal to the flourishing of Arabic sciences and philosophy in the Renaissance. In the context of the present study, three kinds of humanist contribution are of particular importance. First, humanist scholars such as Franciscus Calphurnius, Konrad Gesner, and Jacob Milich studied the lives and works of famous Arabic authors and wrote biographies and bibliographies of them. They thus helped to accommodate the Arabic scientific tradition to the new ideals of the humanist movement. Second, several humanist scientists contributed to the discussion of Arabic scientific and philosophical theories: examples are Francesco Vimercato, the Aristotelian scholar and professor of Greek in Paris; Girolamo Cardano, the mathematician, physician and astrologer; and the aforementioned medical botanists Manardo, Brasavola, and Mattioli. Third, humanist and Arabic traditions were intertwined in the Arabic–Latin and Hebrew–Latin translation movement. The movement involved translators who entertained humanist ideals and wrote a classicizing Latin, notably Paolo Ricci and Jacob Mantino. Other humanist scholars invested great effort in revising medieval Latin translations of Arabic texts in classical style. Examples are Miguel Ledesma, who disguised his revision of Avicenna's *Canon* as a translation; Jean Bruyerin Champier, who turned the medieval translation of Averroes's *Colliget* into up-to-date humanist Latin; and Jacobus Sylvius, who transformed the medieval style of Mesue's *Opera* into humanist syntax and contemporary vocabulary. Among the patrons of the translation movement were several members of the Italian nobility who shared humanist ideals: Giovanni Pico della Mirandola, Cardinal Domenico Grimani, Alberto Pio prince of Carpi, and the later bishop Ercole Gonzaga. Grimani in particular was closely linked to the humanist movement. He was famous for his exquisite collection of Greek manuscripts and was, for this reason, visited by Erasmus. Two Hebrew–Latin translators of Averroes worked directly for Grimani: Elia del Medigo and Abraham de Balmes. In the prefaces, del Medigo and Balmes state that their work was motivated by Grimani's interest in metaphysics and logic. Here, admiration for the Greek and admiration for Averroes go hand in hand. The example of Grimani thus undercuts antagonistic descriptions of Arabic and humanist traditions of the Renaissance—as does that of Pico della Mirandola, the fine Latin stylist and patron of Averroes translations.

As all this evidence shows, the active reception of Arabic sciences and philosophy was not confined to a limited milieu of academic culture, but involved many intellectual currents and milieus of the time, including the humanist movement. The Arabic heritage was an essential part of Renaissance culture. This, however, is only one side of Renaissance attitudes toward Arabic traditions. Other attitudes were much more hostile.

## SUPPRESSION

The polemics against Arabic traditions in the West were not purely rhetorical. They influenced the scientific discussion and formed part of a struggle for basic beliefs, ideological concepts, Christian orthodoxy, intellectual leadership, personal glory, printing success, university reforms, and academic positions. As a result, the discussion of medical, philosophical, and astrological theories was significantly strained by nonscientific interests. The course of the sciences was changed by these attitudes—a development that was accompanied by gains and losses. The term *suppression* implies that these losses were the result of conscious actions, and one may doubt that suppression of Arabic traditions existed in the Renaissance. But it did. In the present context the term *suppression* means no more and no less than conscious opposition to scientific theories for nonscientific reasons. In this regard, the term deserves its negative connotation. My first step will be to show, by mustering the evidence for opposition clearly driven by nonscientific motives, that such suppression of Arabic traditions indeed took place in the Renaissance. Often, however, such ideological beliefs were to some extent mixed with scientific motives. My second step will be to try to delineate these motives.

In medicine, the humanist anti-Arabic polemics were particularly aggressive, partly because so many humanists were also physicians by profession and were engaged in a reform of medicine. They were confronted with a discipline that was heavily dominated by Arabic sources, at the university as well as in practice. The subfield of *materia medica* or medical botany, with its many plant and drug names, was most hotly debated between humanists and their opponents, because here terminological problems, for which humanists had a keen eye, were particularly pressing. As the example of the drug senna has shown, there existed a current of humanist hard-liners who propagated a complete return to ancient *materia medica* and to Dioscorides in particular. Scholars such as Jean Ruel and Leonhart Fuchs wanted more than a mere restitution of ancient sources; they aimed at a Hellenization of modern *materia medica*. Fuchs cherished an ideal of linguistic purification, in which a work of medical botany consisted of Greek plant names and Greek sources only. To this end, they did not shy away from textual manipulations that were meant to deceive their readers and make Arabic traditions disappear under a Grecizing disguise. Hence, suppression of Arabic traditions in Renaissance medicine was a historical reality. In the field of *materia medica,* it resulted in the disappearance of scientific material, in the temporary removal of Arabic authorities from the medical curricula of several universities, and in the academic promotion of anti-Arabic polemicists like Leonhart Fuchs. As the new, Grecizing medical statutes of the University of Tübingen put it in 1538, in an anonymous voice that is clearly that of Fuchs:

> Since there is nobody who would not know that the Arabs have copied almost everything from the Greeks, the Arabs will be used only to a very minimal degree for the teaching of this field, because it is advisable to scoop the precepts of a science from the sources and not from turbid ditches.[9]

In the field of philosophy, the polemics had a humanist and religious coloring. The prime target of the polemics was Averroes, who was attacked not only as a commentator who lacked a knowledge of Greek but also as an allegedly irreligious philosopher. Hence, adherents of Averroes had to face both humanist criticism and the pressure exerted by Church officials. Humanist charges against Averroes have a long tradition: Petrarca, Salutati, Valla, Barbaro, Ficino, Lefèvre d'Étaples, Champier, and Vives—they all criticized Averroes as an incompetent interpreter of Aristotle, and often also as an enemy of religion. Some humanists, in particular those who translated Greek commentators into Latin, loudly called for a replacement of Averroes by his Greek predecessors: "Make the decision to only read Simplicius, day and night . . . and take Averroes out of your hands!"—urges Giovanni Faseolo, the translator of Simplicius's *De anima*.[10] And Agostino Nifo, in 1508, explains in exquisitely irrational rhetoric: "I prefer to err with the Greeks in expounding Greek authors rather than to be right with the Barbarians *(i.e., Averroes)*, who do not know the languages except in dreams."[11] It is clear that a current of Renaissance intellectuals worked actively toward relegating Averroes to a marginal role in university education, and that the belief in the superiority of Greek helped to motivate this activity. In fact, the availability of the Greek Aristotle and the Greek commentators in Latin proved a great challenge to Averroes's position at the universities. Averroes was not marginalized, as his enemies had hoped, but he lost his dominating position, even at the University of Padua, as is testified by the steeply rising number of quotations from the Greek commentators in university lectures of the early sixteenth century.

In the field of intellect theory, Alexander of Aphrodisias turned out to be the principal and, in fact, sole rival of Averroes, because Themistius and Simplicius were often interpreted as agreeing with Averroes on the unicity thesis. From Pietro Pomponazzi onward, a good number of Aristotelian philosophers of the sixteenth century followed Alexander's theory of the complete dependency of the intellectual soul on the body, among them Simone Porzio, Jacopo Zabarella, and Cesare Cremonini. While this development can be viewed as resulting from a philosophical competition for the better argument, there is a period in the history of Renaissance Aristotelianism that is heavily influenced by nonphilosophical factors. In the 1490s, the Paduan philosophers Nicoletto Vernia and Agostino Nifo publicly turned their backs on Averroes's intellect theory, thus prompting a critical phase for the survival of Averroes as a philosophical authority at Padua. Nicoletto Vernia, in his treatise "against the perverse opinion of Averroes about the unicity of the intellect"—and thus against his own previous opinion—invokes the Greek commentators against Averroes.[12] And Agostino Nifo, in 1508, explicitly justifies his turning against the "great drunk" Averroes with his own reading of Aristotle in Greek.[13] But this piece of self-styling does not tell the truth about Nifo. Neither Nifo nor Vernia, in the factual detail of their texts, employs to any significant degree the help of the Greek commentators, or of the Greek Aristotle, for their new anti-Averroist position. Rather, they take their inspiration and ammunition from Thomas Aquinas, Albertus Magnus and, in the case of Nifo, Marsilio Ficino. The fiction of a Greek turn has a second purpose: it helps to cover up the more important cause of the changing attitude toward Averroes, which is pressure from the Church.

Openness toward pagan philosophical sources had been a trademark of the medieval universities since 1255, when the entire Aristotelian corpus was declared mandatory reading for all students of the Paris arts faculty. There were, of course, condemnations, notably that of 1277, and orthodox pressure on individual thinkers. But Church interference with the arts faculty of a medieval university, rather than with the theological faculty, was a rare phenomenon. Hence, it is all the more remarkable that Bishop Pietro Barozzi and the inquisitor Marco da Lendinara of Padua took drastic action against Averroist philosophers in 1489 by threatening the excommunication of anyone discussing Averroes's unicity thesis in teaching. In the following two decades, the pressure on Averroist philosophy professors was tangible: in 1491, Nicoletto Vernia was declined a benefice by Barozzi; in 1492, Vernia composed a straightforward recantation, a treatise against Averroes, which met with Barozzi's explicit approval; some years later, before his death in 1499, Vernia appealed to three theologians for a testimonial of orthodoxy; and, finally, in his testament Vernia asserted before God that he had never believed in Averroes's opinion on the intellect. Sometime before 1493, the young Agostino Nifo was accused of heresy by unnamed adversaries because of his teaching on the soul, but apparently won the protection of Barozzi himself. But Church pressure left its traces in Nifo's subsequent writings on intellect theory, which bear obvious signs of self-censorship. As a probable repercussion of these Paduan events, the Bologna philosopher Alessandro Achillini hid an Averroist thesis beneath a halfhearted refutation of it. In 1513, the Lateran Council condemned all partisans of the unicity thesis as heretics. Hence, between 1489 and 1513, Church repression was a real factor in the history of the Renaissance reception of Averroes. It influenced the philosophical discussion of Averroes's intellect theory, and could also have dire personal and professional repercussions, as the case of Nicoletto Vernia shows. The pressure did not last long, however. It was only a specific group of clerics around Bishop Barozzi who took action against Averroists. Other clerics opposed the idea that Church authorities should interfere with the teaching of the arts faculty. Later philosophers such as Prassicio, Genua, and Vimercato, who were writing after 1513, returned to the traditionally sober discussion of the unicity thesis, which brackets all questions of orthodoxy. The Averroist intellect theory thus survived the Church pressure in the years after 1489. It declined only later in the sixteenth century, and then for other reasons: the changing commentary culture in the arts faculties favored philological commentaries and textbooks, and philosophical alternatives, especially in intellect theory, arose during this time.

In the field of astrology, Giovanni Pico della Mirandola's *Disputationes adversus astrologiam divinatricem* was the founding document for more than a century of controversies about the relative value of Greek and Arabic traditions in astrology. As Albert Pigghe, one of the many followers of Pico, put it: All students of astrology should "give up the fables of Albumasar and the <other> Punics *(i.e., Arabs)* and read with great diligence our Ptolemy."[14] The polemics against Arabic astrology were more bookish and theoretical in character than the polemics in medicine and philosophy, which have a more direct bearing on curricula, teaching contents, and academic careers. This is because astrology was less institutionalized than philosophy and medicine, two disciplines at the heart of Renaissance education. Nevertheless, the astrological discourse too was much influenced by a nonscientific bias that led to the disappearance of Arabic astrological theories, namely, the idea that the only true astrology can come from Ptolemy's *Tetrabiblos,* and that everything else is follies, fictions, deceptions, fables, and superstitions of the Arabs. In the "war" *(bellum),* which Pigghe and others fought against the supporters of Arabic traditions, many doctrines and even entire branches of the discipline of astrology came under attack. As a consequence, interrogations and lots disappeared almost completely from astrological handbooks, and revolutions of the years of the world received only marginal treatment in handbooks. If they were treated, they tended to become "Ptolemized," that is, based on the last conjunction of sun and moon before the spring equinox. In natal astrology, the force of the humanist paradigm is tangible when Johannes Schöner includes in his handbook a lengthy treatment of anniversary horoscopes drawn from Arabic sources, but assures his readers that Ptolemy is always given the preeminence. A further consequence of the Greek–Arabic antagonism created by Pico and his followers was the valorization of a highly theoretical form of astrology—for Ptolemy's *Tetrabiblos* presents a purist's theory of astrology, which cannot be put into practice without a great deal of additional information. Hence, for some Renaissance scholars, the attack on Arabic astrologers served the special purpose of claiming intellectual leadership for theoreticians of astrology against practitioners. In sum, therefore, the attempts to suppress Arabic traditions left their clear traces in the doctrinal development of astrology—as they did in medicine and philosophy.

## SCIENTIFIC AND NONSCIENTIFIC MOTIVES

Ideological motives certainly played an important role in the attacks on Arabic traditions, but they are intermixed, to a certain extent, with scientific reasons. This melange of motives is difficult to disentangle. In what follows, I will first summarize the main charges of anti-Arabic polemics, then make some observations about their specific character, and finally set out to distinguish the layers of motives involved.

Three charges against Arabic traditions dominate in the Renaissance polemics: linguistic corruption, plagiarism, and irreligion. As to the first, it was a basic belief of many Renaissance scholars that many intellectual defects follow from uncultivated language, or from language that breaches with classical practice. The scholastic Latin practiced in many universities was a prime example, in humanist eyes, of such linguistic and intellectual corruption. But the Latin translations of Arabic texts were considered to be even more problematic: they were written in scholastic Latin, and, on top of that, in Arabicized scholastic Latin, which was colored by Arabicizing syntax and contained many Latin transliterations of Arabic terms. The cross-cultural transmission, humanists further argued, resulted in many errors and terminological corruptions, which are particularly detrimental in the field of medicine. The second charge, plagiarism, was made by physicians, philosophers, and astrologers alike, holding that the Arabs have stolen everything from the Greeks, and live like parasites on the work of others. This polemic rests on the *ad fontes* idea, that an intellectual renewal will come about through a return to the ancient sources of contemporary textual culture. This idea is sometimes accompanied by the stronger claim that the true and superior knowledge lies with the ancient sources, and not with subsequent cultures. The third reason, irreligion, was voiced by clerical authorities, but also by many humanists since Petrarca, especially since a renewal of Christian religion was dear to the hearts of many humanist intellectuals. This charge was directed almost exclusively at Averroes, and took its origin from Averroes's presumed denial of personal immortality, and, in later centuries, from the alleged determinism of his philosophy and from the denial of God's knowledge of the sublunar world.

The anti-Arabic polemics were not simply an extension or segment of the broad humanist current of invectives against scholasticism.[15] Of course, antischolastic and anti-Arabic discourses were connected and shared many features, for example, an irritation with traditional university teaching. But the anti-Arabic polemics are a phenomenon in their own right. There are two main reasons for this. First, Latin translations of Arabic texts were, as mentioned, doubly "barbaric," due to their medieval and Arabicizing Latin. Second, Arabic writers exerted an influence over a broader spectrum of topics in the Renaissance university curriculum than did Latin authors of medieval Christianity. The latter were prominently read in logic, grammar, and theology at late medieval and Renaissance universities, but rarely in the other disciplines. The curriculum in medicine, *mathematica,* and the remainder of philosophy was dominated by Greek and Arabic authorities. Hence, it was inevitable that humanist reformers of the sciences and philosophy attacked the Arabic authorities, as the most distinctive non-Greek element in scientific and philosophical university education in Europe.

It is noteworthy that the religion of Islam did not play any significant role in the Renaissance polemics against Arabic sciences and philosophy. In the present study, only three authors among the many critics of Arabic traditions are cited with anti-Islamic polemics as part of their critique: Champier, Vives, and Reisch. Symphorien Champier, in his biography of Avicenna, correctly relates that Avicenna describes himself as a Muslim in his *Metaphysics,* but then adds a manipulated sentence in which the high authority of Avicenna allegedly denies the prophethood of Muhammad, whom Champier then abuses as selfish and vain.[16] Juan Luis Vives, the Spanish humanist, attacks Averroes and Avicenna and everything Arabic as smacking of the "hallucinations" of the Quran and the "blasphemies" of Muhammad.[17] While we do not know much about the background of Champier's polemics, in the case of Vives it is clear that the context is missionary activities, to which Vives contributed with his late work *De veritate fidei christianae.* The German schoolbook author Gregor Reisch, finally, peppers his polemics against Averroes with a denigrating remark about the "dishonesty" of Muhammad, whom Averroes follows.[18] But these are single voices in the much greater chorus critical of Arabic traditions. The absence of Islam in anti-Arabic polemics is also noteworthy in view of the fact that the Christian orthodoxy of Averroist positions was such a prominent topic of Renaissance philosophy. Many considered Averroes to be an enemy of religion, and of Christian religion at that, and many praised Thomas Aquinas for his "triumph" over Averroes on the question of the unicity of the intellect. But Averroes's alleged irreligion was hardly ever related to his Islamic faith. Hence, religion was a major topic of the polemics, but Islam was not.

Orientalizing tones are also very rare in the sources surveyed in this study. Again, Symphorien Champier is among the very few who discusses the Oriental culture of Arabic scientists. He explains that Arabia has always produced excellent physicians and thinkers, and then continues to praise the country for its "plenitude and abundance," and its sacred embellishments such as thyme and myrrh, silver, gold, and precious herbs, where the people possess great fortunes and wear the finest clothes.[19] Champier thus invokes the traditional description of Arabia associated with the land of Queen Saba and implicitly attributes the great scientific productivity to the country's richness. But this image of the cultural background of Arabic science remains exceptional. As was shown above, reliable information on the lives of Arabic scientists, not to speak of their culture, was extremely scarce in the Renaissance, and the predominant but false view turned most of the Arabic scientists into Andalusians. As such, they were not in any way linked to the Ottoman Empire, which was the main object of the imagination of the Orient in the Renaissance.[20] The scarcity of Orientalizing tones in our sources contrasts instructively with the dramatic rise in such tones in the seventeenth century. New attitudes toward the Orient emerge in the sixteenth century, resulting from the academic study of Arabic and Islam and from eyewitness accounts by travelers to the Orient who report on customs and intellectual life. Generally speaking, the Orient of the early modern period differs from its medieval predecessor by being closer and more visual. The rising number of Arabic philologists and travelers who know the language and see the countries has the effect of drawing the Orient closer to Europe. And the many illustrations of the Arab world in print increase its visualization.[21] But these are developments that, in the Renaissance, do not yet motivate Orientalizing descriptions of the Arabic scientists and philosophers.

Let us now try to disentangle scientific motives from external ones. In one sense, the anti-Arabic polemics, the charges of linguistic corruption, plagiarism, and irreligion, were obviously unjustified: it is clear that medieval Arabic–Latin translations were great linguistic achievements in themselves, that the Arabs, as impressive scientists, did not plagiarize, and that Averroes conceived of himself as a proper believer. But polemical utterances may also serve to express, in condensed and biased format, more complex scientific beliefs. They can, for example, indicate a belief in the superiority of Greek over Arabic commentators on Aristotle; in the accumulation of textual mistakes and errors in the transmission from Greek to Arabic to Latin; in the benefits of philological competence for the progress of the sciences; in the technical superiority of Greek sciences over their Arabic and medieval Latin successors; and in the considerable dependency of Arabic authors on Greek material. What do the examples of senna, of Averroes's unicity theory, of Vives's critique of Averroes's *Metaphysics* commentary, and of the handbook tradition in astrology tell us about the validity of these beliefs?

In the field of *materia medica,* it was the principal conclusion of the above analysis in Chapter 4 that the humanist program of reductionism was a failure, whereas the philological turn of medical humanism was beneficial to the sciences and their development. Humanist physicians and botanists, from Leoniceno onward, successfully demonstrated that philological skills, and knowledge of the Greek in particular, were indispensable for any serious progress in medical botany in the Renaissance. The Arabic and Latin traditions in this discipline had a common major source, which was written in Greek: Dioscorides's *Materia medica,* which listed the medical properties of about a thousand plants, animals, and minerals. As a result, the stock of simple drugs that appeared in late medieval pharmacological texts rested on a fusion of originally Greek and Arabic sources. It was essential to understand this fusion, even in the case of the Oriental plant senna. The name *senna* was not mentioned in Dioscorides and Theophrastus. But determining whether ancient authors had mentioned the plant under a different name, required close textual comparison, which indirectly also gave impulses for observation and testing. Mere Latin competence, as in the medieval Synonyma literature, was not enough to solve such questions. The humanists, therefore, were justified in demanding competence in Greek from their colleagues in pharmacology—as were, in fact, those scholars who recognized the value of Andrea Alpago's competence in Arabic and on Arabic drug names. On the other hand, the reductionist idea of identifying all non-Greek plant names with Greek plant names, or of making them subspecies of Greek plants, created numerous philological, scientific, and empirical problems. The basic belief in the superiority of Greek medicine, in *antiqua et vera medicina,* as Manardo put it, was particularly unjustified in the field of medical botany, where Arabic pharmacognosists had added significantly to the work of their Greek predecessors. Arabic scholars had considerably increased the number of simple drugs, and had greatly developed the genre of recipes, that is, the description of composite medicines, which is largely absent from ancient sources, especially in Hippocratic medicine. This is why opposition to the radical form of medical humanism also came from the apothecaries, who were not willing to give up Oriental drugs or dispense with the ample recipe literature of the Arabs. There had been progress in Arabic medical botany—"For who would deny that the centuries always make progress?" asked Guillaume Postel in 1539/1540, with respect to Arabic medicine.[22] In view of this progress, it proved impossible even to make the minimal humanist gesture of avoiding the "barbaric" names of Arabic drugs and authors. Those humanists who tried to do so were obliged to resort to textual manipulations. More moderate humanist scholars, such as Giovanni Manardo, Antonio Musa Brasavola, and Pietro Mattioli, steered a different course and incorporated Arabic sources and plants into their classically inspired texts. Eventually, their approach won the day. In sum, therefore, there is evidence in this discipline, first, of the deterioration of humanism into ideology, and second, of the validity of the claim that the Arabic–Latin tradition of *materia medica* stood in need of being anchored to its Greek sources.

In the field of philosophy, I will not say much on the issue of the Christian orthodoxy of Averroes's intellect theory. Church pressure is obviously a factor external to philosophy itself, as recognized by attackers and defenders of Averroes alike. Most Averroist philosophers of the Renaissance— even Luca Prassicio, who defends Averroes on the immortality of the soul— were ready to admit that Averroes's position is false from a Christian point of view. The layers of motives are more complicated when we come to other forms of opposition to Averroes's philosophy. One charge against the unicity thesis was that it deviates from the true sense of Aristotle's Greek text. This was claimed by some Renaissance scholars, for instance by Agostino Nifo. Despite these claims, arguments from the Greek text appear very rarely, if ever, in the texts of Renaissance Aristotelians, with Francesco Vimercato being the great exception. One reason for this is the very limited knowledge of Greek on the part of most philosophy professors in the period surveyed. Moreover, the unicity thesis was difficult to attack on philological grounds only, that is, as disagreeing with Aristotle's text, since Aristotle's intellect theory itself has separabilist tendencies that blur the neat picture of Aristotle's hylomorphism: several passages stress the separability of the intellect from the body. From a philosophical point of view, too, there was much that recommended Averroes's theory to Renaissance thinkers: it offered an explanation of how human beings can acquire knowledge of universal intelligibles and how two persons can think the same concept. At the same time, it safeguards the individuality of thinking by assigning a central role in intellection to the faculty of imagination, without turning the soul into something perishable, as did Alexander of Aphrodisias. Hence, given these exegetical and argumentative strengths, it is clear that the validity of Renaissance polemics against the unicity thesis as "brainless" and "non-Aristotelian" rested on thin ice.

The critics of Averroes were much more justified in criticizing the reliability of Averroes's commentaries in general as a guide to understanding Aristotle, especially if Averroes is compared with the Greek commentators on Aristotle. Humanist scholars with knowledge of Greek, such as Juan Luis Vives, were able to compare the Greek texts of Aristotle and the ancient commentators with the Arabic–Latin tradition. They were the first to see that the accumulated errors of transmission and translation from Greek into Arabic, sometimes via Syriac, severely impaired the reliability of Averroes's commentary as a guide to the text of Aristotle. The transmission into the Latin world added further problems. Averroes, with his enormous expertise as an Aristotelian scholar, was able to circumvent a good number of textual distortions, especially in vocabulary, but he often fell victim to the many small corruptions of syntax that distort Aristotle's argumentation, such as when Averroes reads "we" instead of "he," that is, Plato, as Vives points out. On the other hand, there is not one ancient commentator who could compete with Averroes's overall expertise on Aristotelian philosophy, as the author of almost forty commentaries on the philosopher's works. For many works of Aristotle, such as on most books of the *Metaphysics* and on the entire zoology, Averroes could not rely on any predecessor in commenting. On other books, such as *De caelo* or the *Topics*, the ancient commentary tradition is extremely meager. Hence, there was hardly any justification for the humanist charge that Averroes was entirely dependent on his Greek predecessors. The humanist opposition to Averroes, then, was based on a mixture of good intuition and mere ideology. This helps to explain the mixed reception of Averroes in the Renaissance: he was admired, for instance by Zabarella, as the greatest expert on Aristotelian philosophy, but lost his dominating position as an expert on Aristotle's very text because of his distance from the Greek text.

The reformers of astrology, who demanded a return to Ptolemy and his *Tetrabiblos,* unfortunately did not say much to explain their belief in the superiority of Ptolemy's astrology. But it can be observed that their opposition rested on two main differences between Ptolemy's astrology and the Arabic–Latin tradition of the discipline: Ptolemy's astrology is, on the one hand, less extensive and detailed than Arabic astrology, and, on the other hand, more theoretical. Hence, for the reformers, Ptolemy's *Tetrabiblios* could serve a twofold purpose: it provided an authoritative warrant to reduce the tasks and techniques of astrology to a more limited and less ambitious size, and it functioned as a theoretical touchstone with which to test the validity of astrological doctrines. Both tendencies can be observed in Renaissance astrological controversies. There is clear evidence, therefore, that Ptolemy was preferred not only for ideological but also for technically astrological reasons. Pico, for instance, correctly observed that the theory of lots had been growing continuously while making its way through the cultures: Ptolemy recognized only one lot, the lot of fortune, other late antique astrologers added more, and Arabic authorities discussed over fifty different lots. Pico rejected the Arabic theory as inflationary theory-building; he also criticized the theory of lots for theoretical reasons that are in agreement with Ptolemy's principles, namely, that the doctrine presupposes the influence of a point on the ecliptic where there is no natural body or quality. As a result of such argumentative techniques, the hu-

manist reformers were successful in expelling two doctrines from handbook literature that are simplistic and weak in theoretical respects: interrogations and lots. In other areas of astrology, however, the humanist
belief in the superiority of Ptolemy prevented the reformers from recog-
nizing the strengths of Arabic astrological doctrines. Such is the case with
anniversary horoscopes and the Alcocoden technique for predicting the
length of life. These latter are genuine technical alternatives to the prorogation techniques offered by Ptolemy, whose treatment of the topic is curt
and partly obscure. Here the main disadvantage of Ptolemy's *Tetrabiblos*
comes to the fore: even according to ancient standards, the text is much too
theoretical to serve as a handbook for astrological practice. Hence, the belief in the superiority of Ptolemy's astrology followed, on one side, an astrological intuition, namely, that the many astrological doctrines needed to be
tested as to their theoretical validity. On the other side, it was a mere humanist prejudice, which blinded the followers of Ptolemy to the technical
disadvantages of the *Tetrabiblos*. If seen from the unhistorical perspective
of present-day knowledge of sources, the humanist prejudices verge on
the ridiculous. Much of Arabic astrology is un-Ptolemaic, but perfectly
Greek in origin. It derives from other Greek authors such as Dorotheus
of Sidon.

As all this shows, Renaissance opposition to Arabic traditions in philosophy and the sciences was motivated by an amalgam of scientific arguments, mere partisanship, and outright ideology. Let us turn to the
arguments first. A good number of medical, philosophical, and astrological reasons were involved. The humanists were right in claiming that
knowledge of Greek was essential for progress in medical botany, philosophy, and astrology, given that these disciplines were widely practiced in
Arabic and Latin as Dioscoridean medical botany, Aristotelian philosophy, and Ptolemaic astrology. The three Greek founding figures of scientific disciplines had determined many of the principles, doctrines, and
practices of the subsequent centuries. Ignorance of their original texts,
therefore, could be damaging to scientific practice. We have seen examples
of such developments: much confusion in plant identification; the obscure
presentation of pre-Socratic philosophy in Averroes; or the proliferation of
astrological details beyond any foundation through astrological principles. Hence, there was good scientific intuition in the humanist resistance
to specific Arabic doctrines, which belies trends in modern scholarship
that portray humanism as a curse to the progress of the sciences (as did, for
instance, Lynn Thorndike, George Sarton, and John Herman Randall).[23]
Moreover, we have also met with examples, rightly criticized by humanists,

of unreflecting partisanship in favor of Arabic and scholastic traditions: astrologers who continue to use interrogations and lots without responding to the massive theoretical objections against them; Averroist philosophers like Luca Prassicio, who is avowedly ignorant of the Greek Aristotle or the Greek commentators; or the many colleagues of Pomponazzi in Padua and Ferrara, who would turn their back on him, as Pomponazzi complains, if he did not defend Averroes. There was very good reason, therefore, to oppose this pro-Arabic bias with scientific arguments.

From another perspective, however, the humanist call for a return to the Greek sources in medical botany, philosophy, and astrology was a very strange and provocative idea. The *ad fontes* idea in itself was not yet provocative, but many humanist scientists claimed more than that, as we have seen. They believed in the superiority of Greek science over the Arabic and medieval successors, and in the intellectual "shipwreck" and "barbaric destruction" of the sciences in Arabic and Latin cultures. As the humanist physician Girolamo Donzellini put it: "When the science of medicine was transported from the Greeks to the Arabs, it was shipwrecked *(naufragium fecit)*, and when the Latins received it from the Arabs, they were very unproductively involved in it for a long time."[24] Or as Philipp Melanchthon remarked on the history of astrology: "In Egypt the barbarism of the Sarracens *(Sarracenica barbaries)* destroyed the academy in Alexandria and the arts *(studia)*." The science of the stars would have perished completely, adds Melanchthon, "if Ptolemy, briefly before the Sarracens' turbulence, had not summed up the entire art in one volume."[25] Historically, these statements are wrong, and the concept of a decadent history of science that is expressed in it does not accord at all with historical data. Many Renaissance contemporaries protested against the humanist charge, pointing to the Arabic achievements in science, and to the numerous things unknown to the Greeks, but known to the Arabs. For example, in 1576, Mathias de l'Obel wrote as follows: "To the great disadvantage of the sick, senna became known to the Greeks only very late *(i.e., in the Byzantine era)*; we owe senna together with its name to the Arabs."[26]

As modern historians, we have gotten accustomed to a positive image of the Renaissance idea of a return to the ancients because we recognize its productivity. But we tend to overlook the violent, disastrous, and reactionary elements that the idea also implies. Humanist scholars invaded the medical botany, philosophy, and astrology of their time—disciplines that were scientifically vibrant and active—and tried to turn the clock back. This attempt acted as a catalyst for many new trends, but it also put a lot of strain on the development of the sciences. Not only did it result in outright suppression, as described; at times it also produced pure fiction, as in the fictitious botany of the Greek plant *colutea,* which Jean Ruel and Leonhart Fuchs used to replace the Arabic plant senna; in Agostino Nifo's autobiographical fiction of a Greek turn against Averroes, where his true sources are Albertus Magnus, Thomas Aquinas, and Marsilio Ficino; and in Giovanni Pico's explaining away of Pseudo-Ptolemy's theory of conjunctions with a desperately far-fetched interpretation of the *Centiloquium.* A good number of humanists, in fact, came to realize that the radical form of humanist science was an impossible endeavor and began to compromise. Even Leonhart Fuchs quoted Pseudo-Mesue on senna in a late work of his; Girolamo Cardano composed a treatise on interrogations, even though "in other places, I have often condemned the branch of interrogations";[27] and Francesco Giuntini explained that in his youth he had followed Ptolemy's very restrictive theory of lots, but that he now adopts again the (Arabic) theory, since Ptolemy is a friend, but *magis amica veritas.* In addition to these explicit reversals, we have met with numerous examples of Arabic theories that continue to exert their attraction and survive under a new, Grecizing garment.

The entry of humanism into the history of Arabic sciences in Europe was not a mere side-episode of history, but a dramatic intervention. It is true that the radical forms of humanism failed, and that many Arabic and medieval traditions survived and developed. But medicine, philosophy, and astrology, the three areas of learning most clearly influenced by Arabic sources, were heavily transformed by the humanist impact—with considerable gains and damages, and many damages to Arabic traditions in particular. These traditions lost their dominating position in many areas and subareas, in spite of the great success of some Arabic theories that we can also observe in the Renaissance.

~

The humanist movement was not at all inimical to Arabic sciences as such. We have seen many examples of humanist scholars who contributed to the flourishing of Arabic sciences and philosophy in the Renaissance. Moreover, and very notably, humanists did not oppose Arabic sciences because they were Oriental or because they originated from Islamic culture. Rather, they opposed them partly for scientific reasons, partly as a result of ideological beliefs in linguistic purism and in Greek superiority, and partly because Arabic authors were an obstacle—an obstacle to the humanists' project of

renewing Europe through Greece and Rome. The Arabic obstacle lay athwart
a prominent cultural road: the university practice of medicine and philosophy, and, to a lesser degree, also of astrology. This institutional aspect
explains much of the controversy's acerbity. One might say, therefore, that
the Renaissance opposition to and suppression of Arabic sciences and philosophy took place mainly for European reasons, not for Oriental ones. Humanists opposed Arabic traditions because they entertained strong beliefs
about antiquity, not because they entertained strong beliefs about Arabic or
Islamic culture.

The fact remains, however, that Arabic traditions were attacked for
being Arabic, that is, they were attacked on the basis of a cultural and
linguistic labeling. Renaissance humanists are the inventors of cultural
clichés that persist even today—that Arabic science amounts to plagiarism;
that it is nothing more than Greek thought in Arabic garb; that the Arabs
were mere transmitters of science from antiquity to medieval Europe; in
other words, as Epicurus is said to have claimed, that "only the Greeks are
able to philosophize."[28] This is the sad inheritance of the humanist polemics against Arabic sciences.

In the sixteenth century, and even more so in the seventeenth, European culture developed an enormous dynamic that changed Europe
lastingly and laid the foundation for the global dominance of Western
culture. It would be wrong to believe that Islamic culture declined in the
same period. Rather, it is Western culture that underwent an extraordinary development, in a notoriously complex process. The process involved
technical inventions, scientific breakthroughs, the exploitation of overseas
countries, the development of military techniques, individual achievements in science and politics, social and religious institutions, and ideas
such as the notion of progress over antiquity, the concept of mechaniza-
tion, and the valorization of private economic success—to mention only a
few prominent factors. The humanists around 1500 stood on the threshold
of this development. They played an important role in that they redefined
the relation between Europe and Arabic science. Today, we all too readily
believe that it was the West by itself that in the seventeenth century developed a dynamic unparalleled in history. One source of this belief is the
very fact that, in the Renaissance, humanist hard-liners had largely
erased the contribution of Arabic science to European culture. They
worked on a purist construction of the Western tradition as essentially
Greek, and, at best, also Roman and Christian. But, in fact, the development of European civilization in later centuries rested on ancient and
medieval traditions that involved many more cultures and languages,
many of them Oriental. European culture prospered in part because of

Arabic medicine, philosophy, and astrology, and because of the other Arabic sciences such as Arabic mathematics, astronomy, physics, magic, and alchemy that had long permeated European thought in the Middle Ages. It is one of the tasks of historical research to point out the historical falsity of clichés such as the image of Arabic science drawn by humanist polemicists, especially if they survive even today. Many medieval and many Renaissance scholars knew better and were fully aware that Arabic scholars, too, are able to philosophize.

# APPENDIX

## The Availability of Arabic Authors in Latin Editions of the Renaissance

he subject of this Appendix is the textual transmission of Arabic sciences and philosophy in the Renaissance, based on the evidence provided by the printed editions of Latin translations. In order to arrive at a comprehensive picture of this transmission, the survey identifies the Arabic authors and texts printed in the Renaissance, attempts to differentiate between authentic and pseudepigraphic works, refers to the medieval and Renaissance Latin translations and translators, points to differences between the medieval and the Renaissance transmission, and names the Renaissance revisers of Latin texts of Arabic origin. The results of the inquiry are presented alphabetically under the Latin names of the Arabic authors, so that the Appendix may also be used as a reference chapter on the Latin reception of Arabic sciences and philosophy. Each entry ends with a list of printed editions of the author's works until the year 1700 AD. The inventory lists the date and place of the edition and offers brief quotations from the title page. These data make it possible to identify the edition in the many printed and electronic catalogs. My primary interest is in the texts of the title pages because these pages, as an early form of advertisement, are part of the popular discourse about Arabic authors, which has been the subject of this study, and because the pages transport information relevant for the history of transmission. Whenever the title pages tell something about the activity of translators or revisers, longer quotations are given.

The present inventory aims at reliability; it cannot aspire to completeness. As a rule, quotations from title pages are given only for those editions that I have seen myself, either in the original or in a reproduction. An exception is the section on Averroes, which also contains quotations from the bibliography of Aristotle editions by Edward Cranz and Charles Schmitt.[1] When I have not seen an edition that is attested by several independent catalogs, a simple reference is given that consists only of the edition's date and place. The quotations follow the original Latin titles in spelling, but not in capitalization and punctuation. Underlining indicates that the edition presents a new textual format, that is, either a new translation or a revision of an older translation. The notes do not offer full bibliographies on the various authors and works mentioned, but only refer to select secondary sources that I considered useful for the reader.

The inventory is limited in a number of respects: its focus is philosophical and scientific literature, and hence it excludes religious and historical texts; it does not list excerpts or citations, but only complete works; it lists only those commentaries on Arabic authors that cite the text in full; it does not take account of vernacular editions of Arabic texts; it includes oriental Christian and Jewish writers only if their works were composed in Arabic; it does not list anonymous works of Arabic provenance, such as *Secretum secretorum, Turba philosophorum, Tabula smaragdina,* or the author 'Jafar Indus' who has not been identified;[2] it only considers authors printed, and the reader is reminded of the fact that Arabic authors and their works were still also distributed in manuscript, particularly in alchemy, magic, and astrology (see below s.v. 'Alkindi'). Because of these restrictions, a few well-known names are missing: Abulfeda (Abū l-Fidā') and Barhebraeus (Abū l-Farağ ibn al-'Ibrī), as historians;[3] Abraham ibn Ezra as an author writing in Hebrew; Avicebron (Solomon ibn Gabirol), Avempace (Ibn Bāğğa),[4] Azarchel (az-Zarqālī), and Alcoarismi (al-Ḥwārizmī), as authors who did not reach print. The name of the famous mathematician al-Ḥwārizmī lived on in the common Renaissance book title *algorismus.*[5] Azarchel was well-known in the Renaissance for the invention of the Saphea *(ṣafīḥa),* a universal astrolabe, which is valid for any geographical latitude.[6]

With the exception of the named authors and traditions, the inventory gives a comprehensive picture of the presence of Arabic sciences and philosophy in the Latin West. The wealth of material testifies impressively to the liveliness of the Arabic scientific tradition in the Renaissance.

## AUTHORS CONTAINED IN THE APPENDIX

Abenguefit (Ps.-Serapion)
al-Abharī
Albategnius
Albohali
Albubater
Albucasis

Albumasar
Alcabitius
Alfarabi
Alfraganus
Algazel
Alhazen

Alkindi
Alpetragius
Avenzoar
Averroes
Avicenna
Costa ben Luca
Ebenbitar (Ibn al-Bayṭār)
Geber filius Affla
Pseudo-Geber
Haly filius Abbas
Haly filius Abenragel
Haly Rodoan
Pseudo-Haly (Aḥmad ibn Yūsuf )
Ibn Buṭlān
Ibn Ǧazla
Ibn al-Ǧazzār

Ibn Muʿāḏ (Abhomad)
Pseudo-Ibn Sīrīn
Ibn Ṭufayl
Iesus Haly
Iohannitius
Isaac Israeli
Isḥāq ibn ʿImrān
al-Isrāʾīlī (Pseudo-Almansor)
Maimonides
Messahalah
Mesue
Omar Tiberiadis
Rhazes
Serapion
Thebit ben Corat
Zahel

## ABENGUEFIT (IBN WĀFID, PSEUDO-SERAPION)

ʿAbd ar-Raḥmān ibn Wāfid lived in Andalusia, where he died probably in 1074 AD; he was vizier in Toledo and founder of the royal botanical garden in the valley of the Tagus. His theoretical treatise on agriculture was translated into Castilian, and this translation served as a source for the Renaissance scholar Gabriel Alonso de Herrera and his *Agricultura general*.[7] A treatise on ophthalmology of Abenguefit is lost, but the "Book on Simple Medicines" *(Kitāb al-Adwiya al-mufrada)* survives in Arabic.[8] In the later twelfth century, Gerard of Cremona translated the short introductory section of this comprehensive work into Latin.[9] It is in tune with the great interest of Renaissance scholars in *materia medica* that Abenguefit's introduction was accessible in a good number of prints; it often accompanied the *Opera omnia* of Pseudo-Mesue, the most comprehensive handbook of pharmacology of Renaissance Latin culture.

As has recently been discovered,[10] Abenguefit's "Book on Simple Medicines" received a second Latin translation toward the end of the thirteenth century: by Simon of Genoa, subdeacon to Pope Boniface VIII until the latter's death in 1303, and Abraham ben Šem Tob of Tortosa, the translators of Albucasis as well.[11] This Latin version, which covers the entire book, is the very well-known *Liber aggregatus in medicinis simplicibus* by (Pseudo-)Serapion. The second longer part of the book, which follows upon a short theoretical introduction, lists more than five hundred simple drugs. For later medieval and Renaissance medicine, Pseudo-Serapion was a very influential authority on simples. For this reason, he became a major target of humanist pharmacologists, as we saw in Chapter 4 when surveying the Renaissance discussion of the drug senna. There was no new translation of Pseudo-Serapion in the Renaissance. In 1552, a humanist revision of the text appeared that was the work of the Milanese physician Niccolò Mutoni, who converted the text into a manual of simple medicines, in which Pseudo-Serapion's drug descriptions are turned into humanist Latin and augmented by passages cited from other authors, notably Galen and Disocorides.

(1473).    *Liber Serapionis aggregatus in medicinis simplicibus,* Milan: Lib. aggreg.[12]

(1479).    *Liber Serapionis aggregatus in medicinis simplicibus,* Venice: Lib. aggreg.

(1497).    *Practica Joannis Serapionis dicta Breviarium. Liber Serapionis de simplici medicina,* Venice: Lib. aggreg.

(1497).    *Mesue cum expositione Mondini . . . Incipit subtilissimus Abhenguefit libellus,* Venice: De simpl.

(1503).    *Practica Joannis Serapionis dicta Breviarium. Liber Serapionis de simplici medicina,* Venice: Lib. aggreg.

(1525).    *Practica Joannis Serapionis. Index operum . . . Practica Joannis Serapionis aliter Breviarium nuncupata. Liber Serapionis de simplici medicina,* Lyon: Lib. aggreg.

(1527).    *Divi Mesue et nova quedam . . . opera . . . subtilissimus Habenguefit De simplicibus medicinis libellus,* Venice: De simpl.

(1530).    *Practica Joannis Serapionis. Necessarium ac perutile opus totius medicine practice . . . Verumetiam librum copiosum de simplici medicina,* Venice: Lib. aggreg.

(1531).    *In hoc volumine continentur . . . Ioannis Serapionis arabis De simplicibus medicinis opus praeclarum et ingens,* Strasbourg: Lib. aggreg.

(1531).    *Tacuini sanitatis Elluchasam Elimithar . . . Albengnefit De virtutibus medicinarum et ciborum,* Strasbourg (with Ibn Buṭlān): De simpl.

(1532).    *Tacuini sex rerum non naturalium . . . Albengnefit De virtutibus ciborum et medicinarum,* Strasbourg (with Ibn Buṭlān): De simpl.

(1533).    *Opera divi Ioannis Mesue . . . Subtilissimus Abhenguefit De simplicibus medicinis libellus,* Lyon: De simpl.

(1535).     *Opera divi Ioannis Mesue ... Subtilissimus Abhenguefit De simplicibus medicinis libellus,* Lyon: De simpl.

(1541).     *Opera divi Ioannis Mesue ... Subtilissimus Abhenguefit De simplicibus medicinis,* Lyon: De simpl.

(1549).     *Mesue et omnia quae cum eo imprimi consueverunt ... Abenguefit De simplicibus medicinis,* Venice: De simpl.

(1552).     *Ioannis Serapionis De simplicium medicamentorum historia libri
septem ex Arabum ac Graecorum, praesertim Pauli Aeginetae,
Dioscoridis et Galeni commentariis quam accuratissime excerpti,
interprete Nicolao Mutono medico Mediolanensi,* Venice: Lib.
aggreg.

(1558).     *Mesue qui Graecorum et Arabum postremus medicinam practicam illustravit ... De virtutibus simplicium medicinarum Abenguefit,* Venice: De simpl.

(1561/62).  *Mesuae graecorum ac arabum clarissimi medici opera ...
De virtutibus simplicium medicinarum Abenguefit,* Venice: De
simpl.

(1570).     *Mesuae medici clarissimi opera ... De virtutibus simplicium medicinarum Abenguefit,* Venice: De simpl.

(1581).     *Ioannis Mesuae medici clarissimi opera ... ,* Venice: De simpl.

(1589).     *Ioannis Mesuae Damasceni medici clarissimi opera ... ,* Venice:
De simpl.

(1602).     *Ioannis Mesuae Damasceni medici clarissimi opera ... De virtutibus simplicium medicinarum atque ciborum Albengnefit,*
Venice: De simpl.

(1623).     *Ioannis Mesuae Damasceni medici clarissimi opera ... De virtutibus simplicium medicinarum atque ciborum Albengnefit,* Venice:
De simpl.

## AL-ABHARĪ

Atīraddīn al-Abharī, who died in Mosul in 1262 or 1265 AD, was the author of a compendium of philosophy and of a logical introduction, which
count among the most successful scientific books in the Arabic world of
the fourteenth and fifteenth centuries AD; they were often commented and
super-commented upon.[13] The compendium, which bears the title *Hidāyat
al-ḥikma* ("Guidance in Philosophy"), presents mainstream Avicennian
philosophy in three parts: logic, natural philosophy, and metaphysics.[14]
The logical introduction, *Kitāb al-Īsāġūġī,* is an adaptation of Porphyry's

*Isagoge,* but also covers other topics of the Aristotelian *Organon,* such as the syllogism and the theory of demonstration.[15] Al-Abharī was not available in Latin until the Franciscan friar Tommaso Obicini of Novara (d. 1632), who spent ten years of his life in Palestina and, after his return to Italy, promoted the study of Arabic in his convent in Rome,[16] produced a Latin translation of the "Introduction," which he published on facing pages with the Arabic in 1625. In the dedicatory letter he argues that missionaries in Arabic countries will not be successful if they do not know the *termini orientales* in philosophy and theology.

(1625).          *Īsāġūġī. Isagoge id est breve introductorium arabicum in scientiam logices cum versione latina ac Theses sanctae fidei, R.P.Fr. Thomae Novariensis . . . linguarum orientalium magistri opera studioque editae,* Rome.


## ALBATEGNIUS

The astronomer Muḥammad ibn Ġābir ibn Sinān al-Battānī (d. 929 AD), a descendant of a family that adhered to the Sabian religion, but himself a Muslim, who lived for the most part of his life in Raqqa in Syria, was the first Arabic scholar to fully adopt and apply the Ptolemaic planetary models and to recognize the essential role of observations as a test of astronomical theory.[17] By comparing his own very accurate observations with those of Ptolemy, Albategnius was able to make a number of important corrections. His main work is a *Zīġ,* an astronomical handbook with text and tables, later called *az-Zīġ aṣ-ṣābi'* ("The Sabian Handbook"),[18] which was translated into Latin by Plato of Tivoli (fl. 1135–1145 AD). Hermann of Carinthia recommends a translation of Albategnius's work by Robert of Ketton, but we have no other indication that a second translation ever existed.[19] Plato of Tivoli's text, without the tables, was printed twice in the Renaissance, with annotations by Regiomontanus. The treatise was also available in manuscript and was used and frequently referred to by many astronomers of the time: Regiomontanus, Copernicus, Tycho Brahe, Kepler, Riccioli, and the young Galilei.[20] Bernardino Baldi was much intrigued by the exactness of Albategnius's observations and discussed them extensively in his *Vite de matematici.*

(1537).          *Continentur in hoc libro Rudimenta astronomica Alfragrani, item Albategnius astronomus peritissimus De motu stellarum . . . cum demonstrationibus geometricis et additionibus Ioannis de Regiomonte,* Nuremberg.

(1645).     *Mahometis Albatenii De scientia stellarum liber cum aliquot ad-
            ditionibus Ioannis Regiomontani ex biliotheca Vaticana tran-
            scriptus,* Bologna.


## ALBOHALI

Abū ʿAlī Yaḥyā ibn Ġālib al-Ḥayyāṭ (fl. around 800 AD) was a pupil of Messahalah and thus belonged to the earliest phase of Arabic astrology. His
treatise *Kitāb Aḥkām al-mawālīd* ("On Judgements made from Nativities")
is largely based on the book by Messahalah on the same topic: on the prediction of the course of life on the basis of a birth chart. Most of its forty-
five chapters discuss the different themes of these predictions with respect
to the twelve places of the horoscope.[21] This treatise was translated into
Latin in 1136 by Plato of Tivoli.[22] A second translation was made in 1152
or 1153 by "Iohannis Toletanus."[23] It is this second translation that was
printed twice in the Renaissance by Joachim Heller in Nuremberg, with a
dedicatory epistle to Philipp Melanchthon, who had been Heller's teacher
at Wittenberg University.[24] In addition, texts by Abū ʿAlī al-Ḥayyāṭ on astrological judgments were included in the twelfth-century compilation
*Liber novem iudicum,* together with *Iudicia* by Alkindi, Omar, and Zahel.
Here his name is transcribed as "Alchait" or "Albenait."[25]

(1509).     *Liber novem iudicum in iudiciis astrorum. Clarissimi auctores
            Mesehella, Aomar, Alkindus, Zael, Albenait, Dorotheus, Jergis,
            Aristoteles, Ptholemeus istius voluminis,* Venice: Iudicia.

(1546).     *Albohali arabis astrologi antiquissimi ac clarissimi De iudiciis na-
            tivitatum liber unus,* Nuremberg: De nativit.

(1549).     *Albohali arabis astrologi antiquissimi ac clarissimi De iudiciis na-
            tivitatum liber unus,* Nuremberg: De nativit.

(1571).     *Albohazen Haly filii Abenragel... de iudiciis astrorum libri
            octo... Compendium duodecim domorum coelestium ex claris-
            simis et vetustissimis authoribus, scilicet Messahalla, Aomare, Al-
            kindo, Zaele, Albenait, Dorotheo, Iergi, Aristotele et Ptolemaeo
            collectum,* Basel: Iudicia.


## ALBUBATER (ACHASITH)

The astrologer Abū Bakr al-Ḥasan ibn al-Ḥasīb from Kufa was a contemporary of Albohali; he lived in the first half of the ninth century AD in
Baghdad and wrote on the same topic: genethlialogy, or nativities. Two

parts of his fourfold magnum opus on genethlialogy are extant in Arabic
and Latin: the *Kitāb al-Mawālīd* ("On Nativities") and the *Kitāb Taḥāwīl sinī
l-mawālīd* ("On the Revolutions of the Years of the Nativities"), that is, on
anniversary horoscopes.[26] The latter text was translated by Plato of Tivoli
in the twelfth century as *De revolutionibus nativitatum,* but had very restricted manuscript circulation and was not printed in the Renaissance;
Plato of Tivoli refers to ibn al-Ḥasīb as "Alkasem filius Achasith."[27] But
even the formidable Renaissance bio-bibliographer of astrology Simon de
Phares (late fifteenth century) does not know either "Achasit" or his book
on anniversary horoscopes, but only *De nativitatibus:*[28] the *Kitāb al-Mawālīd,*
a book of great size on the manifold themes of prediction discussed with
respect to the twelve astrological places. "On Nativities" was translated
into Hebrew, and later from Hebrew into Latin by Salio, canon of Padua,
with the help of a Jew "David," probably in Barcelona or Toledo in 1218.[29]
It is this text by Albubater that was printed in the Renaissance.

(1492).          Venice.

(1501).          *Albubather et Centiloquium divi Hermetis,* Venice.

(1540).          *Albubatris astrologi diligentissimi Liber genethliacus sive De nati-
                 vitatibus,* Nuremberg.


### ALBUCASIS

Abū l-Qāsim Ḥalaf az-Zahrāwī was the greatest authority on surgery in
the Latin Middle Ages and, via quotations in the work of Guy de Chauliac
(d. 1368), also in the early Renaissance, until humanist surgery eclipsed
his influence.[30] We do not know much more about az-Zahrāwī other than
that he was a practicing physician in Cordoba at the time of the Caliph
ʿAbd ar-Raḥmān III, who reigned 912–961. The Latin image of Albucasis
as an author on surgery and pharmacology rests on the fact that only two
of the thirty books of his medical summa, apparently his sole work, were
translated into Latin in the Middle Ages. The original title of this work is
*Kitāb at-Taṣrīf li-man ʿaǧiza ʿan at-taʾlīf* ("The distribution <of medical knowl-
edge> to those who are incapable of collecting it").[31]

   Book 30 on surgery was translated in the twelfth century by Gerard of
Cremona, book 28 on simple medicines at the end of the thirteenth
century by Simon of Genoa and the Jewish scholar Abraham ben Šem Tob
of Tortosa, perhaps from Hebrew.[32] Book 28 became known under the title
*Liber servitoris;* it contains concrete instructions for the preparation of
simple medicines, which were often quoted in Arabic sources.[33] The *Liber
servitoris* is recommended as one of six reference works for the use of apoth-

ecaries by Saladin of Asculo around 1450.[34] The treatise was often printed with the *Opera Mesue*.

Another translation of the *Taṣrīf* appeared 1519 in an Augsburg edition; it covers the first two books on theoretical medicine. The editor is Paolo Ricci, the translator of Averroes and Christian convert from Judaism (on Ricci see Chapter 2), but from his dedicatory epistle it is not clear who had translated the two books. Ricci does not claim to have translated them himself.[35]

## Liber servitoris

(1471).          *Incipit Liber servitoris Liber xxviii Bulchasin Benaberazerin*, Venice.

(1478).          Venice.

(1478–1479).     Pavia (with Nicolaus Salernitanus).

(1479).          *Incipit Liber servitoris Liber xxviii Bulchasin Benaberacerin*, Venice (with Mesue).

(ca. 1482–1483). *Antidotarius Nicolai . . . Liber servitoris*, s.l. (Strasbourg) (with Nicolaus Salernitanus).

(1484).          *Incipit Liber servitoris Liber xxviii Bulchasin Benaberacerin*, Venice (with Mesue).

(1489–1491).     *Incipit Liber servitoris Liber xxviii Bulchasin Benaberazerin*, Venice (with Mesue).

(1495).          *Mesue cum expositione Mondini . . . Libellus Bulcasis sive servitoris*, Venice.

(1497).          *Mesue cum expositione Mondini . . . Libellus Bulcasis sive servitoris*, Venice.

(1502).          *Mesue cum expositione Mondini . . . Libellus Bulcasis sive servitoris*, Venice.

(1508).          *Mesue cum expositione Mondini . . . Libellus Bulcasis sive servitoris*, Venice.

(1510).          *Mesue cum expositione Mondini . . . Libellus Bulcasis sive servitoris*, Lyon.

(1510).          *Mesue cum expositione Mondini . . . Libellus Bulcasis sive servitoris*, Venice.

(1515).          *Mesue cum expositione Mondini . . . Libellus Bulcasis sive servitoris*, Lyon.

(1519).          *Mesue cum expositione Mondini . . . Libellus Bulcasis sive servitoris*, Lyon.

(1525).     *Mesue cum expositione Mondini . . . Libellus Bulcasis sive servitoris,* Lyon.

(1527).     *Divi Mesue et nova quedam . . . opera . . . Bulcasis sive servitoris medicorum profundissimi libellus copiosissimus,* Venice.

(1533).     *Opera divi Ioannis Mesue,* Lyon.

(1535).     *Opera divi Ioannis Mesue,* Lyon.

(1538).     Venice.

(1541).     *Opera divi Ioannis Mesue,* s.l.

(1549).     *Mesue et omnia quae cum eo imprimi consueverunt,* Venice.

(1558).     *Mesue qui Graecorum ac Arabum postremus medicinam practicam illustravit,* Venice.

(1561/1562).     *Mesuae Graecorum ac Arabum clarissimi medici opera,* Venice.

(1568/1570).     *Mesuae medici clarissimi opera,* Venice.

(1581).     *Ioannis Mesuae medici clarissimi opera,* Venice.

(1602).     *Ioannis Mesuae Damasceni medici clarissimi opera . . . Liber servitoris,* Venice

(1623).     *Ioannis Mesuae Damasceni medici clarissimi opera . . . Liber servitoris,* Venice

### Chirurgia, Liber theoricae

(1500).     *Cyrurgia parva Guidonis. Cyrurgia Albucasis cum cauteriis et aliis instrumentis,* Venice.

(1519).     *Liber theoricae necnon practicae Alsaharavii . . . qui vulgo Acararius dicitur,* Augsburg.

(1532).     *Albucasis chirurgicorum omnium primarii Libri tres,* Strasbourg: Chirurgia.

(1541).     *Methodus medendi certa, clara et brevis . . . autore Albucase,* Basel: Chirurgia.

(1544).     *Chirurgia Albucasis,* Strasbourg: Chirurgia.

## ALBUMASAR

Abū Maʿšar Ǧaʿfar ibn Muḥammad ibn ʿUmar al-Balḫī (787–886 AD) was the most influential and best-known astrologer of both the Arabic and Latin worlds. He was a native of Balḫ in Khurasan in Persia, but lived for most of his life in Baghdad. His predecessors (Messahalah, Albohali, Albubater, and others) had established Arabic astrological literature on the basis of translations from Greek, Sanskrit, and Persian. Albumasar composed general reference works and practical manuals of astrology, which became standard accounts of the astrological fields of anniversary horoscopes and conjunctions.[36]

His most important and acclaimed work is the *Introductorium maius in astronomiam,* in Arabic *Kitāb al-Mudḥal al-kabīr ilā ʿilm aḥkām an-nuǧūm* ("The Large Book Introducing the Science of Making Judgments from the Stars"). It was translated twice into Latin, by John of Seville (probably in 1133), and in 1140 by Hermann of Carinthia.[37] In the first of its eight chapters, Albumasar gives a philosophical justification of astrology, by placing it within the framework of Aristotelian physics. The Renaissance editions print Hermann of Carinthia's version of the text.[38] An epitome of the *Introductorium maius,* composed by Albumasar himself, was translated into Latin by Adelard of Bath, but not printed in the Renaissance.[39]

The second major work of Albumasar available in Latin translation was *De magnis coniunctionibus,* in Arabic *Kitāb al-Milal wa-d-duwal* ("The Book of Religions and Dynasties"), a major work of historical astrology.[40] The twelfth-century translator is not named, but much evidence points to John of Seville. The version that became canonical in the Middle Ages seems to be the result of a revision made in Toledo on the basis of a different Arabic manuscript.[41] The Renaissance printed editions print this canonical version.[42] Theories of conjunctions of the planets, in particular of Saturn and Jupiter, and their significance for world history were very popular among Renaissance astrologers, as was discussed in Chapter 6, and Albumasar figured as a prime authority in this discipline.

Numerous Latin manuscripts and three Renaissance editions survive of another text of general astrology, the *Flores Albumasaris,* whose title apparently derives from the Arabic *Kitāb an-Nukat* ("Book of Anecdotes"). This text also traveled under longer Arabic titles. It contains a long list of chapters on various issues of the discipline of revolutions of the years of the world, that is, on anniversary horoscopes that concern general matters of the world.[43] It was translated partially in the twelfth century, probably by John of Seville.[44]

The three works mentioned constitute the main body of *Albumasar Latinus,* and it is this body that was the basis of his fame in the Renaissance. There are, however, several other texts associated with the name of Albumasar, the authenticity of which is still in question.[45] One clearly authentic work by Albumasar reached the Latin West via the intermediary of Byzantine Greek. The first five of the nine books *De revolutionibus nativitatum,* a work on anniversary horoscopes for individual persons, titled in Arabic *Kitāb Taḥāwīl sinī l-mawālīd* ("On the Revolutions of the Years of the

Nativities"), were translated into Greek in the tenth century, and thence into Latin in the thirteenth century, perhaps in the year 1262, most likely by Stephen of Messina.[46] The text was available in the Renaissance in an edition of 1559, where the text is attributed to Hermes. Another partial Latin translation of the same work, but from Arabic, was produced in the twelfth century, probably by John of Seville, but did not reach print.[47]

The *Apomasaris apotelesmata* printed in 1577 are not by Albumasar, but by Pseudo-Ibn Sīrīn (see below).

(1488 or 1506). *Albumasar Flores astrologiae*, Venice.

(1488). *Flores Albumasaris*, Augsburg.

(1489). *Albumasar De magnis coniunctionibus*, Augsburg.

(1489). *Introductorium in astronomiam Albumasaris Abalachi*, Augsburg.

(1495). *Flores Albumasaris*, Augsburg.

(1506). *Introductorium in astronomiam*, Venice.

(1515). *Albumasar De magnis coniunctionibus, annorum revolutionibus ac eorum profectionibus octo continens tractatus*, Venice.

(1559). *In Claudii Ptolemaei Quadripartitum ... praeterea Hermetis philosophi De revolutionibus nativitatum libri duo incerto interprete*, Basel (under the name of Hermes).

## ALCABITIUS

Abū ṣ-Ṣaqr ʿAbd al-ʿAzīz ibn ʿUtmān ibn ʿAlī al-Qabīṣī was the author of the most widely read elementary introduction to astrology in the Middle Ages and the Renaissance. The exact dates of his life are unkown: he was a student of al-ʿImrānī (d. 955 AD); at least four of his works are dedicated to Sayf ad-Dawla, the Hamdanid ruler of Aleppo (945–967), which indicates that he lived and worked in close relation to the court in Syria. Apart from his introduction to astrology, Alcabitius is known as the author of about a dozen mathematical, astronomical, and astrological treatises.[48]

His *Kitāb al-Mudhal ilā ṣināʿat aḥkām an-nuǧūm* ("Book of the Introduction to the Art of Judicial Astrology") was translated into Latin as *Introductorius ad magisterium iudiciorum astrorum* in the twelfth century by John of Seville;[49] this was the beginning of a reception much more successful than that in the Arabic world. There exist several medieval Latin (but no Arabic) commentaries on the *Introductorius*.[50] As was discussed in Chapter 1, the

*Introductorius* was the most often printed astrological work of Arabic provenance. It was subject to at least two commentaries in the sixteenth century, one by the French astrologer Pierre Turrel, who inserted comments *(additiones)* into Alcabitius's text, printed around 1520, and one, much more substantial, by the German professor of mathematics Valentin Nabod (1523–1593), printed in 1560.[51] An effort to improve the Latin text of the *Introductorius* was made in the early sixteenth century by Antonio de Fantis, who seems to have compared at least one Latin manuscript and perhaps also the fourteenth-century commentary by John Danko of Saxony. His text appeared twice in 1521 in Venice, with the printers Peter Liechtenstein and Melchior Sessa.[52] Antonio de Fantis's Venice editions also contain a two-page treatise *De coniunctionibus planetarum* ascribed to Alcabitius, of which no Arabic equivalent has yet been found; it was translated into French and Spanish.[53]

It adds to the historical significance of Alcabitius that his *Introductorius* appears as mandatory astrological reading in the university curricula of Bologna, Ferrara, and Kraków. From indirect evidence it is clear that he was read also at many other Renaissance universities (see Chapter 1).

(1473).    *Introductorium Alchabitii Arabici ad scientiam iudicialem astronomiae*, s.l.[54]

(1482).    *Libellus ysagogicus Abdilazi,* Venice.

(1485).    *Libellus ysagogicus Abdilazi,* Venice.

(1491).    *Libellus ysagogicus Abdilazi,* Venice.

(1502).    *Alchabitius cum comento,* Venice.

(1503).    *Alchabitius cum comento: Cum gratia et privilegio. Libellus isagogicus Abdilazi,* Venice.

(1508).    Frankfurt an der Oder.

(1512).    *Alchabitius cum commento noviter impresso: Libellus isagogicus Abdilazi,* Venice.

(ca. 1520).    *Alkabitius astronomie iudiciarie principia tractans,* Lyon (with Pierre Turrel's comments).

(1521).    *Alcabitii ad magisterium iudiciorum astrorum Isagoge commentario Ioannis Saxonii declarata,* Paris.

<u>(1521).</u>    *Preclarum summi in astrorum scientia principis Alchabitij opus . . . pristino candori nuperrime restitutum ab excellentissimo doctore Antonio de Fantis Taruifino qui notabilem eiusdem auctoris libellum De planetarum coniunctionibus nusquam antea impressum addidit,* Venice (Sessa).

(1521).    *Preclarum summi in astrorum scientia principis Alchabitii opus ad scrutanda stellarum magisteria isagogicum . . . restitutum ab . . . Antonio de Fantis,* Venice (Liechtenstein).

(1560).    *Enarratio elementorum astrologiae, in qua praeter Alcabicii, qui Arabum doctrinam compendio prodidit, expositionem atque cum Ptolemaei principiis collationem, reiectis sortilegiis et absurdis vulgoque receptis opinionibus de verae artis praeceptorum origine et usu satis disseritur* (Cologne, 1560) (*Introductorius* 1–4 with commentary by Valentin Nabod).

## ALFARABI

Very few facts are known about the life of Alfarabi (Abū Naṣr Muḥammad ibn Muḥammad al-Fārābī), the most important representative of the Baghdad Aristotelian philosophers. His origins lie in Fāryāb in northeast Iran or, more likely, in Fārāb in southern Kazakhstan. About his ethnic descent, whether Persian, Turkish, or Arab, nothing certain is known. He lived for a long time in Baghdad, but spent the last eight years of his life in Syria and Egypt. He died in 950/951 in Damascus.[55]

Alfarabi has written on many topics of the philosophical tradition, but his emphasis lies on logic and on the theory of the religious community. The Latin West—in contrast to the Hebrew tradition—knew only a small part of his oeuvre.[56] His encyclopedic magnum opus on "The Principles of the Opinions of the Inhabitants of the Perfect City" was not translated into Latin. Of his numerous logical works, only one is extant in a complete Latin translation: his "Guide to the Book of Rhetoric" (*Ṣadr li-kitāb al-ḫaṭāba*), translated by Hermannus Alemannus around 1250 as *Didascalia in rhetoricam Aristotelis ex glosa Alpharabi*.[57] In 1481 and 1515, two editions appeared of excerpts from this translation (paragraphs 39–56 of the modern edition) under the title *Declaratio compendiosa super libris Rhetoricorum Aristotelis*; these excerpts had been rearranged by the editor Lancillotus de Zerlis, physician of Verona, so that they formed a table of contents of Aristotle's *Rhetoric*.[58]

Another two works by Alfarabi reached print: *De intellectu (Risāla fī l-ʿaql)*, 1508 in an edition of Avicenna's philosophical works, and again in 1638. Abraham de Balmes, the prolific translator of works of Averroes (d. 1523), produced a new Latin translation of *De intellectu* from Hebrew; this translation is extant in manuscript.[59] Abraham remarks that Averroes's *Epitome of logic* contains much material from Alfarabi "of Baghdad" (*babilonius*), who was venerated as the "greatest philosopher" (*maximus philosophus*) of his time.[60] The 1638 edition also printed *De scientiis (Iḥṣāʾ al-ʿulūm*, "Enumeration of the Sciences"). This is a comprehensive work on the division and order of the sciences, in which Alfarabi subsumes the traditional Islamic sciences among the theoretical sciences of the Aristotelian tradition.[61] *De scientiis* was translated twice in twelfth-century Toledo, once literally by Gerard of Cremona and once, in an abbreviated fashion, by Dominicus Gundisalvi. It was the latter text that was printed in 1638. Alfarabi's *De scientiis* much influenced the discussion of the classification of the sciences in the twelfth and thirteenth centuries, but apparently not in the Renaissance.[62]

The two early prints of his Epitome of the *Rhetoric* indicate that Alfarabi's few Renaissance readers conceived of him as a commentator of Aristotle. This is also how the bio-bibliographers Foresti da Bergamo, Hartmann Schedel, and Johann Staindel refer to him. They note that "some of his commentaries exist among the Latins."[63] They also address him as *astronomus*, and Konrad Gesner presents him as the author of *De compositione astrolabii*,[64] a common title, which is often associated with (Pseudo-)Messahalah. In Arabic, a commentary by Alfarabi on the *Almagest* and a treatise against astrology survive.[65]

The very short printing history of Alfarabi should not obscure the fact that the Renaissance reception of Alfarabi as an Aristotelian philosopher profited much from the boom of Averroism. Through the works of Averroes, Alfarabi's philosophical positions were better known than in the Middle Ages.[66]

(1481).   *Declaratio compendiosa per viam divisionis Alfarabii super libris rethoricorum Aristotilis,* Venice: Didascalia in rhetoricam.

(1508).   *Avicenne perhypatetici philosophi . . . opera . . . Alpharabius De intelligentiis,* Venice, repr. Frankfurt am Main, 1961: De intellectu.

(1515).   *Rhetorica Aristotelis cum . . . Egidii de Roma . . . commentariis . . . necnon Alpharabii compendiosa declaratione,* Venice: Didascalia in rhetoricam.

(1638).   *Alpharabii vetustissimi Aristotelis interpretis opera omnia,* Paris, repr. Frankfurt am Main, 1969: De scientiis, De intellectu.

## ALFRAGANUS

Aḥmad ibn Muḥammad ibn Kaṯīr al-Farġānī (fl. ca. 830–860) was the author of a compendium of Ptolemaic astronomy, which was the most widely read piece of Arabic astronomy in the Latin West. The exact dates of Alfraganus's life are unknown and were already a matter of dispute in the later Renaissance: Jacob Christmann (in 1590) conjectured on the basis of Alfraganus's observations that he lived around 950 AD,[67] whereas Jacob Golius and Gerard Voss[68] in the seventeenth century followed the bio-bibliographer Barhebraeus, who placed him in the lifetime of the Caliph al-Ma'mūn (d. 833), that is in the ninth century—as is still assumed today.[69]

Alfraganus is also known as an author of treatises on the astrolabe, on sundials, and of a lost commentary on the astronomical tables by al-Ḫwārizmī,[70] but the Latin world knew him by his main work, the astronomical compendium titled *Kitāb Ǧawāmiʿ ʿilm an-nuǧūm* ("Book of the Summaries of the Science of the Stars"), also known as the "Thirty Chapters." It was translated into Latin in 1135 by John of Seville,[71] and later in the century by Gerard of Cremona.[72] Both translations were widely distributed in manuscript,[73] but only John's version was printed in the Renaissance.[74] Alfraganus was a very influential figure in medieval astronomy; it was mainly through his work that the West became acquainted, in a nonmathematical manner, with the basic features of Ptolemaic astronomy. The principal university textbook on this topic, John of Sacrobosco's *Sphaera*, much depended on Alfraganus.[75] His doctrinal and institutional influence in the Renaissance remains to be assessed by future research.

The 1537 edition of Alfraganus's compendium contains the introductory passages from a lecture by Regiomontanus on Alfraganus, which he had held at the University of Padua in the spring of 1464.[76] In 1590, a new Latin version of the *Ǧawāmiʿ* appeared, which was produced by Jacob Christmann (d. 1613), professor of Hebrew in Heidelberg on the basis of the Hebrew version (by Jacob Anatoli). The dedicatory epistle to elector Johann Casimir is a programmatic essay about the usefulness of academic studies of the Arabic.[77] Alfraganus was again translated in the seventeenth century by Jacob Golius, professor of Arabic in Leiden for over forty years and one of the founding figures of Arabic studies in Europe. Golius printed both the Arabic text and his Latin translation and added long notes in an appendix; the notes cover only the first nine chapters.[78] Alfraganus is thus one of the very few Arabic scientists who has a place both in the history of the Arabic influence on Renaissance culture and in the history of Arabic scholarship in Europe.

(1493). *Brevis ac perutilis compilatio Alfragani astronomorum peritissimi totum id continens quod ad rudimenta astronomica est opportunum*, Ferrara.

(1537).     *Continentur in hoc libro Rudimenta astronomica Alfragrani, item Albategnius astronomus peritissimus De motu stellarum . . . Item Oratio introductoria in omnes scientias mathematicas Ioannis de Regiomonte Patavii habita cum Alfraganum publice praelegeret,* Nuremberg.

(1546).     *Alfragani astronomorum peritissimi compendium id omne quod ad astronomica rudimenta spectat complectens,* Paris.

(1590).     *Muhamedis Alfragani arabis chronologica et astronomica Elementa . . . additus est commentarius . . . autore M. Iacobo Christmanno,* Frankfurt am Main.[79]

(1669).     *Muhammedis filii Ketiri Ferganensis, qui vulgo Alfraganus dicitur, Elementa astronomica arabice et latine . . . opera Jacobi Golii,* Amsterdam, repr. Frankfurt am Main, 1986.


## ALGAZEL

Abū Ḥāmid al-Ġazālī (ca. 1056–1111 AD) was one of the most influential theologians of Islam. Professor of Muslim law and theology at the prestigious college Niẓāmīya in Baghdad, he abandoned his position and left the city in 1095, in order to live a religious life that is not compromised by serving political authorities.[80] A central theme of his entire work is the theologian's view of and attitude toward philosophy. The main fruit of his thought on this topic is the *Tahāfut al-falāsifa* ("Incoherence of the Philosophers"), completed before he left Baghdad, a thorough criticism and partial adoption of a series of philosophical positions—which prompted Averroes to a famous response, the *Tahāfut at-tahāfut (Destructio destructionum)*.

The only work by al-Ġazālī available in Latin in the twelfth and thirteenth centuries was a text written in about the same years, the *Maqāṣid al-falāsifa* ("The Aims of the Philosophers"), a description of the philosophers' logic, physics, and metaphysics based on the Persian *Dānešnāme* by Avicenna; it was translated by Dominicus Gundisalvi and Magister Iohannes in the twelfth century. The preface in which Algazel explains that he is only going to expose the doctrines of the philosophers in order to refute them later was also translated into Latin, but had a very restricted manuscript circulation and seems to have been known only to Roger Bacon.[81] As a consequence, Algazel was generally thought to be a straightforward follower of Avicenna *(sequax Avicennae)* by the scholastics.[82] The Renaissance theologian Konrad Wimpina continues this tradition when describing Algazel, in 1493, as a *collega Avicennae* in all doctrines, so that the refutation of Avicenna's errors automatically applies to Algazel as well.[83] The Latin *Maqāsid* was printed in 1506, without the preface. The logic part of the *Maqāsid* reached the West again, when Ramon Llull (d. 1315/1316) incorporated excerpts from an Arabic compendium based on the *Maqāsid* into his own compendium of logic, which he originally wrote in Arabic and then translated himself into Latin under the title *Compendium logicae Algazelis*.[84] In the Renaissance, the *Maqāsid* was newly translated into Latin, this time from Hebrew, together with the Hebrew commentary by Moses Narboni (which dates 1344–1349). This anonymous translation survives in Ms. Vat. lat. 4554, and apparently never reached print.[85]

The translations of Averroes's *Tahāfut at-tahāfut* in 1328 and again in 1528 (see below s.v. 'Averroes'), which cited Algazel's text, did not lead to a general revision of the Western image of Algazel: even Agostino Nifo, the sole Latin commentator of the *Destructio destructionum* and the foremost Western expert on Algazel,[86] continued to address Algazel as "the abbreviator of Avicenna."[87] Nevertheless, as in the case of Alfarabi, the Renaissance interest in Averroes also led to a new interest in Algazel, especially among the Italian Aristotelians surveyed in Chapter 5.[88]

In 1650, Algazel's *Tarǧama ʿaqīda ahl as-sunna* ("Interpretation of the Creed of the People of the Sunna," *Interpretatio confessionis fidei orthodoxorum*) was printed in Arabic with Latin translation by the Oxford professor of Arabic Edward Pocock Sr. (d. 1691) in his comprehensive commentary on a chapter of Barhebraeus's history of the Arabs.

(1506).     *Logica et philosophia Algazelis arabis,* Venice, repr. Frankfurt am Main, 1969, Hildesheim, 2001.

(1650).     *Specimen historiae Arabum sive Gregorii AbulFarajii Malatiensis . . . narratio . . . opera et studio Edvardi Pocockii,* Oxford.[89]

See also below the editions of *Destructio destructionum* s.v. 'Averroes.'

## ALHAZEN

Al-Ḥasan ibn al-Ḥasan ibn al-Haytam was one of the foremost scientists of Arabic culture; he is particularly well-known for his contribution to mathematical and experimental optics and for his cosmographical interpretation of Ptolemy's mathematical astronomy. Alhazen was born in Basra in Iraq in 965 AD. According to some sources, he made an offer to the Fatimid ruler in Cairo to regulate the flow of the Nile, but did not succeed with this enterprise. He continued to live in Cairo, where he died around 1040 AD. Alhazen is credited with the composition of almost one hundred treatises, mainly in the fields of optics, mathematics, astronomy, and physics.[90]

Many of these treatises have survived in Arabic (his summaries of Aristotle and Galen, however, are lost), but only three were translated into Latin. Among these is his major and most comprehensive work, the *Kitāb al-manāẓir* ("Optics"), known in Latin as *Perspectiva* or *De aspectibus*. The circumstances of the translation in the late twelfth or early thirteenth century are as yet unknown.[91] Alhazen's *Perspectiva*, which is particularly noteworthy for its theories of reflection and refraction, was well-distributed in medieval Latin manuscripts and was much used by Roger Bacon, Pecham, and Witelo. Friedrich Risner's 1572 edition of the work—which, like all manuscripts, lacks chapters 1–3 of book 1[92]—initiated a new phase in the reception of the book among Kepler, Snell, Beeckman, Fermat, and others.[93] The *Perspectiva* played an important part in the development of modern optics and only ceased to be cited in the last decades of the seventeenth century.[94] Friedrich Risner (d. 1580 or 1581) was an assistant and collaborator of Peter Ramus and a well-known mathematician himself;[95] he published the text on the basis of two manuscripts and introduced chapter divisions, for which he created his own, sometimes misleading titles.[96] The same volume, to which Risner gave the title *Opticae thesaurus*, also contains Witelo's *Perspectiva* and Pseudo-Alhazen's *De crepusculis*, which is in fact a treatise by Ibn Muʿāḏ (see below s.v.).

The only other optical treatise by Alhazen available in Latin was his *Maqāla fī l-marāya al-muḥriqa bi l-qutūʿ* ("On Paraboloidal Burning Mirrors"), *Liber de speculis comburentibus*, which treats the reflection from the concave surface of paraboloidal mirrors.[97] It was not printed in full, but in 1548 Antonius Gogava published an abbreviation that contained the first four propositions.[98]

Of the about twenty astronomical works written by Alhazen, one reached the Latin West: the *Maqāla fī hayʾat al-ʿālam* ("On the Configuration of the World"). This important and ambitious work is an attempt to establish a physical theory of the world that is in accordance with the mathematical theory of Ptolemy's *Almagest*. Alhazen's treatise was translated at least twice, once anonymously under the title *Liber Aboali* in the twelfth or thirteenth century, and once into Spanish by Abraham Hebraeus for King Alfonso X (d. 1284) and thence into Latin under the title *Liber de celo et mundo* by an unkown translator.[99] In the Renaissance, Abraham de Balmes (d. 1523), the prolific translator of Averroes, produced a new Latin version of the text, by translating it from the Hebrew version of Profacius Judaeus, that is, of Jacob ben Makir, who was also the translator of Avenzoar and Azarchel (d. 1303–1306).[100] Abraham de Balmes's Latin version is extant in manuscript Vat. lat. 4566.[101] The medieval versions of the treatise did not reach print either, but had a significant influence, in manuscript form, on Renaissance astronomy, especially via Peuerbach's *Theoricae novae planetarum.*[102]

(1548).     *Claudii Ptolemaei Pelusiensis mathematici operis quadripartiti . . . item, De sectione conica orthogona . . . deque speculo ustorio libelli duo hactenus desiderati restituti ab Antonio Gogava Graviensi,* Louvain.[103]

(1572).     *Opticae thesaurus: Alhazeni arabis libri septem . . . , eiusdem liber De crepusculis et nubium ascensionibus,* Basel, repr. New York, 1972.


## ALKINDI

Ya'qūb ibn Ishāq al-Kindī (d. between 861 and 866 AD), the first great philosopher and scientist of Arabic culture, was received very differently in the Middle Ages than in the Renaissance.[104] Whereas in the twelfth and thirteenth centuries, Alkindi's philosophical treatises *De quinque essentiis, De somno et visione,* and *De intellectu* were read and quoted by the scholastics in Toledan translations, these texts found hardly any readers after 1300—in contrast to the medical, meteorological, and astrological works. Alkindi was the leading figure of a circle of translators from Greek into Arabic in Baghdad,[105] and a very prolific author on almost all scientific subjects of the Graeco-Arabic tradition and of Islamic culture, with the exception of alchemy. About forty titles have survived in Arabic, which is only a small section of his entire oeuvre;[106] some are extant only in Hebrew or Latin. There exist a good number of apparently pseudo-epigraphic treatises in Latin.

Apart from the three philosophical treatises mentioned, the Latin translations of the twelfth century also covered the following authentic works—and it is as an author of these works that Alkindi was known in the Renaissance: *De medicinarum compositarum gradibus* on the composition of compound drugs, and *De aspectibus* on optics,[107] both translated by Gerard of Cremona; *De radiis,* a treatise on the mathematical foundation of astrology and magic, and the meteorological treatise *De mutatione temporum,* which is based on an Arabic combination of two letters by Alkindi,[108] both by anonymous translators;[109] and the comprehensive handbook of judicial astrology called *al-Arba'ūna bāban* ("The Forty Chapters"), which was translated twice into Latin, by Robert of Ketton and by Hugo of Santalla. Robert's translation was not printed in the Renaissance, but Hugo's was printed twice as part of the astrological compendium *Liber novem iudicum* ("Book of the Nine Judges"), a successful medieval handbook of astrology that was compiled in the mid-twelfth century, probably by Hugo of Santalla himself.[110]

Alkindi was best known in the Renaissance for the first of these treatises, the short medical tract *De gradibus* (*Kitāb fī ma'rifat quwā l-adwiya al-murakkaba*, "Knowledge of the Powers of Composite Drugs"). This treatise presents a mathematical theory for the degrees of the effects of composite medicines, which is based on the ancient theory of four degrees for simple medicines.[111] It was often printed as an appendix to the *Opera Mesue*, together with other medical works of Arabic origin (by Albucasis and Abenguefit). It is by virtue of this treatise that Girolamo Cardano ranked Alkindi among the "twelve men most outstanding in the disciplines."[112] Averroes had criticized Alkindi's theory of degrees in his *Colliget*, chapter 5.58, and perhaps it was for this reason that Jacob Mantino produced a new translation of Averroes's chapter from the Hebrew (see below s.v. 'Averroes' and Chapter 3).

The case of Alkindi is a reminder that the printing history of an author may not truly reflect his influence, since three of his works, *De radiis*, *De aspectibus*, and Robert of Ketton's translation of the *Iudicia* were copied in manuscript in the fifteenth to seventeenth centuries.[113] The influence of the manuscript tradition is tangible in Konrad Gesner's presentation of Alkindi as the author not only of *De gradibus* but also of *De radiis*.[114] It is also tangible in Konrad Wimpina's 1493 treatise on the errors of the philosophers, which attacks Alkindi, "the most knowledgable astrologer," for advocating astrological determinism.[115]

### De medicinarum compositarum gradibus

(1497).     *Mesue cum expositione Mondini super Canones universales...*
            *Alchindi authoris prestantissimi in suum de medicinarum com-*
            *positarum gradibus investigandis libellum prefatio,* Venice.

(1501).     *Alchindus de gradibus medicinarum compositarum,* Salamanca.

(1527).     *Divi Mesue et nova quedam ... opera ... Alchindi medicorum*
            *perspicacissimi de medicinarum compositarum gradibus investi-*
            *gandis libellus,* Venice.

(1531).     *Tacuini sanitatis Elluchasem Elimithar ... Iacobus Alkindus De*
            *rerum gradibus,* Strasbourg (with Ibn Butlān).

(1532).     *Tacuini sex rerum non naturalium ... Iacobus Alkindus De*
            *rerum gradibus,* Strasbourg (with Ibn Butlān).

(1533).     *Opera divi Ioannis Mesue . . . Alchindi medicorum perspicacissimi De medicinarum compositarum gradibus investigandis libellus,* Lyon.

(1535).     *Opera divi Ioannis Mesue . . . Alchindi medicorum perspicacissimi De medicinarum compositarum gradibus investigandis libellus,* Lyon.

(1538).     Venice (Mesue).

(1541).     *Opera divi Ioannis Mesue . . . Alchindi medicorum perspicacissimi . . . libellus,* Lyon.

(1549).     *Mesue et omnia quae cum eo imprimi consueverunt . . . Alchindus De . . . gradibus,* Venice.

(1558).     *Mesue qui Graecorum et Arabum postremus medicinam practicam illustravit . . . Alchindus De . . . gradibus,* Venice.

(1561/1562).     *Mesuae Graecorum ac Arabum clarissimi medici opera . . . Alchindus De . . . gradibus,* Venice.

(1570).     *Mesuae medici clarissimi opera . . . Alchindus De . . . gradibus,* Venice.

(1575).     Venice.

(1579).     *De dosibus . . . opuscula illustrium medicorum . . . Alchindi De gradibus medicinarum,* Venice.

(1581).     *Ioannis Mesuae medici clarissimi opera . . . ,* Venice.

(1583).     Basel.

(1584).     *Opuscula illustrium medicorum de dosibus . . . Alchindi De gradibus medicinarum,* s.l. (Heidelberg).

(1589).     *Ioannis Mesuae Damasceni medici clarissimi opera . . . ,* Venice.

(1602).     *Ioannis Mesuae Damasceni medici clarissimi opera . . . Alchindus de investigandis compositarum medicinarum gradibus,* Venice.

(1623).     *Ioannis Mesuae Damasceni medici clarissimi opera . . . Alchindus de investigandis compositarum medicinarum gradibus,* Venice.

### De mutatione temporum

(1507).     *Astrorum iudices Alkindus, Gaphar De pluviis imbribus et ventis ac aeris mutatione,* Venice.

(1540).     *Alkindus De temporum mutationibus sive de imbribus nunquam antea excussus, nunc vero per D. Io. Hieronymum a Scalingiis emissus,* Paris.

## Iudicia

(1509).     *Liber novem iudicum in iudiciis astrorum. Clarissimi auctores Mesehella, Aomar, Alkindus, Zael, Albenait, Dorotheus, Jergis, Aristoteles, Ptholemeus istius voluminis,* Venice.

(1571).     *Albohazen Haly filii Abenragel ... de iudiciis astrorum libri octo ... Compendium duodecim domorum coelestium ex clarissimis et vetustissimis authoribus, scilicet Messahalla, Aomare, Alkindo, Zaele, Albenait, Dorotheo, Iergi, Aristotele et Ptolemaeo collectum,* Basel: Iudicia.

## ALPETRAGIUS

Nothing is known about the astronomer Alpetragius (Nūr ad-Dīn al-Biṭrūǧī) apart from what can be deduced from his *Kitāb fi l-hay'a* ("On Astronomy"): Alpetragius refers to his teacher Ibn Ṭufayl who had recently died; this places the work not much after Ibn Ṭufayl's death in 1185 AD. The name 'al-Biṭrūǧī' indicates that his family came from Pedroches (Biṭrawš) near Cordoba.[116] In 1217 at Toledo, Michael Scot completed his Latin translation of the text; it became known under various titles, of which *De motibus celorum* was the most common.[117] Alpetragius has acquired fame for his courageous, but astronomically problematic attack against the basic principles of Ptolemaic astronomy; he aimed at excluding eccentric circles and epicycles from his theory in order to make it more Aristotelian—an intention he shared with his Andalusian contemporary Averroes.[118]

Michael Scot's translation was not printed in the Renaissance. Instead, there appeared in 1531 in Venice a new translation made from the Hebrew by Calo Calonymos ben David of Naples (d. after 1526),[119] who had produced two translations of works by Averroes before (see below s.v. 'Averroes'). Calo Calonymos's translation is closer to the Arabic than Michael Scot's: as is obvious from the *apparatus criticus* of the Latin edition, the Renaissance version shares with the Arabic manuscripts many phrases and sentences that are omitted in Michael's text.[120] Alpetragius's critical attitude toward Ptolemy was also known to Renaissance reformers of Ptolemy's astronomy, such as Giovan Battista Amico, who quotes Alpetragius from the 1531 edition.[121]

(1531).     *Spherae tractatus Ioannis de Sacrobusto ... Alpetragii arabi theorica planetarum nuperrime latinis mandata literis a Calo Calonymos hebreo neapolitano ubi nititur salvare apparentias in motibus planetarum absque eccentricis et epicyclis,* Venice.

## AVENZOAR

Avenzoar (Abū Marwān ibn Zuhr) was the most prominent descendant of a family of physicians in Andalusia. He was personal physician of the Almohad Caliph ʿAbd al-Muʾmin and was appointed vizier; he died in 1162 in Sevilla. Ibn Zuhr's most important work was the *Taysīr fi l-mudāwāt wa-t-tadbīr* ("Facilitation for Therapy and Diatetics").[122] Averroes, his much younger contemporary and friend, praises the book at the end of his *Kullīyāt (Colliget).*[123] The *Taysīr,* a manual mainly on pathology, was translated at least twice into Hebrew and later from Hebrew into Latin as *Liber theisir,* probably by John of Capua around 1280.[124] Not much later, in 1299, Profacius Judeus translated Ibn Zuhr's regimen *(al-Aġḏiya)* from Hebrew into Latin under the title *De regimine sanitatis.*[125]

In the Renaissance, Avenzoar's *Theisir* was printed several times, but apparently no attempts were made to improve upon the medieval version of the text. This was different with his *De regimine:* in 1618, there appeared an edition of *Abohaly Abenzoar De regimine sanitatis liber* which was prepared by Johann Georg Schenck of Grafenberg (d. ca. 1620).[126] The title page claims that Schenck "translated" the work. A comparison of the beginning of the translations suggests that Schenck merely revised the work stylistically.

A short tract *De cura lapidis* attributed to "Alguazir abuale zor filius abmeleth filii zor" appeared in a number of prints around 1500 that collect brief medical works, in particular aphorisms, by Greek and Arabic authors. This is apparently a work by Avenzoar's father, likewise a distinguished physician: Abū l-ʿAlāʾ Zuhr ibn ʿAbd al-Malik (d. 1131); the Arabic original has not yet been found.[127]

### Liber theisir

| | |
|---|---|
| (1490/91). | *Abumeron Avenzohar . . . In nomine domini amen. Incipit liber Theicrisi,* Venice. |
| (1496). | Venice. |
| (1497). | *Abhumeron Abynzoar, Colliget Averrois . . . Incipit liber Theisir,* Venice. |
| (1514). | *Abhomeron Abynzohar, Colliget Averroys . . . Incipit liber Theicir,* Venice. |
| (1530). | *Colliget Averrois: Habes in hoc volumine studiose lector gloriosi illius senis Abhomeron Abinzoar librum Theysir,* Venice. |

(1531).     *Abhomeron: geminum de medica facultate opus ... alterum Ab-homeron Abynzohar, Colliget Averroys reliquum,* Lyon.

(1542).     *Colliget Averrois ... Theizir Abynzoar morbos omnes ... conti-nens,* Venice.

(1549).     *Colliget Averrois ... Theizir Abynzoar morbos omnes ... conti-nens,* Venice.

(1553).     *Averrois Cordubensis Colliget libri septem ... Abimeron Abynzo-ahar,* Venice.

(1576).     *De febribus opus sane aureum ... Arabes: Avicenna, Rasis, Abim-eron Avenzoar, Averrois, Isaac, Serapion, Haly Abatis, Actuarius,* Venice (excerpt on fevers).

### De cura lapidis

(1497).     *Contenta in hoc volumine: liber Rasis ad Almansorem ... libellus Zoar De cura lapidis,* Venice.

(1500).     *Hoc in volumine hec continentur: Aphorismi Rabi Moysi ... liber Zoar De cura lapidis,* Venice.

(1500).     *Contenta in hoc volumine: liber Rasis ad Almansorem ... libellus Zoar De cura lapidis,* Venice.

(1508).     *Contenta in hoc volumine: liber Rasis ad Almansorem ... libellus Zoar De cura lapidis,* Venice.

### De regimine sanitatis

(1618).     *Abohaly Abenzoar Arabis medici praestantissimi et vetustissimi De regimine sanitatis liber ... annis amplius quadringentis de-sideratus nunc vero recens e Schenckiana bibliotheca medica erutus, translatus et publicatus opera Joannis Georgii Schenckii a Grafenberg,* Basel.

## AVERROES

Averroes (Abū l-Walīd Muḥammad ibn Rušd), born in Cordoba in 1126 AD, was an author with many faces: Muslim jurist and theologian, physician and Aristotelian philosopher. The Latin West knew him as a commentator on Aristotle and as a physician, but his juridical work and his treatises on the relation between philosophy and religion remained unknown.[128] The medieval Hebrew world had an image of Averroes similar to that of the scholastics, but arrived at different conclusions on his doctrines; the reason was that a different set of commentaries had been translated into Hebrew. In the Islamic world, Averroes was rarely read for many centuries; the revival of interest in Averroes in the late nineteenth century was based mainly on the theological-philosophical treatises, and not on the commentaries.[129]

Averroes's grandfather and father had been Qadis and jurists, and he himself was educated in the same profession and became Qadi first in Seville, later in Cordoba. He was held in high esteem by the Almohad ruler Abū Yaʿqūb Yūsuf, who appointed him personal physician in 1182. With the next ruler, Yaʿqūb al-Manṣūr, Averroes fell into disgrace, was subjected to an inquisitorial prosecution in 1196–1197 and exiled to Lucena, close to Cordoba, for a short period. He was restored to grace and died at the court of the sultan in Marrakesh in 1198.[130]

Not much more than twenty years after Averroes's death, Western scholars in Sicily and Italy began to translate Averroes's commentaries on Aristotle from Arabic into Latin. Among these translators, Michael Scot, who had started his career as translator in Toledo,[131] is the most famous. As has recently been shown,[132] Michael Scot can be credited with the translation of six and probably even seven commentaries by Averroes (Comm. mag. Phys., Comm. mag. Cael., Comm. mag. An., Comm. mag. Metaph., Comm. med. Gen., Comp. Parv. nat., and, probably, Comm. med. De animalibus), while five translations stem from William of Luna at Naples (Comm. med. Isag. of Porphyry, Comm. med. Cat., Comm. med. Int., Comm. med. An. pr. and Comm. med. An. post.), and another three from Hermannus Alemannus in Toledo in 1240 and 1256 (Comm. med. Poet., Comm. med. Rhet. fragm., Comm. med. Eth. Nic.). One commentary translation is anonymous (Comm. med. Meteor. 4), one probably derives from Theodore of Antioch (Comm. mag. Phys. Prooem.).

In the 1480s, a new wave of Latin translations of Averroes's commentaries began. This translation movement has been discussed in Chapter 3 above; it counts among the great accomplishments of Renaissance philology. In contrast to the medieval translations of Averroes's works, all the new translations of Averroes were made from the Hebrew, and some of the translators were Jews: del Medigo, Balmes, Mantino, and Calo Calonymos; Burana was a Christian, and Ricci a convert from Judaism. Several commentaries were translated two or even three times. The translators attempted not only to make new commentaries accessible but also to improve upon the medieval Latin versions of the commentaries and upon the work of their colleagues. The following table lists the translators of Averroes's commentaries in chronological sequence:

| Elia del Medigo (d. 1493) | Comp. Meteor. + Comm. med. Meteor. (fragm.)<br>Comm. mag. Metaph. Prooem 12 (two times)<br>Quaest. in An. pr.<br>Comp. An. (fragm.) (Ms. Vat. lat. 4549)[133]<br>Comm. med. Animal. (Ms. Vat. lat. 4549)[134] |
|---|---|
| Anonymous | Comm. med. An. (Ms. Vat. lat. 4551) |
| Giovanni Burana<br>(d. before 1523) | Comp. An. pr.<br>Comm. med. An. pr.<br>Comm. med. An. post.<br>Comm. mag. An. post. |
| Abraham de Balmes<br>(d. 1523) | Comp. Org.<br>Quaesita logica (Int., An. pr., An. post.)<br>Comm. mag. An. post.<br>Comm. med. Top.<br>Comm. med. Soph. El.<br>Comm. med. Rhet.<br>Comm. med. Poet.<br>Comp. Gen.<br>Comp. An.<br>Comp. Parv. nat.<br>Comm. med. Phys. (Ms. Vat. lat. 4548)[135] |
| Paolo Ricci (d. 1541) | Comm. med. Cael.<br>Comm. mag. Metaph. Prooem. 12 |
| Vitalis Nisso (d. ?) | Comp. Gen. |
| Jacob Mantino (d. 1549) | Comm. med. Animal.<br>Comp. Metaph.<br>Comm. med. Isag.<br>Comm. med. Cat.<br>Comm. med. Int.<br>Comm. med. Top. 1–4<br>Comm. med. Poet.<br>Comm. med. Phys.<br>Comm. mag. Phys. Prooem.<br>Comm. mag. An. 3.5+36<br>[retroversion, not from an Arabic-Hebrew, but from a Latin–Hebrew version][136]<br>Comm. mag. An. post. (fragm.) |

Apart from Averroes's commentaries on Aristotle (and Porphyry's *Isagoge*), the Latin Middle Ages also had access to a number of independent philosophical treatises, all of which are colored by Averroes's Aristotelianism: *De substantia orbis,* translated in the early thirteenth century by Michael Scot; *Destructio destructionum (Tahāfut at-tahāfut),* a comprehensive reply to Algazel's theological critique of the philosophical tradition, translated from Arabic into Latin by Calonymus ben Calonymus in Arles in

1328;[137] *De separatione primi principii,* translated from the Arabic by Alfonso Dinis (d. 1352) and Abner de Burgos (d. ca. 1347) in Valladolid;[138] *De animae beatitudine,* perhaps by the same two translators. This corpus of philosophical treatises available in Latin was further enlarged in the Renaissance, as the following table of Hebrew–Latin translations shows:

| | |
|---|---|
| Elia del Medigo (d. 1493) | Epitome of Plato's Republic[139]<br>Tractatus de intellectu speculativo (Ms. Vat. lat. 4549)[140]<br>De substantia orbis (lemmata, with his own commentary) (Ms. Vat. lat. 4553) |
| Abraham de Balmes (d. 1523) | Quaesita naturalia, De substantia orbis 6–7 (Ms. Vat. ottob. lat. 2060)[141]<br>Liber modorum rationis de opinionibus legis [Kašf ʿan manāhiǧ] (Ms. Vat. ottob. lat. 2060, Ms. Milan Ambros. G. 290)[142] |
| Calo Calonymos ben David (d. after 1526) | Destructio destructionum[143]<br>Epistola de connexione intellectus abstracti cum homine[144]<br>Libellus de viis probationum in opinionibus legis, 1527 [Kašf ʿan manāhiǧ, quotations in Calo's De mundi creatione]<br>Libellus de differentia inter legem et scientiam in copulatione [Faṣl al-maqāl, quotations in Calo's De mundi creatione][145]<br>De qualitate esse mundi [one of the Physical Questions] (Ms. Vat. ottob. lat. 2016)[146] |
| Jacob Mantino (d. 1549) | Epitome of Plato's Republic |

The *Epitome of Plato's Republic,* a collection of those statements of Plato's political theory that accord with the principles of Aristotelian demonstrative science, was translated twice in the Renaissance. Del Medigo's translation, however, did not reach print, while Mantino's seems to have found readers, since it was printed four times. The new Hebrew–Latin translation of the *Destructio destructionum* covered the entire twenty disputations and thus supplanted the medieval translation that had comprised only fourteen.[147]

The Renaissance reception of Averroes is characterized by the fact that Averroes was read not only as a commentator but also as a philosopher in his own right, who was esteemed by scholars as well as by the aristocratic patrons of new Averroes translations (see Chapters 3 and 5 above). Several Renaissance works facilitated reading Averroes's commentaries: In his commentaries on Aristotle's *De anima, Physics, De caelo,* and *Metaphysics,* Agostino Nifo often inserts super-commentaries on Averroes's text.[148] Nifo also wrote commentaries on Averroes's *Destructio destructionum, De be-*

*atitudine animae,* and *De substantia orbis.* Several sixteenth-century commentaries on Averroes's *De substantia orbis* exist in addition to Nifo's: by Pietro Pomponazzi, Giovanni Battista Confalonieri, Giovanni Francesco Beati, Mainetto Mainetti, and Nicolò Vito di Gozze.[149] Marcantonio Zimara composed two often-printed reading tools, a commentary on the differences between Aristotle and Averroes in 1508 and an alphabetical lexicon on Aristotle and Averroes in 1537.[150] As a result of the ernomous efforts of printers and editors, Averroes's commentaries, in medieval and Renaissance translations, were readily available in many printed editions. The printing history of Averroes, as described above in Chapter 1, reached its highpoint with the monumental 1550/1552 Giunta edition of the collected works of Aristotle and Averroes, which facilitates the study of Averroes until today.[151]

As was said above, Averroes's theological-philosophical treatises,[152] as well as his juridical work, were not translated into Latin, nor were his many commentaries on Galen. His main medical opus, however, the *Kitāb al-Kullīyāt fī t-tibb* ("Book on the Generalities of Medicine"), was available: it was translated in 1285 by the Italian Jew Bonacosa, and became known under the title *Colliget.* This text encompasses the entire field of theoretical medicine; it defends Aristotle's physiological doctrines against Galen.[153] In the Renaissance, two "translations" appeared, of which the first, by Jean Bruyerin Champier, is in fact a reworking of Bonacosa's text; it covers three of the original seven books (2, 6, and 7). The second translation, by Jacob Mantino, is only of chapters 57–59 of the fifth book; Mantino adopts many phrases and sentences from Bonacosa, but corrects the older translation against a Hebrew manuscript (for an analysis of the two Renaissance versions see Chapter 3 above). Commentaries exist, in manuscript, on Averroes's *Colliget* by the Ferrara professor Pietro Mainardi (of 1500) and by the Paduan professor Matteo Corti (of 1527).[154]

A particularly popular work in the Arabic world was Averroes's commentary on Avicenna's medical poem *Cantica* (*al-Urǧūza fī t-tibb,* "Poem on Medicine"); the commentary was translated together with Avicenna's text by Armengaud of Blaise around 1300. In 1527 Andrea Alpago published corrections of the *Cantica* (see below s.v. 'Avicenna'), which do not, however, extend to Averroes's commentary.

Three short medical treatises by Averroes, which had already circulated in late medieval manuscripts, were first printed around 1497–1500 in Bologna: Averroes's *Maqāla fī t-Tiryāq* (*De tyriaca* or *De theriaca*); his treatise 'On Poisons' (*De venenis*), which has not been identified in the Arabic; and *De concordia inter Aristotilem et Galienum de generatione sanguinis,* which is an alternative title for Averroes's middle commentary on *De partibus animalium.*

The editor Baptista de Avolio, professor of philosophy and the arts in Bologna, dedicated the Bologna edition to Cardinal Grimani, who is praised for his knowledge of the liberal arts and for his admiration of Averroes, the best interpreter of Aristotle. The theriac, an antidote against poison, was of particular interest to sixteenth-century pharmacology since it was thought to be the most prestigious of all composite drugs and since it was of classical origin.[155] There is no indication that Andrea Alpago was involved in the correction of *De tyriaca*.[156] A fourth medical text became available only in the Renaissance: Elia del Medigo produced a translation of the *Libellus de spermate vel de humano semine,* which was printed in 1560.[157]

In the list of editions below, the first printing of a new translation is underlined and its translator is named in brackets.

### Medical Works: Colliget, Commentary on Avicenna's Cantica, De theriaca

(1482).           *Incipit liber de medicina Averoys qui dicitur Coliget,* Venice (in fact: Ferrara): Colliget.

(1483).           Venice: Comm. Cant.

(1484).           *Incipit translatio Canticorum Avicenne cum commento Averoys,* Venice: Comm. Cant.

(1490–1495).     *Cantica Avicenne cum comento Averois,* Venice: Comm. Cant. (with Avicenna's *Canon*).

(1490/1491).     *Abumeron Avenzohar . . . Incipit liber de medicina Averrois qui dicitur Colliget,* Venice: Colliget.

(1496).           Venice: Colliget.

(1497).           *Abhumeron Abynzoar, Colliget Averrois,* Venice: Colliget.

(ca. 1497–1500). *Index Averrois librorum in hoc opusculo contentorum. Primus de venenis. Secundus de tyriaca. Tertius de concordia inter Aristotilem et Galienum de generatione sanguinis,* s.l. (Bologna): De venenis, De tyriaca, De concordia.

(ca. 1502).       s.l. (Lyon): De venenis, De tyriaca (with Regimen santiatis).

(1506).           Paris: De venenis, De tyriaca (with Regimen santiatis).

(1514).           *Abhomeron Abynzohar, Colliget Averroys,* Venice: Colliget.

(1517).           *Regimen sanitatis Magnini . . . Item Averrois De venenis,* Lyon: De venenis, De tyriaca.

(1520/1522).     *Primus Avicenne Canonis,* Venice: Comm. Cant. (with Avicenna's *Canon*).

(1523).           Venice: Comm. Cant. (with Avicenna's *Canon*).

(1524).        *Regimen sanitatis Magnimi . . . Item Averrois De venenis,* Paris: De venenis, De tyriaca.

(1530).        *Colliget Averrois . . . Habes et Averrois librum Colliget,* Venice: Colliget.

(1531).        *Abhomeron: geminum de medica facultate opus . . . alterum Abhomeron Abynzohar, Colliget Averroys reliquum,* Lyon: Colliget.

(1537).        *Collectaneorum de re medica Averrhoi . . . sectiones tres . . . a Ioanne Bruyerino Campegio . . . nunc primum latinitate donatae,* Lyon: Colliget (Bruyerin).

(1542).        *Colliget Averrois totam medicinam . . . complectens,* Venice: Colliget.

(1549).        *Colliget Averrois totam medicinam . . . complectens,* Venice: Colliget.

(1550/1552).   *Aristotelis . . . omnia . . . opera . . . Averrois . . . commentarii aliique ipsius in logica, philosophia et medicina libri,* Venice: Colliget (Bruyerin/Mantino), Comm. Cant., De theriaca.

(1553).        *Averrois Cordubensis Colliget libri septem cum quibus . . . impressimus translationem . . . a Iacob Mantino . . . factam. Addidimus . . . sectiones Collectaneorum . . . a Iohanne Bruyerino Campegio . . . latinitate donatas,* Venice: Colliget, Comm. Cant., De theriaca.

(1560).        *Aristotelis . . . omnia . . . opera . . . Tomus nonus . . . Colliget et Avicennae Cantica cum eiusdem Averrois expositione,* Venice: Colliget, Comm. Cant., De theriaca, De spermate (del Medigo).

(1562).        *Aristotelis omnia . . . opera . . . Decimum Volumen. Averrois Cordubensis Colliget libri septem,* Venice, repr. Frankfurt am Main, 1962: Colliget, Comm. Cant., De theriaca.

(1574?/1575).  *Aristotelis omnia quae extant opera,* Venice: Colliget, Comm. Cant., De theriaca.

(1576).        *De febribus opus sane aureum . . . Arabes: Avicenna, Rasis, Abimeron Avenzoar, Averrois, Isaac, Serapion, Haly Abatis, Actuarius,* Venice (excerpt on fevers).

## Commentaries on Aristotle

(1472–1474).   Padua: Comm. mag. Phys. + Prooem., Comm. mag. Cael., Comm. med. Gen., Comm. med. Meteor. IV, Comm. mag. An., Comp. Parv. nat., Comm. mag. Metaph.

(1481).   Venice: Comm. med. Poet.

(1483).   Venice: Comm. med. Isag., Comm. med. Cat., Comm. med. Int., Comm. med. An. pr., Comm. med. An. post., Comm. mag. Phys. + Prooem., Comm. mag. Cael., Comm. med. Gen., Comm. med. Meteor. 4, Comm. mag. An., Comp. Parv. nat., Comm. mag. Metaph., Comm. med. Eth. Nic.

(1483).   Venice: Comm. med. Isag., Comm. med. Cat., Comm. med. Int., Comm. med. An. pr., Comm. med. An. post.

(1488).   Venice: Comp. Meteor. + Comm. med. Meteor. (fragm.), Comm. mag. Metaph. Prooem. 12 (del Medigo).

(1489).   Venice: Comm. med. Isag., Comm. med. Cat., Comm. med. Int., Comm. med. An. pr., Comm. med. An. post., Comm. mag. Phys. + Prooem., Comm. mag. Cael., Comm. med. Gen., Comm. med. Meteor. 4, Comp. Meteor. + Comm. med. Meteor. (fragm.) (del Medigo), Comm. mag. An., Comp. Parv. nat., Comm. mag. Metaph., Comm. med. Eth. Nic.

(1495–1496).   Venice: Comm. med. Isag., Comm. med. Cat., Comm. med. Int., Comm. med. An. pr., Comm. med. An. post., Comm. mag. Phys. + Prooem., Comm. mag. Cael., Comm. med. Gen., Comm. med. Meteor. 4, Comp. Meteor. + Comm. med. Meteor. (fragm.) (del Medigo), Comm. mag. An., Comp. Parv. nat., Comm. mag. Metaph., Comm. med. Eth. Nic.

(1497).   Venice: Quaest. in An. pr. (del Medigo) (with Laurentius Maiolus).

(1501).   Venice (107.693):[158] Comm. med. Isag., Comm. med. Cat., Comm. med. Int., Comm. med. An. pr., Comm. med. An. post., Comm. mag. Phys. + Prooem., Comm. mag. Cael., Comm. med. Gen., Comp. Meteor. + Comm. med. Meteor. (fragm.) (del Medigo), Comm. med. Meteor. 4, Comm. mag. An., Comp. Parv. nat., Comm. mag. Metaph., Comm. med. Eth. Nic.

(1507/1508).   Venice (107.753): Comm. med. Isag., Comm. med. Int., Comm. med. An. pr., Comm. med. An. post., Comm. mag. Phys., Comm. mag. Cael., Comm. med. Gen., Comp. Meteor. + Comm. med. Meteor. (fragm.) (del Medigo), Comm. med. Meteor. 4, Comm. mag. An., Comp. Parv. nat., Comm. mag. Metaph., Comm. med. Eth. Nic.

(1511).      *Hoc opere contenta: De prooemio ... Averois in Phisico auditu prooemium emendatum, ... Averois in quattuor De celo et mundo libros paraphrasis ... de hebraicis latebris in latinum splendorem conversa, Averois in duodecimo Metaphisice prooemium quoque de hebraico decerptum exe<m>plari,* Milan: <u>Comm. med. Cael.</u> (Ricci), <u>Comm. mag. Metaph. Prooem. 12</u> (Ricci), Comm. mag. Phys. Prooem.

(1515).      *Rhetorica Aristotelis cum ... Egidii de Roma ... commentariis ... Addita eiusdem Aristotelis Poetica cum magni Averroys in eandem summa,* Venice (107.828): Comm. med. Rhet. (fragm.), Comm. med. Poet.

(1516).      Venice (107.838): Comm. med. Isag., Comm. med. Cat., Comm. med. Int., Comm. med. An. pr., Comm. med. An. post., Comm. mag. Phys. + Prooem., Comm. mag. Cael., Comm. med. Gen., Comp. Meteor. + Comm. med. Meteor. (fragm.) (del Medigo), Comm. med. Meteor. 4, Comm. mag. An., Comp. Parv. nat., Comm. mag. Metaph., Comm. med. Eth. Nic.

(1520).      *Physica Aristotelis cum comment<ariis> Averrois,* Lyon (107.866): Comm. mag. Phys. + Prooem.

(1520/1521).      *Libri physicorum octo cum ... epitomatis ... Averroeque eius exactissimo interprete ... ,* Lyon/Pavia (107.870): Comm. med. Isag., Comm. med. Int., Comm. med. An. pr., Comm. med. An. post., Comm. mag. Phys. + Prooem., Comm. mag. Cael., Comm. med. Gen., Comp. Meteor. + Comm. med. Meteor. (fragm.) (del Medigo), Comm. med. Meteor. 4, Comm. mag. An., Comp. Parv. nat., Comm. med. Eth. Nic.

(1520).      *Aristotelis Stagyritae libri quatuor De coelo ... Averroe ... interprete,* Pavia (107.870A): Comm. mag. Cael., Comm. med. Gen.

(ca.1520).      *Accipe lector studiose Aristotelem ... ac eius fidelissimum interpretem Averroem ... ,* Venice (107.875): Comm. mag. Phys. + Prooem., Comm. mag. Cael., Comm. med. Gen.

(1521).      Rome: <u>Comm. med. Anim.</u> (Mantino).

(1521).      *Habes hoc in enchiridion ... omnia quae ad Aristotelis logicen pertinent opera ... ,* Pavia (107.876B): Comm. med. Isag., Comm. med. Cat., Comm. med. Int., Comm. med. An. pr., Comm. med. An. post.

(1521).        *Aristotelis Stagyritae Ethicorum libri decem cum Averrois Cor-
               dubensis exactissimis commentariis,* Pavia (107.876C): Comm.
               med. Eth. Nic.

(1521).        *Aristotelis Stagyritae libri Metaphysicorum duodecim . . . Averro-
               eque eius fidelissimo interprete,* Pavia (107.876D): Comm. mag.
               Metaph.

(1521).        Rome: Comp. Metaph. (Mantino).

(1522).        Venice: Quaest. in An. pr. (Medigo) (with Egidius).

(1523).        Bologna: Comp. Metaph. (Mantino).

(1523).        *Index illorum que in hoc volumine continentur: libri Posteriorum
               analiticorum cum magnis commentariis Averroys, libri Thopi-
               corum paraphrases Averroys . . . ,* Venice (107.887): Comp.
               Org., Quaesita logica, Comm. mag. An. post., Comm. med.
               Top., Comm. med. Soph. El., Comm. med. Rhet., Comm.
               med. Poet. (Balmes).

(1524).        *Libri Physicorum octo . . . Averroeque eius interprete,* Lyon
               (107.889): Comm. mag. Phys. + Prooem.

(1524).        Venice: Comp. An. pr., Comm. med. An. pr. (Burana).

(1529/"1520"). *Physica Aristotelis cum com<mentariis> Averrois,* Lyon
               (107.906): Comm. mag. Phys. + Prooem.

(1529).        *Aristotelis De celo et mundo cum com<mentariis> Averrois,* Lyon
               (107.909): Comm. mag. Cael., Comm. med. Gen.

(1529).        *Metaphysic<a> Aristotelis cum commentariis Averrois,* Lyon
               (107.910): Comm. mag. Metaph.

(1530).        *Ethica et Politica Aristotelis cum commentariis Averrois,* Lyon
               (107.915): Comm. med. Eth. Nic.

(1530).        *Aristotelis Stagyritae libri tres De anima . . . eiusdemque Parva
               naturalia cum Averroi Cordubensi . . . interprete,* Lyon (107.916):
               Comm. mag. An., Comp. Parv. nat.

(1530).        *Libri Meteororum Aristotelis cum com<mentariis> Averrois,*
               Lyon (107.920): Comp. Meteor. + Comm. med. Meteor.
               (fragm.) (del Medigo), Comm. med. Meteor. 4.

(1530).        *Logica Aristotelis cum com<mentariis> Averrois,* Lyon (107.921):
               Comm. med. Isag., Comm. med. Cat., Comm. med. Int.,
               Comm. med. An. pr., Comm. med. An. post.

(1539).        Venice: Comm. med. Isag., Comm. med. Cat., Comm. med.
               Int., Comm. med. An. pr., Comm. med. An. post.

(1540).  *Aristotelis libri tres De anima . . . , digressiones omnes Averrois ac notatu digna in eosdem,* Venice (107.999): Comm. mag. An. (digress.)

(1540).  *Aristotelis Stagiritae De physico auditu libri octo . . . Averrois Cordubensis digressiones omnes in eosdem,* Venice (108.001): Comm. mag. Phys. (digress.)

(1541).  *Aristotelis . . . De coelo libri . . . Averrois digressiones in eosdem,* Venice (108.030): Comm. mag. Cael. (digress.), Comm. med. Gen. (digress.), Comp. Meteor. + Comm. med. Meteor. (fragm.) (del Medigo), Comm. med. Meteor. 4.

(1541).  *Aristotelis . . . Metaphysicorum libri quatuordecim . . . Averrois . . . digressiones omnes in eosdem,* Venice (108.031): Comm. mag. Metaph. (digress.)

(1542).  *Ethicorum libri decem cum Averrois Cordubensis exactissimis commentariis,* Lyon (108.037): Comm. med. Eth. Nic.

(1542).  *Aristotelis Stagyritae libri tres De anima . . . eiusdemque Parva naturalia cum Averroi Cordubensi . . . interprete,* Lyon (108.038): Comm. mag. An., Comp. Parv. nat.

(1542).  *Aristotelis Stagyritae libri quatuor De coelo . . . cum epitomis . . . ac Averroe . . . interprete,* Lyon (108.039): Comm. mag. Cael., Comm. med. Gen.

(1542).  *Libri Metaphysicorum duodecim cum . . . epitomatis . . . Averroeque . . . interprete,* Lyon (108.040): Comm. mag. Metaph.

(1542).  *Meteororum libri quatuor cum Averrois Cordubensis exactissimis commentariis,* Lyon (108.041): Comp. Meteor. + Comm. med. Meteor. (fragm.) (del Medigo), Comm. med. Meteor. 4.

(1542).  *Aristotelis Stagyritae libri Physicorum octo cum . . . epitomatis . . . Averroeque eius exactiss&lt;imo&gt; interprete,* Lyon (108.042): Comm. mag. Phys. + Prooem.

(1542).  Lyon: Comm. med. Isag., Comm. med. Cat., Comm. med. Int., Comm. med. An. pr., Comm. med. An. post.

<u>(1542).</u>  *Averrois Cordubensis Epithoma totius Metaphysices Aristotelis. Prohemium duodecimi libri Metaphysices. Eiusdem Paraphrases in libris quatuor De coelo et duobus De generatione et corruptione Aristotelis,* Venice: Comp. Metaph. (Mantino), Comm. mag. Metaph. Prooem. 12 (Ricci), Comm. med. Cael. (Ricci), <u>Comp. Gen.</u> (Nisso)

(1543).        *Libri Physicorum octo cum ... epitomatis ... Averroeque eius exactiss<imo> interprete,* Lyon (108.066): Comm. mag. Phys. + Prooem.

(1547).        *Aristotelis Stagiritae Metaphysicorum libri quatuordecim ... Averrois Cordubensis digressiones omnes in eosdem,* Lyon (108.128): Comm. mag. Metaph. (digress.).

(1550/1552).   *Aristotelis Stagiritae omnia quae extant opera ... Averrois Cordubensis in ea opera omnes qui ad nos pervenere commentarii aliique ipsius in logica, philosophia et medicina libri,* Venice (108.193): Comp. Org. (Balmes), Quaesita logica (Balmes), Comm. med. Isag. (Mantino), Comm. med. Cat. (Mantino), Comm. med. Int. (Mantino), Comm. med. An. pr. (Burana), Comm. med. An. post. (Burana), Comm. mag. An. post. (Burana, fused with Balmes),[159] Comm. med. Top. (Balmes), Comm. med. Top. 1–4 (Mantino), Comm. med. Soph. El. (Balmes), Comm. med. Rhet. (Balmes), Comm. med. Poet. (Mantino), Comm. med. Phys. (Mantino), Comm. mag. Phys., Comm. mag. Phys. Prooem. (ant. transl./Mantino), Comm. med. Cael. (Ricci), Comm. mag. Cael., Comp. Gen. (Nisso), Comm. med. Gen., Comp. Meteor. + Comm. med. Meteor. (fragm.) (del Medigo), Comm. med. Meteor. 4, Comm. med. Anim. (Mantino), Comm. mag. An., Comm. mag. An. 3.5+36 (Mantino), Comp. Parv. nat., Comp. Metaph. (Mantino), Comm. mag. Metaph., Comm. mag. Metaph. Prooem. 12 (Ricci/Mantino), Comm. med. Eth. Nic.

(1552).        *Averoys compendium necessarium ex libris Aristotelis De generatione et corruptione, De anima, De sensu et sensato, De memoria et reminiscentia, deque somno et vigilia conversum ex arabico in latinum sermonem ab accuratissimo interprete Abraham de Balmes,* Venice: Comp. Gen., Comp. An., Comp. Parv. nat. (Balmes).

(1554).        *Aristotelis Stagiritae De physico auditu libri octo ... Averrois Cordubensis digressiones omnes in eosdem,* Venice (108.290A): Comm. mag. Phys. (digress.).

(1556).        *Aristotelis Stagiritae Metaphysicorum libri quatuordecim ... Averrois Cordubensis digressiones omnes in eosdem,* Lyon (108.319): Comm. mag. Metaph. (digress.).

(1557).        *Aristotelis Stagiritae Metaphysicorum libri quatuordecim ... Averrois Cordubensis digressiones omnes in eosdem,* Lyon (108.342): Comm. mag. Metaph. (digress.).

(1560).   *Aristotelis Stagiritae omnia quae extant opera . . . Averrois Cor-*
*dubensis in ea opera omnes qui ad nos pervenere commentarii,*
Venice (108.423): Comp. Org. (Balmes), Comp. An. pr.
(Burana), Quaesita logica (Balmes), Comm. med. Isag.,
Comm. med. Cat., Comm. med. Int., Comm. med. An. pr.
(Burana), Comm. med. An. post. (Burana), Comm. mag. An.
post. (Burana), Comm. med. Top. (Balmes), Comm. med.
Soph. El. (Balmes), Comm. med. Rhet. (Balmes), Comm.
med. Poet. (Balmes), Comm. mag. Phys. + Prooem., Comm.
med. Cael. (Ricci), Comm. mag. Cael., Comp. Gen., Comm.
med. Gen., Comp. Meteor. + Comm. med. Meteor. (fragm.)
(del Medigo), Comm. med. Meteor. 4, Comm. med. Anim.
(Mantino), Comm. mag. An., Comp. Parv. nat., Comp.
Metaph. (Mantino), "Comm. med. Metaph. 1–7" (a compilation rather than a translation by del Medigo), Comm. mag.
Metaph., Comm. mag. Metaph. Prooem. 12 (Ricci/Man-
tino), Comm. med. Eth. Nic.

(1562).   *Aristotelis omnia quae extant opera . . . Averrois Cordubensis in*
*ea opera omnes qui ad haec usque tempora pervenere commen-*
*tarii, nonnulli etiam ipisus in logica, philosophia et medicina*
*libri,* Venice, repr. Frankfurt am Main 1962 (108.456):
Comp. Org. (Balmes), Quaesita logica (Balmes), Comm.
med. Isag. (Mantino), Comm. med. Cat. (Mantino),
Comm. med. Int. (Mantino), Comm. med. An. pr.
(Burana), Comm. med. An. post. (Burana), Comm. mag.
An. post. (Balmes/Burana/fragm. Mantino), Comm.
med. Top. (Balmes/1–4 Mantino), Comm. med. Soph. El.
(Balmes), Comm. med. Rhet. (Balmes), Comm. med. Poet.
(Mantino), Comm. med. Phys. (Mantino), Comm. mag.
Phys., Comm. mag. Phys. Prooem. (ant. trans./Mantino),
Comm. med. Cael. (Ricci), Comm. mag. Cael., Comp.
Gen., Comm. med. Gen., Comp. Meteor. + Comm. med.
Meteor. (fragm.) (del Medigo), Comm. med. Meteor. 4,
Comm. med. Anim. (Mantino), Comm. mag. An., Comp.
Parv. nat., Comp. Metaph. (Mantino), Comm. mag.
Metaph., Comm. mag. Metaph. Prooem. 12 (Ricci/Man-
tino), Comm. med. Eth. Nic.

(1564).   *Aristotelis Stagiritae . . . De physico auditu libri octo . . . Aver-*
*rois . . . in eosdem digressiones,* Padua (108.479C): Comm.
mag. Phys. (digress.).

(1574).          *De anima libri tres, cum Averrois commentariis,* Venice
                 (108.595): Comm. mag. An.

(1574?/1575).    *Aristotelis omnia quae extant opera ... Averrois Cordubensis in*
                 *ea opera omnes qui ad haec usque tempora pervenere commen-*
                 *tarii ...,* Venice (108.599): exactly the same set of commen-
                 taries as in the edition of 1562 (108.456).

### Destructio destructionum

(1497).          *Eutyci Augustini Niphi Philothei Suessani in librum Destructio de-*
                 *structionum Auerroys commentationes,* Venice.

(after 1500).    *Eutyci Augustini Niphi Philothei Suessani in librum Destructio de-*
                 *structionum Auerroys commentationes,* s.l. (Venice).

(1508).          Venice (with Aristotle's moral philosophy).

(1517).          *Eutici Augustini Niphi Philothei Suessani in librum Destructio de-*
                 *structionum Averrois commentationes,* Venice.

(1521).          *Eutyci Augustini Niphi Philothei Suessani in librum Destructio de-*
                 *structionum Averrois commentarii,* Pavia.

(1527).          *Subtilissimus liber Averrois qui dicitur Destructio destructionum*
                 *philosophie Algazelis nuperrime traductus ... cui additus est*
                 *libellus seu epistola Averrois De connexione intellectus abstr-*
                 *acti cum homine ab ... Calo Calonymos hebreo Neapolitano,*
                 Venice.

(1529).          *Destructio destructionum Averrois. Eutici Augustini Niphi Phi-*
                 *lothei Suessani in librum Destructio destructionum commentarii,*
                 Lyon (versio antiqua).

(1542).          *Eutyci Augustini Niphi Philothei Suessani in librum Destructio de-*
                 *structionum Averrois commentarii,* Lyon (versio antiqua).

(1550/1552).     *Aristotelis ... omnia ... opera ... Averrois ... commentarii ali-*
                 *ique ipsius in logica, philosophia et medicina libri* (Calonymos),
                 Venice.

(1560).          *Aristotelis ... omnia ... opera ... Averrois ... commentarii*
                 (Calonymos), Venice.

(1562).          *Nonum volumen: Averrois Cordubensis ... Destructio destruc-*
                 *tionum philosophie Algazelis* (Calonymos), Venice, repr. Frank-
                 furt am Main, 1962.

(1573).          *Aristotelis omnia ... opera ... Averrois Cordubensis ... De-*
                 *structio destructionum philosophie Algazelis* (Calonymos),
                 Venice.[160]

## De substantia orbis

(ca. 1473–1474). s.l. (Padua).

(1482). *Incipit primus liber Aristotelis Stragerite . . . De auditu physicorum . . . Averois De substantia orbis,* Venice.

(1483). *De anima liber primus . . . Liber Averrois De substantia orbis,* Venice.

(1486). *Incipit expositio . . . Gaetani Thienensis super libros De anima . . . Incipit expositio super libro De substantia orbis Joannis de Gandavo,* Vicenza (with John of Jandun).

(1489). *Aristotelis Stagyrite peripateticorum principis De celo et mundo liber primus . . . Averrois Cordubensis De substantia orbis liber,* Venice.

(1493). *Gaietanus super libros de anima . . . Item De substantia orbis Joannis de Gandavo cum questionibus eiusdem,* Venice (with John of Jandun).

(1495/1496). *Aristotelis Stagirite . . . De physico auditu liber primus et Averrois Cordubensis commentaria,* Venice.

(1505). *Gaietanus super libros de anima . . . Item De substantia orbis Joannis de Gandavo cum questionibus eiusdem,* Venice (with Jean den Jandun).

(1507). Frankfurt am Main.

(1508). *Eutychi Augustini Niphi philothei Suessani commentationes in librum De substantia orbis,* Venice (with Nifo).

(1514). Venice (with John of Jandun).

(1525). *Averrois libellus De substantia orbis nuper castigatus et duobus capitulis auctus diligentique studio expositus per Joannem Baptistam Confalonerium Veronensem,* Venice (with Confalonieri).

(1529). *Aristotelis De celo et mundo cum com\<mentariis\> Averrois . . . necnon eiusdem opusculum de substantia orbis,* Lyon.

(1542?). Padua (chapter 7 with Beati).

(1542). *Aristotelis Stagyritae libri quatuor De coelo et mundo . . . Averroe fidissime interprete necnon eiusdem opusculum De substantia orbis,* Lyon.

(1546). *Augustini Niphi . . . commentationes in librum Averrois De substantia orbis,* Venice (with Nifo).

(1550/1552). *Aristotelis . . . omnia . . . opera . . . Averrois . . . commentarii aliique ipsius in logica, philosophia et medicina libri. Nonum volumen:*

Sermo de substantia orbis castigatus ac duobus capitulis auctus ab Abrahamo de Balmes hebraeo latinitate donatis (versio antiqua/<u>Balmes</u>), Venice.

(1552; 1551 in colophon). *Ioannis de Ianduno in libros Aristotelis De coelo et mundo quae extant quaestiones . . . adiecimus Averrois sermonem De substantia orbis,* Venice: Giunta (with John of Jandun).

(1552). *Ioannis de Ianduno in libros Aristotelis De coelo et mundo quaestiones . . . adiecimus Averrois sermonem De substantia orbis,* Venice: Scoto (with John of Jandun).

(1559). *Augustini Niphi Suessani philosophi clarissimi commentationes in librum Averrois De substantia orbis,* Venice (with Nifo).

(1562). *Nonum volumen: Averrois Cordubensis Sermo de substantia,* Venice, repr. Frankfurt am Main, 1962.

(1564). Venice (with John of Jandun).

(1570). Bologna (with Mainetti).

(1573). *Aristotelis omnia . . . opera . . . Averrois Cordubensis Sermo de substantia orbis,* Venice.

### Epitome of Plato's Republic

<u>(1539).</u> *Averrois Paraphrasis super libros De republica Platonis nunc primum latinitate donata Iacob Mantino medico hebraeo interprete,* Rome.

(1552). *Francisci Philelphi De morali disciplina . . . Averrois Paraphrasis in libros De republica Platonis,* Venice.

(1560). *Tertium volumen: Aristotelis Stagiritae libri moralem totam philosophiam complectentes cum Averrois Cordubensis . . . in Platonis libros De republica Paraphrasi,* Venice, repr. Frankfurt am Main, 1962.

(1578). Venice.

### De animae beatitudine, Epistola de connexione intellectus abstracti cum homine

(1501). *Aristotelis . . . Secretum secretorum . . . Averrois magni commentatoris de anime beatitudine,* Bologna: De an. beat.

(1508). *Euthici Augustini Niphi Philothei Suessani in Averroys de animae beatitudine,* Venice: De an. beat.

(1524). *Suessanus De beatitudine animae. Euthici Augustini Niphi Philothei Suessani in Averrois De animae beatitudine,* Venice: De an. beat.

(1527). *Subtilissimus liber Averois qui dicitur Destructio destructionum philosophie Algazelis nuperrime traductus . . . cui additus est libellus seu epistola Averois De connexione intellectus abstracti cum homine ab . . . Calo Calonymos hebreo Neapolitano,* Venice: Epist. de conn.

(1550/1552). *Aristotelis . . . omnia . . . opera . . . Averrois . . . commentarii aliique ipsius in logica, philosophia et medicina libri,* Venice: De an. beat., Epist. de conn.

(1560). *Aristotelis . . . omnia . . . opera . . . Averrois . . . commentarii,* Venice: De an. beat., Epist. de conn.

(1562). *Nonum volumen: Averrois Cordubensis . . . De animae beatitudine seu Epistola de intellectu,* Venice, repr. Frankfurt am Main 1962: De an. beat., Epist. de conn.

(1574?–1575). *Aristotelis omnia . . . opera . . . Averrois . . . commentarii, nonnulli etiam ipsius in logica, philosophia et medicina libri,* Venice: De an. beat., Epist. de conn.


## AVICENNA

Avicenna (Abū ʿAlī al-Ḥusayn ibn ʿAbdallāh ibn Sīnā), who was born near Buḫārā on the Silk Road and died in 1037 AD in Hamaḏān after spending the greater part of his life at the courts of local rulers in Persia, was a towering medical and philosophical authority both of the Arabic and the Latin worlds.[161] He significantly influenced Arabic philosophy for many centuries, from the eleventh to at least the sixteenth century,[162] and his *Canon medicinae (Qānūn fī ṭ-ṭibb),* a comprehensive account of medical knowledge, was the most important and influential handbook of medicine in the Orient and Occident for an equally long period of time.[163]

In the twelfth and thirteenth centuries, the *Canon* and several large parts of Avicenna's principal philosophical work *aš-Šifāʾ* ("The Cure") were translated into Latin. The book *aš-Šifāʾ* is organized in four sections: logic, natural philosophy, mathematics—this section was not translated—and metaphysics. A number of independent translators of the twelfth and thirteenth centuries were involved, all of whom worked in Spain: Avendauth, Dominicus Gundisalvi, Alfred of Shareshill, Michael Scot, Juan Gonzalves de Burgos together with a certain 'Salomon,' and Hermannus

Alemannus. Among these translations, some were very influential in scholasticism and widely distributed in Latin manuscripts: *De anima* (in the translation of Avendauth and Gundisalvi), *Metaphysics* (Gundisalvi), *De animalibus* (Michael Scot), and the meteorological *De congelatione et conglutinatione lapidum* (Alfred of Shareshill).[164] In the Renaissance, Andrea Alpago translated five philosophical treatises from Arabic into Latin (printed 1546), among them important tracts on the soul and its afterlife. However, the philosophical interest in Avicenna was less intense in the Renaissance than in the thirteenth and fourteenth centuries, as is obvious from the very few printed editions of his philosophical work. In Renaissance psychology, for example, Avicenna remained a major authority on the external senses, but he has ceased to significantly influence the discussion on the intellect.

In Renaissance medicine, in contrast, Avicenna reached the peak of his influence. The *Canon medicinae (Qānūn fī ṭ-ṭibb)*, translated by Gerard of Cremona in the twelfth century, was firmly established as the major textbook for university teaching in many countries, despite the widespread predilection for Greek authors, and it was commented on by a great number of scholars.[165] Its printing history in the Renaissance is impressive (see Chapter 1 above on editions and curricula). The *Canon* is divided into five books: on medical theory, on simple medicinal drugs, on specific ailments and their treatment, on general diseases that affect the entire body, and on composite drugs.

Several scholars of the later fifteenth and sixteenth centuries produced new Latin versions of Avicenna's *Canon* or parts of it: Girolamo Ramusio (d. 1486) and Andrea Alpago (d. 1522) translated directly from the Arabic, Jacob Mantino (d. 1549) and Jean Cinqarbres (d. 1587) from the Hebrew; Miguel Jerónimo Ledesma (d. 1547) published a humanist revision of the medieval translation without recourse to the Arabic or Hebrew, and Andrea Gratiolo (fl. ca. 1540–1580) composed a new Latin version by comparing older Latin translations. No other text of Arabic origin received an equal amount of philological attention in the Renaissance. The motives, techniques and accomplishments of these translators have been surveyed above (Chapter 3). Andrea Alpago also translated two short medical treatises by Avicenna *(De removendis nocumentis,* and *De syrupo acetoso)* and two sections from Arabic commentaries on the *Canon* by Ibn an-Nafīs and Quṭb ad-Dīn aš-Šīrāzī, all of which were first published in 1547. And he proposed substantial corrections to the medieval translations of Avicenna's *Cantica* and *De viribus cordis* (first published 1527), which were reprinted several times. Andrea Alpago clearly was the most productive and accomplished Arabic–Latin translator of the Renaissance. In the seventeenth century, new Latin translations of the *Canon* appeared, which have a context different from their sixteenth-century predecessors, since they were produced directly from the Arabic by European scholars of oriental languages: Peter Kirsten (printed 1609), Vopiscus Plemp (1658), Pierre Vattier (1659), and Georg Welsch (1674). These scholars could already work with the Arabic text of the *Qānūn* printed in 1593 by the Stamperia Orientale Medicea, the Medici press for Arabic works founded in 1584.[166]

Avicenna's medical didactic poem *Cantica (al-Urǧūza fī ṭ-ṭibb)*, translated in prose by Armengaud of Blaise in 1283 or 1284, either from Arabic or Hebrew,[167] received over twenty editions in the Renaissance,[168] which indicates that it counted among the popular medical texts of the time. This is underlined by the fact that, in addition to Alpago's corrections, two new Arabic–Latin translations were written in the first half of the seventeenth century: in elegiac couplets by Jean Faucher, physician in Beaucaire, printed posthumously by Faucher's son Guillaume in 1630,[169] and in prose by the Dutch physician, mathematician, and Orientalist Antonius Deusing, printed in 1649 with a long dedication to Jacob Golius, professor of Arabic at Leiden.[170]

The following inventory does not list the commentaries on the *Canon*, which cite Avicenna's text in full, because it would have been too difficult, in view of the sheer number of commentaries, to differentiate them from commentaries that cite only abbreviated lemmata. The reader is referred to Siraisi's impressive list of printed commentaries and lectures on Avicenna's *Canon* after about 1500, which comprises forty-three items by thirty-two different authors, and to her long list of commentaries extant only in manuscript.[171]

## Philosophical Works

(ca. 1485).   *Incipit opus egregium de anima qui Sextus naturalium Avicene dicitur*, Pavia.

(1495).   *Metaphysica Avicenne sive eius Prima philosophia*, Venice, repr. Frankfurt am Main, 1966.

(ca. 1500).   *Avicenna De animalibus per magistrum Michaelem Scotum de arabico in latinum translatus*, Venice.

(1508).   *Avicenne perhypatetici philosophi ac medicorum facile primi opera in lucem redacta*, Venice, repr. Frankfurt am Main, 1961.

(1514).   *Flores Avicenne collecti . . . super decem et novem libris De animalibus*, Lyon.

(1546).   *Avicennae philosophi praeclarissimi ac medicorum principis Compendium de anima, De mahad . . . , Aphorismi de anima,*

*De diffinitionibus et quaesitis, De divisione scientiarum,* Venice, repr. Farnborough, 1969.

## Canon and Other Medical Works[172] (De viribus cordis sive De medicinis cordialibus, Cantica, De removendis nocumentis, De syrupo acetoso).

| | |
|---|---|
| (1472). | s.l. (Italy): Canon 3. |
| (1473). | *Liber canonis,* Milan: Canon. |
| (1473). | *Liber canonis,* s.l. (Strasbourg): Canon. |
| (1476). | *Liber canonis,* Padua: Canon, De med. cord. |
| (1479). | *Liber canonis,* Padua: Canon, De med. cord. |
| (1482). | Bologna: Canon 4.1, 3–5. |
| (1482/1483). | Pavia: Canon, De med. cord. |
| (1482/1483). | Venice: Canon, De med. cord. |
| (1483). | Venice: Cantica. |
| (1484). | *Incipit translatio Canticorum Avicenne,* Venice: Cantica. |
| (1486). | *Liber canonis,* Venice: Canon, De med. cord. |
| (1489/1490). | *Liber canonis,* Venice: Canon, De med. cord. |
| (1490). | *Liber canonis,* Venice: Canon, De med. cord. |
| (1490/1495). | *Gentilis Fulginatis expositiones cum textu Avicenne,* Venice: Canon, De med. cord., Cantica. |
| (1498). | *Primus canonis Avicenne principis cum explanatione Jacobi de partibus,* Lyon: Canon. |
| (1500). | *Liber Avicenna,* Venice: Canon, De med. cord. |
| (ca. 1503). | *Hic merito inscribi potens vite liber corporalis . . . ,* Venice: Canon, De med. cord., Cantica. |
| (ca. 1505). | Venice: Canon 3. |
| (1505). | *Liber canonis Avicenne revisus et ab omni errore mendaque purgatus,* Venice: Canon, De med. cord., Cantica. |
| (1506). | Pavia: Canon 1.1–2.4 and 4.1. |
| (1507). | *In hoc volumine . . . continentur infrascripti codices . . . Textus duarum primarum fen primi Avicenne . . . ,* Venice: Canon 1.1–2.4 and 4.1, Cantica. |
| (1507). | *Liber canonis Avicenne revisus et ab omni errore mendaque purgatus,* Venice, repr. Hildesheim, 1964: Canon, De med. cord., Cantica. |

(1508).    *Flores Avicenne,* Lyon: Canon.

(1510).    Pavia: Canon 1, 2.

(1511/1512).    Pavia: Canon 3.

(1514).    *Flores Avicenne collecti super quinque Canonibus... necnon super decem et novem libris De animalibus cum Canticis eiusdem,* Lyon: Canon, Cantica, De animal.

(1515).    *Articella nuperrime impressa... Textus duarum primarum fen primi Canonis Avicenne... Cantica Avicenne,* Lyon: Canon 1.1–2, 4.1, 3–5, Cantica.

(1519).    Lyon: Canon 1.1–2, 4.1, 3–5, Cantica.

(1520/1522).    *Primus Avicenne Canonis,* Venice: Canon, De med. cord., Cantica.

(1522).    *Avicenna: Liber Canonis totius medicine,* Lyon: Canon, De med. cord., Cantica.

(1523).    *Praesens maximus codex est totius scientie medicine principis Aboali Abinsene,* Venice: Canon, Cantica.

(1525).    *Articella nuperrime impressa... Textus duarum primarum fen primi Canonis Avicenne... Cantica Avicenne,* Lyon: Canon 1.1–2, 4.1, 3–5, Cantica.

(1527).    *Principis Avicennae libri Canonis necnon De medicinis cordialibus et Cantica ab Andrea Bellunensi ex antiquis Arabum originalibus... correcti,* Venice: Canon (corr. Alpago), De med. cord. (corr. Alpago), Cantica (corr. Alpago).

(1527).    *Liber principis... Aboali Abinceni aliter Avicenne vulgo dicti De viribus cordis cum commentariis Jacobi Lupi Bilbilitani,* Toulouse: De med. cord.

(1530).    *Avicennae quarta fen primi libri... nunc primum M. Iacob Mantini medici hebrei opera latinitate donata,* Venice: Canon 1.4 (Mantino).

(1531).    *Avicennae arabis medicorum... facile principis quarta fen primi... nunc primum M. Iacob. Mantini medici hebraei latinitate donata,* Ettlingen: Canon 1.4.

(1532).    *Avicennae arabis medicorum... facile principis quarta fen primi... interprete Iacob Mantino medico hebreo,* Paris: Canon 1.4.

(1532).    *Avicennae arabis inter omnes medicos... facile principis quarta fen primi... per M. Iacob Mantinum... latinitate donata,* Haguenau: Canon 1.4.

(1534).      *Articella nuperrime impressa . . . ,* Lyon: Canon 1.1–2, 4.1, 3–5, Cantica.

(1538).      Bruges: Canon 1.1.3.29 (Mantino) (with Cornelis van Baersdorp).

(ca. 1540).      Venice?: Canon 1.1 (Mantino).

(1544).      *Avicennae Liber canonis, De medicinis cordialibus et Cantica cum castigationibus Andreae Alpagi Bellunensis,* Venice: Canon, De med. cord. (corr. Alpago), Cantica (corr. Alpago).

(1547).      Venice: De rem. nocum., De syr., Expositiones super Canonem (Alpago).

(1547).      *Avicennae primi libri fen prima nunc primum per magistrum Iacobum Mantinum . . . translata,* Padua: Canon 1.1.

(1547/1548).      *Prima primi Canonis Avicenne sectio Michaele Hieronymo Ledesma Valentino medico et interprete et enarratore,* Valencia: Canon 1.1 (Ledesma).

(1549).      *Prima fen quarti Canonis Avicennae de febribus,* Paris: Canon 4.1.

(1555).      s.l. (Paris): Canon 1.4.

(1555).      *Avicennae liber Canonis, De medicinis cordialibus et Cantica,* Venice: Canon, De med. cord. (corr. Alpago), Cantica (corr. Alpago), De rem. nocum., De syr.

(1556).      *Avicennae medicorum Arabum principis liber Canonis,* Basel, repr. Teheran, 1976: Canon, De med. cord. (corr. Alpago), Cantica (corr. Alpago), De rem. nocum., De syr.

(1562).      *Avicennae liber Canonis, De medicinis cordialibus, Cantica,* Venice: Canon, De med. cord. (corr. Alpago), Cantica (corr. Alpago), De rem. nocum., De syr.

(1562).      *Cantica Avicennae in usum scholae Witebergensis seorsim edita,* Wittenberg: Cantica (corr. Alpago).

(1564).      *Avicennae principis et philosophi sapientissimi libri in re medica omnes,* Venice: Canon, De med. cord. (corr. Alpago), Cantica (corr. Alpago), De rem. nocum., De syr.

(1570).      *Avicennae medicorum arabum facile principis libri tertii fen secunda . . . ad fidem codicis hebraici latinus factus interprete Iohanne Quinquarboreo Aurilacensi,* Paris: Canon 3.2 (Cinqarbres).

(1576).      *De febribus opus sane aureum . . . Arabes: Avicenna, Rasis, Abimeron Avenzoar, Averrois, Isaac, Serapion, Haly Abatis, Actuarius,* Venice (excerpt from Canon on fevers).

(1572).     Paris: Canon 3.1.4 (Cinqarbres).

(1580).     *Principis Avicennae liber primus . . . Andrea Gratiolo Salodiano interprete,* Venice: Canon 1 (Gratiolo).

(1582).     *Avicennae liber Canonis De medicinis cordialibus Cantica,* Venice: Canon, De med. cord. (corr. Alpago), Cantica (corr. Alpago), De rem. nocum., De syr.

(1586).     Paris: Canon 3.1.5 (Cinqarbres).

(1595).     *Avicennae arabum medicorum principis Canon medicinae,* Venice: Canon, De med. cord. (corr. Alpago), Cantica (corr. Alpago), De rem. nocum., De syr.

(1608).     *Avicennae arabum medicorum principis Canon medicinae,* Venice: Canon, De med. cord. (corr. Alpago), Cantica (corr. Alpago), De rem. nocum., De syr.

(1609 or 1610).  *Liber secundus de Canone Canonis . . . studio . . . Petri Kirsteni . . . Arabice per partes editus et ad verbum in latinum translatus,* Breslau: Canon 2 (Kirsten).

(1611).     *Avicennae summi inter medicos nominis fen I libri I Canonis,* Vicenza: Canon 1.1.

(1630).     *Cantica Avicennae carmine elegiaco παραφραστικῶς ex arabico latine reddita a Ioanne Fauchero medico Bellocarensi,* Nîmes: Cantica (Faucher).

(1636).     *Avicennae summi inter arabes medici fen I libri I Canonis . . . editio correctior,* Padua: Canon 1.1.

(1647).     *Avicennae summi inter medicos nominis fen I libri I Canonis,* Venice: Canon 1.1.

(1648).     *Avicennae arabum medicorum summi fen I libri I Canonis . . . nova editio castigatior,* Padua: Canon 1.1.

(1649).     *Canticum . . . de medicina seu breve, perspicuum et concinne digestum institutionum medicarum compendium,* Groningen: Cantica (Deusing).

(1658).     *Clarissimi et praecellentissimi doctoris Abuali Ibn-Tsina . . . Canon medicinae interprete et scholiaste Vopisco Fortunato Plempio,* Louvain: Canon 1.2 and 4.1 (Plemp).

(1658).     *Avicennae quarti libri Canonis fen prima de febribus nova editio,* Padua: Canon 4.1.

(1659).     *Avicennae quarti libri Canonis fen prima de febribus,* Padua: Canon 4.1.

<u>(1659).</u>         *Abugalii filii Sinae . . . De morbis mentis tractatus,* Paris: Canon
                3.1 (Vattier).

<u>(1674).</u>         *Georgii Hieronymi Velschii Exercitatio de vena medinensi ad
                mentem Ebnsinae . . . specimen exhibens novae versionis ex ara-
                bico,* Augsburg: Canon 4.3.21–22 (Welsch).


## COSTA BEN LUCA

The Christian scholar Qusṭā ibn Lūqā—apart from being praised by his
contemporaries for the quality of his Graeco-Arabic translations—is
credited with a great number of treatises in many fields, especially in
mathematics, philosophy, and medicine. He was born in Baʿlabakk (Leb-
anon), lived for some time in Baghdad, and died around 912 or 913 AD in
Armenia.[173] At least three of his works were translated into Latin. The first
is a treatise on the psychological functions of the material *spiritus,* which is
contrasted with the immaterial soul: *Risāla fī l-Faṣl bayna r-rūḥ wa-n-nafs*
("Treatise on the Difference between Spirit and Soul"). The Latin translation, which is addressed to Archbishop Raymond of Toledo, was produced
by John of Seville and Limia *(De differentia spiritus et animae).* Some manuscripts transport an abbreviated and stylistically revised version of John's
translation.[174] The second treatise is preserved only in Latin: *De physicis ligaturis.* Some Latin manuscripts attribute it to Costa ben Luca (some as-
cribe it to Dioscorides, who is quoted in the text). In all probability, it was
translated by Constantine the African in the eleventh century; later manuscripts erroneously name Arnald of Villanova as translator or Galen as author.[175] This treatise, which was the work best known in the Renaissance,
discusses the medical effects of various substances that are used as amu-
lets; Costa ben Luca argues that most of these effects rely on self-suggestion
rather than on occult powers. The third translation is of the *Kitāb al-ʿamal
bi-l-kura an-nuǧūmīya* ("On the Use of the Celestial Globe"), *De sphaera solida.*
The text was rendered into Latin by Stephanus Arnaldi (or Arlandi) in the
early fourteenth century; apparently, it did not reach print.[176]
    Costa ben Luca's *De differentia spiritus et animae* is extant in more than
150 Latin manuscripts; with respect to distribution, it was the second
most influential philosophical text translated from Arabic into Latin in
the Middle Ages[177] (after the *Liber de causis,* which survives in at least 237
Latin manuscripts). The main reason for the enormous distribution of the
work was its inclusion in the Corpus vetustius of Aristotle's works on
natural science in the early thirteenth century,[178] and its appearance, from

1255 onward, on the curriculum of the arts faculty in Paris (see Chapter 1 above). In the Renaissance, Costa ben Luca has ceased to be a philosophical classic. *De differentia* was printed only once and under the name of Constantinus Africanus (1536). *De physicis ligaturis* was much better known; it was often printed among the works of Arnald of Villanova, Galen, and Agrippa von Nettesheim.[179]

## De differentia spiritus et animae

(1536).          *Constantini Africani . . . opera*, Basel: De spiritu, De ligaturis.

## De physicis ligaturis

(1504).          *Hec sunt opera Arnaldi de Villa nova . . . De phisicis ligaturis . . . Incipit liber Costa ben Luce De phisicis ligaturis translatus a magistro Arnaldo de Villa nova de greco in latinum*, Lyon.

(1505).          *Hec sunt opera Arnaldi de Villa nova . . . De physicis ligaturis*, Venice.

(1509).          *Hec sunt opera Arnaldi de Villa nova . . . De physicis ligaturis*, Lyon.

(1514).          Lyon (with Arnald of Villanova).

(1515–1516).   *Quarta impressio . . . continens omnes Galeni libros . . . Secundum Galeni volumen . . . Liber de incantatione*, Pavia.

(1520).          *Arnaldi de Villa nova . . . opera . . . Incipit liber Costa ben Luce De phisicis ligaturis translatus a magistro Arnaldo de Villa nova de greco in latinum*, Lyon.

(1528).          *Sexta impressio . . . omnes Galeni libros continens . . . Secundus . . . tomus . . . De incantatione, adiuratione et colli suspensione*, Lyon.

(1532).          *Arnaldi de Villa nova . . . opera . . . De physicis ligaturis*, Lyon.

(1536).          *Constantini Africani . . . opera*, Basel: De spiritu, De ligaturis.

(1541–1545).   *Galeni operum omnium sectio octava . . . De incantatione*, Venice.

(1542).          *Operum Galeni tomus sextus . . . De incarnatione* (sic) *et amuletis et colli suspensione etc incerto interprete, spurius qui Graece non habetur*, Basel.

(1546).          *Claudii Galeni . . . opuscula aliquot . . . Liber Galeno ascriptus De incantatione, adiuratione et suspensione*, Paris.

(1548). *In Caii Plinii secundi Naturalis Historiae ... I et II cap. libri xxx commentarius ... Item De incantatione et adiuratione collique suspensione epistola incerti authoris,* Würzburg.

(1549). *Cl. Galeni ... opera ... tomus vi ... De incantatione et amuletis et colli suspensione etc incerto interprete, spurius, Graece non habetur,* Basel.

(ca. 1550). *Henrici Cornelii Agrippae ab Nettesheym ... opera ... De incantatione et adiuratione collique suspensione epistola authoris incerti,* Lyon.

(1562–1563). *Galeni omnia quae extant ... Spurii libri ... De incantationibus et amuletis,* Venice.

(1567). *Henrici Cor. Agrippae ab Nettesheym De occulta philosophia libri III quibus accesserunt ... Epistola de incantatione et adiuratione collique suspensione,* Paris.

(1576–1577). *Galeni omnia quae extant opera ... septima classis ... De incantatione, adiuratione et suspensione liber Galeno falso ascriptus Graece non habetur,* Venice.

(1585). *Arnaldi Villanovani ... opera omnia ... De physicis ligaturis,* Basel.

(1586). *Arnaldi Villanovani ... Praxis medicinalis ... Arnaldi Villanovani De physicis ligaturis tractatus translatus de graeco in latinum,* Lyon.

(ca. 1600). *Henrici Cornelii Agrippae ab Nettesheym ... opera ... De incantatione et adiuratione collique suspensione epistola authoris incerti,* Lyon.

(1625). *Galeni opera ... Spurii libri ... De incantatione, adiuratione et suspensione liber Galeno falso adscriptus graece non habetur,* Venice.

(ca. 1630). *Henrici Cornelii Agrippae ab Nettesheym ... opera ... De incantatione et adjuratione collique suspensione epistola authoris incerti,* Lyon (i.e., Strasbourg).

(1679). *Operum Hippocratis Coi et Galeni ... tomus x ... Galeno ascriptus liber de incantatione, adjuratione et suspensione,* Paris.

## EBENBITAR (IBN AL-BAYṬĀR)

The botanist Ibn al-Bayṭār, who was born in Malaga at the end of the twelfth century and who died in Damascus in 1248 AD, is the author of the

largest and most famous work on simple medicines written in Arabic, the
*Kitāb al-Ǧāmiʿ li-mufradāt al-adwiya wa-l-aġḏiya* ("The Comprehensive Book
of Simple Medicines and Food"), a collection of excerpts of over 260 authors.[180] Ibn al-Bayṭār was not translated into Latin in the Middle Ages. In
1583, however, an edition of a treatise by "Ebenbitar" on lemons *(De limonibus)* appeared in a Latin version by Andrea Alpago (d. 1522). Alpago,
otherwise known as the translator of Avicenna, produced this translation
sometime between about 1487 and 1513 in the Near East. It was dedicated
to Benedetto da Monselice, who was a professor at the University of Padua
around 1480; Alpago at this time was a student in Padua. In the dedicatory epistle, Alpago recalls that Benedetto da Monselice had asked him
whether he could translate for him an Arabic treatise on lemons.[181] As the
editor of the 1583 edition explains, Alpago's manuscript had reached
Benedetto in Venice, who had given it to Leonardo Butironi, the father of
the editor, before 1513.[182] *De limonibus* is a translation of a treatise originally written by Ibn Ǧumayʿ (d. 1198 AD), which Ibn al-Bayṭār incorporated
into his *Kitāb al-Ǧāmiʿ*. The original title of the treatise by Ibn Ǧumayʿ was:
*Maqāla fī l-Līmūn wa-šarābihī wa-manāfiʿihī* ("Treatise on Lemon, on Its Juice
and on Its Benefits").[183] A second Latin edition appeared in 1602, a third
together with a commentary in 1757. "Ebenbitar" was also known to the
Latin West as an authority cited many times in Andrea Alpago's Latin lexicon of Arabic loan words in Avicenna's *Canon medicinae*, which was first
printed in 1527.

(1583).        Venice.

(1602).        *De limonibus tractatus Embitar arabis per Andream Bellunensem
               latinitate donatus,* Paris.

## GEBER FILIUS AFFLA

The astronomer Ǧābir ibn Aflaḥ of Seville appears to have been an important figure in the development of the Andalusian current that revolted
against Ptolemaic astronomy,[184] but hardly anything is known about his
life. Since his son was acquainted with Maimonides (d. 1204 AD), it seems
that he lived in the early or mid-twelfth century.[185] He is the author of a
comprehensive work called *Iṣlāḥ al-Maǧisṭī* ("Correction of the *Almagest*"),
in which he criticizes Ptolemy from a theoretical standpoint; but in contrast to his younger contemporaries Alpetragius and Averroes who start
from Aristotelian principles, his critique is based on Ptolemaic premises.
The *Iṣlāḥ* was translated into Latin by Gerard of Cremona,[186] and referred
to as *Astronomia Gebri* or *Liber Geber super Almagesti*.[187] The translation was

well distributed in about thirty manuscripts,[188] and Geber's critique of Ptolemy, as well as his trigonometry influenced late medieval and Renaissance astronomers.[189] Regiomontanus's *De triangulis* of the early 1460s much relied on Geber filius Affla without acknowledgment, which prompted Cardano to accuse Regiomontanus of plagiarism.[190] Cardano counts Geber among the twelve outstanding men in the scientific disciplines.[191] In 1534, the text was printed by Johann Petreius, together with Peter Apian's treatise on trigonometry.

(1534).     *Instrumentum primi mobilis a Petro Apiano ... Accedunt iis Gebri filii Affla Hispalensis astronomi vetustissimi pariter et peritissimi libri ix De astronomia,* Nuremberg.

## PSEUDO-GEBER

"Ğābir ibn Ḥayyān" and "Geber" are famous names of medieval alchemy and natural philosophy, but the historical connection between the Arabic and the Latin "author" is rather thin. Arabic bibliographers know of hundreds of titles attributed to "Ğābir," and it is unlikely that the entire corpus was produced by only one person; the texts date from the middle of the ninth to the middle of the tenth century AD.[192] As far as is known today, only three Arabic works of the corpus were translated into Latin in the Middle Ages: *Kitāb as-Sabʿīn* ("The Seventy Books"), *Liber de septuaginta,* and *Kitāb ar-Raḥma* ("The Book of Compassion"), *Liber misericordiae,* and extracts from the *Kitāb al-Mulk* ("Book of the Kingdom"), *Liber regni.* A number of treatises by a Latin Pseudo-Geber draw on the doctrines of the *Liber de septuaginta* and the *Liber misericordiae* without being translations themselves. The five "Geberian" works most often printed in the Renaissance are of this kind, that is, they are of occidental origin. This group of texts consists of a major work central to medieval alchemy, the *Summa perfectionis,* which in fact was composed by the thirteenth-century writer Paul of Taranto, and a series of opuscula that are designed as commentary literature on the Pseudo-Geberian *Summa: De investigatione perfectionis, De inventione veritatis, De fornacibus construendis,* and *Testamentum.*[193] The three real translations of Ğābir were not available in print.

The following list of editions breaks with the rules of this Appendix by listing works not of direct, but only of indirect Arabic origin.[194] This exception is made because the title pages and prefaces of the printed works of Geber Latinus put much emphasis on the Arabic provenance of Geber's works. In alchemy, more than in other sciences, the Arabic character of a work added to its authority. Emphasis on the Arabic origin is particularly common in vernacular German editions (e.g., "von Geber erstlich Arabisch beschriben," 1625). Geber is an important "Arabic" figure of Renaissance culture, not least because his name was associated with positive values of Arabic science. Just as Avicenna, he was thought to be a prince or king.[195] In the case of Avicenna, the legend of his kingship originates in a mistranslation of an Arabic honorary title.[196] As has been shown recently, the origins of Geber's legend lie in the fourteenth century, when the colophon of his *Liber regni* was misunderstood as meaning "the book of king Geber," the alleged son of Muhammad's daughter and a Baghdadian prince.[197]

(1473).          s.l. (Padua?).

(ca. 1486–1490). *Incipit liber Geber,* s.l. (Rome).

(ca. 1525).      *Gebris philosophi perspicacissimi Summa perfectionis magisterii,* Rome.

(1529).          *Geberi philosophi ac alchimistae maximi de alchimia libri tres,* Strasbourg.

(1531).          *Geberi philosophi ac alchimistae maximi de alchimia libri tres,* Strasbourg.

(1541).          *In hoc volumine de alchemia continentur haec: Gebri arabis philosophi solertissimi . . . De investigatione perfectionis metallorum liber unus. Summae perfectionis metallorum . . . libri duo. Quae sequuntur omnia nunc primum excusa sunt. Eiusdem de inventione veritatis seu perfectionis metallorum liber unus. De fornacibus construendis liber unus,* Nuremberg.

(1542).          *Geberis philosophi perspicacissimi Summa perfectionis magisterii,* Venice.

(1545).          *Alchemiae Gebri arabis philosophi solertissimi libri,* Nuremberg and Bern.

(1548).          *De alchemia dialogi duo quorum prior genuinam librorum Gebri sententiam . . . retegit,* Lyon.

(1561).          *Verae alchemiae . . . doctrina . . . Gebri opera,* Basel.

(1572).          *Artis chemicae principes Avicenna atque Geber,* Basel.

(1595).          *De lapidis physici conditionibus liber quo . . . Gebri et Raimundi Lullii methodica continetur explicatio,* Cologne.

(1598).          *Gebri arabis philosophi ac alchimistae acutissimi de alchemia traditio summae perfectionis,* Strasbourg.

(1668).          *Gebri arabis chimia sive traditio summae perfectionis,* Leiden.

(1679).         *Ginaeceum chimicum . . . Geberi arabis Summa perfectionis mag-*
                *isterii . . . Eiusdem Liber investigationis,* Lyon.

(1682).         *Gebri regis Arabum chymia,* Danzig.


## HALY FILIUS ABBAS

The Persian physician 'Alī ibn al-'Abbās al-Maǧūsī stemmed from a Zoroastrian family, and was perhaps himself a Zoroastrian. It is likely that he
spent most of his life in the city of Šīrāz. He lived in the later tenth century
AD. His sole work, the substantial *Kitāb Kāmil aṣ-ṣinā'a aṭ-ṭibbīya* ("The Complete Book of the Medical Art"), also called *al-Kitāb al-Malakī* ("The Royal
Book"), a full-fledged encyclopedia of the entire field of theoretical and
practical medicine,[198] was dedicated to the Buyid ruler 'Aḍud ad-Dawla,
before the latter seized power in Baghdad and left Šīrāz in 977-978 AD.[199]

The Latin reception of this work is a complicated story. It was "translated" twice—and both translations were printed in the Renaissance—once
in a free Latin adaptation by Constantine the African in the later eleventh
century under the title *Pantegni,* and once by Stephen of Antioch as *Liber
regalis* or *Regalis dispositio,* probably in 1127; the latter translation, however,
was much less widely transmitted. Constantine the African adapted the
ten books of the theoretical part into Latin, but for the second, practical
part of the *Pantegni,* Constantine preferred, for the most part, different
Arabic sources, for instance treatises by Ibn al-Ǧazzār (see below), while
retaining the overall structure of the *Practica* by al-Maǧūsī.[200] The *Pantegni*
was very influential in the twelfth and thirteenth centuries as the first
major medical summa of the Graeco-Arabic tradition available in the
Latin West and as a source for natural philosophers and encyclopedists.
Its textual format is not stable, however, and it has been demonstrated
that the Renaissance editions of 1515 and 1536/1539 of the *Pantegni* were
based on different textual traditions. The Lyon edition of 1515, which attributes the *Pantegni* to Isaac Israeli, is based on manuscripts of lesser
quality and presents a text that has been stylistically revised. The Basel
edition of 1536/1539 does not print the *Practica Pantegni* as a whole piece,
but combines parts of it with other translations by Constantine the African; the prologue has been rewritten in humanist style, but the text itself
seems to be of higher quality than in the Lyon edition since it shares readings with a large group of manuscripts.[201]

The small number of Renaissance editions—if compared to Avicenna,
Mesue, and Rhazes, but also to Avenzoar, Albucasis, and Serapion—reflects

the fact that the *Pantegni* had not been successful as a university text-book,[202] and that its influence had declined since the late thirteenth century.

(1492).     *Incipit tabula omnium librorum Haly Abatis . . . Incipit pro-logus . . . in libro medicine quod dicitur Regalis dispositio,* Venice (trans. by Stephen of Antioch).

(1515).     *Omnia opera Ysaac . . . Pantechni decem libri theorices et decem practices,* Lyon (trans. by Constantine the African).

(1523).     *Haly filius Abbas. Liber totius medicine necessaria continens . . . unde et Regalis dispositionis nomen assumpsit,* Lyon (Steph.).

(1536/1539).  *Constantini Africani . . . opera,* Basel (Constant.).

(1576).     *De febribus opus sane aureum . . . Arabes: Avicenna, Rasis, Abim-eron Avenzoar, Averrois, Isaac, Serapion, Haly Abatis, Actuarius,* Venice (excerpt on fevers).


## HALY FILIUS ABENRAGEL

Abū l-Ḥasan ʿAlī ibn abī r-Riǧāl was translated later than most other Arabic astrologers, in the second half of the thirteenth century, but his influence was all the more impressive. The exact dates of his life are not known. Abenragel was the personal astrologer of al-Muʿizz ibn Bādīs, the fourth ruler of the Zirid dynasty in Ifrīqiya (ruled 1016–1062), and held high positions at the court in Qayrawān; it seems that he died some years after 1041. Abenragel's oeuvre does not count many titles: apart from his astrological summa, we only know of a didactic poem on astrology and two lost treatises (astronomical tables and a *Kitāb ar-Rumūz wa-l-uṣūl,* "Book of Riddles and Roots"). The summa is a very comprehensive work on the various branches of the astrological art: general concepts of astrology (book 1), interrogations (2–3), nativities (4–5), anniversary horoscopes (6), elections (7), and historical and political astrology, meteorology (8). Its title is: *Kitāb al-Bāriʿ fī aḥkām an-nuǧūm* ("The Outstanding Book on Making Judgments from the Stars").[203]

In 1254, Jehuda ben Mose and Alvaro of Oviedo produced an Old Castilian translation of Abenragel's major work; Alvaro also began a si-multaneous Latin translation, which was not completed.[204] In 1256, Egidio de Tebaldis and Pietro de Regio, two Italian scholars who were serving Alfonso X of León and Castille as notaries and ambassadors,[205] translated the entire book from Castilian into Latin[206] under the title

*De iudiciis astrorum.* It was further translated into several vernacular languages.

There are a number of indications that Haly filius Abenragel counted among the most widely read Arabic astrologers in the Renaissance, equaled only by Alcabitius and Albumasar: he was often printed and often cited as a major reference work on astrology,[207] and his *De iudiciis astrorum* was the sole astrological text of Arabic provenance that was subjected to a substantial revision in the Renaissance. In 1551, Antonius Stupa published a humanist revision of the text, which was based on Latin witnesses only. Stupa's principal aim was to extinguish the Gallicisms and Hispanicisms of the medieval Latin version, which go back to the intermediary Old Castilian. Modern scholars have censured this edition for its many mistakes in content, which, it is said, are a result of Stupa's linguistic castigation of the text. This censure, however, has not yet been backed up with concrete analysis of specific passages.[208]

In 1563, the last chapter of *De iudiciis astrorum,* which deals with comets, was printed separately and in a revised version, which is a thorough reworking of the chapter in humanist Latin.

(1485).　　　*Preclarissimus liber completus in iudiciis astrorum quem edidit Albohazen Haly filius Abenragel,* Venice.

(1503).　　　s.l. (Venice).

(1520).　　　*Haly de iuditiis. Preclarissimus in iuditiis astrorum Albohazen Haly filius Abenragel,* Venice.

(1523).　　　*Haly de iuditiis. Preclarissimus in iuditiis astrorum Albohazen Haly filius Abenragel,* Venice.

(1525).　　　*Omar De nativitatibus . . . De revolutionibus nativitatum ex Abenragele de fridariis seu temporaria potestate planetarum,* Venice.

(1551).　　　*Albohazen Haly filii Abenragel libri de iudiciis astrorum, . . . de extrema barbarie vindicati ac latinitati donati per Antonium Stupam Rhoetum Praegalliensem,* Basel.

(1563).　　　*Meteororum . . . loci . . . digesti . . . a M. Marco Frytschio . . . Tractatus Albohazenhalii filii Abenragelis De cometarum significationibus per xii signa zodiaci,* Nuremberg.

(1571).　　　*Albohazen Haly filii Abenragel scriptoris arabici de iudiciis astrorum libri octo . . . a barbarie vindicati et puritati linguae donati per Antonium Stupam Rhaetum Praegalliensem,* Basel.

# HALY RODOAN

The physician and astrologer Abū l-Ḥasan ʿAlī ibn Riḍwān lived in Cairo from 998 AD to about 1061 AD. Born and raised in poor circumstances, he was an autodidact whose career culminated in his appointment as chief physician of Egypt by the Fatimid ruler. His medical writings are characterized by fierce opposition against deviations from the teachings of Galen. In 1049–1050 he had a famous controversy with the physician Ibn Buṭlān (see below). Of his many writings, about twenty have survived in Arabic manuscripts, most of them on medical issues.[209] Only one medical work of his was known in Latin, his "Commentary on Galen's *Ars parva*" *(Šarḥ aṣ-ṣināʿa aṣ-ṣaġīra li-Ġālīnūs)*, through the twelfth-century translation by Gerard of Cremona *(Tegni Galeni cum commento)*.[210] This commentary, which was integrated into the widely read collection of medical treatises called *Articella*, influenced the Latin discussion of the scientific principles of the medical art.[211] It regularly appears in early prints of the *Articella*.

The other book by which Haly Rodoan was known in the Latin West was astrological: his "Commentary of the *Tetrabiblos* of Ptolemy" *(Šarḥ al-maqālāt al-arbaʿ li-Baṭlūmiyūs)*.[212] The *commentum* was translated into Spanish by an unknown translator, and thence, on order of King Alfonso X, into Latin by Egidio de Tebaldis, the Italian notary who also translated Abenragel.[213] Haly Rodoan's commentary had a very tangible influence on Renaissance astrological literature and especially on the commentary tradition on Ptolemy's *Tetrabiblos,* for example, on the commentaries by Cardano and Nabod.[214]

## Tegni Galeni cum commento

(ca. 1476).   *Vita brevis, ars vero longa ... Galieni principis medicorum Microtegni cum comento Hali liber incipit,* Padua (in *Articella*).

(1483).   *In hoc praeclaro libro ... Galieni principis medicorum micro Tegni cum commento Hali Rodoham liber incipit,* Venice (in *Articella*).

(1487).   *In hoc praeclaro libro ... Galieni principis medicorum micro Tegni cum commento Hali Rodoham liber incipit,* Venice (in *Articella*).

(1491).   *Gregorius a Vulpe Vincentinus ... Galieni principis medicorum migro Tegni cum commento Hali Rodoam liber incipit,* Venice (in *Articella*).

(1493).   Venice (in *Articella*).

(1500).          *Articella . . . Galieni principis medicorum migro Tegni cum com-*
                 *mento Hali Rodoam liber incipit,* Venice (in *Articella*).

(1501).          Pavia.

(s.d., ca. 1516).  *Index eorum omnium que in hac arte parva Galeni pertrac-*
                 *tantur . . . Item additiones Haly Rodoam admodumque acute ac*
                 *docte,* s.l. (Lyon).

(1519).          Venice.

(1523).          *Articella . . . Galeni liber qui Techni inscribitur . . . cum commento*
                 *Haly,* Venice.

(1557).          *Plus quam commentum in parvam Galeni artem Turisani . . . ad-*
                 *ditis quibusdam . . . ea autem sunt: Hali qui eandem Galeni artem*
                 *primus exposuit,* Venice.

### Quadripartitum Ptolemei cum commento

(1493).          *Liber quadripartiti Ptholemei . . . cum commento Haly Heben*
                 *Rodan,* Venice.

(1519).          *Quadripartitum Ptolomei. Que in hoc volumine continentur hec*
                 *sunt. Liber quadripartiti Ptolomei . . . cum commento Haly Heben*
                 *Rodan,* Venice.

## PSEUDO-HALY (AḤMAD IBN YŪSUF)

In Latin manuscripts and prints, a certain "Haly" is credited with the
authorship of a commentary on Pseudo-Ptolemy's *Centiloquium,* an astrological treatise consisting of one hundred aphorisms. This "Haly,"
however, is not identical with either Haly filius Abenragel or Haly
Rodoan. Rather, the true commentator on the *Centiloquium* is Abū Gaʿfar
Aḥmad ibn Yūsuf ibn Ibrāhīm ibn ad-Dāya, who lived around 900 in
Cairo.[215] Aḥmad ibn Yūsuf was a mathematician, astronomer, and astrologer. Two mathematical works of his, *On Ratio and Proportion* and *On
Similar Arcs,* were translated into Latin by Gerard of Cremona in the twelfth
century.[216] But only his commentary on the *Centiloquium* was printed in
the Renaissance. Pseudo-Ptolemy's *Centiloquium* itself is certainly older
than Aḥmad ibn Yūsuf's commentary, since it is quoted already in the
ninth century by aṣ-Ṣaymarī.[217] Moreover, there are indications that the
Greek text of the *Centiloquium* is not a Byzantine translation from the
Arabic, but a work of Greek antiquity of the centuries after Ptolemy.[218]
The *Centiloquium* was translated from Arabic into Latin (or revised) four
or five times in the twelfth century. The commentary by Aḥmad ibn

Yūsuf is not part of the earliest translation of the *Centiloquium* by Adelard of Bath, but was translated (or revised) several times together with
the *Centiloquium* in the course of the twelfth century. The confusion of
Aḥmad ibn Yūsuf with "Haly" probably originated with the version of
Plato of Tivoli.

(1484).        *Liber quadripartiti Ptolomei . . . Incipit liber centum verborum*
               *Ptholomei . . . Expositio Haly super primo verbo,* Venice.

(1493).        *Liber quadripartiti Ptholomei. Centiloquium eiusdem . . . In-*
               *cipit liber centum verborum Ptholomei cum commento Haly,*
               Venice.

(1519).        *Quadripartitum Ptolomei . . . Centiloquium eiusdem . . . In-*
               *cipit liber Ccentum verborum Ptolomei cum commento Haly,*
               Venice.


## IBN BUṬLĀN

Abū l-Ḥasan al-Muḥtār ibn ʿAbdūn ibn Buṭlān (whose Latinized name
was "Elluchasem Elimithar") was a Christian physician, who was born
in Baghdad. In 1049 AD, he left Baghdad for extensive travels in Syria,
Byzantium, and also Egypt, where he was involved in a controversy with
Haly Rodoan. Later, he became a monk in a monastery in Antioch,
where he died in 1066 AD.[219] Apart from a number of smaller works, Ibn
Buṭlān produced an influential reference work on dietetics and hygiene,
the *Taqwīm aṣ-ṣiḥḥa* ("Almanac of Health"), translated as *Tacuinum sani-*
*tatis* in Sicily at the court of Manfred of Hohenstaufen who reigned 1254
to 1266 AD; the name of the translator is not known.[220] Ibn Buṭlān's
medical Almanac is remarkable for a format borrowed from astronomical literature: the text is organized in the form of tables. In the Arabic
world, this format was adopted by other authors such as Ibn Ǧazla; in
the Latin West it inaugurated a textual tradition of its own, the *Tacuina*
*sanitatis,* which were often richly illustrated and designed for the usage
of noble laymen.[221] The title of the Renaissance German translation of
the work *Schachtafelen der Gesuntheyt,* published by the Strasbourg physician Michael Herr in 1533, refers to the checkered appearance of the
text.[222]

(1531).        *Tacuini sanitatis Elluchasem Elimithar medici de Baldath de sex*
               *rebus non naturalibus,* Strasbourg.

(1532).        *Tacuini sex rerum non naturalium . . . Tacuini aegritudinum et*
               *morborum ferme omnium,* Strasbourg.[223]

## IBN ĞAZLA

Abū ʿAlī Yaḥyā ibn ʿĪsā ibn Ğazla ("Byngezla" in Latin), a Christian scholar who converted to Islam in 1074 AD, was a physician in Baghdad, where he died in 1100 AD. His main work, the *Taqwīm al-abdān fī tadbīr al-insān* ("Almanac of the Bodies on Human Diatetics") is very similar in format and content to Ibn Buṭlān's almanac. In forty-four tables and accompanying canons, it presents descriptions of 352 maladies and their remedies.[224] The work was translated into Latin by the Jewish scholar Farağ ibn Sālim of Agrigent (Sicily) for Charles I of Naples; the translation was finished in 1280 AD.[225] In 1532 it was printed in Latin as *Canones tacuinorum* (the canons) and *Tacuinorum tabulae* (the tables). In 1533, it appeared together with Ibn Buṭlān in a German translation by Michael Herr.

(1532).    *Tacuini aegritudinum et morborum ferme omnium corporis humani cum curis eorundem Buhahylyha Byngezla autore,* Strasbourg.

## IBN AL-ĞAZZĀR

The name of Ibn al-Ğazzār seems not to have been known in the Latin West, but several of his medical works were available in Latin translations of the eleventh century, traveling under the name of their translator and redactor Constantinus Africanus. Ibn abī Ḥālid ibn al-Ğazzār, a physician and historian and a pupil of Isaac Israeli, lived in Qayrawān, the capital of Ifrīqiya, where he died around 980 AD.[226] He is known to have written many medical works, most of which are not extant. Best known in the Arabic and the Latin worlds was his *Zād al-musāfir wa-qūt al-ḥāḍir* ("Provisions for the Traveler and Nourishment for the Settled"), which was known in Constantinus's Latin translation as *Viaticum peregrinantis.* This is a succinct survey of diseases and their treatment, which are discussed *a capite ad calcem;* the flowery title means no more than that Ibn al-Ğazzār aims at a broader audience for his brief and readable book.[227] In Renaissance editions, the work is credited either to Constantine the African or Isaac Israeli.[228] It is a general feature of Constantine's translations that they are not literal; he adds or omits phrases and sentences and has a very particular style. This is also true for the *Viaticum* and the other short medical works of Ibn al-Ğazzār that he translated: *De oblivione, De coitu,* and (perhaps) *De stomacho.*[229]

Two treatises by Ibn al-Ğazzār appear as parts of the *Pantegni,* Constantine the African's Latin adaptation of the medical magnum opus of Haly Abbas: *De elephantia* (*Pantegni* chapter 4, 2–4) and *De gradibus* (book 9). *De elephantia* ("On Leprosy") was twice printed as an independent treatise in the Renaissance.[230] Constantine's *De gradibus* is a Latin adaptation of Ibn al-Ǧazzār's influential treatise on simple drugs called *Kitāb al-I'timād fī l-adwiya al-mufrada* ("The Book of Reliability on Simple Drugs"). In 1233, this work received a second Latin translation by Stephanus of Saragossa, resident in Lérida ("Stephanus de Ceza<rau>gusta civis Ilerdensis") under the title *Liber fiduciae de simplicibus medicinis,* of which a Renaissance edition does not seem to exist.[231]

### Viaticum

(1505). *In hoc volumine continentur Introductorium iuvenum Gerardi de Solo . . . commentum eiusdem super Viatico cum textu,* Venice (comm. Gerardus).

(1510/1511). *Opera parva . . . Arasi . . . additus est Constantini monachi Viaticus,* Lyon.

(1515). *Omnia opera Ysaac . . . Viaticum Ysaac quod Constantinus sibi attribuit,* Lyon.

(1536). *Constantini Africani . . . opera,* Basel.[232]

### De oblivione, De coitu, De stomacho, De elephantia

(1515). *Omnia opera Ysaac . . . Item libelli eiusdem (ut ferunt) Constantini De obliuione,* Lyon: De oblivione, De stomacho, De elephantia.

(1536). *Constantini Africani . . . opera,* Basel: De stomacho, De coitu.

(1541). *Methodus medendi . . . autore Albucase . . . Constantini Africani De humana natura liber unus. Item eiusdem De elephantia,* Basel: De elephantia.

For *De gradibus* and *De elephantia,* see also: Haly filius Abbas, *Pantegni.*

## IBN MU'ĀD (ABHOMAD, PSEUDO-ALHAZEN)

Not much is known about the Andalusian juridical scholar, mathematician, and astronomer Abū 'Abdallāh Muḥammad ibn Mu'ād al-Ǧayyānī, of Jaen in Southern Spain, who spent some time in Egypt. The only certain date of his life is 1079, the year in which he observed a solar eclipse. In all likelihood, Ibn Mu'ād is the author of a work on spherical trigonometry and a treatise on Euclid's notion of ratio,[233] and also a treatise "On the Dawn," which has survived in Hebrew and in a Latin translation by Gerard of Cremona.[234] The Latin title is *De crepusculis* or *De crepusculo matutino et vespertino* ("On the Morning and the Evening Twilight"); ibn Mu'ād in this treatise attempts to determine the height of the atmosphere on the basis of the angle of depression of the sun at daybreak and nightfall.[235] Since the fourteenth century, *De crepusculis* has been wrongly attributed to Alhazen,[236] and this is the attribution that was adopted by the printers of the sixteenth century. Some sixteenth-century manuscripts refer to the author as 'Abhomad.'[237]

Ibn Mu'ād also composed astronomical tables, which were calculated for Jaen. They were translated into Latin by Gerard of Cremona.[238] The introductory *regulae* of the *Tabulae Jahen* were printed without the tables themselves in 1549 by Joachim Heller.[239]

(1542).          *Petri Nonii Salaciensis De crepusculis . . . Item Allacen arabis ve-
                 tustissimi De causis crepusculorum liber unus,* Lissabon.

(1549).          *De elementis . . . Messahalae . . . Item iisdem de rebus scriptum
                 cuiusdam Saraceni continens praeterea praecepta ad usum tabu-
                 larum astronomicarum utilissima,* Nuremberg.

(1572).          *Opticae thesaurus: Alhazeni arabis libri septem . . . eiusdem liber
                 De crepusculis et nubium ascensionibus,* Basel.

(1573).          *Petri Nonii Salaciensis . . . de crepusculis liber unus cum libello Al-
                 lacen De causis crepusculorum,* Coimbra.

(1592).          *Petri Nonii Salaciensis . . . de crepusculis liber unus cum libello Al-
                 lacen De causis crepusculorum,* Basel.


### PSEUDO-IBN SĪRĪN

Abū Bakr Muḥammad ibn Sīrīn was a famous transmitter of hadiths in the first century of Islam; he died in 728 AD. In later centuries, he became well-known as one the first interpreters of dreams. It is unlikely, however, that he is the author of the various Arabic texts on dream interpretation that were ascribed to him.[240] Parts of this Arabic tradition reached the Latin West via a Greek compilation; in the Greek tradition, this compilation was known as the *Oneirocriticon* of Sirim or Achmet son of Sereim. In 1175/1176, Leo Tuscus in Constantinople translated the text from Greek into Latin, under the title *De somniis et oraculis*.[241] A second Latin translation from the Greek was produced in the sixteenth century by the classical scholar Johannes Leunclavius. Leunclavius, or Löwenklau in German (1541–1594), was the author of a famous work, *Historiae Musulmanae Turcorum,* and a translator of several Hellenistic and Byzantine authors.[242]

The translation bears the title *Apomasaris Apotelesmata,* and thus seems to indicate that the author is the famous astrologer Albumasar. This possibility was already ruled out by Leunclavius himself in his preface to the edition, on the wrong assumption that the author of the treatise must be a Christian.[243] Modern research has shown that the *Apotelesmata* ultimately derive from various Arabic sources on dream interpretation, some of them connected with the name of Ibn Sīrīn.[244] A German translation of the Latin text, called *Traumbuch Apomasaris,* was integrated into the very influential compendium of domestic economics by Johann Coler (d. 1639) and printed several times in the seventeenth century. The medieval Latin translation by Leo Tuscus was edited in 1603 by Nicolas Rigault (d. 1654) together with the Greek text in a facing column.[245]

(1577).    *Apomasaris Apotelesmata sive De significatis et eventis insomniorum . . . Ioanne Leunclaio interprete,* Frankfurt am Main.

(1603).    *Artemidori Daldiani et Achmetis Sereimi filii Oneirocritica,* Paris (transl. by Tuscus).


## IBN ṬUFAYL

The Andalusian philosopher, physician, and poet Abū Bakr ibn Ṭufayl was born near Granada around 1116 AD. He worked as a physician in Granada, and in 1147 AD moved to Marrakesh, where he was eventually appointed personal physician to the caliph Abū Yaʿqūb Yūsuf. Averroes was introduced to the caliph by Ibn Ṭufayl and later became his successor as court physician. Ibn Ṭufayl died in 1185 AD.[246] He has written a number of poems, among them a very long didactic poem on the various branches of medicine, but his fame rests on the philosophical novel *Ḥayy ibn Yaqẓān.* It is titled after the protagonist of the story, who grows up on an island without any contact with human civilization; his intellect gradually acquires philosophical knowledge, which, in the final development, results in a union with God.[247] The text was not known in the Latin Middle Ages, but was translated into Hebrew by an unknown translator and commented upon by Moses Narboni.[248] In the late fifteenth century, an anonymous translator, probably a Christian convert from Judaism, produced a very literal Latin version from the Hebrew text; it is extant in one manuscript.[249] It is likely that the translator is one of the learned Hebrew scholars in the service of Pico della Mirandola, since Pico says about a text by Ibn Ṭufayl with the title *Quo quisque pacto per se philosophus evadat* around 1493: "We have turned it from Hebrew into Latin in the past year."[250]

When in 1671 the Arabic text appeared in print together with a Latin translation, Ibn Ṭufayl's philosophical novel, which received several subsequent translations in the vernaculars, became a widely read piece of enlightenment culture (as indicated by the reference to *ratio humana* on the title page of the 1671 edition). The Latin translation, which is difficult to read because of its literalness,[251] was a joint product of Edward Pocock, the accomplished Oxford Arabist (d. 1691), and his son, Edward Pocock Jr.[252]

(1671).      *Philosophus autodidactus sive epistola Abi Jaafar Ebn Tophail de Hai Ebn Yokdhan, in qua ostenditur quomodo ex inferiorum contemplatione ad superiorum notitiam ratio humana ascendere possit, ex arabica in linguam latinam versa ab Eduardo Pocockio A.M.,* Oxford.


## IESUS HALY

Šaraf ad-Dīn ʿAlī ibn ʿĪsā is the author of the most important work on ophthalmology to reach the Latin West. ʿAlī ibn ʿĪsā, a Christian, lived and practiced in Baghdad, but we know not much more about his life than that he was the pupil of ʿAbd Allāh ibn aṭ-Ṭayyib (d. 1043), the commentator on Galen, and a physician at the hospital al-ʿAḍudīya in Baghdad. His comprehensive *Taḏkirat al-kaḥḥālīn* ("Promptuary for Oculists"), apparently his only surviving work, has three parts: on the anatomy and physiology of the eye, on external eye diseases, on nonvisible eye diseases; it ends with a list of simple medicines. Altogether, it proposes treatments for 130 eye ailments. It was of great influence on subsequent Arabic oculists.[253] In 1271, the treatise was translated into Latin by Dominicus Marrochinus, a Dominican friar;[254] the quality of the translation is doubtful.[255] In Renaissance editions, the common title for the work is *De oculis*. It is always followed by a treatise titled: *Canamusali de Baldac De oculis*. Despite its title, this treatise does not seem to be a translation of a text by ʿAmmār ibn ʿAlī al-Mawṣilī, an important oculist who was a contemporary of ʿAlī ibn ʿĪsā,[256] but a Latin compilation by a person called "David Armenicus."[257]

(1499).      *Cyrurgia Guidonis de Cauliaco . . . Jesu Hali De oculis. Canamusali de Baldac De oculis,* Venice.

(1500).      *Cyrurgia parva Guidonis . . . Tractatus de oculis Jesu Hali. Tractatus De oculis Canamusali,* Venice.

(1513).      *Cyrurgia Guidonis de Cauliaco . . . Jesu Hali De oculis. Canamusali de Baldac De oculis,* Venice.

## IOHANNITIUS

The Latin West had only a very dim and reduced image of Ḥunayn ibn Isḥāq, the great translator and mediator between Greek and Arabic cultures, the accomplished physician and influential promoter of Hellenistic learning. Ḥunayn was a Nestorian Christian, who lived from 808 to 873 AD. He was a student of the leading physician of his time, Yūḥannā ibn Māsawayh (Mesue). The number of his translations from the Greek exceeds one hundred; among these, the translations of Galen and Hippocrates are perhaps the most influential. He is known to have written numerous works of his own, on medicine, but also on other sciences, and on wisdom literature, many of which are extant in Arabic.[258] Of his original works, only two were translated into Latin: a short introductory treatise on humoral pathology, *Kitāb al-Mudḫal fī ṭ-ṭibb* ("Introduction to Medicine"), and a work on ophthalmology, *Kitāb al-ʿAšr maqālāt fī l-ʿayn* ("Ten Treatises on the Eye").

The first work, the *Isagoge Iohannitii*, reached an enormous distribution in Latin manuscripts because, since the twelfth century, it was copied as the first treatise of the *Articella*, a twelfth-century collection of medical treatises, which in the early thirteenth century became part of medical university education in many parts of medieval Europe. The *Isagoge* was first adopted into Latin with many changes in 1075–1085; it was later subjected to a substantial revision; circumstantial evidence points to Constantine the African as the translator.[259] The Renaissance printing history of the *Articella*, and with it the *Isagoge*, probably reflects changes in the medical curriculum in the sixteenth century: the early tide of editions of the *Articella* breaks off in 1534, most probably as a result of the new interests of medical humanism. Ḥunayn's *Isagoge* continues to be a model for other brief introductions to medicine, such as the *Tabulae Isagogicae in universam medicinam ex arte Humain idest Ioannitii Arabis* by Fabio Paolino (printed in *Canon* editions of 1595 and 1608).

The second work, Ḥunayn's "Ten Treatises on the Eye," was not available to the Latins in a translation proper, but Constantine the African based his own work *De oculis* largely on this Arabic source. Via Constantine, Iohannitius's physiological optics and pathology of the eye were of considerable influence in the West. In the Renaissance, this treatise was once printed under the name of Constantine and many times among the works of Galen, often with the additional information: "spurious, translated from the Greek by Demetrius, the Greek exemplar is still missing." The Demetrius version is not a separate translation, but a reworking of Constantine's *De oculis*;[260] the Latin style suggests that the Pseudo-Galenic treatise is the product of a humanist redactor.[261]

## Isagoge

| | |
|---|---|
| (ca. 1476). | *Vita brevis, ars vero longa . . . Incipiunt Isagoge Iohannitii ad Tegni Galieni primus liber medicine,* Padua. |
| (1483). | *In hoc praeclaro libro . . . Primo est liber Johannitii qui dicitur Isagoge in greco,* Venice. |
| (1487). | *In hoc praeclaro libro . . . Primo est liber Joannitii qui dicitur Isagoge in greco,* Venice. |
| (1491). | *Gregorius a Vulpe Vincentinus . . . Primo est liber Joannitii qui dicitur Isagoge in greco,* Venice. |
| (1493). | Venice. |
| (1497). | Leipzig. |
| (1500). | *Articella . . . Primo est liber Joannitii qui dicitur Isagoge in greco,* Venice. |
| (1502). | *Liber Hysagoge Joannici,* Venice. |
| (1507). | *In hoc volumine . . . continentur infrascripti codices: Liber hysagoge Joannitii,* Venice. |
| (1513). | Venice. |
| (1515). | *Articella nuperrime impressa . . . Hysagoge Joannitii,* Lyon. |
| (1519). | Lyon (in *Articella*). |
| (1523). | *Articella . . . Joannitii liber grece Isagoge appellatus,* Venice. |
| (1525). | *Articella nuperrime impressa . . . Hysagoge Joannitii,* Lyon. |
| (1527). | *Thaddei Florentini expositiones . . . in subtilissimum Joannitii Isagogarum libellum,* Venice. |
| (1534). | *In hoc opusculo continentur . . . Isagoge sive introductio Ioannitii in artem parvam Galeni,* Strasbourg. |
| (1534). | *Articella . . . Hysagoge Ioannitii,* Lyon. |
| (1557). | *Plus quam commentum in parvam Galeni artem Turisani Florentini . . . Ioannitii ad eandem introductio,* Venice. |

## De oculis

| | |
|---|---|
| (1515). | *Omnia opera Ysaac in hoc volumine contenta . . . liber De oculis Constantini,* Lyon. |
| (1541–1545). | *Galeni operum omnium sectio octava . . . De oculis . . . Galeni de oculis a Demetrio translatus,* Venice. |
| (1542). | *Operum Galeni tomus sextus . . . De oculis therapeuticon Demetrio graeco interprete,* Basel. |

(1546).     *Claudii Galeni . . . opuscula aliquot . . . Liber de oculis,* Paris.

(1549).     *Cl. Galeni . . . opera . . . tomus vi . . . De oculis therapeuticon Demetrio Graeco interprete, spurius, graecum exemplar desideratur,* Basel.

(1562–1563).     *Galeni omnia quae extant . . . Spurii libri . . . De oculis therapeuticon Demetrio Calcondylo interprete,* Venice.

(1576–1577).     *Galeni omnia quae extant opera . . . septima classis . . . De oculis, spurius, Demetrio graeco interprete . . . graecum exemplar desideratur,* Venice.

(1586).     *Galeni librorum septima classis . . . De oculis, spurius, Demetrio graeco interprete . . . graecum exemplar desideratur,* Venice.

(1609).     *Galeni opera . . . Libri septimae classis . . . De oculis, spurius, Demetrio graeco interprete . . . graecum exemplar desideratur,* Venice.

## ISAAC ISRAELI

Isḥāq ibn Sulaymān al-Isrā'īlī, the Jewish physician and philosopher, was a native of Egypt who emigrated to Qayrawān in North Africa around 905–907, where he was the companion and student of Isḥāq ibn 'Imrān and eventually was appointed court physician to the first Fatimid ruler. He probably died around 932, and certainly before 955–956.[262] His medical writings were translated into Latin by Constantine the African in the late eleventh century.[263] While Constantine made himself the author of what were in fact his translations of Isḥāq ibn 'Imrān, Ibn al-Ġazzār, and al-Maǧūsī (Haly filius Abbas), the Latin translations of Isaac Israeli alone traveled under the name of their real author. Constantine the African translated three works of his, all of which were printed in 1515: *Kitāb al-Ḥummayāt (De febribus),* an important work on fevers much esteemed by both Arabic and Latin authors, *Kitāb al-Bawl (De urinis),* and *Kitāb al-Aġḏiya (De dietis universalibus et particularibus),* perhaps the most comprehensive Arabic work in the field of hygiene.

The last one was the only work by Isaac Israeli that received philological attention in the Renaissance: it was printed four times, twice in a stylistically revised version produced by the German neo-Latin poet and physician Johannes Posthius (1537–1597);[264] Posthius explains in the dedication to the 1570 edition that the work counts among the most important on dietetics, but that it is difficult to find and that it is marred by an erroneous translation in bad Latin style. The dedication is missing in the second, posthumous edition of 1607, which praises Posthius's "first translation."[265]

Two philosophical texts by Isaac Israeli were translated by Gerard of Cremona in the twelfth century: *Kitāb al-Usṭuquṣṣāt (De elementis)*, a treatise on Aristotle's and Galen's theories of the elements and on their number and properties; *Kitāb al-Ḥudūd wa-r-rusūm* ("On Definitions and Descriptions"), *De differentia inter descriptiones rerum et definitiones earum,* on the traditional definitions of the discipline of philosophy.[266] This second treatise also exists in an anonymous revision, which stems from Dominicus Gundisalvi, as its style indicates.[267] The title *On Definitions* recalls the title of a text by al-Kindī, whose philosophical writings are amply used by Isaac Israeli in other works as well.[268] The translations by Constantine the African and Gerard of Cremona cover almost the entire oeuvre of Isaac Israeli, which makes him a rare case in the history of Arabic–Latin translations. The bio-bibliographers of the Renaissance knew all these works;[269] they add the title *De stomacho,* probably referring to a treatise with this title by Constantine the African, which, however, seems to originate from Ibn al-Ġazzār.

(1487).          *Eximii Isaac medicine monarce De particularibus dietis libellus,* Padua: Diet. part.

(1515).          *Omnia opera Ysaac in hoc volumine contenta . . . liber De definitionibus, liber De elementis, liber Dietarum universalium . . . liber Dietarum particularium, liber De urinis . . . liber De febribus,* Lyon: De febribus, Diet. univ., Diet. part., De urinis, Elem., Defin.

<u>(1570).</u>       *Isaac Iudaei Salomonis Arabiae regis adoptivi filii De diaetis universalibus et particularibus libri II . . . Liber omnibus philosophiae et medicinae immo sanitatis studiosus . . . superiori seculo ex Arabica lingua in Latinam conversus, nunc vero opera D. Ioannis Posthii Germershemii sedulo castigatus et in lucem editus,* Basel: Diet. univ., Diet. part.

(1576).          *De febribus opus sane aureum . . . Arabes: Avicenna, Rasis, Abimeron Avenzoar, Averrois, Isaac, Serapion, Haly Abatis, Actuarius,* Venice: De febribus.

(1607).          *Thesaurus sanitatis de victus salubris ratione . . . sive De dietis universalibus et particularibus libri II ab Isaaco Iudaeo Salomonis Arabiae regis adoptivo filio lingua arabica conscripti et nunc primum latinitate donati opera Ioannis Posthii,* Antwerp: Diet. univ., Diet. part.

## ISHĀQ IBN ʿIMRĀN

Just as Isaac Israeli (d. ca. 932 AD) and Ibn al-Ġazzār (d. ca. 980), Isḥāq ibn ʿImrān was a physician active in Qayrawān in North Africa, who was translated into Latin by Constantine the African in the later eleventh century. Ibn ʿImrān originally came from Baghdad, but was invited to Qayrawān by the last Aġlabid ruler Ziyādat Allāh, who after a few years, sometime before 909 AD, ordered Ibn ʿImrān to be killed, in consequence of a slander.[270] Ibn ʿImrān is known to have written a couple of medical works, of which the most significant is the *Maqāla fī l-mālīḫūliyā* ("Treatise on Melancholy"), which continues and enlarges the doctrine of medical melancholy as expounded by the Greek author Rufus of Ephesus. The treatise was adopted into Latin by Constantinus Africanus: about half of the text is a literal translation with occasional omissions, but at the middle of the second chapter the Latin text departs from the Arabic original.[271] In the Latin manuscript tradition, *De melancholia* was attributed to Constantine, with the effect that Isḥāq ibn ʿImrān remained unknown in the West as a medical author.[272]

(1536).        *Constantini Africani . . . opera . . . Constantini Africani medici de melancholia libri duo,* Basel.[273]

## AL-ISRĀʾĪLĪ (PSEUDO-ALMANSOR)

It is only in the 1970s that the Arabic source of the widely distributed astrological text *Almansoris iudicia seu propositiones* has been identified.[274] It is a treatise with the title *Fuṣūl fī ʿilm an-nuǧūm li-l-Isrāʾīlī ḫadima bihā l-Ḥākim bi-Amrillāh* ("Chapters on the Science of the Stars by al-Isrāʾīlī with which he renders a service to al-Ḥākim bi-Amrillāh"). The unknown Jewish astrologer obviously dedicated his work to the sixth and particularly cruel Fatimid caliph, al-Ḥākim bi-Amrillāh Abū ʿAlī al-Manṣūr. This fixes the date of the treatise between 996 AD and 1021 AD, when al-Ḥākim reigned. When Plato of Tivoli translated the text in 1136,[275] he probably chose the title *Capitula stellarum oblata regi magno sarracenorum Almansor astrologo filio Abrahae Judaei:* "Chapters on the stars dedicated to the great king of the Arabs Almansor by the astrologer Ibn Abraham Judaei." There are many variants of this title in the Latin manuscript tradition;[276] some of the manuscripts and editions have *oblata . . . ab Almansore,* thus mistaking *Almansor* for the author. The Renaissance bio-bibliographical tradition (Simon de Phares, Gesner, and Baldi) ascribes the work to *Almansor.*[277] The treatise itself is a loosely ordered collection of 150 astrological aphorisms, which cover different branches of astrological science: catarchic astrology *(electiones)*, nativities, and years of the world.[278]

(ca. 1492).  *Centiloquium divi Hermetis . . . Almansor scripsit ter quinquagita (sic) polorum praecepta quibus est scire quid astra velint,* Venice.

(1493).  *Liber quadripartiti Ptholemei . . . Centum quinquaginta propositiones Almansoris,* Venice.

(1501).  *Albubather et Centiloqium divi Hermetis . . . Almansoris Iudicia seu propositiones. Incipiunt capitula stellarum oblata regi magno Saracenorum ab Almansore astrologo,* Venice.

(1519).  *Quadripartiti Ptolomei . . . Centum quinquaginta propositiones Almansoris,* Venice.

(1533).  *Iulii Firmici Materni . . . astronomicôn libri VIII . . . Almansoris astrologi propositiones ad Saracenorum regem,* Basel (with Firmicus Maternus).

(1551).  *Iulii Firmici Materni . . . astronomicôn libri VIII . . . Almanzoris astrologi Propositiones ad Saracenorum regem,* Basel (with Firmicus Maternus).

(1641).  *Astrologia aphoristica Ptolomaei, Hermetis, Ludovici De Rigiis, Almansoris, Hieronymi Cardani et autoris innominati,* Ulm.

(1674).  *Astrologia aphoristica Ptolomaei, Hermetis, Ludovici De Rigiis, Almansoris, Hieronymi Cardani et autoris innominati,* Ulm.

## MAIMONIDES (RABI MOYSES)

Abū ʿImrān Mūsā ibn Maymūn, Mose ben Maymon (Rabi Moyses in medieval Latin, Moses Maimonides in modern languages) is the author of the most important work of Jewish medieval philosophy, the "Guide of the Perplexed." This work was written in Arabic, as were all his other major works, with the exception of the Hebrew *Mishneh Torah.* Maimonides was born into a Jewish family in Cordoba in 1138 (not 1135); during his childhood, the family was forced to leave Cordoba, when the new intolerant Almohad rulers seized power in Andalusia. The family lived for some time in the Maghreb, until they moved to Egypt in 1165. Maimonides worked as personal physician to the eldest son of Saladin, and was also the spokesman of the Jewish community in Cairo; he died in 1204.[279]

His magnum opus, the *Dalālat al-ḥā'irīn* ("Guide of the Perplexed"), was translated into Hebrew as *Mōrē ha-nebūkīm* by Samuel ibn Tibbon and again by Jehuda al-Ḥarisi, soon after it had been finished (Hebrew versions exist for almost all of Maimonides' scientific writings). Al-Ḥarisi's version was the source for the medieval Latin translation, which an unknown translator produced in the early thirteenth century. Its title is *Dux neutrorum vel dubiorum*.[280] The "Guide of the Perplexed" is a philosophical-theological summa, which covers a wide range of topics in physics, metaphysics, epistemology, political philosophy (especially with respect to the religious community), prophecy, and negative theology. As the main work of Jewish Aristotelianism, it defends the place of philosophy within a religious society. A partial translation of the "Guide of the Perplexed" (book 2, preface and chapter 1), which was based on the Hebrew version of Ibn Tibbon, was known in the Latin West under the title *Liber de uno deo benedicto*.[281] In 1629, Johann Buxtorf junior (1599–1664), like his father a professor of Hebrew in Basel, published a new Latin translation of the entire "Guide" from Ibn Tibbon's version, in order to promote Hebrew studies.[282] Maimonides and his *Dux neutrorum* had a very tangible influence on the scholastic discussion in the thirteenth century. In the later fourteenth century the influence declined. A Renaissance author clearly interested in Maimonides was Pico della Mirandola.[283]

In the Renaissance, Maimonides was valued primarily as a medical author. At least six of about a dozen medical works of his were translated into Latin in the Middle Ages: John of Capua, in all likelihood, translated around 1300 the *Kitāb tadbīr aṣ-ṣiḥḥa* ("The Diatetics of Health"); this tract was written for the use of the son of Saladin, as is expressed in the Latin title: *Liber de regimine sanitatis ad Soldanum regem*.[284] In 1305, or more probably 1307 in Barcelona, Armengaud of Blaise translated the *Kitāb as-Sumūm wa-l-mutaḥarriz mina l-adwiya al-qattāla* ("On Poisons and on Protection against Deadly Medicines"), which was known under the Latin title *Tractatus de medicinis contra venena;* it was not printed.[285] Three further tracts were translated in the Middle Ages, but did not reach print: *De asmate* (translated from Arabic by Armengaud probably in 1294, and again by an unknown translator of the same century),[286] *De haemorrhoidibus*,[287] and *De coitu*.[288]

Maimonides's best-known medical work is his *Kitāb al-Fuṣūl* ("Book of Aphorisms"), a collection of about 1,500 quotations either drawn from or freely based on the writings of Galen. Maimonides here chooses a traditional literary format, which had also been adopted by Mesue and Rhazes. The circumstances of the Latin translation, which was probably produced in the thirteenth century, have not yet been elucidated.[289] In 1579 a humanist revision of the *Aphorismi* appeared.

## Aphorismi

(1489). *Prologus. Incipiunt Aphorismi excellentissimi Raby Moyses,* Bologna.

(1497). *Liber Rasis ad Almansorem . . . Afforismi Rabi Moysi,* Venice.

(1500). *Aphorismi Rabi Moysi,* Venice.

(1500). *Liber Rasis ad Almansorem . . . Aphorismi Rabi Moysi,* Venice.

(1508). *Liber Rasis ad Almansorem . . . Aphorismi Rabi Moysi,* Venice.

(1579). *Aphorismi Rabi Moysis medici antiquissimi . . . ab interitu vindicati et iam primum in lucem editi,* Basel.[290]

## De regimine sanitatis

(ca. 1480/1481). *Tractatus Rabi Moysi quem domino et magnificho Soldano Babilonie transmisit,* Florence.

(1514). *Consiliorum consummatissimi . . . doctoris domini Joannis Matthei de Gradi . . . repertorium. Additis antiquissimi medici Rabbi Moysi De regimine vite quinque tractatibus ad Sultanum inscriptis,* Venice.

(1517). *Consilia Joannis Matthei de Gradi . . . additis antiquissimi medici Rabbi Moysi De regimine vite quinque tractatibus ad Sultanum inscriptis,* Pavia.

(1518). *Tractatus Rabbi Moysi De regimine sanitatis ad Soldanum regem,* Augsburg.

(1521). *Consilia Jo. Mat. de Gradi . . . additis antiquissimi medici Rabbi Moysi De regimine vite quinque tractatibus ad Sultanum inscriptis,* Venice.

(1535). *Consummatissimi . . . Ioannis Matthei de Gradi . . . consilia . . . Additis antiquissimi medici Rabbi Moysi De regimine vitae quinque tractatibus ad Sultanum inscriptis,* Lyon.

(1553). *Averrois Cordubensis Colliget . . . Tractatus Rabi Moysi Abemmaimon quem Soldano Babyloniae transmisit,* Venice.

## Dux dubitantium

(1520). *Rabi Mossei Aegyptii Dux seu director dubitantium aut perplexorum,* Paris, repr. Frankfurt am Main, 1964.

(1629). *Rabbi Mosis Majemonidis Liber mōre nevūkīm ... nunc vero nove ad linguae hebraicae cognitionem uberius propagandam eiusque usum et amplitudinem evidentius Christianorum scholis declarandam in linguam latinam perspicue et fideliter conversus a Johanne Buxtorfio filio,* Basel.

## MESSAHALAH

Māšā'allāh, one of the earliest and most influential Arabic astrologers, was a Jewish scholar from Basra, who was one of the expert astrologers appointed by the caliph to cast a horoscope for the foundation of Baghdad in 762. He died around 815. The Arabic bio-bibliographers credit him with numerous works, few of which have survived in Arabic. They treat all parts of the astrological science and draw on Sassanian, Greek, and Indian theories.[291] At least fifteen treatises, or parts of treatises, were translated into Latin in the Middle Ages (underlining indicates the treatises that were printed in the Renaissance):[292]

*Libellus de nativitatibus 14 distinctus capitulis* (translated by Hugo of Santalla;[293] corresponds to no. 1 in the tenth-century index of Arabic books by Ibn an-Nadīm); *De compositione astrolabii* (no. 5); *Epistola in pluviis et ventis* (related to no. 7; translated by "magister Drogo");[294] a series of abstracts: *De cogitationibus,*[295] *De occultis,*[296] *Liber iudiciorum, De interpretationibus* (probably from no. 9); *Liber super annona* or *De mercibus* (probably no. 13);[297] *De nativitatibus* (no. 14). Neither attested nor extant in the Arabic are: *De revolutionibus annorum mundi,*[298] *Epistola in rebus eclipsis lunae* (translated by John of Seville),[299] *De significationibus planetarum in nativitate, De septem planetis, De receptione* (perhaps translated by John of Seville),[300] *De scientia motus orbis* (translated by Gerard of Cremona),[301] and *Septem claves.* Many of these treatises had a remarkably wide distribution in Latin manuscript.[302]

The astrological treatises printed in the Renaissance treat the following topics: revolutions of the years of the world, with special respect to kings and kingdoms *(De revolutionibus),* eclipses of the moon and sun, conjunctions of the planets *(Epistola),* nativities *(De significationibus),* and interrogations *(De receptione).* Gerard of Cremona's translation concerned a treatise with a character more astronomical than astrological—which is a typical choice of Gerard's:[303] *De scientia* presents general principles of Aristotelian cosmology and Ptolemaic astronomy with respect to the art of astrology.[304] The two Nuremberg editions of 1549 were produced by Joachim Heller and thus testify to the interest in Messahalah's work in the circle of Philipp Melanchthon, who had been Heller's teacher. As in the case of the earlier edition of Albohali,

Heller mentions that the basis of his edition is a manuscript that originally belonged to the library of the Hungarian king Matthias Corvinus.[305]

In the Latin Middle Ages, Messahalah's name was attached to a very popular treatise on the construction and use of the astrolabe, which seems to be a thirteenth-century Western compilation; this compilation was based on a twelfth-century Latin translation of a treatise by Ibn aṣ-Ṣaffār (d. 1035 AD).[306] It was printed several times in the later Renaissance in the *Appendix matheseos* of the popular encyclopedia *Margarita philosophica* by Gregor Reisch.

Messahalah was also known in the Renaissance as one of the astrological authorities of the *Liber novem iudicum* (see s.v. 'Abohali,' 'Alkindi,' 'Omar,' and 'Zahel'), which does not, however, seem to transport authentic material by Messahalah.

## Messahalah

(1493).    *Liber quadripartiti Ptholemei ... Messahallach de receptionibus planetarum. Eiusdem de interrogationibus. Epistola eiusdem cum duodecim capitulis. Eiusdem de revolutionibus annorum mundi,* Venice: De recept., Epist. in rebus eclips., De revol. ann. mundi.[307]

(1504).    *Messahalah De scientia motus orbis,* Nuremberg: De scientia.

(1519).    *Quadripartiti Ptolomei ... Messahallach de receptionibus planetarum. Eiusdem de interrogationibus. Epistola eiusdem cum duodecim capitulis. Eiusdem de revolutionibus annorum mundi,* Venice: De recept., Epist. in rebus eclips., De revol. ann. mundi.

(1533).    *Iulii Firmici Materni ... astronomicôn libri VIII ... Messahalah de ratione circuli et stellarum et qualiter in hoc seculo operentur liber I,* Basel: Epist. in rebus eclips.

(1549).    *Messahalae antiquissimi ac laudatissimi inter Arabes astrologi libri tres,* Nuremberg: De revol. ann. mundi, Epist. in rebus eclips., De sign. plant., De cogitat., De recept.

(1549).    *De elementis et orbibus coelestibus liber ... Messahalae,* Nuremberg: De scientia.

(1551).    *Iulii Firmici Materni ... astronomicôn libri VIII ... Messahalah de ratione circuli et stellarum et qualiter in hoc seculo operentur liber I,* Basel: Epist. in rebus eclips.

## Pseudo-Messahalah: De compositione astrolabii

(1512).    *Margarita philosophica ... Tractatus de compositione astrolabii Messehalath,* Strasbourg.

(1515).     *Margarita philosophica ... Tractatus de compositione astrolabii Messehalath*, Strasbourg.

(1535).     *Margarita philosophica ... De compositione astrolabii Messahalath*, Basel.

(1583).     *Margarita philosophica ... De compositione astrolabii Messahalath*, Basel.

## MESUE

The only text by the true Mesue translated into Latin, at present knowledge, was the *Kitāb an-Nawādir aṭ-ṭibbīya* ("The Book of Medical Anecdotes"), which was known in Latin as *Aphorismi Iohannis Damasceni*. Yūḥannā ibn Māsawayh, a Nestorian Christian who was born after 786 and died in 857, was a medical celebrity of his day: he was director of the hospital in Baghdad and personal physician to several caliphs and was influential in commissioning the translation of Greek medical works into Arabic. Arabic biobibliographers credit him with about forty works, of which only a few have survived complete in Arabic manuscripts.[308] The aphorisms, a collection of sentences or short paragraphs on manifold topics of the medical art, are inspired by a Greek paradigm, the aphorisms of Hippocrates. Ibn Māsawayh's aphorisms were translated twice into Latin in the Middle Ages, once by an anonymous translator in the late eleventh or early twelfth century and once as the sixth book of Rhazes's *De secretis medicinae* by the Dominican Giles of Santarem (d. 1265).[309] The identical origin of the two Latin versions was first realized, apparently, in the Renaissance, by the anonymous compiler of a paraphrase printed in Strasbourg in 1528, who combined material from both translations; the compiler is possibly Alban Thorer, the author of a very similar paraphrase printed in 1542.[310] In 1649, Antonius Deusing of Groningen, the Arabic–Latin translator of Avicenna's *Cantica*, published a new Latin version of the *Aphorismi*, dedicated to the Arabic professor Jacob Golius.

    The great fame that Mesue acquired in the Middle Ages and in the Renaissance did not rest on this short text, but on a comprehensive, fourfold work of theoretical and practical pharmacology, printed in the Renaissance as the *Opera Mesue*. These texts were clearly not written by Yūḥannā ibn Māsawayh, since they quote later authors up to the late tenth century. The *Opera* of Pseudo-Mesue consist of four parts: *Canones universales* (rules for the preparation of drugs), *De simplicibus* (on laxative drugs), *Grabadin* (a pharmacopeia), and *Practica* (a therapeutics). The textual format in which the *Opera* are transmitted is the product of an anonymous Latin compiler of the thirteenth century, who took Avicenna's *Canon* as a model for arranging the material. The first two parts of Pseudo-Mesue's *Canones universales* are probably direct translations from the Arabic, but the original is not known. In general it is clear for linguistic reasons that much material of the *Opera Mesue* is of Arabic origin. The first three parts of the *Opera (Canones, De simplicibus,* and *Grabadin)* are connected by cross-references; they quote a very similar set of Greek and Arabic authors. It is thus possible that they are the work of an as yet unidentified Arabic author of the eleventh or twelfth century.[311]

The *Opera* of Pseudo-Mesue were a great printing success in the Renaissance, equaled only by Avicenna's *Canon* and Rhazes's *Liber ad Almansorem.* These three texts have in common that they play an important role in the medical education of Renaissance universities and that they appear in the curricula of medical faculties. The great attention payed to Pseudo-Mesue's *Opera* is also due to fact that the *Opera* constitute a very comprehensive handbook of pharmacology, and pharmacology was the most thriving area of humanist medicine. Several authors of the sixteenth century wrote commentaries on parts of the *Opera* and published them in print: Angelus Palea and Bartholomaeus Urbevetanus, Johann Dantz, Andrea Marini and Giovanni Costeo.[312] Among them is also a humanist protagonist of anti-Arabism, Giovanni Manardo; his literal commentary on Pseudo-Mesue's *De simplicibus* first appeared around 1521 (it is discussed in Chapter 4 above). These commentaries are a clear indication of the fact that Pseudo-Mesue reached the peak of his influence in the Latin West in the sixteenth century.

The textual transmission of the *Opera* in the Renaissance is marked by an important break: In 1542, the humanist physician Jacobus Sylvius (alias Jacques Dubois) published a thorough revision of *Canones, De simplicibus,* and *Grabadin* on the basis of Latin sources only. His changes to the Latin translation of the thirteenth century concern Latin vocabulary and syntax, as well as the arrangement of the material. Sylvius's technique of revision (which is analyzed in Chapter 3) is daring, but not unsuccessful, since it profits significantly from his medical expertise.

### Aphorismi

| | |
|---|---|
| (1481). | Milan. |
| (1489) | *Prologus. Incipiunt Aphorismi . . . Raby Moyses . . . Amphorismi Iohannis Damasceni,* Bologna. |
| (1497). | *Liber Rasis ad Almansorem . . . Afforismi Damasceni,* Venice. |
| (1500). | *Aphorismi Rabi Moysi. Aphorismi Joannis Damasceni,* Venice. |

(1500).          *Liber Rasis ad Almansorem* . . . *Aphorismi Damasceni,* Venice.

(1502).          *Liber Hysagoge Joannici* . . . *Liber aphorismorum Damasceni,* Venice.

(1507).          *In hoc volumine* . . . *continentur infrascripti codices* . . . *liber Aphorismorum Damasceni,* Venice.

(1508).          *Liber Rasis ad Almansorem* . . . *Aphorismi Damasceni,* Venice.

(1515).          *Articella nuperrime impressa* . . . *Aphorismi Joannis Damasceni,* Lyon.

(1519).          Lyon (in *Articella*).

(1525).          *Articella nuperrime impressa* . . . *Aphorismi Joannis Damasceni,* Lyon.

(1528).          *Alexandri Benedicti* . . . *De hystoria corporis humani* . . . *Aphorismi Damasceni,* Strasbourg (paraphrase).

(1534).          *Articella nuperrime impressa* . . . *Aphorismi Joannis Damasceni,* Lyon.

(1542).          *Alexandri Aphrodisei* . . . *opusculum* . . . *Ioannis Damasceni* . . . *eiusdem Aphorismorum libellus eodem Torino paraphraste,* Basel (paraphrase).

(1543).          *Iani Damasceni Decapolitani* . . . *curandi artis libri vii* . . . *Iani Damasceni Decapolitani* . . . *liber primus Albano Torino paraphraste. Aphorismus primus,* Basel (paraphrase).

(1579).          *Aphorismi Rabi Moysis* . . . *denique Ioannis Damasceni Aphorismi utilissimi ad filium,* Basel (paraphrase).

(1649).          *Canticum* . . . *Avicennae* . . . *cui adjecti Aphorismi medici Johannis Mesuaei Damasceni ex Arabico Latine reddita ab Antonio Deusingo,* Groningen (Deusing).


*Opera*

(1471).          *In nomine dei* . . . *principium verborum Ioannis filii Mesue,* s.l. (Padua).

(1471).          *In nomine dei* . . . *principium verborum Ioannis filii Mesue,* Venice.

(1473).          s.l. (Milano).

(1475).          Naples.

(1478).          Naples.

(1478).          *Incipit liber de consolacione* . . . *Johannis heben Mesue,* Lyon.

(1478).        Pavia.

(1479).        *In nomine dei ... principium verborum Joannis filii Mesue,*
               Venice.

(1479).        Milano.

(ca. 1482–1483). *Antidotarius Nicolai ... Johannis Mesue Grabadin,* s.l. (Stras-
               bourg) (Grabadin, with Nicolaus Salernitanus).

(1484).        *In nomine dei ... principium verborum Joannis filii Mesue,*
               Venice.

(1489/1490/1491). *Mesue cum additionibus Francisci de Pedemontium,* Venice.

(1495).        *Mesue cum expositione Mondini super Canones universales,*
               Venice.

(1497).        *Mesue cum expositione Mondini super Canones universales,*
               Venice.

(1502).        *Mesue cum expositione Mondini super Canones universales,*
               Venice.

(1505).        *Textus Mesue noviter emendatus,* Venice.

(1508).        *Mesue cum expositione Mondini super Canones universales,*
               Venice.

(1510).        *Mesue cum expositione Mondini super Canones universales,*
               Lyon.

(1510).        *Mesue cum expositione Mondini super Canones universales,*
               Venice.

(1511).        *Canones universales divi Mesue,* Lyon.

(1513).        *Canones universales divi Mesue,* Venice.

(1515).        *Mesue cum expositione Mondini super Canones universales,*
               Lyon.

(1517).        *Generales divi Mesue Canones,* Pavia.

(1519).        *Mesue cum expositione Mondini super Canones universales,*
               Lyon.

(1523).        *Domini Mesue vita ... Canones universales divi Mesue,* Lyon.

(1525).        *Mesue cum expositione Mondini super Canones universales,* Lyon.

(1527).        *Divi Mesue et nova quedam ultra ea que secum associari con-
               sueverunt opera,* Venice.

(1531).        *Domini Mesue vita ... Canones universales divi Mesue,* Lyon.

(1533).        *Opera divi Ioannis Mesue,* Lyon.

(1535).  *Opera divi Ioannis Mesue*, Lyon.

(1538).  Venice.

(1540).  *Textus Mesue*, s.l. (Lyon).

(1541).  *Opera divi Ioannis Mesue*, s.l. (Lyon).

(1542).  *Ioannis Mesuae Damasceni De re medica libri tres Iacobo Sylvio medico interprete*, Paris.

(1544).  *Ioannis Mesuae Damasceni De re medica libri tres Iacobo Sylvio medico interprete*, Paris.

(1545).  *Universales Ioannis Mesue praestantissimi et celeberrimi medici Canones*, Basel.

(1548).  *Ioannis Mesuae Damasceni De re medica libri tres Iacobo Sylvio medico interprete*, Lyon (printer: apud Guilelmum Rouillium sub scuto Veneto).

(1548).  *Ioannis Mesuae Damasceni De re medica libri tres Iacobo Sylvio medico interprete*, Lyon (printer: apud Ioan. Tornaesium et Gulielmum Gazeium).

(1549).  *Mesue et omnia quae cum eo imprimi consueverunt*, Venice.

(1550).  *Ioannis Mesuae Damasceni De re medica libri tres Iacobo Sylvio medico interprete*, Lyon.

(1551).  *De morbis internis curandis liber unus*, Lyon: only Grabadin.

(1553).  *Ioannis Mesuae Damasceni De re medica libri tres Iacobo Sylvio medico interprete*, Paris.

(1558).  *Mesue qui Graecorum ac Arabum postremus medicinam practicam illustravit*, Venice.

(1560).  Lyon.

(1561).  *Ioannis Mesuae Damasceni De re medica libri tres Iacobo Sylvio medico interprete*, Paris.

(1561/1562).  *Mesuae Graecorum ac Arabum clarissimi medici opera*, Venice.

(1566).  *Ioannis Mesuae Damasceni De re medica libri tres Iacobo Sylvio medico interprete*, Lyon.

(1568/1570).  *Mesuae medici clarissimi opera*, Venice.

(1581).  *Ioannis Mesuae medici clarissimi opera*, Venice.

(1589).  *Ioannis Mesuae Damasceni medici clarissimi opera*, Venice.

(1602).  *Ioannis Mesuae Damasceni medici clarissimi opera*, Venice.

(1623).  *Ioannis Mesuae Damasceni medici clarissimi opera*, Venice.

(1630).          *Iacobi Sylvii ... Opera medica ... Ioannis Mesuae Damasceni De re medica libri tres Iacobo Sylvio medico interprete*, Geneva.

(1634).          *Iacobi Sylvii ... Opera medica*, Geneva.

(1635).          Geneva.[313]


## OMAR TIBERIADES

Together with Messahalah, the Persian astrologer ʿUmar ibn al-Farruḫān at-Ṭabarī belonged to the earliest generation of astrologers active in Baghdad in the first decades of the Abbasid empire. Omar was involved in drawing up the horoscope for the foundation of Baghdad in 762, and apparently continued to have good relations to the court. He still lived in 812.[314] Of his various astrological writings, at least two were translated into Latin: the first is the *Kitāb al-Mawālīd* ("On Nativities"), a work in three books with some appendices, which was translated by John of Seville in the first half of the twelfth century.[315] One of the appendices, a short treatise beginning "Dixit Kankaf Indus . . . ," which discusses the Mighty Fardārs, that is, periods of 360 solar years, was printed in 1549 as an appendix to the rules of the *Tabulae Jahen (De diversarum gentium eris)* by Ibn Muʿāḏ (see above s.v.).[316] Second, Hugo of Santalla in the twelfth century translated the *Muḫtaṣar masāʾil al-Qaysarānī* ("Abridgment of the Caesarean (?) Interrogations"), a book of 138 chapters on astrological judgments; this translation was split up and incorporated into two compilatory works, the *Liber trium iudicum,* and its expansion, the *Liber novem iudicum,* dating from the mid-twelfth century.[317] Chapters 81–85 of the "Abridgment" in Hugo's translation, which concern weather forecasting, also appear separately in Latin manuscripts.[318]

The treatise *De interrogationibus,* which Luca Gaurico (in 1503 and 1525) prints as the fourth chapter of *De nativitatibus* in a translation by "Salomon" dated 1217, is a translation by Salio of Padua of an abbreviated version of Omar's *Iudicia.*[319]

(1503).          *Omar Tiberiadis astronomi preclarissimi Liber de nativitatibus et interrogationibus,* Venice: De nativit., De interr.[320]

(1509).          *Liber novem iudicum in iudiciis astrorum. Clarissimi auctores Mesehella, Aomar, Alkindus, Zael, Albenait, Dorotheus, Jergis, Aristoteles, Ptholemeus istius voluminis,* Venice: Iudicia.

(1525).          *Omar De nativitatibus et interrogationibus nuper castigatus et in ordinem redactus per d. Lucam Gauricum,* Venice: De nativit., De interr.

(1533).     *Iulii Firmici Materni ... astronomicôn libri VIII ... Omar De nativitatibus libri III,* Basel: De nativit. (with Firmicus Maternus).

(1549).     *De elementis ... Messahalae ... Item iisdem de rebus scriptum cuiusdam Saraceni,* Nuremberg: Kankaf Indus.

(1551).     *Iulii Firmici Materni ... astronomicôn libri VIII ... Omar De nativitatibus libri III,* Basel: De nativit. (with Firmicus Maternus).

(1571).     *Albohazen Haly filii Abenragel ... de iudiciis astrorum libri octo ... Compendium duodecim domorum coelestium ex clarissimis et vetustissimis authoribus, scilicet Messahalla, Aomare, Alkindo, Zaele, Albenait, Dorotheo, Iergi, Aristotele et Ptolemaeo collectum,* Basel: Iudicia.

## RHAZES

Abū Bakr Muḥammad ibn Zakarīyā᾽ ar-Rāzī is one of the foremost representatives of Arabic science. As a medical author, he was very influential in the Arabic and Latin Worlds and also in the Renaissance. Alchemical treatises of his were also known in both cultures. His philosophical work was controversially discussed in the Orient because of its critical attitude toward prophetic religion, but was not available in Latin (apart from some fragments).[321] Rhazes was born in Rayy (Persia), probably in 865. For part of his life, he was the director of the hospital in Baghdad, but he seems to have returned often to his hometown Rayy, where he also headed the hospital and where he died in 925. One of his main works on medicine is dedicated to the Samanid ruler of Rayy, Manṣūr ibn Isḥāq.[322] In the history of Arabic medicine, Rhazes stands signally apart because of his diagnostic abilities and his most comprehensive learning.

Rhazes has left an enormous oeuvre, especially in medicine, of which a significant portion has reached the Latin West, primarily through the Toledan translations of Gerard of Cremona in the twelfth century. In contrast to many other authors, apparently the entire medieval Latin corpus of Rhazes's medical works reached print in the Renaissance. The following translations are firmly attributed to Gerard of Cremona:[323] (1) *al-Kitāb al-Manṣūrī fī ṭ-ṭibb* ("The Book of Medicine <dedicated to> Manṣūr"), *Liber ad Almansorem,* a handbook of theoretical and practical medicine in ten books, which was by far the most often printed of Rhazes's works.[324] The translation exists in two recensions, of which only the later one appears to be by Gerard;[325] (2) the *Kitāb at-Taqsīm wa-t-tašḡīr* or *Taqsīm al-᾽ilal* ("The

Division of Causes"), *Liber divisionum,* a diagnostic guide for distinguishing diseases, symptoms, causes, and so on;[326] (3) the *Kitāb al-Mudḫal ilā ṭ-ṭibb* ("Introduction to Medicine"), *Introductio in medicinam,* an entirely theoretical work, which gives an outline of the basic principles of Galenic medicine.[327]

It seems probable that the following four translations were also made by Gerard:[328] (4) *Kitāb al-Aqrābāḏīn* ("Pharmacopeia"),[329] *Antidotarium,* a treatise on the composition of drugs;[330] (5) *Awǧāʿ al-mafāṣil* ("Ailments of the Joints"), *De aegritudinibus iuncturarum,* one of many short tracts written by Rhazes on the symptoms and treatments of specific diseases.[331] Of the same kind are the treatises (6) *De aegritudinibus puerorum,* which is extant only in Hebrew and Latin,[332] and (7) *Kitāb al-Ḥaṣā fī l-kulā wa-l-matāna* ("On the Stones in Kidney and Bladder"), *Tractatus de praeservatione ab aegritudine lapidis.*[333]

In the second half of the thirteenth century, Giles of Portugal, a Dominican in the convent of Santarem, produced a translation of (8) the *Maqāla fī sirr ṣināʿat aṭ-ṭibb* ("Treatise on the Secret Art of Medicine"), *De secretis medicinae,*[334] which was also printed under the title *Aphorismi.*[335] Rhazes's fame as a medical practicioner rests mainly on the clinical observations that he noted, among thousands of excerpts collected from earlier medical authors, in his multivolume work (9) *al-Ḥāwī* ("The Comprehensive Book"), translated into Latin as *Continens* by Faraǧ ibn Sālim in Sicily; the translation was finished in 1279.[336] Apart from these works, there are further medical treatises attributed to Rhazes extant in manuscripts and early prints, the authenticity of which is doubtful or remains to be investigated.[337]

As is evident from the list of printed editions, Rhazes's most successful piece of work in the West was book 9 of the *Liber ad Almansorem,* which contains a pathology *a capite ad calcem.* This book served as a manual for practical medicine in many medieval and Renaissance universities and appeared in a number of university statutes.[338] It was often commented upon in the Latin West, especially by Montpellier physicians in the fourteenth, and by Italian physicians in the fifteenth century, such as Cristoforo Barzizza (Padua), Syllanus de Nigris (Pavia), Giovanni Arcolano (Ferrara), Giovanni Matteo Ferrari da Grado (Milan), and Marco Gatinaria (Pavia).[339] In the sixteenth century, Leonardo Giacchini (Pisa) and Giambattista da Monte (Padua) wrote commentaries on *Liber ad Almansorem* 9, and Andreas Vesalius, at the end of his medical studies in Louvain in 1537, published a paraphrase of the same book.[340] The 1544 edition presents a humanist revision of the *Opera* of Rhazes by the Swiss scholar Alban Thorer (d. 1550); the preface to this edition is a programmatic statement on the value of Arabic medicine.

Rhazes continued to be reputed a first rank medical author until the late eighteenth century because of his treatise on smallpox, which was first translated in the Renaissance from a Greek intermediary: (10) the *Kitāb al-Ǧudarī wa-l-ḥasba* ("On Smallpox and Measles"), *De pestilentia,* which gives a detailed description of the symptoms and the development of the diseases. The first Greek–Arabic translation of the treatise was produced by Giorgio Valla of Piacenza (d. 1500), a relative of Lorenzo Valla;[341] it was first published in 1498. The Greek text appeared in 1548. In 1549, the Strasbourg physician Guinter of Andernach published a new Greek–Latin translation,[342] which was followed by another Greek–Latin translation in 1556 by Niccolò Macchelli; in the dedicatory epistles, Guinter of Andernach and Macchelli both claimed to be the first to translate the treatise into Latin. The first Arabic–Latin translation was made by Thomas Hunt and appeared in London, 1747, under the title *De variolis et morbillis.*[343]

The name of Rhazes is attached to many alchemical treatises in Latin, most of which do not seem to be authentic.[344] There is certainty, however, about two of Rhazes's alchemical works, the short *Sirr al-asrār* ("The Secret of Secrets"),[345] and the comprehensive *Kitāb al-Asrār* ("Book of Secrets"), which in three books treats the substances, the instruments, and the operations of alchemy.[346] The latter work was translated into Latin in the twelfth or early thirteenth century as *Liber Ebu Bacchar er Raisy.* It received several reworkings in the course of the thirteenth century, one by Paul of Taranto, who used much of Rhazes's work when compiling the Pseudo-Geberian *Summa perfectionis.*[347]

## Continens

| | |
|---|---|
| (1486). | *Incipit prologus libri Elhaui, idest totum continentis Bubikir Zacharie Errasis filii,* Brescia. |
| (s.d., ca. 1495). | *Continens Rasis ordinatus et correctus per . . . Hieronymum Surianum,* s.l. |
| (1506). | Venice. |
| (1509). | *Continens Rasis ordinatus et correctus per . . . Hieronymum Surianum,* Venice. |
| (1515). | *Liber Helchavy idest continens artem medicine,* Venice. |
| (1529). | *Continens Rasis,* Venice. |
| (1542). | *Habes candide lector Continentem Rasis,* Venice. |

Liber ad Almansorem and other Medical Works
(Lib. divis.; De aegrit. iunct.; De aegrit. puer.; Aphorismi; Antidot.;
De lapide; Introd. in med.; De section. et vent.)

(1472).    s.l. (Milan): Lib. ad Almans. 9 (comm. Ferrari da Grado).

(1476).    s.l. (Padua): Lib. ad Almans. 9 (comm. Syllanus).

(1480).    Padua: Lib. ad Almans. 9 (comm. Arcolano).

(1481).    Milan: Lib. ad Almans.; Lib. divis.; De aegrit. iunct.; De
           aegrit. puer.; Aphorismi; De section. et vent.

(1481).    Milan: Lib. ad Almans. 9 (comm. Ferrari da Grado).

(1483).    *Incipit nonus liber Almansoris cum expositione . . . Sillani de Nigris,* Venice: Lib. ad Almans. 9 (comm. Syllanus, comm.
           Tussignanus).

(1489).    *Prologus. Incipiunt Aphorismi excellentissimi Raby Moyses,* Bologna: Aphorismi.

(1490).    *Incipit clarificatorium Iohannis de Tornamira super nono Almansoris cum textu ipsius Rasis,* Lyon: Lib. ad Almans. 9
           (comm. Tornamira).

(1490).    *Almansoris liber nonus cum expositione Sillani,* Venice: Lib. ad
           Almans. 9 (comm. Syllanus, comm. Tussignanus).

(1493).    *Practica Joannis Arculani. Expositio noni libri Almansoris,*
           Venice: Lib. ad Almans. 9 (comm. Arcolano).

(1494).    *Cristofori Barzizii . . . introductorium . . . Rasis Zacharie filii in
           nonum regis Almansoris filii,* Pavia: Lib. ad Almans. 9 (comm.
           Barzizza).

(1497).    *Almansoris liber nonus cum expositione Syllani,* Venice: Lib. ad
           Almans. 9 (comm. Syllanus, comm. Tussignanus).

(1497).    *Practica Joannis Arculani. Expositio noni libri Almansoris,*
           Venice: Lib. ad Almans. 9 (comm. Arcolano).

(1497).    *Liber Rasis ad Almansorem,* Venice: Lib. ad Almans.; Lib.
           divis.; De aegrit. iunct.; De aegrit. puer.; Aphorismi; Antidot.; De lapide; Introd. in med.

(1497).    *Illustrissime et excellentissime princeps . . . in nonum Almansoris,*
           Pavia: Lib. ad Almans. 9 (comm. Ferrari da Grado).

(1500).    *Liber Rasis ad Almansorem,* Venice: Lib. ad Almans.; Lib.
           divis.; De aegrit. iunct.; De aegrit. puer.; Aphorismi; Antidot.; De lapide; Introd. in med.

(1500). Venice: De aegrit. puer. (with Ketham).

(1501). *Incipit clarificatorium Iohannis de Tornamira super nono Almansoris cum textu ipsius Rasis,* Lyon: Lib. ad Almans. 9 (comm. Tornamira).

(1502). *Practica ... Joannis Matthei de Gradi noviter correcta ... Pars prima commentarii textualis in nonum Almansoris,* Venice: Lib. ad Almans. 9 (comm. Ferrari da Grado).

(1504). *Practica Geraldi de Solo super nono Almansoris,* Lyon: Lib. ad Almans. 9 (comm Geraldus).

(1504). Venice: Lib. ad Almans. 9 (comm. Arcolano).

(1505). *In hoc volumine continentur Introductorium iuvenum Gerardi de Solo ... commentum eiusdem super nono Almansoris cum textu,* Venice: Lib. ad Almans. 9 (comm. Geraldus).

(1507). *In hoc volumine ... continentur infrascripti codices ... textus noni Almansoris,* Venice.

(1508). *Liber Rasis ad Almansorem,* Venice: Lib. ad Almans.; Lib. divis.; De aegrit. iunct.; De aegrit. puer.; Aphorismi; Antidot.; De lapide; Introd. in med.

(1510/1511). *Opera parva Abubetri filii Zacharie filii Arasi ... Liber ad Almansorem decem tractatus continens ... ,* Lyon: Lib. ad Almans.; Lib. divis.; De aegrit. iunct.; De aegrit. puer.; Aphorismi; Antidot.; De lapide; Introd. in med.

(1517). *Almansoris liber nonus,* Venice: Lib. ad. Almans. 9 (comm. Syllanus, comm. Tussignanus).

(1518). *Sillanus super nono Almansoris,* Venice: Lib. ad Almans. 9 (comm. Syllanus, comm. Tussignanus).

(1519). Lyon: Lib. ad Almans. 9 (comm. Ferrari da Grado).

(1521). *Practica Jo. Matthei de Gradi cum tabula. Praxis in nonum Almansoris,* Venice: Lib. ad Almans. 9 (comm. Ferrari da Grado).

(1521). *Practica Joannis de Tornamira cum tabula,* Venice: Lib. ad Almans. IX (comm. Tornamira).

(1524). Venice: Lib. ad Almans. 9.

(1527). *Practica Joannis Matthei de Gradi. Praxis in nonum Almansoris,* Lyon: Lib. ad Almans. 9 (comm. Ferrari da Grado).

(1531). *In hoc volumine continentur ... Serapionis ... De simplicibus medicinis ... Rasis filii Zachariae de eisdem opusculum perutile . . . De simplicibus ad Almansorem,* Strasbourg: Lib. ad Almans. 8 (?).

(1533).     *Opus medicinae practicae . . . Galeatii de Sancta Sophia in nonum tractatum libri Rhasis ad regem Almansorem,* Haguenau: Lib. ad Almans. 9 (comm. Galeazzo di Santa Sofia).

(1534).     *Rhasis philosophi tractatus nonus ad regem Almansorem,* Paris: Lib. ad Almans. 9.

(1534).     *Articella nuperrime impressa . . . Textus noni ad Almansorem,* Lyon.

(1540).     *Io. Arculani . . . opera . . . commentarii in Razis arabis nonum librum ad regem Almansorem,* Basel: Lib. ad Almans. 9 (comm. Arcolano).

(1542).     *Ioannis Arculani commentaria in nonum librum Rasis ad regem Almansorem,* Venice: Lib. ad Almans. 9 (comm. Arcolano).

<u>(1544).</u>   *Abubetri Rhazae Maomethi . . . opera exquisitiora,* Basel, repr. Bruxelles 1973: Lib. ad Almans.; Lib. divis.; De aegrit. iunct.; De aegrit. puer.; Aphorismi; Antidot.; De lapide; Introd. in med. (Thorer).

(1554).     *Ioannis Baptistae Montani Veronensis in nonum librum Rhasis ad Mansorem regem Arabum expositio,* Venice: Lib. ad Almans. 9 (comm. da Monte).

(1557).     *Practica Ioannis Arculani . . . nonum librum Rasis ad Almansorem regem accuratius explicat,* Venice: Lib. ad Almans. 9 (comm. Arcolano).

(1560).     *Practica Ioannis Arculani,* Venice: Lib. ad Almans. 9 (comm. Arcolano).

(1560).     *Practica seu commentaria in nonum Rasis ad Almansorem Ioannis Matthaei Gradii,* Venice (comm. Ferrari da Grado).

(1562).     *Ioannis Baptistae Montani . . . in nonum librum Rhasis ad regem Almansorem lectiones,* Basel: Lib. ad Almans. 9 (comm. da Monte).

(1564).     *Leonardi Iacchini Emporiensis . . . in nonum librum Rasis . . . ad Almansorem regem . . . commentaria,* Basel: Lib. ad Almans. 9 (comm. Giacchini).

(1576).     *De febribus opus sane aureum . . . Arabes: Avicenna, Rasis, Abimeron Avenzoar, Averrois, Isaac, Serapion, Haly Abatis, Actuarius,* Venice (excerpt on fevers).

(1577).    *Leonardi Iacchini Emporiensis . . . in nonum librum Rasis arabis medici ad Almansorem regem . . . commentaria,* Lyon: Lib. ad Almans. 9 (comm. Giacchini).

(1579/1580).    *Leonardi Iacchini Emporiensis in nonum Rasis ad Almansorem . . . commentarii,* Basel: Lib. ad Almans. 9 (comm. Giacchini).

(1622).    *Leonardi Iacchini Emporiensis . . . in nonum librum Rasis . . . ad Almansorem regem . . . commentaria,* Lyon: Lib. ad Almans. 9 (comm. Giacchini).

## De pestilentia

<u>(1498)</u>.    *Georgio Valla Placentino interprete hoc in volumine continentur Nicephori logica . . . Rhazes De pestilentia,* Venice (trans. Valla).

(1528).    *Rhazes philosophus de ratione curandi pestilentiam e graeco in latinum versus per Georgium Vallam Placentinum,* Paris (Valla).

(1529).    *Pselli de victus ratione . . . Rhazae cognomento experimentatoris De pestilentia liber Georgio Valla Placentino interprete,* Basel (Valla).

<u>(1548)</u>.    *Alexandru Trallianou . . . Biblia dyokaideka . . . Rhazae De pestilentia libellus ex Syrorum lingua in graecam translatus,* Paris (Greek text with comments by Jacques Goupyl).

<u>(1549)</u>.    *Alexandri Tralliani . . . libri duodecim. Razae De pestilentia libellus. Omnes nunc primum de graeco accuratissime conversi . . . per Ioannem Guinterium Andernacum,* Strasbourg (Guinter of Andernach).

(1552).    *Alexandri Tralliani . . . libri duodecim. Razae De pestilentia libellus,* Venice (Guinter).

(1555).    *Alexandri Tralliani . . . libri duodecim. Razae De pestilentia libellus,* Venice (Guinter).

(1556).    *Alexandri Tralliani medici libri duodecim,* Basel (Guinter).

<u>(1556)</u>.    *Razae libellus De peste de graeco in latinum sermonem versus per Nicolaum Macchellum,* Venice (Macchellus).

(1560).    *Alexandri Tralliani medici libri duodecim,* Lyon (Guinter).

(1575).    *Alexandri Tralliani medici libri duodecim,* Lyon (Guinter).

## SERAPION

Yūhannā ibn Sarābiyūn was a Syrian physician of the Abbasid period, who lived in the second half of the ninth century AD.[348] He wrote all his works in Syriac, among them two compendia of medicine, a longer and a shorter one, which were translated into Arabic. In the twelfth century, Gerard of Cremona produced a Latin version of the "Small Compendium" *(al-Kunnāš aṣ-ṣaġīr)*, which traveled under the title *Practica* or *Breviarium*. The Latin text of the *Practica* is a valuable document since the Syriac text seems to be lost, and of the Arabic only fragments exist.[349] In 1550, there appeared a new *translatio*, as the title page claims, of the *Practica* by Andrea Alpago, the accomplished Arabic–Latin translator of Avicenna (d. 1522). The new text is clearly a reworking of Gerard's translation, but it is unclear whether Alpago also consulted an Arabic manuscript or not. There is one chapter of the *Practica* of which the Arabic has been edited: chapter 5.21 on poisonous serpents. Alpago makes two sensible changes to Gerard's translation, one of which requires knowledge of the Arabic text.[350] The 1550 edition also contains Alpago's index to transliterated drug names, with Latin equivalents, in Serapion's *Practica*.[351]

The second major text associated with the name of "Serapion" bears the title *Liber aggregatus in medicinis simplicibus*. This work is not by Ibn Sarābiyūn, but originated in a completely different time and context: it was written in the eleventh century by the Andalusian scholar Ibn Wāfid (Abenguefit, see the lists of editions above s.v.) and was translated into Latin at the end of the thirteenth century. Its original title was *Kitāb al-Adwiya al-mufrada* ("Book on Simple Medicines").[352] The name "Serapion" was attached to the Latin text apparently to enhance its authority. It became one of the most widely read reference books on simple drugs in the later Middle Ages and the Renaissance and was subject to much humanist criticism.

(1497).        *Practica Joannis Serapionis dicta Breviarium. Liber Serapionis de simplici medicina,* Venice.

(1503).        *Practica Joannis Serapionis dicta Breviarium. Liber Serapionis de simplici medicina,* Venice.

(1525).        *Practica Joannis Serapionis. Index operum . . . Practica Joannis Serapionis aliter Breviarium nuncupata. Liber Serapionis de simplici medicina,* Lyon.

(1530).        *Practica Joannis Serapionis. Necessarium ac perutile opus totius medicine practice,* Venice.

(1543). *Iani Damasceni Decapolitani . . . therapeutice methodi, hoc est curandi artis libri vii partim Albano Torino Vitodurano paraphraste, partim Gerardo Iatro Cremonensi metaphraste,* Basel: Practica (paraphrase).

(1550). *Serapionis . . . Practica . . . quam postremo Andreas Alpagus Bellunensis medicus, philosophus, idiomatisque arabici peritissimus in latinum convertit,* Venice.

(1576). *De febribus opus sane aureum . . . Arabes: Avicenna, Rasis, Abimeron Avenzoar, Averrois, Isaac, Serapion, Haly Abatis, Actuarius,* Venice: Practica, excerpt on fevers.

## THEBIT BEN CORAT

Thābit ibn Qurra was a famous astronomer, mathematician, physician, and magician, as well as translator from Greek, who was born in Ḥarrān in 826 and died in Baghdad in 901. His principal fame in the Middle Ages and the Renaissance rested on a misattribution: "Thebit's" very influential theory of trepidation, which explained the movement of precession (in other terms: the movement of the stars of the eighth sphere) as an oscillatory motion of accession and recession, appears in a treatise *De motu octavae spherae,* which derives from a not yet identified Arabic text written around 1080, that is, after Thebit's death.[353] This treatise was not printed, even though "Thebit's" theory continued to be discussed in the Renaissance, for instance by Benardino Baldi (see Chapter 2).[354] Today historians of science highly value the true Thebit's work on infinitesimal mathematics and his attempt to mathematize astronomy.[355]

Only a small part of his large oeuvre[356] was translated into Latin:[357] Apart from *De motu* and another treatise that seems to be wrongly attributed to him (*De anno solis, Fī sanat aš-šams bi-l-arṣād,* "On the Solar Year on the basis of Observations"),[358] medieval Latin translations of a number of short tracts exist: *De his quae indigent expositione antequam legatur Almagesti* (*Tashīl al-Maǧistī,* "Explanation of the Almagest"); *De recta imaginatione spherae et circulorum eius diversorum,* on the construction of a celestial sphere and its arcs of reference; and *De quantitate stellarum et planetarum et proportione terrae* on celestial and terrestial measurements; *Liber carastonis* (*Risāla fī al-qarasṭūn,* "Treatise on the Roman Balance") on the science of weights; and *De figura sectore* (*Risāla fī šakl al-qaṭṭāʿ,* "Treatise on the Figure of Secants") on spherical geometry.[359] Of these translations, *De motu, De his quae indigent,* and *De figura sectore,* are apparently the work of Gerard of Cremona;[360] *De figura sectore* is also extant in two anonymous

Latin translations.[361] A widely copied treatise on talismans and on the underlying astrological and physical principles of this magical art was translated twice, as *De imaginibus* by John of Seville,[362] and as *Liber prestigiorum* by Adelard of Bath.[363] John of Seville's translation was printed in 1559.

In view of the fact that some of these works had an extremely wide distribution in medieval manuscripts (e.g., *De quantitate*), the Renaissance reception of Thebit is meager, if the number of printed editions is taken as indicative. Several tracts were not printed at all: *De motu, De his quae indigent, Liber carastonis,* and *De quantitate stellarum.* As an introductory author to Ptolemaic astronomy, Thebit was much less read in the Renaissance than in the Middle Ages; this role had been taken over by other authors, and by John of Sacrobosco in particular.

(1503). *Textus Spere materialis Joannis de Sacrobusto ... Verba Thebit acutissimi astronomi de imagine tocius mundi atque sperici ... per eundem* (sc. Conradum Noricum) *addita,* Leipzig: De recta imag.

(1509). *Textus Spere materialis Joannis de Sacrobusto ... Verba Thebit acutissimi astronomi de imagine totius mundi atque corporis sperici ... per eundem addita,* Leipzig: De recta imag.

(1518). *Sphera cum commentis ... Thebit de imaginatione sphere,* Venice: De recta imag. (with Sacrobosco).

(1559). Frankfurt am Main: De imagin.

## ZAHEL

Sahl ibn Bišr was an influential ninth-century astrologer. He first served at the court in Merv in Khorasan and later left for Baghdad, where he served the vizier; he died around 845 AD.[364] He has left a comprehensive astrological oeuvre, which remains largely unexplored; it incorporates much material from the Greek astrologer Dorotheus of Sidon.[365] Several treatises by Zahel were translated into Latin in the twelfth century, but the relation of these texts to the Arabic original has not yet been investigated fully. Five of these translations travel as one corpus in the Latin manuscript tradition:[366] *Liber introductorius, De 50 praeceptis, De interrogationibus* (these three treatises probably derive from Zahel's *Kitāb al-Aḥkām,* "Book of Judgments," on interrogational astrology),[367] *De significatore temporis* or *De temporibus* (a translation of the *Kitāb al-Awqāt,* "On Times," that is, the best time for beginning an activity),[368] and *Liber electionum* (a translation of the

*Kitāb al-Iḫtiyārāt ʿalā l-buyūt al-itnay ʿašar,* "On Elections according to the Twelve Houses").[369] In addition, in 1138 Hermann of Carinthia produced a translation of a treatise on predictions pertaining to whole nations and regions, which is titled *Fatidica* or *Pronostica;*[370] the Arabic original seems to be the *Kitāb Taḥāwīl sinī l-ʿālam* ("On the Revolutions of the Years of the World").[371] Finally, Hugo of Santalla translated—perhaps with the help of Hermann of Carinthia—astrological judgments, *iudicia,* by Zahel and incorporates them into the *Liber trium iudicum* (see above, s.v. 'Alkindi' and 'Omar'); such judgments also appear in the *Liber novem iudicum,* but in a partially different translation.[372] Of these Latin versions, the most successful in terms of manuscript tradition was *De temporibus.*[373] In the Renaissance, Zahel was an important authority on interrogations and elections.

(1493).      *Liber quadripartiti Ptholemei . . . Zahel de interrogationibus. Eiusdem de electionibus. Eiusdem de temporum significationibus in iudiciis,* Venice: Introd., De praec., De interr., De temp., De election.

(ca. 1508).   Frankfurt an der Oder: Introd.

(1509).      *Liber novem iudicum in iudiciis astrorum. Clarissimi auctores Mesehella, Aomar, Alkindus, Zael, Albenait, Dorotheus, Jergis, Aristoteles, Ptholemeus istius voluminis,* Venice: Iudicia.

(1509).      *Meseallach et Ptholemeus de electionibus,* Venice: De election.

(1513).      Paris: De election.[374]

(1519).      *Quadripartiti Ptolomei . . . Zahel de interrogationibus. Eiusdem de electionibus. Eiusdem de temporum significationibus in iudiciis,* Venice: Introd., De praec., De interr., De temp., De election.

(1533).      *Iulii Firmici Materni . . . astronomicôn libri VIII . . . Zahelis arabis De electionibus liber I,* Basel: De election. (with Firmicus Maternus).

(1551).      *Iulii Firmici Materni . . . astronomicôn libri VIII . . . Zahelis arabis De electionibus liber I,* Basel: De election. (with Firmicus Maternus).

(1571).      *Albohazen Haly filii Abenragel . . . de iudiciis astrorum libri octo . . . Compendium duodecim domorum coelestium ex clarissimis et vetustissimis authoribus, scilicet Messahalla, Aomare, Alkindo, Zaele, Albenait, Dorotheo, Iergi, Aristotele et Ptolemaeo collectum,* Basel: Iudicia.

# ABBREVIATIONS

| | |
|---|---|
| BEA | *The Biographical Encyclopedia of Astronomers,* ed. T. Hockey et al., 2 vols. (New York, 2007). |
| Cont. Eras. | *Contemporaries of Erasmus,* ed. P. G. Bietenholz, 3 vols. (Toronto et al., 1985–1987). |
| DBI | *Dizionario biografico degli italiani* (Rome, 1960–). |
| DSB | *Dictionary of Scientific Biography,* ed. C. C. Gillespie et al., 18 vols. (New York, 1970–1980). |
| EI$^2$ | *The Encyclopaedia of Islam,* new edition, 12 vols. (Leiden, 1960–2004). |
| EJud$^2$ | *Encyclopaedia Judaica, Second Edition,* 22 vols. (Detroit, MI, 2007). |
| Enc. Iran. | *Encyclopaedia Iranica,* ed. E. Yarshater (London, 1985–). |
| Enc. Ren. | *Encyclopedia of the Renaissance,* 6 vols., ed. P. F. Grendler (New York, 1999). |
| Enc. Ren. Phil. | *Encyclopedia of Renaissance Philosophy,* ed. M. Sgarbi, forthcoming. |
| Forcellini | *Lexicon totius Latinitatis,* ed. F. Corradini et al., 6 vols. (Prato, 1858–1875). |
| GW | *Gesamtkatalog der Wiegendrucke* (Leipzig, 1925–). |
| IA | *Index Aureliensis: catalogus librorum sedecimo saeculo impressorum* (Baden-Baden, 1965–). |
| Klebs | A. C. Klebs, 'Incunabula scientifica et medica,' Osiris 5 (1938): 1–359 (repr. Hildesheim, 1963). |
| LexMA | *Lexikon des Mittelalters,* 11 vols (Munich, 1980–1999). |

| | |
|---|---|
| MBW Pers. | *Melanchthons Briefwechsel. Kritische und kommentierte Gesamtausgabe. Bände 11–14: Personen,* ed. H. Scheible (Stuttgart-Bad Cannstatt, 2003–). |
| MPL | *Patrologiae cursus completus: Series latina,* ed. J.-P. Migne, 221 vols. (Paris, 1844–1855). |
| NDB | *Neue deutsche Biographie* (Berlin, 1953–). |
| ODNB | *Oxford Dictionary of National Biography* (Oxford, 2004). |
| Rout. Enc. Philos. | *Routledge Encyclopedia of Philosophy,* 10 vols., ed. E. Craig (London, 1998). |
| SEP | *Stanford Encyclopedia of Philosophy,* ed. E. N. Zalta, http://plato.stanford.edu. |
| ThLL | *Thesaurus linguae Latinae* (Leipzig, 1900–). |
| Ueberweg I | *Philosophie in der Islamischen Welt, Band 1, 8.–10. Jahrhundert, herausgegeben von Ulrich Rudolph unter Mitarbeit von Renate Würsch* (Zurich, 2012). |
| Verf.-Lex.[2] | *Die deutsche Literatur des Mittelalters: Verfasserlexikon,* 2nd ed., ed. K. Ruh and B. Wachinger (Berlin, 1978–2008). |
| < > | added |
| [ ] | deleted |
| *(italics)* | explanatory remarks |

# NOTES

## PREFACE

1. Petrarca, *Epistolae rerum senilium,* ep. 12, 2, in: *Opera* (1965), 2:1009: "Vix mihi persuadebitur ab Arabia posse aliquid boni esse"; the entire passage is quoted and translated in Burnett, 'Petrarch and Averroes,' 49. For context see Chapter 5 below, section 'The Public Image.'

2. Fuchs, *Paradoxorum medicinae libri tres* (1535), sig. A6r: "A fontibus autem haurient si ex Graecis, minime autem Arabibus, eius artis praecepta didicerint, quod in illis nihil non purum, nihil non eruditum, nihil porro quod non summo iudicio excultum depromptumque sit, inveniri queat, in his vero nihil ferme quod non rancidum putidumve sit offendi possit." On Fuchs, see Chapter 4 below.

3. Pomponazzi, *Questio de immortalitate anime* (1955), 93: "Ego magis abhorreo opinionem Averrois quam diabolum." For discussion, see Chapter 5 below.

4. Fries, *Defensio medicorum principis Avicennae* (1533), f. 40v: "Conspicimus interire doctrinalem medicinam et ab orco revocari nugamenta pernecantia. Avicenna damnatur et Apuleii mendacia adprobantur."

5. Postel, *Grammatica arabica* (ca. 1539/1540), sig. D2v: "Quid? Quod videas dilucide et clare dict[u]um apud Aben Sina uno aut altero folio tantum, quod vix Galenus cum suo Asiatismo quinque aut sex libris maximis absolvat?"

6. Prassicio, *Questio* (1521), sig. A1v: "Ego autem . . . nil certi nilque veri de immortalitate animae reperio nisi ea quae ex Averroi excerpi possunt." Cf. Chapter 5 below.

7. Klein-Franke, *Die klassische Antike,* 42 ("statt einer versinkenden Tradition nachzuweinen").

8. Schmitz, 'Der Anteil,' 233 ("Verhaftung in arabistischer Tradition").

9. Baader, 'Medizinisches Reformdenken,' 267.

10. Thorndike, *Science and Thought,* 10: "Another problem is, whether there was not a falling off in civilization in general and in scientific productiveness in particular after the remarkable activity of the twelfth and thirteenth centuries—in short, whether instead of a renaissance something of a backsliding did not set in with Petrarch."

11. Thorndike, *History,* 5:3. Cf. Sarton, *The Appreciation,* 174: "Hence, we might say that the main peculiarity of the Renaissance was its negative orientation; its creative activities concerned the past." Cf. Randall, *The Making,* 212: "For natural science humanism was an almost unmitigated curse." These trends in scholarship are surveyed by Cochrane, 'Science and Humanism,' 1039–1041, and Lindberg, 'Conceptions of the Scientific Revolution,' 13–15. For the historical origin of the cliché of the anti-scientific bias of humanism, see Grafton, *Defenders of the Text,* 1–5, and, in greater detail, Grafton, 'The New Science,' 203–207.

12. Pagel, 'Medical Humanism: A Historical Necessity,' 385n: "Nevertheless, the polemical attitude of the humanists was admittedly legitimate—methodical return to the ancient texts was the only way out of terminological chaos. It was a salutary reaction and meant to benefit mankind."

13. Document 6497 "On the contribution of the Islamic civilisation to European culture" (1991) and Document 9626 "Cultural co-operation between Europe and the south Mediterranean countries" (2002) of the Parliamentary Assembly of the Council of Europe. The texts are accessible on the website of the Parliamentary Assembly s.v. 'Official documents': www.assembly.coe.int.

14. Gouguenheim, *Aristote au Mont Saint-Michel.*

15. See Hasse, 'Pseudowissenschaft vom Abendland'; Bataillon, 'Sur Aristote et le Mont-Saint-Michel'; Büttgen et al., *Les Grecs, les Arabes et nous: Enquête sur l'islamophobie savante.*

16. An example of such an attitude is Brague, 'Das islamische Volk': "Früher wurde dieser arabische Einfluss *(sc., auf den Westen)* vernachlässigt. Jetzt wird er übertrieben." In line with this, Brague tries to minimize the intellectual contribution of Islamic theology, which he describes—against the historical evidence—as not being concerned with the rational investigation of divine things; see Brague, 'Wie islamisch ist die islamische Philosophie?' 167 (French version, 113). The articles and books by Serafín Fanjul on "the myth of Muslim Spain" are further examples of the tendency to downplay Arabic influence.

17. Flasch, *Meister Eckhart: Die Geburt der "Deutschen Mystik" aus dem Geist der arabischen Philosophie;* Belting, *Florence and Baghdad: Renaissance Art and Arab Science.*

18. Compare Grafton's methodical criticism of Rose's history of Renaissance mathematics: Grafton, 'Review of Rose, *The Italian Renaissance.*'

19. Examples of anti-Arab polemics, attacking the alchemical author Geber, can be found in the writings of Symphorien Champier, which criticize the science of alchemy; see Copenhaver, *Symphorien Champier,* 229–235.

20. Jacquart, 'The Influence,' 979.

21. See Schacht, 'Ibn al-Nafîs, Servetus and Colombo'; Ullmann, *Die Medizin,* 172–176; Jacquart and Micheau, *La médecine,* 202–203. On Ibn an-Nafîs' theory itself see Fancy, *Science and Religion in Mamluk Egypt.*

22. Kennedy, 'Late Medieval Planetary Theory'; Swerdlow, 'The Derivation,' 503–504, for an important argument; Saliba, 'The Astronomical Tradition,' 258–290, esp. 265–272, in Saliba, *A History of Arabic Astronomy,* art. 14

(with further literature); Hugonnard-Roche, 'The Influence,' 298–303; Ragep, 'Tusi and Copernicus'; Saliba, *Islamic Science and the Making of the European Renaissance* (only chapter 6 is on the Renaissance); Ragep, 'Copernicus and His Islamic Predecessors.' On the seventeenth and eighteenth centuries, see Kunitzsch, 'Late Traces of Arabic Influence.'

23. See Di Bono, 'Copernicus,' on Amico and Fracastoro; and Blåsjö, 'A Critique,' which suggests parallel development rather than direct influence.

24. Arabic criticism of Ptolemy, and Averroes's criticism in particular, was transported in several textual genres, notably in commentaries on *Metaphysics, De caelo* and *Physics*. See Barker, 'Copernicus and the Critics of Ptolemy'; Shank, 'Setting up Copernicus?'; Hasse, 'Averroes' Critique of Ptolemy'; Sylla, 'Astronomy at Cracow University.'

25. On the history of Arabic philology in the Renaissance, the most comprehensive study remains Fück, *Die arabischen Studien*. See also Dannenfeldt, 'The Renaissance Humanists'; Klein-Franke, *Die klassische Antike;* Piemontese, 'Venezia'; Jones, 'Thomas Erpenius'; Jones, 'Piracy'; Hamilton, *William Bedwell;* Hamilton, *'Nam tirones sumus,'* 105–117; Hamilton, *Europe and the Arab World;* Tinto, *La Tipografia;* Nuovo, 'Il Corano arabo ritrovato'; Daiber, 'The Reception'; Toomer, *Eastern Wisedome;* Bobzin, 'Agostino Giustiniani'; Bobzin, *Der Koran im Zeitalter;* Bobzin, 'Islamkundliche Quellen'; Hamilton, *Arab Culture,* esp. 79–94; Burman, *Reading the Qur'ān;* Hamilton, *The Forbidden Fruit;* Hamilton, 'The Long Apprenticeship'; Glei, *'Arabismus latine personatus';* Hamilton, *The Arcadian Library,* 297–313; Glei, *Frühe Koranübersetzungen;* Bobzin, *Ließ ein Papst den Koran verbrennen?;* Hasse, 'Contacts with the Arab World,' 287–289; and, for further literature, the survey by Bobzin, 'Geschichte der arabischen Philologie.'

26. Brentjes, *Travellers from Europe;* Hamilton, *The Arcadian Library,* 13–103; MacLean, *The Rise of Oriental Travel.*

27. For Leo Africanus, see Chapter 2 below.

28. See also Bobzin, 'Geschichte der arabischen Philologie,' 176–179.

29. Lichtenberg, *Sudelbücher,* F 578: "Alle Unparteilichkeit ist artifiziell. Der Mensch ist immer parteiisch und tut sehr recht daran."

## NOTE ON TERMINOLOGY

1. See Gutas, 'The Study of Arabic,' 16–19.

2. As is common practice in other disciplines; cf., for example, Grendler, *Universities,* xvi.

3. See Kristeller's 1988 article 'Humanism,' which sums up his many earlier publications on the issue (the quotation is on p. 133). For the opposite view see, for example, Garin, *Il rinascimento italiano,* or idem, 'Die Kultur der Renaissance,' 431–534. For a recent discussion, see Hankins, 'Humanism, Scholasticism,' the articles in Mazzocco, *Interpretations of Renaissance Humanism,* and Paganini, 'Umanesimo e "Denkstil." '

## 1 · INTRODUCTION: EDITIONS AND CURRICULA

1. The calculation is made on the basis of the catalog by Schum, *Beschreibendes Verzeichniss*. See, in general, Lehmann, *Mittelalterliche Bibliothekskataloge*, 2:1–5, and Döbler, 'Der Katalog,' for further literature.
2. This category includes the Hebrew–Latin tradition, such as the Old Testament.
3. Gutas, *Greek Thought, Arabic Culture*.
4. Endress, 'Reading Avicenna in the Madrasa.'
5. Hasse, 'The Social Conditions,' 79–84.
6. Champier, *De medicine claris scriptoribus* (1506?), f. 33v: "Girardus Cremonensis, vir in medicinis studiosus et exercitatus. Erat enim vir in omni scientia et lingua doctissimus... Transtulit Canones medicine in Toleto de arabico in latinum quos princeps Aboali edidit."
7. Hasse, *Latin Averroes Translations*.
8. Carion, *Chronica* (1533), s.v. 'die vierde Monarchi': "Das Almagestum Ptolemei hat er erstlich aus Sarracenischer sprach in Latin bringen lassen und dadurch die schöne kunst Astronomia, die inn gantzem Europa kein mensch lange zeit gelernet hat, widder auff bracht"; Blancanus, *Clarorum mathematicorum chronologia* (1615), 58: "Federicus Secundus imperator primus Almagestum ex arabo in latinum converti curavit adeoque astronomiam omnem excoluit."
9. Jean Bruyerin Champier, dedicatory letter to Averroes, *Collectaneorum... sectiones tres* (1537), sig. A3v: "Cum vero in Hispania imperaret Alphonsus literarum, et maxime mathematicarum, sitientissimus, Mauris adhuc Bethicam obtinentibus, facile fuit ut partim ob viciniam, partim ob frequens populorum commercium comportarentur in Hispaniam citeriorem Auerrhoi aliorumque Maurorum lingua conscripti libri, ubi ab Hispano quopiam Latinitate utcumque donati sint, aut florentibus iam in Lutecia Parisiorum philosophiae medicinaeque studiis, ex Hispania in Gallias deportati, illic transferri potuerunt."
10. Milich, *Oratio de Avicennae vita* (1550), sig. B7v: "E(t)si enim pulsa lingua graeca, veteres autores etiam exulabant, tamen aliqui studiosiores requirebant fontes. Quare et in linguam Arabicam pleraque Hippocratis Galeni et Ptolemaei opera a Sarracenis in linguam Arabicam sunt conversa, ex qua lingua non multo post nostrorum imperatorum beneficio Lotharii ac Friderici secundi in latinam linguam translata sunt."
11. Girolamo Donzellini, preface to Leonardus Iacchinus (Leonardo Giacchini), *In nonum librum Rasis... ad Almansorem* (1564), sig. B4r: "A Graecis ad Arabas delata *(sc., medicinae scientia)*, naufragium fecit, ac Latini ab Arabibus illam recipientes, diu admodum infoeliciter in illa versati sunt. Deus tandem nostram sortem miseratus, cum una cum linguarum peritia, scientias in lucem reuocaret, hanc etiam diuinam artem illustrauit, excitatis aliquot viris, qui e limpidis Graecorum fontibus illam docerent." The passage forms part of

Donzellini's letter of dedication, which is dated to 1563. On Donzellini, see DBI s.v. 'Donzellini (Donzellino, Donzellinus), Girolamo' (by Jacobson Schutte).

12. Champier in an undated letter to Sébastien Monteux, printed in Monteux, *Annotatiunculae* (1533), f. 54v: "A Plinii enim tempore complures lingua externa ac arabica scripserunt, quo quidem barbarorum feritas Italiam, Galliam, Germaniam ac Hispaniam invasit et magis literarum decus quam urbium ornamenta sustulit, ex qua nobili studiorum strage medicinae vilitas orta est inconstantissimaque omnium artium facta."

13. The Appendix lists 133 editions for Averroes, but 19 of them are listed more than once.

14. Some of Roger Bacon's texts, however, circulated in manuscript. For further examples of scholastic authors famous today, but poorly represented in early prints, such as John Scotus Eriugena, John of Salisbury, Siger of Brabant, and Marsilius of Padua, see Kraye, 'The Role of Medieval.'

15. See Hasse and Bertolacci, *The Arabic, Latin and Hebrew Reception of Avicenna's Metaphysics,* and Hasse, *Avicenna's* De Anima.

16. For examples, see Hasse, 'Arabic Philosophy and Averroism,' 121–129 (on Avicenna's doctrines of prophecy and spontaneous generation), and Leinsle, *Das Ding und die Methode,* chapter 2 (on the subject of metaphysics and on essence and existence, as discussed by Jesuit authors). See also d'Alverny, *Avicenne en occident,* art. 13 ('Avicenne et les médecins de Venise').

17. The phrase *princeps medicorum* appears on several title pages of Avicenna editions; see the Appendix s.v. 'Avicenna.' On Avicenna's epithet as *princeps,* see Hasse, 'King Avicenna.'

18. On Thebit's works, see EI² s.v. 'Thābit b. Kurra' (by Rashed and Morelon), on his Latin transmission, the partially outdated Carmody, *Arabic Astronomical,* 116–129 (to be supplanted by the 'New Carmody' in preparation by Charles Burnett and David Juste).

19. As historians of medicine have observed before, see, for example, Keil, in LexMa s.v. 'Mesue,' and Schmitz, *Geschichte der Pharmazie,* 382.

20. Siraisi, *Avicenna in Renaissance Italy,* 127–174.

21. Lucchetta, *Il medico e filosofo,* document 12, p. 90.

22. Siraisi, *Avicenna in Renaissance Italy,* 139–141.

23. Ibid., 143–156. See also the literature on the development of Arabic studies cited in the endnotes to the Preface.

24. These editions are listed by Burnett, Yamamoto, and Yano, *Al-Qabīsī,* 192–194, and in the Appendix.

25. On Ratdolt, who also published several editions of Albumasar, see NDB s.v. 'Ratdolt, Erhard' (by Künast), for further literature.

26. On the medieval commentary tradition on Alcabitius, see Arnzen, 'Vergessene Pflichtlektüre,' 100–107, 110–112.

27. On Turrel, see Thorndike, *History,* 5:307–312.

28. On Nabod, see Thorndike, *History,* 6:119–123.

29. Nabod, *Enarratio elementorum astrologiae, in qua praeter Alcabicii, qui Arabum doctrinam compendio prodidit, expositionem atque cum Ptolemaei principiis collationem, reiectis sortilegiis et absurdis vulgoque receptis opinionibus de verae artis praeceptorum origine et usu satis disseritur* (Cologne, 1560).

30. Nabod, *Enarratio* (1560), sig. δ1v–δ2r.

31. On de Fantis and his editions, see Scarabel, 'Une "édition critique" latine,' 9–18, and Burnett, Yamamoto, and Yano, *Al-Qabīsī*, 220–221.

32. Haly filius Abenragel, *Libri de iudiciis astrorum* (1551), sig. A3v: "in meliorem latinae linguae sermonem transcribi." For a full quotation and translation of this passage of the *Epistola nuncupatoria*, see Burnett, 'The Second Revelation,' 189–190.

33. On the printing history of Averroes, see Cranz, 'Editions of the Latin Aristotle'; Schmitt, 'Renaissance Averroism,' in *The Aristotelian Tradition*, art. 8, 121–141; and Burnett, 'Revisiting.' See also the prefaces of Roland Hissette to his critical editions of Averroes, *Super libro Praedicamentorum*, 39*–53* and Averroes, *Super libro Peri Hermeneias*, 19*–24*.

34. See Hissette, 'Des éditions,' 108–109.

35. As Vernia mentions in the preface to the second volume, the request came from Marco Sanudo, son of the governor of Padua (in 1480): "Marcus Sanutus eius filius, dum philosophiae incumberet videretque Averoim commentatorem corruptum esse, me rogavit ut eum quoad possem emendarem." And Vernia adds: "I tried, to the best of my abilities, to most carefully correct the extant <texts> by the commentator, since I had found in them many corruptions" ("conabor item pro virili mea residuum commentatoris diligentissime emendare, cum in eo plurimas corruptiones invenerim"). Quoted after Hissette, 'Le corpus averroicum,' 302, 305. For the date of Sanudo's pledge, see Hissette, 'Le corpus averroicum,' 306. See also Hissette, 'Des éditions,' 105–122.

36. For references to these commentaries see Chapter 5, note 175. For further Renaissance commentaries on Averroes, in particular, on *De substantia orbis* and *Colliget*, see the Appendix and Martin, 'Humanism,' 70–75.

37. See the references to Hebrew super-commentaries on Averroes in Steinschneider, *Die hebraeischen Übersetzungen*, §22ff. (logic), §49ff., 54, 59, 60b, 72 (natural philosophy), §86 (metaphysics). Steinschneider does not record a super-commentary on the Long commentary on *De anima*. For super-commentaries on the Middle and Long commentary on the *Physics*, see Zonta, 'Aristotle's *Physics*,' 204–205; for super-commentaries of the fifteenth century, see Zonta, 'The Autumn of Medieval Jewish Philosophy.'

38. See Lohr, *Latin Aristotle Commentaries*, 2:504–512.

39. On this famous edition, see Schmitt, 'Renaissance Averroism,' in *The Aristotelian Tradition*, art. 8, pp. 121–142, Burnett, 'The Second Revelation,' and Burnett, 'Revisiting.'

40. Cf. Schmitt, 'Renaissance Averroism,' 132, and Grendler, *Universities*, 261, mistakenly; and Hissette, 'Jacob Mantino utilisateur,' 157, correctly.

41. DBI s.v. 'Bagolino, Gerolamo' (by Vasoli); Lohr, *Latin Aristotle Commentaries*, 2:28.

42. Schmitt, 'Renaissance Averroism,' 141–142.

43. Aristotle and Averroes, *Omnia . . . opera* (1550/1552), vol. 1, f. 5r: "quippe qui et multa accuratissime scribendo, et summa cum omnium admiratione in patavina academia publice interpretando, omnibus fuerat admirabilis."

44. Ibid., f. 20v: "Haec nos Averrois opera ne diutius studiosorum desideria nimis longa expectatione torqueremus, nunc tandem emisimus, alia quoque non pauca, quae in linguam nostram nondum venerunt, aliquando edituri. Nuper enim Bernardus Naugerius, vir eloquentissimus, qui politioribus literis ita delectatur, ut philosophiam ab Averroe subtiliter expressam admiretur, cum pro Repub. Veneta legationem Constantinopoli ageret, librorum qui in ea urbe apud Iudaeos et Arabas medicos inveniuntur hunc misit indicem." On Bernardo Navagero (1507–1565), see DBI s.v. 'Navagero, Bernardo' (by Santarelli).

45. Burnett, 'Revisiting.'

46. On logic at the University of Padua and on Tomitano, see Grendler, *Universities*, 250–253. See also Lohr, *Latin Aristotle Commentaries*, 2:461–464 (on Tomitano) and 302–304 (on Pavese).

47. The last appearance of the Giunta edition marked a turning point in the eyes of Charles Schmitt; see Schmitt, 'Renaissance Averroism,' in *The Aristotelian Tradition*, art. 8, 140: "With the 1575 edition, the era of Averroes' major influence in Italy came to a close."

48. As noted already by Schmitt, 'Renaissance Averroism,' 140n89: "A search through the BM and BN catalogues fails to reveal any seventeenth-century Averroes editions."

49. Lucchetta, *Il medico e filosofo*, document 12, p. 90: "rettulerunt et relationem suam fecerunt singulatim cum vidissent nonnulla opera eiusdem magistri Andree, translat[at]a ex arabo in latinum, illaque sumopere comendarunt tamquam utilia et necessaria." Cf. also document 11, pp. 88–90.

50. Jacquart, 'La Réception'; Siraisi, *Taddeo Alderotti*, 107; Jacquart and Micheau, *La médecine*, 191–192.

51. Siraisi, *Taddeo Alderotti*, 107; Grendler, *Universities*, 319–323; Jacquart and Micheau, *La médecine*, 193–196 (with a useful table).

52. Siraisi, *Avicenna in Renaissance Italy*, 186; Grendler, *Universities*, 323–324.

53. Jacquart and Micheau, *La médecine*, 191–192; Grendler, *Universities*, 323–324; Siraisi, *Avicenna in Renaissance Italy*, 85, 178.

54. Illgen, *Die abendländischen Kommentatoren*, 5–7 (=2: 117–119), and Sezgin, *Geschichte*, 3:282.

55. Liess, *Geschichte der medizinischen Fakultät*, 292 (1555/1560), 299 (1571).

56. Schmitz, *Geschichte der Pharmazie*, 497–501.

57. Goltz, *Studien*, 357. On Cordus's *Dispensatorium*, see Friedrich and Müller-Jahncke, *Geschichte der Pharmazie*, 2:199–203.

58. Siraisi, *Avicenna in Renaissance Italy*, 84, 123.

59. Stübler, *Geschichte*, 32–41; Nutton, 'John Caius,' 377–378.

60. Cited from Stübler, *Leonhart Fuchs*, 171: "Et quum nemo sit, qui nesciat Arabes omnia ferme sua e Graecis transscripsisse, parcissime deinceps ad doctrinam

studii huius adhibebuntur, quod consultius sit artis praecepta a fontibus quam a turbidis rivulis haurire. Posthabitis itaque quam maxime fieri potest Arabibus ac aliis ineptis ac barbaris autoribus unus deinceps Hippocrates et Galenus . . . enarrabuntur."

61. The reintroduction of Arabic authors was noted by Nutton, 'John Caius,' 377–378, and Siraisi, *Avicenna in Renaissance Italy*, 78.

62. Liess, *Geschichte der medizinischen*, 93-98 ("Wiederbelebung der arabischen Medizin") and 299: "veteremque medicinam, quam non immerito catholicam appellare licebit . . . explicabit."

63. Nauck, *Zur Geschichte*, 92: "das allain die alten Classici authores, alls Hipocrites, Gallenus, Dioscordes, Avicena unnd Rosis publice gelesen werden solten."

64. Eisenlohr, *Sammlung*, 337.

65. Siraisi, *Avicenna in Renaissance Italy*, 78.

66. Friedensburg, *Geschichte*, 273-279.

67. Wegele, *Geschichte*, 2:249.

68. Siraisi, *Avicenna in Renaissance Italy*, 83.

69. Nutton, 'John Caius,' 378n27; Siraisi, *Avicenna in Renaissance Italy*, 77.

70. Siraisi, *Avicenna in Renaissance Italy*, 78.

71. Ibid., 85-89.

72. Denifle and Chatelain, *Chartularium*, 1:278 (no. 246). The three sources in fact written by Arabic authors are the following: "librum primum metheorum cum quarto in Ascensione; . . . librum de causis in septem septimanis; . . . librum de differentia spiritus et animae in duabus septimanis."

73. Malagola, *Statuti*, 274.

74. Denifle and Chatelain, *Chartularium*, 3:145 (no. 1319).

75. Lhotsky, *Die Wiener Artistenfakultät*, 252-253.

76. See Lorenz, 'Libri ordinarie legendi.'

77. Baudry, *La querelle*, 68: "Statuit eadem facultas ut textui Aristotelis fides adhiberi debeat non secundum expositionem Wycklef, Occan, suorumve sequatium vel aliorum suspectorum, sed secundum quod eum exponunt et intelligunt suus commentator Averroys, ubi contra fidem non militat, vel dominus Albertus magnus, vel sanctus Thomas de Aquino, vel Egidius de Roma, vel aliquis alter quem placebit facultati concorditer acceptare."

78. Kristeller, 'Paduan Averroism,' 112; Monfasani, 'Aristotelians,' 249-256. Cf. Kraye, 'The Philosophy of the Italian Renaissance,' 16: "In Italian universities the study of philosophy was propaedeutic to medicine rather than, as in Oxford and Paris, theology. This encouraged an atmosphere in which philosophy could operate as an autonomous discipline, guided solely by rational criteria." On the establishing of faculties of theology in Italy, see Grendler, *Universities*, 353-357.

79. Hasse, 'Der mutmaßliche arabische Einfluss.'

80. On this development, see Schmitt, 'Renaissance Aristotelianism,' 159-193, esp. 163, 179.

81. Kraye, 'The Philosophy of the Italian Renaissance,' 19–26, 40–41; Grendler, *Universities*, 271–279; Bianchi, *Studi sull'aristotelismo*, 180–183; Bianchi, 'Continuity and Change,' 56–61.

82. Lorenz, 'Libri ordinarie legendi'; North, 'The *Quadrivium*'; Grendler, *Universities*, 408–412.

83. Grendler, *Universities*, 418.

84. For instance, in the Bolognese statutes of 1405; see Malagola, *Statuti*, 276: "legatur tractatus astrolabii Mes[sa]chale."

85. Kunitzsch, 'On the Authenticity,' 42–62, in *The Arabs and the Stars*, art. 10.

86. Thorndike, *The Sphere*, 43.

87. Malagola, *Statuti*, 276: "In tertio anno, primo legatur Alkabicius, quo lecto legatur Centiloquium Ptolomei cum commento Haly."

88. As maintained by Grendler, *Universities*, 411.

89. Malagola, *Statuti*, 264: "Item statuerunt, ordinaverunt et firmaverunt quod doctor electus vel eligendus per dictam Universitatem ad salarium ad legendum in astrologia, teneatur iudicia dare gratis scolaribus dicte Universitatis infra unum mensem postquam fuerint postulata, et etiam singulariter iudicium anni in scriptis ponere ad stationem generalium Bidellorum." The English translation is from Thorndike, *University Records*, 282.

90. Stelcel, *Codex diplomaticus*, 47: 'Magistri Martini autem dicti Rex Ptolomeum in Quadripartito, Alcabicium, Centiloquium verborum Ptolomei, Albumasar et alios libros spectantes ad astrologiam, iudicium quoque correctum et a senioribus in eadem facultate revisum et approbatum, universitati singulis annis praesentabit." For context in Kraków, see Markowski, 'Die mathematischen,' and Stopka, 'Les programmes.'

91. On the discrepancy between the official records and the actual teaching of astrology, see North, 'The *Quadrivium*,' 352–358.

92. For references to astrological teaching at Italian universities other than Padua and Bologna, see Grendler, *Universities*, 422–429. See also Rutkin, 'Astrology,' esp. n. 26, for further literature on astrology at early modern universities. For an example of lectures in astrology at an Italian university, see Rutkin, 'The Use and Abuse,' 141–145. Rutkin analyzes Filippo Fantoni's lectures on Ptolemy's *Tetrabiblos* at the University of Pisa in 1585–1586.

93. As convincingly argued by Azzolini, *The Duke and the Stars*, 26–50.

94. Beaujouan, 'Motives and Opportunities,' 223.

95. Eamon, 'Astrology and Society,' 147–148.

96. Lemay, 'The Teaching of Astronomy,' 200–206.

97. North, *God's Clockmaker*, 325.

98. Graf-Stuhlhofer, *Humanismus*, 144–149.

99. Reisch, *Margarita philosophica* (1973), lib. 7, tr. 2, 293–328.

100. On astrological medicine in the Renaissance, see Müller-Jahncke, *Astrologisch-magische Theorie*, 135–259, Hirai, 'The New Astral Medicine,' and, for brief information mainly on English authors, Chapman, 'Astrological Medicine.' For astrological medicine in general, see Akasoy and Burnett, *Astro-medicine*.

101. Cardano, *De subtilitate* (1559), lib. 16, 570: "Huic Mahometus Moisis filius Arabs, Algebraticae, ut ita dicam, artis inventor, succedit. Ob id inventum ab artis nomine cognomen adeptus est."

## 2 · BIO-BIBLIOGRAPHY: A CANON OF LEARNED MEN

1. On the format of Renaissance biography, see, most informatively and pertinently, Cochrane, *Historians,* 52–56, 393–422, esp. 393–400. Cf. also Petoletti, 'Les Recueils *De viris illustribus*,' Ijsewijn, 'Die humanistische Biographie' (on the diversity of Renaissance writings in this genre), the miscellaneous articles in Buck, *Biographie und Autobiographie,* and in *Biographie zwischen Renaissance und Barock,* ed. Berschin, and McLaughlin, 'Biography and Autobiography,' 37–65. On medical biography, see Siraisi, *History,* 106–133, and Rütten, *Geschichte der Medizingeschichtsschreibung,* esp. the articles by Strohmaier on Arabic medical biography, by McVaugh on historical accounts of surgery in medieval Latin, by Nutton on Latin biographies of Galen, and by Siraisi on Latin biographies of Hippocrates.

2. On the Latin biography of Avicenna in the Renaissance, see d'Alverny, 'Survivance,' in *Avicenne en occident,* art. 15; Siraisi, *Avicenna,* 161–164 ("is a subject that has yet to be fully studied"); Hasse, 'King Avicenna.' On the depictions of Averroes, see Polzer, 'The "Triumph of Thomas" Panel,' and Zahlten, 'Disputation mit Averroes.'

3. An interesting case not treated here is the biography of Mesue; see *Domini Mesue vita* in the editions of the *Opera* of Pseudo-Mesue of Lyon, 1523, and Lyon, 1540, and Sylvius's dedicatory epistle in his edition of the *Opera,* here quoted from the Paris, 1544 edition (sig. A2r) on the question of whether Mesue was writing in Greek or in Arabic; Sylvius argues for the Arabic: "Alii arabice omnino scripsisse existimant, quae mihi sententia magis arridet, cum et versio vetus impolita quidem, sed (ut coniectare licet) fidelis, voces arabicas non paucas servarit, non substituerit." Cf. also Konrad Gesner on the identity of Mesue, as discussed in the present chapter.

4. Bonatti, *Tractatus astronomie* (1491), tr. 7, pars 1, ch. 8, sig. A1v: "sicut dixit ille reverendus praedecessor noster Albumasar, qui fuit melior in hac arte quam unquam latini habuissent, qui studuit Athenis ubi studium tunc vigebat, ut ipse idem fatetur in quodam suo libro." Cf. Bonatti, *Tractatus astronomie,* tr. 8, pars 4, ch. 6, sig. F6v: "sicut fuit ille reverendissimus predecessor Albumasar Tricas qui licet studuerit Athenis fuit tamen latinus, prout ipse idem in tractatu Agiget confitetur."

5. Burnett, 'The "Sons of Averroes,"' 261–262, esp. n. 7: "Nam dicitur filios suos fuisse cum imperatore Federico, qui temporibus nostris obiit."

6. Zamora, *De preconiis hispanie* (1955), tr. 7, 176–177: "Et sicut de Hispania fuit Aristoteles philosophorum praecipuus, ita et Averroyz commentator eius eximius, et exempla quae ipse ponit in Commento super Phisicis istud probant;

fuit etenim Cordubensis . . . Accedit ad Hispaniam decorandam Avicenna philosophus qui prae ceteris philosophis plures libros composuit, omnes scientias circuiens et perscrutans, triviales et quadriviales, caelicas et mundanas. Verumtamen, ut aliqui assueverant, quater viginti philosophi Cordubae congregati omnes libros illos composuerunt et Avicenni, cuiusdam regis filio, ascripserunt, ut ex hoc in maiori auctoritate libri haberentur et filii regis nomen in maiori memoria et reverentia haberetur." This passage is briefly discussed by Rico, 'Aristoteles Hispanus,' 160–162.

7. A gloss on f. 7v of a *De animalibus* manuscript (MS Venice, Bibl. Marciana 2666 [L. VI, lvi]); printed in d'Alverny, *Avicenna Latinus: Codices,* 116: "Iurgen est versus Persiam et est christianorum et in ea fuit Avicenna, ut hic dicit, quare patet quod non fuit Yspanus nec de Barbaria, immo videtur fuisse de terra Corasceni quae est circa Baldach, ut potest haberi per eum in secundo Canonis . . . Quare manifestum est ipsum non fuisse Ispanum nec Africum ex praedictis."

8. See d'Alverny, 'Survivance,' in *Avicenne en occident,* art. 15, 81–83.

9. Hasse, 'King Avicenna,' 230–231. That Algazel was the uncle and educator of the prince Avicenna is a legend that may also have derived from a misunderstanding of medieval incipits, perhaps incipits of Algazel's logic. The legend is found in the library catalog of Richard of Fournival of the mid-thirteenth century, quoted in Kischlat, *Studien,* 128: 'Abuhanudin Algazelin avunculi et nutritoris principis Albohali Aviscenni liber logicorum.'

10. See the often printed incipit of the *Canones,* that is, the first part of the *Opera Mesue,* which makes Mesue a descendant of a king of Damascus: "In nomine dei misericordis cuius nutu sermo recipit gratiam et doctrina perfectionem. Principium verborum Joannis filii Mesue filii Hamech filii Hely filii Abdela regis Damasci" (here quoted from *Mesue . . . Canones universales* [1497], f. 2r).

11. On the kingship of Geber, see Appendix s.v. 'Pseudo-Geber.' On Isaac Israeli, see the title page of the 1607 edition: "ab Isaaco Iudaeo Salomonis Arabiae regis adoptivo filio."

12. On Foresti's sources in general, see Pianetti, '"Fra" Jacopo Filippo Foresti,' 147–151; Krümmel, *Das "Supplementum Chronicarum,"* 159–222.

13. Antoninus Florentinus, *Cronica* (1491).

14. On the *Liber de vita et moribus,* see Prelog, 'Die Handschriften'; on its Arabic sources Hasse, 'Plato arabico-latinus,' 45–52.

15. On Colonna's *De viribus illustribus,* see Ross, 'Giovanni Colonna' (with a few specimens of a critical text). For a general survey of medieval literary history, see Lehmann, 'Literaturgeschichte im Mittelalter,' and Godman, 'Literaturgeschichtsschreibung,' 177–197, as well as Lehmann's more specialized articles: 'Der Schriftstellerkatalog,' and 'Staindel-Funde'; for medical historiography, see Heischkel, 'Die Geschichte,' 204–206, and the articles in Gadebusch Bondio and Ricklin, *Exempla medicorum;* for philosophical historiography, see the articles in Ricklin and Gadebusch Bondio, *Exempla docent.*

16. See McVaugh, 'Historical Awareness.'

17. Salutati, *De nobilitate legum et medicinae,* 70: "Fuit et post hos omnes cunctorum que scripta repperit congregator Rhasis, qui librum maximum scripsit, quem quod omnia medicine contineat nominant 'Continentem.' Sequitur autem in tempore cunctorum que prius scripta fuerant digestor et ordinator egregrius Avicenna, quem regem nescio qua tamen auctoritate iactatis; vir plane mirabilis et qui lumen maximum attulerit medicine. Fuerunt et alii, ut commentator Averoys, Mesue, Rabi Moyses, et alii multi quos operosius quam ingeniosius est referre."

18. On Tortelli, see Schullian and Belloni, *Giovanni Tortelli,* xiv–xvii, and LexMa s.v. 'Tortelli,' with further literature (by Cortesi).

19. Tortelli uses the term *Poenus.* For further examples of the usage of the term for Arabic authors, see Index s.v. 'Poenus.'

20. Tortelli, *Liber de medicina et medicis* (1954), 15–16: "Ex quibus omnibus longo tempore post Rasis quidam Poenus lingua arabica opus grande ad medicinae artem faciens utiliter compilavit. Sed distinctius et scientius post eum in eadem lingua Avicenna Hispalensis effecit, qui cum reliquorum scripta viderit tum maxime Galienum secutus est, quem in suo libro primo de infrigidandis medicorum principem appellavit."

21. Ibid., 16–17: "Dinus vero adeo utilia in Avicennam commentaria scripsit ut sine illis nonnulli asserunt medici Avicennae doctrinam stare non posse."

22. Pianetti, '"Fra" Jacopo Filippo Foresti,' 100–106; DBI s.v. 'Foresti, Giacomo Filippo' (by Megli Frattini); Cochrane, *Historians,* 377–386; Krümmel, *Das "Supplementum Chronicarum,"* 56–73.

23. Foresti, *Supplementum* (1486), f. 218r: "Albaterius natione arabs medicus," "Johannes medicus cognomento Serapion," and "Isaach Benimiram medicus," f. 219r: "Rasis medicus," f. 224v: "Avicena quoque hispalensis medicus," "Averois quoque medicus," and "Zour etiam medicus," and Foresti, *Supplementum* (1506), f. 299r: "Mesue medicus." Cf. the shorter entries in Foresti, *Supplementum* (1483), f. 107r on "Albateribus," "Joannes," and "Isaach"; f. 108r on "Rasis"; and f. 115v on "Avicenna," "Averois," and "Zoar."

24. Foresti, *Supplementum* (1486), f. 224v: "Averois quoque medicus cognomento commentator, philosophus quidem clarissimus, et ipse apud Cordubam Hispaniae urbem hoc tempore, ut ex libro eius qui de celo tractat habetur, floruit et multa composuit et adeo excellenter in omnes Aristotelis libros scripsit ut commentatoris cognomentum promeruerit. Deinde de substantia orbis et de sectis composuit librumque pulchrum de medicina collegit atque de Tyriacha et de diluviis ac alia multa edidit. Avicene enim medici aemulus et inimicissimus fuit." The title *De sectis,* which recalls Galen's treatise *On sects* in medicine, does not appear on the long list of Averroes's medical writings in Cruz Hernández, *Abū-l-Walīd Muḥammad Ibn Rušd,* 50–51.

25. Pietro d'Abano, *Conciliator,* diff. 9, f. 14vb: "Tempore autem gratiae 1149 floruit commentator, ut 2 innuit coeli et mundi."

26. Averroes, *Commentum magnum super libro De celo,* 2.111, 480.

27. We know from his own notes that Foresti acquired two volumes of Averroes for the library of his convent: "Commentator supra physicam Aristotelis, Commentator supra Logicam Aristotelis. Magna volumina"; see Krümmel, *Das "Supplementum Chronicarum,"* 157.

28. Foresti, *Supplementum* (1506), f. 273v: "Alpharabium natione arabem, philosophum et astronomum insignissimum hisdem temporibus fuisse et viguisse tradunt, ac multa praeclarissima volumina eum scripsisse ferunt. E quibus etiam quaedam apud latinos habentur commentaria."

29. See Appendix s.v. 'Alfarabi.'

30. Foresti, *Supplementum* (1506), f. 273v: "Avedadis similiter *(just as Alfarabi)* arabs philosophus et medicus hisdem temporibus *(sc., 982 AD)* floruit et nonnulla edidit memoratu digna, e quibus in Aristotelis libros multa habentur commentaria, quae a nostris habentur multo in pretio."

31. Tortelli, *Liber de medicina et medicis* (1954), 15–16: "Eodem tempore Evax rex Arabum medicinae doctissimus volumen magnum ad Neronem scripsit in quo non solum herbarum virtutes sed lapidum etiam species, nomina, colores et quibus in regionibus valeant comperiri monstravit. De quo Plinius in quinto et vicesimo Naturalis Historiae dixit: Evax Dionysius et Metrodorus herbarum effigies pinxere atque subscripsere effectus."

32. Foresti, *Supplementum* (1506), f. 176v: "Evax arabum rex insignis philosophus et medicus ac rhetor. . . ." He gives a reference to Pliny's *Naturalis Historia,* book 25 (ch. 8); this reference is to a sentence that appears in the early editions of Pliny, but in fact is a medieval interpolation: "Evax rex arabum qui de simplicium effectibus scripsit"; see Halleux, 'Damigéron,' 327–330.

33. Pingree, 'The Diffusion,' 59–64.

34. Gesner, *Bibliotheca universalis* (1545), f. 224v.

35. Pingree, 'The Diffusion,' 60–61.

36. Foresti, *Supplementum* (1506), f. 81v.

37. Foresti, *Supplementum* (1486), f. 218r: "Albaterius natione arabs medicus quidem clarissimus tempestate hac floruit et multa conscripsit ac transtulit et inter cetera Galleni magni libros omnes in linguam arabam convertit Serapione medico teste c. 97."

38. See Appendix s.v. 'Iohannitius.'

39. Champier, *De medicine claris scriptoribus* (1506?), f. 21v: "Albatarich sive Albatarius natione arabs philosophus et medicus clarissimus floruit anno christi M.lxx. et multa ac varia conscripsit et transtulit et inter cetera Galeni libros omnes in linguam suam arabicam convertit. Et de ipso testatur Serapion capitulo de ligno aloes." Tiraqueau, *De nobilitate* (1616), 163: "Albatharich, Serapioni eiusdem operis, c. 85. Albatenius, qui primus Galenum in arabicam linguam vertit, ut dicit Serapio" (quoted from Steinschneider, *Vite,* 26n5).

40. Serapion, *Practica . . . De simplici medicina* (1497), f. 125ra (ch. 197): "Galenus in translatione Albatarich: Algaloyan est lignum a loco indum," and f. 111rb (ch. 85): "Albatarich: nominatur *(scil. molochia)* a triplex matinum et venditur in Babilonia in fasciculis." Cf. also f. 109ra (ch. 67): "Galenus in translatione

Batarie: lisacos est semen"; f. 122va (ch. 181): "Galenus in translatione[m] Athabarich: Lacca est gummi arboris"; f. 124va (ch. 194): "Et ita transtulit Albatarich in translatione sua in medicinis simplicibus"; f. 143va (ch. 329): "Galenus in translatione Albatharich: Non est essentia eius secundum nomen suum in graeco."

41. On al-Biṭrīq and his son Yaḥyā ibn l-Biṭrīq, see Dunlop, 'The Translations,' Ullmann, *Die Medizin,* 326, and Arnzen, *Aristoteles'* De anima, 149–150. Ḥunayn ibn Isḥāq attributes a (later?) Arabic translation of this treatise by Galen to Ḥubayš (Bergsträsser, *Ḥunayn ibn Isḥāq,* no. 53).

42. As has long been pointed out by Steinschneider, 'Die toxikologischen Schriften,' 366 (with references in Serapion); Steinschneider, *Die arabischen Übersetzungen,* (339)–(340). The exact references in Ibn al-Bayṭār are given by Ullmann, *Die Medizin,* 47–48. On Ibn al-Bayṭār, see Appendix s.v. 'Ebenbitar.'

43. Volaterranus, *Commentariorum urbanorum . . . libri* (1530), f. 247v.

44. Simon de Phares, in Boudet, *Le Recueil,* 1:320–321: "Albaterius de la nacion de Arabie, expert praticien es sciences de medicine et astrologie. Cestui, selon ce que recite Serapion, traslata les livres de Galien de langue grecque en langue arabicq."

45. Bartolotti, *De antiquitate medicinae* (1954), 61.

46. Mexía, *Le Vite di tutti gl'imperadori* (1561), 755–756: "Fiorirono ancora ne' tempi di quest' Imperadore nell'arte della Medicina quel grande ed eccellente Albetenio che scrisse commenti sopra Aristotele e tradusse Galeno nella lingua Arabica, e Serapione che scrisse de' semplici, ed altresi Rhasi detto ancora Almancor similmente Arabo ilquale abbreviò tutti i libri de' suoi antecessori e ridusse la sostanza in un libro intitolato Contenente i Medici." On Mexía, see Cont. Eras. s.v. 'Mexía, Pero' (by Deutscher).

47. Boudet, *Le Recueil,* 2:159.

48. On this Pseudo-Avicennian treatise, which was composed in twelfth-century Andalusia in Arabic and translated into Latin around 1230, see Moureau, 'Some Considerations.'

49. Foresti, *Supplementum* (1486), f. 220v: "Avicena quoque Hispalensis medicus omnium clarissimus eminentissimi ingenii vir, de cuius vita nil certum ferme habetur, his temporibus universo orbi clarissimus habetur, quem Mesue et Zoar medici Aboalim vocant, sed vulgares Bythinie regem fuisse dicunt, et ab Averoi medico venenatum tradunt, sed antequam periret ipse illum Averoim interfecisse. Qui cum in omni doctrina eruditissimus esset, librum excellentissimum composuit, in quo omnem logicam omnemque naturalem philosophiam multis voluminibus complecti voluit; methaphysicam deinde accuratissime enucleavit. Distinctiusque ac clarius prae ceteris medicis post alios medicos in lingua arabica, postquam omnium medicorum scripta vidisset, medicinam omnem in quinque libris descripsit, in quibus se Galieni interpretem appellat. Scripsit quoque et de viribus cordis, de tyriacha, de dilu[i]viis, canticorum etiam librum edidit. De alchimia ad Assem philosophum insuper et liber quidam qui de cholica tractat eidem ascribitur." Cf. the

shorter entry in Foresti, *Supplementum* (1483), f. 115v, which already mentions the mutual murder of Avicenna and Averroes.

50. On this tradition, see also d'Alverny, *Avicenne en occident,* art. 15, 79–83.

51. See Chandelier, 'Comment Averroès empoisonné Avicenne.' The first source is Jacquier's *Flagellum haereticorum fascinariorum* (1581), p. 92: "sicut a quibusdam narratur quod Averrois intoxicavit Avicennam, ponendo toxicum super folia libri quem Avicenna solitus erat contractare." This treatise by the Dominican inquisitor Nicolas Jacquier (d. 1472) dates to 1458, but was printed in 1581. The second source is a long note in a manuscript of Avicenna's *Canon* dating to the 1470s (MS Nuremberg, Stadtbibliothek, Cent. VI 2, f. 6r, quoted after Chandelier): "Dicitur etiam quod propter hoc Avicenna posuerit Averroim in turrim . . . Et ad marginem sexternorum posuit *(i.e. Averroes)* venenum quod collegit in superficie turris, postquam Avicenna vidit commentum istud valde pulcrum *(i.e. on his Cantica)*, delectabatur in legendo et in vertendo folia tetigit ad linguam ad humectandum digito ut mores est Italicis et sic infectus in lingua fuit intoxicatus quare fecit comburi Averroim et ipse postea veneno obiit." Chandelier is able to attribute the handwriting of the note to the young Heidelberg student of medicine Peter Lutz of Schweigern.

52. Boudet has suggested that Foresti's source on Arabic and Italian physicians was Michael of Carrara, a physician who also lived in Bergamo; see Boudet, *Le Recueil,* 2:159.

53. Tortelli, *Liber de medicina et medicis* (1954), 15–16.

54. On Foresti's influence on later world chronicles and collections of biographies, see Krümmel, *Das "Supplementum Chronicarum,"* 344–372.

55. Volaterranus, *Commentariorum urbanorum . . . libri* (1530), f. 247v: "Medicinae insuper peritissimi secundum graecorum famam arabes visi. In primis Albatenius qui in Aristotelem commentarios edidit ac Galeni libros in arabum vertit sermonem, teste Ioanne Serapione, qui eodem fere tempore fuit anno MLXX. Scripsitque de simplicibus Rasis, quem et Almansorem vocant, arabs et ipse, qui cunctorum ante se scriptorum commentarios ingenti volumine coegit, quod medici continentem uocant, quod cuncta contineant. Post annos deinde LXX Avicenna emicuit arabs item genere, patria Cordubensis, e stirpe regia, qui praeter medicinam, metaphysicam valde quidem posteris probatam scripsit. Huius coaevi Averrois, Aristotelis commentator et medicus, Zour quoque qui sapiens cognominatur. Ioannes Mesuae anno MCLX de pharmacis solventibus opus edidit, et alia quaedam quae per mortem absolvere non licuit." On Volaterranus (Raffaele Maffei of Volterra), see Cont. Eras. s.v. 'Maffei, Raffaele' (by d'Amico).

56. Bartolotti, *De antiquitate medicinae* (1954), 61: "Mauritani quoque aliique ex barbaris suos habuere medicos ut Rasim, Isach, Avenzoar, Alzaravium, Halyabatem, Avicennam, <Albatenium qui Galeni libros in arabicum sermonem vertit et Aristotelis libros commentatus est>, Ioannem Mesue <Volaterranus: Avicenna fuit post Rasim 70 annos et Ioannes Mesue fuit anno Domini 1160>, Averroem et reliquos" (angle brackets indicate later additions by Bartolotti himself).

57. Schedel, *Liber chronicarum* (1493), f. 108r (Evax, under the year 84 AD), 181r (Alfarabi, 974 AD: "pauca apud latinos habentur commentaria"), 192v (Serapion, Isaac Israeli, without heading), 193r (Rhazes, without heading), 202r (Avicenna, Averroes, Avenzoar, 1144 AD), and 203v (Mesue, 1160 AD).

58. On these additions, see Lieberknecht, *Die* Canones *des Pseudo-Mesue,* 32.

59. On Schedel's medical interests see Schnell, 'Arzt und Literat,' 53–57, with reference to the more than forty biographies of physicians in the *Liber chronicarum,* among which the vita of Avicenna is the longest. There is extensive literature on Schedel and his chronicle; see Verf.-Lex² s.v. 'Schedel, Hartmann' (by Hernad and Worstbrock).

60. Schedel, *Liber chronicarum* (1493), f. 202r: "Aaboali Abinceni sunt nomina arabica priorem et avum significantia. Abon enim et abin apud arabes idem est quod filius. Mos enim (ut omnes asserunt) apud arabes fuit ut scriptores in titulis nomen prioris et avi ponerent et suum tacerent, ut maioribus honores deferrent."

61. Cf. Hugh of Siena, *Expositio super prima et secunda fen primi canonis Avicenna,* opening, edited in Lockwood, *Ugo Benzi,* 214: "Antiqua enim arabum consuetudo est nomen patris et avi ponere. Certum autem est apud arabes *abon* significare idem quod apud nos *filius,* et *abin* est suus genitivus." Cf. Jacopo da Forli, *Expositio cum questionibus* (1488), f. 1r: "Nos autem dicimus: *Avicenna* cum hoc tamen non fuerit nomen auctoris, quod vero fuerit ignoratur. Mundinus vero refert se habuisse a fide digno arabe quod illa nomina *aben, abin* sunt nomina indeclinabilia synonima filium importantia; sensus igitur est: Incipit etc. quem princeps abohali, idest filius aly abinsceni, idest filii sceni etc. Mos autem apud veteres fuit ut quicumque liberum componeret, patris et avi nomen in titulo scriberet et proprium nomen auctoris reticeret." Hugh is dependent on Jacopo, who in turn quotes Mondino de' Liuzzi; see d'Alverny, 'Survivance,' in *Avicenne en occident,* art. 15, 86n33.

62. Schedel, *Liber chronicarum* (1493), f. 202r: "Fuit enim a nativitate christi anno M. cl ut asserit Conciliator. Et Egidius Romanus in suis quodlibetis, quaestio de unitate intellectus, dicit se vidissee filios Averrois in curia Friderici imperatoris.' On the testimony of Giles, see Burnett, 'The "Sons of Averroes,"' 261–262, esp. n7. Schedel's reference to the *Conciliator* most likely denotes the following passage: Pietro d'Abano, *Conciliator* (1985), diff. 9, ppt. 2, f.14vb: "Tempore autem gratiae 1149 floruit Commentator, ut 2 innuit Coeli et mundi."

63. Verf.-Lex² s.v. 'Schedel, Hartmann' (by Hernad and Worstbrock), 617: 'scheut auch gänzlich Fabulöses nicht.'

64. The German version of Schedel's biography is considerably shorter: the discussion of the letters with Augustine is much abbreviated, the alleged contemporaries Algazel, Alfarabi, and Averroes are omitted, as is any mention of Avicenna's philosophical works; see Schedel, *Das buch der Cronicken* (1493), f. 202r.

65. Schedel, *Liber chronicarum* (1493), f. 202r: "Dedit tamen magnam operam exercitio medicinae faciendo, ut a prioribus intellexi, in sua urbe construi

hospitale habitaculum, ubi innumeri ponebantur infirmi, quos ipse visitabat. Homo fuit colore brunus et iocundus quia pulsando et cantando librum fecit canticorum. Et ut fertur satis iuvenis mortuus est. Non enim attigit quinquagesimum annum."

66. Printed by d'Alverny, 'Survivance,' in *Avicenne en occident,* art. 15, 85, and in Siraisi, *Avicenna in Renaissance Italy,* 161–162n97.

67. Schedel, *Liber chronicarum* (1493), f. 202r: "Dicunt autem reperiri epistolas mutuas inter utrosque quod absurdum est."

68. Ibid.: "Mirabile enim esset cum in volumine epistolarum Augustini ad viles personas epistolae scriptae sunt et epistolae ad principem habitae non nominentur. Avicenna quoque conculas asportatas de sancto Iacobo nominat, qui tempore Augustini non erat ubi nunc est. Christianos item nominat et eorum consuetudines de carne et corallo cum christiani pauci nominis erant et non manifestos ritus observantes tempore Augustini. Creditur igitur fuisse tempore Averrois aut parum ante."

69. Ibid.: "Creditur igitur fuisse tempore Averrois aut parum ante. Algazel etiam et Alfarabius videntur fuisse contemporanei Avicennae, qui fuerunt tempore Averrois. Declinante ergo imperio Romano et arabibus Hispaniam occupantibus floruit."

70. Boudet, *Le Recueil,* 2:85–120. On Simon's library, of which twenty-five volumes have been retraced in the Bibliothèque nationale, see Boudet, *Lire dans le ciel.*

71. The texts are in Boudet, *Le Recueil,* 1:285 (Evax), 312 (Alfarabi), 320 (Isaac Israeli), 320–321 (Albaterius), 373–374 (Serapion), 375 (Rhazes), 390 (Avicenna), 392 (Avenzoar), 395 (Averroes), and 397–398 (Mesue). On Foresti as a source of Simon de Phares, see Boudet, *Le Recueil,* 2:142, 158–160.

72. The spelling of the names is that of the Appendix. The spellings in Boudet's index to the *Recueil* (2:407–491) are: Abulcasis, Alchabitius, Mahomet Alcoarithmi, Alchindi, and Aomar Tiberiadis.

73. The absence of these Arabic astronomers has been pointed out by Boudet, *Le Recueil,* 2:177.

74. As argued in DSB s.v. 'Simon de Phares' (by Poulle).

75. Simon de Phares, *Recueil,* 279–280: "Albumazar le Grant florit a Athenes en ce temps, selon ce qui plaist a aucuns, et estoient avecque lui Plusardonus, ung nommé Hermes et Guellius, car alors en Athenes avoit grant excercite de la science de astrologie. Cestui Albumazar fut precepteur des dessusdicts, et fut cellui qui fist le *Grant Introductoire* et le *Traicté des grandes conjunctions* que noz ignorans calumpniateurs s'efforcent calumpnier pour une cause qu'il ne fait <seullement> que reciter <après autres>."

76. Boudet, *Le Recueil,* 1:279–280, notes.

77. Bonatti, *Tractatus astronomie* (1491), tr. 7, pars 1, ch. 8, sig. A1v and tr. 8, pars 4, ch. 6, sig. F6v.

78. Naukler, *Chronica* (1675), 851: "Avicenna Hispalensis medicus, cum in omni doctrina excelleret, librum composuit, in quo omnem logicam omnemque naturalem philosophiam complecti voluit, metaphysicam accuratissime

enucleavit. Post alios medicos in lingua arabica medicinam omnem descripsit in quinque libris, in quibus se Galeni interpretem appellat. Averrois medicus, cognomento commentator, apud Cordubam Hispaniae floruit. Commentarios in Aristotelis libros scripsit, librum quoque de medicina composuit, Avicennae inimicissimus fuit." On Naukler's chronicle, see Theuerkauf, 'Soziale Bedingungen' (with a brief discussion of Naukler's view of Islamic culture, 330–331).

79. Johann Staindel, *Suppletio virorum illustrium,* manuscript addition in Inc. 65 of the Staatl. Bibl. Passau, ff. 142v–179r. On Staindel, see Verf.-Lex² s.v. 'Staindel, Johann' (by Schnith). On the *Suppletio* and its sources—first, Foresti, but also Diogenes Laertius and Jerome—see Lehmann, 'Staindel-Funde.' Lehmann and Schnith wrongly give "Inc. 117" instead of "Inc. 65" as the Passau signature.

80. Staindel, *Suppletio,* MS Passau, f. 175v: "Alpharabius natione arabs insignis philosophus et scientia clarus multa scripsit de quibus pauca apud latinos habentur commentaria: de ortu scientiarum liber 1, de divisione scientiarum liber 1. Et plura alia que nos nondum vidimus. Claruit tempore Ottonis Imperatoris." See Appendix s.v. "Alfarabi" for these two treatises, of which the first probably is not by Alfarabi.

81. Champier, *De medicine claris scriptoribus* (1506?), ff. 11r–41r.

82. On Champier's role in this debate, see Copenhaver, *Symphorien Champier,* 67–81; Klein-Franke, *Die klassische Antike,* 29–33.

83. In Chapter 4, section 'The Early 1530s: Cordus, Champier, and Brasavola.'

84. Copenhaver, *Symphorien Champier,* 45–52. Cf. the older account of Champier's life and works in Thorndike, *History,* 5:111–126.

85. Champier, *De medicine claris scriptoribus* (1506?), f. 3r: 'reges qui in medicinis claruerunt'; f. 17r: 'Tractatus secundus de philosophis qui in medicinis claruerunt'; f. 26r: 'Tractatus tertius de viris ecclesiasticis qui in medicinis claruerunt et in ea arte scripserunt'; f. 31r: 'Tractatus quartus de italis qui in medicinis claruerunt et in ea doctrina scripserunt'; f. 36r: 'Tractatus quintus de claris medicis qui in gallia, hispania, anglia et germania claruerunt.'

86. Ibid.: *Evax arabum rex* (f. 14r), *Sabid arabum rex* (14r), *De Avicenna rege* (15r), and *Mesues regis damasci nepos* (15v); as philosophers: *Rasis medicus* (21v), *Hali filius Abbas* (21v), *Albatarich sive Albatarius natione arabs philosophus* (21v), *Isaach Beimiram medicus* (21v), *Johannes filius Serapionis medicus* (22r), *Abymeron Abynzoar medicus* (22r), *Averrois quoque medicus* (22v), and *Rabby Moyses iudaeus* (23r).

87. A treatise *De sublimationibus* does not appear either in Ruska, 'Pseudepigraphe Rasis-Schriften,' or in Schmitt and Knox, *Pseudo-Aristoteles Latinus.* Copenhaver mentions, but does not identify the treatise in *Symphorien Champier,* 139.

88. *Sabid rex Arabum* had already been mentioned by Ficino; see Ficino, *Opera* (1962), 645.

89. Champier, *De medicine claris scriptoribus* (1506?), f. 14r: "Ipsa enim Arabia semper summos in medicinis viros varii atque mirabilis ingenii produxit, uti fuerunt et ipse Evax in herbaria arte mirabilis, noster quoque Avicenna Cordube princeps eminentissimi ingenii vir, et ipse Sabid insignis medicus arabum

rex. Dicam igitur de hac regione quippiam cum gentes habeat quarum potentiam et divitias ceterae omnes admirantur obstupescuntque copiam et uber-
tatem. Nec illud praetermittam quod, cum plurimum sacris oblectetur, nulla
est sui pars quae non tota fragret thymiamate aut mirrha."

90. See, for instance, Petrus Comestor (d. 1187), *Historia scholastica,* ch. 26: "De
regina Saba" of the book "Historia libri III regum," MPL 198, cols. 1369–
1370, and Foresti da Bergamo's entry, "Sabba regina . . . apud Ethiopiam seu
Arabiam regina" (for the year 4179), which contains a description of her
country, its riches, gold, silver, iron, precious stones, and fragrant wood
(*Supplementum cronicarum* (1506), f. 81v).

91. In a similar vein, Jacob Milich explained the excellence of Arabic medicine by the
fact that the Arabs lived in countries where plants with extraordinary medical
effects were growing. See Milich, *Oratio de Avicennae vita,* sig. B4r–v: "Multa enim
in natura et in arte deprehenderunt recentiores Arabes, quae Graecis ignota
fuerunt, qua in re et ab ipsa regionum natura, in quibus fuerunt, multi adiuti
sunt. Sunt enim feracissimae remediorum Syria, Arabia, Aegyptus, Cyrenaica,
Africa, Hispania et haec quidem loca perlustrarunt Arabici scriptores omnia."

92. Aristotle, *Metaphysics,* ch. 1,1, 981b17–25.

93. Champier, *De medicine claris scriptoribus* (1506?), f. 15r–v: "Methaphisicam composuit in qua se machometistam confitetur et illam Machometi legem, si lex
dici possit, videtur contemnere, cum tractatu nono circa principium de felicitate loquens ita ait: 'Tu autem scis delectationes corporum et gaudia quid
sunt. Lex enim nostra quam dedit Machomet ostendit dispositionem felicitatis et miseriae quae sunt secundum corpus.' Haec ibi. Et postea dicit eodem
capite: 'Et alia est p<ro>missio quae apprehenditur intellectu et argumentatione demonstrativa et prophetia approbat et haec est felicitas animarum vel
miseria postquam sunt exutae a corporibus suis. Sapientibus vero theologis
multo magis cupiditas fuit ad consequendum hanc felicitatem corporum quam
dedit Machomet.' Haec ille. Vide, dicit Machometum non fuisse prophetam et
non intellectualem felicitatem arabibus et suis, sed solum corporalem et volup-
tuosam dedisse. Fuit semper ille pestilentissimus dracho carnalis et hebes, qui
non salutem patriae sed gloriam quaerebat." Champier's text has "permissio,"
Avicenna's text "promissio."

94. Avicenna, *Liber de philosophia prima* (1977–1983), ch. 9,7, p. 507, lines 94–97, 97–03
(Arabic = *al-Ilāhīyāt,* ed. Anawati and Zayed, p. 423, lines 5–7 and 8–11, = *al-
Ilāhīyāt,* ed. Marmura, p. 347, line 19, p. 348, line 2 and lines 3–6).

95. See Daniel, *Islam and the West,* ch. 5.4, 148–152; Akbari, *Idols in the East,*
248–279.

96. See Rahman, 'Ibn Sīnā,' 500–501, and *Prophecy in Islam,* 52–55 (on Avicenna
and Alfarabi).

97. See Gutas, *Avicenna,* 254–261 (=288–296), esp. the table on p. 259 (=293).

98. Avicenna, *Liber de philosophia prima* (1977–1983), ch. 9,7, p. 507: "Sapientibus
vero theologis multo maior cupiditas fuit ad consequendum hanc felicitatem
quam felicitatem corporum."

99. On Champier's anti-Islamic polemics, see Copenhaver, *Symphorien Champier,* 136, 141, and Martin, *Subverting Aristotle,* ch. 6. Champier was accused of being the author of the infamous "De tribus impostoribus" (Moses, Jesus, Muhammad); see Niewöhner, *Veritas sive varietas,* 305–306.

100. Voss, *De philosophia* (1658), 114–116 (according to Rauchenberger, *Johannes Leo,* 122); Hottinger, *Bibliothecarius quadripartitus* (1664), 246–291. Hottinger's text was reprinted by Fabricius, *Bibliotheca graeca* (1708–1728), 13:259–298. The manuscript used by Hottinger is preserved in Florence, Bibliotheca Medicea Laurentiana, MS Plut. 36, cod. 35, Ms 184, ff. 31r–69v. See Codazzi, 'Il trattato dell'arte metrica,' 180–184, for information on this manuscript and an edition of Leo's metrical treatise contained in it. A proper assessment of *De viris illustribus* has to await a critical edition, the main task of which will be to trace Leo's Arabic sources.

101. On Leo Africanus's life, see Davis, *Trickster Travels,* esp. chapters 2, 3, and 9, and Rauchenberger, *Johannes Leo,* 27–101.

102. Leo Africanus, *De viris* (MS Florence), f. 41r–v (missing in Fabricius, *Bibliotheca* [1708–1728], 13:273, after: "Adeo quod aegroti praepositum et complexionem semper cognoverat"): "Dixit ibnu Giulgiul in Vita philosophorum quod quadam die uenit ad mesuah quidam agricola: Iuxta domum eius exclamans cui dixit Mesuah quid tibi est o agricola. Respondit agricola, membrum meum mihi fortiter dolet."

103. In Leo Africanus's table of contents at the end of the treatise (*Tabula secundum ordinem descriptorum,* f. 69r–v), the names of physicians have the following spellings: De Mesue chaliphe medico, De Mesuah christiano id est mesue, De Rasi, De Ezaragui, De Auicenna, and De Ibnu zohar. The philosophers: De Farabio, De Gazali, De Ibnu saig, De Ibnu Ttophail, and De Aueroi.

104. Conflated with Haly filius Abbas, physician. The confusion was pointed out by Ullmann, *Die Medizin,* 305.

105. Conflated with Ibn al-Ġazzār, physician.

106. The Latin names of the physicians are De Abhulhesen ibnu telmid, De Ettabarani, De Ibnu zor, and De Ibnu elchatib.

107. The Latin names of the theologians and philosophers are De Esciari, De bachillani, De Ibnu elchathib Rasi, De Ephtheseni uel Ettephtheseni, and De Abubahar ibnu chalson.

108. If the identification of "Abulhasan ibnu haidor" with ʿAlī ibn Mūsā ibn ʿAbd Allāh ibn Muḥammad ibn Haydūr at-Tādilī is correct. See Ullmann, *Die Natur- und Geheimwissenschaften,* 344.

109. The Latin names of the scientists are De Abulhusein essophi, De Esseriph, De Ibnu elaitar, De elmuhaied ettosi, and De Ttograi.

110. Cf., for instance, Codazzi, 'Il trattato,' 184–187, on Leo's biography of the famous grammarian al-Ḥalīl in *De arte metrica.*

111. On the Islamic bio-bibliographical tradition, see Humphreys, *Islamic History,* 132, 188–192, and Auchterlonie, *Arabic Biographical Dictionaries.*

112. Leo Africanus, *De viris* (MS Florence), on Mesue (f. 32v: "Dixit etiam ibnu Giulgiul In vita philosophorum"), Alfarabi (f. 35r), Avicenna (f. 38r, 39r), Pseudo-Mesue (f. 41r), ibn Bāǧǧa (f. 46r), and Abū Bakr ibn Zuhr (f. 46v). Cf. Daiber, 'The Reception,' 67.

113. Leo Africanus quotes the historian of al-Andalus Ibn Ḥayyān (d. 987–98) on Rhazes (f. 36r: "Dixit ibnu haijan lusitaniae chronista"); Ibn Ḥallikān (d. 1282) on Avicenna (f. 39v: "Dixit ibnu challican chronicista"); the annal-istic historian of Baghdad ʿAbd ar-Raḥmān ibn ʿAlī ibn al-Ǧawzī (d. 1200) on Mesue (f. 32v: "Dixit Geuzi historiographus de ciuitate bagdad"), Algazel (f. 42r, 42v, 43r, cf. Griffel, 'Toleranzkonzepte,' 136) and Ibn Chatib (f. 62v); "ibnu Fadlilla," who is probably the Mamluk author Ibn Faḍl Allāh al-ʿUmarī (d.1349) on Tūsī (f. 53r: "Et dixit ibnu Fadlilla"), cf. EI² s.v. 'Ibn Faḍl Allāh al-ʿUmarī' (by Salibi). I have not identified: "ibn chatir in chronicis suis uniuer-salibus" (f. 39v on Avicenna), "ibnu Abididunia in mirabilibus mundi" (f. 40v on aṭ-Ṭabarī), "ibnu elhussein in chronicis Cilicie" (f. 45v on al-Idrīsī), or "Masendrani historiographus tartarorum" (f. 53r on Tūsī).

114. Cf. Codazzi, 'Il trattato,' 184 ("biografia di al-Khalīl sensibilmente diversa da quella data concordemente dalle altre fonti"); Ullmann, *Die Medizin,* 305 ("Hinzu kommt, daß die Nachrichten des Leo Africanus außerordentlich unzuverlässig sind").

115. See Appendix s.v. 'Mesue.'

116. Leo Africanus, *De viris* (MS Florence), f. 33v (on al-Ašʿarī), f. 43r (on Algazel). There are further references to the Latin tradition: Leo expresses the belief that the *librum Elmalihi* by "Ibnu Telmid" (which, in fact, is *al-Kitāb al-Malakī,* the "Royal Book" by Haly filius Abbas) has been translated into Latin; and he mentions that Idrīsī's book on geography was translated into Latin (which it was not) because King Roger of Sicily found that there was nothing equivalent in Latin (f. 34r, on "Ibnu Telmid," and f. 45v, on Idrīsī).

117. Ibid., f. 37v: "Errahis est in Arabico uocabulum commune uel equiuocum et inprimis significat superiorem, nobilem, ducem ac principem: et prouenit ab hac dictione Rase, quae in Arabico est idem quod apud nos caput. Auicenna uero uocatus fuit Rahis, quia erat nobilis et honoratus apud dominos et populos uirtutibus suis. Eoque medicorum maxima pars, quando eum in medicina aut citare aut aduocare uolunt, aiunt: medicorum princeps hoc dixit, absque alio aliquo epytheto et intelligitur de Auicenna dixisse."

118. Ibid. (immediately following): "Et quod dicitur inter latina lambentes quod Rex chordube fuerit, mendacium est, quoniam non inuenitur apud scrip-tores eius uitae latina lingua. Non enim extendissem in hoc sermonem meum, nisi uidissem quod scripserat Iacobus de forliuio: qui multa tandem dicens nihil dixit. Et tamen ei inerat ratio incognita: In culpa magis est qui librum Auicenne in latinum transtulit, et interpretatus est, quia consider-auit hanc dictionem rahis nil aliud nisi principem significare et ille, ut dicam, miser ac nescius Arabice grammatices uocabula communia non in-telligebat neque cognoscebat."

119. Jacopo da Forli, *Expositio cum questionibus* (1488), f. 1r.
120. See Schedel, *Liber chronicarum* (1493), f. 202r, Volaterranus, *Commentariorum urbanorum . . . libri* (1530), f. 247v, and Hasse, 'King Avicenna,' 234.
121. Ficino, *Opera* (1562), 645, on kings who have practiced medicine: "Quales fuerunt Sabor et Giges Medorum reges; Sabid rex Arabum; Mitridates Persarum rex; rex Egyptiorum Hermes; Mesues regis Damasci nepos. Volunt nonnulli et Avicennam fuisse Cordube principem."
122. On the history of this legend, see d'Alverny, *Avicenne en occident*, art. 15 and 16, and Hasse, 'King Avicenna.'
123. Ibn Ḥallikān relates the true story of Avicenna's death; he also reports the prison story as stemming from Kamāl ad-dīn ibn Yūnus. See Ibn Ḥallikān, *Biographical Dictionary*, 1:444.
124. On the Arabic sources of Avicenna's biography, see Enc. Iran. s.v. 'Avicenna' (by Gutas).
125. Leo Africanus, *De viris* (MS Florence), on Averroes (f. 48r: "Dixit Ibnu habdul Malech Marochi Et Ibnu habbar bettice historiographus," 49r, 50r).
126. Ibid., on Averroes (f. 48r, 48v, 49r, 50r, 51v), and also on Abraham Ibnu Sahal (f. 67v).
127. Ibid., f. 48r.
128. Ibid., f. 50r: "Dixit ibnu el habbar quod Auerois fuit etiam legista secundum Malich et theologus secundum Esciarim, hoc est secundum opinionem eius. At secundum opinionem ueram imitabatur Aristotelem et multi nobiles ac ciuitatis cordube doctores et uiri prestantes ei inuidebant, prout ibnu zoar medicus et sui praeceptoris filius, qui decreuerunt eum heresi incusare."
129. Ibid., f. 52r–v: "Postremo mortuus est Auerois in ciuitate Marrochi et fuit sepultus extra portam Coriariorum tempore Maumedis mansoris pontificis ac regis marrochi anno sexingentesimo tertio de elhegira. Dixit interpres se uidisse sepulchrum et epitaphium eius."
130. See Munk's article in *Dictionnaire des sciences philosophiques* (of 1847): 'Ibn-Roschd,' esp. 159–160 ("Ces détails, ainsi que les autres fables débitées par Léon, ont été répétés par Brucker et par une foule d'autres écrivains, sans qu'on se soit aperçu de ce qu'il y a de fabuleux et d'impossible dans les récits de Léon Africain, qui a fait d'énormes anachronismes"), and, a few years later, Renan, *Averroès et l'Averroisme*, 9–10.
131. On Leo's contacts with these scholars, see Rauchenberger, *Johannes Leo*, 97–99, 115–116, Burman, *Reading the Qur'ān*, 165–168, and, most comprehensively, Davis, *Trickster Travels,* index. See also Bobzin, *Der Koran im Zeitalter,* 84–88 (on Viterbo) and 277–363 (on Widmannstetter). On Jacob Mantino, see Chapter 3 below, section 'Averroes: Preface to *Metaphysics* 12.'
132. Avicenna, *Liber canonis* (1505), reverse of title page: "Vita Avicennae quam quidam illius arabs discipulus scriptam reliquit." The colophon of the entire edition reads: 'Impressus et diligentissime correctus mandato et impensis heredum quondam nobilis viri D. Octaviani Scoti civis Modoetiensis Venetiis anno salutis MCCCCCV die 24 Januarii per presbyterum Bonetum

Locatellum Bergomensem." On Locatello, see Norton, *Italian Printers, 1501–1520,* 139–140, and DBI s.v. 'Locatello, Boneto' (by Ruggerini).

133. d'Alverny, 'Survivance,' in *Avicenne en occident,* art. 15, 97: "Elle reflète une source orientale"; Siraisi, *Avicenna in Renaissance Italy,* 162: "may possibly be based on an Arabic source."

134. Ǧūzǧānī's text is in Gohlman, *The Life of Ibn Sina,* 82–91.

135. See Nuovo, 'Il Corano arabo ritrovato'; cf. Bobzin, 'Islamkundliche Quellen,' 55–57, Bobzin, *Der Koran im Zeitalter,* 182–185, and Bobzin, *Ließ ein Papst den Koran verbrennen?* On the printers Paganino and Alessandro Paganini, see Norton, *Italian Printers, 1501–1520,* 145; cf. also 105–106, 116–117.

136. Calphurnius, *Francisci Calphurnii Vindocinensis Epigrammata* (ca. 1517), sig. a1–c4r. I have consulted the digital copy of the university library of Salamanca.

137. Avicenna, *Liber Canonis totius medicinae . . . Una cum eius vita a domino Francisco Calphurnio non minus vere quam eleganter excerpta* (1522), f. 1v: "Sed quoniam multis non est in liquido quanto et quo tempore vixerit, id placuit ex antiquis annalibus excerptum eliminare. Ille enim ante maturos dies occidit, quia vix octavum et quadragesimum annum expleverat cum funesta morte preventus interiit. Habuit Averroim philosophum et medicum contemporaneum, qui semper in eius aliorumque medicorum gloriam liventibus oblatrantis lingue iaculis obluctabatur. Habuit etiam convictitios Algazellem et Alfarabium medicos illustres, quos itidem nitebatur maledicus ille Averrois in Avicennam commovere . . . Tradunt alii arabem illum abolayque principem fuisse, qui post evecti patris exequias sese totum disciplinis tradidit enutriendum . . . Verum quia diversos Avicenne mortis scriptores diversa flectit opinio ea causa nihil quod omni parte audeam asserere scribo, attamen reiectis multorum opinionibus facile mihi persuadeo non a vero discedere si cum domino Symphoriano Camperio Lugdunensi de Avicenne dicam interitu . . . potu venefico illum interemit."

138. Ibid., f. 2v: "prout fuit ille impius Averrhous sine lege infidelis rusticus paganus et gentilis qui unitatem intellectus contra Aristotelis mentem . . . somniavit."

139. See d'Alverny, 'Survivance,' in *Avicenna en occident,* art. 15, 99–100; Siraisi, *Avicenna in Renaissance Italy,* 163; Palmer, 'Nicolò Massa.' Not worth consulting is Goehl, 'Die Vita Avicennae.'

140. Avicenna, *Principis Avicennae libri Canonis* (1527), sig. +0r: "Notandum praeterea quod Avicenna non fuit Hispanus, sicut aliqui de ipso scribunt, immo fuit Persicus ex civitate dicta Bochara."

141. Nutton, 'Wittenberg Anatomy,' 14.

142. Milich, *Oratio de Avicennae vita,* sig. B3r–C3r.

143. For brief information see DSB s.v. 'Gesner, Konrad' (by Pilet) and MBW Pers. s.v. 'Gesner, Konrad' (with further literature). See, in general, Fischer et al., *Conrad Gessner.* Schipperges, *Ideologie,* 24–25, gives a succinct description of Gesner's entries on Arabs.

144. Gesner, *Bibliotheca universalis* (1545), f. 224v (Evax), 578v (Rhazes), f. 454r–v (Serapion), f. 297v–298v (Haly filius Abbas), f. 467r (Isaac Israeli), f. 110r–111r (Avicenna), f. 100r–102r (Averroes), f. 1r (Avenzoar) and f. 631r (Zoar, father of Avenzoar), f. 437r–v and f. 409v–411r (Mesue); the physicians: f. 298v (Haly Rodoan), f. 20v (Albucasis), f. 350v and 463r (Johannitius), f. 219v (Ibn Buṭlān), f. 372r (Jesus filius Haly); the astrologers and astronomers: f. 1r (Abraham ibn Ezra), f. 17r (Albategnius), f. 20v (Albubater), f. 20v–21r (Albumasar), f. 21r (Alcabitius), f. 29r–v (Alfraganus), f. 30r (Almansor), f. 30v (Alpetragius), f. 30v (Alfarabi), f. 298v (Haly filius Abenragel), f. 336v (Omar Tiberiadis), f. 511r (Messahalah), f. 607r (Thebit), f. 630v (Zahel), f. 30r (Alhazen), the philosophers f. 29v (Algazel), f. 351r (Alkindi), f. 219v (Avempace), f. 514r (Maimonides); f. 266v (Geber).

145. Ibid., f. 110r–v.

146. Ibid., f. 102r: "In commentariis suis in Aristotelem Graecos maxime imitatus est, ut Alexandrum Aphrodisiam et Themistium. A nonnullis stimulus medicorum vocatur, nam et Galeno locis non paucis contradicit."

147. Gadebusch Bondio, 'Exempla medicorum,' 381–388; Nutton, 'John Caius,' 383.

148. Occasionally, Gesner quotes from the editions. He cites Abenragel's and Haly filius Abbas's own prefaces (f. 298v), and the editors' prefaces for editions of Haly Rodoan (298v) and Ibn Buṭlān (219v).

149. See the Appendix s.v. 'Mesue' and 'Serapion.'

150. Gesner, *Bibliotheca universalis* (1545), f. 409v–411r: "Recentiores quidam eundem cum Mesue aut ei contemporaneum fuisse putarunt, sed falso: cum Mesuas itidem Damasco natus et educatus plurius saeculis post, puta anno Christi 1563 *(mistake for 1163)* sub Friderico Barbarosa vixerit."

151. Ibid., f. 454r–v.

152. Ibid., f. 370r: "Ianus Damascenus: vide in Ioanne Damasceno. Nam videtur de industria aliquis mutasse nomen, ut novus titulus appareret, in opere illo medico sex librorum, quod Henricus Petrus Basileae nuper excudit."

153. Ibid., ff. 219v (Ibn Buṭlān), 336v (Omar), 372r (Johannitius).

154. Ibid., f. 426r: "Ioannes Hispalensis Alchabitii Isagogen ad magisterium astrorum barbaro stilo latinum fecit."

155. Ibid., f. 20v–21r: "Albumasar astrologus, alio nomine Iaphar, libros 8 scripsit de magnis coniunctionibus, annorum revolutionibus, ac eorum profectionibus. Chartae sunt 30 excusae in 4 Augustae Vindelicorum, 1489. Titulus alias huiusmodi est: 'Hic est liber individuorum superiorum, in summa, de significationibus super accidentia quae efficiuntur in mundo, de praesentia eorum respectu ascendentium inceptionum coniunctionalium, et aliorum.' Argumenta singulorum tractatuum, quamvis obscuro et barbaro sermone scripta, addidimus tamen ut ex eis de totius operis stilo iudicare liceret. Quis autem latinus interpres sit, non invenio . . . Et haec usque ad nauseam, adscripsimus tamen. Inseruntur autem passim depictae zodiaci et planetarum imagines. Albumasaris flores astrologiae, excusi

Augustae Vindelicorum, 1488, in 4, chartis quinque, cum figuris Zodiaci et planetarum."

156. Gaurico, *Oratio de inventoribus,* in *Opera omnia* (1575), 1:1–8, here 5: "Profuerunt autrem posteris in scribendo Aegyptii, Chaldaei et Arabes: Hermes, Almansor qui Centiloquia scripsere, item Dorotheus, astrologiae columen, Messala, Thebitius, Albategnius, Aboasares eiusque discipulus Sadam, item Alchabitius, Ioannes de Saxonia eius commentator, Albubater, Alchindus, Alfraganus, Alpetragius, Iergis, Omar, Hales, Abenrageles, Hales Rodoan Ptolemaei commentator, Hales Israelita, Albumasar, Abraham Avenezre, Abraham Iudeus, Gamaliel quem Christi discipulum fuisse inquiunt, Aboalisar Cinator et reliqui quos non facile enumerare possemus."

157. Brunfels, *Catalogus illustrium medicorum* (1530), 51–52. 'Albaterius' (al-Biṭrīq) is misspelled 'Albatenius.'

158. Reinhold, *Tabulae Directionum* (1554), sig. ε3v.

159. Cf. Commandino's preface to the edition of Euclid, *Elementorum libri 15* (1572); Clavius's prolegomena to the edition of Euclid's elements, first edited in 1574; see Clavius, *Opera mathematica* (1611–1612), 1:4–5, where Clavius mentions the Arabs' contribution to arithmetic: "Quam mirum in modum postea Pythagoras eiusque successores nec non Aegyptii, Graeci denique atque Arabes amplificarunt"; Tycho Brahe, *De disciplinis mathematicis oratio* (held in 1574), in Brahe, *Opera omnia* (1913), 1:143–173. Ramus's historical outline of mathematics in *Scholae mathematicae* (1569) does not cover authors between Theon of Alexandria and the Emperor Maximilian. See the survey of such histories of mathematics in Rose, *The Italian Renaissance,* 259–260, and, less informatively, Sarton, *The Appreciation,* 163–165. On Ramus, Commandino, and Clavio, see also Nenci's introduction to the edition of Baldi's *Le vite* (1998), 31–36.

160. Regiomontanus, *Oratio introductoria in omnes scientias mathematicas,* published in 1537, in the edition of Alfraganus's *Rudimenta,* sig. a4r: "Oratio Iohannis de Monteregio habita Patavii in praelectione Alfragani." The Arabs appear on sig. b2r. I am grateful to Pietro D. Omodeo for drawing my attention to this passage.

161. On Baldi, see Rose, *The Italian Renaissance,* 243–253; DSB s.v. 'Baldi, Bernardino' (by Drake); Cochrane, *Historians,* 399–400; Moyer, 'Renaissance Representations,' 471–484; Nenci's introduction to his edition of Baldi's *Le vite,* 13–21; Nenci, *Bernardino Baldi.*

162. In the nineteenth century, fifty-four lives were published by Narducci (Italian mathematicians) and Steinschneider (Arabic mathematicians): Narducci, 'Vite inedite,' and Narducci, 'Vita di Pitagora'; Steinschneider, *Vite.* Steinschneider's edition is stuffed with copious notes, which reveal that he read Baldi as a colleague, trying to assess the reliability of the Italian's information on Arabic scholars. A modern edition of the seventy-two *Vite* of medieval and Renaissance scholars with full identification of the sources was published by Elio Nenci: Baldi, *Le vite* (1998).

163. Baldi, *Cronica de matematici* (1707).

164. Rose, *The Italian Renaissance*, 262. Dannenfeldt, 'The Renaissance Humanists,' 103, lists Baldi among Italian scholars who knew Arabic. On Raimondi, see Siraisi, *Avicenna in Renaissance Italy*, 146–153, and the Appendix s.v. 'Avicenna.'

165. Rose, *The Italian Renaissance*, 262. Rose was misled by Steinscheider's notes, which cite much Arabic material. Cf. Rose's comments on Baldi's ambiguous attitude toward Arabic authors as excellent astronomers and *barbari*.

166. Occasionally, Baldi draws on Damiano da Goes, Francesco Barozzi, Girolamo Cardano, Francesco Maurolico, and Friedrich Risner (astronomers), and on Hartmann Schedel, Pedro Mexía, André Tiraqueau, Francesco Barocci, Francesco Taraffa, and Konrad Gesner (historians and biographers). Steinschneider's and Nenci's indices under these names lead to Baldi's references to these scholars.

167. As already noted by Moyer, 'Renaissance Representations,' 476.

168. As described by Biliński, 'Vite dei matematici,' 312–313. Nenci's edition presents the *vite* according to the chronology established by Baldi.

169. The original spellings are as following: *Messala* (Messahalah); *Albategno* (Albategnius); *Alì, che talhora cognominossi Alboazeno* (Haly filius Abenragel); *Arzahele* (Azarchel); *Punico, sì come fu contemporaneo di Ali Abenragele;* "Punicus" is a Renaissance Latin term for Arabs; Baldi's entry in fact discusses Haly Rodoan, who appears again in the next entry of the manuscript (as already realized by Steinschneider, *Vite*, 56–57); *Abenrodano* (Haly Rodoan); *Almansore* (Almansor); *Alhazen* (Alhazen); *Alchindo* (Alkindi); *Alpetragio* (Alpetragius); *Gebro* (Geber filius Afflah); *Alfagrano* (Alfraganus); *Tebitte* (Thebit ben Corat); and *Albumasaro* (Albumasar).

170. Baldi, *Le vite* (1998), 76: "Ch'egli *(i.e. Albategno)* fosse medico et eccellente viene affermato da Fra' Filippo nel *Supplemento,* ancorchè da lui sia stroppiato il nome e distorto in Albaterio." At the end of his entry on Albategnius, Baldi criticizes Foresti's date of 1063 AD for Albategnius's time of flourishing.

171. Mexía, *Le Vite di tutti gl'imperadori* (1561), 756: "Albetenio che scrisse commenti sopra Aristotele e tradusse Galeno nella lingua Arabica."

172. At first sight, Alhazen seems the Arabic author most favored by Baldi (as emphasized by Rose, *The Italian Renaissance*, 263), but although Alhazen is praised highly, the entry consists mainly of quotations from Friedrich Risner's preface to the 1572 edition of Alhazen's optics and does not reveal many of Baldi's own thoughts about Arabic science.

173. Baldi, *Le vite* (1998), 75–82.

174. Ibid., 126–133.

175. Ibid., 141–148. On the problematic attribution of this theory to Thebit, see Appendix s.v. 'Thebit ben Corat.'

176. This is also apparent in his remark that Messahalah was probably the Graeco-Arabic translator of Ptolemy's *Planispherium*. Baldi, *Le vite* (1998), 73: "Affaticossi parimente Messala intorno a l'Astrolabio, o *Planisferio* di Tolomeo, e per

quanto mi credo fu egli che da la lingua greca lo converse ne l'Araba." Baldi's historical sense also appears when he dates the flourishing of the arts and in particular of the mathematical disciplines among the Arabs to 1100 AD; for this, his immediate source is Friedrich Risner's above-mentioned preface to Alhazen: Baldi, *Le vite* (1998), 118: "Il Risnerio, seguendo le congietture, tiene potersi credere ch'egli fiorisse intorno mille e cento anni dopo la natività di Cristo, nel qual tempo fra' barbari e saraceni fiorivano gli studi di tutte l'arti buone, e particolarmente quelli de le mathematiche."

177. Champier, *De medicine claris scriptoribus* (1506?), f. 28v; Gesner, *Bibliotheca universalis* (1545), f. 186r-v.

178. Ibid., f. 33v; Gesner, *Bibliotheca universalis* (1545), f. 274r.

179. As Nenci has shown, the "Abalachius filius Albumasaris Astrologus 568" is part of Giuntini's two-page *Catalogus doctorum virorum quorum ad absolvendum Astrologiae speculum annotationes lucubrationesque nos iuvarunt*, which is printed at the beginning of Giuntini's *Speculum astrologiae*. The "Abalachius" entry, however, appears only in the enlarged version of the *Catalogus*, which accompanies not the 1573, but the 1581 edition of the *Speculum*, sig. †9r. See Nenci's commentary in Baldi, *Le vite* (1998), 69n9.

180. Baldi, *Le vite* (1998), 69–70: "Hebbe Albumasaro un figliolo detto Abalachio anch'egli matematico nobile, come afferma Francesco Giuntino. Fu Albumasaro di natione Spagnuolo, e fra gl'astrologi illustri di Spagna è connumerato da Damiano da Goa. Fiorì egli secondo il Reinoldo nella *Tavoletta historica* ottocento e quarantaquattro anni dopo Cristo, onde s'inganna il Giuntino, che vuole ch'egli vivesse del cinquecento quaranta, percioché se Albategnio, il quale epitomò l'*Almagesto* ampliato d'Albumasaro fiorì dell'ottocento ottanta in circa, come habbiamo provato nell'historia sua, chiaro è che Albumasaro fiorisse o ne' tempi suoi, o poco dopo lui; erra dunque il Giuntino."

181. See DSB s.v. 'Abū Ma'shar' (by Pingree) and Burnett, 'Abu Ma'shar,' 18.

182. Damiano de Goes, *Hispania* (1542), sig. D4r-v, "Avicenna medicus praeclarus et philosophus. Averrois quoque medicus et philosophus. Rasis Almansor. Mesalac medicus astrologus. Habramus astrologus. Albumasar astrologus."

183. Reinhold, *Tabulae Directionum* (1554), sig. ε3v. The half-page table is titled: "Philosophi et artifices insignes." Identified by Nenci in Baldi, *Le vite* (1998), 70n11.

184. Rudolph of Bruges's preface is printed in Ziegler et al., *Sphaerae atque astrorum coelestium ratio* (1536), 227–232, here 229: "Ex quibus et Ionica lingua duo volumina collegit, primam Theoricen, alterum Tetrastin appellans, arabice dicta Almagesti et Alarba, quorum Almagesti quidem Albeten commodissime restringit, Tetrastin vero Albumazar non minus commodissime ampliavit." See Steinschneider's comment in Baldi, *De le Vite* (1874), 27n8 (451–452).

185. Giuntini, *Speculum astrologiae* (1573), sig. *5r: "Albumasar Astrologus qui Iaphar dicitur 540 *(sc., post Christum)*."

186. For an introduction to the problem of chronology in the late Renaissance, see Cochrane, *Historians,* 380–382, and, most pertinently, Grafton, *Joseph Scaliger, 2: Historical Chronology,* 1–10.

187. See Gaurico, *Calendarium ecclesiasticorum novum* (1552); Baldi, *Le vite* (1998); Biancani (=Blancanus), *Clarorum mathematicorum chronologia* (1615); Voss, *De universae mathesios* (1650). On Baldi, Biancani, and Voss, see Biliński, 'Vite dei matematici,' 318–320.

188. On the medieval genre, see Krüger, *Die Universalchroniken,* and von den Brincken, 'Die Rezeption'; on the sixteenth century, see, in particular, Klempt, *Die Säkularisierung,* and Haeusler, *Das Ende der Geschichte,* 142–173.

189. I have consulted the following editions: Franck, *Chronica* (1536); Carion, *Chronica* (1537); Egenolff, *Chronica* (1534); Sleidanus, *De quatuor summis imperiis* (1556); Freigius, *Historiae synopsis* (1602); Neander, *Chronicon* (1586); and Reusner, *Isagoges historicae* (1600).

190. On Golius, see Jones, 'Piracy,' 98. On Pocock, see Daiber, 'The Reception,' 68–74.

191. Cf., for instance, Verf.-Lex² s.v. 'Schedel, Hartmann' (by Hernad and Worstbrock), where Foresti, Schedel's principal source, is grouped among the "Historiographie des italienischen Humanismus."

192. On the medieval character of Foresti's chronicle, see Cochrane, *Historians,* 382–386 (under the heading "The Rejection of Humanism").

193. As shown by Martin, 'Providence.'

194. Pico, *Disputationes* (1946/1952), lib. 1, 72 (adopted by Pigghe, *Adversus ... astrologiae defensio* [1519], f. 18r): "Erat enim grammatices artis professor et a scribendis historiis, ut narrat Avenrodan, ad astrologiam se convertit." This Pico picked up from a passage in Haly Rodoan's commentary on Ptolemaeus, *Liber quadripartiti* (1493), f. 2va: "Tamen Albumasar et alii qui fecerunt cronicas, ipsi qui non intellexerunt nisi textum et res accipiunt per auditum quia non exquirunt que est veritas vel que non, crediderunt quod hic Ptholemeus fuerit unus ex regibus Alexandrie." Cf. the remark by Ibn an-Nadīm that Albumasar was a student of the hadiths, that is, the transmitted sayings attributed to Muhammad, before he turned to philosophy and astrology; see Burnett, 'Abu Ma'shar,' 18.

195. See Clubb, *Giambattista Della Porta,* 172–185; Tomkins, *Albumazar.*

196. Reisch, *Margarita philosophica* (1973), ch. 8.10: "Albumasar impietatem impietati accumulans ait: Qui deo supplicaverit hora qua luna cum capite draconis iovi coniungitur, impetrat quicquid petierit."

197. Melton, *Astrologaster* (1620), 37: "Albumazar also is as divellish as the rest, heaping impietie on impietie ..." As noted by Hugh Dick in the commentary to his edition of Tomkins, *Albumazar,* 162.

## 3 · PHILOLOGY: TRANSLATORS' PROGRAMS AND TECHNIQUES

1. On the Renaissance Arabic–Hebrew–Latin translation movement, see Wolfson, 'The Twice-Revealed Averroes'; Wolfson, 'Revised Plan'; d'Alverny, 'Avicenne et les médecins,' in *Avicenne en occident,* art. 13, 177–198; Schmitt, 'Renaissance Averroism,' in *The Aristotelian Tradition,* art. 8, 121–142; Siraisi, *Avicenna in Renaissance Italy,* 133–143; Coviello and Fornaciari, *Averroè: Parafrasi,* x–xv (on del Medigo and Mantino); Tamani, 'Le traduzioni'; Tamani, 'Traduzioni ebraico-latine'; Burnett, 'The Second Revelation'; Hasse, 'The Social Conditions'; Tamani, 'I traduttori ebrei'; Di Donato, 'Traduttori di Averroè.' See also the more specific literature cited in the footnotes to the various translators discussed below.

2. This is true even of the sixteenth century; see Ivry, 'Remnants.'

3. I have inspected the manuscript Vienna, Österreichische Nationalbibliothek, Cod. hebr. 114, ff. 1–89v. Cf. Steinschneider, *Die hebraeischen Übersetzungen,* 95. See the brief remarks on the three translations by Minio-Paluello, *Opuscula,* 150–154. A first comparative study on these translations by Silvia Di Donato is forthcoming in the proceedings of the conference *From Corduba to Cologne,* October 2010, in the series *Miscellanea Mediaevalia.*

4. The translators of the *Theologia* were Moses Arovas, Pier Nicola Castellani, and Jacques Charpentier. A manuscript of the Arabic *Theology* was found in Damascus by Francesco Rosi, and brought to Cyprus, where he engaged the Jewish physician Moses Arovas to translate it into Latin; this translation was revised by Pier Nicola Castellani and published with a dedication to Pope Leo X in Rome in 1519: *Sapientissimi philosophi Aristotelis Stagiritae Theologia sive mistica phylosophia secundum Aegyptios noviter reperta et in latinum castigatissime redacta.* Later in the century, Jacques Charpentier produced a linguistic revision of the text, which was printed several times. See Fenton, 'The Arabic and Hebrew Versions,' 241 (with remarks on the difficult identification of these translators), and Kraye, 'The Pseudo-Aristotelian *Theology.*'

5. On the translators of the Quran, Juan de Segovia, Juan Andrés, Flavius Mithridates, Johannes Gabriel Terrolensis, and Guillaume Postel, see Bobzin, *Der Koran im Zeitalter,* 64–88, 470–495, and Burman, *Reading the Qur'ān,* 133–197.

6. A negative example is Pfeiffer's *History.*

7. This is how Fück and Bobzin explain why they pass over the medieval Arabic–Latin translation movements in their histories of Arabic philology and instead concentrate on the translation of the Quran and the activities of the missionaries. Cf. Fück, *Die arabischen Studien,* 4 ("Aber diese Übersetzungen . . . vermittelten noch keinen Einblick in das eigentlich islamische Wesen . . . und gaben ebensowenig Anlaß zu einem selbständigen Studium der arabischen Sprache"); Bobzin, 'Geschichte der arabischen Philologie,' 156 ("Aber erst der Missionsgedanke wurde zur treibenden Kraft, eine

systematische Beschäftigung mit dem Arabischen einzuleiten"). Fück and Bobzin are aware of the fact that the exclusion of scientific texts and the inclusion of religious texts in such histories is difficult to justify.

8. Pfeiffer starts his seminal history of classical scholarship with a broad definition (Pfeiffer, *History ... From the Beginnings*, 3: "Scholarship *("Philologie" in the German edition)* is the art of understanding, explaining, and restoring the literary tradition"), but later concentrates on the celebrated invention of "pure scholarship" (170) in Alexandria; this criterion recurs when he evaluates the achievements of Renaissance philologists (Pfeiffer, *History ... From 1300*, 37 on Poggio and Valla on the "reintroduction of textual criticism," 40 on Valla). Cf. Pfeiffer's remarks on the uncritical attitude of the Middle Ages toward the text (*History ... From 1300*, 47). Cf. also Kenney's opinion that the activities of medieval scholars with respect to classical sources were "essentially anti-philological" (*The Classical Text*, 3). Cf. the review of Grafton, 'The Origins.'

9. Among the many studies on this subject, I refer the reader to Kenney, *The Classical Text*, 4–10; Grafton, *Defenders of the Text*, 47–75; Grafton, *Joseph Scaliger*, 1:9–100, and, for further literature and information, to the article by Reeve, 'Classical Scholarship,' with bibliography, 295–296.

10. The philological nature of the work of the Andalusian translators is particularly apparent when they revise earlier translations; see Burnett, 'The Strategy of Revision.' On the translators' techniques in general, see Burnett, 'Translating from Arabic.'

11. On the Greek-Latin translators, see, among others, Minio-Paluello, *Opuscula;* Brams and Vanhamel, *Guillaume de Moerbeke;* and the overview article by Dod, 'Aristoteles Latinus.'

12. It is not a prerequisite of this kind of philological activity that the scholar be aware of the historical distance between himself and the original text, as a narrow definition of philology may imply; cf. Gumbrecht, *Die Macht der Philologie*, 11–13 (English title: *The Powers of Philology*).

13. Previous literature on the philological and philosophical problems posed by this preface include Freudenthal and Fränkel, *Die ... Fragmente*, esp. 67–69; Steinschneider, *Die hebraeischen Übersetzungen*, 171–176; Moraux, *Alexandre d'Aphrodise*, 16–17; Bouyges, *Averroès ... Notice*, lxxix–lxxx, cxi–cxii; Martin, *Averroès: Grand Commentaire*, 24–43 (comments on the Arabic text); Genequand, *Ibn Rushd's Metaphysics*, 7–8 (on the authenticity of the quotations from Alexander), to be consulted together with the review by Gutas, 'Charles Genequand: Ibn Rushd's Metaphysics,' 122–126. From a doctrinal point of view, a rewarding study is Ivry, 'Averroes and the West,' 153–158.

14. That Michael Scot was the translator of the Long commentary on the *Metaphysics* is shown in Hasse, *Latin Averroes Translations*.

15. On Elia del Medigo, see Cassuto, *Gli ebrei*, 282–326; Kieszkowski, 'Les rapports,' 41–77; Roth, *The Jews in the Renaissance*, 112–116; Geffen, 'Insights into the Life,' 72; Puig, 'Continuidad medieval,' 47–54; Idel, 'Jewish Mystical

Thought,' 30–32; DBI s.v. 'Del Medigo, Elia' (by Bartòla); Licata, *La via della ragione,* 23–81; SEP s.v. 'Elijah Delmedigo' (by Ross); and in particular, Bartòla, 'Eliyahu del Medigo.'

16. On del Medigo's philosophical works in Latin, some of which are extant only in Latin-Hebrew translations, see Bland, 'Elijah del Medigo'; Mahoney, 'Giovanni Pico,' 128–138; Licata, *La via della ragione,* 83–93; and Engel, *Elijah Del Medigo's Theory.*

17. The compendium was translated together with a chapter of the Middle commentary on the *Meteorology.* See Bartòla, 'Eliyahu del Medigo,' 259; Steinschneider, *Die hebraeischen Übersetzungen,* 135.

18. Bartòla, 'Eliyahu del Medigo,' 261–265. These two translations were never printed, but still exist in manuscript. The relation to Pico is unclear.

19. Bartòla, 'Eliyahu del Medigo,' 268; Steinschneider, *Die hebraeischen Übersetzungen,* 98–99; Kieszkowski, 'Les rapports,' 45, 64.

20. Coviello and Fornaciari, *Averroè: Parafrasi* (1992); on the relations to Pico, cf. the introduction, xix–xxiii.

21. Zonta, *Il Commento medio,* 15–18. Whether this text is related to Pico, is unclear; see Bartòla, 'Eliyahu del Medigo,' 275; Kieszkowski, 'Les rapports,' 50–51.

22. For brief information and further literature on Grimani, see Cont. Eras. s.v. 'Grimani, Domenico' (by Chambers). See also Paschini, *Domenico Grimani;* Lowry, 'Two Great Venetian Libraries,' esp. 146–164.

23. Letter to Grimani, in Kieszkowski, 'Les rapports,' 76: "Et quia bene ratiocinatus fuisti, volui transducere (tibi) prohemium Commentatoris, quod fecit in XIIo Metaphysicae, in quo ponit ordinem librorum et multa bona," and 77: "Id etiam quod dicit in prohemio praedicto multum facit in hoc."

24. Ibid., 76: "Et quamvis alias transduxi dignissimo domino Johanni comiti Mirandolano, tamen illam transductionem non habeo, et forte in nulla sententia variatur."

25. Ibid., 77: "Quia ego scio quod ex te potes multa bona et difficilia cognoscere, in hoc autem non solum sententiam suam servavi, immo et verba meliore modo, quo fieri possit. Valete."

26. See Del Medigo's preface to *De primo motore,* in John of Jandun, *Questiones de physico auditu* (1506), f. 121vb. I owe the reference to this passage to Silvia Di Donato.

27. Bartòla, 'Eliyahu del Medigo,' 274n78. The translation of the prooemium can be dated to after 1487, when del Medigo finally settled in Padua—and before he left again for his home country, where he died in 1493.

28. Tamani, 'I libri ebraici del cardinal Domenico Grimani,' 5–52. Pico's Hebrew library is more difficult to trace than Grimani's since his manuscripts and prints do not contain an ex libris. Of the 123 Hebrew manuscripts mentioned in the inventory of 1498, only 4 are Averroes manuscripts. See Tamani, 'I libri ebraici.'

29. Pico recalls this in a letter of 1489 to Pizzamano: Paschini, *Domenico Grimani,* 10.

30. On Ricci, see the fundamental study by Roling, *Aristotelische Naturphilosophie*, as well as Roling, 'Meditatoris fungi munere,' and Roling, 'Prinzip, Intellekt und Allegorese.' Cf. also Secret, *Les Kabbalistes chrétiens*, 87–99; Cont. Eras. s.v. 'Ricius, Paulus' (by Bietenholz).

31. Pietro Pomponazzi in Bologna lectures of 1514 quotes "Paulus Israelita" on the authenticity of the prooemia by Averroes to *Physics*, book 3 and book 8. In a different redaction of the lectures, the same reference is to a professor in Pavia, which strongly suggests the identification of "Paulus Israelita" and "Ricci"; these references are in Nardi, *Studi*, 131, 143. Pomponazzi's reference is to the following passage in Ricci's 1511 edition, f. 7r: "Secundum vero quod in tertio De auditu prooemii loco legitur est Averrois in sexagesimo primi Phisicorum commento digressio ... Tertium autem quod octavi De auditu prooemium creditur ... omnino totumque prooemium fictitium est, quod interpretis aut scriptoris dolus vel inscitia effinxit."

32. See Cont. Eras. s.v. 'Poncher, Étienne' (by de la Garanderie). From 1502 to 1511, Poncher was also chancellor of the Duchy of Milan (with interruptions), which may explain Ricci's acquaintance with Poncher, who was known as a supporter of the humanist movement.

33. This translation is not Ricci's own, as is sometimes stated. Cf. Roling, *Aristotelische Naturphilosophie*, 8, 87n. On Theodore's and Mantino's versions of the preface, see Burnett and Mendelsohn, 'Aristotle and Averroes on Method,' 53–60. On Averroes's preface, see also Glasner, *Averroes' Physics*, 41–43.

34. Ricci, *De Prooemio et eius partibus questio*, in Averroes, *Hoc opere contenta ...* (1511), f. 1v: "Hunc demum cordubensem Averroim de caelo, de mundo deque totius machinae partibus disputantem fors obtulit, quem utique tuae celsitudini tum oppificis celebritate, tum subiecti nobilitate, sensus claritate, voluminis in latinorum bibliotheca novitate, eius denique magna in parte cum theologicis quibus insistis concinnitate haud incongrue destinandum fore duxi."

35. Ibid., f. 7r: "Nec solum eiuscemodi tria prooemia, sed universa in Averoi latina editio, ut quandoque aliquibus ex conphilosophis nostris patefeci, crebris corruptelis erroribusque abundat. Haec admonuisse volui ne quando aliqua huius prooemii et aliorum subsequentium, quae de hebraeorum bibliotheca excerpsi a latina editione discriminari conspexeris, in querulam et criminatoriam vocem prorumpas. Ubique enim hebraei et nonnulli arabicae linguae adsunt eruditi, quos si veri explorator fueris consule pro viribusque examina unde omnia quidem sedulo ac adamussim tradita esse dignosces. Sin autem minus eiuscemodi periculum feceris nec his nostris editionibus mens tua acquieverit, tunc non me argue interpretem, sed te corripe ipsum." See Burnett, 'The Second Revelation,' 193, for a translation of the entire passage.

36. Grendler, *The Universities*, 31–32. The closure lasted until 1517.

37. Ricci, *In apostolorum simbolum ... dialogus* (1514), sig. A2r, titled: "Petri Pomponetii unici nostro aevo philosophiae moderatoris ad discipulos et conphilosophantes epistolium."

38. Nardi, *Studi,* 131, 143.

39. Nifo composed an *Expositio in IV libros De caelo et mundo* which was printed several times (Lohr, *Latin Aristotle Commentaries,* 2:285). On Vernia, see Mahoney, 'Philosophy and Science,' in *Two Aristotelians,* art. 1, 136–142. As a guide to the reception of *De caelo* in the sixteenth century, one may consult Lohr, *Latin Aristotle Commentaries,* v. 2, and Cranz, *A Bibliography,* 170–171.

40. On Mantino's life and works, the substantial article by Kaufmann of 1893 remains the main reference: Kaufmann, 'Jacob Mantino'; cf. Roth, *The Jews in the Renaissance,* 77–79; Carpi, 'Sulla permanenza'; Kottek, 'Jacob Mantino'; DBI s.v. 'Martino, Giacobbe' (by Saracco). Mantino's technique of translation is discussed by Hissette, 'Guillaume de Luna,' 148–158; Coviello and Fornaciari, *Averroè: Parafrasi,* x–xv; Burnett and Mendelsohn, 'Aristotle and Averroes on Method,' 55–60.

41. The prefaces are reprinted in Kaufmann, 'Jacob Mantino,' 220–227.

42. Ibid., 220 (to Pope Leo X): "Hanc cum diu a latinis desideratam nuper offendissem nostris, hoc est Hebraicis, exarctam litteris, incredibili gaudio sum affectus."

43. Ibid. (to Pope Leo X): "facturum me arbitratus gratissimum iis qui scientiam, quae de natura est, profitentur. Nam ut rerum naturalium cognitionem sine Aristotele adipisci difficile est, ita Aristotelem sine Averroi profiteri, meo quidem iudicio, non est valde probandum." Cf. ibid., p. 222 (to Gonzaga): "Huius equidem expositores atque commentatores innumerabiles prope dixerim extant, inter quos unus tantum Averroys Cordubensis machometanus sententiis ipsius Aristotelis maxime accedere nullus prorsus potest ambigere."

44. Ibid., 222 (to Gonzaga): "Et quamvis in his libris vertendis et iam conversis a nobis latinam eloquentiam non profiteamur, fateor enim me eam non esse assecutum, illam tamen traductionem quae pridem foede et barbare latinis data fuit atque obscure non imitabimur, sed pro viribus conabimur sententiam integram autori reddere et intelligibilem. Quapropter hoc epithoma metaphysicae Averroys prelo transferre decrevi, cum in longa commentatione ipsius quam latini habent multa inculta et mutilata appareant, propter depravatam traductionem, idque profecto est familiare omnibus priscis traductionibus Averroys fuitque causa ut multi hac aetate doctrinam Averroys damnent."

45. Ibid., 224: "Et quamvis pro illustrissimo patrono meo reverendo Hercule Gonsaga, principe proculdubio liberalissimo omnique virtute ornatissimo, Averrois philosophiam sim interpretatus, non omittam tamen his negotiis impeditus tibi morem gerere."

46. Aristotle and Averroes, *Omnia . . . opera* (1550/1552), vol. 1, f. 7v, 8r, 9r; Aristotle and Averroes, *Omnia . . . opera* (1562), vol. 1 (part 2), f. 319r (on Comm. mag. Post. anal.).

47. The dedication is printed in Zedler, *Averroes' Destructio destructionum,* 57–58; cf. her comments, 26–27.

48. On Ercole Gonzaga (1505–1563), see the long and informative article by Brunelli in DBI s.v. 'Gonzaga, Ercole.' For the context of the history of his family, see Brinton, *The Gonzaga,* 163–164, 173–174. On his presidency of the Council of Trent, see the (not fully reliable) book by Pescasio, *Cardinale Ercole Gonzaga.*

49. According to a report by Ercole Gonzaga's secretary to his mother, Isabella d'Este; see Oliva, 'Note sull'insegnamento,' 264: "Lo Ex.te Mag. Petro Pomponatio vien ogni dì alle ventidue vel circha a levare il S.re di casa et li fa compagnia al Studio, dove a quella hora legge et la lectione sua è il Metheoro d'Aristotele molto delettevole a sentirla"; mentioned by Grendler, *The Universities,* 166.

50. Nardi, *Studi,* 213.

51. Ibid., 53.

52. Kaufmann, 'Jacob Mantino,' 223: "et aquam claram ex illo magno fonte praeclarissimi philosophi Petri Pomponatii patricii Mantuani bibis."

53. Avicenna, *Quarta fen primi libri* (1530), sig. a2v: "quia vestro florentissimo studio patavino a puero fuerim semper addictus."

54. Kaufmann, 'Jacob Mantino,' 36–37. There is evidence that Mantino was living in Padua in 1533; see Carpi, 'Sulla permanenza.'

55. On Pico's attitude toward humanism, see Breen, 'Giovanni Pico della Mirandola,' esp. 392–394.

56. On Grimani's intellectual profile, see Lowry, 'Two Great Venetian Libraries.'

57. Roth, *The Jews in the Renaissance,* 113.

58. Kieszkowski, 'Les rapports,' 89: "Et ideo secundum literas latinas nihil deficit, sed quia nescio an ita sit in lingua arabum sicut in nostra, ideo dixi forte et dubitavi de hoc."

59. The Arabic term for "exposition" is *talḫīṣ* (Averroes, *Tafsīr mā baʿd aṭ-ṭabīʿa* [1938–1948], 1393), which has the technical meaning "exposition of subjects" in Averroes's philosophical language; see Gutas, 'Aspects of Literary Form,' 41–43.

60. Kieszkowski, 'Les rapports,' 78.

61. On the adoption of this classical method by Arabic–Latin translators, see Burnett, 'Translating from Arabic.'

62. For the Hebrew text, see Kieszkowski, 'Les rapports,' 78, for the Arabic: Averroes, *Tafsīr mā baʿd aṭ-ṭabīʿa* (1938–1948), 1393.

63. Kieszkowski, 'Les rapports,' 82. Cf., 81: "considerationem esse necessariam in principiis entis" instead of "considerationem principiorum entis esse necessariam." The Arabic phrase is *naẓara fī.*

64. Kieszkowski, 'Les rapports,' 84.

65. Paolo Ricci translates: "nullum . . . sermonem comprobare potest"; see Aristotle and Averroes, *Omnia . . . opera* (1562), vol. 8, f. 288ra and f. 288rb.

66. Kieszkowski, 'Les rapports,' 79.

67. Ibid., 85.

68. Coviello and Fornaciari, *Averroè: Parafrasi,* xiii: "Più in generale la lingua di Elia del Medigo, nella sua 'irregolarità' sintattica e grammaticale segue comportamenti costanti. In primo luogo il suo latino si modella sull'uso del

volgare parlato; frequenti sono quindi gli anacoluti e le costruzioni 'ad sensum.'" At the same time, they explain the transposition of a noun to the beginning of the sentence as an imitation of Hebrew syntax (xiv).

69. Kieszkowski, 'Les rapports,' 79.

70. Gildersleeve and Lodge, *Latin Grammar*, §617; Kühner and Stegmann, *Ausführliche Grammatik*, 2:2, §193, 11, p. 289.

71. Aristotle and Averroes, *Omnia... opera* (1562), vol. 8, f. 286rb.

72. Kieszkowski, 'Les rapports,' 80, 82, 84, 87, 88.

73. Ibid., 87.

74. Aristotle and Averroes, *Omnia... opera* (1562), vol. 8, f. 289rb.

75. This is said in answer to a question raised by Steinschneider, *Die hebraeischen Übersetzungen*, 173, and Bartòla, 'Eliyhau del Medigo,' 275n79.

76. For instance, the reading "qui... est" (והם) (*Omnia... opera,* f. 286raC: "qui quidem principium est et prima substantia"), where del Medigo and Mantino have "et de" (וב, in Arabic: *wa-fī*), or the missing negation "quae substantiae sint..." (*Omnia... opera,* f. 289raB), where del Medigo writes "quae sunt aliae a formis..." and Mantino "quae non sint formae..." (with the Arabic *ġayr* and the Hebrew manuscript d). See Averroes, *Tafsīr mā baʿd aṭ-ṭabīʿa* (1938–1948), 1394, 1402.

77. Aristotle and Averroes, *Omnia... opera* (1562), vol. 8, f. 286ra.

78. For example, a concessive relation, Aristotle and Averroes, *Omnia... opera* (1562), vol. 8, f. 286va: "licet enim idem facere aggressus est in dictione signata per literam Aleph maiorem, de hoc tamen in duabus hic dictionibus sermonem consumat."

79. Hasse, 'Abbreviation,' on the abbreviation techniques of Constantine the African, Hermann of Carinthia, Michael Scot, Theodore of Antioch, and Paolo Ricci.

80. Averroes, *Tafsīr mā baʿd aṭ-ṭabīʿa* (1938–1948), 1395–1396.

81. Aristotle and Averroes, *Omnia... opera* (1562), vol. 8, f. 286va (Ricci), f. 286vb (Mantino); cf. Kieszkowski, 'Les rapports,' 81.

82. Cf. ThLL s.v. 'condecens.'

83. Ricci, in Averroes, *Hoc opere contenta...* (1511), f. 7r: "Nec solum eiuscemodi tria prooemia, sed universa in Averoi latina editio, ut quandoque aliquibus ex conphilosophis nostris patefeci, crebris corruptelis erroribusque abundat."

84. The Venice 1542 edition of *Averrois Epithoma totius Metaphysices Aristotelis: Prohemium duodecimi libri Metaphysices* prints Ricci's version, according to Bouyges, *Averroès... Notice,* lxxx.

85. One such unclassical expression adopted from del Medigo is *finis intentus.* Cf. Cf. ThLL s.v. 'intentus.'

86. Averroes, *Tafsīr mā baʿd aṭ-ṭabīʿa* (1938–1948), 1394; Kieszkowski, 'Les rapports,' 79 (del Medigo); Aristotle and Averroes, *Omnia... opera* (1562), vol. 8, f. 286ra (Ricci), f. 286rb (Mantino).

87. Averroes, *Tafsīr mā baʿd aṭ-ṭabīʿa* (1938–1948), 1395; translation by Gutas, review of C. Genequand, *Ibn Rushd's Metaphysics,* 124.

88. Kieszkowski, 'Les rapports,' 80; Aristotle and Averroes, *Omnia...opera* (1562), vol. 8, f. 286vb (Mantino).

89. Kieszkowski, 'Les rapports,' 80 (Hebrew); Averroes, *Tafsīr mā baʿd aṭ-ṭabīʿa* (1938-1948), 1395.

90. Hissette, 'Jacob Mantino tributaire,' and 'Jacob Mantino utilisateur.'

91. Ricci's translation fuses two sentences, so that it is unclear which of the two readings he is translating: "Nec absonum est huc quoque afferre ea quae ad summariam aliarum huius scientiae dictionum intelligentiam Alexander praeposuit" (Aristotle and Averroes, *Omnia... opera* (1562), vol. 8, f. 286va).

92. Kieszkowski, 'Les rapports,' 80.

93. *fī-hi iḥtimāl* is the reading of the sole Arabic manuscript. Freudenthal and Fränkel, *Die... Fragmente,* 69, propose instead the reading *fī-hi iǧmāl* (which they translate as "ist summarisch ausgedrückt"). This reading is adopted by Bouyges in his edition, but rejected by Gutas, review of C. Genequand, *Ibn Rushd's Metaphysics,* 142n11.

94. On Abraham de Balmes, the most pertinent studies are Pierro, 'Abramo di Meir de Balmes,' Tamani, 'Le traduzioni' (a survey of all of his translations), and Tamani, 'Una tradizione' (on the *Quaesita naturalia*). Cf. also the old study by Ferorelli, 'Abramo de Balmes.' On Balmes's Latin grammar of Hebrew, which was printed in 1523, see Campanini, 'Peculium Abrae.'

95. Averroes, *Libri Posteriorum analiticorum* (1523), sig. AA3v: "mea namque natalitia exacta sunt in litio civitate in agro salentino... quae ubi olim rudiae fuerunt est, ex reliquiis rudiarum nacta originem, unde Ennius fuit oriundus, cuius etiam moris fuit verborum leporem negligere modo aureas exposuerit sententias" (in the dedication to Cardinal Grimani).

96. According to Gedalyah ibn Yahya; see Tamani, 'Le traduzioni,' 629n21.

97. Di Donato, 'Traduttori di Averroè,' 37. On Alberto Pio's interest in Aristotle, the Greek commentators, and Averroes and on his relation to Pomponazzi, see Schmitt, 'Alberto Pio,' in *The Aristotelian Tradition,* art. 6, 43-64 (on Balmes, 60).

98. Averroes, *Libri Posteriorum analiticorum* (1523), sig. AA2v: "Hos ergo, dive Grimane, libros ita nostris posteris tibi inscriptos tradimus, ut per te illos habuisse censeant, quando te iubente eos edidimus, tuoque nomini inscripsimus et te favente publicos facimus." In 1521-1523, Antonio Grimani, Domenico's father, was doge of Venice.

99. To the best of my knowledge; Cranz, *A Bibliography,* 198, lists the 1523 volume, which, however, only contains the paraphrase of the *Topics,* not the Middle commentary.

100. Averroes, *Libri Posteriorum analiticorum* (1523), sig. AA2r.

101. Ibid.: "Gaudeant ergo *(scil. the partisans of the Greek Aristotle)* ventulo illi suo, nos Averroym imitemur, Averroy credamus in maiore parte suorum dictorum praecipue in logica, ubi nullum fidei scandalum est metuendum."

102. Ibid.: "Sed missa isthaec faciamus, ad hos futiles viros redeamus qui potius ventositate quam philosophia inflati ut graecizare se ostentent in nostrum

Averroym oblatrant, dicentes quod Aristotelis mentem non habuerit, quia a graeca littera sua sententia saepe recedit."

103. Ibid., sig. AA3r: "Nec credat aliquis quod eloquentiae ornatu et orationis fucu ea scripserimus, sed nostris communibus usitatis verbis et modis passim venientibus ipsa tradimus, cum maluerim de abusu eloquentiae quam de mutata sententia aut serie autoris increpandum esse, tum quia mea prima ineunte aetate meis hebraeis litteris meis talmudisticis gymnasiis immissus sum, quibus veritas praeponitur spreta eloquentia, eo magis quia linguarum disparitas contrarias eloquii normas saepe parire solet. Romani enim dicacitatem, hebraei vero veritatem semper praehonorandam censuerunt."

104. Aristotle and Averroes, *Omnia . . . opera* (1550/1552), vol. 1, f. 15v: "<Averrois> Expositio in octo libros Topicorum, Abramo de Balmes interprete, mendosa obscura nunc castigatissima et ad maximam claritatem restituta legitur, cui una illam super quatuor primos libros a Iacob Mantino nuper translatam adiecimus, quam super reliquos quatuor morte praeventus explere non valuit." And f. 8r: "Constitui deinceps Topicorum et Elenchorum libros cum media Averrois expositione ab Abramo de Balmes et Iacobo Mantino latinitate donata. Nam cum Iacobi interpretatio super quatuor primos libros Topicorum dumtaxat haberetur, quae vero Abrami erat tota quidem legeretur, sed obscuritate et erratis referta maximam lectori afferret difficultatem, nobis, qui ob Mantini dilucidi interpretis interitum residuis carebamus, visum est id saltim quod ab eo conversum fuit Abrami interpretationi prius a nobis expurgatae et Aristotelis contextui aptare, ut illius et candorem et veritatem interpretationum iudicio tuo perpenderes."

105. Averroes, *Talḫīṣ kitāb al-ǧadal* (1979), 1.1, 29.

106. Aristotle and Averroes, *Omnia . . . opera* (1562), vol. 1, 3, f. 3v–4r.

107. Hebrew: Lasinio, 'Studii sopra Averroe,' 139.

108. Aristotle and Averroes, *Omnia . . . opera* (1562), vol. 1, 3, f. 4r.

109. Ibid.

110. Cf. Forcellini s.v. 'universalis.'

111. Aristotle and Averroes, *Omnia . . . opera* (1562), vol. 1, 3, lib. 1, f. 22v (Arabic, p. 53, l. 2). Compare the following list of Balmes's and Mantino's renderings of עניין/*ma'nā* in book 1: f. 4r: transtulit Aristoteles hoc nomen ad hanc rem (Balmes)—ideo Aristoteles usus est hoc nomine ad significandum (Mantino) (=Arabic, p. 30, ll. 2–3); f. 6r: distinguendo res ratione dictionum—dum dividit res per dictiones (=Arabic, p. 33, ll. 8–9); f. 6v: illae enim iuvant hanc rem—et ideo conducunt ad hoc negotium (Arabic, p. 34, l. 5); 8v: non quatenus significant rem illi inexistentem—non quatenus significent aliquod existens (Arabic, p. 36, l. 8); f. 11v: nomen unius dicitur de tribus significatis—hoc nomen idem primo quidem tria significat (Arabic, p. 40, ll. 5, 6); f. 12v: et sic est oratio significans rei rationem—tunc erit oratio significans ipsam rem (Arabic, p. 41, l. 15); f. 19v: quod hanc rem iuvat—conducet autem ad hoc negotium (Arabic, p. 50, l. 6); f. 21v: acuitas significat dipositionem

cultri et dipositionem vocis—quod quidem significat aliquid in gladio et aliquid in voce (Arabic, p. 51, l. 10); f. 22v: si consideremus an oppositum unius rerum dicatur—si oppositum unius illorum significatorum dicatur (Arabic, p. 52, ult. l.); f. 23v: consideremus alteram duarum rerum—consideremus alteram duarum rerum (Arabic, p. 55, ll. 4,5); f. 23v: una harum duarum rerum—una illarum duarum rerum (Arabic, p. 56, l. 1); f. 25v: alterius duorum significatorum—unum quoddam significatum ex illis significatis (Arabic, p. 57, l. 15; p. 58, l. 1: "significatum" used repeatedly); f. 26r: quoddam illorum significatorum—illarum rerum quae significantur (Arabic, p. 58, l. 8); f. 26r: admittendo rem—acceptare id; de hac re—de ea re; aliam rem—aliam rem; significat rem—significat illam rem (Arabic, p. 58, ll. 15–16).

112. For example, Aristotle and Averroes, *Omnia . . . opera* (1562), vol. 8, ch. 4.9, f. 75va: "Sed quia hoc non potest demonstrari, sed tantum homo utitur in defensione eius sermonibus veris et valde probabilibus *(fī ġāya aš-šuhra)* quos nullus potest negare." Similar passages are in ch. 4.5, f. 70va–b.

113. Aristoteles Latinus, *Topica* (1969), 1, 1, p. 5.

114. Aristoteles Latinus, *Topica* (1969), 1, 10–11, pp. 15–18, 198–200; Aristotle and Averroes, *Omnia . . . opera* (1562), vol. 1, 3, f. 13v–16v.

115. Reisch, *Margarita philosophica* (1517/1973), 2.5.1, pp. 98–99. Of course, the tradition had long modified one or the other term used by Boethius, as in the case of the four predicables enumerated in *Topics* 1,5 (1,4): Boethius and the twelfth-century translator had written: *terminus, proprium, genus, accidens;* but Reisch (p. 99) uses *definitio* instead of *terminus,* as do both Balmes and Mantino (Aristotle and Averroes, *Omnia . . . opera* (1562), vol. 1, 3, f. 8v–10r).

116. Aristoteles Latinus, *Topica* (1969), 1, 12, p. 18 and p. 200; Aristotle and Averroes, *Omnia . . . opera* (1562), vol. 1, 3, ch. 1, 7 + 10, f. 12v and f. 17r.

117. Aristoteles Latinus, *Topica* (1969), 1, 13, pp. 19, 201; Aristotle and Averroes, *Omnia . . . opera* (1562), vol. 1, 3, ch. 1, 11, f. 18r.

118. Aristotle and Averroes, *Omnia . . . opera* (1562), vol. 1, 3, ch. 1, f. 4r–4v.

119. Di Donato, 'Il *Kitāb al-Kašf . . .* nella traduzione.' Cf. also Di Donato, 'Il *Kašf . . .* confronto.'

120. Di Donato, 'Il *Kitāb al-Kašf . . .* nella traduzione,' 22, 31.

121. Minio-Paluello, *Opuscula,* 152: "<Balmes> non sembra molto pratico del linguaggio tecnico latino corrispondente alla terminologia greco-araba-ebraica. Il Mantino, ugualmente ebreo, è più accurato del de Balmes nel rendere accuratamente anche nei dettagli il testo da cui traduce, ed è più del de Balmes padrone del linguaggio tecnico latino."

122. Aristotle and Averroes, *Omnia . . . opera* (1550/1552), vol. 1, f. 7v: "Averrois expositiones . . . nuper a Iacob Mantino et translatae et longe dilucidiores factae"; f.8r: "ob Mantini dilucidi interpretis interitum"; f. 9r: "ex Iacobi Mantini dilucidissima novaque translatione"; f. 13r: "et ob id Iacob Mantinus ea dilucidissime ut sui moris est interpretatus esset"; Aristotle and

Averroes, *Omnia . . . opera* (1562), vol. 1, 2, f. 319r: "Hucusque doctissimi Mantini candide lector aurea super hoc primo Posteriorum pervenit translatio. Cetera vero morte preventus perficere haud potuit."

123. These translations are surveyed by Siraisi, *Avicenna in Renaissance Italy*, 133–143. See also her discussion of the prefaces on pp. 165–170. Cf. also d'Alverny, 'Avicenne et les médecins,' in *Avicenne en occident*, art. 13, 182–188.

124. On Ramusio, see Jacquart, 'Arabisants,' in *La science médicale*, art. 11, 399–415, and the older, but more extensive study by Lucchetta, 'Girolamo Ramusio.'

125. Ramusio, MS Paris, Bibl. Nat. Arabe 2897, f. 349r: "Ad Flavium Ramusius. Inter scribendum arabice multa vidi que latino Abuali omnino repugnant, multa annotavi que latinus dimisit aut incorrepta scripsit, multa perlegi que non ita ad unguem latina facere potui ob imperitiam arabice lingue, quare in dies adiscens in dies melius corrigo meque ac scripta mea castigo, dum igitur hunc primum Abuali arabice scripsi nullum ordinem fere servavi, nam ad utiliora prius semper accessi posteriora relinquens minus utilia, et ut ordinem reperias, Flavi optime, aduerte: primo vidi usque ad anathomiam 1, de pulsu urina et egestione a charta 50, quarta primi a charta 107, doctrina sexta de virtutibus a charta 160, fen secunda a charta 174, fen tertia a charta 272. Reliquum quarte primi et de anathomia invenies latinum meum correctum, nec admireris si omnia non ita dilligenter castigata sint, cum sensim omnia perceperim, et ut citius adiscerem arabice scripsi et litteris omnia labefeci. Spero in *(inde: Lucchetta)* futuro anno omnia repetere et dilligentissime omni studio perscrutari et in optimam formam redigere, quare diis magnis et Catte gratias ago etc." Cf. the edition in Lucchetta, *Girolamo Ramusio*, 42n, and the French translation in d'Alverny, 'Survivance,' in *Avicenne en occident*, art. 15, 95.

126. Ramusio, MS Paris, Bibl. Nat. Arabe 2897, f. 160v: "Volui prius videre que in gymnasio Patavino publice leguntur"; see the longer quotation in Jacquart, 'Arabisants,' 402n11.

127. The main research tool on Alpago is Lucchetta, *Il medico e filosofo;* for her arguments for the date 1487, see p. 16. Cf. Veit, 'Der Arzt Andrea Alpago.'

128. For an analysis of the six letters with political content that are still exant, see Veit, 'Andrea Alpago and Schah Ismāʿīl.'

129. See Siraisi, *Avicenna in Renaissance Italy*, 134–135. Note Siraisi's reference to the frequent comparisons of Alpago's and Mantino's versions in Oddo Oddi's commentary (135, 192–194); see Oddi, *In primam totam fen primi libri* (1575). For the section discussed below on the perceiving faculties, Oddi's commentary does not go into a detailed comparison of the two translations.

130. Avicenna, *Principis Avicennae libri Canonis* (1527), sig. A2r: "Quidnam vel ille sibi voluerit, in non paucis huius libri sui locis, vel quid iuxta eius decreta sibi periclitandum foret, dum de vita hominum contendant, tanta scilicet erratorum caligine obductus habebatur, quamobrem patruus olim meus Andreas . . . hunc librum sibi recognoscendum proposuit . . . Redditum sibi

esse nunc Avicennam, quem vere se habere gloriari possint medici, cuius antea falsa quaedam fere et inani imagine efferebantur."

131. Lucchetta, *Il medico e filosofo*, doc. 12, p. 90: "rettulerunt et relationem suam fecerunt singulatim cum vidissent nonnulla opera eiusdem magistri Andree, translat[at]a ex arabo in latinum, illaque sumopere comendarunt tamquam utilia et necessaria." Further down in the document, Alpago's work is referred to as "optima transla[c]tione et correctione locorum in libris Avicene." Cf. also doc. 11, 88–90.

132. Lucchetta, *Il medico e filosofo*, doc. 13, p. 91.

133. Avicenna, *Quarta fen primi* (1530), sig. A1v–A2r "Sed quoniam Avicenna in scribendo gentilitio ac sibi peculiari arabico idiomate usus est, quod a latinis hominibus non ita facile comparatur, eius operum traductio maximis ac multis erroribus scatet, quos Andreas Belunensis, aetatis nostrae medicus insignis et arabica latinaque lingua pariter eruditus, magna ex parte laudabiliter emendaverit, alieno tamen semine campum penitus expurgare non potuit, sed adhuc plurima relicta sunt quae veluti nebula quadam veritatem lectionis obducant" (for a reprint of the preface, see Kaufmann, 'Jacob Mantino,' 224–225).

134. Avicenna, *Quarta fen primi* (1530), sig. A2r: "Cum autem tres potissimum Avicennae partes in gymnasiis publice legantur, . . . ea omnia in latinam orationem vertere proponens a quarta fen primi libri interpretari exorsus sum quia haec pars maiorem caeteris in universali medendi arte utilitatem afferre videtur."

135. "Hagenoae" is Haguenau/Alsace, not The Hague, as in Siraisi, *Avicenna in Renaissance Italy*, 135, 363.

136. Kaufmann, 'Jacob Mantino,' 212. From 1539 to 1541 Mantino was also a teacher of medicine at the Sapienza in Rome.

137. Avicenna, *Primi libri fen prima* (1547), sig. A1v: "cum superioribus diebus primam, ut vocant, fen primi libri Avicennae ex hebraico, quem quidem nihil penitus ab ipso arabico sensu discrepare norunt omnes vel mediocriter eruditi, in latinum transtulerim."

138. As has been pointed out by Burnett, 'The Second Revelation,' 194–195.

139. See Zonta, *La filosofia antica*, 96–116. See also Tamani, 'Le versione ebraiche del *Canone*.'

140. On Ledesma, see García Ballester, 'The Circulation,' 188–190 (Spanish version, *Los moriscos*, 24–26), and Siraisi, *Avicenna in Renaissance Italy*, 138–139.

141. Avicenna, *Prima primi Canonis Avicenne sectio* (1547/1548), sig. A2v.

142. The first nonreligious work to be printed was a geographical treatise in 1585; on the very few previous editions—a psalter, a book of hours, a confession of faith, and a Quran—see Pinto, 'Una rarissima opera,' and in general Endress, 'Die Anfänge.'

143. Avicenna, *Prima primi Canonis Avicenne sectio* (1547/1548), sig. A2v–A3r: "Item Andreas Bellunensis novus interpres, atqui is aliquando Gentilis aut Nicoli

aut alterius cuiuspiam sententiam verius sequitur quam veritatem. Quibus praeter peculiaria nostra in vestigandis linguarum proprietatibus studia, adde consultum fuisse socium arabicae linguae non minus quam rei medicae peritum."

144. García Ballester, 'The Circulation,' 188–189. Siraisi, *Avicenna in Renaissance Italy,* 139, is more careful: "'very old' (and perhaps Arabic?) manuscript."

145. On Gratiolo, see d'Alverny, 'Avicenne et les médecins,' in *Avicenne en occident,* art. 13, 182–184; Siraisi, *Avicenna in Renaissance Italy,* 141–142, 160–161; and Lohr, introduction to Philoponus, *Commentaria . . . in libros Posteriorum* (1542/1995), xv.

146. Lohr, introduction to Philoponus, *Commentaria . . . in libros Posteriorum* (1542/1995), xv.

147. Siraisi, *Avicenna in Renaissance Italy,* 134–135, 192–194.

148. There is much information on Oddi in Siraisi, *Avicenna in Renaissance Italy,* index s.v. Graziolo was also supported by Bernardino Paterno, Niccolò Sanmichele, and Girolamo Fracastoro: Avicenna, *Principis Avicennae liber primus* (1580), sig. B4r; cf. Siraisi, *Avicenna,* 141.

149. Avicenna, *Principis Avicennae liber primus* (1580), sig. B3r–v: "qui proculdubio multas per se illustres sententias obscuravit, multa non intellexit, omnia barbarie foedavit, praeter rem voces quasdam in orationis cursu inculcans, quae omnino a propria ipsius autoris oratione absunt."

150. On Donzellini (ca. 1513–1587), see DBI s.v. 'Donzellini (Donzellino, Donzellinus), Girolamo.'

151. Avicenna, *Principis Avicennae liber primus* (1580), sig. B3v–B4r.

152. Ibid., sig. B3v: "Caeterum in hoc autore vertendo, omnibus eius interpretibus consultis, multorum emendationibus inspectis, praesertim Andreae Bellunensis . . . , verum ac germanum (ni fallor) scriptoris sensum sum expiscatus, quam latinis ac perspicuis verbis, quantum dicendo assequi potui, expressi."

153. This chapter is put into the context of Avicenna's medical teaching by Siraisi, *Avicenna in Renaissance Italy,* 29.

154. See Hasse, *Avicenna's De Anima,* 40–41, 48–49, 53–54, 133, 135, 144, and Sudhoff, 'Die Lehre von den Hirnventrikeln,' 174–177, 184–189.

155. Jacquart, 'Arabisants,' in *La science médicale,* art. 11, 399–415, esp. 405–407. Cf. the older study of Ramusio's translation by d'Alverny, 'Survivance,' in *Avicenne en occident,* art. 15, 91–97.

156. A parallel case of extreme literalness of translation are the Greek–Latin lecture notes, preserved in manuscript, of a contemporary of Ramusio: Antonio Urceo, that is, Codro (1446–1550), who was teaching Greek in Bologna beginning in 1485. In order to demonstrate the specific character of the Greek language, Codro translates Isocrates into a syntactically hybrid Latin. For specimens of this Latin, see Raimondi, *Codro,* 192–196 (I owe this reference to Ludwig Braun).

157. Ramusio, MS Paris, Bibl. Nat. Arabe 2897, f. 169v and f. 170v.

158. On Ramusio's Latin poetry, which has not survived, see Lucchetta, 'Girolamo Ramusio,' 48. For a specimen of Ramusio's Latin style, see his letter to Flavius quoted note 128 above.

159. On faculty theory in the Middle Ages and the Renaissance, see Hasse, 'The Soul's Faculties,' and Park, 'The Organic Soul.'

160. Avicenna, *Liber canonis* (1507), 1.1.6.5, f. 24va–25rb.

161. Avicenna, *Qānūn* (1877), 1:71–72.

162. Hasse, *Avicenna's* De Anima, 167.

163. Avicenna, *Liber canonis* (1507), 1.1.6.5, f. 25ra.

164. Ramusio, MS Paris, Bibl. Nat. Arabe 2897, f. 169v.

165. Avicenna, *Liber canonis* (1507), f. 25ra.

166. Ramusio, MS Paris, Bibl. Nat. Arabe 2897, f. 170r.

167. Avicenna, *Liber Canonis medicinae* (1527), f. 22ra (Alpago): "sicut hominem volare et montem smaragdinum"; Avicenna, *Primi libri fen prima* (1547), sig. F6r (Mantino): "ut hominem evolantem exempli causa, et montem smaragdinum."

168. Cf. the following two examples: Gerard: "ex quibus cum in suis impedimentis accidit laesio"; Ramusio: "cum accidit ei nocumentum in operationibus eius"; Alpago: "in quarum operationibus cum accidit laesio"; Mantino: "quarum actiones cum laeduntur" (Avicenna, *Liber canonis* (1507), f. 25ra; MS Paris, Bibl. Nat. Arabe 2897, f. 170v; Avicenna, *Liber Canonis medicinae* (1527), f. 22ra; Avicenna, *Primi libri fen prima* (1547), sig. F6v). Gerard: "virtus vero comprehensiva"; Ramusio: "sed virtus remansiva"; Alpago: "reliqua vero virtus"; Mantino: "restat modo virtus" (f. 25rb; f. 171v; f. 22rb; sig. F7r).

169. Wright, *A Grammar,* 2:318.

170. Avicenna, *Qānūn* (1877), 1:72.

171. Ramusio, MS Paris, Bibl. Nat. Arabe 2897, f. 171r.

172. Wright, *A Grammar,* 2:16.

173. Ramusio, MS Paris, Bibl. Nat. Arabe 2897, f. 170r: "Ergo *quod* iudicat in eis et comprehendit eas virtus alia," and f. 170v: "Et medicus *quod* considerat in virtutibus...."

174. Wright, *A Grammar,* 2:57ff.

175. Avicenna, *Liber Canonis medicinae* (1527), f. 22ra (Alpago); Avicenna, *Primi libri fen prima* (1547), sig. F7r (Mantino).

176. Avicenna, *Liber canonis* (1507), f. 25ra.

177. Ramusio, MS Paris, Bibl. Nat. Arabe 2897, f. 170v.

178. Alpago's emendations are printed in the margin of Avicenna, *Liber Canonis medicinae* (1527), f. 21vb–22rb.

179. Avicenna, *Qānūn* (1877), 1:72.

180. For a more negative evaluation of Alpago's technique of translation with respect to vocabulary, see Goltz, *Studien,* 347–348.

181. Avicenna, *Primi libri fen prima* (1547), sig. F6v–F7r. I use the copy of Vienna, Österreichische Nationalbibliothek, shelfmark 68.M.19.

182. Mantino's *tamen* translates אמנם, Mantino's *nam* translates ש (I am grateful to David Wirmer for this information). On the 1491 Venice print of the *Canon* in Hebrew (GW 3113, Klebs 132.1), see Tamani, *Il Canon medicinae.*

183. On *complexio* and *temperatura* in sixteenth-century medicine, see Siraisi, *Avicenna in Renaissance Italy,* 296–305. On the general phenomenon, see Nutton, 'The Changing Language.'

184. Compare the following two sentences, the first about *ḫayāl,* the second about *mutaḫayyila:* Avicenna, *Primi libri fen prima* (1547), sig. F5v: "ad eam quae sensus communis et imaginaria vocatur, et est una et eadem virtus apud medicos, sed apud doctos philosophos sunt duae virtutes," and sig. F6r: "Nam si ipsa utetur virtute existimativa vitali, de qua paulo post fiet mentio, vel quod de se ipsa excitetur ad suam actionem, tunc vocamus eam imaginariam." It is possible that Mantino's choice reflects the intermediary Hebrew, from which he was translating.

185. For a contemporary source, cf. Reisch, *Margarita philosophica* (1517, repr. 1963), 434, which is put in context by Park, 'The Organic Soul.'

186. Cf. Albertus Magnus, *De homine* (1896), qu. 37, p. 323.

187. Avicenna, *Primi libri fen prima* (1547), sig. F6v: "quia imaginatio sensilia retinet, haec autem iudicat de sensibilibus <u>res</u> non sensatas"; ibid.: "huius vero actio est iudicare de re sensata, per <u>aliquod</u> exi[g]ens extra sensatum"; ibid.: "ita existimativa in eis iudicat de illis <u>formis</u> quae ad existimativam deferuntur."

188. Avicenna, *Qānūn* (1877), 1:71–72.

189. Avicenna, *Prima primi Canonis Avicenne sectio* (1547/1548), f. 110r.

190. Ibid., sig. A2v–A3r: "atqui is aliquando Gentilis aut Nicoli aut alterius cuiuspiam sententiam verius sequitur quam veritatem."

191. Avicenna, *Qānūn* (1877), 1:72.

192. Avicenna, *Prima primi Canonis Avicenne sectio* (1547/1548), f. 112r.

193. Avicenna, *Qānūn* (1877), 1:72.

194. Avicenna, *Prima primi Canonis Avicenne sectio* (1547/1548), f. 112v.

195. Avicenna, *Liber canonis* (1507), f. 25rb.

196. Cf. for a different conclusion, see Siraisi, *Avicenna in Renaissance Italy,* 139: "Ledesma succeeded in producing what seems to be the only independent new translation of a substantial section of the *Canon* directly from Arabic to appear in the sixteenth century."

197. Avicenna, *Liber primus de universalibus . . . praeceptis* (1580), f. 106r–v.

198. Ibid., f. 106r: "posterior vero in iis se exercet quae iam in imaginatione reposita sunt usu videlicet componendi et dividendi." Cf. Gerard's version (Avicenna, *Liber canonis* [1507], f. 25ra): "et ista est illa quae se exercet in eis quae in imaginatione recondita sunt exercitio componendi et dividendi."

199. Avicenna, *Liber primus de universalibus . . . praeceptis* (1580), f. 106v: "quemadmodum imaginativa est thesaurus earum, quae ad sensum pertinent ex speciebus atque imaginibus sensibilibus." Cf. Mantino's translation (Avicenna, *Primi libri fen prima* (1547), sig. F7r): "quemadmodum imaginatio est thesaurarium rerum quae ad sensum deferuntur ex formis sensatis."

200. Avicenna, *Liber primus de universalibus . . . praeceptis* (1580), f. 106v: "quae scilicet rem, quae in thesauro cogitativae reposita est, in memoriam revocat." Cf. Mantino's translation (Avicenna, *Primi libri fen prima* (1547), sig. F7r): "et quae reminiscitur rei quae a custodia thesauri cogitativae occultata est."

201. Avicenna, *Qānūn* (1877), 1:72; Avicenna, MS Paris, Bibl. Nat. Arabe 2897, f. 171r (Ramusio); Avicenna, *Primi libri fen prima* (1547), sig. F7r (Mantino); Avicenna, *Liber primus de universalibus . . . praeceptis* (1580), f. 106v.

202. Avicenna, *Prima primi Canonis Avicenne sectio* (1547/1548), sig. A2v.

203. McVaugh, *Arnaldi de Villanova,* 61–62n15, convincingly argues for the year 1285. The text had a very rapid diffusion after its translation. The early date of 1255 that is transported in some secondary literature is the result of a misreading. Cf. Steinschneider, *Die hebraeischen Übersetzungen,* 672, and LexMA s.v. 'Bonacosa' (by Premuda).

204. Steinschneider, *Die hebraeischen Übersetzungen,* 675.

205. Burnett, 'The Second Revelation,' 194.

206. Aristotle and Averroes, *Omnia . . . opera* (1562), 10:175–177.

207. Averroes, *Collectaneorum . . . sectiones tres* (1537), sig. A2v. Coronné read the text with the help of a Jewish physician as he himself remarks in a letter to Symphorien Champier in 1533: "inveni certe hoc est Avicennam, non arabem quidem illum in sua lingua principio scriptum et editum, sed hebraeum antiquissimum in hanc proximam linguam transfusum . . . ad hanc felicitatem nactus sum hominem hebraeum qui locis difficilioribus mihi praeit et tanquam lucerna accensa praelucet, totos tres menses illum iam audii" (printed in Monteux, *Annotatiunculae* (1533), f. 34r). Siraisi, *Avicenna in Renaissance Italy,* 136, has drawn attention to this letter, which she quotes from: Avicenna, *Libri tertii fen primae tractatus quartus* (1572).

208. Averroes, *Collectaneorum . . . sectiones tres* (1537), sig. A3r–v: "Consideranti namque mihi Galeni principis nostri universa paene opera Latinitate donata, simul ac graecorum aliorum posteriorum, diligentia virorum doctorum, venit in mentem me quoque posse medicinam adiuvare si hanc quam aggredior materiam, aliis vel intentatam vel corruptissime depravatam, quam possem diligentissime tractarem. Laboravi equidem hac in re vehementer quoniam codices quos typis excusos habemus depravatissimi essent, unde aut nihil aut parum adiumenti ab iis capere potuimus. Verum ab hinc menses aliquot inciderat in manus nostras codex vetustissimus, qui tres quatuorve sectiones horum collectaneorum complectebatur quique referebat eam tempestatem qua in Gallias immigrarunt Arabum atque Mauritanorum tum philosophia, tum medicina."

209. To the best of my knowledge; see the extensive article in EJud[2] s.v. 'Incunabula' and Steinschneider, *Die hebraeischen Übersetzungen,* 672–675, on the two Hebrew translations of the *Kitāb al-Kullīyāt.* Cf. Endress, 'Averrois Opera,' 376–377.

210. For the humanist usage of the term *codex vetustissimus,* see the examples in Rizzo, *Il lessico filologico,* 147–168.

211. Averroes, *Collectaneorum . . . sectiones tres* (1537), sig. B1r: "Sed illud repetendum me multum laboris insumpsisse multumque vigiliarum collocasse in his convertendis et instaurandis, cum a nullo praeterquam a codicis mei vetustate, quae fidelius scripturam reddidit, quam formis excusi libri, iuvarer."

212. Ibid.: "Cum igitur medicinae candidati a lectione horum collectaneorum, ob tam incultum, tamque horridum sermonem abhorrerent . . . officium meum esse putavi ut . . . hunc materiae campum ingrederer et pro dignitate excultiorem faceremus. In quo gloriari non verebimur nos primos esse qui huiuscemodi barbarorum interpretationem susceperimus, quae (ut speramus) ad eloquendum aptior et ad medicandum commodior utiliorque iudicabitur."

213. Copenhaver, *Symphorien Champier,* 77.

214. Averroes, *Collectaneorum . . . sectiones tres* (1537), sig. B1v: "Lugduni ex aedibus vestri Symphoriani Campegii libertatis medicae assertoris, MDXXXVII, Idibus Ianuarii."

215. Copenhaver, *Symphorien Champier,* 65. Cf. Cont. Eras. s.v. 'Tournon, François de" (by Griffiths).

216. Averroes, *Collectaneorum . . . sectiones tres* (1537), sig. A4r.

217. Bürgel, 'Averroes "contra Galenum,'" 326, 328, 330, with apparatus criticus on pp. 338–339. The paragraph numbering is Bürgel's.

218. Omissions with Bonacosa: Aristotle and Averroes, *Omnia . . . opera* (1562), vol. 10, f. 192va, after "aquam non nutrire asserit" (sentence on natural philosophy, Bürgel §2); f. 192va: "hoc est, naturali animalique" (nutritive faculty, Bürgel §5); f. 193va, after "huic reluctantur" (sentence on progress, Bürgel §13); f. 193vb, after "motum obtineat" (on Galen's method, Bürgel §19); f. 194ra, after "eius summitate" (epiglottis, Bürgel §22).

219. Additions with Bonacosa: Aristotle/Averroes, *Omnia . . . opera* (1562), vol. 10, f. 192b, "quod mehercle" until "claudicaverit" (logical excursus, Bonacosa f. 28vb, Bürgel §7); f. 193ra, "Quod si obstinate" until "non nervorum" (on muscles, Bonacosa f. 29ra, Bürgel §8); f. 193rb, "Ars vero anatomices" until "absoluta" (on anatomy, Bonacosa f. 29rb-va, Bürgel §11).

220. Bruyerin also shares Bonacosa's transposition of a reference to swallowing and the exchanging of the two paragraphs 24 and 25. See Aristotle and Averroes, *Omnia . . . opera* (1562), vol. 10, f. 192vb, "ac deglutionis" (Bonacosa f. 28va, Bürgel §5); f. 194ra, paragraph beginning with "Uva columella" and paragraph beginning with "Epiglottidem vocis" (Bonacosa f. 30rb, Bürgel §§24–25).

221. Bonacosa's text reads: "Item non est impossibile . . . quod illa malitia non penetret ad pulmonem per illam colligantiam, quia una rerum, ex qua aegrotant membra, est colligantia," which means: "It is not impossible . . . that this malice does not pass over to the lungs by way of contact, because one of the things because of which the limbs get sick is contact." Bruyerin with full right eliminates the *non*. See Aristotle and Averroes, *Omnia . . . opera* (1562), vol. 10, f. 29vb (Bonacosa), f. 193va (Bruyerin), Bürgel §15: "Porro id quoque usuvenire potest quod . . . affectus ad pulmones per consensum pertingat . . . etc." The

*non* is not only in the 1562 edition of the *Colliget;* I have also checked Averroes, *Colliget* (1497), f. 53v.

222. Averroes, *Kitāb al-Kullīyāt* (1987), ed. Fórneas Besteiro, 1:91, 95 (with apparatus criticus, 2:47–50).

223. Aristotle and Averroes, *Omnia . . . opera* (1562), vol. 10, f. 28rb (Bonacosa), f. 192rb (Bruyerin).

224. Ibid., f. 30ra (Bonacosa), f. 194ra (Bruyerin).

225. Ibid., f. 193rb ("hoc est partes nobiliores ab ignobilioribus distinguens ac separans"); f. 194ra ("quae nonnullis spiritalis fistula dicitur"); f. 194ra ("Uva, columella aliis nominata").

226. The following passages in Bonacosa's translation do not have a counterpart in Bruyerin's: Aristotle and Averroes, *Omnia . . . opera* (1562), vol. 10, f. 28rb ("motus arteriarum qui venit a corde"); f. 28va ("sed alii dicunt quod stat virtuti <nutritivae> sicut motus pulsus"; the reference to the nutritive faculty is missing, I have added it from the Arabic); f. 29ra ("concedit quod motus palpebrarum fit per musculos").

227. Ibid., f. 174v: "Ioannes Bruyerinus Campegius, vir doctus ac in dicendo perpolitus, tres collectaneorum Averroys sectiones, tribus Colliget libris secundo scilicet, sexto ac septimo respondentes latinitate donavit easque in lucem protulit. Nos autem studiosorum commodo eas post antiquam librorum Colliget translationem una cum ipsis imprimendas curavimus."

228. Ibid., sig. A1v: "Collectaneorum item sectiones tres, tribus Colliget libris, secundo scilicet, sexto et septimo respondentes, a Ioanne Bruyerino Campegio elegantissime latinitate donatae, post antiquam translationem ob studiosorum commodum appositae sunt."

229. Kaufmann, 'Jacob Mantino,' 222: "Quas deo duce teque favente transferre spero in latinum idioma, cum primum librum Colliget Averroys (ut incepi) verti perfecerim; est enim ille liber totus perverse depravateque translatus, quod facile iudicari poterit si nostra cum prima examinetur traductione."

230. On Alkindi's theory and Averroes's critique, see McVaugh, *Arnaldi de Villanova,* 53–74; Langermann, 'Another Andalusian Revolt?'

231. Aristotle and Averroes, *Omnia . . . opera* (1562), vol. 10, f. 121va (Bonacosa), 121vb (Mantino).

232. Ibid.

233. Cf. McVaugh, *Arnaldi de Villanova,* 308: "Preparing any sort of edition of the Latin *Colliget* is complicated by Bonacosa's often over-free or inaccurate translation, and by the fact that all the manuscripts of the translation I have seen, including B, seem to reflect a tradition corrupted very early" (B is MS Cesena, Biblioteca Malatestiana D. XXV. 4). I have checked the text of an incunabula edition of the *Colliget* to exclude that the sentences were added in the sixteenth century. They are already in Averroes, *Colliget* (1497), f. 86r. Cf. again McVaugh, *Arnaldi de Villanova,* 307: "As it proved, the printed version of 1562 is very close to that of the best manuscripts."

234. This confirms the impression of McVaugh, *Arnaldi de Villanova,* 308: "the edition of Venice, 1562, where the material . . . is printed side by side with a more accurate retranslation by Iacobus Mantinus."

235. See the list of editions in the Appendix s.v. 'Mesue' and in Lieberknecht, *Die* Canones *des Pseudo-Mesue,* 201–202.

236. On Sylvius, see O'Malley, *Andreas Vesalius,* 47–53, 246–251; DSB s.v. 'Dubois, Jacques' (by O'Malley); Baader, 'Jacques Dubois,' and Lieberknecht, *Die* Canones *des Pseudo-Mesue,* 100–102.

237. Mesue, *Opera* (1548), sig. A2v: "Quanquam eum semper arabica familia praedicavit, graeca vero fere aspernata est, sed verbis tantum."

238. Ibid., sig. A3v: "Haec autem omnia Mesues ad vivum exsecutus est, is profecto ad Galeni canonem medicamentorum compositorum scriptor idoneus fuerit, sed remediorum etiam idem pollentium et saepe iisdem constantium copia passim illusit Mesues. Id autem qui Mesuae vitio vertunt, Galenum et caeteros omnes medicos illustres accusant."

239. Ibid., sig. A2v: "Mesues operibus ipsis scripta communione iunxit et prima scripsit medicamenta purgantia maxime clementia et parte maxima graecis prioribus incognita magno mortalium incommodo, quippe qui elleboro, peplio, colocynthide et similiter valentibus medicamentis purgabantur quanto cum periculo, abunde docet Hippocrates lib. 5 aph. 1."

240. Ibid., sig. A4v: "Malui autem hac epistola Mesuae defensionem et laudes meritas inscribere quam tuas."

241. Ibid., sig. A1v: "Quam vero non sit facile rebus longa et obscura caligine mersis et velut illuvie sordida obsitis dare aut verius reddere splendorem, norunt omnes."

242. Ibid., sig. A2r–v: "Quae fortasse aliquando etiam expoliemus, si hunc laborem doctis non displicere intellexerimus."

243. Goltz, *Studien,* 344–346.

244. Lieberknecht, *Die* Canones *des Pseudo-Mesue,* 101.

245. The following sums up the results of the work of Ullmann, *Die Medizin,* 304–306, and Lieberknecht, *Die* Canones *des Pseudo-Mesue,* 1–34. A major step in research in the past was Steinschneider, *Die hebraeischen Übersetzungen,* 717–721.

246. On this development of terminology, see Siraisi, *Avicenna in Renaissance Italy,* 296–305; Nutton, 'The Changing Language.' For another specimen of Sylvius's humanist technique of revision, his edition of Gatinaria's commentary on Rhazes, see Baader, 'Jacques Dubois,' 149–151.

247. Goltz, *Studien,* 345: "Auch änderte er oft die Reihenfolge der Rezeptbestandteile, wofür es keinen ersichtlichen Grund gibt."

248. Mesue, *Opera* (1561), f. 2va, reprinted in Lieberknecht, *Die* Canones *des Pseudo-Mesue,* Appendix, p. 2 (versio antiqua): "Dotatur enim omne duplici, ut aiunt philosophi, virtute, scilicet elementari et coelesti. Hac quidem communi, hac vero propria. Et enim calefactivum et frigidativum, calidum et frigidum omne; solutivum autem non quia calidum, nec quia frigidum, sed quia coelesti virtute dotatum, sic ipsius mixtionem regulante. Et ob hoc quidem

solutivum hoc, illud vero provocativum, aliud vero aliter et aliter. Et hoc quia a coelesti virtute tale supra complexionem fertur." Cf. Lieberknecht's comments, *Die* Canones, 48–49.

249. See the examples collected by Goltz, *Studien,* 345–346, and, with respect not to Mesue, but to Gatinaria, Baader, 'Jacques Dubois,' 150.

250. Cf. Forcellini s.v. 'sativus.' The great influence of Pliny the Elder on late medieval and Renaissance medicine is emphasized by Nutton, 'Changing Language,' 188–189.

251. As stated in Mantino's dedication; see Avicenna, *Quarta fen primi* (1530), sig. A2v.

252. Note also that Giles of Viterbo was the dedicatee of Calo Calonymos's *De mundi creatione;* see Di Donato, 'Traduttori di Averroè,' 39.

253. See Perfetti's introduction (pp. xvii–xviii) and notes to Pomponazzi, *Expositio super primo et secundo* (2004).

254. See Perfetti, *Aristotle's Zoology,* 85–120.

255. Aristotle and Averroes, *Omnia ... opera* (1562), vol. 1, 3: (1) *In primum librum Posteriorum Resolutorium.* (2) *Contradictionum solutiones in Aristotelis et Averrois dicta in primum librum Posteriorum Resolutorium.* (3) *In novem Averrois Quesita Demonstrativa argumenta.*

256. Pavese, *In prologum Averrois super Analytica posteriora* (1552). On Tomitano and Pavese, see Lohr, *Latin Aristotle Commentaries,* 2:302–304, 461–464. See also Burnett, 'Revisiting.'

257. Martin, 'Humanism,' 75; Lohr, *Latin Aristotle Commentaries,* II;31, 225. Martin also draws attention to the Naples professor Francesco Storella (d. 1575), who composed two brief treatises in which he analyzes the new Hebrew-Latin Averroes translations of the Renaissance from a philological point of view; see Martin, "Humanism," 77.

258. On this issue, see the very informative volume edited by Di Liscia, Kessler, and Methuen, *Method and Order in Renaissance Philosophy of Nature,* with additional literature on the disputed question of whether Galilei's scientific method was influenced by the Aristotelian theory of *regressus,* as John Hermann Randall *(The School of Padua)* had surmised, or, as is much more likely, by the tradition of geometrical method. For one of many replies to the problematic Randall thesis, see the convincing article by Jardine, 'Problems of Knowledge,' 708–711.

259. Siraisi, *Avicenna in Renaissance Italy,* 135, 192–194.

260. Santorio, *Commentaria in primam fen* (1625), 15: "Sed Bellunensis, quem ut plurimum sequimur, distinctius de patria et de nomine Avicennae mentionem facit." Cf. 17, when citing the beginning of the *Canon:* "Prooemium Avicennae secundum Bellunensem." On Santorio, see Siraisi, *Avicenna in Renaissance Italy,* 206–211, esp. 208.

261. Examples are Leonhard Rauwolf and Prospero Alpini. See Dannenfeldt, *Leonhard Rauwolf,* 23, and Brentjes, *Travellers from Europe,* art. 1, 446. See Chapter 4, final section.

4 · *MATERIA MEDICA*: HUMANISTS ON LAXATIVES

1. See Kristeller, 'Humanism,' 114–118.

2. Leoniceno, *De Plinii et plurium aliorum medicorum in medicina erroribus* (1509). On Leoniceno, Manardo, and Brasavola, see Nutton, 'The Rise of Medical Humanism.' Cf. also Cont. Eras. s.v. 'Leoniceno' (by Lowry).

3. Manardo, *Epistolarum medicinalium libros XX* (1540). This is the first edition to contain all twenty books.

4. Champier's 1522 edition of Avicenna's *Liber canonis totius medicine* (Lyon) opens with several folios of critical examination of Avicenna's philosophical and medical errors (see Siraisi, *Avicenna in Renaissance Italy*, 71–73, 166). But the following publication was more sensational: *Symphonia Galeni ad Hippocratem, Cornelii Celsi ad Avicennam . . . Item, Clysteriorum Campi contra Arabum opinionem, pro Galeni sententia, ac omnium graecorum medicorum doctrina, a domino Symphoriano aurato equite ac Favergiae domino digesti, contra communem Arabum ac Poenorum traditionem summa cum diligentia congesti ac in lucem propagati* (Lyon?, 1528?). That Champier's book was understood as a provocation directed against the followers of Arabic medicine is apparent from the *Defensio medicorum* of Lorenz Fries, f. 40r. On Champier's controversies with Fries, Unger, and Fuchs, see Copenhaver, *Symphorien Champier*, 76–78.

5. Fuchs, *Errata recentiorum medicorum LX numero* (1530).

6. Fuchs, *Paradoxorum medicinae libri tres* (1535), f. 27v: "Ego sane arabum autorum lectionem nunquam ita perniciosam esse antea, ut hodie, credidi, in dies enim magis atque magis quantum incommodare possint animadverto, quapropter mitius quam par erat etiam illos olim a me tractatos esse fateor. Iam multo acerbius tractandos arbitror, ut hoc pacto posteris saltem consulatur, ne arabica illa lupanaria incidant improvidi. Et certe me illorum hostem futurum esse omnium acerrimum publice omnibus denuntio, maiorique conatu quam hactenus unquam coepi, illorum castra oppugnare pergam, neque desinam, modo vita dei benignitate supersit, donec castris penitus exuero. Quis enim has pestes in vitam humanam diutius grassari pateretur? Nemo sane, nisi qui homines, christianos praesertim, omnes ablatos iri cupiat. Ad fontes potius pergamus, illicque aquam puram nullis inquinamentis perturbatam hauriamus."

7. Most of these studies have the disadvantage that they do not counterbalance the discussion of the polemics with a concrete analysis of the medical theory of the polemicists. Cf. Schipperges, *Ideologie*, 14–26; Baader, 'Medizinisches Reformdenken,' and 'Medizinische Theorie'; Copenhaver, *Symphorien Champier*, 67–81; Klein-Franke, *Die klassische Antike*, 17–52; Dilg, 'The Antarabism'; Siraisi, *Avicenna in Renaissance Italy*, 65–76; briefly Friedrich and Müller-Jahncke, *Geschichte der Pharmazie*, 97–99; and Pormann, 'The Dispute.' The only survey I am aware of that tests the rhetoric against the technical detail is given by Goltz, *Studien*, 334–365. On medical botany in Renaissance universities, see Grendler, *The Universities*, 342–351.

8. Leoniceno, *De Plinii et plurium aliorum . . . erroribus* (1509), f. 74r: "Quare diligentius considerantibus doctrina Avicennae cum tot expositoribus, qui
partim alii ab aliis, partim iidem a se ipsis in variis locis discordant, chaos
quoddam videri possit, qua confusionis obscuritate nihil possit esse humanae
vitae periculosius." For context, see Mugnai Carrara, *La biblioteca*, 29–30.
Champier, *Apologetica epistola*, in S. Monteux, *Annotatiunculae* (1533), f. 23r:
"Magnum certe ex Avicennae erroribus generi humano imminet periculum."
Champier, *Castigationes* (1532), f. 3r: "Multi enim medicorum . . . noxia illa
Arabum, Poenorum et Mahumetensium in medicinis placita pro salutaribus
amplectentes multos mortalium . . . morti dederunt." Cf. also Siraisi, *Avicenna
in Renaissance Italy*, 69–70.

9. Leoniceno, *De Plinii et plurium aliorum . . . erroribus* (1509), f. 9v: "quemadmodum in eupatorio, centaurea maiore, argemonio ac plerisque aliis plantis
contingit, pro quibus diversae ab iis, quas eisdem vocabulis antiqui designa-
runt, quotidie a nostrae aetatis medicis recipiuntur. Quibus tamen non adeo
Plinius quam arabes auctores, qui et ipsi alius aliter de eadem re tradiderunt,
errandi occasionem praestiterunt." Champier, letter to Nicole de Lande
(1508): "ut vocabula arabica quae ab Hali Abbate tradita sunt quae successu
temporum credo esse in multis locis corrupta" (quoted from Copenhaver,
*Symphorien Champier*, 68n).

10. Fuchs, *Paradoxorum medicinae libri tres* (1535), praef., f. a6r: "praeter incon-
ditam ac plane ineptam, adde etiam obscuram dicendi rationem." See
Chapter 3 above on Ledesma, Gratiolo, Bruyerin Champier, and Mantino.

11. Fuchs, *Paradoxorum medicinae libri tres* (1535), ch. 1.1, f. 2r: "Unde discant velim
studiosi Arabibus in citandis Graecis parum fidei adhibendum esse, quod
nonnumquam illis tribuant, quae scripsisse nusquam comperimus. Id autem
cur fecerint, haec in primis ratio est, quod singula ex iis libris qui e graeca in
arabicam linguam translati fuerunt, transcripserint, quos certe genuinum
autorum sensum minime referre in confesso est, ut quibus multa praeter
Graecorum mentem per interpretes adiecta, multa etiam omissa sint." On
Valla, Ficino, and Vives, see Chapter 5, section 'The Public Image.'

12. Manardo, *Epistolarum medicinalium libri XX* (1542), epist. 3,4, p. 28: "Constat Avicennam totam fere suum de medicina librum ex aliis medicis, tum graecis, tum
barbaris transcripsisse"; Fuchs, *Paradoxorum medicinae libri tres* (1535), praef., f.
a6r: "Arabes quotquot fuerunt nihil aliud studuerunt quam ut fucorum instar
alieno labore partis fruerentur seque alienis atque adeo furtivis vestirent
plumis"; Barbaro, *Epistolae* (1943), 1, p. 92: "Et, hercule, si conferas eius viri *(sc.,
Averrois)* scripta cum graecis, invenies singula eius verba singula esse furta ex
Alexandro, Themistio, Simplicio." For the context of Averroism, see Chapter 5.

13. 'Materia medica' as a term more appropriately describes the field than 'pharmacology'; see Dietrich, 'Islamic Sciences,' 50–51.

14. Jacquart, *La science médicale*, esp. art. 11; Goltz, *Studien*, 334–365. Cf. Siraisi,
*Avicenna in Renaissance Italy*, 12: "Moreover, a challenge for future studies of
Renaissance medicine remains to trace the full history of efforts to come to

grips with, to interpret, and finally in large part to discard technical vocabulary of Arabic origin (although often unrecognizably distorted) in pharmacological botany and anatomy." Cf. also Jacquart's conclusion with respect to humanism (in 'The Influence,' 979): "When the Renaissance humanists wanted to eliminate Arabic terms from medical vocabulary, they also swept the table of nuances which reflect the historic evolution of concepts."

15. See the list of Arabic sources on senna in Dietrich, *Disocorides triumphans,* no. 4.115, 629.

16. Pseudo-Mesue, *De simplicibus* (1581), f. 71vb–72ra; Pseudo-Serapion, *Liber aggregatus in medicinis simplicibus* (1525), f. 131rb; Averroes, *Colliget* (1572), f. 119v. Cf. Avicenna, *Liber canonis* (1507), ch. 2.2.282, f. 117rb. Albuzale (Abū ṣ-Ṣalt), *De medicinis simplicibus* (2004), no. 5.11, 121 (Albuzale's text was not printed in the Renaissance).

17. Postel, *Grammatica Arabica* (ca. 1539/1540), praef., sig. D2v: "Quam multa autem, quae foelicissime nobis hoc saeculo succedunt, Arabibus solum, non Galeno debemus? Nolo recitare omnium medicinarum temperamentum sacharum, rhabarbarum (nec enim hodie quod nunc habemus antiquorum est, sed a traditione poenorum), turbit, sene, manna, quae praestantissimae sunt medicinae, et composita illa scherab, giulab, sufus, lohoc, roob, et huiusmodi quae ita nobis sunt in usu et singulari commoditate ut sine maximo detrimento tolli non possint."

18. *European Pharmacopoeia,* 7th ed., 0206. I am grateful to my Würzburg colleague in pharmaceutical biology Martin J. Müller for advice on the botany and pharmacognosy of senna.

19. Ainslie, *Materia Indica,* 1:391: "Mr. G. Hughes of Palamcotah, a few years ago, succeeded perfectly in cultivating the true senna of Arabia, in the Southern part of the Indian peninsula"; Flückiger, *Pharmakognosie,* 634; Reinhardt, *Kulturgeschichte der Nutzpflanzen,* 2:313–314; Bornkessel, 'Senna-Ernte,' 172.

20. Wichtl, *Teedrogen und Phytopharmaka,* 546–550; Tyler et al., *Pharmacognosy,* 64–66.

21. *Ortus sanitatis de herbis et plantis . . .* (ca. 1497), ch. 427.

22. Anonymous, *Antidotarium Nicolai* (1471), s.v. For context, see Goltz, *Mittelalterliche Pharmazie,* 137, 168.

23. Saladin of Asculo, *Compendium aromatariorum* (1919), 6:112.

24. Pseudo-Mesue, *Opera* (1549), f. 64ra (versio antiqua).

25. See the lists printed in Dressendörfer, '*In apotecis,*' 82–86, and in Jacquart and Micheau, *La médecine,* 226.

26. Pegolotti, *Libro di divisamenti di paesi* (1936), 34, 70, 99, 108, 123, 138. For information on Pegolotti, see Evans's introduction.

27. Ashtor, *Levant Trade,* 69. Still valuable on the various goods traded in the Levant (though not on senna) is Heyd, *Geschichte des Levantehandels.* For a description of nineteenth-century trade with Senna, see Flückiger, *Pharmakognosie,* 628–631.

28. Ashtor, *Levant Trade,* 450, 479. For further literature, see LexMA s.v. 'Levantehandel' (by Balard).

29. I am not aware of any study on the discussion of senna apart from Thorndike, *History*, 5:460 (on Symphorien Champier and Brasavola), and Greene, *Landmarks*, 2:607–608 (on Ruel).

30. On Barbaro's work not being a commentary, see Ramminger, 'A Commentary?'

31. See Ramminger, 'Zur Entstehungsgeschichte.' Cf. Riddle, 'Dioscorides,' 46 (with wrong, i.e., Venetian date "1516" of the edition); Godman, *From Poliziano*, 217. For further literature on Barbaro's *Corollarium*, see Greene, *Landmarks*, 2: 553–568; Cont. Eras. s.v. 'Barbaro, Ermolao (I)' (by Lowry).

32. Barbaro, *In Dioscoridem Corollariorum libri quinque* (1530), ch. 4.789, f. 65r: "Sunt apud graecos inter quos Isacius qui peplium pro senna modo appellata intelligant. Ego diversum puto genus, cum senna siliquas et in iis semen ferat, peplis non ferat." Up to now, all attempts to identify 'Isacius' have been of no avail. One is inclined to think of Isaac Israeli, who is, however, not a Greek author, but an Arabic-writing Jew. Isaac Israeli's *Omnia opera* (1515), which contain the *De dietis universalibus et particularibus,* do not discuss either *peplis* or *senna.* Johann Ramminger (Munich), who prepares an edition of the *Corollarium,* has been so kind as to search for other references to 'Isacius' in Barbaro's oeuvre—without success.

33. Barbaro, *In Dioscoridem Corollariorum libri quinque* (1530), preface to book 2, f. 22v. For a partial translation and interpretation of the preface to book 1, see Godman, *From Poliziano*, 219–220.

34. Described by Riddle, 'Dioscorides,' 44–113.

35. See the analysis of Marcello Virgilio Adriani's intellectual, literary, and political motives in Godman, *From Poliziano*, 212–231.

36. Marcello Virgilio, *Pedacii Dioscoridae Anazarbei de medicina materia* (1518), preface, sig. AA3r: "Antiqua enim et graeca antiquorum et graecorum rationibus tradere et confirmare praecipuum nobis consilium est."

37. Ibid., opening of the preface: "Auxisse postremo et paulo minus quam inextricabiles tam multas difficultates fecisse novam posteriorum scriptorum medicinam, quae mixtis cum antiquis graecorum vocibus barbaris appellationibus suis, quid, qualeve hoc aut illud in antiquis scriptoribus fuerit aegre cognoscendum reliquit." The text of the preface is also edited in Riddle, 'Dioscorides,' 36–37.

38. Marcello Virgilio, *Pedacii Dioscoridae Anazarbei de medicina materia* (1518), f. 262r (de buglosso): "Nec laboramus pro re, cunctis enim nota est, sed pro medicae materiae antiquis nominibus et historia."

39. Johannes Lonicerus, *Nova Scholia,* prefatory letter, in Walther Ryff (=Gualtherus Rivius), *Pedanii Dioscorides Anazarbei de medicinali materia libri sex Joanne Ruellio Suessionensi interprete* (1543), praef.: "Marcelli versionem, quia romana est et elegans, reliqui, quorsum enim attinet de integro Dioscoridem latinum facere, cum proprie nobis et romane loquitur Virgilii Marcelli eloquentissimi viri vigiliis" (quoted from Riddle, 'Dioscorides,' 75). On Ryff's controversial edition, see Stübler, *Leonhart Fuchs,* 244–252, and Riddle, 'Dioscorides,' 75–79. Johannes Lonitzer was the father of Adam Lonitzer, the

author of the often printed *Kreuterbuch* (first edition Frankfurt am Main, 1557).

40. Marcello Virgilio, *Pedacii Dioscoridae Anazarbei de medicina materia* (1518), f. 188r–v: "Quam nostra aetas senam officinis omnibus et vulgo notam vocat, antiquitas ut opinati aliqui sunt delphinion appellavit. Posterioris et quo nunc medicina utitur nominis incerta ratio nobis et causa; antiquioris notissimi nautis in mari pisces auctores sunt."

41. For various modern attempts to identify the plant (with Delphinium Ajacis L., Delphinium peregrinum L., Delphinium Consolida L.), see Berendes, *Des Pedanios Dioskurides . . . Arzneimittellehre,* 310–311. Cf. Aufmesser, *Etymologische . . . Erläuterungen,* 107, who, without discussion, identifies *delphinion* with larkspur (Delphinium Ajacis).

42. The chapter on delphinion does not seem to have been part of the earliest versions of the Greek text; see Wellmann's critical apparatus to the Greek edition: Dioscorides, *De materia medica* (1958), lib. 3, ch. 73 RV, 84.

43. Marcello Virgilio, *Pedacii Dioscoridae Anazarbei de medicina materia* (1518), f. 188v: "Demirabitur aliquis etiam neque immerito in multiplici senae proprietate nihil a Dioscoride in delphinio dictum quod ad faciendam homini medicinam praeterquam in scorpionis ictu pertineat. Nos ex figura siliquarum senae quae totam delphinorum formam referunt, et posteriorum quorundam auctoritate usi, senam delphinium et sosandrum antiquis dictam fuisse testati sumus. Ingenue tamen fatemur pro comperto aliud nihil de ea nos habere."

44. On Marcello Virgilio's reply, see Godman, *From Poliziano,* 231–234.

45. On this work, which originated from marginal notes on Manardo's copy of Pseudo-Mesue, see Herczeg, 'Johannes Manardus,' 92–94. Not worth consulting is Spina and Guido, 'Il commento di Giovanni Manardo.' For further information on Manardo, see Riddle, 'Dioscorides,' 50–54; Nutton, 'The Rise of Medical Humanism,' 8–11; cf. also Greene, *Landmarks,* 2:584–597 (on botany); Cont. Eras. s.v. 'Manardo' (by Cavagna); DBI s.v. 'Manardi (Manardo), Giovanni' (by Mugnai Carrara).

46. Manardo, *Epistolarum medicinalium libri XX* (1542), epist. 3,4, pp. 28–29: "Constat Avicennam totam fere suum de medicina librum ex aliis medicis, tum graecis, tum barbaris transcripsisse. Ex quo fit ut non sit ad intelligendum ipsum, sicubi difficilis videtur, via melior quam ad ipsos recurrere autores, a quibus id de quo dubitatur assumpsit. Quisquis autem hoc melius facere norit, is utique melius eum intelligat necesse est. Et qui ab hac via recesserunt, multipliciter semper lapsi sunt. Primus, quod sciam, hanc intelligendi Avicennae viam nobis monstravit Nicolaus Leonicenus, vir et graece et latine doctissimus ac medicinae sicuti bonarum omnium artium cultor eminentissimus. Hoc idem Hermolaus Barbarus aevi nostri splendor et gloria aliquando mihi Venetiis indicavit, cum de communibus studiis in eius bibliotheca colloqueremur. Tantorum ego ducum castra sequutus, quoties scrupulus aliquis in Avicenna incidit, ad antiquiores praesertim Graecos me confero." For context, see Siraisi, *Avicenna in Renaissance Italy,* 69.

47. Manardo, *Epistolarum medicinalium libri XX* (1542), epist. 6,5, pp. 69–70: "Utinam, mi compater, multi hoc tempore medici invenirentur, qui tecum *(sc., Ioannus Paulus Castilionus)* sentirent, nec tantum nomini et quorundam autoritati quantum veritati et rationibus crederent; facile enim sperare possemus, ut antiqua et vera medicina, quae iam caput exerere occoepit, exuta plene barbarie, detersaque omni caligine ac labe, levigata ac splendenti facie se totam speciosam et castam veris amatoribus ostenderet. Sed magna medicorum pars et hi maxime qui celebriores vulgo habentur tam male initiati sunt ut praedicare per compita et fora non vereantur se malle cum Avicenna, Serapione, Mesue ac reliquis, quorum verba pro oraculis hactenus habuere, non recte sentire, quam cum recentioribus, qui medicinam a graecis fontibus hauriunt, recte."

48. A similar observation is made by Schipperges, 'Diaetetische Leitlinien,' 256.

49. Manardo, *In Ioannis Mesue Simplicia medicamenta* (1549), 548: "De sena. Averrois hanc quoque reponit inter medicinas novas. Avicenna de ea non scribit ex professo, meminit tamen eius in fine capitis de fumo terrae, et capite de quintana. Scribit Serapion, sed non affert (ut fere aliis in medicinis consuevit) Dioscoridis et Galeni de ea verba. Quidam delphinion, alii pelecinon, Isacius peplion. Ego aliquando dubitavi an esset empetron, aliquando analypon. Parum autem refert scire, quo nomine a veteribus sit nuncupata, modo eius vires non ignoremus, quae certe nobiles sunt et efficaces."

50. Manlius de Bosco's only comment on Pseudo-Mesue's teaching on senna concerns the passage on pods *(folliculi);* see Manlius de Bosco, *Luminare maius* (1536), f. 12v: "Hic vero possumus certificari quod intentio Mesue est quod sene sit foliculus plantae et melior pars plantae est foliculus et in foliis eius est debilitas et propter id dico quod per senem intelligitur foliculus et non folia eius ut faciunt quidam."

51. Averroes, *Colliget* (1562), f. 119v; Avicenna, *Liber canonis* (1507), ch. 2.2.282, f. 117rb; Pseudo-Serapion, *Liber aggregatus* (1525), f. 131rb.

52. Pseudo-Mesue, *Opera* (1549), f. 64ra. I here follow the wording of Manardo's quotations of the text.

53. Manardo, *In Ioannis Mesue Simplicia medicamenta* (1549), 549: "Experientia ipsa ostendit, folia in purgandi vi esse potentiora. Nascitur in Apulia, sed praestantior est quae ex Aegypto affertur."

54. Ibid.: "Averrois dicit siccam in secundo, Serapion paucae caliditatis et siccitatis, cui ego credo."

55. Ibid.: "Experientia contrarium ostendit, saporque subamarus et adstringens, sed nec illud dixit Serapion."

56. Ibid.: "Utinam dixisses quo loco et sub quo nomine; sed puto te hic, sicuti et alibi plerunque hallucinatum."

57. Ibid.: "De empetro dixit Dioscorides decoquendum in iure vel aqua dulci, et nos scimus ius in quo vulgaris haec bullierit vel per noctem maduerit facile purgare. Serapion decoctam plus conferre ait quam contritam."

58. Ibid.: "Bilem absolute dixit Serapion, non addendo ustam, et quod valet melancholiae, rhagadis, nervorum quassationibus, capillorum defluvio, dolori

capitis antiquo, scabiei, et morbo comitali. Dioscorides de empetro quod purgat pituitam, bilem et aquosa. Averrois negat senam posse purgare pituitam. Ego experientia didici quod potest, et quod multum valet in morbo gallico vocato."

59. Ibid.: "Serapion dat tritae drachmam unam, decoctae quinque. Ego scio utranque decoctam excedi tuto posse, unciamque decocti ex aqua mediocriter purgare."

60. Wichtl, *Teedrogen und Phytopharmaka*, 552. For a table listing the percentage of the different hydroxyanthracen derivatives in Alexandria and Tinnevelley senna leaves and pods, see Kabelitz and Reif, 'Anthranoide in Sennesdrogen,' 5087–5088.

61. Wichtl, *Teedrogen und Phytopharmaka*, 547, 549.

62. See Hoppe, *Drogenkunde*, 236–240, for a list of some species of cassia.

63. See Ainslie, *Materia Indica*, 1:390; Targioni Tozzetti, 'Sulla coltivazione della Sena,' 23, and Flückiger, *Pharmakognosie*, 629–630 (with partially outdated terminology on cassia and senna).

64. Dietrich, *Dioscurides triumphans*, 2:688: "Die Pflanze, den Arabern unbekannt, ist nicht zu bestimmen." For various nineteenth-century attempts of identification (Crithmum maritimum L., Frankenia pulverulenta L.), see Berendes, *Des Pedanios Dioskurides . . . Arzneimittellehre*, 468–469. Aufmesser, *Etymologische . . . Erläuterungen*, 188, identifies it with sea-heat (Fanklinia pulverulenta), without discussion.

65. This is Manardo's free rendering of Averroes, *Colliget* (1562), f. 119v: "non est potens trahere humores viscosos."

66. Krause, *Euricius Cordus*, 67–77; Dilg, *Das Botanologicon*, 9–10; Greene, *Landmarks*, 1:360–367. For further literature on Cordus, see Kühlmann et al., *Humanistische Lyrik*, 1084–1086, and MBW Pers. s.v. 'Cordus, Euricius.'

67. Cordus, *Botanologicon* (1534), 12: "Fateor, cum animadverterem plerosque immo omnes fere pharmacopolas in tanta errorum caligine versari, ut ne dimidiam quidem partem eorum, quae illis scribimus, simplicium vel noscant vel habeant, dolui vehementer, et qualemcunque huic rei opem pro virili mea ferre volui." On the relationship between physicians and apothecaries, see Dilg, *Das Botanologicon*, 94–99.

68. See, for example, his invective against the *Luminare maius:* Cordus, *Botanologicon* (1534), 153.

69. Ibid., 16: "Hoc de me non vanus et citra iactantiam affirmare ausim, humanioribus studiis, quae tamen infra mediocritatem attigi, me saepissime fuisse adiutum, ut ea quae in medicina legi faciliore captu deprehenderim acriori etiam iudicio intellexerim, immo et hodie medicam materiam exactius quam vitilitigatores illi, cognoscam atque accommodatius opportuniusque ea uti calleam."

70. Ibid., 65: "Ralla: Frustra meditor quid tibi respondeam. Non leviter deiicienda stat Dioscoridis autoritas. Cordus: Imo veluti Marpesia cautis." See also 132.

71. On Cordus's methodological standpoint, see Dilg, *Das Botanologicon*, 86–93.

72. Cordus, *Botanologicon* (1534), 169: "Megobacchus: Quin et senam, sunt qui delphinium esse putent. Cordus: Non tamen undequaque respondente historia. Quod si maxime fuerit, non tamen veteres de eo quicquam scripsisse crediderim. Nam adiecticium hoc in Dioscoride caput hinc arguitur, quod in antiquissimis id codicibus non inveniatur, neque Galenus neque Aegineta neque Plinius delphinii usquam meminit, quantum ego observarim. Megobacchus: Quandocumque et a quibuscunque reperta sit, quomodocunque etiam vocetur sena multis laudibus predicata, quia multis viribus praedita est medicina. Cordus: Hinc non immerito et sosandros dicta, quod hominem scilicet sospitet."

73. Cordus, *Botanologicon* (1534), index et quasi epilogus, s.v. "sene," sig. N4r–v: "Sene forte veteribus ignota, quamvis delphinion apud Dioscoridem credatur, quod tamen caput in antiquissimis codicibus non legitur, neque apud Galenum neque Aeginetam invenitur."

74. Ibid., 74.

75. Ibid., 163: "Sunt et aliae plures Arabum medicinae, quarum nec ille nec alius quispiam ex veteribus Graecis, quod ego sciam, meminit, earum complura nomina in schedulam quandam notata penes me in pera, ut reor, habeo. Megobacchus: Ea nobis recensere ne pigeat. Cordus: Non invenio, memoriter aliquot enumerabo. Sunt igitur illa, ambra, anacardion (seu anacardus), behen, bellirici, bel, camphora, casia fistula (quam ita iam vocant), coton (latine gossipium), culcul, dadi, dend, doronici, emblici, fagre, feleng, faufel, et ipsa de qua iam dubitas galanga, kadi, kanabel, lingua avis, moschus, manna, macis (si non machir est), mirobalani citrinae, chaebulae et indae (nisi quis eas inter palmulas Dioscoridis reputet), michad, mummia, musa (seu maum), nersin, nux muscata, nux metel, nux mechil, et forte nux hende, nux indica (nisi haec sit quam Galenus regiam nucem appellat et septem uncias continere dicit), nux vomica (modo et illa non sit, ut quidam volunt, thitymali caryitis fructus), phel, piperella (licet viticis semen ita etiam dicatur), ramech, ribes, sadervam, sandalum, sel, sene, tamarindi, turbith, viret, zanarbum, zeduaria, zelim, zurumbeth."

76. For a general introduction to Champier's role in the medical disputes of the time, see Copenhaver, *Symphorien Champier,* 67–81, with further literature.

77. Two letters by Champier and one by Manardo are published in Monteux, *Annotatiunculae* (1533), ff. 26r–33v. In 1533, Champier sent a copy of his newly published *Hortus Gallicus* to Manardo.

78. In his *Cribratio medicorum* of 1534, Champier repeatedly refers to Manardo as "amicus noster"; see Champier, *Cribratio medicamentorum* (1534), for example, 52, 85. Whether Champier read the annotations on Pseudo-Mesue in a printed version or in manuscript is an open question. Manardo's preface to the annotations dates from 1521, and in 1522 he refers to the work as "under the anvil" (Manardo, *Epistolarum medicinalium libri XX* (1542), epist. 11,1, p. 232: "Epistolae meae medicinales quas sub praelo, et annotationes in Mesuen, quas sub incude habeo, me in respondendo fecerunt tardiorem"), but it seems that the

earliest edition extant is Basel (1535). On editions of Manardo's works, see Herczeg, 'Johannes Manardus,' 88–91, 93.

79. Fries, *Defensio medicorum principis Avicennae* (1533), f. 40r: "Vidimus nuper Lugdunensis medici invectionem in arabes medicos, Avicennam praesertim, quem libello 'Symphonia Galeni ad Hippocratem' inscripto stolidis convitiis dilaniavit; oblitus quanta laude ipsum extulerit in libro quem antea de claris medicinae scriptoribus edidit. Prosilierunt etiam medicinales epistolae, Avicennam livida manu caedentes." On *De medicine claris scriptoribus* (1506?), see Chapter 2 above.

80. Champier, *Castigationes* (1532), ch. 49, ff. 63v–64v. That important medicines were unknown to the ancients is expressed openly in a passage on sugar on f. 55r: "Si cui autem res incredenda videatur de re tam nobili antiquos medicos non scripsisse, consideret quam multa sint nostro aevo pretiosissima antiquitati penitus ignota, veluti casia nostra fistula, ambra, moschus, camphora, iubetum, aliaque non parum multa."

81. Champier, *Ioanni Manardo,* in S. Monteux, *Annotatiunculae* (1533), f. 31r: "Item tempore Galeni ignotae erant medicinae, quas nostri benedictas vocant, ut sunt manna roris, cassia fistula, rhabarbarum, sene, et fortasse laudabilis agaricus." On the two kinds of agaricum, see Lieberknecht, *Die* Canones *des Pseudo-Mesue,* 56. Champier proceeds to argue: if Galen had known rhubarb, he would not have recommended *scammonium;* if he had known our senna, he would not have used *veratrum;* if he had known *manna* and *cassia,* he certainly would have praised them. Champier, *Ioanni Manardo,* f. 31r: "Si sene nostrum expertus fuisset, non fuisset veratro usus."

82. Champier, *Cribratio medicamentorum* (1534), lib. 3, 43–47, lib. 6, 100–101, esp. 47: "Averrhous ille apostata medicus quinto Colliget inquit: nulla medicinarum purgativarum est sine veneno aliquo in aliqua parte sui etc. Quae omnia istorum dogmata intelligenda sunt de medicinis venenosis. Ignorabant autem antiqui medicinas benedictas, ut sunt cassia, rheubarbarum, manna, sene."

83. Ibid., lib. 3, 52–53.

84. In his quotation of Manardo, Champier leaves out the references to Dioscorides, thus taking out the mark of Manardo's implicit argument in favor of *empetron.* Champier uses the Dioscorides quotation in a brief note later in the book; see ibid., 85: "Sene vero apud Mesuem educit melancholiam, et choleram adustam. Dioscorides vero loquens de epetro *(sic),* quod purgat pituitam, bilem, et aquosa."

85. On the dating, see Meyer, *Geschichte der Botanik,* 4:238, and Thorndike, *History,* 5:445n. On Brasavola, see Nutton, 'The Rise of Medical Humanism,' 11–17. Cf. the older accounts of Thorndike, *History,* 5:445–471; Greene, *Landmarks,* 658–701; DBI s.v. 'Brasavola, Antonio, detto Antonio Musa' (by Gliozzi).

86. For a comparison, see Dilg, *Das Botanologicon,* 53–54.

87. Brasavola, *Examen omnium simplicium medicamentorum* (1537), 65.

88. Ibid., 65–66: "Sen. Res mira, herba adeo digna decantari, et tot nominibus insignita Graecis ignota fuit? Bra. Fuit quidem, ut videtur. Sen. Nonne eorum

tempore extabat aut deinceps terra peperit? Bra. Et tunc extabat, sed fortassis aut ab his non experta vel non visa. Sen. Quomodo non visa a Dioscoride, qui fere universum orbem circuivit? Bra. Dioscorides militavit quidem, sed universum orbem perlustrasse non constat, illuc saltem non ivit, ubi nascebatur, aut mox etiam ad propinquiores provincias delata est, ut de pluribus aliis rebus factum constat, quas antiqui ignorarunt aut data opera reliquerunt. Certum vero est centesimam partem herbarum in universo orbe constantium non esse descriptam a Dioscoride, nec plantarum a Theophrasto aut Plinio, sed in dies addiscimus et crescit ars medica."

89. Ibid., 66: "plures conati sunt eam apud Dioscoridem invenire," followed by the refutation.

90. Ibid., 66–67: "Adeo tenuis coniectura est, ut nec filo inhaereat: nam et ita bovem capum esse probarem, quia bulliuntur ut nutriant."

91. Ibid., 67: "Hi adhuc peius sentiunt qui Delphinium putarunt: nam sena folia non sunt delphino similia nec opus est ad folliculos fugere, quoniam de foliis Disocorides loquitur."

92. Marcello Virgilio, *Pedacii Disoscoridae Anazarbei de medicina materia* (1518), f. 188v: "Animadvertendum et illud in delphinii historia dicere Dioscoridem folia eius delphinorum figura esse, cum tamen aliter sensus indicet et quod delphinis simile in delphinio est siliqua tantum sit. Eamque ob causam non male forsan suspicetur aliquis periisse aliquid eo loco quod ad has siliquas pertineret, facili librariorum lapsu cum plura sigillatim et per se similiter desinentia scripturi sunt."

93. Manardo, *In Ioannis Mesue Simplicia medicamenta* (1549), f. 549: "Nascitur in Apulia: sed praestantior est quae ex Aegypto affertur"; Brasavola, *Examen* (1537), 67: "ut ligustica sena ostendit, quamvis ea inefficacior sit quae ex partibus Aegypti defertur"; and 69: "tamen potius laudarem, ut Venetiis senam emeres, quae Alexandria et orientalibus partibus defertur, magis quam ex Apulia aut Liguria; experientia enim ostendit semiunciam Aegyptiacae tantum fere proficere, quantum uncia nostratis."

94. Brasavola, *Examen* (1537), 67–68: "Sen. Quomodo id novisti? Bra. In primis Arabum autoritate, quae non est adeo temere despicienda, ut nonnulli recentes faciunt, praesertim ubi Graecis non opponuntur; postea ipsa quotidiana experientia, quae est omnium rerum magistra."

95. Camerarius, *De plantis epitome . . . Matthioli* (1586), 538–539.

96. For the pictorial tradition before Camerarius, see Fuchs, *De historia stirpium* (1542), 447; Lonitzer, *Kreuterbuch* (1679), 104; Mattioli, *Commentarii* (1565), 782.

97. Flückiger called this kind of senna "Cassia obovata" (*Pharmakognosie*, 627, 631). Meyer, Trueblood, and Heller, *Commentary*, 436, identify Fuchs's senna with "Cassia italica" growing in "Egypt, Sudan, and India" (131). On this kind of *Cassia*, cf. Hoppe, *Drogenkunde*, 1:237, and Andrews, *The Flowering Plants*, 2:114–117.

98. Targioni Tozzetti, 'Sulla coltivazione della Sena,' 17–26.

99. Nutton, 'The Rise of Medical Humanism,' 11–17.

100. DSB s.v. 'Ruel, Jean' (by Jovet and Mallet); Riddle, 'Dioscorides,' 34.

101. Pseudo-Mesue, *Opera* (1549), f. 64r (quoted above in Chapter 3, section 'Mesue: *Opera*' [Table 3]).

102. Theophrastus, *De historia plantarum* (1552), ch. 3.17, 72: "Colutea quoque Liparae propria traditur. Arbos magnitudine praestans, fructum in siliqua ferens, magnitudine lentis, qui oves mirum in modum pinguefacit. Nascitur semine et fimo praecipue ovillo. Tempus serendi cum Arcturus occidit; serendum semine praemadefacto, cum iam in aqua pullulare inceperit. Habet folium non absimile foenograeco; germinat primo unicaulis, triennio maxime; quo quidem tempore baculos decidunt, tunc enim praestantiores esse videntur." The Greek text is easily accessible in Theophrastus, *Enquiry into Plants* (1980/1990), vol. 1, ch. 3.17, 264–266.

103. Pseudo-Serapion, *Liber aggregatus in medicinis simplicibus* (1525), f. 131rb: "Sene quando desiccatur reponitur et ipse habet vaginas oblongas et obtortas in quibus sunt semina ordinate distincta, et ille vaginae habent pendiculum subtile quo adhaerent ramulis, et quando agitantur vaginae a vento, cadunt et pastores colligunt eas."

104. Ruel, *De natura stirpium* (1537), ch. 1.70, 147: "Mauritani sermone suo vernaculo sene nominant, duoque eius faciunt genera, suae spontis et hortense; nam et semine nascitur in fimo praecipue ovillo. Addunt ubique siliquas oblongas, quibus semina in ordinem utriusque digesta clauduntur, tenui pendere pediculo. Ob id facile ventorum impetu deturbari, haerentesque pertinacius perticis decuti. Caducas opiliones legunt, saginandis ovibus perquam utiles."

105. In the words of Ruel, ibid., 147: "Arbor magnitudine praestans, fructum ferens in siliqua magnitudine lentis, qui mirum in modum oves saginet. Nascitur semine et fimo praecipue ovillo. Serendi tempus, cum Arcturus occidit. Mandari solo debet praemaceratum, cum iam in aqua germinare coeperit. Folium mittit foenograeco non absimile. Arbor primum unicaulis emicat, trima concidendis scipionibus et baculis idonea."

106. In classical Latin, *opiliones* usually meant "shepherds" and thus was more specific than the term *pastores;* see ThLL s.v. 'opilio': "pastor pecoris, imprimis ovium." The term *pertica* was used mainly by authors on agriculture; see ThLL s.v. 'pertica.'

107. The edition first appeared in 1539 in Paris, two years after Ruel's death. On Johannes Zacharias Actuarius, see LexMA s.v. 'Zacharias, Johannes' (by Hohlweg), and, most pertinently, Bouras-Vallianatos, 'Ioannes Zacharias Aktuarios,' forthcoming.

108. Ruel, *De natura stirpium* (1537), ch. 1.70, 147: "Demiror hanc arborem Theophrasto celebratam a Plinio, Dioscoride, Galeno, Pauloque praetermissam ... Sed quae posteritas graecorum medicorum memoriae mandavit, paucis exponemus," etc.

109. Actuarius, *De medicamentorum compositione* (1556), 183: "Sene a barbaris appellatum siliquosus ille fructus est, qui citra noxam drachmae pondusculo

sumptus, pituitam et bilem trahit. Quae his deinceps succedunt, perquam moderate purgant: nimirum casia *(sic)* nigra et manna. Casia quidem, tribus aut quatuor drachmis sumpta, vix moveret alvum; manna vero etiam maiore modo quam casia sumenda est, et flavam bilem simplici ratione pellit. Casia vero ipsa flavam bilem, sed magis retorridam et quae iam in atram degeneravit, extrahit, quin etiam pituitam educit mediocrem."

110. Pseudo-Mesue, *Opera* (1549), f. 64r: "decoquatur cum iuribus gallorum . . . et aperit oppilationes viscerum."

111. Pseudo-Serapion, *Liber aggregatus in medicinis simplicibus* (1525), f. 131rb: "et confert . . . dolori capitis antiquo et scabiei et bothor *('pimples' from Arabic butur)* et pruritui et epilepsie . . . potio decoctionis eius plus confert quam ipsum tritum."

112. Ruel, *De natura stirpium* (1537), 147: "Sed quae posteritas graecorum medicorum memoriae mandavit, paucis exponemus. Actuarius siliquosum quendam fructum a barbaris vocari tradit, qui citra noxam denarii pondusculo potus pituitam et bilem deiicit. Secundum has reliquos humores modestissime purgat, retorridam atramque bilem earumque suffusiones ex gallinaceo iure depellit. Vetusto capitis dolori, scabiei, comitialibus, impetigini, succurrit. Sed fervefacti ius potius quam triti farina propinatur. Interaneorum obstructiones explicat."

113. Greene, *Landmarks,* 2:607–608.

114. Ruel, *De natura stirpium* (1537), ch. 3.49, 537: "Hallucinantur graviter qui delphinium esse senam Mauritanis dictam existiment."

115. Ibid.: "Nanque delphinium folio vestitur oblongo, tenui, repando, sena vero orbiculari et velut foeni graeci rotundo . . . Utrumque tamen siliquatur, sed sena tumidum ostendit folliculum membraneum, in quo semen atrum, durum, lentis magnitudine continetur. . . ." The round leaves similar to fenugreek are drawn from Theophrastus, the parchment-like pods are Ruel's own addition to the monography on colutea (147).

116. Ibid., 537: "Quid veteribus fuerit sena inter arbores demonstravimus."

117. Mondella, *Epistolae medicinales* (1557), epist. 15, 348. The letter's header is: "Aloisius Mundella medicus, Vincentio Metello. De Sena."

118. Ibid.: "Duo potissimum sunt quae me ab illius opinione removent. Primum quod Theophrastus coluteam arborem esse dicit eamque magnam; quodque nullam illi vim, ut quandoque consuevit, adscribat quae alvum solvat et atrum humorem educat, quod senae proprium et peculiare esse Arabum autores affirmant."

119. Ibid.

120. Amatus Lusitanus, *In Dioscorides de materia medica libros quinque enarrationes* (1554), 340. On Lusitanus and his commentary, see Riddle, 'Dioscorides,' 61–64, on his life EJud² s.v. 'Amatus Lusitanus' (by Leibowitz).

121. Pseudo-Mesue, *Opera* (1549), f. 64ra: "Sene est folliculus plantae . . . Melior pars plantae eius est folliculus, deinde folia."

122. Pseudo-Serapion, *Liber aggregatus in medicinis simplicibus* (1525), f. 131rb.

123. This is how Mattioli later explained the origin of Ruel's *colutea* theory; see Mattioli, *Commentarii* (1565), 782: "caducas opiliones colligunt. Haec Serapio. Cuius postremis verbis fortasse deceptus Ruellius senam esse arborem existimavit, quam Theophrastus lib. iii. cap. xvii. de plantarum historia coluteam appellat."

124. Saladin de Asculo, *Compendium aromatariorum* (1919), 100 ("De herbis, ut mezereon, sene, laureola"); Manardo, *Epistolae medicinales* (1577), epist. 3.2, 13b ("herbae, quam senam vocant"); Brasavola, *Examen omnium simplicium* (1537), 65 ("herba adeo digna").

125. On Bock, see DSB s.v. 'Bock, Jerome' (by Stannard).

126. Bock, *Kreutterbuch* (1577), ch. 49, f. 339r: "Ob wol der Senet noch bey den Gelehrten im zanck stehet / under welches Capitel er gehoere inn Dioscoride / so ist er doch sonst so gar bekandt worden inn Germania / das ich sein nicht hab moegen vergessen."

127. Ibid., f. 339r–v: "Mich wundert das Dioscorides / Plinius / Galenus und Aegineta dises gewaechss haben verschwigen / so es doch vom alten Theophrasto bedacht ist / da er spricht lib. i. capit. xviii. Colutea / oder wie Gaza schreibt / Coletia sey gemeyn zuo Lipara / zwischen Sicilien und Italien / bey dem Tyrrhenischen Meer / welches ich gern auff das Sene verstande / von wegen der artlichen beschreibung unnd wuerckung des selben. Das aber etliche Sene fuer Delphinium oder Analypon / oder Peplis halten / lass ich bleiben / ich nenne mit dem Ruellio Senam / Coluteam Theophrasti."

128. Ibid., f. 339r: "Erstlich soll man wissen / das dis gewaechs den Winter Frost nicht mag leiden / wie das Kuerbsen unnd Melonen geschlecht. Die bletter an den runden holzechten stengelen seind anzusehen wie am spitzen Klee / oder an dem kleinen Foeno graeco doch groesser. Die stengel tragen an ihren gypffeln schoene weisse bluemlein / als Violen / darauss werden schwartze krumme Schoettlein / nit anders dann wie droben von den Pfrimmen gemelt ist. Der eingeschlossen sam ist etwas breyt unnd grawfarb / von welchem ich hab etwan junge stoecklein gezielet. Sonst kompt der Senet zu uns uberfluessig auss Italia / da er von ihm selber ungepflantzet wachset. Dess gleichen inn Gallia / sagt Ruellius lib. i. Cap. lxx. Der best Senet aber ist der klein zart / so man auss Alexandria pflegt zu bringen / tregt bleychgaele bluemlein / ist sonst dem andern vast gleich."

129. Ibid., f. 339v: "würt allein in leib zuo purgieren erwoehlet / erstlich die schotten mit dem samen / darnach die bletter."

130. On Fuchs see Stübler, *Leonhart Fuchs;* Greene, *Landmarks,* 1:271–303; Riddle, 'Dioscorides,' 78–80; Krafft, 'Fuchs (Fuchsius, Füchsel), Leonhart'; Brinkhus et al., *Leonhart Fuchs;* MBW Pers. s.v. 'Fuchs, Leonhard.'

131. Fuchs, *De historia stirpium* (1542, repr. 1999), preface, sig. a5v: "Atque hoc quidem nomine laudem meritus est non vulgarem, potuissetque multorum evitare reprehensionem, nisi singulis fere herbis sua ex Dioscoride nomina accersere voluisset, quasi vero Dioscorides omnium regionum stirpes

descripsisset cognitasque habuisset, cum tamen constet quamvis prope-
modum terram suas privatim ferre herbas."

132. Ibid., preface, sig. a5r: "Quamvis vero ille non magnam admodum herbarum
notitiam habuisse videatur, ut qui professione medicus non fuerit."

133. Ibid., preface, sig. a6v: "Quare ubi potuimus habere dictionem graecam per
omnia germanicae appellationi respondentem, hanc potius quam falsam
aliquam ex Dioscoride illis imposuimus."

134. Ibid., preface, sig. a6r: "Caeterum cum multarum stirpium quas neque Dioscorides neque alii veteres cognoverunt historias huic operi inseru-
erimus, nam cum magna ex parte sint vulnerariae atque adeo in quotidiano
multorum maxime chirurgorum usu praetereundas minime putavimus,
illarum certe usitatis atque adeo barbaris nobis utendum fuit nomenclaturis, quandoquidem latinis destitueremur."

135. Ibid. The fifty-seven exotic plants are listed in Meyer, Trueblood, and Heller,
*Commentary*, 129–133.

136. Fuchs, *De historia stirpium* (1542, repr. 1999), preface, sig. a4v.

137. Ibid., ch. 148 ("De Colytea"), 445: "Differunt autem inter se magnitudine,
siquidem primum genus maius est altero, utpote quod quadriennio, Theophrasto loco paulo ante citato teste, se in arborem efferat. Siliquae etiam
illius magis turgido spiritu distenduntur, sena vero siliquas profert lunatas, nec ita praetumidas. Semen denique colyteae rotundum, lentis ob-
tinens similitudinem; senae autem oblongum, cordisque humani instar
acuminatum."

138. Cf. the quotation from Theophrastus in n. 102 above, and Ruel, *De natura
stirpium* (1537), 147: "Colutea magni Alexandri aetate non temere alibi quam
in Lipara nascebatur, iam tamen in Gallia provenit, omnibus nota *bagenaulde*
nomine. Namque frutex est quadriennio se in arborem efferens, ramis exilibus, folio foenograeci, membraneo folliculo, pellucente, praetumido et
veluti quodam spiritu distendente turgido, ita ut digitis si prematur, crepi-
tans dissiliat, in quo semen atrum, durum, latum, lentis magnitudine, pisi
gustu."

139. See the depictions in Wichtl, *Teedrogen und Phytopharmaka*, 551, 554.

140. For a much earlier picture, dating between 1379 and 1393, see Anonymous,
*De simplici medicina* (1960), s.v. 'Sene.' Senna is also depicted in editions of the
*Hortus sanitatis*, for instance, in *Ortus sanitatis de herbis et plantis . . .* (ca. 1497),
ch. 427.

141. Flückiger, *Pharmakognosie*, 633.

142. Camerarius, *De plantis epitome . . . Matthioli* (1586), 538–539; Jakob Theodorus
(=Tabernaemontanus), *Neuw vollkommentlich Kreuterbuch* (1613), 2:230–232.
Cf. Bauhin, *Pinax* (1623), 397: "I. Senna Alexandrina sive foliis acutis . . .
II. Senna Italica sive foliis obtusis."

143. Bock is the less probable source because the wording is closer to Ruel. See
Fuchs, *De historia stirpium* (1542, repr. 1999), 445: "Forma: Frutex est ramis
exilibus, folio foenigraeci"; Ruel, *De natura stirpium* (1537), 147: "Nanque

frutex est quadriennio se in arborem efferens, ramis exilibus, folio foeni-graeci"; Theophrastus, *De historia plantarum* (1644), 248: "Habet folium non absimile foenograeco"; Bock, *Kreutterbuch* (1577), f. 339r: "Die bletter an den runden holzechten stengelen seind anzusehen wie am spitzen Klee / oder an dem kleinen Foeno graeco, doch groesser."

144. Cf. Meyer, Trueblood, and Heller, *Commentary*, 125: "The sources of most of the plants figured remain unknown."

145. Fuchs, *De historia stirpium* (1542, repr. 1999), 147: "Cum non forma tantum, immo etiam sapore duo haec colyteae genera sibi similia sunt, facile etiam colligitur facultate inter se minime differre. Quare utrunque citra moles-tiam atram bilem retorridamque educit, caputque ac cerebrum, et instru-menta sensuum a noxiis purgat humoribus. Quid multa? Adversus omnia vitia ab atra bile nata valet." The translation is from Meyer, Trueblood, and Heller, *Commentary*, 437 (slightly changed).

146. Compare Ruel, *De natura stirpium* (1537), 147: "reliquos humores modestis-sime purgat, retorridam atramque bilem ... depellit," and Actuarius, *De me-dicamentorum compositione* (1556), 183: "Casia vero ipsa flavam bilem sed magis retorridam et quae iam in atram degeneravit extrahit."

147. Pseudo-Mesue, *Opera* (1549), f. 64ra: "solutione educit cum facilitate melan-choliam et choleram adustam, et mundificat cerebrum, cor et hepar et splenem et membra sensuum et pulmonem."

148. It is true that Fuchs says in an appendix to the preface that he had the Latin but not the Greek version of Theophrastus with him when writing the article on *colutea*; nevertheless, his quotations are from Ruel's *De natura stirpium*.

149. See Stübler, *Leonhart Fuchs*, 38, 53–54, 107.

150. See Stübler, *Leonhart Fuchs*, 80–81; Krafft, 'Fuchs,' 139a.

151. Fuchs, *New Kreuterbuch* (1543), ch. 169, sig. m3v: "Von Senet. Cap. CLXIX. Namen. Senet nennen wir hie die kreüter so vom Theophrasto Coluteae ge-nent werden / darumb das sie einander / fürnemlich mit den blettern gleich seind. Geschlecht. Der kreüter daruon wir hie handlen / seind zwey ge-schlecht. Eins würt fürnemlich vom Theophrasto beschriben / und Colutea geheyssen. Ist in den Apotecken unbekant / bey vns nent mans welsch Linsen. Das ander würdt von dem Actuario Sena genent."

152. Ibid., sig. m4v: "Die blumen seind auch geel / aber anderst gestalt dann der welschen Linsen / dann sie seind groesser / an der farb bleychgeeler. Ein yede blum hat fünff bletter / die thun sich etwas von einander / vnd machen doch keinen volkomnen stern / inn welcher mitte wechst ein kleins krumbs hül-ßlin herauß / das würdt nach abfallen der blumen immerzu groesser."

153. See the description and pictures in Wichtl, *Teedrogen und Phytopharmaka*, 546 and 551.

154. Lonitzer, *Kreuterbuch* (1679), 104: "Senet / welches dieser zeit sehr gemein ist / und Sena genant wird / ist zweyerley Geschlecht. Das eine wird vom Theophrasto beschrieben und Kolutea genannt. Wir nennens Welsche Linsen / oder Schaffs Linsen / und ist in den Apotecken unbekant. Ital. la

Colutea. Gall. Baguenaudier. Das ander ist das gemeine Senet / welches allenthalben Sena genennet wird. Ital. la Sena. Gall. Sene."

155. An excellent inquiry into the qualities of Mattioli's commentary is Nutton, 'Mattioli and the Art of the Commentary.' On Mattioli, see also Greene, *Landmarks,* 2:798–806; Riddle, 'Dioscorides,' 92–97; Goltz, *Studien,* 350–351; Palmer, 'Medical Botany,' 152–157; Ferri, *Pietro Andrea Mattioli;* DBI s.v. 'Mattioli (Matthioli), Pietro Andrea' (by Petri).

156. See Riddle, 'Dioscorides.'

157. Mattioli, *Commentarii* (1565), 3.70, 781: "Quod certe manifesto satis argumento esse potest delphinii historiam plane fabulosam esse vel saltem Dioscoridi adscriptam et illegitimam."

158. Ibid., 784: "Quae tamen etsi Theophrasti aevo Liparae peculiaris fuerit, hoc tamen tempore pluribus in locis sponte natam vidimus, praesertimque in Tridentino agro apud Ananios, ubi ego primum eam quam plurimis ostendi."

159. Ibid., 781–782: "Siquidem (ut pluries expertus sum) prius sata *(i.e., than in May),* facile frigore deperditur, serius vero hyemem similiter non tolerat."

160. Ibid., 782: "Unde facile Ruellius, qui numquam forsitan in Hetruria fuerat, ubi passim legitima sena seritur, praesertim in Florentino agro, magno errore putavit, senam non herbam esse sed arborem coluteam vocatam, ratus easdem plantas esse."

161. Ibid., 783: "Et quamvis asserat Fuchsius in suis de plantarum historia commentariis has plantas facultate inter se minime differre, eum tamen in errore versari putaverim, quod certo sciam coluteae semen non minus quam genista vomitiones ciere."

162. Ibid., 784: "Caeterum e sena ea magis praestat quae ex Alexandria Aegypti vel ex Syria affertur, quamvis e nostrate adhuc in agro virente et rosarum diluto ego quotannis Syrupum parare consueverim qui tuto largeque omnes purgat humores."

163. Ibid.: "In quorum sententiam ego quoque pedibus eo; nam etsi quibusdam, praesertimque mulieribus, quandoque contingat ut epoto senae diluto eorum intestina torminibus vexentur, id numquam ex senae facultate provenire existimavi, sed ex crassis, pituitosisque excrementis, ab hoc pharmaco eo impulsis, quae sua crassitie intestinorum meatus adeo replent, ut in eis non parvam extensionem faciant, dum viam ad exitum quaerunt. Neque tamen me unquam aliquem invenisse memini, qui diceret se epota sena ventriculi dolore laborasse."

164. Ibid.: "Cui sane, ut iam millies periculum fecimus, non minor deiectoria vis inest quam foliis. Primum vero genus, quo Venetiis omnes fere officinae scatent, non modo minorem huiusmodi vim habet, sed prope nullam. Ut hinc plane colligere liceat Mesuem hic non esse sine limitatione damnandum, quandoquidem is, ut arbitror, de folliculis decerptis intelligat, non autem de caducis, quibus tantum utuntur ii, qui Mesui falso reclamant. In quorum numero ego quoque quandoque fui, verum cum postea integrum

senae campum sevissem, ut folliculos decerptos virentes et succo praegnantes, mox siccatos experirer, rem aliter facto periculo se habere facile comperi."

165. See Palmer, 'Pharmacy,' 110, and 'Medical Botany,' 152–154, 154: "So that although historians of botany continue to declare that the Renaissance 'turned men's eyes to living plants in the field and away from the pages of the classics' (Morton, *History of Botanical Science,* 116), as far as Italy is concerned they are wide of the mark."

166. In relating practical details and personal experiences, Mattioli was not alone; other contemporary botanists such as Brasavola, Valerius Cordus, and William Turner did the same. See Nutton, 'Mattioli and the Art of the Commentary,' 139–141.

167. Mattioli, *Commentarii* (1565), 784, line 9.

168. Mattioli does not, however, refer to Marcello Virgilio by name save as an authority on the value of the oldest Dioscorides manuscript—the one that does not contain *delphinion*.

169. Mattioli, *Commentarii* (1565), 781: "Quorum gravis error ita evidens est ut iam ab omnibus explodatur. Idque non iniuria quod in delphinio (demus eius caput legitimum esse) folia, non siliqua, quae delphinorum formam repraesentent, requirantur; foliaque sint tenuia, longa, divisa; flores vero purpurei ... Constat itaqua sena Mauritano nomine dicta foliis glycyrrhiza crassis, suppinguibus, fabarum sapore, caule cubitali, a quo ramuli exeunt in lori modum flexiles; flores ei emicant aurei, veluti brassicae, purpureis intercursantibus venulis."

170. As already observed by Bock, *Kreutterbuch* (1577), ch. 49, f. 339r: "Erstlich soll man wissen / das dis gewaechs den Winter Frost nicht mag leiden." Mattioli, *Commentarii* (1565), 782–783: "Sed eius error hinc praeterea apertissime diluitur quod colutea siliquas proferat colore in purpuram primum vergentes, mox albicantes, flatu turgentes, quae si premantur, crepitum magnum edunt; semine eius *(scil. coluteae)* parvo, rotundo, lentis effigie. Sena vero siliquatur folliculis lunatis, qui nullo spiritu tumido distenduntur, quibus semen includitur vinacearum instar. Adde etiam quod colutea arbor sit multis perdurans annis, sena vero non nisi paucis mensibus vivat."

171. Mattioli, *Commentarii* (1565), 781–782: "Siquidem (ut pluries expertus sum) prius sata facile frigore deperditur, serius vero hyemem similiter non tolerat. Meminit senae Serapio in suo de simplicium medicamentorum volumine, ubi sic habet ex auctoritate Abohanifa: sena asservatur sicca, siliquas lunatas et oblongas profert, in quibus semina in ordinem digesta clauduntur. Siliquae tenui pediculo pendent, ob idque facile ventorum impetu deturbantur. Caducas opiliones colligunt. Haec Serapio."

172. Fuchs, *De compendorum miscendorumque medicamentorum ratione* (1556), 1, ch. 79, 272: "Mesue scribit tergere, discutere, citra molestiam melancholiam et bilem torridam purgare a cerebro, sensoriis, pulmone, corde, iecore et liene, corpus efficere floridum, obstructionesque viscerum solvere, aliaque

posse quae apud illum legere quivis potest." The first edition appeared 1555 in Basel; see Stübler, *Leonhart Fuchs,* 50–52.

173. Daléchamps, *Historia generalis plantarum* (1586/1587), lib. 2, ch. 51, p. 217: "Res autem nova potius videtur, veteribus graecis et latinis scriptoribus indicta."

174. l'Obel and Pena, *Nova stirpium adversaria* (1576), 406: "Sero et magno aegrotantium incommodo innotuit Graecis sena, quam una et appellationem Arabibus debemus."

175. In Dodonaeus, Senna appears after "Asarum" and "Aloe" as the third of a long list of "herbae purgantes"; see Dodonaeus, *Stirpium historiae pemptades sex* (Antwerp, 1616, first printed in 1583), lib. 2, ch. 4, 360–361.

176. Camerarius, *De plantis epitome . . . Matthioli* (1586), 539: "Forma: Oblongiora et in acumen vergentia habet folia, omnibus suis partibus tenerior est vulgari. Locus: Ex Alexandria Aegypti ad nos convehitur. Facultates: Valentius praestat ea universa quae vulgo usitata."

177. Jacob Theodorus (=Tabernaemontanus), *Neuw vollkommentlich Kreuterbuch* (1613), lib. 2, 230–232 (lib. 2, 903–904 in the 1731 edition, repr. 1963).

178. C. Bauhin, *Pinax* (1623), 397.

179. Cesalpino, *De plantis libri XVI* (1583), lib. 3, ch. 27–40, 110–118: Cassia, Siliqua sylvestris, Tamarindus, Anagyris, Laburnum, Cytisus, Ebenus, Genista, Aspalathus, Spina, Emerus, and Sena (ch. 39, 118: "Sunt inter arbores quae semina ferant in siliquam"), Barba Iovis. This classification is adopted with modifications by Bauhin, *Pinax* (1623), lib. 11, sec. 1, 389–397.

180. Morison, *Plantarum historiae . . . distributio nova* (1715), 2:201–202: "Legumina siliquis propendentibus donata."

181. Hermann, *Horti academici Lugduno-Batavi catalogus* (1687), 556–558. Paul Hermann describes the following plants: Senna Italica, Senna Occidentalis odore Opii viroso, Eadem hirsuta, Senna Occidentalis odore Opii minus viroso, Senna Orientalis fruticosa Sophera dicta, Senna Orientalis hexaphylla, Senna Orientalis tetraphylla Absus dicta, Senna Occidentalis siliqua multiplici, and Eadem siliqua singulari.

182. Ray, *Historia plantarum,* 3 vols. (1686/1688/1704), 2:1742–1743.

183. Tournefort, *Institutiones rei herbariae* (1700), 1:618–619: "Sectio V. De arboribus et fruticibus flore rosaceo, cuius pistillum abit in siliquam. Genus I. Senna . . . Genus II. Poinciana . . . Genus III. Cassia." Cf. the French wording of the first edition, *Elemens de Botanique* (1694), 491–492: "Sections V. Des arbres et des arbrisseaux qui ont les fleurs en rose et le fruit en gousse. Genre I. Senna . . . Genre II. Poinciana . . . Genre III. Cassia."

184. Tournefort's description of senna is the following: "Senna est plantae genus, flore A ex quinque ut plurimum petalis B in orbem positis constante, cuius pistillum C abit deinde in siliquam D, E planam admodum, ut plurimum incurvam, ex duabus membranis F, G compositam, quas inter nidulantur semina K gigartis similia, tenuibus veluti dissepimentis H, I inter se distincta" (*Institutiones rei herbariae* (1700), 1:618). The capital letters refer to the illustration.

185. Johann Jacob Dillenius, for example, identified a cassia plant with quadrangular pods with a senna plant with flowers of six leaves described in Hermann's Leiden catalog. See Dillenius, *Hortus Elthamensis* (1732), 72–73: "Cassia siliqua quadrangulari t. LXIII . . . Est Senna Orientalis hexaphylla: Tala Zeylonensibus Herm. Hort. Lugd. B. p. 557. Raj. Hist. Plant. Tom. II. p. 1743. N. 3. ubi ex Horto L. Bat. describitur . . . Ceterum observari velim, me Tournefortium hic sequi, et a Cassiae genere non excludere eas species, quarum siliquae pulpa destituuntur."

186. Linnaeus, *Species plantarum* (1753), 376: "5 phyllus. Petala 5. Antherae supernae 3 steriles; infimae 3 rostratae. Legumen."

187. Ibid., 377: "8. Cassia foliolis trijugatis quadrijugatisque subovatis . . . Habitat in Aegypto."

188. For a guide to nineteenth-century botanical classification of senna, see *Index Londinensis*, 2:83–88 (s.v. cassia acutifolia, cassia angustifolia, cassia obovata, senna).

189. On "Cassia italica" (an alternative term being "Cassia obovata"), see Andrews, *The Flowering Plants*, 2:114–121, esp. the table (114), on the difference between "Cassia italica" and "Cassia senna": "(dd) Whole plant more or less glabrous: (e) Leaflets obovate; pod with an undulate crest on each side. C. italica. (ee) Leaflets elliptic to lanceolate, acute at the apex; pod without undulate crests on its sides. C. senna." See also Hoppe, *Drogenkunde*, 237.

190. l'Obel and Pena, *Nova stirpium adversaria* (1576), 406; Morison, *Plantarum historiae . . . distributio nova* (1715), 2:201: "at rem novam veteribus graecis et latinis scriptoribus indictam, et a mauritanis primo propositam, a Serapione sub sene, a Mesue qui domesticam et silvestrem facit sub senna censemus"—which is a literal quotation from Bauhin, *Pinax* (1623), lib. 11, sec. 1, 397.

191. Cesalpino, *De plantis libri XVI* (1583), lib. 3, ch. 39, 118.

192. Daléchamps, *Historia generalis plantarum* (1586/1587), 218.

193. On Costeo, see Lieberknecht, *Die Canones des Pseudo-Mesue*, 103, and Siraisi, *Avicenna in Renaissance Italy*, 140.

194. Costeo, in Mesue, *Opera* (1570), f. 86r–v: "Verum (ut dicam quod sentio) si folliculos tales haberemus quales Mesue eligi docet, completos ut ipse ait et seminibus amplis compressis, colore ad viridem et subnigrum accedente, cum astrictione subamaros, tales auem sunt qui cum ad perfectam magnitudinem devenere colliguntur antea quam vesica vetustate resiccetur, colore vivido et sapore proprio exuatur, dehiscatque ac sponte decidat; si, inquam, tales haberemus, hos ipsos tum ob perfectiorem aetatem ut docte Brasavolus ait, tum ob densiorem substantiam ac vim contentorum seminum excellere ego certe semper contenderim. Atqui raro habemus tales . . . non est ergo Mesue erroris notandus, qui folliculum foliis praeferat."

195. l'Obel and Penna, *Nova adversaria stirpium* (1576), 406: "Eapropter qui adversum Mesuem tuetur non esse corrigenda, imo siccitate sua atque quadam astrictione ventriculum iuvare, nescit quid commentetur."

196. Daléchamps, *Historia generalis plantarum* (1586/1587), 218–219.

197. Jacob Theodorus (=Tabernaemontanus), *Neuw vollkommentlich Kreuterbuch* (1613), lib. 2, 230–232, for example, 231: "Es werden aber die Senetblätter fürnemlich gebraucht / das grobe verbrannte melancholische Geblüt auss dem Leib zuführen / darnach die Gall und den zähen Schleim damit zu expurgieren" (lib. 2, 903–904 in the 1731 edition, repr. 1963).

198. Morison, *Plantarum historiae . . . distributio nova* (1715), 201–202.

199. Ibid., 202: "Laudant Arabum plerique folliculos Senae, at nostri Europei folia magis commendant: folliculi etenim nisi maturi flatus gignunt et torsiones ventris adferunt; saepe autem illos ante maturitatem colligi contingit." This is a literal quotation from Dodonaeus, *Stirpium historiae pemptades sex* (1616), lib. 2, ch. 3, 361.

200. Ray, *Historia plantarum* (1688), 2:1742.

201. On temperament and humors in the Renaissance, see Siraisi, *Avicenna in Renaissance Italy*, 296–314, on Galenism in general, 17–18; see also Temkin, *Galenism*, ch. 4 ("Fall and Afterlife"), and Wear, 'Medicine in Early Modern Europe,' 320–361.

202. Sennerus, *Dissertatio botanico-medica inauguralis de senna* (Altdorf, 1733), 16: "Speciales Sennae facultates, quas in singularibus partium humani corporis affectibus purgando potissimum, tum et alterando exerit, longum foret, si enarrare vellem, praesertim ex Arabum Medicorum testimoniis, quae summatim quasi percensuit Symphor. Campegius Cribrat. Medicam. Lib. 3. P.M. 52 et seq. *(on this book by Champier, see section 'The Early 1530s' above).* Ego tamen illorum quam aliorum scriptorum experimenta praecipua saltem breviter commemorabo. Joh. Mesue l.c. collaudat Sennam, quod clementer purget melancholiam a cerebro, sensoriis, pulmone, corde, hepate, liene."

203. Tundermann, *Meletemata De Sennae Foliis* (Dorpat, 1856). He refers to Heerlein, Bley, Diesel, Lassaigne, Feneulle, Buchheim, and Casselmann.

204. Dietrich, *Dioscorides triumphans,* 39. Leclerc, *Histoire de la médecine arabe,* 2:232–233, gives a list of fifty-three new drugs which he finds particularly important, among them senna.

205. Dietrich, 'Islamic Sciences,' 57–58, quoting from Abū l-'Abbās an-Nabātī (fl. 1200): "If Dioscorides omitted the description of a drug because it was well known in his time, both in name and appearance, and widespread, this is not to be counted as a failing on his part."

206. Cf. Dilg, 'The Antarabism,' 288: "the downfall of Arabism was, at least in Germany, definitely sealed with the end of the Renaissance humanism," and Sarton, 'Arabic Science,' 316–317: "The decline of Arabic influence began in the fourteenth century and was almost complete by the middle of the sixteenth." Cf., for the same position, Schipperges, *Ideologie,* 22–23. My findings agree with what Siraisi has shown for the history of academic teaching of Avicenna: "These arguments *(i.e., of the humanist polemics against the Arabs)* were genuinely persuasive and their force was widely recognized; yet they did not prevail. The *Canon* survived in numerous university medical curricula,

continued to be taught seriously in many universities, including leading centers of medical education, and continued to be consulted by practicioners" (Siraisi, *Avicenna in Renaissance Italy,* 76; cf. 105); similarly Temkin, *Galenism,* 127–128. They also agree with Schmitz, 'Der Anteil,' 242, who points out that "die Humanisten den Primat der aristotelisch-scholastischen und spätarabistischen Medizin und Pharmazie nicht zu brechen vermochten"—which is a fact Schmitz deplores.

207. On the apothecaries' interest in Arabic medicine, see Dilg, 'The Antarabism,' 285, and Goltz, *Studien,* 357. On dispensatoria in the sixteenth century, see Friedrich and Müller-Jahncke, *Geschichte der Pharmazie,* 119–123, 194–216.

208. Paglia and d'Orvieto, *In Antidotarium Joannis filii Mesuae* (1543), 50: "plurimum aromatarii hodierni sunt idiotae, et nec curant studere nisi his e quibus magis lucrum accipiunt. In quorum officinis libri . . . non habent locum."

209. Pagel, 'Medical Humanism: A Historical Necessity,' 385n: "Nevertheless, the polemical attitude of the humanists was admittedly legitimate—methodical return to the ancient texts was the only way out of terminological chaos. It was a salutary reaction and meant to benefit mankind."

210. The rise of botanical illustrations is often described as epoch-making. See Morton, *History of Botanical Science,* 123–124, Mägdefrau, *Geschichte der Botanik,* 29, Greene, *Landmarks,* 2:798, and Dilg, *Das Botanologicon,* 108 ("epochemachende Leistung"). But cf. Palmer, 'Medical Botany,' 154: "So that although historians of botany continue to declare that the Renaissance 'turned men's eyes to living plants in the field and away from the pages of the classics' . . . , as far as Italy is concerned they are wide of the mark."

211. As convincingly argued by Kuhn, *The Structure,* 170. The development of *materia medica* was dramatic, but not revolutionary; the humanists did not, in the end, conquer or convert the field. For a sensible historical concept of "revolution," see Porter, 'The Scientific Revolution.'

212. Jacquart, *La science médicale,* art. 10, 277–290. Cf. also Jacquart's introduction, xii–xiii.

213. As has been shown for the example of mineral drugs; see Goltz, *Studien,* 358–361.

214. See Dannenfeldt, *Leonhard Rauwolf,* 1–12 (also on Belon), and Brentjes, *Travellers from Europe,* art. 1, 439–448; cf. also art. 5, 384–390. For context, see Hamilton, *The Arcadian Library,* 31. On Rauwolf in particular, see Dannenfeldt, *Leonhard Rauwolf,* esp. 252–279, where the 364 plants named or described by Rauwolf are listed.

215. On the *Interpretatio,* see Lucchetta, *Il medico e filosofo,* 39–47, and Veit, 'Der Arzt Andrea Alpago,' 308–312. A selection of the longer entries of the *Interpretatio* with Italian translation is printed in Vercellin, *Il Canone,* 52–141.

216. See Jacquart, *La science médicale,* art. 11, 408–415, for a comparative study of medieval and Renaissance *Synonyma,* which demonstrates the quality of Alpago's lexicon.

217. Serapion, *Practica... quam postremo Andreas Alpagus Bellunensis... in latinum convertit* (1550), sig. +5v–+6v. The index is titled: *Index arabicorum nominum medicamentorum simplicium per ordinem alphabeti cum translatione latina.*

218. Alpini, *Historiae Aegypti Naturalis* (1735), 85–86: "Jacobus Manus Salodiensis, qui ante me septennio magna cum laude Medicinam apud Nationem Venetam fecerat, vir in lingua Arabica etiam doctissimus, quem ego Cairi offendi, retulit aliquando mihi se habuisse Arabicum Dictionarium omnium nominum quae usui medicinae facerent, ipsumque Latinitate donavisse, additis ab ipso nonnullis observationibus, ut studiosorum animos ad maiora excitaret, tum ut nomina Arabica quae in Serapionis et Avicennae indice ex Bellonense depravata leguntur, emendarentur, quando illorum ne verbum quidem a nostris prolatum Arabes intelligant." For context and an English translation of the passage, see Brentjes, *Travellers from Europe,* art. 1, 446, and Siraisi, *History,* 249.

219. Rauwolf also once refers to Alpago's lexica; see Dannenfeldt, *Leonhard Rauwolf,* 23.

220. Postel, *Grammatica Arabica* (ca. 1539/1540), sig. D2v.

221. Siraisi, *Avicenna in Renaissance Italy,* 148.

222. See Hamilton, *William Bedwell,* 119, for the text, and 71–73, for context.

223. Jones, 'Thomas Erpenius.'

224. Christmann, *Alphabetum Arabicum* (1582), sig. A3r: "Arabes multa simplicia, quorum permagnus est usus, invenere; iidem doses, quas Graeci non sine crudelitate et fastidio aegris propinare sunt soliti, optime temperarunt... Quam enim multa medicamenta simplicia ac composita nominibus Arabicis circumferuntur? Quorum tamen proprietatem pauci intelligimus. Optandum sane esset ut ex usu linguae Arabicae ea descripta haberemus, quorum appellationes adhuc in medicina retinentur."

225. Erpenius, *Orationes tres* (1621), 79: "utilissima et hactenus vobis Graecisque incognita simplicia." See the English translation in Jones, 'Thomas Erpenius,' 22–23.

226. As is very obvious from the six reference works that Saladin of Asculo (Saladino Ferro) recommends to apothecaries around 1450: texts by Avicenna, Serapion, Simon of Genoa, Albucasis (Liber servitoris), Mesue, and Nicolaus of Salerno. See Saladin of Asculo, *Compendium aromatariorum* (1488), sig. A1v–A2r: "Libri necessarii ispi aromatario. Dico quod sex libri sunt necessarii cuilibet aromatario, scilicet duo de simplicibus, ut secundus Avicennae et Serapio de simplicibus. Tertius est liber de synon<y>mis Simonis Ianuensis. Quartus est liber Servitoris, in quo continentur praeparationes omnium fere necessariorum ad aromatarios pertinentium. Quintus est liber Mesue, idest, Iohannis Damasceni, qui dividitur in tres libros. Nam primus tractat de consolatione medicinarum et de quibusdam simplicibus solutivis famosis. Secundus est eius antidotarium, et isti sunt necessarii aromatario. Tertius vero est in practica, et iste non est necessarius aromatario, sed medico. Sextus est antidotarium Nicolai de Salerno, licet sunt duo

antidotaria Nicolai, scilicet magnum quod non est in usu propter eius pro-
lixitatem, licet sit optimum, et antidotarium parvum quo omnes communiter utuntur."

227. Siraisi, *Avicenna in Renaissance Italy,* 158.

228. On Walther Ryff (Gualtherus Rivius), see Riddle, 'Disocorides,' 76–78.


### 5. · PHILOSOPHY: AVERROES'S PARTISANS AND ENEMIES

1. The prefaces to the 1550/1552 Giunta edition give a prominent historical place to Averroes by claiming that no other philosopher supplanted more carefully and reliably what had been left out by Aristotle: Aristotle and Averroes, *Omnia . . . opera* (1550/1552), vol. 1, 1, f. 2v: "Ac iam constet inter omnes qui proximis saeculis sunt philosophati, eas philosophiae partes, quae ab Aristotele sunt omissae, ab alio hactenus nemine vel diligentius inspectas, vel fundamentis solidioribus fuisse constitutas." On this important edition, see Chapter 1.

2. See Francesco Robortello, preface to Averroes, *Paraphrasis in libros de republica Platonis* (1552), sig. a2r: "Averroes . . . inter philosophos maxime insignis, et optimus Aristotelis interpres semper est habitus."

3. Agostino Nifo's edition of and commentary on Averroes's *Destructio destructionum,* which appeared in 1497, is explicitly written for the use of students and natural philosophers. Cf. Nifo, *In librum Destructio destructionum* (1497), sig. A1v: "Elucubratus sum proximis diebus quoddam opusculum quod destructio destructionum inscribitur, ut et profitendo et edendo aliquid adolescentibus studiosis, ac ad rerum physicarum cognitionem aspirantibus prodesse possim." On this treatise, see Mahoney, 'Philosophy and Science,' in *Two Aristotelians,* art. 1, 179–200.

4. Benedetto Rinio, preface to Avicenna, *Liber Canonis* (1556), sig. A3v: "Excitavit hic Latinorum studia occasionemque pluribus dedit acrius naturam investigandi, qui nunc ea de causa celebrantur. Tantum enim Averrois doctrina eruditionis plena fuit, ut sectatores etiam invenerit, qui hodiernis temporibus adhuc illius placita defendunt acerrime." On Rinio and his edition of the *Canon,* see Siraisi, *Avicenna in Renaissance Italy,* 139–140, and 167 (on the preface).

5. Schmitt, 'Renaissance Averroism,' in *The Aristotelian Tradition,* art. 8, 121–142; Burnett, 'The Second Revelation,' 185–198.

6. See Mahoney, 'Philosophy and Science,' in *Two Aristotelians,* art. 1, 136–161. The term *fidelissimus ille Aristotelis interpres* comes from his *Quaestio an coelum sit ex materia et forma constitutum vel non* of 1483 (154).

7. See the passages collected by Mahoney, in ibid., 181 ("elegantissimus philosophus"), and 'Agostino Nifo's Early Views,' in *Two Aristotelians,* art. 7, 453–454.

8. Nifo, *Super tres libros de anima* (1503), 2, comm. 97, sig. h5va: "Ecce igitur quomodo verba latinorum qui sequuntur Avicennam et Alfarabium

philosophiam perhypatheticorum non sapiunt; propter quod dicta huius philosophi semper inveniuntur convenientia cum verbis Aristotelis, idcirco consuevi vobis dicere Averroem nihil aliud esse nisi arabem Aristotelem."

9. See Lohr, *Latin Aristotle Commentaries*, 2:504–512: Cranz, *A Bibliography*, 241. On Marcantonio Zimara, see also Nardi, *Saggi*, 322–355.

10. Prassicio, *Questio de immortalitate anime intellective secundum mentem Aristotelis a nemine verius quam ab Averroi interpretati* (1521).

11. Patrizi, *Discussiones peripateticae* (1999), 66: "si quidem vera est Averrois omnium interpretum, ut multi putant, Aristotelicissimi sententia." As pointed out by Martin, 'Rethinking,' 18.

12. As Craig Martin has shown. See Pomponazzi, *Dubitationes in quartum Meteorologicorum* (1563), f. 16v: "quae tamen non est mens Aristotelis. Averroem ego excuso quoniam habuit textus corruptos. Sed inculpandi sunt Latini, qui meliores textus habent et clariores." On Pomponazzi's discussion of Averroes's meteorological theories, see Martin, *Renaissance Meteorology*, 49–50.

13. Zabarella, *De propositionibus necessariis*, 2.2, in *Opera logica* (1603), 380: "Averroes . . . excusandus est, quia Arabs fuit et Graecae linguae ignarus, eo praesertim quod etsi in appellatione aliquid erroris commisit, in re tamen ipsa non erravit. Imo adeo egregie se gessit ut nemo profundius ac melius quam ipse mentem Aristotelis hac in re penetravit, nam et Graecorum errores detexit atque impugnavit et totam rei veritatem pulcherrime explanavit." Discussed also in Martin, 'Rethinking,' 15.

14. On Averroes's views on immortality, see Taylor, 'Personal Immortality'; de Libera, *Averroès: L'intelligence*, n230.

15. An example is Giles of Rome (Aegidius Romanus), *Errores philosophorum* (1944), 15–16, a text written in the climate of the Parisian condemnations of 1277: "Commentator autem omnes errores Philosophi *(sc., Aristotelis)* asseruit, immo cum maiori pertinacia et magis ironice locutus est contra ponentes mundum incepisse quam Philosophus fecerit. Immo sine comparatione plus est ipse arguendus quam Philosophus, quia magis directe fidem nostram impugnavit ostendens esse falsum cui non potest subesse falsitas, eo quod innitatur primae veritati."

16. Petrarca, *Invective contra medicum* (1950), 2:53: "At tu, miser, erroneus post ydolum tuum confragosis anfractibus delectare, venturus ad finem impietati debitum, ad quem tuus venit Averrois"; Petrarca, *De sui ipsius et multorum ignorantia*, in *Prose* (1955), 750–752.

17. Petrarca, *Epistolae rerum senilium*, ep. 13, 6 [5] *(Donato Apennigenae Grammatico)*, in *Opera* (1965), 2:1017 (here the text reads: "Averroim").

18. Petrarca, *Epistolae sine titulo*, ep. 18, in *Opera* (1965), 2:812 (=*Ep. sen.* 15 [14], 6): "Extremum quaeso ut, cum primum perveneris quo suspiras, quod cito fore confido, contra eum illum rabidum Averroim, qui furore actus infando contra dominum suum Christum contraque catholicam fidem latrat, collectis undique blasphemiis eius, quod ut scis iam coeperamus, sed me ingens semper et nunc solito maior occupatio nec minor temporis quam scientiae retraxit

inopia, opusculum unum scribas..." (translation from: Cassirer, Kristeller, and Randall, *The Renaissance Philosophy of Man,* 143). The quotation is also in Di Napoli, *L'immortalità,* 65n46.

19. Petrarca, *Epistolae rerum senilium,* ep. 12, 2, in *Opera* (1965), 2:1009. See Burnett, 'Petrarch and Averroes.' On the targets of Petrarca's anti-Aristotelian polemics in *De sui ipsius et multorum ignorantia,* see Kristeller, 'Petrarch's "Averroists."'

20. Novati, *Epistolario di Coluccio Salutati,* 3:191: "Irreligiosissimum tamen Averroym non sine motu cachinnationis admiror, qui cum de Deo et animae aeternitate pessime senserit, ad quem refertur cuncta religio, illius muliercule crediderit iuramento, quae se iactavit ex emisso contra naturam semine in livelli balneo concepisse, nisi forsitan ipsam timuisse putaverit quod ipse penitus deridebat" (letter to Baruffaldi, probably of 1397).

21. Valla, *Dialecticarum disputationum libri tres* (1541), 5: "Nam Avicenna et Averrois plane barbari fuerunt, nostrae linguae prorsus ignari et graecae vix tincti. Quorum etiam si magni viri fuerint, ubi de vi verborum agitur quae plurimae sunt in philosophia quaestiones, quantula debet esse auctoritas?"

22. Martin, 'Rethinking,' 13.

23. Barbaro, *Epistolae* (1943), 1:92: "Et, hercule, si conferas eius viri (scil. Averrois) scripta cum graecis, invenies singula eius verba singula esse furta ex Alexandro, Themistio, Simplicio." The translation is taken from Kraye, 'Philologists and Philosophers,' 156–157n9. The preceding sentence is: "Utor expositoribus graecis, latinis, arabibus; praecipue vero graecis, unde omnis et excitata et consummata philosophiae cognitio est, Iamblicho, Porphyrio, Alexandro, Themistio, Simplicio, Philopono, caeteris huius modi; post hos Averroi, quem ut multis ante se ita nemine post se inferiorem fuisse comperio." The letter is addressed to Arnoldo di Bost in Ghent.

24. Alexander of Aphrodisias, *Enarratio De anima ex Aristotelis institutione* (1495, repr. 2008), sig. a3r: "Fere qui Averoim exacte legerit et suis quaeque locis singulatim singula contulerit, eius doctrinam ab optimis auctoribus prodisse comperiet. In eius enim interpretationibus Alexandrum, Themistium et Simplicium licet videre." On Donato's translation, see Cranz, 'Alexander Aphrodisiensis,' 85–86, and Kessler, introduction to Alexander of Aphrodisias, *Enarratio De anima* (2008), xxxi–xxxvi.

25. Giglioni, 'Introduction,' 6–7.

26. Simplicius, *Commentarii in libros de anima Aristotelis* (1543), f. 35v (introductory letter to the second book): "Verum et inter hos *(scil. Averroem et Simplicium)* ita sese res habet, ut quicquid boni in his praesertim De anima libris Arabs ille dixerit, de hoc sumpserit" (quoted after Nardi, *Saggi,* 397).

27. Ibid.: "Simplicium unum vobis die noctuque versandum proponite;... et illum *(sc., Averroem)* iam de manibus deponite."

28. Ficino, *Opera* (1962), 1537: "Totus enim ferme terrarum orbis a Peripateticis occupatus in duas plurimum sectas divisus est, Alexandrinam et Averroicam. Illi quidem intellectum nostrum esse mortalem existimant, hi vero unicum

esse contendunt. Utrique religionem omnem funditus aeque tollunt, prae-sertim quia divinam circa homines providentiam negare videntur, et utro-bique a suo etiam Aristotele defecisse." For context, see Hankins, *Plato in the Italian Renaissance,* 274–276, and Kristeller, *Die Philosophie,* 15–16.

29. Ficino, *Theologia platonica* (1964–1970), vol. 3, ch. 15.1, p. 8: "Averrois, hispanus patria, lingua arabs, Aristotelis doctrinae deditus, graecae linguae ignarus, aristotelicos libros in linguam barbaram e graeca perversos potiusquam con-versos legisse traditur, ut non mirum sit, si in quibusdam rebus reconditis brevissimi scriptoris mens eum latuerit." On Ficino's attitude toward Aver-roes, see Allan, 'Marsilio Ficino on Saturn,' 81–97.

30. Lefèvre, *The Prefatory Epistles* (1972), 375 (a quotation of the year 1516).

31. According to Rice, 'Humanist Aristotelianism,' 139.

32. Vives, *De causis corruptarum artium* (1990), ch. 5.3, 191–198 (=pp. 490–504).

33. For a different view, see Monfasani, 'Aristotelians,' 266–267: "True, Pe-trarch attacked the Averroists, but Coluccio Salutati did not echo him, and Petrarch's anti-Averroistic strictures hardly resonated with Quattrocento humanists. Neither Leonardo Bruni nor George of Trebizond nor Lorenzo Valla, just to mention some of the more important humanists who dealt with philosophy, made Averroism a special target."

34. Nifo, *Expositiones in Aristotelis libros Metaphysices* (1559, repr. 1967), 1.3, p. 57: "Magno miratu dignum est quonam pacto vir iste tantam fidem lucratus sit apud latinos in exponendis verbis Aristotelis, cum vix unum verbum recte ex-posuerit." Cf. 1.5, p. 59: "Et errat errore translatoris."

35. Reisch, *Margarita philosophica* (1973), ch. 11.30, p. 472: "Discipulus: Cur Aver-roim et maledictum et somniantem appellaveris? Magister: Quia hic cath-olicam irridens sinceritatem ad Mahumetis caninam perfidiam rediit et dum Aristotelis lucubrationes in multis non indocte exposuisset, tamen eum locum qui de perpetuitate animae rationalis est, non tam obtenebravit quam etiam rationibus et argumentationibus tele araneae paulo fortioribus depravare co-natus est, intantum ut diceret unum et eundem numero omnibus hominibus esse intellectum."

36. In the absence of Ockamist theology in Italy, Scotists and Thomists domi-nated Italian theology, which was a source of hostility toward Averroes; this is very clearly pointed out by Monfasani, 'Aristotelians,' 256–266.

37. On Trombetta, see Poppi, *La filosofia,* 63–85, 258–266. See also Mahoney, *Two Aristotelians,* art. 9.

38. See the dedication and letters accompanying the edition of 1498, in Trom-betta, *Tractatus singularis* (1498), f. 1v: "iucundius tamen erit, ut qui praecipites et deliri sunt in via Averroys ex ipso aliquid fructus et utilitatis percipiant." Cf. f. 23vb: "licet contrarium ipse Averroys profiteatur, qui ex ruditate ingenii a vera philosophia et a veritate plurimum declinavit et praeceps in multos er-rores lapsus in hoc et in aliis plurimis non habuit intentionem praeceptoris sui *(sc., Aristotelis).* Qui ulterius propter quasdam rationes logicales et pueriles voluntarie se brutum et bestiam fecit."

39. Barozzi, *Edictum contra disputantes de unitate intellectus* (1890/1891), 279: "Mandamus ut nullus vestrum sub poena excommunicationis latae sententiae quam si contrafaceritis ipso facto incurratis, audeat vel praesumat de unitate intellectus quovis quaesito colore publice disputare; et si hoc ex Aristotelis sententia fuisse secundum Averroin hominem doctum quidem sed scelestum. A quo et Avicenna, hispalensem medicorum praestantissimum Bithiniae, ut multi putant, regem veneno enectum et quem ab Avicenna priusquam moreretur interfectum tradunt, in controversiam deducatur." On Barozzi and his decree, see Di Napoli, *L'immortalità,* 185–188; Gios, *L'attività pastorale,* 291–303; Monfasani, 'Aristotelians,' 250–251, 264–266; Grendler, *Universities,* 283–285. On Barozzi, see also the recent articles in Nante, Cavalli, and Gios, *Pietro Barozzi.*

40. Ragnisco, 'Documenti inediti,' 280–281: "disputavi ac tenui quod opinio unitatis intellectus Averrois fuerit opinio Aristotelis et post multos annos, dum vidissem et graecos et arabes doctissimos, repperi non solum dictam opinionem alienam esse a fide nostra et veritate, sed etiam ab intellectu Aristotelis ... Deum testor quod numquam credidi tali opinioni."

41. Printed in Vernia, *Contra perversam Averroys opinionem* (1505), f. 2v.

42. The decree is printed in Trombetta, *Tractatus singularis contra Averroystas* (1498), f. 30rb: "Nullis itaque in paduana et vincentina diocesibus nunc degentibus aut in futurum sub excommunicationis poena trina monitione canonica praemissa in posterum licebit praefatam Averrois opinionem tanquam veram et certam asserere."

43. See Bianchi, *Pour une histoire,* 128–130, esp. 128n4: "dixit quod non placet secunda pars bullae praecipiens philosophis, ut publice persuadendo doceant veritatem fidei."

44. Grendler, *Universities,* 289–290, with n. 70.

45. Mansi, *Sacrorum Conciliorum nova et amplissima collectio,* 32, col. 842: "hoc sacro approbante Concilio damnamus et reprobamus omnes asserentes animam intellectivam mortalem esse aut unicam in cunctis hominibus ... omnesque huiusmodi erroris assertionibus inhaerentes veluti damnatissimas haereses seminantes per omnia ut detestabiles et abominabiles haereticos et infideles catholicam fidem labefactantes, vitandos et puniendos fore decernimus. Insuper omnibus et singulis philosophis ... mandamus ut ... teneantur eisdem *(scil. auditoribus suis)* veritatem religionis christianae omni conatu manifestam facere et persuadendo pro posse docere ac omni studio huiusmodi philosophorum argumenta ... pro viribus excludere atque resolvere." See Di Napoli, *L'immortalità,* 220–225; Monfasani, 'Aristotelians,' 269; Grendler, *Universities,* 289–290 (with English translation). On the difficult interpretation of the bull, see Bianchi, *Pour une histoire,* 119–124.

46. Foresti, *Supplementum* (1486), f. 224v: "sed vulgares Bythiniae regem fuisse dicunt et ab Averoi medico venenatum tradunt, sed antequam periret, ipse illum Averoim interfecisse." Cf. Foresti, *Supplementum* (1483), f. 115v. Compare the very similar wording of Barozzi's decree of 1489 previously cited.

47. On the tradition of depicting Thomas triumphing over Averroes, see Polzer, 'The "Triumph of Thomas" Panel,' and Zahlten, 'Disputation mit Averroes.' I am grateful to Damian Dombrowski for advice on this pictorial tradition in the Renaissance.

48. On this panel painting, see Guerrini, 'Dentro il dipinto.'

49. Polzer, 'The "Triumph of Thomas" Panel,' 49–54. Filippino Lippi's painting of the Triumph scene does not depict Averroes, at least not explicitly, but an old man in occidental clothing lying below the feet of Thomas Aquinas, with a face twisted in pain, holding a scroll saying *Sapientia vincit malitiam* ("Wisdom defeats malice"). In the foreground are heretics of Christian history: Arius, Sabellius, Appollinaris of Laodicea, Euchites, Mani, and Photius. See Nelson, 'Filippino Lippi', 415, 515 (the Triumph scene), 519 (the miracle scene), 543 and 576n138: "la figura rappresenta probabilmente un generico 'infidele.'"

50. A woodcut with Averroes below Aquinas's classroom desk is on f. 1r of Thomas Aquinas, *In decem libros Ethicorum Aristotelis profundissima commentaria,* which was printed in Venice either in 1526 (according to the Explicit) or in 1531 (according to the title page). Cf. the references to images of "Averroè sconfitto" in *Manoscritti e stampe venete,* 181, in editions of *Parva naturalia* of 1523, and *De substantia orbis* of 1525, which I was not able to identify.

51. For a different view, cf. Monfasani, 'Aristotelians,' 260: "So, if there were no institutional or educational imperatives driving the local religious authorities in Italy to correct the Averroists, a tolerant climate of opinion was theoretically possible and historically the case."

52. As has recently been shown by Martin, *Subverting Aristotle,* chs. 5–8.

53. For recent discussions of the term "Averroism," see Imbach, 'L'Averroïsme latin'; Hayoun and de Libera, *Averroès et l'averroïsme;* Niewöhner and Sturlese, *Averroismus;* Bianchi, 'Les aristotélismes'; Kuksewicz, 'Some Remarks,' 93–121; Brenet, *Transferts du sujet,* esp. 21–22; Coccia, *La trasparenza,* 20–53; Calma, *Études,* 11–20, 367–373; Hasse, 'Averroica Secta'; and Martin, 'Rethinking,' 3–19.

54. Thomas Aquinas, *De unitate intellectus* (1976), ch. 1, pp. 294–295, lin. 307–312: "Sed quia ex quibusdam verbis consequentibus Averroyste accipere volunt intentionem Aristotilis fuisse quod intellectus non sit anima que est actus corporis, aut pars talis anime, ideo etiam diligentius eius verba sequentia consideranda sunt."

55. The exceptions I am aware of occur in two manuscripts of Averroes's Long commentary on the *Metaphysics:* MS Paris BN 15453, f. 347v and f. 354r ("Explicit liber Metaphysice Aristotelis cum commento Averroiste"), and repeatedly in MS Paris BN 6504.

56. See Hankins, 'Humanism, Scholasticism,' esp. 36–39, on 'Averroism.'

57. On the occurrence and meaning of the term *Averroista,* see Hasse, 'Averroica Secta,' and Martin, 'Rethinking.' Important early occurrences of *Averroista* not mentioned in these articles are in Peter John Olivi; see Piron, 'Olivi et les averroïstes.'

58. Pietro Pomponazzi, commentary on the *Physics* of 1518, as quoted in Nardi, *Studi,* 239n: "Alterum notandum est quod multi Averroistae, de quorum numero ego sum, <dicunt> quod secundum ipsum <Averroem> deus continue conservat mundum et quod deus est causa efficiens totius mundi et finalis, ut apparet expresse ab ipso in Destructionibus."

59. As shown by Martin, 'Rethinking.'

60. Lull writes several treatises against Averroes and Averroists toward the end of his life, especially in 1309–1311. See Imbach, 'Lulle face aux Averroïstes.' A broad range of errors attributed to Averroes is discussed in Giles of Rome's *Errores philosophorum* of about 1270; many passages of this text are quoted in a Renaissance treatise of similar format, Konrad Wimpina's *Tractatus de erroribus philozophorum in fide Christiana* (1493).

61. One such historian with a misleading notion of "Averroism" is the great scholar René Antoine Gauthier, for example, in his article 'Notes sur les débuts,' which is countered with good arguments by Bazán, 'Was There Ever?' 31–53. Calma, *Études,* 368–369, argues for a definition of Averroism "plus élargie."

62. As has long been pointed out by Schmitt, 'Renaissance Averroism,' in *The Aristotelian Tradition,* art. 8, 123.

63. Van Steenberghen, *Introduction à l'étude,* 531–554 ("L'averroïsme latin"), esp. 553: "L'averroïsme latin proprement dit est né au XIVe siècle avec Jean de Jandun ou peu avant lui"; and Van Steenberghen, *Maître Siger de Brabant,* 394–395. Kristeller, 'Paduan Averroism,' 117: "I think that it will be more appropriate to speak henceforth not of 'Paduan Averroism,' but rather of Italian secular Aristotelianism."

64. As did William of Alnwick in Bologna in 1322–1323: "It is strange that some persons put so much effort on maintaining the sinful and arrogant opinion of Averroes on the unicity of the intellect" ("Et mirum est quod aliqui homines tantum laborant ad tenendum iniquam et frivolam opinionem Averrois de unitate intellectus"); see Maier, 'Wilhelm von Alnwicks Bologneser Quaestionen,' 275.

65. That Averroes never enjoyed greater influence in Europe than in the sixteenth century—given the interest in his entire oeuvre, the new translations, and enormous editions—is surmised with good reasons by Bianchi, 'Continuity and Change,' 60.

66. For a different opinion, cf. Calma, *Études,* 11–21.

67. On commentaries and editions of Averroes, see Chapter 1 and the Appendix; on Renaissance translations, see Chapter 3.

68. Hasse, 'Averroica secta,' 324–331.

69. On John of Jandun's philosophical psychology, see Brenet, *Transferts du sujet.*

70. Vernia, *Utrum anima intellectiva,* MS Venice, Biblioteca Marciana, Cod. Lat. 6, 105, ff. 156r–160v, here f. 156rb.

71. Ibid., f. 157bis va: "Ex quibus sequitur Io(hannes) Gan(davensis) male dixisse."

72. Pico della Mirandola, *Conclusiones* (1973), p. 34, art. 3: "non solum in hoc, sed ferme in omnibus quesitis philosophiae doctrinam Averrois corrupit omnino et depravavit."

73. Nifo, *In librum Destructio destructionum* (1497), dub. 7.3, f. 86va: "Paucitas enim exercitii huius hominis in libris Averrois fecit istum hominem errare"; here quoted from Mahoney, *Two Aristotelians,* art. 9, 8n. On Nifo's attitude toward John of Jandun, see also articles 8 and 12 in Mahoney's volume.

74. Nifo, *Super tres libros de anima* (1503), sig. n5r. See also sig. o1r: "haec expositio non est consona litterae."

75. Nifo, *De sensu agente,* in *In librum Destructio desctructionum* (1497), f. 124r: "Et scio quod multis fuit occasio errandi propter eius famam, intantum quod homo non putabatur Averroista nisi qui erat Gandavensis."

76. On John of Jandun's theory of intelligible species, see Mahoney, 'Themes and Problems,' 282–286; Spruit, *Species intelligibilis,* 1:326–336; Brenet, *Transferts du sujet,* 135–194.

77. Nifo, *In librum Destructio destructionum* (1497), dub. 3.18, f. 47vb. See Mahoney, *Two Aristotelians,* art. 9, 9–10.

78. Trombetta, *Tractatus,* art. 1, f. 12va: "Presupponendo ex ista opinione quod intellectus possibilis non habet intentionem intellectam novam nisi per hoc quod recipit eam a fantasmate, ut communiter tenent omnes recte sententientes de opinione Commentatoris... Et hi quidem sicut sententiam Commentatoris pervertunt, ita et philosophi, et quod pernitiosissimum est, contra fidem et veritatem pestilentissimos excitant errores." See Mahoney's comments in *Two Aristotelians,* art. 9, 22–24, and Spruit, *Species intelligibilis,* 2:74–76.

79. On this controversy, see Mahoney, 'Albert the Great,' 558–559.

80. Nifo, *Questio de speciebus intelligibilibus,* here quoted from Mahoney, 'Albert the Great,' 558–559.

81. On Zimara on intelligible species, see Spruit, *Species intelligibilis,* 2:103–110. On Zimara on the active intellect, see Mahoney, 'Albert the Great,' 557, and Mahoney, *Two Aristotelians,* art. 8. On Zimara on intellect theory, see Zimara, *Solutiones contradictionum,* in Aristotle and Averroes, *Omnia... opera* (1562), 8, contr. 14, f. 420vb: "Defensores Averrois in quaestione de unitate intellectus... devenerunt ad hoc ut dicerent animam intellectivam esse veram formam dantem verum esse substantiale homini ad intentionem Averrois..."; and f. 423va: "Et illud maximo argumento est quod omnes Latini nostri, sicut Albertus cognomento Magnus... et beatus doctor, insuper subtilissimus Scotus et Egidius Romanus... voluerunt Averrois sententiam fuisse animam intellectivam non esse formam substantialem hominis. Taceo Greogorium Ariminensem, Iohannem de Gandavo, Gaetanum et Paulum Venetum et multos praeclaros Averroistas ex viventibus, qui tenuerunt hanc fuisse Averrois sententiam."

82. On Nifo's treatise *De immortalitate,* see Boulègue, 'A propos de la thèse.'

83. Nifo, *De immortalitate* (1518), ch. 4, f. 1vb: "Postea Pomponatius... de mente Averrois animam in homine dupliciter constituit... Que quidem opinio non

Averrois sed Gandavensis est . . ."; and f. 2ra: "Inconsulte ergo Pomponatius protulit opinionem Averrois cum non sit omnium qui Averroem interpretantur, sed Gandavensis."

84. Ibid., ch. 4, f. 1vb–2ra.

85. Ibid., ch. 18, f. 4va: "Potest ergo secundum Averroem intellectus esse forma informans humanum corpus et separata; informans quidem quia illud inesse formaliter constituit; separata vero quia ab eo non dependet." Ibid., f. 4vb: "Secundo sequitur hominem formaliter intelligere secundum utramque opinionem, quoniam intelligit per intellectivam que est informans essentialiter corpus humanum, non autem per formam, que humano corpore assistit, ut putat Gandavensis; quod haec sit opinio Averrois . . . nemo est qui dubitat."

86. Prassicio, *Questio de immortalitate anime* (1521), sig. C1vb.

87. Ibid., sig. C4va.

88. Ibid., sig. D1rb.

89. The combined mentioning of Roger (Bacon) and Siger is inherited from Nifo, *De immortalitate* (1518), ch. 4, f. 4vb: "quam opinionem Suggerius et Rogerius uterque Bacconitanus ad Averrois mentem tradunt." Cf. Nifo, *De anima* (1503), f. 83: "ut Rogerius et Suggerius uterque Bacconitanus, Thomeque coetanei," here quoted from Nardi, *Sigieri,* 43–44.

90. Prassicio, *Questio de immortalitate anime* (1521), sig. B2vb–B3ra: "mirandum est de Augustino acceptante animam intellectivam esse intelligentiarum infimam et ponente aliam compositionem in illa, quod non est aliud nisi apostetare in doctrina Averrois . . . Quam positionem revera Ioannes Candavensis mirum in modum ebibit ac pergustavit. Modo que demencia quisve melancolicus humor seu spiritus Averroim arripuisset hanc publicam fatuitatem excogitare intellectum seu anima intellectivam ut intelligentiarum infimam per essentiam secundum remque separatam et incorruptibilem uniri formaliter, ut tenet Augustinus, comiscerique cum re corruptibili utpote cum sensitivo et vegetativo . . . Ex quibus verbis clare constat secundum Averroim animam nullo modo nobis formaliter et univoce uniri, ut dicit Augustinus, qui cum Suggerio et Rogerio in hac re condemnandi sunt."

91. On Thomas Wylton, see Brenet, *Les possibilités.*

92. On this concept, see Etzwiler, 'Baconthorpe and the Latin Averroism,' esp. 266–269.

93. Prassicio, *Questio de immortalitate anime* (1521), sig. B3ra–vb.

94. John of Jandun, *Super libros Aristotelis de anima* (1587, repr. 1966), lib. 3, qu. 5, col. 242: "Sed indubitanter, salva pace illorum, nunquam illud dixit Commentator nec fuit eius intentio, sicut patet scientibus commenta eius." In col. 245, John refers to Siger of Brabant: "Et debes scire quod istam solutionem huius rationis qualiter homo intelligit quantum ad aliquid posuit reverendus doctor philosophiae magister Remigius (?) de Brabantia in quodam suo Tractatu de intellectu, qui sic incipit: 'Cum anima sit aliorum cognoscitiva.'" For context, see Mahoney, 'Themes and Problems,' 275.

95. Calma, *Études,* 16–19.

96. Etzwiler, 'Baconthorpe and the Latin Averroism,' 258: "Sequitur de secundo articulo, an scilicet Commentator voluit salvare quod intellectus esset forma informans hominem . . . Ubi dicit singulariter unus doctor, Wiltonensis, quod sic."

97. Cited by Nardi from the *Quaestio de remanentia elementorum in misto* (sic), see Nardi, *Studi,* 18–19: "Ideo teneo quod opinio Averrois devitat maiores difficultates quam alia; nec me pudet amore veritatis me ipsum retractare. Unde qui dicunt me aliis adversari ut contradicam, mentiuntur. Oportet enim in philosophia haereticum esse, qui veritatem invenire cupit."

98. Renan, *Averroès et l'Averroïsme,* ch. 3,1, p. 322: "L'Université de Padoue mérite une place dans l'histoire de la philosophie, moins comme ayant inauguré une doctrine originale que comme ayant continué plus longtemps qu'aucune autre école les habitudes du moyen âge. La philosophie de Padoue, en effet, n'est autre chose que la scolastique se survivant à elle-même et prolongeant sur un point isolé sa lente décrépitude."

99. P. O. Kristeller and F. E. Cranz tend to marginalize the philosophical liveliness of Renaissance Averroism; cf. Cranz, 'Editions of the Latin Aristotle,' 120: "Again it is to be noted that Averroes appears not so much as the exponent of a specific philosophic position but rather as part of the general learning of the times." Cf. Kristeller, 'Paduan Averroism,' 113–115.

100. As also pointed out by Wisnovsky, *Avicenna's Metaphysics,* 94–96. Cf. Shields, 'Some Recent Approaches,' 158–172, for a survey of recent interpretations of Aristotle's hylemorphism.

101. Averroes, *Talḫīṣ mā baʿd aṭ-ṭabīʿa* (1958), 154.

102. For summaries of the eight most important psychological works by Averroes, see the exemplary study by Wirmer, in Averroes, *Über den Intellekt* (2008), 329–364. Further important studies on Averroes's intellect theory are Davidson, *Alfarabi, Avicenna, and Averroes,* ch. 6–8; Geoffroy's introduction to Averroes, *La beatitude de l'âme;* Taylor, 'Intelligibles in Act in Averroes,' and Taylor's introduction to Averroes, *Long Commentary;* and Black, 'Models of the Mind.'

103. Averroes, *Commentarium magnum De anima* (1953), 3.5, p. 391, lin. 127–140, and p. 393, lin. 176–195 (on Themistius), p. 395, esp. lin. 228–235 (on Alexander). For an English translation of the commentary on 3.5, see Averroes, *Long Commentary,* 303–329; for a French translation, see de Libera, *Averroès: L'intelligence,* 57–81; and for a German translation, see Averroes, *Über den Intellekt,* 158–231. On the Arabic fragments of the commentary, see Sirat and Geoffroy, *L'Original arabe.*

104. Averroes, *Commentarium magnum De anima* (1953), 3.5, p. 387, lin. 23, and p. 388, lin. 56.

105. Ibid., p. 404, lin. 500, and p. 405, lin. 520.

106. Ibid., p. 400, lin. 379–393, and p. 401, lin. 419–423, and p. 412, lin. 724–728.

107. Ibid., p. 393, lin. 176–181.

108. Ibid., p. 390, lin. 86–104; p. 406, lin. 566–569.

109. Hasse, 'The Attraction of Averroism,' and 'Aufstieg und Niedergang des Averroismus.'

110. Pomponazzi, *Questio de immortalitate anime* (1955), ed. Kristeller, 93. See this chapter, section 'Pietro Pomponazzi.'

111. Averroes, *Commentarium magnum De anima* (1953), p. 411, lin. 713, and p. 412, lin. 723. Cf. Achillini, *Quolibeta* (1551), III, dub. 2, 3, f. 10rb ("tunc non posset discipulus addiscere a magistro"); Vimercato, *De anima rationali* (1574), 47a.

112. Averroes, *Commentarium magnum De anima* (1953), 3.5, p. 400, lin. 441–446, and p. 403, lin. 431–440. Cf. Paul of Venice, *Summa philosophie naturalis* (1503/1974), f. 88vb ("non esset intellectus reflexus supra se nec universalium cognitivus"); Achillini, *Quolibeta* (1551), 3, dub. 2, 3, f. 10ra ("non esset cognoscitivus universalium"); Prassicio, *Questio* (1521), sig. D2va: "ubi ait *(sc., Aristoteles):* utrum autem contingit aliquid separatorum intelligere ipsum existentem non separatum a magnitudine ... Quae verba solus Averroes summus Aristotelis interpres recte interpretatus est."

113. Thomas Aquinas, *De spiritualibus creaturis,* art. 9, ad 6, p. 403: "Ad sextum dicendum quod in hac ratione praecipuam vim videtur Averroes constituere: quia videlicet sequeretur, ut ipse dicit, si intellectus possibilis non esset unus in omnibus hominibus, quod res intellecta individuaretur et numeraretur per individuationem et numerationem singularium hominum, et sic esset intellecta in potentia et non in actu."

114. On the unicity thesis in the Renaissance see Renan, *Averroès et l'Averroïsme,* ch. 3; Nardi, *Saggi;* Di Napoli, *L'immortalità;* Schmitt, 'Renaissance Averroism,' in *The Aristotelian Tradition,* art. 8, 121–142; Schmitt, *Aristotle and the Renaissance,* ch. 1 and 2; Kessler, 'The Intellective Soul'; Hasse, 'Arabic Philosophy and Averroism,' 113–121.

115. Mahoney, 'Nicoletto Vernia,' in *Two Aristotelians,* art. 3, 144–163, and 'Philosophy and Science,' art. 1, 201–202; Kessler, 'Nicoletto Vernia,' esp. 281; Lohr, 'Metaphysics and Natural Philosophy,' 280–287.

116. Cf. Schmitt, *Aristotle and the Renaissance,* 37: "Consequently, the establishment in 1497 of the famous Paduan chair to teach Aristotle from the Greek text was perhaps more the product of wishful thinking than of taking advantage of a genuine pedagogical opportunity." For warnings not to overrate the influence of Greek around 1500 see Wilson, *From Byzantium,* 161–162, and Nutton, 'Hellenism Postponed.'

117. On Paul of Venice, see Nardi, *Saggi,* 75–93; Bottin, 'Paolo Veneto'; Kessler, 'The Intellective Soul,' 488–491; Lohr, *Latin Aristotle Commentaries,* 1.2:52–56.

118. Paul of Venice, *Summa philosophie naturalis* (1503/1974), ch. 5.37, f. 88va: "Secundo notandum quod nulla istarum fuit opinio Aristotelis, scilicet quod unicus est intellectus in omnibus hominibus iuxta impositionem Commentatoris tertio De anima"; and f. 88vb: "Quarta conclusio: Intellectus non numeratur numeratione individuorum, sed est unicus omnibus hominibus ...

ergo oportet dare unicum intellectum in omnibus hominibus secundum opinionem et intentionem Aristotelis." For a different, non-Averroist reading of the *Summa,* see SEP, s.v. 'Paul of Venice' (by Conti).

119. Paul of Venice, *Summa philosophie naturalis* (1503/1974), ch. 5.37, f. 88va–89ra.

120. Ibid., f. 88vb: "Si enim *(sc., intellectus humanus)* esset materialis, non esset intellectus reflexus supra se nec universalium cognitivus."

121. Ibid., 5.36, f. 88ra, quarta conclusio; cf. also f. 89ra, ad tertium.

122. See Kuksewicz, 'Paul de Venise,' 301.

123. Paul of Venice, *Commentum de anima* (1481), ch. 3.27, f. Z8ra (quoted after Kuksewicz, 'Paul de Venise,' 302): "Dicendum ergo secundum opinionem fidei quod intellectus plurificatur ad plurificationem individuorum speciei humanae."

124. Angelo d'Arezzo, *Scriptum super libro Profirii de quinque praedicabilibus,* quoted from Grabmann, 'Der Bologneser Averroist,' 267: "Propter quod est sciendum quod secundum intentionem Commentatoris et Aristotelis intellectus est unus numero in omnibus hominibus licet hoc sit contra fidem."

125. Gaetano disagrees with Averroes's unicity thesis in his *De anima* commentary. In his *Quaestio de perpetuitate intellectus,* he defends the idea that the human soul with its active and potential intellects is immortal, a position not shared by Averroes either; see Di Liscia and Ebbersmeyer, 'Gaetano,' esp. 170, 177.

126. Di Napoli, *L'immortalità,* 99. On Gaetano, see Kessler, 'The Intellective Soul,' 490–492; Lohr, *Latin Aristotle Commentaries,* 1.1:94–95.

127. Gaetano da Thiene, *Super libros de anima* (1505), 3.36, f. 73va: "Ad argumenta facta contra alias vias ex fundamentis suis facile potest responderi, quod brevitatis causa relinquo, et hic pauca congregare volui de isto quaesito difficili ad iuniorum introductionem."

128. Cf. Tignosi, *In libros Aristotelis de anima commentarii* (1551), 405: "A magistro Paulo doctore meo hic movetur dubium." On Tignosi's early education and biography in general, see Sensi, 'Niccolò Tignosi,' esp. 362–369; cf. also Thorndike, *Science and Thought,* 161–179, 308–365; Lohr, *Latin Aristotle Commentaries,* 1.2:44–45.

129. Tignosi, *In libros Aristotelis de anima commentarii* (1551), 351: "expositione huius praeclarissimi Commentatoris."

130. Ibid., 350.

131. This was shown by Mahoney, 'Albert the Great,' 543–563.

132. Sensi, 'Niccolò Tignosi,' 406.

133. Tignosi, *In libros Aristotelis de anima commentarii* (1551), 409: "Dicendum ergo secundum opi[o]nionem fidei, quod intellectus plurificatur ad multiplicationem individuorum specie<i> humanae." This position is borrowed from the late commentary of *De anima* by Paul of Venice, *Commentum de anima* (1481), ch. 3.27, sig. Z8ra.

134. Tignosi, *In libros Aristotelis de anima commentarii* (1551), 409: Quae omnia *(i.e. the arguments against Averroes)* maximam reciperent instantiam et contrarium

forte sustentaretur non difficulter, sed fides christianorum alium morem postulat, qui secundum conscientiam verissimus est. Nam plurificatio animae intellectivae apud loquentes naturaliter solum cum magna difficultate tenetur et caetera." This again is borrowed from Paul's late commentary on *De anima;* see Kuksewicz, 'Paul de Venise,' 302.

135. Tignosi, *In libros Aristotelis de anima commentarii* (1551), 350: "Averroes subtilissimus commentator dixit intellectum humanum nequaquam uniri corpori ut forma, sed per phantasmata intellecta in actu."

136. Ibid., 352: "Beatus Thomas in suo tractatu quem fecit contra Averroem probat Aristotelem voluisse intellectum esse vere formam substantialem hominis"; Tignosi concludes this passage as follows: "Haec sunt quae vera colliguntur, et habebitur Aristotelis sententia vera" (353).

137. Ibid., 352: "Si quis tamen mere naturaliter loqueretur, istis obiectis facile responderet sustentans opinionem Averrois et Ioannis, sed pro nunc ista sufficiant."

138. Ibid., 338: "Ad quod, ut bene notavit Gaietanus, respondet Commentator praemittendo quod intellecta speculativa constituuntur per duo subiecta, quorum unum est generatum et aliud non, et id declarat per similitudinem . . . Ubi videtur quod per intellectum speculativum intelligit intentiones causatas in intellectu materiali a forma imaginata, tamen lumine intellectus agentis concurrente. Patet igitur quomodo intellectio est contingens, quamvis intellectus sint aeterni."

139. On Vernia, see Di Napoli, *L'immortalità,* 181–193; Vasoli, *Studi,* 241–256; Kessler, 'The Intellective Soul,' 492–494; Monfasani, 'Aristotelians,' 250–251; Mahoney, *Two Aristotelians;* Rout. Enc. Philos. s.v. 'Vernia, Nicoletto' (by Mahoney); De Bellis, *Nicoletto Vernia;* Lohr, *Latin Aristotle Commentaries,* 1.2:47–48.

140. Further evidence of his views on the unicity thesis can be drawn from Vernia's annotations on a copy of John of Jandun's *De anima,* printed in 1473 and now at the Biblioteca Universitaria at Padua. As Mahoney has shown, the annotations suggest that Vernia understood Averroes and Aristotle to hold that there is but one substance that is both possible and active intellect (Mahoney, 'Nicoletto Vernia's Annotations,' in *Two Aristotelians,* art. 4, 591)— which is reminiscent of Vernia's idea, set forth in the *Questio utrum anima intellectiva,* that possible and active intellect are eternally united.

141. See the description of the manuscript (MS Venice, Biblioteca Marciana, Cod. Lat. VI, 105, ff. 156r–160v) in Pagallo, 'Sull'autore (Nicoletto Vernia?)'; Mahoney, 'Nicoletto Vernia,' in *Two Aristotelians,* art. 3, 145–149; and Pagallo, 'L'animus averroisticus,' 142–146.

142. Gios, *L'attività pastorale,* 294n8; Monfasani, 'Aristotelians,' 250–251n21. Monfasani also notes that Vernia, when he was giving up teaching in 1498, was in need of a retirement income and that the preface to Cardinal Grimani and the anonymous poem that accompany *Contra perversam* "are blatant appeals for a benefice."

143. See this chapter, section 'The Public Image.'

144. On the 1513 bull breaking with a long tradition of the noninterference of the ecclesiastical authorities with the Italian universities, see Bianchi, *Pour une histoire,* 128. Barozzi clearly paved the way for this "coupure nette."

145. As has been noted before by Monfasani, 'Aristotelians,' 251n21: "Vernia's declaration of his belief toward the end of the treatise is formulistic and reads as if it were prepared to satisfy an inquisitor."

146. Vernia, *Contra perversam* (1505), f. 9vb–10ra: "Dico secundum sacrosanctam Romanam ecclesiam et veritatem quod intellectiva anima est forma substantialis corporis humani, dans ei esse formaliter et intrinsece a sublimi deo creata et in humano corpore infusa, multiplicataque in ipsis secundum multitudinem eorum . . . et non tantum credo haec omnia dicta ex fide, sed physice dico omnia possunt probari et etiam cum Aristotele potest sic dici."

147. Averroes, *Commentarium magnum De anima* (1953), 3.5, p. 397.

148. Ragnisco, 'Documenti inediti,' 281: "dum vidissem et graecos et arabes doctissimos."

149. Mahoney, 'Philosophy and Science,' in *Two Aristotelians,* art. 1, 201: "One remarkable thing that emerged from the analysis of the writings of Vernia was his shift from following Averroes as the true guide to Aristotle to accepting the Greek Commentators . . . as the more accurate interpreters of the Philosopher"; and 202: "That Vernia and Nifo both knew and cited Simplicius . . . marks a major turning-point in late medieval and Renaissance Aristotelianism." See also art. 3, 162, and Kessler, 'Nicoletto Vernia,' esp. 281. But cf. Monfasani, 'Aristotelians,' 250n21, who points out that Vernia was reading Alexander and Simplicius "in a rather wilful manner" in order to use them for his purpose of self-justification.

150. Vernia, *Contra perversam* (1505), f. 9ra.

151. Thomas Aquinas, *De unitate intellectus* (1976), ch. 2, 302: "etiam Graeci et Arabes hoc senserunt quod intellectus sit pars vel potentia seu virtus animae quae est corporis formae."

152. Vernia, *Contra perversam* (1505), f. 11rb: "sic intellectus prout est aliquid naturae animae est individuus, prout autem emittit actiones intelligendi est in virtute universali et sic recipit universalia quae non sunt in eo sicut accidens in subiecto neque sicut forma in materia et ideo per ipsum non individua<n>tur." I read 'individuantur' with Albertus, *De intellectu* (1890), 1.7, p. 488b.

153. Vernia, *Contra perversam* (1505), f. 11va: "Maxime cum ad modum golie Averroym Christi nominis inimicum gladium quo iuguletur ad certamen deferentem penitus extinxit." The theologians who approved of the orthodoxy of Vernia's treatise were Antonio Trombetta, OFM, Maurice O'Fihely, and Vincenzo Merlino, OP; see Monfasani, 'Aristotelians,' 251n21, 264–265.

154. Albertus, *De intellectu* (1890), 1.7, p. 488.

155. Albertus Magnus, *De anima* (1968), 3.2.7, p. 186: "Propter quod etiam formalis intellectus, qui est forma speculata, non habet transmutationem ex intellectu possibili, sed ex phantasmate, in quo est, sicut diximus. Et sic

satisfacit Theophrasto quaerenti qualiter esse possit quod intellectus possibilis sit separatus et intransmutabilis et similiter agens, et tamen speculativus sit transmutabilis et temporalis, sicut id quod exit de potentia ad actum. Et in veritate in ista solutione bene satisfacit et verum dicit Averroes."

156. Ibid., 3.2.12, p. 194: "Et in hac sententia convenit nobiscum Averroes in commento de anima."

157. Ibid., 3.2.12, p. 194: "Et sic soluta est prima quaestio de tribus inductis quaestionibus, scilicet quomodo potest esse quod agens est substantia stans et similiter possibilis, et tamen speculativus sit factus et mutabilis."

158. As has been noted before, see de Libera, *Métaphysique et noétique*, 384.

159. A brief account of Albertus's complex attitude toward Averroes is in Hayoun and de Libera, *Averroès et l'averroïsme*, 86–90.

160. Nifo, *In tres libros Aristotelis de anima* (1559), 3 t. c. 5, col. 645–646: "Sapit enim haec positio aliquo modo unitatem animae intellectivae posse esse diversorum corporum, propterea mihi non placet." Quoted from Kessler, 'Nicoletto Vernia,' 286n35.

161. Cf. Garin, *La cultura filosofica*, 293: "Esponente caratteristico dell'aristotelismo averroizzante, e noto per la sua spregiudicatezza e miscredenza, Nicoletto Vernia è figura di insegnante ancora da illustrare nel suo peso effettivo"; Nardi, *Saggi*, 95–114 (chapter: "La miscredenza e il carattere morale di Nicoletto Vernia").

162. For an introduction to his life and works, see Matsen, *Alessandro Achillini*, 21–41. Cf. the older account in Thorndike, *History*, 5:37–49. See also Lohr, *Latin Aristotle Commentaries*, 2:5–6.

163. Vimercato, *De anima rationali* (1574), p. 36b: "nonnullis unicum in hominibus omnibus intellectum, ut Alexander Achillinus, et multo antea, iuxta ipsius Achillini et aliorum quorundam sententiam, commentator Averroys ... ponentibus." See the references to Pomponazzi and Zimara, in Nardi, *Saggi*, 231–233. See also Renan, *Averroès et l'Averroïsme*, 360 ("le champion de l'averroisme"), and Nardi, *Saggi*, 179–223, 226–279, esp. 180, 274–279, and in DBI, vol. 1, s.v. "Achillini": "L'Achillini nella storia dell'averroismo rappresenta, insomma, la corrente che è stata detta *sigieriana*."

164. Achillini therefore did not adopt the position of Siger of Brabant by maintaining both the unicity of the intellect and its being the substantial form of the human being, as argued by Nardi, *Saggi*, 182, 204, followed by Kessler, 'The Intellective Soul,' 495. Cf. more correctly Kristeller, 'Paduan Averroism,' 114: "Achillini did not accept the unity of the intellect."

165. Achillini, *Quolibeta* (1555) 3, dub. 2, f. 10ra: "Utrum intellectum possibilem habeat omnis homo. Respondeo per duo dicta: Primum opinio philosophi est quod sic. Secundum illa opinio non est vera." This passage has been misunderstood by Garin, who read the question as meaning "whether every human being has its own possible intellect" rather than "the same possible intellect" (cf. Garin, *Storia*, 504). The line of argumentation is the following:

after having stated the question and the double conclusion, Achillini proceeds to demonstrate the first conclusion, which is that Aristotle affirms the thesis that every human being has a possible intellect; this Achillini does by pointing out four *viae* that argue for the unicity of the intellect ("Si intellectus possibilis esset multiplicatus . . . etc.").

166. Achillini, *Quolibeta* (1551), 3, dub. 3 fin., f. 12vb: "Sed quia in hoc quaesito Commentator ad suum falsum fundamentum de unitate intellectus consequenter loquitur, posito enim intellectu uno non multiplicato, ponit accidentia non inhaerere illi. Nos autem oppositum illius fundamenti tenemus. Ideo non oportet nos concordare in conclusione sequente ex illo"; and 3, dub. 4 fin., f. 14vb: "Hoc autem quolibetum tertium hoc dicto claudimus quod non errat Commentator in hoc quaesito, an intellectus possibilis det esse homini, sed in alio huic circumstanti errat, ut prius patuit etc."

167. Ibid., 3, dub. 2, f. 10ra–11rb.

168. Paul of Venice, *Summa philosophie naturalis* (1503/1974), f. 88vb: "non esset intellectus reflexus supra se nec universalium cognitivus."

169. As noted by Matsen, *Alessandro Achillini,* 27.

170. Achillini, *De elementis* (1551), 2.5, f. 123ra: "Sed dices: quomodo stat opinio Aristotelis cum fide? Quia secundum rationem naturalem aut intellectus est unus, ut intellexit Averrois Aristotelem, aut intellectus est plurificatus incipiens esse, ut intellexit Aphrodisieus Alexander. Istarum autem nulla est conformis fidei. Responsio: haec est circumstantia propter quam dixi oportere relinquere Philosophum et inter duas illas opiniones falsas probabiliorem eligendo: illa est opinio Averrois." Garin has drawn attention to this passage in his *Storia,* 503–504. For a summary of the contents of *De elementis,* see Nardi, *Saggi,* 240–248.

171. On Nifo, see Di Napoli, *L'immortalità,* 203–214; Zambelli, 'I problemi'; Lohr, *Latin Aristotle Commentaries,* 2:282–287; Perfetti, *Aristotle's Zoology,* 85–120; Mahoney, *Two Aristotelians;* Rout. Enc. Philos. s.v. 'Nifo, Agostino' (by Mahoney), with further literature; Kessler, *Die Philosophie,* 157–171; De Bellis, *Nicoletto Vernia.*

172. On this work, see Zambelli, 'I problemi,' esp. 133–140; Mahoney, *Two Aristotelians,* art. 1, 179–200; and Kuhn, 'Die Verwandlung.'

173. Mahoney, 'Philosophy and Science,' in *Two Aristotelians,* art. 1, 173–174; Cranz, 'Editions of the Latin Aristotle,' 120. The title page of the 1497 edition of Nifo's commentary on Averroes's *Destructio destructionum* also announces the 1495–1496 edition of the works of Aristotle and Averroes by the same publisher, Octaviano Scoto in Venice: *Destructiones destructionum . . . , Eiusdem Augustini questio de sensu agente, Omnia Aristotelis Opera tam in logica quam in philosophia naturali et morali et metaphysica cum sui fidelissimi interpretis Averroys Cordubensis commentariis* (1497) (GW 2340 and 3106).

174. Nifo, *In Averroys De animae beatitudine* (1508). See the critical edition of the Latin version of Averroes's text by Marc Geoffroy and Carlos Steel and the introductory chapter on Nifo, in Averroes, *La bèatitude de l'âme* (2001), 94–111.

175. Super-commentaries on Averroes appear in Nifo's commentaries on *De anima*, printed in 1503; *Metaphysics* 12, printed in 1505; *Physics*, printed in 1508; *De caelo*, printed in 1517, 1553, and 1567; and complete *Metaphysics*, printed in 1547 and 1559.

176. On Nifo's not knowning Arabic, see the six passages cited in Steel's introduction to Averroes, *La béatitude*, 96: "ego non legi eum in lingua sua," "cum non biberim haec in fonte proprio, nescio si textus est superfluus," and so on.

177. Mahoney, *Two Aristotelians*, art. 1, 181: "Aristoteles transpositus," "sacerdos Aristotelis," "elegantissimus philosophus."

178. Nifo, *Aristotelis physicarum acroasum . . . liber* (1508), f. 218va: "Propterea mihi videtur Averroes magnus temulentus. Hi<c> enim cum litteras grecas ignoraverit non nisi temere grecum auctorem exponere potest." This passage was noted by Mahoney, *Two Aristotelians*, art. 5, 89.

179. Nifo, *De intellectu* (1503), dedication: "Eamque *(sc., questionem meam de intellectu)*, ne labores iuventutis mee perditum irent, imprimendam esse curavissem, nisi emuli affuissent, qui me hereseos accusassent. At malui ad hoc temporis pervenire morando, quam huiuscemodi criminis culpam subire. Iam cessant accusationes. Emulorum iniquitas, sic mea fide postulante, in propatulo est. Ergo suo tribuant commodo si quam utilitatem accepere qui me insidiis prosequuti sunt, discantque inter ea *(1527: interea)* diligentius legere que volunt criminari, ut cautius egisse videantur. Sed valeant isti! Satisque mihi sit Petrum Barotium episcopum Patavinum, christianorum nostre etatis decus et splendorem, te *(1527: et)* cui non minus in fide quam in phylosophia tribuo, et quamplurimos alios tum theologos tum philosophos iudices ac defensores habuisse, qui semper innocentie mee testes eritis. Tractaveram hanc nobilissimam materiam et de fontibus omnium antiquorum phylosophorum exhaustam recenti stilo quod omnes fere commendare visi sunt, preter admodum paucos quorum precipuus fuit Hieronymus Malelavellus . . . Hic hortatus est me ut universum opus in capitula secarem asserens antiqua stilo esse antiquo tractanda. Hac unica huiusce viri ratione persuasus . . . pristinam mutavi sententiam. Placuit quedam tollere, mutare alia, addere plurima. Nihil delevi quod sit contra fidem catholicam; non enim potest destrui quod factum non invenitur."

180. The dedication to Sebastiano Badoer is not mentioned in Mahoney, *Two Aristotelians*, or in Rout. Enc. Phil. s.v. 'Nifo, Agostino' (by Mahoney), or in Spruit, *Agostino Nifo*, or in Kuhn, 'Die Verwandlung,' 291–308, or in Hasse, 'Aufstieg und Niedergang des Averroismus,' 455–461. It was cited almost in full, though with several misprints, by Nardi, *Saggi*, 102n18, and discussed by Di Napoli, *L'immortalità*, 204–205, De Bellis, *Nicoletto Vernia*, 71, and Steel, introduction to Averroes, *La béatitude de l'âme*, 96–97. It is unfortunate that the recent edition of *De intellectu* by Spruit is a reprint of the 1554 edition, which does not contain the dedication.

181. Nifo, *In librum Destructio desctructionum* (1497), dub. 1.23, f. 23rb: "Et debes scire quod haec est opinio Aristotelis et Averroys et est error purus secundum

nos christianos. Aliter est dicendum, ut declaravi in libro de intellectu ubi contradixi his philosophis fortissimis rationibus."

182. See the passages quoted in Nardi, *Sigieri*, 15–17. This is how Nifo describes Siger's standpoint as a middle course between the *Latini* (i.e., Thomas Aquinas and others) and the *Averroici* in *De intellectu* (1554), lib. 3, ch. 16, f. 30rb: "Ecce quomodo mediat inter latinos et Averroicos; ab Averroicis enim accepit intellectus impartibilitatem, immaterialitatem et unitatem, a latinis autem quod sit forma constituens hominem."

183. Nifo, *Super tres libros de anima* (1503), 1.12, sig. b7rb–va: "Debes scire secundum istum philosophum *(scil. Averroem)* et Aristotelem quod animae species sunt multae, attamen principalis est anima intellectiva et haec est una numero in omnibus hominibus. Imaginatur enim hic homo quod quemadmodum una numero int<elligenti>a est in orbe et in qualibet parte orbis, sic una numero anima est in tota spera hominum et in qualibet parte illius, id est, in quolibet individuo illius. Unde sicut orbis lunae est una spera cuius una forma est int<elligenti>a lunae, sic hominum nexus est una spera cuius forma est intellectiva anima."

184. Ibid., sig. b8ra: "Sed debes scire quod tota haec opinio Averrois est falsa secundum nos christicolas et contra ipsam arguemus in tertia demonstratione, deo dante. Pro nunc eam induxi in declaratione eius."

185. Ibid., 3.5, sig. o2vb: "Et debes scire quod omnia quae hic dixi tantum ut expositor scripsi. Ideo post hanc expositionem feci librum separatum, qui de intellectu inscribitur, ubi omnia retracto quae hic Averroys dicit, et ostendo qualiter ea quae ipse dicit nec philosophiam nec quicquam veritatis sapiunt. Tu autem lege eum."

186. Nifo, *In tres libros Aristotelis De anima* (1559), sig. **6v (final paragraph of the *praefatio*): "Quantum igitur inique Alex. Calcidonius collectanea nostra publicaverit, quantum venenose, ex hisce patet. Ego enim publicare illa non destinaveram, nisi nono pressis in anno." For diverging interpretations of this passage, see Nardi, *Saggi*, 286n13, and Mahoney, *Two Aristotelians*, art. 7, 458.

187. Nifo, *In tres libros Aristotelis De anima* (1559), sig. **6v: "Quam quidem licet olim per collectanea nostra aliqua ex parte exposuimus, non tamen mature nec semper secundum verum Aristotelis sensum. Secuti enim sumus in omnibus fere expositiones Averrois quem non arbitror in omnibus verba Aristotelis recte fuisse interpretatum."

188. Nifo, *De intellectu* (1503), 3.27, f. 36r: "Sed nescio profecto quid est quod movit Averoym, Themistium et alios istum errorem pessimum asseverare. Aristotelis enim auctoritas non, argumentorum firmitas minime, famositas opinantium nec, experientia nequaquam."

189. Ibid.: "Taceo autem alias quam plures et singulares, quas ille summus theologus Petrus Barrocius Patavinae urbis dignissimus antistes inquit in libro suo quem *De ratione bene moriendi* inscripsit, quoniam superfluum puto." Barozzi's work *De ratione bene moriendi* (together with another consolatory treatise dated 1481) was printed posthumously in Venice, 1531, and s.l., 1581

(under the title *De modo bene moriendi* on the title page). On this treatise, see Gios, *L'attività pastorale,* 77–79.

190. Nifo, *De intellectu* (1503), 2.21, f. 26v: "Et ne prolixus sim, videtur quod expositor noster in libro *De unitate intellectus contra Averoym* quam pulchre ponderans omnia verba Aristotelis quae dicit in secundo *De anima,* ostendit positionem Averoys ex toto verbis Aristotelis ibi contradicere."

191. A most helpful summary of Nifo's sometimes convoluted argument is given by Spruit, *Agostino Nifo,* 35–107.

192. Nifo, *De intellectu* (1503), 4.10, f. 40r: "Multi in positione Perypateticorum ac Averoys persistentes et ego diu credidimus intellectum potentiae esse intellectum separatum."

193. Ibid., 2.20, f. 25r: "Quemadmodum et nos diu sentire visi sumus, credentes opinionem Averoys falsam esse veram, rationibus naturalibus non convincibilem."

194. Letter by Barozzi of February 23, 1504, quoted from Grendler, *Universities,* 288: "La lectura di theologia secondo la via de Schoto, la quale è come una medicina de li errori *de eternitate mundi, de unitate intellectus, et de hoc quod de nihilo nihil fiat* et altri simili, i quali pullulano da li philosophi: senza la quale el se poteria dire che in quel Studio non se lezesse cossa la quale non se lega anche in Studio de' pagani."

195. See his conclusions to chapters 3 and 4: Nifo, *De intellectu* (1554), 3.31, f. 34va: "rationales animae sint tot numero quot homines," and 4.24, f. 41vb: "Ex his ergo intellectum agentem et intellectum potentiae esse virtutes et facultates animae rationalis ... palam," and V.45, f. 52vb: "Et ideo dicimus aliud esse intellectum, aliud quo intellectum intelligitur. Intellectum enim est quidditas quam mens removet a materia ac a conditionibus materiae; quo intelligimus est species intelligibilis, quae est accidens realiter inhaerens animae mediatae, qua anima transit in intelligibile et fit id." On Nifo's theory of intelligible species, see Mahoney, *Two Aristotelians,* art. 9, 4–17; for the fifteenth-century discussion of the topic, see Spruit, *Species intelligibilis,* 1:385–395. Not correct is HWPh s.v. 'Vernunft; Verstand' (by Leinkauf ), col. 798: "A. Nifo hingegen verteidigt in seinen *De intellectu libri sex* die Position des Averroes in der Deutung des Siger von Brabant: 1) der *intellectus possibilis* ist *einer* für die ganze menschliche Art."

196. Nifo, *De intellectu* (1554), 2.23, 3.32.

197. Ibid., 2.1, f. 15va: "Quoniam eorum qui locuti sunt in quaestionibus rationalis animae nullus in tantum laboravit sicut Averroes, ideo inter reliquos cum eo pugnare mihi honos erit, ubi volo ut sciatis quod omnia quae possunt induci in sustentatione unius viri ego inducam, nec praeterire volo verbum ... quare si ultimum posse eius debile ac frivolum monstrabo esse, tunc quilibet debet acquiescere in credendo opinionem hanc esse falsam et contra rationem."

198. Ibid., 2.15, f. 20va.

199. Ibid., 3.25, f. 32va: "Inducuntur quaestiones adversus Averroim, quas ego saepe saepius consuevi apponere, ut igitur videatur eius positio tantum

quantum potest, primo inducam quaestiones et ex solutionibus earum declarabitur plus Averrois positio. Postmodum inducemus alibi demonstrationes firmas contra eum. Erit ergo prima ambiguitas: . . . si igitur rationalis anima est una numero omnium, ergo erit hoc individuum . . . Altera quaestio est Averrois quoniam eo dato sequitur quod tu intelligeres per intelligere meum . . . Tertia quaestio: . . . ergo erit et unus numero homo tantum."

200. Ibid., 3.26, f. 33ra–b: "Ideo Averroes diceret quod intellectio una numero est in omnibus uno modo et plures altero modo. Est quidem una quoniam id quo homo intelligit scilicet intellectu est unum, est vero plures per accidens, ratione intentionum imaginatarum cum quibus res intellecta coincidit, verbi gratia intellectum lapidis in me et in te est unum in se, sed quoniam diversa individua cogitata ad diversos homines continuantur, pro tanto diversatur, et ideo intelligere meum non est tuum et stat te intelligere et me non intelligere."

201. For Albertus Magnus's influence on *De intellectu,* see Mahoney, *Two Aristotelians,* art. 11, 202–211, esp. 206.

202. See Albertus Magnus, *De unitate intellectus,* p. 22, l. 10–11: "Et ideo dicitur *(i.e., anima)* a quibusdam esse in horizonte aeternitatis et temporis." The phrase also appears in other works by Albertus.

203. Nifo, *De intellectu* (1503), 2.22, f. 27va. Cf. Vernia, *Contra perversam* (1505), f. 11ra: "Est ergo virtutis univeralis virtus animae quae imago dei est creata in orizonte aeternitatis et temporis." Vernia attributes the saying to Isaac, Nifo to Plato.

204. Nifo, *De intellectu* (1503), 2.22, f. 27vb: "Erit ergo individuus prout forma est quaedam hominis, ut vero potestas quaedam lucis spiritualis porro universalis est."

205. See also Perfetti, *Aristotle's Zoology,* 109, on Nifo's continued use of Averroes in spite of his "ideological distancing from Averroes."

206. Thomas Aquinas, *De unitate intellectus* (1976), ch. 3, 306, lin. 336–337: "secundum istorum positionem destruuntur moralis philosophiae principia."

207. Nifo, *De intellectu* (1503), 3.28, f. 36va–37ra.

208. Ibid., f. 37ra: "Quo fit ut sermo meus continuus quem auditoribus meis continue predico verificetur, scilicet quod si Averoys positio esset ut dicit, proculdubio nullum animal infelicius esset me ac quovis homine, cum qua sententia Ficinum nostrum inveni nobiscum una concordem."

209. On Ficino's influence on Nifo, see Mahoney, *Two Aristotelians,* art. 2. On the moral arguments, cf. Zambelli, 'I problemi,' 132–133.

210. Mahoney, *Two Aristotelians,* art. 2, 520–523; Spruit, *Agostino Nifo,* 373–377.

211. Nifo, *De intellectu* (1554), 3.29, f. 34ra–b: "est igitur una propositio confessa omnibus quod motor unus numero in uno tempore non utitur nisi uno tantum moto ac instrumento sibi sufficienter adaequato . . . rursum nullus motor unus numero exercet opera diversa numero eiusdem speciei pro uno et eodem tempore, sive uno aut pluribus instrumentis."

212. Albertus Magnus, *De anima* (1968), ch. 3.2.7, p. 187, l. 85–87; Thomas Aquinas, *De unitate intellectus* (1976), ch. 4, p. 308, l. 57–58, and p. 308, l. 103–107. This has been pointed out by Mahoney, *Two Aristotelians,* art. 11, 208.

213. Nifo, *Aristotelis physicarum acroasum . . . liber* (1508), 4, f. 100vb (?): "Ego vero in pueritia positionem Averrois defendere non desinebam, asserens illam esse indubie mentem Aristotelis. Nunc vero cum graeca verba Aristotelis legerim et diligenter examinaverim, potius assero positionem huius esse deliramentum et nullatenus ad propositum" (quoted after Mahoney, 'Philosophy and Science,' in *Two Aristotelians,* art. 1, 201n175; I was not able to trace the quotation on f. 100vb).

214. Nifo, *De beatitudine animae* (1508), 1.16, f. 5va: "Ego plures rationes scripsi contra hoc *(i.e., John of Jandun's interpretation)* in libro de intellectu, licet antequam graecam linguam gustarem, crediderim illam non tantum fuisse positionem Averroys, sed Aristote<lis>, et ita scripsi pluribus in locis. Nunc vero aliter sentire cogor." For context, see Steel's introduction to Averroes, *La béatitude de l'âme,* 98.

215. Nifo, *Aristotelis physicarum acroasum . . . liber* (1508), f. 218va: "Malo enim errare cum Graecis in expositione Graecorum auctorum quam recte sentire cum Barbaris qui linguae peritiam non habent nisi per insomnia."

216. Mahoney, 'Philosophy and Science,' in *Two Aristotelians,* art. 1, 201–202. Similarly Kessler, 'The Intellective Soul,' 497–498.

217. Cf. Spruit, *Agostino Nifo,* 17: "Equally problematic is the claim that Nifo substituted Averroes with Greek commentators and Scholastic authorities."

218. Nifo, *De intellectu* (1503), 3.31, f. 37v: "Quod ergo ex his accipimus est positio media, quam Avicenna, Algazel et Constabembeluce et alii Arabes dixerunt, quod rationales animae numeratae sint et formae corporum vitales et intrinsecae, quarum intellectus sunt separati secundum rationem et organo."

219. As has been suspected before; see Steel, 'Siger of Brabant,' on Nifo's "reworking and self-censorship" of his commentary on *De animae beatitudine,* finished, as was *De intellectu,* in 1492, but printed only in 1508. See also Gios, *L'attività pastorale,* 302n29: "Anche la conversione del Nifo è dovuta probabilmente all'intervento del Barozzi, del quale cercherà poi di ricuperare il favore pubblicando il trattato antiaverroista *De intellectu*"; and Di Napoli, *L'immortalità,* 206: "Certo, il Barozzi e l'inquisitore di Padova potevano aver influito su tale cambiamento d'opinione."

220. As underlined also by Bianchi, *Pour une histoire,* 135: "L'immense majorité des maîtres de Padoue, de Bologne et des autres universités italiennes restera fidèle à une longue tradition, enracinée dans la culture philosophique latine dès le XIIIe siècle, qui reconnaissait la distinction d'objet et de méthode entre philosophie et théologie."

221. On Pomponazzi's life and works and for further literature, see Nardi, *Studi;* Pine, *Pietro Pomponazzi;* Lohr, *Latin Aristotle Commentaries,* 2:347–362; Kraye, 'Pietro Pomponazzi'; SEP s.v. 'Pietro Pomponazzi' (by Perfetti).

222. Pomponazzi, *Tractatus* (1990), ch. 4–6, pp. 14–40.

223. Ibid., ch. 9, pp. 78–116. On the textual details of Pomponazzi's reading of Averroes, see Brenet, 'Corps-sujet, corps-objet.'

224. See the surveys of this debate in Di Napoli, *L'immortalità,* 277–338, and Pine, *Pietro Pomponazzi,* 124–138. See also Maclean, 'Heterodoxy in Natural Philosophy,' 10–17.

225. Kraye, 'The Immortality,' ch. 2.13–14, and Pine, *Pietro Pomponazzi,* ch. 2. On the influence of the 1513 bull, see Bianchi, *Pour une histoire,* 130–156.

226. These lectures are transmitted in two different reports by students: the Antonio Surian report edited by Kristeller ('Two Unpublished Questions,' 1955, 85–101) and an anonymous report, which is closer to the original wording of Pomponazzi's lectures, edited by Poppi (*Corsi inediti,* 1970, 2:1–93). On the philosophical standpoint of these lectures, see Poppi, *Saggi sul pensiero inedito,* 27–92, and Kessler, 'The Intellective Soul,' 500–504.

227. Pomponazzi, *Questio de immortalitate anime* (1955), 93–94: "Quia ergo non est tanta unitas inter animam intellectivam et corpus quanta est inter materiam et formam materialem, neque tanta separatio quanta est inter nautam et navim, ideo se habet sicut medium, et ambae opiniones *(that is, two different interpretations of Averroes's position)* ad hunc sensum bene dixerunt." Cf. the text of Poppi's recension, *Corsi inediti,* 2:44–45: "Unde dico quod non est unum vere sicut anima quae deducitur de potentia materiae et ipsa materia, non est disgregatio sicut est nauta navi, sed est unum medium essentiale cuius partes essentialiter una ab altera dependet, ut dictum est."

228. See the chapter on the saying 'Aristotele fu un uomo e poté errare' in Bianchi, *Studi sull'aristotelismo,* 101–124.

229. Pomponazzi, *Quomodo anima intellectiva* (1970), ed. Poppi, *Corsi inediti,* 2:41–42: "[Audivistis tres opiniones: vidistis opinionem Alexandri et eius substentationem, et domini, ut dixi, solum argumentum quod est ex parte modi intelligendi inter omnia alia argumenta semper fecit mihi magnam difficultatem, et adhuc non est bene solutum, tamen illa opinio est reicienda; vidistis enim *(probably: etiam)* duas opiniones in via Commentatoris, et opinionem christianorum.] De opinione Commentatoris dico vobis verum, ego credo quod opinio illa sit <in>imaginabilis, fatua ac chimerica. Oh, dixisti quod est opinio Aristotelis! Dico quod verum est, sed dico quod ipse Aristoteles fuit homo et potuit errare." A few paragraphs farther down (p. 46), Pomponazzi adds that "opinio Commentatoris, Themistii et Theophrasti sit opinio Aristotelis infallanter."

230. Pomponazzi, *Questio de immortalitate anime* (1955), ed. Kristeller, 93: "[Remoto lumine fidei ego valde perplexus sum in materia ista. Contra Alexandrum multum valet argumentum illud de universali.] De opinione autem Averrois mihi videtur quod fuerit opinio Aristotelis, tamen nullo pacto possum illi adhaerere, et videtur mihi maxima fatuitas. Dicat autem quisque quicquid vult, ego magis abhorreo opinionem Averrois quam diabolum." Cf. on Averroes's position: "videtur etiam esse Themistii" (94).

231. Ibid., 93.

232. Ibid.

233. Pomponazzi, *Questio de immortalitate anime* (1955), 85: "Nam quemadmodum dicit Commentator commento quinto *(ed. Crawford, p. 402),* si esset immersa materiae, intelligeret cum conditionibus coniunctis materiae seu materialibus. Hoc autem falsum est. Si enim sic esset, non posset universalia cognoscere quae absque quantitate et hic et nunc cognoscuntur, ideo etc." Cf. the text of Poppi's recension, *Corsi inediti,* 2:3–4: "Tertia ratio sumitur ex modo intelligendi, et ista est potissima rationum quae possunt fieri in ista materia, et est ratio Avicennae quam adducit Commentator in c. 19 huius: intellectus enim intelligit quidquid intelligit sine hic et nunc, sine quantitate et sine conditionibus materialibus. Virtus materialis cognoscit cum hic et nunc, nam, ut dicit Commentator, conclusio a materia est terminatio primae materiae in eis; cum enim intellectus intelligit sine hic et nunc etc. et universaliter cognoscit . . . , ideo concludit quod intellectus est immaterialis et immortalis."

234. Passages from these lectures are printed in Nardi, *Studi,* 195–199. See esp. 197: "Quare credo secundum Aristotelem alterum esse tenendum: aut quod anima intellectiva est corruptibilis, aut quod est tantum una ingenita et incorruptibilis," and 198: "tenuit <Scotus> etiam de mente philosophi animam intellectivam generari et corrumpi, quia non videbat eam secundum philosophum posse generari et non corrumpi. Hunc approbo quoad hoc quod tenuerit ad mentem philosophi animam esse genitam et corruptibilem."

235. Pomponazzi, *Tractatus* (1990), 9:110: "Tamen <intellectus humanus> universale in singulari speculatur, per quod differt ab aeternis et quoquo modo convenit cum bestiis."

236. Ibid.: "quare <homines> neque universale simpliciter, ut aeterna, neque singulariter tantum, ut bestiae, sed universale in singulari contemplantur."

237. On the reception of Alexander of Aphrodisias in the Renaissance, see Hankins and Palmer, *The Recovery,* 27–29.

238. See Perfetti's introduction and notes to Pomponazzi, *Expositio super primo et secundo* (2004).

239. From the commentary on the first book of the *Physics* of 1513, quoted in Nardi, *Studi,* 19–20n2: "Commentator erravit neque ipse est deus."

240. Again quoted in ibid.: "Ea que dicit Averroes sunt nuge. Quod si quis dicat: Tu ergo non defendis Averroem?, mihi continget quod Padue et Ferrarie neminem habeo pro amico."

241. On Prassicio, see Di Napoli, *L'immortalità,* 318–320; Lohr, *Latin Aristotle Commentaries,* 2:368. For a fuller analysis of Prassicio's treatise, see Hasse, 'The Attraction,' 141–144. Among Prassicio's colleagues in the Studio of Naples were other avid readers of Averroes such as Agostino Nifo and Pietro d'Afeltro. On the latter, see ERP s.v. 'D'Afeltro, Pietro' (by Burnett).

242. Prassicio, *Questio* (1521), sig. A1v: "Ego autem quantum intueri possum nil certi nilque veri de immortalitate animae reperio nisi ea quae ex Averroi

excerpi possunt. Et dicant quid velint eius aemuli mordaces et malevoli, sunt enim omnes ut oblatrantes a tergo catelli. Et quamvis tam Aristotelis quam Averrois positio de immortalitate animae sit falsa et erronea, tamen quid ipsi senserint iusta vires conatus sum ostendere. Ea vero quae tenet et sentit catholica nostra religio de animae immortalitate deque eius multiplicatione pro hominum numero testor et fateor esse ipsam infallibilem veritatem."

243. Ibid., sig. C2vb: "Quapropter teneo assertive hanc conclusionem esse ad mentem tam Aristotelis quam Averrois pro ultimo eorum nuncupativo testamento, quod anima intellectiva est simpliciter immortalis ex parte utriusque intellectus." Cf. also the conclusion of the treatise on sig. D4rb.

244. Ibid., sig. B4va: "Anima cum unam numero habeat naturam methaphisice constitutam ut saepe diximus, semper habeat propriam eius intellectionem separatam a phantasmate continuam aeternam quae est substantia et non potentia intelligentis. Cum vero ipsam contingat homini uniri, non univoce et per formalem inhaerentiam ut falso Augustinus sompniat, sed magis per assistentiam effectivam ad phantasmata illa actu intellecta afficiendo, statim fit species universalis in intellectu materiali."

245. Ibid., sig. D4ra: "secundum Averroem asserentem ubique animam intellectivam non esse formam corporis univoce, videlicet per formalem inhaerentiam illi unitam, sed tantum aequivoce per assistentiam effectivam seu affectivam ad phantasmata hominis."

246. Ibid., sig. B2vb–3ra. For a full quotation see this chapter, section 'Four Controversies about Averroes's Doctrine.'

247. Averroes, *Commentarium magnum De anima* (1953), 3.5, 406–407: "Quoniam, quia opinati sumus ex hoc sermone quod intellectus materialis est unicus omnibus hominibus et etiam ex hoc sumus opinati quod species humana est aeterna, ... etc."; see Taylor, 'Personal Immortality,' 102–106, and Wirmer's introduction to Averroes, *Über den Intellekt,* 14, 18–19.

248. On Genua, see Nardi, *Saggi,* 386–394; Poppi, Introduzione, 35–37; Lohr, *Latin Aristotle Commentaries,* 2:200–201; Kessler, 'The Intellective Soul,' 523–527; Spruit, *Species intelligibilis,* 2:164–173; Bakker, 'Natural Philosophy,' 169–175.

249. Genua, *In tres libros De anima* (1576), for example, on f. 131va, 139vb, 142ra, 144va, and 155r.

250. Hankins and Palmer, *The Recovery,* 30–31.

251. Simplicius, *Commentarii in libros de anima Aristotelis* (1543), f. 35v: "Simplicium unum vobis die noctuque versandum proponite; ... et illum *(scil. Averroem)* iam de manibus deponite."

252. Genua, *In tres libros De anima* (1576), sig. A1va: "omnium princeps Averroes quem cum alterum Aristotelem semper habuerimus, sic eius expositionem observabimus."

253. Ibid.: "Ioannis de Gandavo opera, ad quem vos hortor omnes studiosissimi iuvenes, quandoquidem et summus sit Aristotelicus, summus et Averroicus."

254. Ibid., 3.5, f. 132vb. This is also emphasized by Bakker, 'Natural Philosophy,' 171, who draws attention to a passage on f. 4rb: "Intellectus humanus dupliciter consideratur sicut bifariam est in nobis. Altero modo quatenus est una de intelligentiis nobis assistens, et sic consideratur non sub ratione animae, sed sub ratione intellectus ac mentis. Altero modo per unionem quam anima intellectiva nobiscum tenet egressa a seipsa, per quam beneficio phantasmatum intelligimus, et sic non rationem mentis et intellectus, sed rationem animae obtinet."

255. Genua, *In tres libros De anima* (1576), 3.5, f. 139rb: "Redeuntes modo ad principale dicimus, cum Aristotele peripatetice sentientes, unicum intellectum in omnibus reperiri hominibus."

256. Ibid., f. 136rb.

257. Ibid., f. 140rb: "Quare hanc indubie tenemus veram conclusionem esse et in philosophia et in theologia quod animae intellectivae creatae sunt numero distinctae . . . nam illa philosophia est magis credenda vera et fidelis quae magis consona est theologicis veritatibus, in quibus summa reperitur veritas; modo philosophia quam Plato professus est maxime proxima ac similis, quinimo Plato dictus fuit Moyses Atticus."

258. Ibid., 3.16–18, f. 152rb–156vb. See the important studies on Genua by Nardi, *Saggi*, 386–394, and Kessler, 'The Intellective Soul,' 523–527. Bakker, 'Natural Philosophy,' 169–175, shows that Genua's intellect theory is Averroist, while Genua adopts the idea from Simplicius that psychology is an intermediate discipline between natural philosophy and metaphysics.

259. As maintained by Kessler, 'The Intellective Soul,' 524.

260. On Porzio, see Lohr, *Latin Aristotle Commentaries,* 2:364–366; Kessler, 'The Intellective Soul,' 519–521; and Kessler's introduction to Alexander of Aphrodisias, *Enarratio De anima* (2008), lxxii–lxxvi.

261. Porzio, *De humana mente* (1551), 3: "multos quippe audies Averroicos, Simplicianos et Themistianos qui autoritate magis nomineque philosophi suam sententiam astruunt quam eam verborum Aristotelis fide confirmare conentur."

262. Ibid., 19–22, esp. 21: "idem fere sentit."

263. Vimercato was counted among the Averroists by the Conimbricenses, *In tres libros De anima* (1600), in 2. lib., ch. 1, quaest. 7, art. 1, col. 108: "Et hoc quidem argumentum permovit etiam ad praedictam intellectus unitatem in Aristotelis doctrina asserendam non paucos e recentioribus Peripateticis, in quibus sunt Thomas Anglicus, Achillinus, Odo, Iandunus, Mirandulanus, Zimara, Vicomercatus, et quidam alii."

264. On Vimercato's life and writings, which include commentaries on *Metaphysics, Physics, De caelo, De generatione, Meteorologica, De anima 3, De partibus animalium, and Nichomachean Ethics* 1–3, see Gilbert, 'Francesco Vimercato' (with a list of printed editions on 213–214); Lohr, *Latin Aristotle Commentaries,* 2:479–481 (with extensive bibliography); Perfetti, *Aristotle's Zoology*, 136–154. On his psychology, see Nardi, *Saggi,* 404–410. Vimercato's principles of translation

are remarkably practical: He was convinced that the usage of nonclassical words such as *ens* and *substantia* in the translation is necessary lest the work be obscure and open to criticism by the enemies of Aristotle, as Charles Schmitt has pointed out; see Schmitt, *Aristotle and the Renaissance*, 79–81.

265. Vimercato, *De anima rationali* (1574), 19a.

266. Ibid., 36a–b.

267. Ibid., 40a ("quorum sane opinioni . . . veritate cogente subscribere sum coactus"), 45b ("Respondeo ego"), 47a ("ut ingenue dicam quod sentio"), and 51b ("Huic opinioni atque sententiae ego quoque subscribo").

268. Ibid., p. 47a: "quisquis exacte perpendit eas *(sc., rationes)* quae unitatem probant (ut ingenue dicam quod sentio), multo caeteris efficaciores atque Aristotelicis principiis conformes magis comperiet."

269. Cf. Themistius, *In libros Aristotelis de anima paraphrasis* (CAG 5.3) (1899), 103, lin. 36—p. 104, lin. 14 (=*On Aristotle On the Soul*, p. 129); Schroeder and Todd, *Two Greek Aristotelian Commentators*, 105.

270. Vimercato, *De anima rationali* (1574), 47b: "Multae aliae sunt non absimiles Commentatoris rationes, sed omnium quas ipse invenerit optima illa esse videtur quam in destructionibus destructionum disputatione primo adduxit contra Algazelem de animarum sive intellectuum infinitudine."

271. See, for example, Bonaventure, *Commentary on the Sentences*, 2. d.1 p. 1 a.1 q.2 sed contra 5 and concl., and Thomas Aquinas, *Commentary on the Sentences*, 2. d.1 q.1 a.5 sed contra 6 and ad 6, in the context of the discussion of the eternity of the world; both texts are easily accessible in Schönberger and Nickl, *Über die Ewigkeit der Welt*, 14, 21, 46, 74.

272. van den Bergh, *Averroes' Tahafut al-tahafut*, 1:161–163.

273. Vimercato, *De anima rationali* (1574), 47b.

274. Ibid., 51b: "Quod si haec *(i.e., other doctrines against the faith such as the eternity of the world)* Aristoteli tribuere non pertimescimus quotidie in scholis docemus, nec ob id fidem nostram labefactari arbitramur, quid causae est ut etiam in his quae ad animam attinent, quod illum sensisse cognoscimus, libere prodere non possimus? . . . Haec dubia non sunt, ubi aperte non involucris quibusdam aut obscure de his, quae nos quoque credimus, disserere his sufficiant, qui unitatem intellectus et si qua alia absurda sint Aristoteli me tribuisse mirabuntur."

275. Ibid., 49a.

276. Ibid., 47a: "Quamobrem ut Aristotelem in eam partem inclinasse credam, quasi compellunt *(sc., has rationes)*, tametsi hanc suam opinionem testimoniis claris et apertis palam facere noluerit, ob eam forte causam quod et vulgo vix credi potuisset et multa quae ad mores vitamque civilem spectant per eam aboleri pertimesceret."

277. Bernardi, *Eversionis singularis certaminis libri XL* (1562), lib. 32, ch. 1, p. 546: "Ad hanc quaestionem respondeo, quod saepius iam diximus, intellectum nostrum ex veritate sanctissimae religionis nostrae esse numeratum ad numerationem singulorum individuorum, sed hoc tamen repugnare naturae

fundamentis . . . Tamen . . . non dubitabimus innixi fundamentis naturae ad quaestionem propositam respondere: ex illo antecedenti, quod videlicet intellectus est aeternus, sequi intellectum esse unum numero in omnibus hominibus, idque ob eas rationes quae in superioribus allatae fuerunt, ad demonstrandum ipsum intellectum non esse numeratum ad numerationem singulorum hominum, quas quidem rationes ex fundamentis naturae, falsis tamen, vim habere, perspicue patere arbitramur" (quoted in Di Napoli, *L'immortalità*, 364–365, but with wrong reference to page 564 in Bernardi).

278. Cf. Hayoun and de Libera, *Averroès et l'averroïsme*, 117 ("le premier choc"), and Poppi, *La filosofia nello studio francescano*, 265: "segna l'inizio della regressione dell'averroismo." For a warning that the significance of the 1497 chair should not be overemphasized, see Schmitt, *Aristotle and the Renaissance*, 37.

279. Renan, *Averroès et l'Averroïsme*, ch. 3,15, p. 414: "L'averroïsme avait résisté, depuis près de trois siècles, aux attaques du platonisme, des humanistes, des théologiens, du concile de Latran, du concile de Trente, de l'inquisition; il expira le jour où apparut la grande école sérieuse, l'école scientifique . . . Cette . . . école vraiment moderne et tout à fait libre enfin de la barbarie du moyen âge, pouvait seule en finir avec un aristotélisme décrépit. La vraie philosophie des temps modernes, c'est la science positive et expérimentale des choses."

280. Nardi, *Saggi*, 453: "L'averroismo volse al tramonto sul finire del secolo XVI e sul cominciare del secolo successivo, perché al tramonto volgeva ormai l'aristotelismo, del quale l'averroismo pretendeva d'essere la più fedele interpretazione. L'aristotelismo a sua volta finiva per interna dissoluzione . . .'; 454: "Ma quello che determinò il crollo definitivo dell'aristotelismo e dell'averroismo, fu il nascere di una nuova filosofia della natura, fondata su un nuovo metodo di ricerca scientifica: la logica dell'esperienza."

281. Hayoun and de Libera, *Averroès et l'averroïsme*, 8: "averroïsme qui, de fait, allait être détrôné à Padoue au début du XVIe siècle."

282. The *Ratio studiorum* admonishes the philosophy professors to be very careful when citing the commentators on Aristotle who are objectionable from the standpoint of faith; to cite Averroes only without praising him and by showing that Averroes had borrowed from other writers ("si quid boni ex ipso *(sc., Averroe)* proferendum sit, sine laude proferat, et, si fieri potest, id eum aliunde sumpsisse demonstret"); to be very critical of the philosophical sects of the Averroists and the Alexandrists; and to always speak favorably of Thomas Aquinas. See *Ratio atque institutio studiorum* (1603), 68–69.

283. Within the *De anima* tradition, the philological commentary of Julius Pacius (Giulio Pace, d. 1635) is of particular importance; see Pacius, *Aristotelis De anima libri tres, graece et latine* (1596). On Pacius, see Schmitt, *Aristotle in the Renaissance*, 43–47; Lohr, *Latin Aristotle Commentaries*, 2:296–297.

284. Schmitt, *Aristotle and the Renaissance*, 46–63; Schmitt, 'The Rise of the Philosophical Textbook.'

285. As suggested by Hayoun and de Libera, *Averroès et l'averroïsme*, 117–118: "Cible privilégiée des humanistes...l'averroïsme entame son déclin—l'irrésistible chute d'Averroès est le chiffre de toutes les nouveautés italiennes...Face à l'arabisme et à l'aristotélisme, il y a, évidemment, l'humanisme; il y a le retour en force, la 'renaissance,' du platonisme."

286. Melanchthon, *Liber de anima, recognitus ab auctore* (1846), ed. Bretschneider, col. 143 (=ed. Nürnberger, 1961, p. 333): "De hoc externo obiecto excitante intellectum in hac nostra infirmitate recte dicitur obiectum esse res omnes quae sensibus percipiuntur. Inde enim procedit intellectus sua vi ad alia."

287. An example is Pietro Pomponazzi who had based his mortality thesis on the Aristotelian sentence that "the soul does not know anything without phantasmata"; Pomponazzi, *Tractatus* (1990), ch. 9, passim, esp. 82. Pomponazzi adopts the principle "nequaquam sine phantasmate intelligit anima" from Aristotle, *De anima*, ch. 3.8, 432a13–14.

288. Melanchthon, *Liber de anima*, ed. Bretschneider, col. 144, ed. Nürnberger, p. 334: "Nec turbemur vulgari dicto: Nihil est in intellectu, quin prius fuerit in sensu. Id enim nisi dextre intelligeretur, valde absurdum esset. Nam universales notitiae et diiudicatio non prius fuerunt in sensu. Sed fatendum est sensuum actione et singularibus obiectis moveri et excitari intellectum ut procedat ad ratiocinanda universalia, et ad iudicandum."

289. Melanchthon, *Liber de anima*, ed. Bretschneider, cols. 143–144, ed. Nürnberger, p. 333: "Vetus contentio est inter Aristotelicos et Platonicos, an sint aliquae in mentibus notitiae nobiscum natae? Sed simplicius et rectius est retinere hanc sententiam, esse aliquas notitias in mente humana, quae nobiscum natae sunt, ut numeros, ordinis et proportionum agnitionem, intellectum consequentiae in syllogismo. Item principia geometrica, physica et moralia." On Melanchthon's doctrine of the inborn notions, see Frank, *Die theologische Philosophie*, 112–126.

290. Melanchthon, *Liber de anima*, ed. Bretschneider, col. 150, ed. Nürnberger, p. 340: "Sunt igitur normae certitudinis iuxta philosophiam tres: experientia universalis, notitiae principiorum, et intellectus ordinis in syllogismo."

291. See Helm, 'Die Galenrezeption'; Park, 'The Organic Soul'; De Angelis, *Anthropologien*. On Vives's *De anima et vita*, in particular, see Casini, *Cognitive and Moral Psychology*, and Del Nero, 'A Philosophical Treatise.'

292. Vives accuses the commentators of being involved in "rebus inanibus" and "magnis absurditatibus" (Vives, *De anima et vita* (1565), sig. a2v,=*De anima et vita* (1974), 84), while Melanchthon castigates the scholastic tradition for its "insulsissimae argutiae" and "cavillationes" (Melanchthon, *De philosophia oratio* [1961], 93).

293. It was mentioned above that Vives in *De causis corruptarum artium* had attacked Averroes for his ignorance of Greek. In *De anima et vita*, first published in 1538, Vives ignores Averroes; only Aristotle is deemed worthy of being attacked as obscure, superficial, and prolix *(obscurus, lubricus, vafer)*. See Vives, *De anima et vita* (1565), 146 (=*De anima et vita* (1974), 440). Averroes is

castigated by Melanchthon for his "perversitas et petulantia" in a different context, namely, because of his anti-Ptolemaic planetary theory; see Melanchthon, *Initia doctrinae physicae* (1846), col. 232. With regard to the intellect, Melanchthon concedes that Averroes's position is not absurd if understood correctly, that is, as saying that the active intellect is God himself who triggers the most noble actions and thoughts in human beings; see Melanchthon, *Liber de anima* (1846), col. 149: "Etsi autem Averrois divinatio deridetur et fortassis ab Aristotele aliena est, tamen si dextre intelligitur, non est absurda. Cum enim ait facientem intellectum esse ipsum Deum cientem excellentiores motus in hominibus, vere dicit." This appears to be a dim reflection of Averroes's original idea that through the speculative intellect's union with the active intellect we reach godlike knowledge.

294. On Zabarella's intellect theory, see Nardi, *Saggi,* 417–421 (on the influence of Simplicius in Zabarella); Kessler, 'The Intellective Soul,' 530–534; Spruit, *Species intelligibilis,* 2:225–236. For his works, see Lohr, *Latin Aristotle Commentaries,* 2:497–504. On Zabarella's philosophical standpoint in general, see Mikkeli, *An Aristotelian Response,* and SEP s.v. 'Giacomo Zabarella' (by Mikkeli).

295. Zabarella, *De mente agente,* in *De rebus naturalibus libri* 30 (1606/1607, repr. 1966), ch. 8, 1020: "Septima <assertio> est: officium abstrahendi non est intellectus agentis, sed est proprium intellectus patibilis; ita tamen ut abstractio ex necessitate praesupponat operam intellectus agentis, qui phantasmata illustret et claras atque conspicuas esse faciat omnes naturas et quidditates quae in phantasmatibus insunt, ut postea patibilis intellectus accipere possit id quod vult et alia dimittere, quod vocatur abstrahere." Cf. ch. 6, 1017.

296. Ibid., ch. 14, 1035: "Quomodo potest idem esse lumine plenus et illuminans, et simul omni lumine, omnique cognitione carere et illuminari a semetipso?"

297. Ibid., ch. 13, 1031: "Maxime autem omnium intelligibilis Deus est, et est primum in genere intelligibilium; ergo nil aliud statui potest intellectus agens nisi solus Deus. Hoc fuit argumentum efficacissimum Alexandri." Cf. ch. 14, 1035: "Praeterea minime conveniens est ulli praeter Deum facultatem attribuere faciendi actu intelligibilia; solum enim primum intelligibile id facere aptum est." On Zabarella's reception of Alexander, see Kessler's introduction to Alexander of Aphrodisias, *Enarratio De anima* (2008), lvi–lxi.

298. Zabarella, *De mente agente* (1606/1607), ch. 16, 1037: "nam eius lumen semper nobis adest immutabile et numquam deficiens; ideo intelligimus quando volumus, dummodo habitum acquisiverimus."

299. Kuhn, *Venetischer Aristotelismus,* 186–243, esp. 198–209, on the active intellect being identical with God and 217–219 on a comparison between Zabarella's and Cremonini's intellect theory.

300. On Suarez's intellect theory, see Castellote Cubells, *Die Anthropologie,* 182–196, esp. 188–193; Kessler, 'The Intellective Soul,' 514–516; Spruit, *Species*

*intelligibilis,* 2: 294–306; for further secondary literature, see Lohr, *Latin Aristotle Commentaries,* 2:441–445.

301. On the occasion of the discussion of Aristotle's definition of the soul in book 2, these authors insert a succinct presentation of Averroes's theory, which is followed by a similarly curt refutation. See Toletus, *Commentaria una cum quaestionibus . . . De anima* (1583), f. 41rb–47vb; Conimbricenses, *In III libros De anima* (1600), col. 107–110. Toletus castigates Averroes as *impius* and *monstrum* and his follower Achillini for his *delirium;* see Toletus, *Commentaria una cum quaestionibus . . . De anima* (1583), f. 41va, 43va, 136rb (on Averroes) and 45rb (on Achillini). Averroes has ceased to be a provocation for those trends of the Second Scholastic, which take their cue less from Aristotle than from Thomas Aquinas or Duns Scotus. An example is the Dominican John of St. Thomas, whose *De anima* commentary (of 1635) refers to the unicity thesis in no more than one sentence; see John of St. Thomas, *In tres libros De anima* (1948), qu. 4. a. 1, p. 106a: "Et ubicumque agit <Thomas> de intellectu possibili ostendendo contra Averroistas quod non datur unus intellectus in omnibus sed in quolibet distinctus, id probat quia operatio intellectus debet attribui cuilibet intelligenti."

302. Francisco Suarez, *De anima* (1856), lib. 1, ch. 12, 557–558.

303. Ibid., ch. 8, p. 741: "Probatur ergo ratione Aristotelis, in 3 De anima, c. 5, nam cum anima nostra ab intrinseco sit intellectualis, ab intrinseco habere debet necessaria ad exercendas intellectuales operationes . . . Habet ergo anima nostra hanc virtutem sive potentiam." The term "power" *(vis)* is carefully chosen by Suarez in order to indicate that active and potential intellect do not exist separately; they are different powers *(vires)* of the same intellective faculty *(potentia)* of the soul (cf. 744–745).

304. Ibid., lib. 4, ch. 2, 715–716: "unde supra conclusimus intellectum indigere specibus et similitudinibus rerum ad cognoscendum."

305. Ibid., 716.

306. Ibid., 719b: "atque ita fit ut anima cum primum phantasiando cognoscit rem aliquam per virtutem spiritualem, quam intellectum agentem vocamus, quasi depingat rem eandem in intellectu possibili."

307. Ibid., 717b: "repraesentari universale abstractum nihil aliud est quam repraesentari rationem communem non repraesentatis individuis conditionibus."

308. As claimed by Renan, *Averroès et l'Averroïsme,* ch. 3,15, p. 414: "un aristotélisme decrepit."

309. Nifo, *Expositiones* (1559, repr. 1967), 1.5, p. 58b.

310. Ibid., 1.3, p. 57a: "Magno miratu dignum est quonam pacto vir iste tantam fidem lucratus sit apud latinos in exponendis verbis Aristotelis, cum vix unum verbum recte exposuerit, potissimum cum ipse tertio libro De coelo et mundo conqueratur de translationibus, asserens translationes, quas ipse exponit esse Alcindi et falsas, veriores autem esse translationes ipsius Isachi, quas non habebat." Nifo here refers to a passage in Averroes's Long

commentary on *De caelo,* where Averroes complains that he is working with the vulgate translation of Yaḥyā ibn al-Biṭrīq, which he attributes to "Al-chindi," and that he does not have "the more faithful versions of Isaac," that is, of Isḥāq ibn Ḥunayn; see Averroes, *Commentum magnum super libro De celo* (2003), 3.35, p. 567: "Et hec intentio est difficilis ad intelligendum ex ista translatione quam modo habemus, et forte diminutio cecidit in hac translatione ex translatore: nos enim non habemus nisi translationem Alchindi, translationes autem veriores sunt Isaac." For the *De caelo* transmission, see Endress, 'Averroes' *De caelo,*' 47.

311. Zabarella, *De propositionibus necessariis,* 2.2, in *Opera logica* (1603), 380: "nemo profundius ac melius quam ipse mentem Aristotelis hac in re penetravit."

312. On this work, see Hidalgo-Serna's introduction to Vives, *De causis* (1990), 22–26. Vives's attack on Averroes had already been noticed by Renan, *Averroès et l'Averroïsme,* 396–398, and Klein-Franke, *Die klassische Antike,* 43–46, but has never been analyzed in textual detail.

313. Vives, *De causis* (1990), 195.

314. Aristotle, *Metaphysics,* 987a29–b2.

315. Aristoteles Latinus quoted by Vives, *De causis* (1990), 193, corrected against the forthcoming critical edition of the Long commentary on the *Metaphysics* by Stefan Georges and myself. Vives writes "erat opinionis" instead of "erat secudum opinionem"; "scilicet quod" instead of "secundum quod"; and "de natura" instead of "de natura universali." Cf. the version of this text in Aristotle and Averroes, *Opera . . . omnia* (1562), vol. 8, f. 7rb, which contains some Renaissance corrections. For the Arabic text, see Averroes, *Tafsīr mā baʿd aṭ-ṭabīʿa* (1938–1948), 62–63.

316. Vives, *De causis* (1990), 193: "Aristoteles, si revivisceret, intelligeret haec?"

317. Ibid.: "Favete linguis viro tanti nominis et alteri Aristoteli."

318. Averroes quoted in ibid., 194, corrected against Hasse and Georges. Vives's quotation differs as follows: dixit (Vives: dicit), consequebatur (Vives: sequebatur), illam (Vives: istam), consequebatur (Vives: consectabatur), ibi (om. Vives).

319. Ibid.: "Pythagoricos distinguit a philosophis Italicis quasi alii fuerint Italici quam Pythagorici, quod nec pueri nostri ignorant . . . Cur Anaxagoram ex Jonia transfers in Italiam?"

320. Ibid.: "Quid ego haec refellam quae apud Aristotelem nulla sunt, et sumam operam supervacaneam, cum haec sint cuilibet ad exibilationem atque explosionem obiecta? Ubi est mentio Democriti? Ubi Herculeorum? Quid (malum) sunt isti Herculei? An quia Hercules Graece ἡρακλῆς dicitur, ideo heraclitici erunt herculei?"

321. Averroes quoted in ibid.: "Istae igitur opiniones praedictae perveniunt ad nos de considerantibus in philosophia usque nunc."

322. Ibid., 195: "Atqui hic est Abenrois, quem aliquorum dementia Aristoteli parem fecit, superiorem Divo Thoma . . . Te nihil est horridius, incultius, obscoenius, infantius."

323. Ibid., 196: "Nam et Abenrois doctrina et metaphysica Avicennae, denique omnia illa Arabica videntur mihi resipere deliramenta Alcorani et blasphemas Mahumetis insanias."

324. On Vives's biography and his *De veritate*, see the very informative article by González, 'Juan Louis Vives,' esp. 19, 52.

325. On Naẓīf ar-Rūmī, see Bertolacci, *The Reception*, 12–15, and Kraemer, *Humanism*, 132–134.

326. On Michael Scot's motives and techniques as the translator of this commentary, see Hasse, *Latin Averroes Translations*, 32–38.

327. The new edition of Alpha by Primavesi reads συγγενόμενος with the Paris recension, Naẓīf and Vives, whereas Bonitz and Ross follow the Florence recension (Primavesi, 'Aristotle, *Metaphysics* A,' 488).

328. On Aristotle's usage of the preposition παρά, see Bonitz, *Index Aristotelicus*, s.v. (561–562). Its interpretation can also divide modern interpreters, such as on Met. 987b8 Bonitz ("neben") and Ross ("after").

329. Averroes, *Tafsīr mā baʿd aṭ-ṭabīʿa* (1938–48), 63.

330. Incidentally, the medieval Latin translations of the *Metaphysics* from Greek were also faulty: The vetustissima of the twelfth century, for instance, writes *circa* for παρά ("beyond"); William of Moerbeke gets this right by writing *praeter*, but infelicitously chooses *ex novo* for ἐκ νέου ("in his youth").

331. Alexander of Aphrodisias, *In Aristotelis Metaphysica Commentaria*, 49–50. The English translation is from Alexander of Aphrodisias, *On Aristotle's* Metaphysics 1, 77.

332. Cranz, 'Alexander Aphrodisiensis,' 93–95; Hankins and Palmer, *The Recovery*, 28.

333. Nifo, *Expositiones* (1559, repr. 1967), 1.5, p. 59a.

334. Averroes, *Tafsīr mā baʿd aṭ-ṭabīʿa* (1938–1948), p. 63, lin. 7–p. 64, lin. 9.

335. Nifo, *Expositiones* (1559, repr. 1967), 1.5, p. 59a.

336. Nifo's commentary on Alpha 5 has many verbal parallels with Sepulveda's translation of Alexander; I have consulted the following edition of Sepulveda's translation: Alexander of Aphrodisias, *Commentaria in duodecim Aristotelis libros de prima philosophia* (1561), 18b: "Deinceps explicat Platonis sententiam."

337. Averroes, *Tafsīr mā baʿd aṭ-ṭabīʿa* (1938–1948), 347, line 16; 354, line 8; 382, line 5; 397, line 6; 448, line 4; 449, line 13; and 467, line 8.

338. Ibid., 424–425.

339. As shown by Endress, 'Die Entwicklung der Fachsprache,' and by Gutas, *Greek Thought, Arabic Culture*.

340. See Tarán and Gutas, *Aristotle: Poetics,* and Schmidt and Ullmann, *Aristoteles in Fes*.

341. Grendler, *Universities*, 288.

342. See the survey article by Emery and Speer, 'After the Condemnation of 1277.' For masters of arts continuing to hold potentially heretic doctrines after 1277, see the exemplary study by Donati, 'Utrum accidens.' For a balanced

introduction to the problem of censorship in the Middle Ages, see Putallaz, 'Censorship.'

343. Renan and other French thinkers believed that the Renaissance Averroists were forerunners of enlightenment rationalism—a claim justly criticized by Kristeller, 'The Myth of Renaissance Atheism.' On Renan, Kristeller, and the secularist historiography of Renaissance Averroism, see Martin, 'Rethinking,' 4–9.

344. Bianchi, *Pour une histoire,* 128–130.

345. Pine, *Pietro Pomponazzi,* 126–127.

346. See Mahoney, 'Philosophy and Science,' in *Two Aristotelians,* art. 1, 201 and 202 on Vernia's and Nifo's citation of Simplicius as "a major turning point in late medieval and Renaissance Aristotelianism." In the understanding of Charles Lohr, "Averroist Aristotelianism," or the "Averroist paradigm for the sciences" of the Middle Ages "broke down" upon Pletho's critique, that is, after "the arrival of the Greeks at the Council of Florence"; it was replaced by a Greek paradigm (Lohr, 'Metaphysics and Natural Philosophy,' 280–287). Or, as Lohr writes in an article of 2000: "Through these Latin translations the Greek commentaries on Aristotle contributed to the Renaissance liberation of science from the one-sided interpretation of the Philosopher which the scholastics had inherited from Averroes" (Lohr, 'Renaissance Latin Translations,' 34). See also Kessler, 'Nicoletto Vernia,' esp. 281. However, I share Kessler's general view of Renaissance Aristotelianism as comprising many currents—among them Greek, Arabic, and Scholastic.

347. On the delayed influence of Greek texts, which came to full fruition only in the course of the sixteenth century, see Wilson, *From Byzantium,* 161–162. On the "thirty almost Greekless years" (161) in humanist medicine between the 1490s and 1525, see Nutton, 'Hellenism Postponed.'

348. Fries, *Defensio medicorum principis Avicennae* (1533), f. 42v: "Nam pauci sunt qui graecae linguae vires intelligant, multi vero qui postquam ἰῶταν depingere norunt, esurientes graeculos sese iactare non verentur."

349. Nardi, *Studi,* 19–20n2: "mihi continget quod Padue et Ferrarie neminem habeo pro amico."

350. In Nifo's *De beatitudine animae* (1508); see Steel's introduction to Averroes, *La béatitude,* 96.

351. Erpenius, *Orationes tres* (1621), 80: "Considerate philosophi quanti momenti sit alterum illum posse sua lingua disertissime docentem audire Aristotelem, ابن راشد (sic) Ibno-Rasjidum, inquam, quem *Averroem* inepte vocatis, quique ita latine balbutit, ut merito conqueramini intelligi eum vix posse et interpretem in ipso desideritis interprete." The English translation comes from Jones, 'Thomas Erpenius,' 23 (slightly changed).

352. Christmann, *Alphabetum Arabicum* (1582), sig. A3r: "ut qui multo rectius saepe Aristotelem sint interpretati quam ipsimet Graeci." Another praise of the Arabic Averroes comes from the Arabist scholar Thomas Greaves, *De linguae Arabicae utilitate* (1639), 11: "Alterum nobis Lyceum aperuit magnus ille

Averroes, in quem ipsius Aristotelis genium animamque migrasse Pythago-
raeus quispiam dejeraret."

353. The substantial and well-researched article 'Ibn-Roschd' in the *Dictionnaire des sciences philosophiques,* vol. 3, of 1847, 157–172, by Salomon Munk (1805–1867), the German Orientalist working in Paris, marks the beginning of European Arabist scholarship on Averroes. It was followed in 1852 by Ernest Renan's *Averroés et l'Averroisme.*

354. On Alpago's student years in Padua, where he acquired his degree *in artibus* in 1481, see Lucchetta, *Il medico e filosofo,* 9–15.

355. On Arabic manuscripts of Averroes's commentaries, see Bouyges, 'Notes,' esp. 9–25. It is difficult to say anything certain about the earliest arrival of Arabic Averroes manuscripts in Arabist collections. It has been argued that Ms. Leiden, Universiteitsbibliotheek Or. 2073 (middle commentaries on the Organon) once belonged to Guillaume Postel's collection (Bouyges, 'Notes,' 11).

356. See von Kügelgen, *Averroes,* 1–19.

## 6 · ASTROLOGY: PTOLEMY AGAINST THE ARABS

1. The basic reference for research on Renaissance and early modern astrology still remain the chapters on astrology in Thorndike, *History,* vols. 5, 6, and 7. Recent excellent studies of different aspects of Renaissance astrology are: Vanden Broecke, *Limits of Influence,* which is strong on the technical aspects of Renaissance astrology; Grafton, *Cardano's Cosmos,* which focuses on the social background and psychological contribution of a principal Italian astrologer; and Oestmann, *Heinrich Rantzau,* which studies an important German astrologer in several respects: astrological techniques, manuscripts and books, instruments, and sociological setting. Important articles are collected in Zambelli, 'Astrologi hallucinati.' For introductions, see Rutkin, 'Astrology,' 541–561, and Hübner, 'Astrologie in der Renaissance.' See also Dooley, *A Companion.*

2. On the 1524 flood prediction, see the fundamental studies by Hellmann, 'Aus der Blütezeit,' Thorndike, *History,* 5:178–233, and Zambelli, 'Fine del mondo.' See also Zambelli, 'Many Ends'; Talkenberger, *Sintflut;* Zambelli, 'Eine Gustav-Hellmann-Renaissance?; Vanden Broecke, *Limits of Influence,* 81–111, and now, with several historical clarifications, Mentgen, *Astrologie,* 113–127, 135–155.

3. See Aston, 'The Fiery Trigon Conjunction,' and Leppin, *Antichrist und Jüngster Tag,* 140–144, for the English and German public, respectively.

4. See, in addition to the evidence of the second part of this chapter, Barnes, *Prophecy and Gnosis,* esp. ch. 4 on apocalyptic astrology, Leppin, *Antichrist und Jüngster Tag,* esp. 140–144, 169–205, 290, and Gindhart, *Das Kometenjahr 1618,* 17–21.

5. Pico, *Disputationes* (1946/1952), 1:56.

6. For information on these critics of astrology, see the relevant chapters in Tester, *A History,* and Knappich, *Geschichte.*

7. On astrology at the court of Paul III, see Thorndike, *History,* 5: h. 8; cf. also 5:175.

8. To cite the central passage from Sixtus's bull of 1586 (*Bullarum,* 8:650): "statuimus et mandamus ut...contra astrologos, mathematicos et alios quoscumque dictae iudiciariae astrologiae artem, praeterquam circa agriculturam, navigationem et rem medicam in posterum exercentes aut facientes iudicia et nativitates hominum, quibus de futuris contingentibus successibus fortuitisque casibus aut actionibus ex humana voluntate pendentibus aliquid eventurum affirmare audent...tam episcopi et praelati, superiores ac alii ordinarii locorum quam inquisitores...diligentius inquirant et procedant." On the condemnations of astrology, see Baldini, 'The Roman Inquisition's Condemnation,' and Baldini and Spruit, *Catholic Church and Modern Science,* 1:440–468. Cf. on the bulls of 1586 and 1631, Mahlmann-Bauer, 'Die Bulle,' esp. 143–170; Ernst, *Religione,* ch. 11, esp. 255–256, 274–279; and Lerch, *Scientia astrologiae,* 249–257.

9. Warburg, 'Heidnisch-antike Weissagung'; Grafton, *Cardano's Cosmos,* 76–77. On Luther, see Ludolphy, 'Luther und die Astrologie.' On Melanchthon's astrological interests and works, see Thorndike, *History,* 5:ch. 17, Caroti, 'Melanchthon's Astrology,' and Brosseder, *Im Bann der Sterne.* On sixteenth-century theological attempts to justify astrology, see Lerch, *Scientia astrologiae,* 191–210.

10. This popular line of reasoning is exhibited, for example, in the standard handbook by Albumasar, *Introductorium maius (Kitāb al-Mudhal al-kabīr),* ch. 1. For the natural philosophical foundations of medieval and Renaissance astrology, see Rutkin, *Astrology.*

11. For an excellent article on the explanation of long-distance effects, see Copenhaver, 'A Tale of Two Fishes.'

12. Pietro d'Abano translated, probably from Old French intermediaries, at least six treatises by ibn Ezra, which form part of an astrological encyclopedia: *Introductorium qui dicitur Principium sapientie, Liber rationum, Liber nativitatum et revolutionum earum, De interrogationibus, De electionibus,* and *Liber luminarum.* See Smithuis, 'Abraham ibn Ezra's Astrological Works,' esp. 248–250. Pietro's translations were more influential than those of the other Latin translators of ibn Ezra: Henry Bate, Arnoul de Quinquempoix, and Ludovicus de Angulo.

13. There is no manuscript copy of the *Disputationes* extant. On the editing process, see Zambelli, 'Giovanni Mainardi,' and Farmer, *Syncretism in the West,* 151–176.

14. Different layers in Pico's work are distinguished by Vanden Broecke, *Limits of Influence,* ch. 3. On the *Disputationes,* see also Zambelli, *L'apprendista stregone,* esp. 47–48, Grafton, *Commerce,* 111–134, Bertozzi, *Nello specchio del cielo,* and Rutkin, 'The Use and Abuse.'

15. Pico, *Disputationes* (1946/1952), 1:522: "Albumasar vel auctor vel inventor huius erroris."

16. Ibid., 2:530: "Tamen nec a Ptolemaeo, nec etiam ab ullo antiquorum tale aliquid umquam excogitatum, sed Arabum esse iuniorum figmentum."

17. Ibid., 5:606: "Ubi de illis vel Ptolemaeus vel antiquus aliquis umquam fecit mentionem? Meracissimae nugae sunt Arabumque figmenta."

18. Nifo, *De nostrarum calamitatum causis* (1505), f. 33va: "Iuniores et Albumasar et alii multa scribunt de his que nos vana et superstitiosa et contra philosophiam et Ptolomei astronomiam reputamus."

19. Ibid., f. 32ra: "Quamobrem Albumasar princeps horum fabulantium erravit de fine christiane religionis ac mahometane."

20. Pigghe, *Adversus prognosticatorum vulgus* (1518), f. 3v–4r: "Sed si qui se pro astrologis gerunt, omnium mathematicarum disciplinarum sunt ignari, Punicis tantum fabulis et superstitionibus imbuti, contra quos praecipue hoc bellum suscepimus."

21. Ibid., f. 19r: "Accessi autem eo solum consilio, si forte invitare possemus astrologiae studiosos, ut relictis fabulis Albumazaris et Punicorum Ptolomeum nostrum legerent diligentius." Cf. the very similar passage at the end of the preface, f. 5v. That *Punici* refers to the Arabs is apparent from phrases like "Albumazaris caeterorumque Punicorum doctrina" (f. 6r).

22. Cardano, *Commentarii in Ptolemaeum,* in *Opera* (1663), 5:95b.

23. Campanella, *Astrologicorum libri VII in quibus astrologia omni superstitione Arabum et Iudaeorum eliminata physiologice tractatur* (1630).

24. Melanchthon, preface to Ptolemaeus, *De praedictionibus astronomicis* (1553), 4: "Habuit vetustas doctrinam de motibus stellarum haud dubie eruditissime extructam *(sic)* ex antiquissimis observationibus, habuit accurate descriptam seriem temporum et veterum Imperiorum. Cum autem Aegypto Sarracenica barbaries et Academiam Alexandrinam et studia delevit, vetera monumenta perierunt. Et tota doctrina de motibus interitura erat, quae tunc dispersa in multos et varios libros, nisi paulo ante Sarracenicos motus Ptolemaeus totam artem in unum volumen contraxisset." I am grateful to Andreas Lerch for drawing my attention to this passage.

25. Some such attacks are mentioned in passing by Thorndike, *History,* 5:298 (on Marstallerus), 6:107–108 (on Pontus de Tyard), 113 (on Hieronymus Wolf), 141 (on van Lindhout), 172 (on Petrus Antonius), 194 (on Rantzau), and 199 (on George of Ragusa).

26. Such as Tommaso Giannotti and Michele da Pietrasanta, see Zambelli, 'Fine del mondo,' 356–360.

27. Abioso, *Opus vaticinans* (1523), f. 103r (quoted from Zambelli, 'Fine del mondo,' 363): "Arabes ergo et Sarraceni perfecerunt quae Graeci non potuerunt perficere in scientiis."

28. Ibid., f. 100r (after Zambelli, ibid.): "Multa ex Ptolemaei doctrina haurire possumus quae Messalus et Albumasar scripserunt, qui Ptolemaei sapientiae non contradicunt, sed potius cumulo quodam et experimentis manifestant quae Ptolemaeus in Quadripartiti libro truncate et brevi sermone dixit."

29. Such as Tommaso Giannotti and Johannes Stöffler, see Zambelli, 'Fine del mondo,' 316nn184, 185.

30. Sebastian Constantinus, *Liber de potestate syderum* (1534), sig. c4r: "Quoniam haec scientia per osservationes *(sic)* et experientias invenitur multa de novo post Ptolomeum. Adinventa sunt unde superiorum coniunctiones secundum doctrinam Ptolomei auctumari debent." On Sebastian Constantinus, see Thorndike, *History,* 5:215–217. Cf. also Guillaume Postel's remark that "we owe astrology and the practical art of medicine to the Arabs"; Postel, *Grammatica Arabica* (ca. 1539/1540), sig. D2r–v: "Taceo genera disciplinarum praeclarissime ab illius linguae authoribus pertracta. Astrologiam et rei medicae praxim illis debemus."

31. On house division in the Renaissance, see the helpful classification in Bezza, 'Representation of the Skies,' 63–65.

32. For an overview of Renaissance handbooks on astrology, see Lerch, *Scientia astrologiae,* 95–151, with a list of handbooks between 1470 and 1610 on 150–151. I have learned too late that the reference works by David Origanus (d. 1628) deserve more attention; see Lerch, *Scientia astrologiae,* 184: "Die frühneuzeitliche Handbuchtradition scheint hier ihren Höhepunkt erreicht zu haben."

33. On Campanella's book, see the introduction by Ernst to Campanella, *Opusculi astrologici* (2003), 12–17.

34. Bonatti, *Tractatus astronomie* (1491), trans. Dykes (2007).

35. Reisch, *Margarita philosophica* (1973), lib. 7, tr. 2, ch. 1–20, 293–319, esp. 293.

36. Zambelli, *The* Speculum, esp. 222–223.

37. Pingree, *From Astral Omens,* 34.

38. Ibid., 46–47.

39. Pico, *Disputationes* (1946/1952), 4.13, 500–506.

40. Ciruelo, *Apotelesmata* (1521), 4.1, sig. p6v: "Quo errore iudicio meo nullus in religione christiana dari posset maior aut periculosior sive loquamur secundum theologos, sive secundum philosophos . . . Corpora celestia nihil aliud significant naturaliter ab eo quod effective causare possunt. Constat autem quod speciem rei absentis causare non possunt."

41. Haly Rodoan in: Ptolemy, *Liber quadripartiti . . . Centiloquium* (1493), f. 53r: "Et cum ita sit, omnes quaestiones et electiones inclusae sunt in verbis Ptholomaei."

42. Cardano, *Commentarii in Ptolemaeum,* in *Opera* (1663), 5, lib. 3, praef., 242: "Cum interrogationes omnino sint sortilegae atque indignae non solum viro Christiano, sed viro bono. Similiter electiones plus inventae sunt causa avaritiae astrologorum quam quod quicquam conferant in utilitatem eligentis." On the *Centiloquium,* see also 356. For context, see Grafton, *Cardano's Cosmos,* 137.

43. Cardano, *De interrogationibus libellus,* in *Opera* (1663), 5:553: "Cum alias saepe interrogationum genus damnaverim, quod fortuitum est et legi nostrae repugnans, multorumque malorum causam praebet, rem autem ipsam

necessariam esse conspicerem, quod multi unum quaesitum scire cupiant, non autem omnia, ideo haec pauca a praescriptis non discedendo subiicere volui."

44. This advantage of interrogations is mentioned explicitly in Pseudo(?)-Albertus Magnus, *Speculum astronomiae,* ch. 10; see Zambelli, *The* Speculum, 238-239.

45. Pico, *Disputationes* (1946/1952), 4.7, 466: "de ea parte qua maxime utilis vitae putatur astrologia."

46. Ibid., 2.2, 106.

47. Cardano, *Commentarii in Ptolemaeum,* in *Opera* (1663), 5, lib. 3, praef., 242.

48. Ciruelo, *Apotelesmata* (1521), 4.1, sig. p7r.

49. Ibid., sig. p7v-q1r: "Caput secundum de electionibus astrologicis quae pertinent ad modum vivendi uniuscuiusque hominis volentis sequi naturam suam."

50. Pitati, *Almanach novum* (1544), f. 17r-24v; Dariot, *Introductio* (1557), 65-71; Campanella, *Astrologicorum libri* (1630), 209-230; Schöner, *De iudiciis nativitatum* (1545), f. 147r-148r: "Quomodo pro ratione cuiusque geniturae sint eligenda idonea tempora actionum aut negotiorum."

51. Dariot, *Introductio* (1557), 66.

52. A helpful introduction to this branch of astrology is in Dykes, *Persian Nativities,* vol. 3, introduction.

53. On the prorogator and anniversary horoscopes as the two most innovative contributions of Dorotheus to astrology, see Pingree, 'Māshā'allāh,' 9.

54. See Dykes, *Persian Nativities,* vol. 1, introduction, esp. ix-xviii. For the text of the *Book of Aristotle,* which is extant only in a Latin translation, see Burnett and Pingree, *The Liber Aristotilis.*

55. See Leopold, *Compilatio* (1489), sig. e6r, at the opening of ch. 5 on revolutions of the years of the world, and Bonatti, *Tractatus astronomie* (1491), e.g., ch. 9.7-8, sig. DD2r-DD3r. See Dykes's translation of Bonatti, *The Book of Astronomy,* ch. 9.7-8, pp. 1394-1399, and index s.v. 'revolutions of nativities.'

56. On prorogations in Renaissance astrology, see Vanden Broecke, *Limits of Influence,* ch. 8.

57. Cardano, *De revolutione annorum,* in *Opera* (1663), 5:561-575.

58. Cardano, *Commentarii in Ptolemaeum,* in *Opera* (1663), 5:364: "Verum dices nonne sufficiebat Ptolemaeo dicere quod tota figura singulo anno per unum signum procedebat quasi per viam revolutionis, ita ut figura revolutioni similis erigeretur? Respondeo quod non ob duas causas: prima ob inaequalitatem domorum, secunda quia non vult quod procedant nisi illa quinque loca principalia, non omnia."

59. Campanella, *Astrologicorum libri* (1630), 208.

60. Schöner, *De iudiciis nativitatum* (1545), lib. 3, praef., f. 95r-v: "Alii inscribunt hanc doctrinae partem 'de revolutionibus,' ego non facile a Ptolemaeo discedo, qui etsi multum tribuit transitibus stellarum, tamen neque introitus solis in puncta cardinalia observat, sed coniunctiones et oppositiones proxime

praecedentes, neque in rebus genethliacis circa discussionem thematis revolutionis occupatus est . . . Ne igitur a Ptolemaei sententia discederemus, initio docuimus quomodo per progressiones et transitus tempora eventuum praevideri debeant. Deinde his attexuimus Arabum et aliorum traditiones, quas etsi non contemnendas censeo, tamen potiores partes Ptolemaeo semper tribuendas esse iudicavimus."

61. To point to some of the correspondences: Schöner, ch. 3.8, quotes Abenragel, chapter 6.4 (on Alfridaria); Schöner, ch. 3.10, is a literal quotation of Abenragel, ch. 6.10; Schöner, ch. 3.11, draws on Abenragel 6.13 (aphorisms); and Schöner, ch. 3.12, quotes Abenragel, ch. 6.7 (on the return of the planets in the same sign in the revolution chart).

62. Giuntini, *Speculum astrologiae* (1573), 11: "Cum fere tota intentio astrologiae principaliter circa duo videatur, scilicet circa nativitates et circa revolutiones, primum ponam tractatum super hominum nativitates . . . deinde de revolutionibus nati faciam duos tractatus."

63. For an example of a seventeenth-century reception of the Arabic doctrine of anniversary horoscopes, see Argoli, *Ptolomaeus parvus* (1652), lib. 3, 172–211.

64. See North, *Chaucer's Universe*, 214–216, for an explanation of the doctrine as presented by Alcabitius.

65. Hasse, 'Avicenna's "Giver of Forms,"' esp. 229–230.

66. On ancient methods of predicting the length of life, see Heilen, *"Hadriani genitura,"* 2:984–1021; see esp. 985–991, on the ruler of the chart method, as opposed to the aphetic method. For the concept of the ruler of the chart itself, see 1057–1075. On the technical side of the Hyleg and Alcocoden method, see Dykes, *Persian Nativities,* 1:340–343, and Hand's postscript to Antonius de Montulmo, *On the Judgement of Nativities,* 73–78.

67. Bonatti, *Tractatus astronomie* (1491), tr. 9, pars 2, and Leopold, *Compilatio* (1489), 100–104.

68. Pico, *Disputationes* (1946/1952), lib. 2, ch. 7, 146–154, esp. 150: "Quid desipis, barbare? Aut quid somnias? Ubi apud Ptolemaeum alchocoden legistis?" and 148–150: "Barbari et nostrates, praeter istos aphetas quos ipsi hylegh vocant, alium esse credunt ex sententia etiam Ptolemaei qui vivendia spacia decernat, quem vocant alcochoden, de qua re, ut dicebam, nihil usquam ipse Ptolemaeus."

69. Latin wording of the *Centiloquium* according to the following edition: Ptolemy, *Liber quadripartiti . . . Centiloquium* (1493), f. 113r: "Non abscindas per directionem solam nisi complete dominationes significatorum fuerint . . . Intendit: . . . Non debemus per ipsum solum significare mortem nati sine directione hileg et alcochoden."

70. Cardano, *Commentarii in Ptolemaeum,* in *Opera* (1663), 5:270–271.

71. Campanella, *Astrologicorum libri* (1630), ch. 4.4, 128–134, esp. 130: "Dominus geniturae, qui est potentissimus planetarum roborat vitam, non autem metitur, ut putant Arabes, nisi sit princeps apheta."

72. Ciruelo, *Apotelesmata* (1521), ch. 3.2–3, sig. k1r–k4r.

73. Schöner, *De iudiciis nativitatum* (1545), sig. B3r–B4r.

74. Giuntini, *Speculum astrologiae* (1573), ch. 1.4, f. 24v–29r. The example is on f. 121r–v.

75. On Alcabitius's theory of lots, see North, *Chaucer's Universe*, 217–218, 527–529. Cf. also Yamamoto and Burnett, in Albumasar, *On Historical Astrology,* 1:593–597.

76. On ancient theories of the lot of fortune, see Heilen, 'Some Metrical Fragments,' 56–57, and, for a fuller treatment, Heilen, *"Hadriani genitura,"* 2:1158–1182.

77. Bonatti, *Tractatus astronomie* (1491), lib. 8, pars 2; Leopold, *Compilatio* (1489), ch. 4.5, f. d6v–e5v.

78. Pico, *Disputationes* (1946/1952), lib. 6, ch. 18, 142: "Nam et cicerum et fabarum et lentis et hordei et caeparum et mille rerum huiusmodi partes apud Arabes omnes et Latinos legis."

79. Ibid., 134: "virtus in caelo naturalis ubi nulla stella, ubi nec lumen nec alia qualitas naturalis."

80. Nabod, *Enarratio* (1560), diff. 5, 471: "Quapropter in examine sortium diutius hic commorari mihi necessarium aut utile minime visum est."

81. Giuntini, *Speculum astrologiae* (1573): pars infirmitatum (121r), pars vitae (122r), pars mortis (123v), pars legis (125r), pars fortunae (126r), pars matrimonii (127r), pars filiorum (127v).

82. Giuntini, *Speculum astrologiae* (1573), f. 320r.

83. Ibid., f. 205r.

84. Ibid., f. 125v: "Quo ad actiones, omissa Ptolomaei opinione et etiam Iulii Firmici, veniam ad opinionem Arabum, qui in hac materia meliores observatores Ptolomaeo et latinis astrologis."

85. Pingree, 'Māshā'allāh,' 9.

86. Bonatti, *Tractatus astronomie* (1491), tr. 8, part 1, f. 225r–302r; Leopold, *Compilatio* (1489), tr. 5, f. e5v–f1r.

87. Bonatti, *Tractatus astronomie* (1491), f. 225v–226r.

88. Pico, *Disputationes* (1946/1952), 5.5, 548.

89. Ptolemy, *Tetrabiblos* (1940), 2.10, 194–201.

90. Pico, *Disputationes* (1946/1952), 5.5, 556.

91. Nifo, *De nostrarum calamitatum causis* (1505), f. 3rb and f. 32rb.

92. Ibid., f. 33va–b: "Nimirum igitur si tot errata et deliramenta scribant hi qui annuas predictiones faciunt, cum e Ptolomei preceptis sint alieni tum propter Ptolomei difficultatem, tum propter ipsas eius ineptas translationes, tum vel maxime propter facilitatem doctrine Albumasaris."

93. On Pigghe, see Vanden Broecke, *Limits of Influence,* 85–91.

94. Pigghe, *Astrologiae defensio* (1519), praef., f. 4r, f. 5r: "primo, prognostica annorum mundi, quae annis singulis fiunt ab his nostris divinatoribus, solum aniles quasdam fabulas esse."

95. Ibid., f. 6v–7r.

96. Ibid., f. 16v, 17v. See Vanden Broecke, *Limits of Influence,* 89–90.

97. Nabod, *Enarratio* (1560), diff. 4, 359: "quod tempus illud adeo exacte supputari vix queat ut perfectioni seu structurae coelestis thematis absque crasso errore serviat."

98. Ciruelo, *Apotelesmata* (1521), ch. 2.3, f. h5v.

99. Ibid., f. h6r.

100. Ibid., f. i1v.

101. Ibid., f. h4r–i1v.

102. Pitati, *Almanach novum* (1544), f. 27r.

103. Cardano, *Commentarii in Ptolemaeum*, in *Opera* (1663), 5:222.

104. Campanella, *Astrologicorum libri* (1630), 78.

105. Ibid., 98.

106. Magini, *Ephemeridum . . . continuatio* (1610), tr. 3, 251: "De revolutionibus seu annis reversionibus." On the genre of ephemerides and their astrological adjuncts, see Lerch, *Scientia astrologiae,* 153–187.

107. Hellmann, *Versuch,* 26–39. On judicia concerning the flood of 1524, see also Hellmann, 'Aus der Blütezeit,' 25–67.

108. Hellmann, *Versuch,* 40. Cf. Thorndike, *History,* 6:99, 143. On early modern English Almanacs, that is, calendars with annual prognostications, see Capp, *Astrology.* On Italian annual predictions, see Casali (who does not seem to know Hellmann's work), *Le spie.*

109. A "distinctly prevailing presence of Arabic elements over Ptolemaic influences" in yearly prognostics is also observed by Bezza, 'Representation of the Skies,' 84.

110. For examples of conflicting prognostica, see Hellmann, *Versuch,* 5.

111. Pingree, *From Astral Omens,* ch. 4, esp. 43–44. See also Pingree, 'Historical Horoscopes'; Kennedy, 'The World Year Concept'; Albumasar, *On Historical Astrology,* 1:582–587.

112. Albumasar, *On Historical Astrology,* ch. 1.1, §11, vol. 1, pp. 10–11; Kennedy and Pingree, *The Astrological History,* vi.

113. Yamamoto and Burnett, in Albumasar, *On Historical Astrology,* 1:xv–xxii.

114. See Kennedy, 'Al-Battānī's Astrological History,' with an Arabic–English edition plus commentary.

115. Yamamoto and Burnett, in Albumasar, *On Historical Astrology,* 1:xv, 585–586. See also the Appendix in this book, s.v. 'Omar Tiberiades' and 'Messahalah.'

116. This text is extant in Latin and Hebrew only. See Sela's Hebrew–English edition, in Abraham ibn Ezra, *The Book of the World,* 235–259.

117. Abraham ibn Ezra, *The Book of the World.* See Smithuis, 'Abraham ibn Ezra's Astrological Works,' 248.

118. For this text, see Sela's edition in Abraham ibn Ezra, *The Book of the World,* 261–269, with comments, 29–31.

119. Kennedy and Pingree, *The Astrological History;* cf. North, 'Astrology and the Fortunes,' 67.

120. North, 'Astrology and the Fortunes,' 70–76; Thorndike, *History*, 3:325–346.

121. Bonatti, *Tractatus astronomie* (1491), tr. 4, f. 72v–76v; Leopold, *Compilatio* (1489), tr. 5, f. e8v.

122. Alcabitius, *Introduction to Astrology*, ch. 4, §2.

123. See Bezold, 'Astrologische Geschichtsconstruction'; Pruckner, *Studien*, esp. 43–72; Smoller, *History*, 61–84; Haeusler, *Das Ende der Geschichte*, 142–155; North, 'Astrology and the Fortunes,' 59–89.

124. See the monograph on Pierre d'Ailly's Christian astrology by Smoller, *History*.

125. On Cusanus, see Roth, 'Die astronomisch-astrologische "Weltgeschichte,"' 1–29.

126. Pitati, *Almanach novum* (1544), f. 32r: "Horum maxima est coniunctio Saturni et Iovis in principio Arietis, quae in 960 annis fit. Secunda autem amborum coniunctio est in initio uniuscuiusque triplic<it>atis in annis videlicet 240. Tertia coniunctio est Saturni et Martis in initio Cancri, quae quibuslibet 30 vel circiter fit. Quarta est coniunctio Iovis et Saturni in unoquoque signo eiusdem triplicitatis. Quinta est introitus Solis in aequinoctii vernalis punctum, videlicet quae in capite Arietis anno quovis fit. Unde astrologi anni initium sumunt. Sexta est coniunctio luminarium in menstruis lunaribus coniunctionibus."

127. Yamamoto and Burnett, in Albumasar, *On Historical Astrology*, 1:609.

128. Albumasar, *On Historical Astrology*, ch. 1.1, §19, 18–19. On Albategnius, see Kennedy, 'Al-Battānī's Astrological History,' 18–19, 80.

129. The influence of conjunctionist theory in later Arabic astrology still awaits proper investigation; for Saturn–Jupiter conjunctions in the tenth-century author al-Anṣārī, see Orthmann, 'Astrologie und Propaganda.'

130. On the technical aspects of the doctrine, see Kennedy, 'The World Year Concept,' 358–365; North, 'Astrology and the Fortunes,' 63–66; Yamamoto and Burnett, in Albumasar, *On Historical Astrology*, 1:582–584; Heilen, 'Lorenzo Bonincontris Schlußprophezeiung,' 312–315; Kennedy, 'Al-Battānī's Astrological History,' 80–82.

131. Messahalah, in turn, uses a different scheme of small, middle, and great conjunctions, which involves Mars; see Yamamoto and Burnett, in Albumasar, *On Historical Astrology*, 1:585n.

132. As can be gathered from the very helpful table in Heilen, 'Lorenzo Bonincontris Schlußprophezeiung,' 328.

133. Albumasar, *On Historical Astrology*, 1:12–13.

134. Kennedy and Pingree, *The Astrological History*, 40, lines 12–14.

135. On the reception of the theory in the fifteenth to seventeenth centuries, see, for general information, Garin, *Astrology*, 1–28, and Geneva, *Astrology*, 132–140. More thorough studies are Aston, 'The Fiery Trigon Conjunction'; Zambelli, 'Fine del mondo'; Ernst, 'From the Watery Trigon'; Leppin, *Antichrist*, 140–144; Mahlmann-Bauer, 'Die Bulle,' 179–182; Heilen, 'Lorenzo Bonincontris Schlußprophezeiung.'

136. North, 'Astrology and the Fortunes,' 81.
137. Pico, *Disputationes* (1946/1952), 5, 520–622. On book 5, see North, 'Astrology and the Fortunes,' 81–83, and Zambelli, 'Introduction,' 24–28.
138. Pico, *Disputationes* (1946/1952), 5.5, 552–556.
139. Pontano, *Commentariorum . . . libri duo* (1531), p. 112, nr. 63: "Quum Saturnus Iupiterque coniunguntur, uter eorum sublimior sit vide ac iuxta illius naturam pronunciato. Idem etiam in caeteris stellis facito." For the Greek text, see Pseudo-Ptolemy, *Fructus sive Centiloquium* (1952), p. 49, nr. 63.
140. Pico, *Disputationes* (1946/1952), 5.6, 558–562.
141. Ibid., 564.
142. Ibid., 5.17, 622.
143. Ibid., 5.6, 562: "princeps istius dogmatis."
144. Riccioli, *Almagestum* (1651), bk. 7.5, ch. 9–10, 670–680. Riccioli's comprehensive *Almagestum novum* became a reference work for many astronomers of the later seventeenth and early eighteenth centuries. Riccioli developed an influential nomenclature for formations on the moon, which also included several names of Arabic scholars, as noted by Brentjes, *Travellers from Europe*, art. 8, 32–33.
145. Riccioli, *Almagestum* (1651), 678a: "Non est hic locus redarguendi astrologiae iudiciariae proscriptam olim temeritatem."
146. Ibid., 678–679.
147. Abraham ibn Ezra, *The Book of the World*, 53: "If you come across Abū Maʿshar's *Book on the Conjunctions of the Planets* you would neither like it nor trust it, because he relies on the mean motion for the planetary conjunctions."
148. On the development of chronology and Scaliger's role in particular see Grafton, *Joseph Scaliger*, vol. 2.
149. Riccioli, *Almagestum* (1651), 677: "Ratio vero potissima et peculiaris, praetermissis quae universaliter afferri solent pro efficacia cuiusvis aspectus ac syzygiae planetarum, est connexio insignium in orbe terrae vicissitudinum ac mutationum cum maximis hisce coniunctionibus."
150. See Smoller, *History*, 66–70. Of the various texts that Pierre d'Ailly has devoted to astrological history, the *Concordantia* is presented here because this is a text that Pico singles out for criticism in his *Disputationes* (5.9).
151. Ciruelo, *Apotelesmata* (1521), sig. h2v: "Eius autem imperii anno 36 fuit prima maxima coniunctio duorum superiorum in ariete, hoc est sex annis fere ante nativitatem Christi, et duravit per 800 fere annos usque ad tempora Karoli Magni regis Francorum et primi imperatoris latinorum, quando Leo papa tertius transtulit imperialem dignitatem e Grecis ad Latinos. Et ex tunc incepit alia maxima coniunctio eorundem Saturni et Iovis quae durabit per alios 800 annos usque ad annos Christi 1600 fere."
152. Cardano, *Segmenta,* in *Opera* (1663), 5:32.
153. Leovitius, *De coniunctionibus* (1564), sig. B1r–4r; f. M1r, f. N2v–N3r.
154. Kepler, *Stella nova* (1606), 29.
155. Origanus, *Novae motuum coelestium ephemerides* (1609), 448–450.

156. Spina, *De maximis coniunctionibus* (1621), ch. 2.2, 58–59.

157. Argoli, *Pandosion* (1653), ch. 73, 328–329.

158. Riccioli, *Almagestum* (1651), 677b.

159. Mahlmann-Bauer has drawn attention to several treatises by German astrologers that contain astrological histories: Nicolaus Winckler, *Bedencken* (1583); Laurentius Eichstadius, *Prognosticon* (1622); Simeon Partlicius, *Mundus furiosus* (1623); David Herlicius, *Groß Prognosticon* (1623); Eberhard Welper, *Conjunctio* (1663); and Johann Henrich Voigt, *Vorstellung* (1682). Of these, I was able to consult the first, which is less detailed on astrological history than the sources treated in the present study.

160. Albumasar, *On Historical Astrology*, ch. 1.1, §26, vol. 1, pp. 22–23.

161. Ciruelo also contributed a *Giudicio* for the year 1524; see Zambelli, 'Fine del mondo,' 311–312.

162. Ciruelo, *Apotelesmata* (1521), sig. g8r: "Tum quinto quia de quattuor primis magnis coniunctionibus agens *(i.e., in his enumeration of conjunctions as in the quotation from Pitati above)* Albumazar solas medias coniunctiones superiorum planetarum curavit, quae sunt pure imaginariae et nullum effectum faciunt in mundo."

163. Ciruelo calculates with the following dates: April 10, 1464–November 20, 1484–June 10, 1504–February 1, 1524. The precise data are (in the Julian calendar): April 8, 1464–November 18, 1484–May 25, 1504–January 31, 1524. The biggest difference concerns the 1504 conjunction, which Ciruelo dates to June 10, 1504, as does Regiomontanus, *Ephemerides* (1474), f. 413v. But Regiomontanus and Ciruelo differ on the 1484 conjunction, which Regiomontanus dates to November 25, 1484. On the 1504 conjunction, see also Heilen, 'Lorenzo Bonincontris Schlußprophezeiung,' 317–326.

164. See the critique of conjunctionism in William of Auvergne, *De fide et legibus*, in *Opera omnia* (1674), 1:ch. 20, p. 55aB, and of Albumasar's astrology in Thomas Aquinas, *Summa contra gentiles* (2005), bk. 3, ch. 86, 87.

165. Ciruelo, *Apotelesmata* (1521), sig. g8v: "Praeter haec multa alia vana, absurda et falsa ponit in eodem volumine de magnis coniunctionibus, quae sine ratione et divinatorie dicta sunt ac cuilibet viro sapienti satis indigna; unde iure optimo Parisienses theologi examinato bene libro illo Albumazar archidivinatorem cognominarunt, librum cum suo auctore damnantes."

166. Ibid., sig. g8v–h1r.

167. For a description of this technique, see North, *Chaucer's Universe*, 213–214.

168. Ciruelo, *Apotelesmata* (1521), sig. h2v.

169. On Cardano's astrology, see Ernst, 'Astri e previsioni,' Ernst, 'Veritatis amor,' and Grafton, *Cardano's Cosmos*, with further literature.

170. Cardano, *Pronostico* (1999), 470: "La coniunctione del 1564 dinota la renovatione de tutte le legge, de la Christiana e de la Machometana, e si refermerà una legge nova, scritta e non udita."

171. Grafton, *Cardano's Cosmos,* 69.

172. Cardano, *In Quadripartitum* (1554), sig. a2v: "Tot Albumasares, Abenrageles, Alchabitios, Abubatres, Zaheles, Messahalacos, Bethenes, Firmicos, Bonatos, boni genii, quid iam superest aut reliquum est a tot impostoribus? a tot nugis?"

173. Cardano, *Commentarii in Ptolemaeum,* in *Opera* (1663), 5:173b: "Igitur nec certa dies, nedum hora, aut coeli figura et status haberi possunt."

174. Ibid., 173b: "Hanc ob causam scientiam talium coniunctionum quae ad hanc usque diem adeo celebrata fuit, quam parvi sit momenti clare quisque intelligit. Sunt tamen quaedam generalia quae ex coniunctionibus magnis habentur."

175. Ibid., 173–174.

176. On this work, see Ernst, 'Veritatis amor,' 45–46.

177. Cardano, *Libelli quinque* (1547), f. 211v–212r: "Saturni et Iovis coniunctio minor fit singulis annis 19, diebus 315, horis 19. Unde processus fit per partes 242, mi. 59, sec. 9, hoc est autem per signa signiferi 8, par. 2, mi. 59, sec. 9 et sic in eadem trigona signorum natura, ut ex Ariete in Sagittarium, partibus ferme tribus procedens. Inde cognoscimus ex ignea in terream, inde in aeream, post in aqueam transire signorum naturam, decem completis revolutionibus sub eadem signorum natura ac qualitate. Manent igitur in eodem signorum trigono annis 198, diebus 236 communibus . . . Annis igitur ante Christum 800, diebus 98 sumpsit initium magna coniunctio, quae singulis 794 annis, 214 diebus renovatur."

178. Cf. Etz, who reckons with the mean figure of 794 years between conjunctions in regard to the precessing vernal equinox ('Conjunctions of Jupiter and Saturn').

179. Albumasar, *On Historical Astrology,* ch. 1.1, §17, 12–13, 14–15; cf. the commentary on p. 583.

180. Cardano, *Libelli quinque* (1547), f. 212v: "Tabula magnarum et mediarum coniunctionum secundum medios motus."

181. Nabod, *Enarratio* (1560), diff. 4, 354. He uses the figure 19y 315d 19h.

182. Kepler, *De trigono igneo* (1606), ch. 7, 27: "Quam Astrologiae partem Joannes Picus Mirandulanus mihi nondum eripuit, etsi plerisque que libri duodecim contra astrologos disputavit . . . subscribo."

183. Ibid., ch. 8, 36: "Non haec dico quod astrologorum inductiones usque ad specialia praedicenda defendam, sed ut obtineam circa tempus magnarum coniunctionum tantas fuisse commotiones naturarum ipsorumque affectuum naturalium in hominibus, ut ex iis decepti sint astrologi putantes res ipsas, quae per illas commotiones gerebantur, ab hoc coelesti principio profectas."

184. Ibid., 34: "Ad eundem modum dico observatum esse a veteribus et hodie observari maximam vim conjunctionum planetarum in ciendis facultatibus rerum sublunarium."

185. On the important role of empiricism in Kepler's astrology, see Krafft, 'Tertius interveniens,' 218–221, esp. 221.

186. Kepler, *Discurs von der grossen Conjunction* (1868), 697–711, here 701. Substantial parts of the text are also easily accessible in Strauss and Strauss-Kloebe, *Die Astrologie des Johannes Kepler,* 82–97, here 87.

187. Kepler, *Discurs von der großen Conjunktion* (1868), 702.

188. As John North has observed; see North, 'Astrology and the Fortunes,' 84n11.

189. Kepler, *Epitome astronomiae Copernicanae* (1953), 6.6, 487–488.

190. Kepler, *De trigono igneo* (1606), ch. 7, 29: "Accipe in tabella rotundis minimeque praecisis numeris, quibus saeculis igneus trigonus inierit."

191. Riccioli, *Almagestum* (1651), 674.

192. Origanus, *Novae motuum coelestium ephemerides* (1609), 448. Origanus uses the figure of 794y 331d in Egyptian years, which is the equivalent of 794y 133d in Julian years.

193. See Grafton, *Joseph Scaliger,* vol. 2; Grafton, *Defenders of the Text,* esp. ch. 3, 4, and 8 (on Annius of Viterbo, Scaliger, and Isaac La Peyrère, respectively); and the older, but still very valuable study of the development of world chronology by Klempt, *Die Säkularisierung.*

194. On the decline of astrology in the later seventeenth and eighteenth centuries, see the very informative article by Rutkin, 'Astrology,' 552–561.

195. An ideological bias toward Greek astrology is sometimes also entertained by modern historians of astrology, such as Krafft, 'Johannes Keplers Bemühungen,' 201, 205, where Arabic astrology is described as excesses over and falsifications of Greek astrology.

## 7 · CONCLUSION

1. I adopt this term from Said, *Orientalism,* 54.

2. Such a thesis would be in line with the views of Gouguenheim, Brague, and Fanjul on the Arabic contribution to medieval Latin culture; for references see the Preface.

3. Cf. Renan, *Averroès et l'Averroïsme,* ch. 3.15, 414: "un aristotélisme décrépit."

4. Cf. Brentjes, *Travellers from Europe,* xv. Cf. also Monfasani, 'Aristotelians,' 266: "Petrarch's anti-Averroistic strictures hardly resonated with Quattrocento humanists."

5. For comments and bibliography see the preface to this study.

6. For an overview on the Arabic influences on the theory of miracles, see, with further literature, Hasse, 'Arabic Philosophy and Averroism,' 121–125.

7. Siraisi, *Avicenna in Renaissance Italy,* ch. 6 and appendix 2.

8. References to these commentaries can be found in the author entries in the Appendix.

9. Cited from Stübler, *Leonhart Fuchs,* 171: "Et quum nemo sit, qui nesciat Arabes omnia ferme sua e Graecis transscripsisse, parcissime deinceps ad doctrinam

studii huius adhibebuntur, quod consultius sit artis praecepta a fontibus quam a turbidis rivulis haurire."

10. Simplicius, *Commentarii in libros de anima Aristotelis* (1543), f. 35v: "Simplicium unum vobis die noctuque versandum proponite; ... et illum *(sc., Averroem)* iam de manibus deponite."

11. Nifo, *Aristotelis physicarum acroasum ... liber* (1508), f. 218va: "Malo enim errare cum Graecis in expositione Graecorum auctorum quam recte sentire cum Barbaris qui linguae peritiam non habent nisi per insomnia."

12. Vernia, *Contra perversam Averroys opinionem de unitate intellectus* (1499).

13. Nifo, *Aristotelis physicarum acroasum ... liber* (1508), f. 218va: "Propterea mihi videtur Averroes magnus temulentus. Hi<c> enim cum litteras grecas ignoraverit non nisi temere grecum auctorem exponere potest."

14. Pigghe, *Adversus prognosticatorum vulgus* (1518), f. 19r: "Accessi autem eo solum consilio, si forte invitare possemus astrologiae studiosos, ut relictis fabulis Albumazaris et Punicorum Ptolomeum nostrum legerent diligentius."

15. On this controversy, see Rummel, *The Humanist-Scholastic Debate.*

16. In a letter to a bishop prefacing his *Canon* edition of 1522, Champier again refers to Islam. See Avicenna, *Liber canonis* (1522), sig. +2r: "Verum multis post seculis Avicenna sub mahomethea spurcissima et nephanda secta medicinam profitente Avicenne nomen apud Hispanos increbuit ..." ("But many centuries later *(i.e., after Galen)* Avicenna professed medicine under the most filthy and impious Mohammedan sect and his fame spread among the Spaniards"). For context, see Copenhaver, *Symphorien Champier,* 141.

17. Vives, *De causis* (1990), 196: "Nam et Abenrois doctrina et metaphysica Avicennae, denique omnia illa Arabica videntur mihi resipere deliramenta Alcorani et blasphemas Mahumetis insanias."

18. Reisch, *Margarita philosophica* (1973), ch. 11.30, 472.

19. Champier, *De medicine claris scriptoribus* (1506?), f. 14r.

20. On Renaissance attitudes toward the Turks, see Bisaha, *Creating East and West;* Meserve, *Empires of Islam;* Vitkus, 'Early Modern Orientalism'; Contadini and Norton, *The Renaissance and the Ottoman World.* Cf. also the articles pertaining to the Ottoman Empire in the collective volumes edited by Schülting, Müller, and Hertel, *Early Modern Encounters,* and by MacLean, *Re-Orienting the Renaissance.*

21. See the Preface for literature on Arabic studies in Europe. On travelers, see Brentjes, *Travellers from Europe,* and the very informative survey in Hamilton, *The Arcadian Library,* 13–103.

22. Postel, *Grammatica Arabica* (1539/1540), sig. D2v: "Quis enim neget saecula proficere semper?"

23. Thorndike, *History,* 5:3; Sarton, *The Appreciation,* 174; Randall, *The Making,* 212. For quotations from these scholars see the Preface.

24. Girolamo Donzellini, preface to Leonardus Iacchinus, *In nonum librum Rasis ... ad Almansorem* (1564), sig. B4r: "A Graecis ad Arabas delata *(sc.,*

*medicinae scientia)*, naufragium fecit, ac Latini ab Arabibus illam recipientes, diu admodum infoeliciter in illa versati sunt."

25. Melanchthon, preface to Ptolemaeus, *De praedictionibus astronomicis* (1553), 4: "Habuit vetustas doctrinam de motibus stellarum haud dubie eruditissime extructam *(sic)* ex antiquissimis observationibus, habuit accurate descriptam seriem temporum et veterum Imperiorum. Cum autem Aegypto Sarracenica barbaries et Academiam Alexandrinam et studia delevit, vetera monumenta perierunt. Et tota doctrina de motibus interitura erat, quae tunc dispersa in multos et varios libros, nisi paulo ante Sarracenicos motus Ptolemaeus totam artem in unum volumen contraxisset."

26. l'Obel, Pena, *Nova stirpium adversaria* (1576), 406: "Sero et magno aegrotantium incommodo innotuit Graecis sena, quam una et appellationem Arabibus debemus."

27. Cardano, *De interrogationibus libellus,* in *Opera* (1663), 5:553: "Cum alias saepe interrogationum genus damnaverim."

28. After Clement of Alexandria, *Stromateis,* 1.15.67: "ὁ δὲ Ἐπίκουρος ἔμπαλιν ὑπολαμβάνει μόνους φιλοσοφῆσαι Ἕλληνας δύνασθαι." This saying was used as a motto by the Dutch Arabist Simon van den Bergh for his English translation of Averroes's *Incoherence of the Incoherence* in 1954. On the tradition of this "unfortunate expression of cultural chauvinism" (Franz Rosenthal) in Arabic studies see Gutas, 'The Study of Arabic Philosophy,' 11–12.


APPENDIX

1. Cranz, *A Bibliography of Aristotle Editions 1501–1600, 2nd ed. with addenda and revisions by Charles B. Schmitt.*

2. Bos and Burnett, *Scientific Weather Forecasting,* 32–34; Burnett, 'Lunar Astrology,' 64.

3. Partial Latin translations of these authors appeared in 1650: Abulfeda, *Chorasmiae . . . descriptio* (London, 1650) by John Greaves, and Barhebraeus, *Specimen historiae arabum,* Oxford, by Edward Pocock Sr. In 1663, Pocock published the complete text: Barhebraeus, *Historia compendiosa dynastiarum* (Oxford). See Fück, *Die arabischen Studien,* 86–97; Daiber, 'The Reception,' 68–74; Toomer, *Eastern Wisedome,* 160–161, 212 (on Pocock Sr.) and 172–174 (on Greaves).

4. Avempace once appears on a sixteenth-century title page as one of several possible authors of the *Liber de causis: Aristotelis Stagiritae Metaphysicorum libri xiiii, Theophrasti metaphysicorum liber, De causis libellus Aristoteli seu Avempacae vel Alpharabio aut Proclo asscriptus* (Ingolstadt, 1577). Note that Abraham de Balmes produced a Hebrew–Latin translation of Avempace's *Epistola expeditionis,* which is extant in Ms. Vatican library, lat. 3897, f. 32–65; see Tamani, 'Le traduzioni,' 619.

5. Cardano praises Alcoarismi as one of the twelve most significant scientists of history; Cardano, *De subtilitate* (1559), lib. 16, 570. Cf. the entries on Alkindi and Geber filius Affla below.

6. Such as on the following title page of Johannes Schöner's treatise on the Saphea: *Sapheae recentiores doctrinae patris Abrusahk Azarchelis summi astronomi a Ioanne Schonero . . . innumeris in locis emendatae correctis erroribus eius qui ex Arabico convertit* (Nuremberg, 1534).

7. Cited under the name 'Abencenif'; see Millás Vallicrosa, 'La traducción castellana,' 281–332. On Ibn Wāfid in general, see Ullmann, *Die Natur- und Geheimwissenschaften*, 443–444; Vernet, *Die spanisch-arabische Kultur*, 52–53.

8. Ullmann, *Die Medizin*, 210 and 273.

9. Burnett, 'The Coherence,' 280, 286n223. Iolanda Ventura (in a paper given in Cordoba, May 26, 2015) has pointed to differences between Ibn Wāfid's Arabic text and the translation of Gerard, who may have reorganized the material.

10. Villaverde Amieva, review of Aguirre de Cárcer, *Ibn Wāfid*, 111–118.

11. On the translators see Jacquart, 'La coexistence,' in *La science médicale*, art. 10, 277–285, and Jacquart and Micheau, *La médecine*, 164–165, 210.

12. Cf. the list of *Liber aggregatus* editions in Straberger-Schneider, *Der Liber aggregatus*, 321–326. Like Straberger-Schneider, I have not found any trace of the alleged editions Basel 1499 and Lyon 1510.

13. Brockelmann, *Geschichte*, 1:608–611, and *Supplementband I*, 839–844.

14. On Abharī and his *Hidāya* see Hasse, 'Mosul and Frederick II,' 145–163.

15. Cf. Kleinhans's description of the Latin edition: "In *Isagoge* auctor agit 100 numeris distinctis sectionibus de significatione vocis, de quinque universalibus, de definitione, conversione, syllogismo, demonstratione, argumentatione" (Kleinhans, *Historia studii linguae arabicae*, 70).

16. Kleinhans, *Historia studii linguae arabicae*, 57–72. See also Hamilton, *The Arcadian Library*, 302.

17. On his life and works, see the comprehensive article in DSB s.v. 'al-Battānī' (by Hartner) and BEA s.v. 'Battānī' (by van Dalen).

18. Sezgin, *Geschichte*, 6:182–187; BEA s.v. 'Battānī' (by van Dalen).

19. Burnett, *Hermann of Carinthia*, 9, 30.

20. As shown by Hartner in DSB s.v. 'al-Battānī'; on Copernicus, see also the literature referred to in the Preface.

21. Ullmann, *Die Natur- und Geheimwissenschaften*, 312–313; Sezgin, *Geschichte*, 7:120–121.

22. DSB s.v. 'Plato of Tivoli,' 32 (by Minio-Paluello).

23. Burnett, 'John of Seville and John of Spain,' 73.

24. Bezzel, 'Joachim Heller (ca. 1520–1580),' 296–298. Heller published the work together with the accomplished mathematician, astrologer, and printer Johannes Schöner (d. 1547), but Heller was the author of the dedication. On these editions, see also Thorndike, *A History*, 5:394–396.

25. Burnett, 'A Hermetic Programme,' 101–102.

26. Ullmann, *Die Natur- und Geheimwissenschaften*, 308–309; Sezgin, *Geschichte*, 7:122–124.

27. DSB s.v. 'Plato of Tivoli' (by Minio-Paluello), 32; Carmody, *Arabic Astronomical*, 137.

28. Boudet, *Le Recueil*, 1:291–292, and the references in 2:410.

29. Steinschneider, *Die hebraeischen Übersetzungen*, 546; Carmody, *Arabic Astronomical*, 137; Burnett, *'Partim de suo,'* 70.

30. Nutton, 'Humanist Surgery,' 75–99, with references not to Albucasis, but to Guy de Chauliac.

31. Translation of the title after Ullmann, *Die Medizin*, 149; cf. Steinschneider, *Die hebraeischen Übersetzungen*, 740.

32. Jacquart and Micheau, *La médecine*, 216–218.

33. Ullmann, *Die Medizin*, 271.

34. Saladin of Asculo (Saladino Ferro), *Compendium aromatariorum* (1488), sig. A1v–A2r. See the quotation in Chapter 4, n. 226.

35. Albucasis, *Liber theoricae necnon practicae Alsaharavii* (1519), sig. A1v: "non poteris ... desistere quid frugiferum hunc authorem squalido et diuturno carcere eruas ac libere proferas in lucem." Ricci's preface is put in context by Roling, *Aristotelische Naturphilosophie*, 6.

36. DSB s.v. 'Abū Maʿshar' (by Pingree); Yamamoto and Burnett, *Abū Maʿšar on Historical Astrology*, 1:xiii–xiv; Burnett, 'Abu Ma'shar,' 17–29. Cf. also Ullmann, *Natur- und Geheimwissenschaften*, 316–324, Sezgin, *Geschichte*, 7:139–151.

37. Burnett, 'Abu Ma'shar,' 21. Cf. Lemay, *Abū Maʿšar al-Balḫī*, 4:306–315.

38. Lemay, *Abū Maʿšar al-Balḫī*, 7:129–136.

39. On the Latin translation, called *Ysagoge minor*, see Burnett et al., introduction to Albumasar, *The Abbreviation* (1994), 3–10.

40. On the different titles and attributions associated with this work in the Arabic tradition, see Yamamoto and Burnett, *Abū Maʿšar on Historical Astrology*, 1:xv–xxii.

41. Burnett, 'The Strategy of Revision,' 51–113, 529–540.

42. Yamamoto and Burnett, *Abū Maʿšar on Historical Astrology*, 2:xxiii.

43. The complicated textual transmission of the *Flores* is best explained in Yamamoto and Burnett, *Abū Maʿšar on Historical Astrology*, 1: xvi–xviii (work no. [1], referred to as *The Report*). Cf. Carmody, *Arabic Astronomical*, 92–94 (no. 3). This text is not identical with another work of general astrology by Albumasar, the *De revolutionibus annorum mundi seu liber experimentorum* (no. 4a in Carmody), as Pingree and Sezgin wrongly believe (Sezgin, *Geschichte*, 7:142–143, no. 3; DSB s.v. 'Abū Maʿshar' [by Pingree], no. 9). The Arabic original of the latter text has not yet been identified.

44. Burnett, 'John of Seville and John of Spain,' 61.

45. Carmody, *Arabic Astronomical*, 94–101.

46. Pingree, preface to: *Albumasaris De revolutionibus nativitatum*, v–vi; Burnett, 'Abu Ma'shar,' 22–23.

47. On this translation see Burnett, 'Abu Ma'shar,' 23–29.

48. Burnett, Yamamoto, and Yano, *Al-Qabīsī*, 1–7; Sezgin, *Geschichte*, 5:311–312; 6:208–210; 7:170–171. Cf. EI² s.v. 'al-Kabisi' (by Pingree). 967 AD is the year of death of Sayf ad-Dawla, not of Alcabitius (confused by Ullmann, *Die Natur- und Geheimwissenschaften*, 332).

49. Burnett, Yamamoto, and Yano, *Al-Qabīsī*, 198–202; Burnett, 'John of Seville and John of Spain,' 78.

50. On commentators on Alcabitius, see Arnzen, 'Al-Qabīsī's astrologische Lehrschrift,' 100–107, 110–112: Cecco d'Ascoli, John Danko of Saxony, John of Stendal, anonymous (John of England?), anonymous (MS Venice), Loys de Langle. On John of Stendal, see also Lorenz, *Studium generale Erfordense*, 155–156.

51. On Nabod, see Thorndike, *A History*, 6:119–123.

52. On de Fantis and his editions, see Scarabel, 'Une "édition critique" latine,' 9–18.

53. Discussed and edited by Burnett, Yamamoto, and Yano, *Al-Qabīsī*, 375–385, esp. 375: "This text in its present form would seem to derive from a Western Christian context." Cf. EI² s.v. 'al-Kabīsī' (by Pingree): "perhaps it is not by al-Kabīsī at all."

54. See the list of editions of Alcabitius in Burnett, Yamamoto, and Yano, *Al-Qabīsī*, 192–194.

55. Reliable accounts of Alfarabi's life are in *Enc. Iran.* s.v. 'Fārābī,' 208–213 (by Gutas) and *Ueberweg I*, 369–374 (by Rudolph).

56. On the Latin translations of Alfarabi's philosophical works, see the recent overview in Schneider, *Al-Fārābī: De scientiis*, 37–46, and, for further literature, *Ueberweg I*, 376 (by Rudolph).

57. The Arabic original of this treatise is lost; it once formed the introduction to Alfarabi's "Commentary on Aristotle's Rhetoric" *(Šarḥ kitāb al-ḫatāba li-Arisṭūtālīs);* see Langhade and Grignaschi, *Deux ouvrages*, 131. On its influence in late medieval philosophy, see Maierù, 'Influenze arabe,' 253–257.

58. Langhade and Grignaschi, *Deux ouvrages*, 142–145.

59. Ms. Vat. lat. 12055; see Tamani, 'Le traduzioni,' 619.

60. In the dedicatory preface to Averroes, *Libri Posteriorum analiticorum* (1523), sig. AA3r.

61. Endreß, 'Die wissenschaftliche Literatur,' 50 (with further literature).

62. Not printed was the anonymous *De ortu scientiarum*, which is often ascribed to Alfarabi, but upon scanty evidence: only one of twenty-seven manuscripts attributes the text to Alfarabi (see Burnett, *The Introduction of Arabic Learning*, 64–65). Not printed either was the Alfarabian treatise *al-Tanbīh 'alā sabīl as-sa'āda* ("Reminder of the Way of Happiness"), *Liber exercitationis ad viam felicitatis.*

63. Foresti, *Supplementum* (1506), f. 273v: "E quibus etiam quaedam apud latinos habentur commentaria." For Foresti, Schedel, and Staindel on Alfarabi see Chapter 2.

64. Gesner, *Bibliotheca universalis* (1545), f. 30v: "Alpharabii cuiusdam liber de compositione astrolabii extare fertur."

65. Sezgin, *Geschichte*, 6:195–196.

66. Cf., for example, Agostino Nifo's discussion of arguments by Alfarabi in: *De intellectu* (1554), ch. 6.20, f. 56v, and 6.40, f. 61r.

67. Alfraganus, *Chronologica et astronomica elementa* (1590), 4–5: "Quo tempore vixerit, ex observationibus illius coniicere licet … Ex quibus omnibus apparet Alfraganum vixisse circa annum Christi 950."

68. Voss, *De universae mathesios natura* (1650), 174: "De aetate quod diximus, idem Golius colligebat ex Albufergio in scriptorum veterum ac recentiorum catalogo, ubi eum recenset inter illos mathematicos, qui vixere circa tempora Almamonis imperatoris; is vero claruit circa annum christi DCCCLXXXIII, ut videre est ex historia saracenica. Quare non assentio Iosepho Blancano, qui in Chronologia mathematicorum ad decimum Christi saeculum refert." The reference is to Blancanus, *Clarorum mathematicorum chronologia* (1615), 56.

69. The most comprehensive account of Alfraganus's life and works is in DSB s.v. 'Al-Farghānī' (by Sabra), 541–545.

70. DSB s.v. 'Al-Farghānī' (by Sabra); Sezgin, *Geschichte*, 6:149–151; BEA s.v. 'Farghānī' (by De Young).

71. Burnett, 'Magister Iohannes Hispalensis,' 241; Thorndike, 'John of Seville,' 27–28; Lemay, *Abū Maʿšar al-Balḫī*, 4:308.

72. As confirmed by Gerard's students; see Burnett, 'The Coherence,' 277.

73. Carmody, *Arabic Astronomical*, 113–115.

74. For a description of the five editions, see Toynbee, 'Dante's Obligations,' 413–417.

75. Thorndike, *The Sphere of Sacrobosco*, 15–19.

76. DSB s.v. 'Regiomontanus' (by Rosen), 349b. On the content of the *oratio*, see Rose, *The Italian Renaissance*, 95–98.

77. Fück, *Die arabischen Studien*, 45–46; DSB s.v. 'Christmann, Jacob' (by Verdonk).

78. On Golius, see Fück, *Die arabischen Studien*, 79–84; on this work in particular, see Sezgin in the introduction to the 1986 reprint (Frankfurt am Main) of the 1669 edition, v–vii.

79. Woepcke, 'Notice,' 120, says that Christmann's version was reprinted in Frankfurt am Main in 1618, but I have not traced this edition.

80. On Algazel's life and works, see Enc. Iran. s.v. 'Ġazālī' (by Böwering) and Griffel, *Al-Ghazālī's Philosophical Theology*, 19–59.

81. Salman, 'Algazel et les latins,' 103–127. It is unlikely that the *Maqāṣid* was a preparatory study to the *Incoherence*, as was assumed by earlier scholarship; see Griffel, *Al-Ghazālī's Philosophical Theology*, 98.

82. Salman, 'Algazel et les latins,' and Lohr, 'Logica Algazelis,' 223–232.

83. Wimpina, *Tractatus* (1493), sig. C2v: "Algasel vero vir sane doctissimus omnibus profecto positionibus haud secus quam collega Avicennae innexus est, unde quecumque de eius positione confutata sunt, censeo ad Algeselis errores perducenda."

84. Lohr, *Raimundus Lullus' Compendium*, 8–39.

85. The incipit and explicit of Vat. lat. 4554 read: "Dixit Moyses filius iosue narbonensis filius david ad laudem dei" (f. 1r); "Explicit liber intentionum philosophorum prestantissimi Doctoris Abughamath algazelis. Laus deo et trino (?)" (f. 69v). On Moses Narboni's commentary, see Harvey, 'Why Did Fourteenth-Century Jews,' 366–368.

86. Nifo, *In librum Destructio destructionum Averroys commentationes* (1497).

87. Nifo, *De intellectu* (1554), ch. 4.9, f. 36v: "Avicenna et suus abbreviator Algazel."

88. Cf. also Griffel, 'Toleranzkonzepte,' 129–138, on Jean Bodin's knowledge of al-Ġazālī.

89. For the Arabic text, which is also contained in al-Ġazālī's *Ihyā'*, see 274–283, for the Latin, 284–292. The translation begins with the words: "Laus Deo qui [omnia] condidit et restituit, qui facit quodcunque voluerit . . ." This was noted by Daiber, 'The Reception,' 69.

90. On Alhazen's life and works, see the comprehensive entry in DSB s.v. 'Ibn al-Haytham' (by Sabra), 189–210; and Sezgin, *Geschichte*, 6:251–261 (astronomy), 5:358–374 (mathematics), 7:288 (meteorology); Schramm, *Ibn al-Haythams Weg;* BEA s.v. 'Ibn al-Haytham' (by Langermann).

91. Sabra, *The Optics*, 2:lxxiii–lxxiv.

92. Ibid. lxxvi–lxxvii.

93. DSB s.v. 'Ibn al-Haytham' (by Sabra), 197.

94. On its influence from 1550 to 1650, see Lindberg, introduction to the 1972 reprint edition of the *Opticae thesaurus* (1572, repr. 1972), xxi–xxv.

95. DSB s.v. 'Risner, Friedrich' (by Lindberg).

96. Lindberg, *Theories*, 65–66; Sabra, *The Optics,* 2:lxxv–lxxvii.

97. Heiberg and Wiedemann, 'Ibn al Haitams Schrift,' 201–237; DSB s.v. 'Ibn al-Haytham' (by Sabra), 195, 206.

98. Heiberg and Wiedemann, 'Ibn al Haitams Schrift,' 231; Vanden Broecke, *The Limits of Influence,* 177–178.

99. DSB s.v. 'Ibn al-Haytham' (by Sabra), 197–198, 205; Langermann, *Ibn al-Haytham's On the Configuration,* 40–41. The first translation is edited by Millás Vallicrosa, *Las traducciones,* 285–312, the second translation by Mancha, 'La version Alfonsi,' 133–197.

100. On Profacius, see the entry on 'Avenzoar.'

101. Steinschneider, *Die hebraeischen Übersetzungen,* 560; Tamani, 'Le traduzioni,' 618–619.

102. Hartner, 'The Mercury Horoscope,' 124–131; Hasse, 'Averroes' Critique of Ptolemy,' 82–83, 87.

103. The text begins on sig. S2r: "Sequitur antiqui scriptoris libellus de speculo comburenti concavitatis parabolae."

104. As was pointed out by Burnett, 'Al-Kindī in the Renaissance,' 13–30.

105. Endreß, 'The Circle of al-Kindī,' 43–76.

106. EI² s.v. 'al-Kindī, Abū Yūsuf Yaʿkūb ibn Isḥāk' (by Jolivet/Rashed); *Ueberweg I*, 92–147 (by Endreß and Adamson).

107. The full title of this treatise is *De causis diversitatum aspectus et dandis demon-strationibus geometricis super eas;* see Rashed, *Œuvres . . . d'al-Kindī,* 1:439.

108. Bos and Burnett, *Scientific Weather Forecasting,* esp. 22–28, and Burnett, 'Al-Kindī in the Renaissance,' 17–20.

109. For an attempt to identify the translators, see Hasse and Büttner, 'Notes,' forthcoming.

110. Burnett, 'Al-Kindī on Judicial Astrology,' 94–98; Burnett, 'A Hermetic Pro-gramme,' 99–118, esp. 101–103 on the genesis of the *Liber novem iudicum.*

111. McVaugh, *Arnaldi de Villanova,* 263–305 (with edition and analysis of the treatise); Ullmann, *Die Medizin,* 302.

112. For a quotation and discussion of this passage (in *De subtilitate,* Nurem-berg, 1550, lib. 16, 316–317), see Burnett, 'Al-Kindī in the Renaissance,' 17–20.

113. For evidence, see Burnett, 'Al-Kindī in the Renaissance,' 20–23, esp. 23, on five manuscripts of the *Liber novem iudicum* (Robert of Ketton's translation) copied in seventeenth-century Oxford.

114. Gesner, *Bibliotheca universalis* (1545), f. 351r: "Iacobi Alkindi philosophi liber De gradibus medicaminum compositorum, impressus Argentorati 1531, chartis 6, cum Tacuinis sanitatis Elluchasem etc.; quis ex Arabico in Latinum verterit, non invenio. Hunc librum refutat Averrois in quinto sui Colliget circa finem. Alkindi eiusdem opinor, licet praenomen non addatur, De radiis stellicis opus in magia extat."

115. Wimpina, *Tractatus* (1493), sig. c3r: "Alkindus astrologiae peritissimus." Wimpina's main source is Giles of Rome's thirteenth-century treatise *Errores philosophorum,* in which astrological topics are less prominent.

116. Carmody, introduction to Alpetragius, *De motibus celorum,* 15–16; BEA s.v. 'Biṭrūjī' (by Samsó).

117. Carmody, introduction to Alpetragius, *De motibus celorum,* 15 (cf. the colo-phon on p. 150).

118. See Sabra, 'The Andalusian Revolt,' 133–153; Vernet and Samsó, 'The Devel-opment,' 267–269.

119. Not Calonymos ben Calonymos of Arles (d. after 1328), as maintained in BEA s.v. 'Biṭrūjī' (by Samsó).

120. See Carmdoy's remarks in his introduction to Alpetragius, *De motibus celorum,* 16, and the critical apparatus to the text. On Michael Scot's abbreviating translation technique see Hasse, *Latin Averroes Translations,* 32–38.

121. See Di Bono, 'Copernicus, Amico, Fracastoro,' 148.

122. Ullmann, *Die Medizin,* 162–163; EI² s.v. 'Ibn Zuhr IV' (by Arnaldez). For a de-tailed overview of its contents, see Colin, *Avenzoar,* 89–141.

123. Translated into French in Jacquart and Micheau, *La médecine,* 143–144.

124. Ibid., 206. Cf. Steinschneider, *Die hebraeischen Uebersetzungen*, 748–749. On John of Capua, see Hasselhoff, 'The Reception,' 270–274.

125. Jacquart and Micheau, *La médecine*, 206. A manuscript containing *De regimine* is described in d'Alverny, *Avicenna latinus: Codices*, 56. On Profacius, whose Jewish name is Jacob ben Makir (d. ca. 1303–1306), see Wickersheimer, *Dictionnaire biographique*, s.v. 'Profacius,' together with Jacquart, *Supplément*, s.v. Older references to an alleged translator of the *Liber theisir* called 'Paravicius' are based on a misreading of the word *patavinus* ("of Padua"); see Weisser, 'Paravicius Immortal?' 239–240.

126. ADB s.v. 'Schenck, Johann Georg.'

127. Colin, *Avenzoar*, 50–51. On Abū l-'Alā' Zuhr see Ullmann, *Die Medizin*, 162, and EI² s.v. 'Ibn Zuhr III' (by Arnaldez).

128. On this nonencounter, see Ivry, 'Averroes and the West,' 142–153.

129. On Averroes's reception in the Islamic world see von Kügelgen, *Averroes und die arabische Moderne*, 4–10, 55–56. On his immediate disciples who were jurists for the most part, see Puig, 'Materials,' 255–258.

130. On Averroes's life and the chronology of his works, see Puig, 'Materials,' 241–260, and Wirmer, *Averroes: Über den Intellekt*, 287–313.

131. Burnett, 'Michael Scot,' 101–126.

132. Hasse, *Latin Averroes Translations*.

133. Puig, 'Eliahu del Medigo, traductor,' 713–729.

134. Bartòla, 'Eliyhau del Medigo,' 263–264. Del Medigo's *Commentatio media* of the *Metaphysics*, books Alpha to Zeta, which was included in the Giunta edition of 1560, is apparently a compilation of several sources (among them Comm. mag. Metaph. and Comm. med. Metaph.), rather than a direct translation of Comm. med. Metaph.; see Zonta, *Il* commento medio, 15–18.

135. Tamani, 'Le traduzioni,' 617.

136. Zonta, 'Osservazioni,' 15–28.

137. Zedler, *Averroes' Destructio*, 24–25.

138. Steel and Guldentops, 'An Unknown Treatise,' 87–89.

139. Edited in 1992 by Coviello and Fornaciari, *Averroè: Parafrasi della 'Repubblica.'*

140. Bartòla, 'Eliyhau del Medigo,' 263.

141. Tamani, 'Una traduzione,' 91–101, with edition of titles and opening paragraphs.

142. Tamani, 'Le traduzioni,' 618.

143. Zedler, *Averroes' Destructio*, 26–29.

144. Geoffroy and Steel, *Averroès: La béatitude de l'âme*, 111–112.

145. Di Donato, 'Il *Kašf*... confronto,' 241–248.

146. Di Donato, 'Traduttori di Averroè,' 39n41.

147. Zedler, *Averroes' Destructio*, 45–46.

148. See Chapter 5, n. 175, above.

149. Martin, 'Humanism,' 72. Martin also draws attention to the genre of a single *Quaestio* devoted to a specific chapter in Averroes and its topic (74–75). Pomponazzi's commentary on *De substantia orbis* exists only in manuscript, while Confalonieri's, Beati's, and Mainetti's texts are printed together with *De substantia orbis* (and thus are listed in this Appendix). Gozze's *Commentaria in sermonem Averrois De substantia orbis* were printed in 1580.

150. These works are: *Contradictiones et solutiones in dictis Aristotelis et Averrois,* first published 1508; *Tabula dilucidationum in dictis Aristotelis et Averrois,* first published 1537. See Lohr, *Latin Aristotle Commentaries,* 2:504–512.

151. On this edition see Chapter 1 above, and Schmitt, 'Renaissance Averroism,' in *The Aristotelian Tradition,* art. 8, 121–142; Burnett, 'The Second Revelation,' 185–198; Burnett, 'Revisiting,' 55–64.

152. With the exception that the short treatise *Ḍamīma* was translated by Ramon Martí as *Epistola ad amicum.* It was included in Ramon Martí's *Pugio fidei* (Paris, 1651, ff. 200–202); the text is printed in Alonso, *Teología de Averroes,* 356–365.

153. Ullmann, *Die Medizin,* 166–167; Bürgel, 'Averroes "contra Galenum,"' 266–340.

154. Martin, 'Humanism,' 71.

155. Palmer, 'Pharmacy,' 108–110.

156. As claimed by Steinschneider, *Die hebräischen Uebersetzungen,* 676. Lucchetta even assumed that Andrea Alpago was the translator of the treatise; cf. Lucchetta, *Alpago,* 66. This misinformation seems to derive from a conflation of the following passage with the preceding sentence in the Giunta edition on Alpago's corrections of the *Cantica:* Aristotle and Averroes, *Aristotelis . . . omnia . . . opera* (1550/1552), vol. 1, f. 13r: "Adiecimus postremo epistolam illam Averrois de theriaca quam se in 7 Colliget, cap. 2, scripsisse testatur, quae cum antea a nobis desideraretur, nunc primum ex scriptis Andree a Cruce Veneti chyrurgi celebris reperta est." For the context of surgery in sixteenth-century Veneto, see Palmer, 'Physicians and Surgeons,' 451–460 (reference to dalla Croce, 455–456).

157. Aristotle and Averroes, *Aristotelis . . . omnia . . . opera* (1560), vol. 11, f. 220r: "Haec sunt verba Aver. de semine ex arabico in latinum translacta *(sic)* ab Eccel. Phylosophiae proffessore D. Helia Cretense."

158. The numbers in brackets are from Index Aureliensis as recorded and augmented by Cranz, *A Bibliography.*

159. See Burnett, 'Revisiting,' 55–64.

160. Cf. the list of editions in Zedler, *Averroes' Destructio,* 55–56.

161. Fundamental for the study of Avicenna's life and philosophical oeuvre is Gutas, *Avicenna and the Aristotelian Tradition.*

162. Gutas, 'The Heritage of Avicenna,' 81–97.

163. Ullmann, *Die Medizin,* 152–154.

164. d'Alverny, 'Notes sur les traductions,' in *Avicenne en occident,* art. 4; and Bertolacci, 'A Community of Translators,' pp. 37–54. On the manuscript

distribution see Kischlat, *Studien,* 51–60. See also the helpful table of translations in Burnett, 'Arabic into Latin: The Reception.'

165. See Siraisi, *Avicenna in Renaissance Italy,* ch. 4 on the *Canon* in Renaissance universities, and ch. 6 on the commentators.

166. Siraisi, *Avicenna in Renaissance Italy,* 143–156. See also the literature on the development of Arabic studies cited in the Preface.

167. McVaugh and Ferre, *The* Tabula antidotarii, 2.

168. For editions of the *Cantica,* see also editions of Averroes's commentary on them s.v. 'Averroes' above. There may well be more editions of this work and of *De medicinis cordialibus* besides those listed below since both of them are difficult to spot.

169. Jahier, *Avicenne: Poème de la médecine,* 102n3, and 104; Burnett, 'Learned Knowledge,' 52. On Deusing see ADB s.v. 'Deusing, Anton' (by Hirsch).

170. Jahier, *Avicenne: Poème de la médecine,* 101–108.

171. Siraisi, *Avicenna in Renaissance Italy,* 373–376 (printed commentaries), 368–372 (commentaries in manuscript).

172. Cf. the list of sixteenth- and seventeenth-century editions of the *Canon* in Siraisi, *Avicenna in Renaissance Italy,* 361–366. The *Canon* 'editions' of Bordeaux 1520 and 1524? listed by Siraisi are in fact lexica, in which excerpts from Avicenna's *Canon* are arranged in alphabetical order by Gabriel de Tarrega. I have not found a reliable trace of the 1528 Lyon reprint of the *Flores Avicenne.*

173. Ullmann, *Die Medizin,* 126–128, 190, 244; Sezgin, *Geschichte,* 3:270–274 (medicine); 5:285–286 (mathematics); and 6:180–182 (astronomy); EI² s.v. 'Kusṭā b. Lūkā' (by Hill).

174. Burnett, 'Magister Iohannes Hispalensis et Limiensis,' 221–242 (n. 3 on the abbreviated version).

175. Pingree, 'The Diffusion,' 68–70, 88, 96; Wilcox and Riddle, *Qusṭā ibn Lūqā's Physical Ligatures,* 21–22.

176. Steinschneider, *Die europäischen Übersetzungen,* 77; Carmody, *Arabic Astronomical,* p. 132; Sezgin, *Geschichte,* 6:181.

177. Kischlat, *Studien,* 25–26, 53.

178. Burnett, 'The Introduction of Aristotle's Natural Philosophy,' 35, 50.

179. See the list of editions in Wilcox and Riddle, *Qusṭā ibn Lūqā's Physical Ligatures,* 49–50. I cannot confirm the inclusion of *De physicis ligaturis* viz. *De incantatione* in the Galen editions of Lyon 1550, Venice 1556, Venice 1586, and Venice 1609. I was not able to consult the Galen editions of Venice 1528, Venice 1550, and Basel 1562.

180. Ullmann, *Die Medizin,* 280–283.

181. Ebenbitar, *Tractatus* (1602), sig. B1r: "Andreas Bellunensis exc. D. Benedicto de Montesilicis physico veneto primario s<alutem dicit>. Memini praeceptor observandissime excell. tuam summopere desiderare, ut si quid de limonibus scriptum esset ab authoribus arabicis, illud latinum facerem, cum ab Avicenna et a Latinis diminute de ipsis limonibus scriptum fuerit.

Quare cupiens tuo desiderio pro immensis beneficiis tuis in me collatis sat-
isfacere, statui sermonem de limonibus arabicum vertere ac ipsum transmit-
tere tuae excell., ut ipse et posteri alii intelligant iuvamenta quam plurima ab
authoribus arabicis diligenter adnotata et experta fuisse, quae a latinis
ignorantur."

182. Lucchetta, *Alpago,* 66–69; the title of the 1602 edition is cited in d'Alverny,
'Avicenne et les médecins,' 196n47, in *Avicenne en occident,* art. 13.

183. Ullmann, *Die Medizin,* 165.

184. Vernet and Samsó, 'The Development,' 265–266.

185. Lorch, 'The Astronomy,' 85–87; DSB s.v. 'Jābir ibn Aflaḥ' (by Lorch); BEA s.v.
'Jābir ibn Aflaḥ' (by Calvo).

186. As confirmed by Gerard's students; see Burnett, 'The Coherence,' 278.

187. Lorch, 'The Astronomy,' 91.

188. This information I have from my colleague David Juste of the Ptolemaeus
Arabus et Latinus project (Bavarian Academy of the Sciences). Cf. Lorch,
'The Astronomy'; Carmody, *Arabic Astronomical,* 163.

189. Lorch, 'The Astronomy,' 101–104.

190. Thorndike, *History,* 5:565; DSB s.v. 'Jābir ibn Aflaḥ' (by Lorch), 38–39.

191. Cardano, *De subtilitate* (1559), lib. 16, 570: "Post sequitur Heber Hispanus cla-
rissimo invento, cum Ptolemaeus ex quinque quantitatibus maximo labore
sextam quaerat, hic in eisdem cum tribus quartam. Multa etiam in melius
mutavit, quae ad statum coeli pertinebant, ut facile intelligas multo minus
aestus maximos frigoribus gelidis obesse ingeniis."

192. Ullmann, *Die Natur- und Geheimwissenschaften,* 198–208.

193. Newman, *The* Summa perfectionis *of Pseudo-Geber,* 72–82.

194. For a list of editions of the *Summa perfectionis* see Newman, 'Arabo-Latin
Forgeries,' 293. I have not found a reliable trace of the 1528 Strasbourg edition of *Geberi philosophi . . . de alchimia libri tres.*

195. See the presentation and discussion of early modern editions of Geber in
Bachmann and Hofmeier, *Geheimnisse der Alchemie,* 181–191, esp. 190–191.
Cf. two examples among the various references to Geber's kingship in edi-
tions: Richardus Anglicus, *Correctorium Alchimiae* (1581), 137: "Des Königes
Gebers auß Hispanien Buch der Heyligkeyt"; and Geber, *Summa perfectio*
(1625), sig. (**)2r: "Geber war ein König in dem glückseligen Arabia."

196. Hasse, 'King Avicenna,' 230–243.

197. Newman, 'Arabo-Latin Forgeries,' 284–288 (with an edition of the *Liber
regni*). Newman draws attention to a fourteenth-century remark by the
monk Ademarus (285): "His illustrious birth was the product of the great
Muhammad's daughter and a Baghdadian prince *(ex filia magni Machometi
et ex principe Baldachitarum),* as he himself writes in his book called *Liber
regni.*" This tradition is followed in the Renaissance by Gesner, *Bibliotheca
universalis* (1545), f. 266v: "In dialogo quodam manuscripto, quo Adenarius
et Guilhelmus Parisienses *(i.e., William of Auvergne)* barbaro sermone collo-
quentes introducuntur de astronomia inferiore, ita legimus: Geber ex filia

magni Machmeth et ex principe Baldachitarum originem clarissimam traxit, ut ipsemet in libro suo qui dicitur Regni scribit."

198. For its content, see Ullmann, *Die Medizin*, 140–146, and the book *Islamic Medicine* by the same author, which summarizes much of the content of the *Kitāb Kāmil aṣ-ṣinā'a.*

199. On al-Maǧūsī's life, see Micheau, "ʿAlī ibn al-ʿAbbās al-Maǧūsī,' 1–15.

200. Fundamental for research on the *Pantegni* is Burnett and Jacquart, *Constantine the African.*

201. Jordan, 'The Fortune of Constantine's *Pantegni,*' 286–290.

202. Jacquart and Micheau, *La médecine,* 174.

203. EI² s.v. 'Ibn Abī l-Ridjāl' (by Pingree); Ullmann, *Die Natur- und Geheimwissenschaften,* 336.

204. As stated in a manuscript note quoted by Menéndez Pidal, 'Cómo trabajaron,' 365: "Hic est liber magnus et completus quem Hali Albenragel summus astrologus composuit De iudiciis astrologiae, quem Juda filius Mosse de praecepto domini Alfonsi illustrissimi regis Castelle et Legionis transtulit de arabico in ideoma maternum et Alvarus dicti illustrissimi regis factura eius ex praecepto transtulit de ideomate materno in latinum." For context see Gonzálvez Ruiz, *Hombres y libros,* 611.

205. On these translators Procter, *Alfonso X of Castile,* 128–130.

206. Vernet, *Die spanisch-arabische Kultur,* 228–231; Hilty, *Aly Aben Ragel,* lxi–lxv.

207. See Thorndike, *History,* vol. V, index, s.v. 'Haly Abenragel'; see also Boudet on Simon de Phares's great estimation for Abenragel, his main source in astrological matters (*Le Recueil,* 1:403n, 2:145, 165–166). Chapter 6 of the present study adds further evidence for the reception of Abenragel in the sixteenth century.

208. Stegemann, *Beiträge,* 10 ("Dabei haben sich ... viele Fehler sachlicher Art infolge sprachlicher Mißverständnisse eingeschlichen"); Carmody, *Arabic Astronomical,* 150 ("Edit. 1551 was 'improved' impressionistically, thus falsifying many details").

209. On Haly Rodoan's life and works, see EI² s.v. 'Ibn Riḍwān' (by Schacht) and Dols, *Medieval Islamic Medicine,* 54–66 Cf. Jacquart and Micheau, *La médecine,* 232–233, and Ullmann, *Die Medizin,* 158–159.

210. As confirmed by Gerard's students; see Burnett, 'The Coherence,' 280. Cf. Schipperges, *Die Assimilation,* 49, 89–90.

211. Cf. Ottosson, *Scholastic Medicine,* 98–126.

212. Sezgin, *Geschichte,* 7:44.

213. Steinschneider, *Vite,* 42, with quotations from Egidio's preface; the full text of the preface is in Muñoz Sendido, *La Escala de Mahoma,* 92–94.

214. Grafton, *Cardano's Cosmos,* 136–137; Thorndike, *History,* 6:121; Vanden Broecke, *The Limits of Influence,* 174–177.

215. There are no reliable dates on Aḥmad ibn Yūsuf save for two events he says to have seen himself in 883/884 and 912/913 AD; see Kunitzsch, review of Sezgin, *Geschichte,* 176. Cf. also Ullmann, *Die Natur- und Geheimwissenschaften,* 327–328;

Sezgin, *Geschichte*, 5:288–289; DSB s.v. 'Aḥmad ibn Yūsuf' (by Schrader); and the introduction by Martorello and Bezza to Aḥmad ibn Yūsuf ibn ad-Dāya, *Commento al Centiloquio*, 22–30.

216. As confirmed by Gerard's students; see Burnett, 'The Coherence,' 276–277.

217. Kunitzsch, review of F. Sezgin, *Geschichte*, 175–176. Lemay, 'Origin and Success,' 91–107, has suggested that Aḥmad ibn Yūsuf was also the author of the *Centiloquium* itself, but this is untenable for chronological reasons.

218. As Maria Mavroudi has shown in a paper delivered on November 6, 2015, at a Warburg Institute conference in London on Ptolemy's Science of the Stars in the Middle Ages. Further indications of Greek origin are mentioned by Kunitzsch, review of Sezgin, 176n4.

219. On his life and works, see EI² s.v. 'Ibn Buṭlān' (by Schacht); Ullmann, *Die Medizin*, 157–158, 192, 224.

220. Jacquart and Micheau, *La médecine*, 209–210.

221. LexMA s.v. 'Tacuina sanitatis' (by Schipperges), with further literature.

222. Wickersheimer, 'Les Tacuini sanitatis,' 85–97; Leber and Starke, epilogue to the reprint edition of *Schachtafelen der Gesuntheyt* (1533, repr. 1988), 3–23.

223. Apart from the title page, the 1532 edition is identical with the 1531 edition.

224. EI² s.v. 'Ibn Djazla' (by Vernet), and, much briefer, Ullmann, *Die Medizin*, 160.

225. Steinschneider, *Die hebraeischen Übersetzungen*, 975; on the translator, see Sirat, 'Les traducteurs juifs,' 178–179.

226. On his life and the date of his death, see Bos, *Ibn al-Jazzār on Sexual Diseases*, 5–6.

227. Ibid., 8. Cf. Ullmann, *Die Medizin*, 147–148.

228. On the Western influence of the *Viaticum*, see Wack, *Lovesickness*.

229. On these treatises, see Burnett and Jacquart, *Constantine the African*, 203–232 (De oblivione), 243–244 (De coitu), 354 (references to De stomacho); on *De coitu*, see also Jacquart and Micheau, *La médecine*, 117–118, and Jacquart and Thomasset, *Sexuality*, 116–120. Cf. Sezgin, *Geschichte*, 3:306–307.

230. Burnett and Jacquart, *Constantine the African*, 243–245.

231. On the treatise, see Ullmann, *Die Medizin*, 268–269. On the second translation, see Volger, *Der* Liber fiduciae, xiii–xiv; for context, see Pingree, 'The Diffusion,' 70. A treatise *De proprietatibus* on the usefulness of magic in medicine was translated into Latin, probably in the twelfth century, but did not reach print.

232. Under the title *De omnium morborum qui homini accedere possunt cognitione et curatione,* see Burnett and Jacquart, *Constantine the African*, 318.

233. Sabra, 'The Authorship,' 84–85; DSB s.v. 'al-Jayyānī' (by Dold-Samplonius and Hermelink); Vernet and Samsó, 'The Development,' 258; BEA s.v. 'Ibn Muʿādh' (by Calvo).

234. As confirmed by Gerard's students; see Burnett, 'The Coherence,' 278.

235. Sabra, 'The Authorship,' 77.

236. This was first corrected by Sabra, ibid., 77–85.

237. Ibid., 81.

238. Vernet and Samsó, 'The Development,' 258; Burnett, 'The Coherence,' 278: "Liber tabularum Iahen cum regulis suis."

239. Hermelink, 'Tabulae Jahen,' 108–112.

240. EI² s.v. 'Ibn Sīrīn' (by Fahd).

241. Haskins, *Studies,* 216–218; Mavroudi, *A Byzantine Book,* 32–41.

242. NDB s.v. 'Löwenklau' (by Metzler); for context, see Schulze, *Reich und Türkengefahr,* 28. On this edition, see also Mavroudi, *A Byzantine Book,* 5–9.

243. Pseudo-Ibn Sīrīn, *Apomasaris Apotelesmata* (1557), sig. a5v–a6r: "Quis ipse fuerit, certo non constat. Existimabat ornamentum illud Germaniae nostrae Ioachimus Camerarius Apomasarem eumdem esse qui Albumasar usurpatur. Sed pace tanti viri dixerim nullo id pacto verisimile videri. Nam Albumasar cognomentum erat eius qui vero nomine Iaphar adpellabatur, ut in eius vita perscriptum legimus. Deinde constat Albumasarem fuisse Arabem et religionem saracenicam sequutum. At noster hic Apomasar, si Graecus non fuit, Arabs tamen et Saracenus haberi nullo modo potest, cum apertissima christianae religionis indicia liber ipse contineat."

244. Fahd, *La divination,* 355–356, with n. 6; Mavroudi, *A Byzantine Book,* 1–62.

245. Mavroudi, *A Byzantine Book,* 9–11.

246. For his life and works, see Conrad, 'The World,' 5–22; Kukkonen, *Ibn Tufayl.*

247. Gutas, 'Ibn Ṭufayl,' 222–241; Kukkonen, *Ibn Tufayl.*

248. Conrad, 'Research Resources,' 281; Hayoun, 'Le commentaire,' 23–98.

249. Genua, Biblioteca Universitaria A.IX.29, ff. 79v–116r; for a study of this translation and specimens of the text, see Bacchelli, 'Pico,' 1–25. On the influence of Ibn Ṭufayl on Yoḥanan Alemanno, who was directly acquainted with Pico, see Idel, 'Jewish Mystical Thought,' 29–30.

250. Pico, *Disputationes* (1946/1952), 1:80, corrected with Bacchelli, 'Pico,' 3: "Scripsit etiam Abubater De natalibus praedictionibus et eodem nomine alius philosophica praecipueque librum 'Quo quisque pacto per se philosophus evadat,' quem anno superiore ex hebraeo vertimus in latinum; sed ille Altasibi, hic vero Tripdis filius."

251. Conrad, 'Research Resources,' 282.

252. Daiber, 'The Reception,' 74–80; Toomer, *Eastern Wisedome,* 218–220.

253. EI² s.v. "ʿAlī b. ʿĪsā' (by Mittwoch); Ullmann, *Die Medizin,* 208.

254. Jacquart and Micheau, *La médecine,* 206.

255. EI² s.v. "ʿAlī b. ʿĪsā' (by Mittwoch).

256. Ullmann, *Die Medizin,* 209–210; Savage-Smith, 'Medicine,' 949.

257. Steinschneider, 'Zur Oculistik,' 406–408; Lindberg, *A Catalogue,* 101.

258. Ullmann, *Die Medizin,* 115–119; Sezgin, *Geschichte,* 3:247–256; *Ueberweg I,* 480–496 (by Gutas).

259. Newton, 'Constantine the African,' 16–47; Jacquart, 'À l'aube de la renaissance,' in *La science médicale,* art. 1, 209–240.

260. This is maintained by Lindberg and Teigen against Hirschberg; see Lindberg, *Theories of Vision*, 232n58, and Hirschberg's article of 1903, 'Über das älteste arabische Lehrbuch,' 1080-1094. I was not able to consult the Galen editions of Venice 1528, Venice 1550, and Basel 1562. I cannot confirm the inclusion of the treatise in the Venice 1625 edition.

261. See the sample passages collected by Hirschberg, 'Über das älteste arabische Lehrbuch,' 1091-1093.

262. Altmann and Stern, *Isaac Israeli*, xvii-xxii; Jacquart and Micheau, *La médecine*, 110-111; Veit, *Das Buch der Fieber*, 27-28; SEP s.v. 'Isaac Israeli' (by Levin).

263. Ullmann, *Die Medizin*, 137-138, 200.

264. On Posthius, see Kühlmann et al., *Humanistische Lyrik*, 1365-1368.

265. Karrer, *Johannes Posthius*, 138, 392-393.

266. This work is often referred to as *De definitionibus*. For the full title, see incipit and explicit in Muckle, 'Isaac Israeli,' 300, 328.

267. Hasse and Büttner, 'Notes,' forthcoming.

268. As shown by Altmann and Stern, *Isaac Israeli*.

269. Foresti da Bergamo, *Supplementum* (1506), f. 285v; Schedel, *Liber chronicarum* (1493), f. 192v; Champier, *De . . . scriptoribus* (1506), f. 21v-22r; Gesner, *Bibliotheca universalis* (1545), f. 467r.

270. Sezgin, *Geschichte*, 3:266; Ullmann, *Die Medizin*, 125-126.

271. Garbers, *Isḥāq ibn ʿImrān*, x, 155.

272. Burnett and Jacquart, *Constantine the African*, 354b (index).

273. On pp. 280-298, incipit: "Constantini Africani medici de melancholia libri duo. Melancholia aliis corporis morbis animum magis conturbat." The treatise is not attributed to Rufus of Ephesus in this edition (as stated by Sezgin, *Geschichte*, 3:266).

274. Sezgin, *Geschichte*, 7:175-176. The author is not Abū ʿAlī Yaḥyā ibn abī Mansūr, as had been supposed by Steinschneider, *Die europäischen Übersetzungen*, 63-64, and Ullmann, *Die Natur- und Geheimwissenschaften*, 307. Carmody, *Arabic Astronomical*, 132-134, attributes the treatise to Rhazes.

275. DSB s.v. 'Plato of Tivoli' (by Minio-Paluello), 32. The translator is well attested in the colophons.

276. For the variant titles, see Vadet, 'Les aphorismes,' 50-55.

277. Simon de Phares, ed. Boudet, *Le Recueil*, 1:286; Gesner, *Bibliotheca universalis* (1545), f. 30r; Baldi, *De le Vite deʾ Matematici* (1874), p. 32 (459).

278. For an analysis of its content, see Vadet, 'Les aphorismes,' 94-130.

279. Among the vast literature, see SEP s.v. 'Maimonides' (by Seeskin), and Bos, *Maimonides: On Asthma*, xxiv-xxx, with further bibliography.

280. Kluxen, 'Literargeschichtliches,' 25-35; Hasselhoff, 'The Reception,' 264-270.

281. Rigo, 'Zur Rezeption,' 29-30.

282. Kluxen, 'Literargeschichtliches,' 24.

283. Hasselhoff, 'The Reception,' 268-269, 277.

284. Ullmann, *Die Medizin,* 168; Hasselhoff, 'The Reception,' 271.

285. On this work, see Ullmann, *Die Medizin,* 338; McVaugh and Ferre, *The* Tabula antidotarii, 3.

286. Bos, *Maimonides: On Asthma,* xxxvii–xxxviii.

287. Ullmann, *Die Medizin,* 168; Hasselhoff, 'The Reception,' 272–273.

288. Ullmann, *Die Medizin,* 196; Hasselhoff, 'The Reception,' 273.

289. Hasselhoff, 'The Reception,' 273–274.

290. See the list of editions in Rosner, *The Medical Aphorisms,* 455–457. I was not able to find the editions of Venice 1553, Venice 1625, and Paris 1689.

291. See DSB s.v. 'Māshā'allāh' (by Pingree); Sezgin, *Geschichte,* 7:102–108; Ullmann, *Die Natur- und Geheimwissenschaften,* 303–306; on Messahalah as the possible translator of Dorotheus fragments, see Pingree, 'Māshā'allāh's Arabic Translation,' 191–209.

292. The principal source of the following list is Pingree in DSB, s.v. 'Māshā'allāh.' Carmody, *Arabic Astronomical,* 22–38, also lists the following titles: *De testimoniis lune* (item 16), *De stationibus planetarum* (item 17), and *De electionibus horarum* (item 18), the authenticity of which is not clear.

293. Haskins, *Studies,* 76.

294. Bos and Burnett, *Scientific Weather Forecasting,* 93; Thorndike, 'The Latin Translations,' 67–68; cf. Ullmann, *Die Natur- und Geheimwissenschaften,* 305.

295. Thorndike, 'The Latin Translations,' 53–54.

296. Ibid., 54–56.

297. Ibid., 68–69.

298. Ibid., 66–67.

299. A variant title is *De ratione circuli et stellarum.* As to John of Seville, see Burnett, 'Magister Iohannes Hispalensis,' 226; Lemay, *Abū Maʿšar al-Balḫī,* 4:308; Thorndike, 'The Latin Translations,' 62–66.

300. Ibid., 50–53.

301. Burnett, 'The Coherence,' 278.

302. See the lists in Carmody, *Arabic Astronomical,* 22–38. The 1509 and 1513 editions of *Meseallach et Ptolemeus de electionibus* do indeed contain Zahel's *De electionibus.* See below s.v. 'Zahel.'

303. On the absence of astrology in Gerard's program of translation, see Burnett, 'The Coherence,' 268–269.

304. Pingree, 'Māshā'allāh: Some ... Sources,' 9–14.

305. Bezzel, 'Joachim Heller,' 298, 327–328.

306. Kunitzsch, 'On the Authenticity,' 42–62, in *The Arabs and the Stars,* art. 10.

307. The title page of this edition confusingly treats *De receptionibus planetarum* and *De interrogationibus* as two treatises. But note the incipit on f. 143r: "Incipit liber Messahallach dictus de receptione planetarum et est de interrogationibus quem transtulit Joannes Hispalensis de arabico in latinum."

308. Ullmann, *Die Medizin,* 112–115; Sezgin, *Geschichte,* 3:231–236.

309. See also below s.v. 'Rhazes.'

310. Jacquart and Troupeau, *Yūḥannā ibn Māsawayh,* introduction: on Mesue, 1–5, and his *Nawādir,* 5–13, on the two translations and on Renaissance editions and paraphrases, 13–104.

311. On Pseudo-Mesue, see, most pertinently, Lieberknecht, *Die* Canones *des Pseudo-Mesue,* 1–34. See also Ullmann, *Die Medizin,* 304–305.

312. On these commentaries see Lieberknecht, *Die* Canones *des Pseudo-Mesue,* 102–104.

313. Cf. the lists of editions in Klimaschewski-Bock, *Die 'Distinctio sexta,'* 279–294, and Lieberknecht, *Die* Canones *des Pseudo-Mesue,* 201–204.

314. See DSB s.v. "'Umar ibn al-Farrukhān al-Ṭabarī' (by Pingree); less informative are Sezgin, *Geschichte,* 7:111, Ullmann, *Die Natur- und Geheimwissenschaften,* 306–307, and Carmody, *Arabic Astronomical,* 38–39.

315. Burnett, 'John of Seville and John of Spain,' 60, 78, with one example of a colophon: "perfectus est universus liber Omar Benfargan Tyberiadis de nativitatibus" (MS Florence, Biblioteca nazionale centrale, con. soppr. J.II.10., late thirteenth century).

316. Pingree, 'The *Liber universus,*' 8–12. Pingree infelicitously called this appendix "Liber universus" on the basis of colophons of *De nativitatibus;* but "universus" here means "the entire book on nativities." See the preceding note and the edition of the treatise in Yamamoto and Burnett, *Abū Maʿšar on Historical Astrology,* 2:342–344.

317. Burnett, 'Al-Kindī on Judicial Astrology,' 95–96. On the *Liber trium iudicum,* see Burnett, 'A Group,' 66–67. On the *Liber novem iudicum,* see Burnett, 'A Hermetic Programme,' 99–118. Cf. above s.v. 'Alkindi.'

318. Bos and Burnett, *Scientific Weather Forecasting,* 23, and Appendix IVB (texts).

319. Burnett, 'De meliore homine,' 295–326. The version by Salio of Padua appears within the translation of Hugo of Santalla in MS Dijon BM, 449, and within the *Liber trium iudicum* and *Liber novem iudicum.*

320. Carmody, *Arabic Astronomical,* 38, and Pingree, 'The *Liber universus,*' 8, mention a 1515 edition by Luca Gaurico, which I have not found.

321. On the sporadic Latin reception of Rhazes's *Doubts against Galen,* see Burnett, 'Encounters,' 973–992.

322. Sezgin, *Geschichte,* 3:274–275; Ullmann, *Die Medizin,* 128–129; Jacquart and Micheau, *La médecine,* 57–61; *Ueberweg I,* 264–267 (by Daiber), with up-to-date bibliography on Rhazes's biography.

323. As confirmed by Gerard's students; see Burnett, 'The Coherence,' 280.

324. Sezgin, *Geschichte,* 3:281; Ullmann, *Die Medizin,* 132.

325. Jacquart, 'Note sur la traduction,' in *La science médicale,* art. 8, 359–374.

326. Sezgin, *Geschichte,* 3:284; Ullmann, *Die Medizin,* 132.

327. Jacquart and Micheau, *La médecine,* 64; Sezgin, *Geschichte,* 3:184.

328. As surmised by Jacquart and Micheau, *La médecine,* 150.

329. On the term *Aqrābāḏīn,* see Ullmann, *Die Medizin,* 295.

330. Sezgin, *Geschichte,* 3:283.

331. Ibid., 288.

332. Ullmann, *Die Medizin,* 133; Jacquart, 'Note sur la traduction,' in *La science médicale,* art. 8, 360 n4.

333. Sezgin, *Geschichte,* 3:288; Ullmann, *Die Medizin,* 134; Jacquart and Micheau, *La médecine,* 66.

334. Jacquart and Troupeau, *Yūḥannā ibn Māsawayh,* 88–104; Jacquart and Micheau, *La médecine,* 208; Sezgin, *Geschichte,* 3:286.

335. For instance in the Bologna 1489 edition, f. 139r: "Liber Rasis de secretis in medicina qui liber aphorismorum appellatur." There does not seem to exist a Latin version of a work that is usually referred to as "Rāzī's Aphorisms": *Kitāb al-Muršid* or *al-Fuṣūl* ("The Guide" or "Aphorisms"). Cf. Sezgin, *Geschichte,* 3:284, Ullmann, *Die Medizin,* 134, Jacquart and Micheau, *La médecine,* 64–65.

336. Jacquart and Micheau, *La médecine,* 207.

337. Such titles are *Tractatus febrium, De flebotomia, Liber experimentorum, Liber de sectionibus et cauteriis et ventosis, Tabula de herbis medicis,* and *De facultatibus partium animalium.* See Jacquart, 'Note sur la traduction,' in *La science médicale,* art. 8, 360.

338. Jacquart and Micheau, *La médecine,* 191–192; Grendler, *The Universities,* 323–324; Siraisi, *Avicenna in Renaissance Italy,* 178.

339. See the lists of commentators in Illgen, *Die abendländischen Kommentatoren,* 5–7 (=117–119), and Sezgin, *Geschichte,* 3:282.

340. O'Malley, *Andreas Vesalius,* 69–72.

341. On Valla, see Cont. Eras. s.v. 'Valla, Giorgio' (by d'Amico and Deutscher).

342. Broemser, 'Johann Winter aus Andernach,' 25.

343. Sezgin, *Geschichte,* 3:283; Ullmann, *Die Medizin,* 133.

344. See Ruska, 'Pseudepigraphe Rasis-Schriften,' 31–94; Halleux, 'The Reception,' 889–893.

345. Sezgin, *Geschichte,* 4:279; Ullmann, *Die Natur- und Geheimwissenschaften,* 213.

346. Sezgin, *Geschichte,* 4:280; Ullmann, *Die Natur- und Geheimwissenschaften,* 212–213. It was translated into German by Ruska, *Al-Razi's Buch,* 83–219.

347. Newman, *The* Summa perfectionis, 64–65. Cf. Ruska, *Übersetzung,* 1–26.

348. Pormann, 'Yūḥannā ibn Sarābiyūn,' 234–239.

349. Ullmann, 'Yūḥannā ibn Sarābiyūn,' 278–296; Pormann, 'Yūḥannā ibn Sarābiyūn,' 233–262, with further literature.

350. Compare Serapion, *Practica* (1550), f. 58r (Alpago's text), and Ullmann, 'Yūḥannā ibn Sarābiyūn,' 285–287 (Arabic text and Gerard's translation): Alpago writes *tumefactio* for *taḥayyuǧ* instead of Gerard's *excitatio* (p. 285), and *festinat* for *bādara* (to rush) instead of Gerard's *incipit,* who apprently read *bādaʾa* (to begin) (p. 287); it is very unlikely that Alpago suggested *festinat* without consulting the Arabic text. See also the different versions of the introductory paragraphs of chapter 5.21 in Pormann, 'Yūḥannā ibn Sarābiyūn' (without Alpago), p. 254; note that Pormann shows that the manuscript

Ullmann was using (Brussels, Bibliothèque royale, MS 19891) has additions and variant readings not shared by either the other manuscripts or Gerard's translation. Cf. also Troupeau, 'Du syriaque au latin,' 275.

351. Serapion, *Practica,* (1550), sig. +5v–+6v, under the title: "Index arabicorum nominum medicamentorum simplicium per ordinem alphabeti cum translatione latina."

352. As discovered by Villaverde Amieva, review of Aguirre de Cárcer, *Ibn Wāfid,* 111–118. Before Villaverde Amieva, scholars had long assumed that Pseudo-Serapion was an Andalusian Arabic author writing in the 1240s; cf. the older (and by now outdated) studies by Straberger-Schneider, *Der Liber aggregatus,* 11–14, and Dilg, 'The Liber aggregatus,' 226–227.

353. Hugonnard-Roche, 'The Influence,' 304n11 (with further literature); North, *Astronomy and Cosmology,* 208–209.

354. See also the references in Thorndike, *A History,* 16:747, s.v. 'Thebit ben Corat.'

355. EI² s.v. 'Thābit b. Kurra' (by Rashed and Morelon) with bibliography and Rashed (ed.), *Thābit ibn Qurra.*

356. Cf. Sezgin, *Geschichte,* 3:260–263 (medicine), 5:264–272 (mathematics), and 6:163–170 (astronomy).

357. Cf. the (partly outdated) list of Latin translations in Carmody, *Arabic Astronomical,* 116–129, and Carmody's edition of the Latin Thebit, *The Astronomical Works.* Morelon surmises that Arabic equivalents exist for two of the astronomical Latin tracts attributed to Thebit: *De his quae indigent* and (Pseudo-Thebit) *De anno solis;* see Morelon, *Thābit ibn Qurra,* xviii. There is no information on the Latin tradition in DSB s.v. 'Thābit ibn Qurra' (by Rosenfeld and Grigorian).

358. Morelon, *Thābit ibn Qurra,* xlvi–liii; EI² s.v. 'Thābit b. Kurra' (by Rashed and Morelon).

359. For context, see Rosenfeld and Youschkevitch, 'Geometry,' 483, and Lorch, *Thabit ibn Qurra: On the sector-figure.*

360. Cf. the list drawn up by Gerard's students in Burnett, 'The Coherence,' 276–278.

361. Lorch, *Thabit ibn Qurra: On the sector-figure,* 30–36. There also exists an anonymous translation of *De his quae indigent,* which survives in a fragment; see Burnett, 'Tābit ibn Qurra the Ḥarrānian,' 33–35.

362. Burnett, 'John of Seville,' 60.

363. Pingree, 'The Diffusion,' 74–75.

364. Ullmann, *Die Natur- und Geheimwissenschaften,* 309–312.

365. Ibid.; Sezgin, *Geschichte,* 7:125–128.

366. On this corpus, see Burnett, 'Arabic into Latin,' 116.

367. Ullmann, *Die Natur- und Geheimwissenschaften,* 309–310.

368. Ibid., 311; Sezgin, *Geschichte,* 7:127.

369. Ullmann, *Die Natur- und Geheimwissenschaften,* 310; Sezgin, *Geschichte,* 7:126–127.

370. Burnett, 'Arabic into Latin,' 115–118.

371. Bos and Burnett, *Scientific Weather Forecasting,* 66. On the Arabic, see Sezgin, *Geschichte,* 7: 127.

372. Burnett, 'A Group,' 66–70.

373. Carmody, *Arabic Astronomical,* 40–46.

374. The 1509 Venice and 1513 Paris editions titled *Meseallach et Ptholomeus de electionibus,* with the printers Peter Liechtenstein and Thomas Kees respectively, do indeed contain Zahel's *De electionibus.*

# BIBLIOGRAPHY

For Latin editions of Arabic authors prior to 1700, see the Appendix. As in the remainder of this study, the orthography of Latin titles is preserved, but capitalization and punctuation are modernized.

## PRIMARY SOURCES

Abioso, Giovanni. *Joannes Abiosus . . . mundo presens dirigit opus et sapientibus vaticinans eventus anni MDXXIII per eclypsim primo martii ac etiam per viginti planetarum coniunctiones in februario* A.D. *MDXXIIII* (Naples, 1523).

Abraham ibn Ezra. *The Book of the World. A Parallel Hebrew-English Critical Edition of the Two Versions of the Text,* ed. and trans. S. Sela. 2 vols. (Leiden, 2010).

Abulfeda (Abū l-Fidā'). *Chorasmiae et Mawaralnahrae, hoc est, regionum extra fluvium Oxum, descriptio,* ed. J. Greaves (London, 1650).

Achillini, Alessandro. *Opera omnia in unum collecta* (Venice, 1551), including *Quolibeta de intelligentiis* and *De elementis.*

Aegidius Romanus (Giles of Rome). *Errores philosophorum,* ed. J. Koch and J. O. Riedl (Milwaukee, 1944).

Aḥmad ibn Yūsuf ibn ad-Dāya (see also Appendix s.v. Pseudo-Haly). *Commento al Centiloquio Tolemaico,* ed. F. Martorello and G. Bezza (Milan, 2013).

Albertus Magnus. *De anima,* in *Opera omnia . . . edenda . . . curavit Institutum Alberti Magni Coloniense,* vol. 7.1, ed. C. Stroick (Münster, 1968).

———. *De homine (Summa de creaturis, secunda pars),* in *Opera omnia,* ed. A. Borgnet. 38 vols. (Paris, 1890–1899), vol. 35.

———. *De intellectu et intelligibili,* in *Opera omnia,* ed. A. Borgnet. 38 vols. (Paris, 1890–99), vol. 9.

———. *De unitate intellectus,* in *Opera omnia . . . edenda . . . curavit Institutum Alberti Magni Coloniense,* vol. 17.1, ed. A. Hufnagel (Münster, 1975).

Albumasar (see also Appendix).

——. *De revolutionibus nativitatum,* ed. D. Pingree (Leipzig, 1968).

——. *On Historical Astrology: The Book of Religions and Dynasties (On the Great Conjunctions),* ed. C. Burnett and K. Yamamoto. 2 vols. (Leiden, 2000).

——. See Burnett, C.

Albuzale (Abū l-Ṣalt). *De medicinis simplicibus,* ed. J. Martínez Gázquez, M. R. McVaugh, A. Labarta, L. Cifuentes, and D. Jacquart, in Arnald of Villanova, *Opera medica omnia,* vol. 17 (Barcelona, 2004).

Alcabitius (see also Appendix).

Alcabitius. *The Introduction to Astrology. Editions of the Arabic and Latin texts and an English translation,* ed. C. Burnett, K. Yamamoto, and M. Yano (London, 2004).

Alexander of Aphrodisias. *Commentaria in duodecim Aristotelis libros De prima philosophia* (Venice, 1561).

——. *Enarratio De anima ex Aristotelis institutione,* ed. E. Kessler (Stuttgart–Bad Cannstatt, 2008).

——. *In Aristotelis Metaphysica Commentaria,* ed. M. Hayduck (Berlin, 1891).

——. *On Aristotle's* Metaphysics, trans. W. E. Dooley (Ithaca, 1989).

Alpini, Prospero. *Historiae Aegypti Naturalis* (Leiden, 1735).

Anonymous. *Antidotarium Nicolai* (Venice, 1471, repr. in D. Goltz, *Mittelalterliche Pharmazie und Medizin,* Stuttgart, 1976).

Anonymous. *De simplici medicina: Kräuterbuch-Handschrift aus dem letzten Viertel des 14. Jahrhunderts im Besitz der Basler Universitäts-Bibliothek,* ed. A. Pfister (Basel, 1960).

Anonymous. *Ortus sanitatis de herbis et plantis* (Strasbourg, ca. 1497).

Antonius de Montulmo. *On the Judgement of Nativities, Part 2,* trans. R. Hand (Berkeley Springs, WV, 1995).

Antoninus Florentinus. *Cronica* (Nuremberg, 1491).

Argoli, Andrea. *Pandosion sphaericum* (Padua, 1653).

——. *Ptolomaeus parvus in genethliacis iunctus Arabibus* (Lyon, 1652).

Aristoteles Latinus. *Topica,* ed. L. Minio-Paluello (Brussels, 1969).

Aristotle and Averroes. *Aristotelis omnia quae extant opera . . . Averrois Cordubensis in ea opera omnes . . . commentarii* (Venice, 1562, repr. Frankfurt am Main, 1962).

——. *Aristotelis Stagiritae omnia quae extant opera . . . Averrois Cordubensis in ea opera omnes . . . commentarii* (Venice, 1550/1552).

Averroes (see also Appendix).

Averroes. *Commentarium magnum in Aristotelis De anima libros,* ed. F. S. Crawford (Cambridge, MA, 1953).

——. *Commentum magnum super libro De celo et mundo Aristotelis,* ed. F. J. Carmody and R. Arnzen. 2 vols. (Leuven, 2003).

——. *Commentum medium super libro Peri Hermeneias Aristotelis,* ed. R. Hissette (Louvain, 1996).

——. *Commentum medium super libro Praedicamentorum Aristotelis,* ed. R. Hissette (Louvain, 2010).

——. *La béatitude de l'âme,* ed. M. Geoffroy, C. Steel (Paris, 2001).

——. *Long Commentary on the De anima of Aristotle,* trans. R. C. Taylor (New Haven, 2009).

——. *Tafsīr mā baʿd aṭ-ṭabīʿa,* ed. M. Bouyges, Bibliotheca Arabica Scholasticorum 5–7. 3 vols. (Beirut, 1938–1948).

——. *Talḫīṣ kitāb al-ǧadal,* ed. C. Butterworth (Cairo, 1979).

——. *Talḫīṣ kitāb an-nafs,* ed. A. L. Ivry (Provo, 2002).

——. *Talḫīṣ mā baʿd aṭ-ṭabīʿa,* ed. U. Amin (Cairo, 1958).

——. *Über den Intellekt: Auszüge aus seinen drei Kommentaren zu Aristoteles' De Anima. Arabisch—Lateinisch—Deutsch,* ed. and trans. D. Wirmer (Freiburg im Breisgau, 2008).

——. See Bergh, S. van den.

Avicenna (see also Appendix).

Avicenna. *al-Ilāhīyāt,* ed. G. C. Anawati and S. Zayed (Cairo, 1960–1968).

——. *al-Ilāhīyāt,* ed. and trans. M. E. Marmura (Provo, 2005).

——. *al-Qānūn fī ṭ-ṭibb.* 3 vols. (Būlāq, 1294/1877, repr. Baghdad, 1970).

——. *Liber de anima seu Sextus de naturalibus,* ed. S. Van Riet. 2 vols. (Louvain, 1968–1972).

——. *Liber de philosophia prima sive Scientia divina,* ed. S. Van Riet. 3 vols. (Louvain, 1977–1983).

Baldi, Bernardino. *Cronica de matematici overo epitome dell'istoria delle vite loro opera di monsignor Bernardino Baldi da Urbino Abbate di Guastalla,* ed. A. A. Monticelli (Urbino, 1707).

——. *De le Vite de' Matematici,* ed. M. Steinschneider (Rome, 1874), (separate print of an article, in *Bulletino di bibliografia e di storia delle scienze matematiche e fisiche* 5 [1872]: 427–534).

——. *Le vite de' matematici,* ed. E. Nenci (Milan, 1998).

Barbaro, Ermolao. *Epistolae, Orationes et Carmina,* ed. V. Branca, Nuova collezione di testi umanistici inediti o rari 5–6. 2 vols. (Florence, 1943).

——. *In Dioscoridem Corollariorum libri quinque* (Cologne, 1530).

Barhebraeus (Abū l-Farağ ibn al-ʿIbrī). *Historia compendiosa dynastiarum,* ed. E. Pocock Sr. (Oxford, 1663).

——. *Specimen historiae arabum,* ed. E. Pocock Sr. (Oxford, 1650).

Barozzi, Pietro. *Edictum contra disputantes de unitate intellectus,* ed. P. Ragnisco, in 'Documenti inediti e rari intorno alla vita ed agli scritti di Nicoletto Vernia e di Elia del Medigo,' *Atti e memorie della R. Accademia di Scienze, Lettere ed Arti in Padova* N.S. 7 (1890/1891): 278–279.

Bartolotti, Gian Giacomo. *De antiquitate medicinae,* ed. D. M. Schullian and L. Belloni (Milan, 1954).

Bauhin, Caspar. *Pinax Theatri botanici . . . sive Index in Theophrasti, Dioscoridis, Plinii et botanicorum qui a seculo scripserunt opera* (Basel, 1623).

Bernardi, Antonio. *Eversionis singularis certaminis libri XL* (Basel, 1562).

Blancanus, Josephus (Biancani). *Aristotelis loca mathematica . . . Clarorum mathematicorum chronologia* (Bologna, 1615).

Bock, Hieronymus. *Kreutterbuch* (Strasbourg, 1577, repr. Frankfurt am Main, 1964).

Bonatti, Guido. *The Book of Astronomy,* ed. and trans. B. N. Dykes (Golden Valley, MN, 2007).

——. *Decem continens tractatus astronomie . . . Incipit liber introductorius ad iudicia stellarum* (Augsburg, 1491).

Brahe, Tycho. *Opera omnia* (Copenhagen, 1913).

Brasavola, Antonio Musa. *Examen omnium simplicium medicamentorum quibus Ferrarienses pharmacopolae in suis officinis utuntur* (Lyon, 1537).

Brunfels, Otto. *Catalogus illustrium medicorum sive de primis medicinae scriptoribus* (Strasbourg, 1530).

Calphurnius, Franciscus. *Francisci Calphurnii Vindocinensis Epigrammata* (Paris, ca. 1517).

Camerarius (the Younger), Joachim. *De plantis epitome utilissima Petri Andreae Matthioli* (Frankfurt am Main, 1586).

Campanella, Tommaso. *Astrologicorum libri VII in quibus astrologia omni superstitione Arabum et Iudaeorum eliminata physiologice tractatur* (Frankfurt am Main, 1630).

——. *Opusculi astrologici,* ed. G. Ernst (Milan, 2003).

Cardano, Girolamo. *Commentariorum in Ptolemaeum de astrorum iudiciis libri IV,* in *Opera omnia.* 10 vols. (Lyon, 1663, repr. Stuttgart–Bad Cannstatt, 1966), vol. 5.

——. *De subtilitate libri XXI* (Lyon, 1559).

——. *In Cl. Ptolemaei Pelusiensis IIII de astrorum iudiciis aut ut vulgo vocant quadripartitae constructionis libros commentaria* (Basel, 1554).

——. *Libelli quinque* (Nuremberg, 1547).

——. *Pronostico generale,* ed. G. Ernst, in *Girolamo Cardano: Le opere, le fonti, la vita,* ed. M. Baldi, G. Canziani, and F. Angeli, 457–475 (Milan, 1999).

Carion, Johann. *Chronica* (Schwäbisch Hall, 1537).

Clavius, Christophorus. *Opera mathematica.* 5 vols. (Mainz, 1611–1612).

Cesalpino, Andrea. *De plantis libri XVI* (Florence, 1583).

Champier, Symphorien. *Apologetica epistola responsiva pro defensione Graecorum in Arabum ac Poenorum errata,* in S. Monteux, *Annotatiunculae* (Lyon, 1533), ff. 20r–23r.

——. *Castigationes seu emendationes pharmacopolarum, sive apothecariorum, ac arabum medicorum Mesue, Serapionis, Rasis, Alpharabii et aliorum iuniorum medicorum* (Lyon, 1532).

——. *Cribratio medicamentorum fere omnium, in sex digesta libros* (Lyon, 1534).

——. *De medicine claris scriptoribus* (Lyon, ca. 1506).

——. *Symphonia Galeni ad Hippocratem, Cornelii Celsi ad Avicennam* (Lyon? 1528?).

Christmann, Jacob. *Alphabetum Arabicum cum Isagoge scribendi legendique Arabice* (Neustadt, 1582).

Clement of Alexandria. *Stromateis I,* ed. and trans. C. Mondésert and M. Caster (Paris, 1951).

Conimbricenses. *In tres libros De anima Aristotelis* (Cologne, 1600).

Constantinus, Sebastianus. *Liber de potestate syderum* (Rome, 1534).

Cordus, Euricius. *Botanologicon* (Cologne, 1534).

Daléchamps, Jacques. *Historia generalis plantarum* (Lyon, 1586/1587).

Damiano de Goes. *Hispania* (Leuven, 1542).

Dariot, Claude. *Ad astrorum iudicia facilis introductio* (Lyon, 1557).

Dillenius, Johann Jacob. *Hortus Elthamensis* (London, 1732).

Dioscorides, Pedanius. *De materia medica libri quinque,* ed. M. Wellmann. 3 vols. (Berlin, 1958).

Dodonaeus, Rembertus. *Stirpium historiae pemptades sex* (Antwerp, 1616).

Dupuis, Guillaume. *Ioannis Mesue medici praestantissimi aloen aperire ora venarum aliaque similia non pauca dicenda adversum Ioannem Manardum et Leonardum Fuchsium aliosque neotericos multos medicos defensio* (Lyon, 1537).

Egenolff, Christian. *Chronica von an und abgang aller Welt wesen* (Frankfurt am Main, 1534).

Erpenius, Thomas. *Orationes tres de linguarum Ebraeae atque Arabicae dignitate* (Leiden, 1621).

Euclid, *Elementorum libri 15 una cum scholiis antiquis a Federico Commandino in latinum conversi* (Pesaro, 1572).

Fabricius, Johann Albert. *Bibliotheca graeca.* 14 vols. (Hamburg, 1708–1728).

Ficino, Marsilio. *Marsilii Ficini Florentini . . . Opera.* 2 vols. (Basel, 1563, repr. Turin 1959, 1983).

——. *Platonic Theology,* ed. M. J. B. Allen, J. Hankins. 6 vols. (Cambridge, MA, 2001–2006).

——. *Theologia platonica de immortalitate animae,* ed. R. Marcel. 3 vols. (Paris, 1964, 1964, 1970).

Foresti, Jacopo Filipo da Bergamo. *Novissime historiarum omnium repercussiones que supplementum supplementi cronicarum nuncupantur* (Venice, 1506).

——. *Supplementum chronicarum* (Venice, 1483).

——. *Supplementum cronicarum* (Venice, 1486).

Franck, Sebastian. *Chronica: Zeitbuch vnnd Geschichtbibell* (Ulm, 1536).

Freigius, Johann Thomas. *Historiae synopsis* (Basel, 1602).

Fries, Lorenz. *Defensio medicorum principis Avicennae,* in S. Monteux, *Annotatiunculae* (Lyon, 1533), ff. 39v–45r.

Fuchs, Leonart. *Apologia contra Hieremiam Thriverum Brachelium . . . qua monstratur quod in viscerum inflammationibus, pleuritide praesertim, sanguis e directo lateris affecti mitti debeat* (Haguenau, 1534).

——. *De componendorum miscendorumque medicamentorum ratione* (Lyon, 1556).

——. *De historia stirpium commentarii insignes* (Basel, 1542, repr. Stanford, 1999).

——. *Errata recentiorum medicorum LX numero* (Haguenau, 1530).

——. *New Kreuterbuch* (Basel, 1543).

——. *Paradoxorum medicinae libri tres* (Basel, 1535).

Gaetano da Thiene. *Super libros de anima* (Venice, 1505).

——. See Di Liscia and Ebbersmeyer.

Gaurico, Luca. *Calendarium ecclesiasticum novum* (Venice, 1552).

——. *Opera omnia quae quidem extant* (Basel, 1575).

——. *Oratio de inventoribus,* in *Opera omnia.* 3 vols. (Basel, 1575), 1:1–8.

Geber (see also Appendix).

Geber. *Summa perfectio, das ist deß königlichen weitberühmbten arabischen philosophi Geber Büchlin* (Strasbourg, 1625).

Genua (Passerus), Marcantonio. *In tres libros De anima* (Venice, 1576).

Gesner, Konrad. *Bibliotheca universalis* (Zurich, 1545).

Giuntini, Francesco. *Speculum astrologiae* (Lyon, 1573).

Hermann, Paul. *Horti academici Lugduno-Batavi catalogus* (Leyden, 1687).

Herr, Michael. *Schachtafelen der Gesuntheyt* (Strasbourg, 1533, repr. Weinheim, 1988).

Iacchinus, Leonardus (Leonardo Giacchini). *In nonum librum Rasis Arabis medici ad Almansorem regem de partium morbis eruditissima commentaria* (Basel, 1564).

Ibn Ḥallikān. *Biographical Dictionary,* trans. W. Macguckin De Slane. 4 vols. (Paris, 1843).

Jacopo da Forli. *Expositio cum questionibus super primo canonis Avicennae principis* (Pavia, 1488).

Jacquier, Nicolas. *Flagellum haereticorum fascinariorum* (Frankfurt am Main, 1581).

John of Jandun. *Questiones de physico auditu noviter emendate* (Venice, 1506).

———. *Super libros Aristotelis De anima* (Venice, 1587, repr. Frankfurt am Main, 1966).

John of St. Thomas. *In tres libros De anima,* in *Cursus philosophicus thomisticus,* vol. 3 (Rome, 1948).

Kepler, Johannes. *De stella nova in pede serpentarii et qui sub ejus exortum de novo iniit trigono igneo* (Prague, 1606).

———. *Discurs von der grossen Conjunction,* in *Opera omnia,* ed. C. Frisch. 8 vols. (Frankfurt am Main, 1858–1871), 7:697–711.

———. *Epitome astronomiae Copernicanae,* in *Gesammelte Werke,* vol. 7, ed. M. Caspar (Munich, 1953).

———. *Mysterium cosmographicum,* in *Gesammelte Werke,* vol. 8, ed. F. Hammer (Munich, 1938).

Lefèvre d'Étaples, Jacques. *The Prefatory Epistles and Related Texts,* ed. E. F. Rice Jr. (New York, 1972).

Leoniceno, Nicolò. *De Plinii et plurium aliorum medicorum in medicina erroribus* (Ferrara, 1509).

Leopold of Austria. *Compilatio de astrorum scientia* (Augsburg, 1489).

Leovitius, Cyprianus. *De coniunctionibus magnis insignioribus superiorum planetarum* (1564).

Linnaeus, Carl. *Species plantarum* (Stockholm, 1753).

——. *Systema, genera, species plantarum uno volumine,* ed. H. E. Richter (Leipzig, 1835).

l'Obel, Mathias de (Lobelius), and Pierre Pena. *Nova stirpium adversaria* (Antwerp, 1576).

Lonitzer, Adam. *Kreuterbuch, kunstliche Conterfeytunge der Bäume, Stauden, Hecken, Kräuter, Getreyd, Gewürtze etc.* (Ulm, 1679, repr. Munich, 1962).

Lusitanus, Amatus (João Rodrigues). *In Dioscoridis . . . de materia medica libros quinque enarrationes eruditissimae* (Strasbourg, 1554).

Magini, Giovanni Antonio. *Ephemeridum coelestium motuum continuatio* (Frankfurt am Main, 1610).

Manardo, Giovanni. *Epistolae medicinales* (Lyon, 1557).

——. *Epistolarum medicinalium libri XX* (Venice, 1542).

——. *In Ioannis Mesue Simplicia medicamenta* (Basel, 1549).

Manlius de Bosco. *Luminare maius* (Lyon, 1536).

Marcello Virgilio. *Pedacii Dioscoridae Anazarbei de medicina materia libri sex. Interprete Marcello Virgilio Secretario Florentino: Cum eiusdem annotationibus* (Florence, 1518).

Mattioli, Pietro Andrea. *Commentarii in sex libros Pedacii Dioscoridis Anazarbei de medica materia* (Venice, 1565).

Melanchthon, Philipp. *De philosophia oratio,* ed. R. Nürnberger, *Melanchthons Werke in Auswahl.* 7 vols. (Gütersloh, 1961), 3:88–95.

——. *Initia doctrinae physicae,* ed. K. G. Bretschneider, *Philippi Melanthonis Opera quae supersunt omnia,* vol. 13 (Halle, 1846, repr. New York, 1963), col. 179–412.

——. *Liber de anima,* partially ed. R. Nürnberger, *Melanchthons Werke in Auswahl.* 7 vols. (Gütersloh, 1961), 3:307–372.

——. *Liber de anima recognitus ab auctore,* ed. K. G. Bretschneider, *Philippi Melanthonis Opera quae supersunt omnia,* vol. 13 (Halle, 1846, repr. New York, 1963), col. 5–178.

Melton, John. *Astrologaster or the Figure-Caster* (London, 1620).

Mexía, Pedro. *Le Vite di tutti gl'imperadori composte dal nobile cavaliere Pietro Messia e da M. Lodovico dolce tradotte* (Venice, 1561).

Milich, Jacob. *Oratio de consideranda sympathia et antipathia in rerum natura . . . Oratio de Avicennae vita* (Wittenberg, 1550).

Minner, Hans. *Thesaurus medicaminum,* ed. U. Schmitz, Quellen und Studien zur Geschichte der Pharmazie 13 (Würzburg, 1974).

Mondella, Luigi (Aloysius Mundella). *Epistolae medicinales* (Lyon, 1557).

Monteux, Sébastien. *Annotatiunculae in errata recentiorum medicorum per Leonardum Fuchsium Germanum collecta* (Lyon, 1533).

Morison, Robert. *Plantarum historiae universalis Oxoniensis seu herbarum distributio nova.* 3 vols. (Oxford, 1715).

Nabod, Valentin. *Enarratio elementorum astrologiae* (Cologne, 1560).

Naukler, Johannes (Nauclerus). *Chronica* (Cologne, 1675).

Neander, Michael. *Chronicon sive Synopsis historiarum* (Leipzig, 1586).

Nifo, Agostino. *Aristotelis physicarum acroasum . . . liber* (Venice, 1508).

——. *De immortalitate anime Libellus* (Venice, 1518).

——. *De intellectu,* ed. L. Spruit (Leiden, 2011).

——. *De nostrarum calamitatum causis* (Venice, 1505).

——. *Expositio subtilissima necnon et collectanea commentariaque in tres libros Aristotelis De anima* (Venice, 1559).

——. *Expositiones in Aristotelis libros Metaphysices* (Venice, 1559, repr. Frankfurt am Main, 1967).

——. *In Averroys De animae beatitudine* (Venice, 1508).

——. *In librum Destructio destructionum Averroys commentationes* (Venice, 1497).

——. *In via Aristotelis de intellectu libri sex* (Venice, 1554).

——. *Liber de intellectu* (Venice, 1503).

——. *Super tres libros de anima* (Venice, 1503).

Oddi, Oddo. *In primam totam fen primi libri Canonis Avicenne dilucidissima et expectatissima expositio* (Venice, 1575).

Origanus, David. *Novae motuum coelestium ephemerides Brandenburgicae annorum LX incipientes ab anno 1595 et desinentes in annum 1655* (Frankfurt am Main, 1609).

Orta, Garcia da. *Colloquies on the Simples and Drugs of India,* trans. C. Markham (London, 1913).

——. *Coloquios dos simples e drogas e cousas medicinais de India* (Goa, 1563).

Pacius, Julius. *Aristotelis de anima libri tres, graece et latine* (Frankfurt am Main, 1596).

Paglia, Angelo and Bartolomeo d'Orvieto. *In Antidotarium Joannis filii Mesuae* (Venice, 1543).

Patrizi, Francesco. *Discussionum Peripateticarum tomi IV* (Basel, 1581, repr. Cologne, ed. Z. Pandžić, 1999).

Paul of Venice. *Summa philosophie naturalis* (Venice, 1503, repr. Hildesheim, 1974), including *Liber de anima,* f. 66r–92v.

Pavese, Giovanni Giacomo. *In prologum Auerrois super Analytica posteriora Aristotelis commentarii* (Padua, 1552).

Pegolotti, Francesco. *Libro di divisamenti di paesi e di misure di mercatanzie,* ed. A. Evans, *La pratica della mercatura* (Cambridge, MA, 1936, repr. New York, 1970).

Petrarca, Francesco. *Invective contra medicum,* ed. P. G. Ricci, Storia e letteratura 32 (Rome, 1950).

——. *Opera.* 4 vols. (Basel, 1554, repr. Ridgewood, NJ, 1965).

——. *Prose,* ed. G. Martellotti (Milan, 1955).

Petrus Comestor. *Historia scholastica,* in J.-P. Migne, ed. *Patrologia latina,* vol. 198 (Paris, 1855), cols. 1051–1722.

Philoponus, Johannes. *Commentaria Ioannis Grammatici Alexandrei cognomento Philoponi in libros Posteriorum Aristotelis* (Venice, 1542, repr. Stuttgart–Bad Cannstatt, 1995).

Pico della Mirandola, Giovanni. *Conclusiones sive theses DCCCC: Romae anno 1486 publice disputandae, sed non admissae,* ed. B. Kieszkowski (Geneva, 1973).

——. *Disputationes adversus astrologiam divinatricem,* ed. E. Garin. 2 vols. (Florence, 1946/1952).

Pietro d'Abano. *Conciliator controversiarum quae inter philosophos et medicos versantur* (Venice, 1565, repr. Padua, 1985).

Pigghe, Albert. *Adversus prognosticatorum vulgus astrologiae defensio* (Paris, 1518).

Pitati, Pietro. *Almanach novum . . . Isagogica in coelestem astronomicam disciplinam* (Tübingen, 1544).

Pomponazzi, Pietro. *Dubitationes in quartum Meteorologicorum Aristotelis librum* (Venice, 1563).

——. *Expositio super primo et secundo De partibus animalium,* ed. S. Perfetti (Florence, 2004).

——. *Questio de immortalitate anime,* ed. P. O. Kristeller, 'Two Unpublished Questions on the Soul of Pietro Pomponazzi,' *Medievalia et Humanistica* 9 (1955): 85–96.

——. *Quomodo anima intellectiva sit forma hominis,* ed. A. Poppi, *Corsi inediti dell'insegnamento padovano.* 2 vols. (Padua, 1966–1970), 2:27–62.

——. *Tractatus de immortalitate animae,* ed. B. Mojsisch (Hamburg, 1990).

Pontano, Giovanni Gioviano. *Commentariorum in centum Claudii Ptolemaei sententias libri duo* (Basel, 1531).

Porzio, Simone. *De humana mente disputatio* (Florence, 1551).

Postel, Guillaume. *Grammatica arabica* (Paris, ca. 1539/1540).

Prassicio, Luca. *Questio de immortalitate anime intellective secundum mentem Aristotelis a nemine verius quam ab Averroi interpretati* (Naples, 1521).

Ptolemaeus, Claudius. *De praedictionibus astronomicis cui titulum fecerunt quadripartitum graece et latine libri IIII Philippo Melanthone interprete* (Basel, 1553).

——. *Fructus sive Centiloquium,* ed. E. Boer (Leipzig, 1952).

——. *Liber quadripartiti Ptholemei* (Venice, 1493).

——. *Tetrabiblos,* ed. and trans. F. E. Robbins (Cambridge, MA, 1940).

Raimundus Martini (Ramon Martí). *Pugio fidei* (Paris, 1651).

Ramus, Petrus. *Scholarum mathematicarum libri unus et triginta* (Basel, 1569).

*Ratio atque institutio studiorum societatis Jesu* (Naples, 1603).

Ray, John. *Historia plantarum.* 3 vols. (London, 1686/1688/1704).

Regiomontanus, Johannes. *Ephemerides* (Nuremberg, 1474).

Reinhold, Erasmus. *Tabulae directionum* (Tübingen, 1554).

Reisch, Gregor. *Margarita philosophica* (Basel, 1517, repr. Düsseldorf, 1973).

Reusner, Elias. *Isagoges historicae libri duo* (Jena, 1600).

Ricci, Paolo. *In apostolorum simbolum Pauli Ricii oratoris philosophi et theologi oculatissimi a priori demonstrativus dialogus* (Augsburg, 1514).

Riccioli, Giambattista. *Almagestum novum astronomiam veterem novamque complectens . . . in tres tomos distributam* (Bologna, 1651).

Richardus Anglicus. *Correctorium Alchimiae, das ist: Reformierte Alchymy* (Strasbourg, 1581).

Ruel, Jean. *De natura stirpium libri tres* (Basel, 1537).

Ryff, Walther (Gualtherus Rivius). *Pedanii Dioscoridis Anazarbei de medicinali materia libri sex Joanne Ruellio Suessionensi interprete* (Frankfurt am Main, 1543).

Saladin of Asculo. *Compendium aromatariorum* (Bologna, 1488).

———. *Compendium aromatariorum,* ed. L. Zimmermann (Leipzig, 1919).

Salutati, Coluccio. *Vom Vorrang der Jurisprudenz oder der Medizin/De nobilitate legum et medicinae,* ed. and trans. P. M. Schenkel (Munich, 1990).

Santorio, Santorio. *Commentaria in primam fen primi libri Canonis Avicennae* (Venice, 1625).

Schedel, Hartmann. *Das buch der Cronicken und gedechtnus wirdigern geschichten* (Nuremberg, 1493).

Sennerus, Ioannes Conradus. *Dissertatio botanico-medica inauguralis de senna* (Altdorf, 1733).

———. *Liber chronicarum cum figuris et ymaginibus ab inicio mundi* (Nuremberg, 1493).

Simon de Phares. *Le Recueil des plus Célèbres Astrologues.* 2 vols., ed. J.-P. Boudet (Paris, 1997–1999).

Sleidanus, Johannes. *De quatuor summis imperiis* (Strasbourg, 1556).

Spina, Giovanni Francesco. *De maximis coniunctionibus Saturni et Iovis annorum 1603 et 1702 . . . libri duo* (Macerata, 1621).

Suarez, Francisco. *Commentaria una cum quaestionibus in libros Aristotelis De anima,* ed. S. Castellote. 3 vols. (Madrid, 1978, 1981, 1991).

——. *De anima,* in *Opera omnia,* ed. M. André and C. Berton, vol. 3 (Paris, 1856).

Themistius. *In libros Aristotelis de anima paraphrasis,* ed. R. Heinze, Commentaria in Aristotelem Graeca 5.3 (Berlin, 1899).

——. *On Aristotle On the Soul,* trans. R. B. Todd (London, 1996).

Theodorus, Jakob (Tabernaemontanus). *Neu vollkommen Kraeuter–Buch* (Basel, 1731, repr. Munich, 1963).

——. *Neuw vollkommentlich Kreuterbuch* (Frankfurt am Main, 1613).

Theophrastus. *De historia plantarum libri IX . . . Theodoro Gaza interprete* (Lyon, 1552).

——. *Enquiry into Plants and Minor Works on Odours and Weather Signs,* ed. A. Hort, Loeb Classical Library 70/79. 2 vols. (Cambridge, MA, 1980/1990).

Thomas Aquinas. *De spiritualibus creaturis,* in *Quaestiones disputatae,* ed. R. Spiazzi et al. 2 vols. (Turin, 1953), 2:363–415.

——. *De unitate intellectus contra Averroistas,* in *Opera omnia iussu Leonis XIII P.M. edita* 43 (Rome, 1976), 289–314 (see also A. de Libera, *Thomas d'Aquin: L'unité de l'intellect*).

——. *Summa contra gentiles,* ed. and trans. K. Albert and P. Engelhardt. 4 vols. (Darmstadt, 2005).

Thrivière, Hieremia. *De missione sanguinis in pleuritide ac aliis phlegmonis tam externis quam internis omnibus cum Petro Brissoto et Leonardo Fuchsio disceptatio ad medicos Parisienses* (Louvain, 1532).

Tignosi, Niccolò. *In libros Aristotelis de anima commentarii* (Florence, 1551).

Tiraqueau, André. *De nobilitate et iure primigeniorum,* in *Opera omnia.* 6 vols. (Frankfurt am Main, 1616), vol. 1.

Toletus, Franciscus. *Commentaria una cum quaestionibus in tres libros Aristotelis De anima* (Cologne, 1583).

Tomkins, Thomas. *Albumazar: A Comedy,* ed. H. G. Dick (Berkeley, 1944).

Tortelli, Giovanni. *Liber de medicina et medicis,* ed. D. M. Schullian and L. Belloni (Milan, 1954).

Tournefort, Joseph Pitton de. *Elemens de botanique: ou Méthode pour connoître les plantes.* 3 vols. (Paris, 1694).

——. *Institutiones rei herbariae. Editio altera, gallica longe auctior.* 3 vols. (Paris, 1700).

Trombetta, Antonio. *Tractatus singularis contra Averroystas de humanarum animarum plurificatione ad catholice fidei obsequium Patavii editus* (Venice, 1498).

Tundermann, Carolus. *Meletemata De Sennae Foliis: Dissertatio Inauguralis* (Dorpat, 1856).

Unger, Bernhard. *Apologetica epistola pro defensione arabum medicorum,* in S. Monteux, *Annotatiunculae* (Lyon, 1533), ff. 17v–19v.

Valla, Lorenzo, *Dialecticarum disputationum libri tres* (Cologne, 1541).

Vernia, Nicoletto. *Contra perversam Averroys opinionem de unitate intellectus et de anime felicitate Questiones divine* (Venice, 1505).

——. See section 'Manuscripts.'

Vimercato, Francesco. *In tertium librum Aristotelis De anima Commentaria. De anima rationali peripatetica disceptatio* (Venice, 1574).

Vives, Juan Luis. *De anima et vita* (Zurich, 1565).

——. *De anima et vita,* ed. M. Sancipriano (Padua, 1974).

——. *De causis corruptarum artium,* ed. E. Hidalgo-Serna (Munich, 1990).

Volaterranus (Raffaele Maffei). *Commentariorum urbanorum octo et triginta libri* (Basel, 1530).

Voss, Gerhard Johannes. *De philosophia et philosophorum sectis* (The Hague, 1658).

——. *De universae mathesios natura et constitutione* (Amsterdam, 1650).

William of Auvergne. *Opera omnia.* 2 vols. (Paris, 1674, repr. Frankfurt am Main, 1963).

Wimpina, Konrad. *Tractatus de erroribus philozophorum in fide Christiana* (Leipzig, 1493).

Zabarella, Jacopo. *De rebus naturalibus libri 30* (Frankfurt am Main, 1606/1607, repr. 1966).

——. *Opera logica* (Cologne, 1603).

Zamora, Juan Gil de. *De preconiis hispanie,* ed. M. de Castro y Castro (Madrid, 1955).

Ziegler, Jacob et al. *Sphaerae atque astrorum coelestium ratio* (Basel, 1536).

SECONDARY SOURCES

Ainslie, W. *Materia Indica: or, some account of those articles which are employed by the Hindoos and other Eastern nations in their medicine, arts, and agriculture.* 2 vols. (London, 1826).

Akasoy, A. and C. Burnett, eds. *Astro-medicine: Astrology and Medicine, East and West* (Florence, 2008).

Akbari, S.C. *Idols in the East: European Representations of Islam and the Orient, 1100–1450* (Ithaca, NY, 2009).

Alessio, M. 'Sulla legenda di Averroè empio.' *Rivista Rosminiana di filosofia e di cultura* 45 (1951): 143–147.

Allan, M. J. B. 'Marsilio Ficino on Saturn, the Plotinian Mind, and the Monster of Averroes.' In *Renaissance Averroism and Its Aftermath: Arabic Philosophy in Early Modern Europe,* ed. A. Akasoy and G. Giglioni (Dordrecht, 2013), 81–97.

Allard, A. 'The Influence of Arabic Mathematics in the Medieval West.' In *Encyclopedia of the History of Arabic Science,* ed. R. Rashed. 3 vols. (London, 1996), 2:539–580.

Alonso, P. M. *Commentario al "De substantia orbis" de Averroes (Aristotelismo y Averroismo) por Alvaro de Toledo* (Madrid, 1941).

———. *Teología de Averroes (Estudios y documentos)* (Madrid, 1947).

Altmann, A., and S. M. Stern. *Isaac Israeli: A Neoplatonic Philosopher of the Early Tenth Century: His Works Translated with Comments and an Outline of His Philosophy* (Oxford, 1958).

Andrews, F. W. *The Flowering Plants of the Anglo-Egyptian Sudan.* 3 vols. (Abroath, 1950–1956).

Antonaci, A. *Ricerche sull'aristotelismo del Rinascimento: Marcantonio Zimara* (Lecce-Galatina, 1971).

Arnzen, R. *Aristoteles' De anima: Eine verlorene spätantike Paraphrase in arabischer und persischer Überlieferung. Arabischer Text nebst Kommentar, quellengeschichtlichen Studien und Glossaren* (Leiden, 1998).

———. 'Vergessene Pflichtlektüre: al-Qābīsī's astrologische Lehrschrift im europäischen Mittelalter.' *Zeitschrift für Geschichte der arabisch-islamischen Wissenschaften* 13 (1999/2000): 93–128.

Ashtor, E. *Levant Trade in the Later Middle Ages* (Princeton, 1983).

Aston, M. 'The Fiery Trigon Conjunction: An Elizabethan Astrological Prediction.' *Isis* 61, no. 2 (1970): 158–187.

Auchterlonie, P. *Arabic Biographical Dictionaries: A Summary Guide and Bibliography* (Durham, 1987).

Aufmesser, M. *Etymologische und wortgeschichtliche Erläuterungen zu* De materia medica *des Pedanius Dioscurides Anazarbeus* (Hildesheim, 2000).

Azzolini, M. *The Duke and the Stars: Astrology and Politics in Renaissance Milan* (Cambridge, MA, 2013).

Baader, G. 'Arabismus und Renaissancemedizin in Österreich im 15. und 16. Jahrhundert.' In *Der Weg der Naturwissenschaft von Johannes von Gmunden zu Johannes Kepler,* ed. G. Hamann and H. Grössing. Sitzungsberichte der Österreichischen Akademie der Wissenschaften, philosophisch-historische Klasse 497 (Vienna, 1988), 160–181.

———. 'Die Antikerezeption in der Entwicklung der medizinischen Wissenschaft während der Renaissance.' In *Humanismus und Medizin,* ed. R. Schmitz and G. Keil. Mitteilung der Kommission für Humanismusforschung 11 (Weinheim, 1984): 51–66.

———. 'Jacques Dubois as a Practitioner.' In *The Medical Renaissance of the Sixteenth Century,* ed. A. Wear et al. (Cambridge, 1985), 146–154.

———. 'Medizinische Theorie und Praxis zwischen Arabismus und Renaissancehumanismus.' In *Der Humanismus und die oberen Fakultäten,* ed. G. Keil et al. Mitteilung der Kommission für Humanismusforschung 14 (Weinheim, 1987), 185–213.

———. 'Medizinisches Reformdenken und Arabismus im Deutschland des 16. Jahrhunderts.' *Sudhoffs Archiv* 63 (1979): 261–296.

———. 'Mittelalter und Neuzeit im Werk von Otto Brunfels.' *Medizinhistorisches Journal* 13 (1979): 186–203.

———. 'Mittelalterliche Medizin im italienischen Frühhumanismus.' In *Fachprosa-Studien: Beiträge zur mittelalterlichen Wissenschafts- und Geistesgeschichte,* ed. G. Keil (Berlin, 1982), 204–254.

Bacchelli, F. 'Pico della Mirandola: Traduttore di Ibn Tufayl.' *Giornale critico della filosofia italiana* 72 (1993): 1–25.

Bachmann, M., and T. Hofmeier. *Geheimnisse der Alchemie* (Basel, 1999).

Baffioni, C. 'Per un' interpretazione del concetto averroistico di "intelletto agente."' *Contributi di storia della filosofia* 1 (1985): 79–95.

Bakker, P. 'Natural Philosophy, Metaphysics, or Something in Between? Agostino Nifo, Pietro Pomponazzi, and Marcantonio Genua on the Nature and Place of the Science of the Soul.' In *Mind, Cognition and Representation: The Tradition of Commentaries on Aristotle's De anima,* ed. J. Thijssen (Aldershot, 2008), 151–178.

Balagna, J. *Arabe et humanisme dans la France des derniers Valois* (Paris, 1989).

Baldini, U. 'The Roman Inquisition's Condemnation of Astrology: Antecedents, Reasons and Consequences.' In *Church, Censorship and Culture in Early Modern Italy,* ed. G. Fragnito (Cambridge, 2001), 79–110.

Baldini, U. and L. Spruit, eds. *Catholic Church and Modern Science. Documents from the Archives of the Roman Congregations of the Holy Office and the Index. Volume I: Sixteenth-Century Documents* (Rome, 2009).

Barker, P. 'Copernicus and the Critics of Ptolemy.' *Journal for the History of Astronomy* 30 (1999): 343–358.

Barnes, R. B. *Prophecy and Gnosis: Apocalypticism in the Wake of the Lutheran Reformation* (Stanford, 1988).

Bartòla, A. 'Eliyhau del Medigo e Giovanni Pico della Mirandola: La testimonianza dei codici Vaticani.' *Rinascimento* 33 (1993): 253–278.

Bataillon, L.-J. 'Sur Aristote et le Mont-Saint-Michel.' *Revue des sciences philosophiques et théologiques* 92 (2008): 329–334.

Bataillon, M. 'L'Arabe à Salamanque au temps de la Renaissance.' *Hespéris* 21, nos. 2–3 (1935): 1–17.

Baudry, L. *La querelle des futurs contingents (Louvain 1465–1475): Textes inédits* (Paris, 1950).

Bazán, B. C. 'Was There Ever a "First Averroism?"' In *Geistesleben im 13. Jahrhundert,* ed. J. A. Aertsen and A. Speer (Berlin, 2000), 31–53.

Beaujouan, G. 'Motives and Opportunities for Science in the Medieval Universities.' In *Scientific Change: Symposium on the History of Science,*

*University of Oxford, 9–15 July 1961,* ed. A. C. Crombie (London, 1963), 219–236.

Belting, H. *Florence and Baghdad. Renaissance Art and Arab Science* (Cambridge, MA, 2011).

Benzing, J. *Die Buchdrucker des 16. und 17. Jahrhunderts im deutschen Sprachgebiet.* Beiträge zum Buch- und Bibliothekswesen 12 (Wiesbaden, 1963).

Berendes, J. *Des Pedanios Dioskurides aus Anazarbos Arzneimittellehre in 5 Büchern: ein Interimskommentar zur Faksimile-Ausgabe des Dioskurides Neapolitanus* (Stuttgart, 1902, repr. Graz, 1988).

Bergh, S. van den. *Averroes' Tahafut al-tahafut: The Incoherence of the Incoherence, Translated from the Arabic with Introduction and Notes.* 2 vols. (London, 1954).

Bergsträsser, G. *Ḥunain ibn Isḥāq über die syrischen und arabischen Galen-Übersetzungen,* Abhandlungen für die Kunde des Morgenlandes 17-2 (Leipzig, 1925, repr. Nendeln, Liechtenstein, 1966).

Berschin, W., ed. *Biographie zwischen Renaissance und Barock: zwölf Studien* (Heidelberg, 1993).

Bertolacci, A. 'Community of Translators: The Latin Medieval Versions of Avicenna's Kitāb al-Shifāʾ (Book of the Cure).' In *Communities of Learning: Networks and the Shaping of Intellectual Identity in Europe 1100–1450,* ed. J. N. Crossley and C. J. Mews (Turnhout, 2011), 37–54.

——. *The Reception of Aristotle's Metaphysics in Avicenna's Kitāb al-Šifāʾ: A Milestone of Western Metaphysical Thought* (Leiden, 2006).

——. See Hasse, D. N.

Bertolotti, A. 'Le tipografie orientali e gli orientalisti a Roma nei secoli XVI e XVII.' *Rivista Europea* 9 (1878): 217–268.

Bertozzi, M. *Nello Specchio del cielo: Giovanni Pico della Mirandola e le "Disputationes" contro l'astrologia divinatoria* (Florence, 2008).

Bezold, F. V. 'Astrologische Geschichtsconstruction im Mittelalter.' *Deutsche Zeitschrift für Geschichtswissenschaft* 8 (1892): 29–72.

Bezza, G. 'Representation of the Skies and the Astrological Chart.' In *A Companion to Astrology in the Renaissance,* ed. B. Dooley (Leiden, 2014), 59–86.

Bezzel, I. 'Joachim Heller (ca. 1520–1580) als Drucker in Nürnberg und Eisleben.' *Archiv für die Geschichte des Buchwesens* 37 (1992): 295–330.

Bianchi, L. 'Continuity and Change in the Aristotelian Tradition.' In *Cambridge Companion to Renaissance Philosophy*, ed. J. Hankins (Cambridge, 2007), 49–71.

——. 'Les aristotélismes de la scolastique.' In *Vérités dissonantes: Aristote à la fin du Moyen Âge*, ed. L. Bianchi and E. Randi, (Fribourg, 1993), 1–37.

——. *Pour une histoire de la "double vérité"* (Paris, 2008).

——. 'Rusticus mendax: Marcantonio Zimara e la fortuna di Alberto Magno nel Rinascimento italiano.' *Freiburger Zeitschrift für Philosophie und Theologie* 45 (1998): 264–278.

——. *Studi sull'aristotelismo del rinascimento* (Padua, 2003).

Biliński, B. '*Vite dei matematici* di Bernardino Baldi nei ritrovati manoscritti Rosminiani.' *Accademia Nazionale dei Lincei: Rendiconti della Classe di Scienze fisiche, matematiche e naturali* 8, no. 59 (1975): 305–321.

Bisaha, N. *Creating East and West: Renaissance Humanists and the Ottoman Turks* (Philadelphia, 2004).

Black, D. L. 'Models of the Mind: Metaphysical Presuppositions of the Averroist and Thomistic Accounts of Intellection.' *Documenti et studi sulla tradizione filosofica medievale* 15 (2004): 319–352.

Bland, K. P. 'Elijah del Medigo, Unicity of Intellect, and Immortality of Soul.' *Proceedings of the American Academy for Jewish Research* 61 (1995): 1–22.

Blåsjö, V. 'A Critique of the Arguments for Maragha Influence on Copernicus.' *Journal for the History of Astronomy* 45 (2014): 183–195.

Bobzin, H. 'Agostino Giustiniani (1470–1536) und seine Bedeutung für die Geschichte der Arabistik.' In *XXIV. deutscher Orientalistentag*, ed. W. Diem and A. Falaturi (Stuttgart, 1990), 131–139.

——. *Der Koran im Zeitalter der Reformation: Studien zur Frühgeschichte der Arabistik und Islamkunde in Europa*. Beiruter Texte und Studien 42 (Stuttgart, 1995).

——. 'Geschichte der arabischen Philologie in Europa bis zum Ausgang des achtzehnten Jahrhunderts.' In *Grundriß der Arabischen Philologie*, vol. 3: *Supplement*, ed. W. Fischer (Wiesbaden, 1992), 155–187.

——. 'Guillaume Postel (1510–1581) und die Terminologie der arabischen Nationalgrammatik.' In *Studies in the History of Arabic Grammar II,* ed. K. Versteegh and M. Carter (Amsterdam, 1990), 57–71.

——. 'Islamkundliche Quellen in Jean Bodins *Heptaplomeres.*' In *Jean Bodins Colloquium Heptaplomeres,* ed. G. Gawlick and F. Niewöhner (Wiesbaden, 1996), 41–57.

——. *Ließ ein Papst den Koran verbrennen? Mutmaßungen zum Venezianer Korandruck von 1537/1538* (Munich, 2013).

Bonitz, H. *Index Aristotelicus* (Berlin, 1870).

Bornkessel, B. 'Senna-Ernte im Sari: Anbau und Bearbeitung von Tinnevelley-Senna in Indien.' *Deutsche Apotheker Zeitung* 131 (1991): 171–174.

Bos, G. *Ibn al-Jazzār on Sexual Diseases and Their Treatment: A Critical Edition of Zād al-musāfir wa-qūt al-ḥāḍir. Provisions for the Traveller and Nourishment for the Sedentary, Book 6* (London, 1997).

——. *Maimonides: On Asthma: Maqālah fī al-rabw: A Parallel Arabic-English Text* (Provo, UT, 2002).

Bos, G., and C. Burnett. *Scientific Weather Forecasting in the Middle Ages: The Writings of Al-Kindī. Studies, Editions, and Translations of the Arabic, Hebrew and Latin Texts* (London, 2000).

Bottin, F. 'Paolo Veneto e il problema degli universali.' In *Aristotelismo Veneto e scienza moderna,* ed. L. Olivieri. 2 vols. (Padua, 1983), 1:459–468.

Boudet, J.-P., ed. *Le Recueil des plus celebres astrologues de Simon de Phares* (Paris, 1997).

——. *Lire dans le ciel: la bibliothèque de Simon de Phares, astrologue de XVe siècle* (Brussels, 1994).

Boulègue, L. 'À propos de la thèse d'Averroès: Pietro Pomponazzi *versus* Agostino Nifo.' In *Pietro Pomponazzi entre traditions et innovations,* ed. J. Biard and T. Gontier (Amsterdam, 2009), 83–98.

Bouras-Vallianatos, P. 'Ioannes Zacharias Aktuarios.' In *Lexikon der byzantinischen Autoren,* ed. M. Grünbart and A. Riehle (forthcoming).

Bouyges, M. *Averroès: Tafsīr mā baʿd aṭ-ṭabīʿa: Notice*. Bibliotheca Arabica Scholasticorum 5, 1 (Beirut, 1952).

——. 'Notes sur les philosophes arabes connus des Latins au Moyen Age: V. Inventaire des textes arabes d'Averroès.' *Mélanges de l'Université Saint-Joseph* 8 (1922): 3–54.

Brague, R. *Au moyen du Moyen Âge: Philosophies médiévales en chrétienté, judaïsme et islam* (Paris, 2006).

——. 'Das islamische Volk ist das belogenste.' Newspaper interview with *Die Presse,* Vienna, April 22, 2008.

——. 'Wie islamisch ist die islamische Philosophie?' In *Wissen über Grenzen: Arabisches Wissen und lateinisches Mittelalter,* ed. A. Speer and L. Wegener. Miscellanea Mediaevalia 33 (Berlin, 2006), 156–178 (French version: *Au moyen du Moyen Âge,* 108–132).

Brams, J., and W. Vanhamel, eds. *Guillaume de Moerbeke: recueil d'études à l'occasion du 700e anniversaire de sa mort (1286).* Ancient and Medieval Philosophy 7 (Leuven, 1989).

Breen, Q. 'Giovanni Pico della Mirandola on the Conflict of Philosophy and Rhetoric.' *Journal of the History of Ideas* 13 (1952): 384–412.

Brenet J.-B. 'Corps-sujet, corps-objet: Notes sur Averroès et Thomas d'Aquin dans le *De immortalitate animae* de Pomponazzi.' In *Pietro Pomponazzi entre traditions et innovations,* ed. J. Biard and T. Gontier (Amsterdam, 2009), 11–28.

——. *Les possibilités de jonction: Averroès–Thomas Wylton* (Berlin, 2013).

——. *Transferts du sujet: La noétique d'Averroès selon Jean de Jandun* (Paris, 2003).

Brentjes, S. *Travellers from Europe in the Ottoman and Safavid Empires, 16th–17th Centuries: Seeking, Transforming, Discarding Knowledge* (Farnham, Surrey, 2010).

Brévart, F. B. 'Eine neue deutsche Übersetzung der lat. "sphaera mundi" des Johannes von Sacrobosco.' *Zeitschrift für deutsches Altertum und deutsche Literatur* 108 (1979): 57–65.

——. *Johannes von Sacrobosco: Das Puechlein von der Spera* (Göppingen, 1979).

Brincken, A. von den. 'Die Rezeption mittelalterlicher Historiographie durch den Inkunabeldruck.' In *Geschichtsschreibung und Geschichtsbe-*

*wußtsein im späten Mittelalter,* ed. H. Patze. Vorträge und Forschungen 31 (Sigmaringen, 1987), 215–236.

Brinkhus, G., and C. Pachnicke, eds. *Leonhart Fuchs (1501–1566): Mediziner und Botaniker* (Tübingen, 2001).

Brinton, S. *The Gonzaga: Lords of Mantua* (London, 1927).

Brockelmann, C. *Geschichte der arabischen Litteratur.* 2 vols. (Leiden, 1943–1949); *Supplement.* 3 vols. (Leiden, 1937–1942).

Brockliss, L. 'Curricula.' In *A History of the University in Europe,* ed. W. Rüegg, vol. 2 (Cambridge, 1996), 563–620.

Broemser, F. 'Johann Winter aus Andernach (Ioannes Guinterius Andernacus), 1505–1574: Ein Humanist und Mediziner des 16. Jahrhunderts.' In *Johann Winter aus Andernach,* ed. K. Schäfer (Andernach, 1989), 5–35.

Brosseder, C. *Im Bann der Sterne: Caspar Peucer, Philipp Melanchthon und andere Wittenberger Astrologen* (Berlin, 2004).

Buck, A. ed. *Biographie und Autobiographie in der Renaissance: Arbeitsgespräch in der Herzog August Bibliothek Wolfenbüttel vom 1. bis 3. November 1982* (Wiesbaden, 1983).

*Bullarum diplomatum et privilegiorum sanctorum romanorum pontificum Taurinensis editio,* vol. 8 (Turin, 1863).

Bürgel, J. C. 'Averroes "contra Galenum": Das Kapitel von der Atmung im Colliget des Averroes als ein Zeugnis mittelalterlich-islamischer Kritik an Galen.' *Nachrichten der Akademie der Wissenschaften in Göttingen: Philologisch-Historische Klasse* 1967, no. 9 (Göttingen, 1968), 267–340.

Burman, T. E. *Reading the Qur'ān in Latin Christendom, 1140–1560* (Philadelphia, 2007).

Burnett, C. 'Abu Ma'shar (A.D. 787–886) and His Major Texts on Astrology.' In *Kayd. Studies in History of Mathematics, Astronomy and Astrology in Memory of David Pingree,* ed. G. Gnoli and A. Panaino (Rome, 2009), 17–29.

——. 'Al-Kindī in the Renaissance.' In *Sapientiam amemus: Humanismus und Aristotelismus in der Renaissance. Festschrift für Eckhard Keßler zum 60. Geburtstag,* ed. P. R. Blum (Munich, 1999), 13–30.

——. 'Al-Kindī on Judicial Astrology: *The Forty Chapters.*' *Arabic Sciences and Philosophy* 3 (1993): 77–117.

——. 'Arabic into Latin: The Reception of Arabic Philosophy into Western Europe.' In *Cambridge Companion to Arabic Philosophy,* ed. P. Adamson and R. Taylor (Cambridge, 2005), 370–404.

——. 'Arabic into Latin in Twelfth-Century Spain: The Works of Hermann of Carinthia.' *Mittellateinisches Jahrbuch* 13 (1978): 100–134.

——. 'The Coherence of the Arabic-Latin Translation Program in Toledo in the Twelfth Century.' *Science in Context* 14 (2001): 249–288.

——. '*De meliore homine.* 'Umar ibn al-Farrukhān al-Ṭabarī on Interrogations: A Fourth Translation by Salio of Padua?' In *Adorare caelestia, gubernare terrena: Atti del Colloquio Internazionale in onore di Paolo Lucentini (Napoli, 6–7 Novembre 2007),* ed. P. Arfé, I. Caiazzo Lacombe, and A. Sannino (Turnhout, 2011), 295–326.

——. 'Encounters with Razi the Philosopher: Constantine the African, Petrus Alfonsi and Ramon Martí.' In *Pensamiento hispano medieval: Homenaje a Horacio Santiago-Otero,* ed. J.-M. Soto Rábanos (Madrid, 1998), 973–992.

——. 'A Group of Arabic-Latin Translators Working in Northern Spain in the Mid-Twelfth Century.' *Journal of the Royal Asiatic Society* s.n. (1977): 62–108.

——, ed. *Hermann of Carinthia:* De essentiis (Leiden, 1982).

——. 'A Hermetic Programme of Astrology and Divination in Mid-Twelfth-Century Aragon: The Hidden Preface in the *Liber novem iudicum.*' In *Magic and the Classical Tradition,* ed. C. Burnett and W. F. Ryan (London, 2006), 99–118.

——. *The Introduction of Arabic Learning into England.* The Panizzi Lectures 1996 (London, 1997).

——. 'The Introduction of Aristotle's Natural Philosophy into Great Britain: A Preliminary Survey of the Manuscript Evidence.' In *Aristotle in Britain during the Middle Ages,* ed. J. Marenbon (Turnhout, 1996), 21–50.

——. 'John of Seville and John of Spain: A *mise au point.*' *Bulletin de philosophie médiévale* 44 (2002): 59–78.

——. 'Learned Knowledge of Arabic Poetry, Rhymed Prose, and Didactic Verse from Petrus Alfonsi to Petrarch.' In *Poetry and Philosophy in the*

*Middle Ages: A Festschrift for Peter Dronke,* ed. J. Marenbon (Leiden, 2000), 29–62.

——. 'Lunar Astrology: The Varieties of Texts Using Lunar Mansions, with Emphasis on *Jafar Indus.*' *Micrologus* 12 (2004): 43–133.

——. 'Magister Iohannes Hispalensis et Limiensis' and Qusta ibn Luqa's De differentia spiritus et animae: A Portuguese Contribution to the Arts Curriculum?' *Mediaevalia. Textos e estudos* 7–8 (1995): 221–267.

——. 'Michael Scot and the Transmission of Scientific Culture from Toledo to Bologna via the Court of Frederick II Hohenstaufen.' *Micrologus* 2 (1994): 101–126.

——. 'Partim de suo et partim de alieno: Bartholomew of Parma, the Astrological Texts in MS Bernkastel-Kues, Hospitalsbibliothek 209, and Michael Scot.' In *Seventh Centenary of the Teaching of Astronomy in Bologna 1297–1997,* ed. P. Battistini et al. (Bologna, 2001), 38–76.

——. 'Petrarch and Averroes: An Episode in the History of Poetics.' In *The Medieval Mind: Hispanic Studies in Honour of Alan Deyermond,* ed. I. MacPherson and R. Penny (London, 1997), 49–56.

——. 'Revisiting the 1552–1550 and 1562 Aristotle-Averroes Edition.' In *Renaissance Averroism and Its Aftermath: Arabic Philosophy in Early Modern Europe,* ed. A. Akasoy and G. Giglioni (Dordrecht, 2013), 55–64.

——. 'The Second Revelation of Arabic Philosophy and Science: 1492–1575.' In *Islam and the Italian Renaissance,* ed. C. Burnett and A. Contadini. Warburg Institute Colloquia 5 (London, 1999), 185–198.

——. 'The "Sons of Averroes with the Emperor Frederick" and the Transmission of the Philosophical Works by Ibn Rushd.' In *Averroes and the Aristotelian Tradition: Sources, Constitution and Reception of the Philosophy of Ibn Rushd (1126–1198),* ed. G. Endreß et al. (Leiden, 1999), 259–299.

——. 'The Strategy of Revision in the Arabic-Latin Translations from Toledo: The Case of Abū Maʿshar's *On the Great Conjunctions.*' In *Les Traducteurs au travail: leurs manuscrits et leurs méthodes,* ed. J. Hamesse (Turnhout, 2001), 51–113, 529–540.

——. 'Ṭābit ibn Qurra the Ḥarrānian on Talismans and the Spirits of the Planets.' *La Corónica* 36 (2007): 13–40.

———. 'Translating from Arabic into Latin in the Middle Ages: Theory, Practice, and Criticism.' In *Éditer, traduire, interpréter: essais de méthodologie philosophique,* ed. S. G. Lofts and P. W. Rosemann (Louvain-la-Neuve, 1997), 55–78.

———. 'The Two Faces of Averroes in the Renaissance.' In *Al-Ufq al-kawnī li-fikr Ibn Rušd,* 1998, ed. M. Maṣbaḥi (Marrakesh, 2001), 87–94.

———. See Bos, G.

Burnett, C., and D. Jacquart, eds. *Constantine the African and ʿAlī ibn al-ʿAbbās al-Maǧūsī: The* Pantegni *and Related Texts.* Studies in Ancient Medicine 10 (Leiden, 1994).

Burnett, C., and A. Mendelsohn. 'Aristotle and Averroes on Method in the Middle Ages and Renaissance: The "Oxford Gloss" to the Physics and Pietro d'Afeltro's Expositio Proemii Averroys.' In *Method and Order in Renaissance Philosophy of Nature,* ed. D. A. Di Liscia, E. Kessler, and C. Methuen (Aldershot, 1997), 53–111.

Burnett, C., and D. Pingree, eds. *The Liber Aristotilis of Hugo of Santalla* (London, 1997).

Burnett, C., and K. Yamamoto. *Abū Maʿšar on Historical Astrology: The Book of Religions and Dynasties (On the Great Conjunctions).* 2 vols. (Leiden, 2000).

Burnett, C., K. Yamamoto, and M. Yano. *Abū Maʿšar: The Abbreviation of the Introduction to Astrology Together with the Medieval Latin Translation of Adelard of Bath* (Leiden, 1994).

———. *Al-Qabīsī (Alcabitius): The Introduction to Astrology.* Warburg Institute Studies and Texts 2 (London, 2004).

Butler, P. 'Fifteenth Century Editions of Arabic Authors in Latin Translation.' In *The Macdonald Presentation Volume: A Tribute to Duncan Black Macdonald* (Princeton, 1933), 63–71.

Butterworth, C. E., and B. A. Kessel, eds. *The Introduction of Arabic Philosophy into Europe* (Leiden, 1994).

Büttgen, P., A. de Libera, M. Rashed, and I. Rosier-Catach, eds. *Les Grecs, les Arabes et nous: Enquête sur l'islamophobie savante* (Paris, 2009).

Calma, D. *Études sur le premier siécle de l'averroïsme latin: Approches et textes inédits* (Turnhout, 2011).

Campanini, S. 'Peculium Abrae. La grammatica ebraico-latina di Avraham de Balmes.' *Annali di Ca' Foscari* 36, no. 3 (1997): 5–49.

Capp, B. *Astrology and the Popular Press: English Almanacs, 1500–1800* (London, 1979).

Carmody, F. J., ed. *Al-Biṭrūjī: De motibus celorum. Critical Edition of the Latin Translation of Michael Scot* (Berkeley, 1952).

——. *Arabic Astronomical and Astrological Sciences in Latin Translation: A Critical Bibliography* (Berkeley, 1956).

——. *The Astronomical Works of Thabit B. Qurra* (Berkeley, 1960).

Caroti, S. 'Melanchthon's Astrology.' In *'Astrologi Hallucinati': Stars and the End of the World in Luther's Time,* ed. P. Zambelli (Berlin, 1986), 109–121.

Carpi, D. 'Sulla permanenza a Padova nel 1533 del medico ebreo Jacob di Shemuel Mantino.' *Quaderni per la storia dell'università di Padova* 18 (1985): 196–203.

Casali, E. *Le spie del cielo: Oroscopi, lunari e almanacchi nell'Italia moderna* (Turin, 2003).

Casini, L. *Cognitive and Moral Psychology in Renaissance Philosophy: A Study of Juan Luis Vives' De anima et vita* (Uppsala, 2006).

Cassirer, E., P. O. Kristeller, and J. H. Randall Jr., eds. *The Renaissance Philosophy of Man: Selections in Translation* (Chicago, 1956).

Cassuto, U. *Gli ebrei a Firenze nell'età del Rinascimento* (Florence, 1965).

Castellote Cubells, S. *Die Anthropologie des Suarez: Beiträge zur spanischen Anthropologie des XVI. und XVII. Jahrhunderts* (Munich, ²1982).

Cerulli, E. 'Petrarca e gli Arabi.' In *Studi in onore di A. Schiaffini. Rivista di cultura classica e medioevale* 7 (1965): 331–336.

Chandelier, J. 'Comment Averroès a empoisonné Avicenne.' In *Mélanges Danielle Jacquart,* forthcoming.

Chapman, A. 'Astrological medicine.' In *Health, Medicine and Mortality in the Sixteenth Century,* ed. C. Webster (Cambridge, 1979), 275–300.

Clubb, L. G. *Giambattista Della Porta, Dramatist* (Princeton, 1965).

Coccia, E. *La trasparenza delle immagini: Averroè e l'averroismo* (Milan, 2005).

Cochrane, H. E. W. *Historians and Historiography in the Italian Renaissance* (Chicago, 1981).

——. 'Science and Humanism in the Italian Renaissance.' *American Historical Review* 81 (1976): 1039–1057.

Codazzi, A. 'Il trattato dell'arte metrica di Giovanni Leone Africano.' *Studi orientalistici in onore di Giorgio Levi Della Vida.* 2 vols. (Rome, 1956), 1:180–198.

Colin, G. *Avenzoar: Sa vie et ses œuvres* (Paris, 1911).

Conrad, L. I. 'Introduction: The World of Ibn Ṭufayl.' In *The World of Ibn Ṭufayl: Interdisciplinary Perspectives on Ḥayy ibn Yaqẓān,* ed. L. I. Conrad (Leiden, 1996), 1–37.

——. 'Research Resources on Ibn Ṭufayl and Ḥayy ibn Yaqẓān.' In *The World of Ibn Ṭufayl: Interdisciplinary Perspectives on Ḥayy ibn Yaqẓān,* ed. L. I. Conrad (Leiden, 1996), 267–293.

Contadini, A., and C. Norton, eds. *The Renaissance and the Ottoman World* (London, 2013).

Copenhaver, B. P. *Symphorien Champier and the Reception of the Occultist Tradition in Renaissance France* (The Hague, 1978).

——. 'A Tale of Two Fishes: Magical Objects in Natural History from Antiquity through the Scientific Revolution.' *Journal of the History of Ideas* 52 (1991): 373–398.

Copenhaver, B. P., and C. B. Schmitt. *Renaissance Philosophy* (Oxford 1992).

Coviello, A., and P. E. Fornaciari, eds. *Averroè: Parafrasi della* Republica *nella traduzione latina di Elia del Medigo* (Florence, 1992).

Cranz, F. E. 'Alexander Aphrodisiensis.' In *Catalogus translationum et commentariorum: Medieval and Renaissance Latin Translations and Commentaries,* ed. F. E. Cranz and P. O. Kristeller (Washington, DC, 1960–), 1: 77–135.

——. *A Bibliography of Aristotle Editions 1501–1600. Second Edition with Addenda and Revisions by Charles B. Schmitt* (Baden-Baden, 1984).

——. 'Editions of the Latin Aristotle Accompanied by the Commentaries of Averroes.' In *Philosophy and Humanism: Renaissance Essays in Honor of Paul Oskar Kristeller,* ed. E. P. Mahoney (Leiden, 1976), 116–128.

——. 'The Publishing History of the Aristotle Commentaries of Thomas Aquinas.' *Traditio* 34 (1978): 157–192.

Cruz Hernández, M. *Abū-l-Walīd Muḥammad ibn Rušd (Averroes): Vida, Obra, Pensamiento, Influencia. Secunda edición* (Madrid, 1997).

Daems, W. F. *Nomina simplicium medicinarum ex synonymariis medii aevi collecta: semantische Untersuchungen zum Fachwortschatz hoch- und spätmittelalterlicher Drogenkunde* (Leiden, 1993).

Daiber, H. 'The Reception of Islamic Philosophy at Oxford in the 17th Century: The Pococks' (Father and Son) Contribution to the Understanding of Islamic Philosophy in Europe.' In *The Introduction of Arabic Philosophy into Europe,* ed. C. E. Butterworth and B. A. Kessel (Leiden, 1994), 65–82.

d'Alverny, M.-T. *Avicenna Latinus: Codices* (Louvain-la-Neuve, 1994).

——. *Avicenne en occident: Recueil d'articles de Marie-Thérèse d'Alverny réunis en hommage à l'auteur* (Paris, 1993), esp. art. 4, 'Notes sur les traductions médiévales d'Avicenne'; art. 13, 'Avicenne et les médecins de Venise'; art. 15, 'Survivance et renaissance d'Avicenne à Venise et Padoue'; and art. 16, 'Avicennisme en Italie'.

d'Alverny, M.-T., and F. Hudry. 'Al-Kindi: De Radiis.' *Archives d'histoire doctrinale et littéraire du moyen âge* 41 (1975): 139–269.

Daniel, N. *Islam and the West: The Making of an Image* (Edinburgh, 1980).

Dannenfeldt, K. H. *Leonhard Rauwolf: Sixteenth-Century Physician, Botanist, and Traveler* (Cambridge, MA, 1968).

——. 'The Renaissance Humanists and the Knowledge of Arabic.' *Studies in the Renaissance* 2 (1955): 96–117.

Davidson, H. A. *Alfarabi, Avicenna, and Averroes, on Intellect: Their Cosmologies, Theories of the Active Intellect, and Theories of Human Intellect* (New York, 1992).

Davis, N. Z. *Trickster Travels: A Sixteenth-Century Muslim between Worlds* (New York, 2006).

De Angelis, S. *Anthropologien: Genese und Konfiguration einer 'Wissenschaft vom Menschen' in der Frühen Neuzeit* (Berlin, 2010).

De Bellis, E. *Nicoletto Vernia e Agostino Nifo: aspetti storiografici e metodologici* (Galatina, 2003).

de Libera, A. *Averroès: L'intelligence et la pensée: Grand commentaire du* De anima *Livre III (429a10–435b25)* (Paris, 1998).

——. *Métaphysique et noétique: Albert le Grand* (Paris, 2005).

——, ed. *Thomas d'Aquin: L'unité de l'intellect contre les Averroïstes, suivi des Textes contre Averroès antérieurs à 1270* (Paris, 1994).

——. See Hayoun, M.-R.

Del Nero, V. 'A Philosophical Treatise on the Soul: *De anima et vita* in the Context of Vives's Opus.' In *A Companion to Juan Luis Vives,* ed. C. Fantazzi (Leiden, 2008), 277–314.

Denifle, H., and A. Chatelain, eds. *Chartularium Universitatis parisiensis.* 4 vols. (Paris, 1889–1897).

Derenbourg, H. 'Léon l'Africain et Jacob Mantino.' *Revue des études juives* 7 (1883): 283–285.

Di Bono, M. 'Copernicus, Amico, Fracastoro and Ṭūsī's Device: Observations on the Use and Transmission of a Model.' *Journal for the History of Astronomy* 26 (1995): 133–154.

Di Donato, S. 'Il *Kašf 'an manāhiğ* di Averroè: confronto fra la versione di Abraham De Balmes e le citazioni di Calo Calonimo nel *De mundi creatione.' Materia Giudaica* 9.1–2 (2004): 241–248.

——. 'Il *Kitāb al-Kašf 'an manāhiğ al-adilla fī 'aqā'id al-milla* di Averroè nella traduzione ebraico-latina di Abraham De Balmes.' *Annali di Ca'Foscari* 41, no. 3 (2002): 5–36.

——. 'La traduzione latina della Risāla al-wadā' d'Avempace.' In *Gli ebrei nel Salento: Secoli IX–XVI,* ed. F. Lelli (Galatina, 2013), 301–314.

——. 'Traduttori di Averroè e traduzioni ebraico-latine nel dibattito filosofico del XV e XVI secolo.' In *L'averroismo in etá moderna (1400– 1700),* ed. G. B. Licata (Macerata, 2013), 25–49.

Dietrich, A. *Dioscurides triumphans: Ein anonymer arabischer Kommentar (Ende 12. Jahrh. n. Chr.) zur Materia medica. Arabischer Text nebst kommentierter deutscher Übersetzung.* 2 vols. Abhandlungen der Akademie der Wissenschaften in Göttingen, Philol.-hist. Klasse, dritte Folge, 173 (Göttingen, 1988).

——. 'Islamic Sciences and the Medieval West: Pharmacology.' In *Islam and the Medieval West: Aspects of Intercultural Relations,* ed. K. I. Semaan (Albany, 1980), 50–63.

Dilg, P. 'The Antarabism in the Medicine of Humanism.' In *La diffusione delle scienze islamiche nel medio evo europeo,* ed. B. Scarcia Amoretti (Rome, 1987), 269–289.

——. 'Arabische Pharmazie im lateinischen Mittelalter.' In *Die Begegnung des Westens mit dem Osten. Kongreßakten des 4. Symposions des Mediävistenver-*

*bandes in Köln aus Anlaß des 1000. Todestages der Kaiserin Theophanu,* ed. O. Engels and P. Schreiner (Sigmaringen, 1993), 299–317.

——. *Das Botanologicon des Euricius Cordus: Ein Beitrag zur botanischen Literatur des Humanismus* (Marburg, 1969).

——. 'Das Theriakbüchlein des Euricius Cordus.' In *Fachprosa-Studien: Beiträge zur mittelalterlichen Wissenschafts- und Geistesgeschichte,* ed. G. Keil (Berlin, 1982), 417–447.

——. 'Die botanische Kommentarliteratur Italiens um 1500 und ihr Einfluß auf Deutschland.' In *Der Kommentar in der Renaissance,* ed. A. Buck (Boppard, 1975), 225–252.

——. 'Die Pflanzenkunde im Humanismus—Der Humanismus in der Pflanzenkunde.' In *Humanismus und Naturwissenschaft,* ed. R. Schmitz and F. Krafft (Boppard, 1980), 113–134.

——. 'The *Liber aggregatus in medicinis simplicibus* of Pseudo-Serapion: An Influential Work of Medical Arabism.' In *Islam and the Italian Renaissance,* ed. C. Burnett and A. Contadini. Warburg Institute Colloquia 5 (London, 1999), 221–231.

——. 'Vom Ansehen der Arzneikunst: Historische Reflexionen in Kräuterbüchern des 16. Jahrhunderts.' *Sudhoffs Archiv* 62 (1978): 64–79.

Di Liscia, D. A., and S. Ebbersmeyer. 'Gaetano da Thiene *Quaestio de perpetuitate intellectus.*' In *Sol et homo. Festschrift zum 70. Geburtstag von Eckhard Keßler,* ed. S. Ebbersmeyer et al. (Munich, 2008), 155–193.

Di Liscia, D. A., E. Kessler, and C. Methuen. *Method and Order in Renaissance Philosophy of Nature* (Aldershot, 1997).

Di Napoli, G. *L'immortalità dell'anima nel Rinascimento* (Turin, 1963).

Döbler, E. 'Der Katalog des Amplonius von 1410/1412.' In *Der Schatz des Amplonius: Die große Bibliothek des Mittelalters in Erfurt,* ed. K. Paasch (Erfurt, 2001), 85–89.

Dod, B. G. 'Aristoteles Latinus.' In *The Cambridge History of Later Medieval Philosophy,* ed. N. Kretzmann et al. (Cambridge, 1982), 45–79.

Dols, M. W. *Medieval Islamic Medicine: Ibn Riḍwān's Treatise "On the Prevention of Bodily Ills in Egypt"* (Berkeley, 1984).

Donati, S. '*Utrum accidens possit existere sine subiecto.* Aristotelische Metaphysik und christliche Theologie in den Physikkommentaren des 13. Jahrhunderts.' In *Nach der Verurteilung von 1277: Philosophie und Theologie an der Universität von Paris im letzten Viertel des 13. Jahrhunderts,* ed. J. A. Aertsen et al. Miscellanea Mediaevalia 28 (Berlin, 2000), 576–617.

Dooley, B., ed. *A Companion to Astrology in the Renaissance* (Leiden, 2014).

Dressendörfer, W. '*In apotecis circa realtum:* Venedig als Einkaufsplatz für Arzneidrogen während des 15. Jahrhunderts.' In *Orbis pictus: Kultur- und pharmaziehistorische Studien. Festschrift für Wolfgang-Hagen Hein,* ed. W. Dressendörfer and W.-D. Müller-Jahncke (Frankfurt am Main, 1985), 73–86.

Dubler, C. E. *La 'Materia Medica' de Dioscórides: Transmisión medieval y renacentista.* 6 vols. (Barcelona, 1953–1959).

———. 'Posibles fuentes árabes de la *agricultura general* de Gabriel Alonso de Herrera.' *al-Andalus* 6 (1941): 135–156.

Dunlop, D. M. 'The Translations of al-Biṭrīq and Yaḥyā (Yuḥannā) b. al-Biṭrīq.' *Journal of the Royal Asiatic Society of Great Britain and Ireland* s.n. (1959): 140–150.

Durling, R. J. *A Catalogue of Sixteenth Century Printed Books in the National Library of Medicine* (Bethesda, 1967).

———. 'A Chronological Census of Renaissance Editions and Translations of Galen.' *Journal of the Warburg and Courtauld Institutes* 24 (1961): 230–305.

Dykes, B. N. *Persian Nativities, Volume 1: Māshā'allāh & Abū 'Ali* (Minneapolis, 2009).

———. *Persian Nativities, Volume 3: Abū Ma'shar's* On the Revolutions of the Years of Nativities (Minneapolis, 2010).

Eamon, W. 'Astrology and Society.' In *A Companion to Astrology in the Renaissance,* ed. B. Dooley (Leiden, 2014), 141–191.

Edwards, W. F. 'The Averroism of Iacopo Zabarella.' *Atti del XII congresso internazionale di filosofia.* 12 vols. (Florence, 1958), 9:91–107.

Eisenlohr, T., ed. *Sammlung der württembergischen Schul-Geseze, 3. Abt.* (Tübingen, 1843).

Emery, K., and A. Speer. 'After the Condemnation of 1277: New Evidence, New Perspectives, and Grounds for New Interpretations.' In *Nach der Verurteilung von 1277: Philosophie und Theologie an der Universität von Paris im letzten Viertel des 13. Jahrhunderts,* ed. J. A. Aertsen et al. Miscellanea Mediaevalia 28 (Berlin, 2001), 3–19.

Endreß, G. 'Averroes' *De caelo*: Ibn Rushd's Cosmology in His Commentaries on Aristotle's *On the Heavens*.' *Arabic Sciences and Philosophy* 5 (1995): 9–49.

——. 'Averrois Opera: A Bibliography of Editions and Contributions to the Text.' In *Averroes and the Aristotelian Tradition: Sources, Constitution and Reception of the Philosophy of Ibn Rushd (1126–1198)*, ed. G. Endreß and J. A. Aertsen (Leiden, 1999), 339–381.

——. 'The Circle of al-Kindī: Early Arabic Translations from the Greek and the Rise of Islamic Philosophy.' In *The Ancient Tradition in Christian and Islamic Hellenism*, ed. R. Kruk et al. (Leiden, 1997), 43–76.

——. 'Die Anfänge der arabischen Typographie und die Ablösung der Handschrift durch den Buchdruck.' In *Grundriß der Arabischen Philologie*, ed. W. Fischer, vol. 1 (Wiesbaden, 1982), 291–296.

——. 'Die arabisch-islamische Philosophie—ein Forschungsbericht.' *Zeitschrift für Geschichte der arabisch-islamischen Wissenschaften* 5 (1989): 1–47.

——. 'Die Entwicklung der Fachsprache.' In *Grundriß der arabischen Philologie*, ed. W. Fischer, vol. 3 (Wiesbaden, 1992), 3–23.

——. 'Die wissenschaftliche Literatur.' In *Grundriß der Arabischen Philologie*, ed. H. Gätje, vol. 2 (Wiesbaden, 1987), 400–506, and in *Grundriß der Arabischen Philologie*, ed. W. Fischer, vol. 3 (Wiesbaden, 1992), 3–152.

——. 'Reading Avicenna in the Madrasa: Intellectual Genealogies and Chains of Transmission of Philosophy and the Sciences in the Islamic East.' In *Arabic Theology, Arabic Philosophy. From the Many to the One: Essays in Celebration of Richard M. Frank*, ed. J. E. Montgomery (Louvain, 2006), 371–422.

Engel, M. *Elijah Del Medigo's Theory of Human Intellect* (PhD diss., University of Cambridge, 2014).

Ernst, G. 'Astri e previsioni: il Pronostico di Cardano del 1534.' In *Girolamo Cardano: Le opere, le fonti, la vita*, ed. M. Baldi, G. Canziani, and F. Angeli (Milan, 1999), 457–475.

——. 'From the Watery Trigon to the Fiery Trigon: Celestial Signs, Prophecies and History.' In *"Astrologi hallucinati": Stars and the End of the World in Luther's Time*, ed. P. Zambelli (New York, 1986), 265–280.

————. *Religione, ragione e natura: Ricerche su Tommaso Campanella e il tardo Rinascimento* (Milan, 1991).

————. 'Veritatis amor dulcissimus: Aspects of Cardano's Astrology.' In *Secrets of Nature: Astrology and Alchemy in Early Modern Europe,* ed. W. R. Newman and A. Grafton (Cambridge, MA, 2001), 39–68.

Etz, D. V. 'Conjunctions of Jupiter and Saturn.' *Journal of the Royal Astronomical Society of Canada* 94 (2000): 174–178.

Etzwiler, J. P. 'Baconthorpe and Latin Averroism: The Doctrine of the Unique Intellect.' *Carmelus* 18 (1971): 235–292.

————. 'John Baconthorpe, "Prince of the Averroists?"' *Franciscan Studies* 36 (1976): 148–176.

*European Pharmacopoeia: 7th edition 2011* (Stuttgart, 2010).

Fahd, T. *La divination arabe: Études religieuses, sociologiques et folkloriques sur le milieu natif de l'islam* (Strasbourg, 1966).

Fancy, N. *Science and Religion in Mamluk Egypt: Ibn al-Nafīs, Pulmonary Transit and Bodily Resurrection* (London, 2013).

Farmer, S. A. *Syncretism in the West: Pico's 900 Theses (1486): The Evolution of Traditional Religious and Philosophical Systems. With Text, Translation and Commentary* (Tempe, 1998).

Fellmann, I. *Das Aqrābādīn al-Qalānisī: Quellenkritische und begriffsanalytische Untersuchungen zur arabisch-pharmazeutischen Literatur.* Beiruter Texte und Studien 35 (Beirut, 1986).

Fenton, P. B. 'The Arabic and Hebrew Versions of the *Theology of Aristotle.*' In *Pseudo-Aristotle in the Middle Ages,* ed. J. Kraye et al. (London, 1986), 241–264.

Ferorelli, N. 'Abramo de Balmes ebreo di Lecce e i suoi parenti.' *Archivio storico per le province napoletane* 31 (1906): 632–654.

Ferre, L. See McVaugh, M. R.

Ferri, S. ed. *Pietro Andrea Mattioli, Siena 1501—Trento 1578: la vita, le opere, con l'identificazione delle piante* (Perugia, 1997).

Fichtner, G. 'Neues zu Leben und Werk von Leonhart Fuchs aus seinen Briefen an Joachim Camerarius I. und II. in der Trew-Sammlung.' *Gesnerus* 25 (1968): 65–82.

Fischer, H. et al. *Conrad Gessner 1516–1565: Universalgelehrter, Naturforscher, Arzt* (Zurich, 1967).

Flasch, K. *Meister Eckhart: Die Geburt der "Deutschen Mystik" aus dem Geist der arabischen Philosophie* (Munich, 2006).

Flückiger, F. A. *Pharmakognosie des Pflanzenreiches* (Berlin, ²1883).

Folh Jou, G. 'Medicamentos empleados por los Arabes y su influencia en la farmacia de Occidente, en especial española, desde el siglo XVI.' In *America y la España del siglo XVI,* ed. F. de Solano and F. del Pino. 2 vols. (Madrid, 1982–1983), 1:233–252.

Fornaciari, P. E. See Coviello, A.

Foust, C. M. *Rhubarb: The Wondrous Drug* (Princeton, 1992).

Frank, G. *Die theologische Philosophie Philipp Melanchthons (1497–1560),* (Leipzig, 1995).

——. 'Philipp Melanchthons Liber de anima und die Etablierung der frühneuzeitlichen Anthropologie.' In *Humanismus und Wittenberger Reform,* ed. M. Beyer and G. Wartenberg (Leipzig, 1996), 313–327.

Fränkel, S. See Freudenthal, J.

Freudenthal, J., and S. Fränkel. *Die durch Averroes erhaltenen Fragmente Alexanders zur Metaphysik des Aristoteles* (Berlin, 1885).

Friedensburg, W. *Geschichte der Universität Wittenberg* (Halle, 1917).

Friedrich, C., and W.-D. Müller-Jahncke. *Geschichte der Pharmazie. Band II: von der Frühen Neuzeit bis zur Gegenwart* (Eschborn, 2005).

Fück, J. *Die arabischen Studien in Europa bis in den Anfang des 20. Jahrhunderts* (Leipzig, 1955).

Gadebusch Bondio, M. '*Exempla medicorum:* quelques remarques sur un chapitre négligé de l'histoire de la médecine.' In *Exempla docent. Les exemples des philosophes de l'Antiquité à la Renaissance,* ed. T. Ricklin and M. Gadebusch Bondio (Paris, 2006), 373–396.

Gadebusch Bondio, M., and T. Ricklin, eds. *Exempla medicorum: Die Ärzte und ihre Beispiele (14.–18. Jh.)* (Florence, 2008).

Garbers, K. *Isḥāq ibn ʿImrān: Maqāla fī l-mālīḫūliyā (Abhandlung über die Melancholie) und Constantini Africani libri duo de melancholia* (Hamburg, 1977).

García Ballester, L. 'The Circulation and Use of Medical Manuscripts in Arabic in Sixteenth-Century Spain.' *Journal for the History of Arabic Science* 3 (1979): 183–199.

——. *Los moriscos y la medicina: Un capítulo de la medicina y la ciencia marginadas en la Espanã del siglo XVI* (Barcelona, 1984).

Garin, E. *Astrology in the Renaissance: The Zodiac of Life* (London, 1983).

——. 'Die Kultur der Renaissance.' In *Propyläen Weltgeschichte,* ed. G. Mann et al. 11 vols. (Frankfurt am Main, 1986), 6:429–534.

——. *Il Rinascimento italiano* (Milan, 1941).

——. *La cultura filosofica del Rinascimento italiano* (Florence, 1961).

——. 'Phantasia e imaginatio fra Marsilio Ficino e Pietro Pomponazzi.' *Giornale critico della filosofia italiana* 64 (1985): 351–361.

——. *Storia della filosofia italiana.* 3 vols. (Turin, 1966).

——. 'Testi minori sull'anima nella cultura del 400 in Toscana.' In *Testi umanistici inediti sul "De anima,"* ed. E. Garin et al. (Padua, 1951), 1–36.

Gätje, H. 'Die "inneren Sinne" bei Averroes.' *Zeitschrift der deutschen morgenländischen Gesellschaft* 115 (1965): 255–293.

Gauthier, R. A. 'Notes sur les débuts (1225–1240) du premier "averroïsme."' *Revue des sciences philosophiques et théologiques* 66 (1982): 321–374.

Geffen, D. 'Insights into the Life and Thought of Elijah Medigo Based on His Published and Unpublished Works.' *Proceedings of the American Academy of Jewish Research* 41 (1973–1975): 69–86.

Genequand, C. *Ibn Rushd's Metaphysics: A Translation with Introduction of Ibn Rushd's Commentary on Aristotle's Metaphysics, Book Lām* (Leiden, 1984).

Geneva, A. *Astrology and the Seventeenth Century Mind: William Lilly and the Language of the Stars* (Manchester, 1995).

Giglioni, G. 'Introduction.' In *Renaissance Averroism and Its Aftermath: Arabic Philosophy in Early Modern Europe,* ed. A. Akasoy and G. Giglioni (Dordrecht, 2013), 1–34.

Gilbert, N. W. 'Francesco Vimercato of Milan: a Bio-Bibliography.' *Studies in the Renaissance* 12 (1965): 188–217.

Gildersleeve, B. L., and G. Lodge. *Gildersleeve's Latin Grammar* (London, ³1968).

Gilson, E. 'Autour de Pomponazzi: problématique de l'immortalité de l'âme en Italie au début du XVIe siècle.' *Archives d'histoire doctrinale et littéraire du moyen âge* 36 (1961): 163–279.

Gindhart, M. *Das Kometenjahr 1618: Antikes und zeitgenössisches Wissen in der frühneuzeitlichen Kometenliteratur des deutschsprachigen Raumes* (Wiesbaden, 2006).

Giorgiantino, M. 'Per l'Averroismo napoletano.' *La Rinascita* 3 (1939): 872–881.

Gios, P. *L'attività pastorale del vescovo Pietro Barozzi a Padova (1487–1507)* (Padua, 1977).

Glasner, R. *Averroes' Physics: A Turning Point in Medieval Natural Philosophy* (Oxford, 2009).

Glei, R. F. '*Arabismus latine personatus.* Die Koranübersetzung von Ludovico Marracci (1698) und die Funktion des Lateinischen.' In *Jahrbuch für Europäische Wissenschaftskultur* 5 (2009/2010): 93–115.

——, ed. *Frühe Koranübersetzungen: europäische und außereuropäische Fallstudien* (Trier, 2012).

Godman, P. *From Poliziano to Machiavelli: Florentine Humanism in the High Renaissance* (Princeton, 1998).

——. 'Literaturgeschichtsschreibung im lateinischen Mittelalter und in der italienischen Renaissance.' In *Mediävistische Komparatistik: Festschrift für Franz Josef Worstbrock zum 60. Geburtstag,* ed. W. Harms and J.-D. Müller (Stuttgart, 1997), 177–198.

Goehl, K. 'Die Vita Avicennae des Sorsanus oder al-Dschusadschani, lateinisch und deutsch.' In *Editionen und Studien zur lateinischen und deutschen Fachprosa des Mittelalters: Festgabe für Gundolf Keil zum 65. Geburtstag,* ed. K. Goehl and J. G. Mayer. Texte und Wissen 3 (Würzburg, 2000), 317–338.

Gohlman, W. E. *The Life of Ibn Sina: A Critical Edition and Annotated Translation* (Albany, 1974).

Goldstein, B. R. *Al-Biṭrūjī: On the Principles of Astronomy. An Edition of the Arabic and Hebrew Versions with Translation, Analysis, and an Arabic-Hebrew-English Glossary.* 2 vols. (New Haven, 1977).

Göllner, C. 'Un coup d'œil sur les éditions vénitiennes du XVe siècle des œuvres d'Ibn Rušd Abul-Valid Muhammed (Averroès).' *Studia et acta Orientalia* 5-6 (1967): 361-364.

Goltz, D. *Mittelalterliche Pharmazie und Medizin: Dargestellt an Geschichte und Inhalt des Antidotarium Nicolai. Mit einem Nachdruck der Druckfassung von 1471* (Stuttgart, 1976).

———. *Studien zur Geschichte der Mineralnamen in Pharmazie, Chemie und Medizin von den Anfängen bis Paracelsus.* Sudhoffs Archiv Beihefte 14 (Wiesbaden, 1972).

González, E. G. 'Juan Louis Vives: Works and Days.' In *A Companion to Juan Luis Vives,* ed. C. Fantazzi (Leiden, 2008), 15-64.

Gonzalez Castrillo, R. *Rhazes y Avicena en la Biblioteca de la Facultad de Medicina de la Universidad Complutense: Descripción de su obra médica impresa y comentarios* (Madrid, 1984).

Gonzálvez Ruiz, R. *Hombres y libros de Toledo (1086–1300)* (Madrid, 1997).

Göpfert, W. *Drogen auf alten Landkarten und das zeitgenössische Wissen über ihre Herkunft.* 2 vols. (Düsseldorf, 1985).

Gorce, M. M. 'Averroisme.' In *Dictionnaire d'histoire et de géographie ecclésiastiques,* vol. 5 (Paris, 1931), cols. 1032-1092.

Gouguenheim, S. *Aristote au Mont Saint-Michel: Les racines grecques de l'Europe chrétienne* (Paris, 2008).

Grabmann, M. 'Der Bologneser Averroist Angelo d'Arezzo (ca. 1325).' In *Mittelalterliches Geistesleben* (Munich, 1926), 2:261-271.

Graf-Stuhlhofer, F. *Humanismus zwischen Hof und Universität. Georg Tannstetter (Collimitius) und sein wissenschaftliches Umfeld im Wien des frühen 16. Jahrhunderts* (Vienna, 1996).

Grafton, A. 'The Availability of Ancient Works.' In *The Cambridge History of Renaissance Philosophy,* ed. C. B. Schmitt et al. (Cambridge, 1988), 767-791.

———. *Cardano's Cosmos: The Worlds and Works of a Renaissance Astrologer* (Cambridge, MA, 1999).

———. *Commerce with the Classics: Ancient Books and Renaissance Readers* (Ann Arbor, 1997).

——. *Defenders of the Text: The Traditions of Scholarship in an Age of Science, 1450–1800* (Cambridge, MA, 1991).

——. *Joseph Scaliger: A Study in the History of Classical Scholarship, vol. 1: Textual Criticism and Exegesis* (Oxford, 1983).

——. *Joseph Scaliger: A Study in the History of Classical Scholarship, vol. 2: Historical Chronology* (Oxford, 1993).

——. 'The New Science and the Traditions of Humanism.' In *Renaissance Humanism,* ed. J. Kraye (Cambridge, 1996), 203–223.

——. 'The Origins of Scholarship.' *American Scholar* 48 (1979): 236–261 (review of R. Pfeiffer, *History of Classical Scholarship from 1300 to 1850*).

——. 'Review of P. L. Rose, *The Italian Renaissance of Mathematics.' Annals of Science* 36 (1979): 628–629.

Greene, E. L. *Landmarks of Botanical History: Edited by Frank N. Egerton.* 2 vols. (Stanford, 1983).

Grendler, P. F. *The Universities of the Italian Renaissance* (Baltimore, 2002).

Griffel, F. *Al-Ghazālī's Philosophical Theology* (Oxford, 2009).

——. 'Toleranzkonzepte im Islam und ihr Einfluß auf Jean Bodins *Colloquium Heptaplomeres.' In Bodins Polimeres: Neue Studien zu Jean Bodins Spätwerk,* ed. R. Häfner (Wiesbaden, 1999), 119–144.

Grignaschi, M. See Langhade, J.

Guerrini, P. 'Dentro il dipinto: il Tommaso d'Aquino di Benozzo Gozzoli.' In *Il ritratto et la memoria: Materiali 2,* ed. A. Gentili, P. Morel, and C. Cieri Via (Rome, 1993), 113–133.

Guido, F. See Spina, G.

Guigues, P. 'Les noms arabes dans Sérapion, "Liber de simplici medicina": Essai de restitution et d'identification de noms arabes de médicaments usités au moyen âge.' *Journal asiatique* 10, ser. 5/6 (1905): 473–546, 49–112.

Gumbrecht, H. U. *Die Macht der Philologie: Über einen verborgenen Impuls im wissenschaftlichen Umgang mit Texten* (Frankfurt am Main, 2003).

Gutas, D. 'Aspects of Literary Form and Genre in Arabic Logical Works.' In *Glosses and Commentaries on Aristotelian Logical Texts,* ed. C. Burnett. Warburg Institute Surveys and Texts 23 (London, 1993), 29–76.

——. *Avicenna and the Aristotelian Tradition: Introduction to Reading Avicenna's Philosophical Works. Second, Revised and Enlarged Edition, Including an Inventory of Avicenna's Authentic Works* (Leiden, 2014).

——. *Greek Thought, Arabic Culture: The Graeco-Arabic Translation Movement in Baghdad and Early 'Abbāsid Society (2nd–4th/8th–10th Centuries)* (London, 1998).

——. 'The Heritage of Avicenna: The Golden Age of Arabic Philosophy, 1000–ca. 1350.' In *Avicenna and His Heritage: Acts of the International Colloquium Leuven—Louvain-La-Neuve September 8–September 11, 1999,* ed. J. Janssens and D. de Smet (Leuven, 2002), 81–97.

——. 'Ibn Ṭufayl on Ibn Sīnā's Eastern Philosophy.' *Oriens* 34 (1994): 222–241.

——. 'Review of Charles Genequand: Ibn Rushd's Metaphysics.' *Der Islam* 64 (1987): 122–126.

——. 'The Study of Arabic Philosophy in the Twentieth Century: An Essay on the Historiography of Arabic Philosophy.' *British Journal of Middle Eastern Studies* 29 (2002): 5–25.

——. See Tarán, L.

Haefeli-Till, D. *Der 'Liber de oculis' des Constantinus Africanus: Übersetzung und Kommentar* (Zurich, 1977).

Haeusler, M. *Das Ende der Geschichte in der mittelalterlichen Weltchronistik* (Cologne, 1980).

Halleux, R. 'Damigéron, Evax et Marbode: L'héritage alexandrin dans les lapidaires médiévaux.' *Studi medievali* 15 (1974): 327–347.

——. 'The Reception of Arabic Alchemy in the West.' In *Encyclopedia of the History of Arabic Science,* ed. R. Rashed. 3 vols. (London, 1996), 3:886–902.

Hamilton, A. *Arab Culture and Ottoman Magnificence in Antwerp's Golden Age* (London, 2001).

——. *The Arcadian Library: Western Appreciation of Arab and Islamic Civilization* (Oxford, 2011).

——. 'An Egyptian Traveller in the Republic of Letters: Josephus Barbatus or Abudacnus the Copt.' *Journal of the Warburg and Courtauld Institutes* 57 (1994): 123–150.

——. *Europe and the Arab World:\Five Centuries of Books by European Scholars and Travellers from the Libraries of the Arcadian Group* (Oxford, 1994).

——. *The Forbidden Fruit: The Koran in Early Modern Europe.* Publication of the 2007 Hadassah and Daniel Khalili Memorial Lecture in Islamic Art and Culture (London, 2008).

——. 'Franciscus Raphelengius: The Hebraist and His Manuscripts.' *De Gulden passer* 68 (1990): 105–117.

——. 'The Long Apprenticeship: Casaubon and Arabic.' In *"I have always loved the Holy Tongue": Isaac Casaubon, the Jews, and a Forgotten Chapter in Renaissance Scholarship,* ed. A. Grafton and J. Weinberg (Cambridge, MA, 2011), 293–306.

——. '*Nam tirones sumus:* Franciscus Raphelengius' *Lexicon Arabico-Latinum (Leiden 1613).*' *De Gulden passer* 66–67 (1988–1989): 557–589.

——. *William Bedwell, the Arabist, 1563–1632* (Leiden, 1985).

Hankins, J. 'Humanism, Scholasticism, and Renaissance Philosophy.' In *Cambridge Companion to Renaissance Philosophy,* ed. J. Hankins (Cambridge, 2007), 30–48.

——. *Plato in the Italian Renaissance* (Leiden, ³1994).

Hankins, J., and A. Palmer. *The Recovery of Ancient Philosophy in the Renaissance: A Brief Guide* (Florence, 2008).

Hartner, W. 'The Mercury Horoscope of Marcantonio Michiel of Venice: A Study in the History of Renaissance Astrology and Astronomy.' *Vistas in Astronomy* 1 (1955): 84–138 (repr. Hartner, *Oriens–Occidens,* 440–495).

——. *Oriens–Occidens: Ausgewählte Schiften zur Wissenschafts- und Kulturgeschichte* (Hildesheim, 1968).

——. 'Tycho Brahe et Albumasar: La question de l'autorité scientifique au début de la recherche libre en astronomie.' *La science au seizième siècle* (Paris, 1960): 135–150 (repr. Hartner, *Oriens–Occidens,* 496–507).

Harvey, S. 'Why Did Fourteenth-Century Jews Turn to Alghazali's Account of Natural Science?' *Jewish Quarterly Review* 91 (2001): 359–376.

Haskins, C. H. *Studies in the History of Medieval Science* (Cambridge, 1924).

Hasse, D. N. 'Abbreviation in Medieval Latin Translations from Arabic.'
      In *Vehicles of Transmission, Translation, and Transformation in Medieval
      Textual Culture,* ed. R. Wisnovsky, F. Wallis, F. C. Fumo, and C.
      Fraenkel (Turnhout, 2011), 159–172.

———. 'Arabic Philosophy and Averroism.' In *Cambridge Companion to
      Renaissance Philosophy,* ed. J. Hankins (Cambridge, 2007), 113–136.

———. 'The Attraction of Averroism in the Renaissance: Vernia, Achillini,
      Prassicio.' In *Philosophy, Science and Exegesis in Greek, Arabic and Latin
      Commentaries,* ed. P. Adamson, H. Baltussen, and M. W. F. Stone
      (London, 2004), 131–147.

———. 'Aufstieg und Niedergang des Averroismus in der Renaissance:
      Niccolò Tignosi, Agostino Nifo, Francesco Vimercato.' In *"Herbst des
      Mittelalters"? Fragen zur Bewertung des 14. und 15. Jahrhunderts,* ed. J. A.
      Aertsen and M. Pickavé. Miscellanea Mediaevalia 31 (Berlin, 2004),
      447–473.

———. 'Averroes' Critique of Ptolemy and Its Reception by John of Jandun
      and Agostino Nifo.' In *Averroes' Natural Philosophy and Its Reception in
      the Latin West,* ed. P. J. J. M. Bakker (Leuven, 2015), 69–88.

———. 'Averroica secta: Notes on the Formation of Averroist Movements in
      Fourteenth-Century Bologna and Renaissance Italy.' In *Averroès et
      les Averroïsmes juif et latin,* ed. J.-B. Brenet (Turnhout, 2007), 307–331.

———. *Avicenna's* De Anima *in the Latin West: The Formation of a Peripatetic
      Philosophy of the Soul 1160–1300.* Warburg Institute Studies and Texts
      1 (London, 2000).

———. 'Avicenna's "Giver of Forms" in Latin Philosophy, Especially in the
      Works of Albertus Magnus.' In *The Arabic, Hebrew and Latin Reception
      of Avicenna's Metaphysics,* ed. D. N. Hasse and A. Bertolacci (Berlin,
      2011), 225–249.

———. 'Contacts with the Arab World.' In *The Oxford Handbook of Neo-Latin,*
      ed. S. Knight and S. Tilg (Oxford, 2015), 279–293.

———. 'Der mutmaßliche arabische Einfluss auf die literarische Form der
      Universitätsliteratur des 13. Jahrhunderts.' In *Albertus Magnus und
      der Ursprung der Universitätsidee,* ed. L. Honnefelder (Berlin, 2011),
      241–258.

———. 'Die humanistische Polemik gegen arabische Autoritäten: Grundsätz-
      liches zum Forschungsstand.' *Neulateinisches Jahrbuch* 3 (2001): 65–79.

——. 'King Avicenna: The Iconographic Consequences of a Mistranslation.' *Journal of the Warburg and Courtauld Institutes* 60 (1997): 230–243.

——. *Latin Averroes Translations of the First Half of the Thirteenth Century* (Hildesheim, 2010).

——. 'Mosul and Frederick II Hohenstaufen: Notes on Aṯīraddīn al-Abharī and Sirāǧaddīn al-Urmawī.' In *Occident et Proche-Orient: Contacts scientifiques au temps des Croisades,* ed. I. Draelants, A. Tihon, and B. van den Abeele (Louvain-la-Neuve, 2000), 145–163.

——. 'Plato arabico-latinus: Philosophy—Wisdom Literature—Occult Sciences.' In *The Platonic Tradition in the Middle Ages: A Doxographic Approach,* ed. S. Gersh and M. J. F. M. Hoenen (Berlin, 2002), 31–65.

——. 'Pseudowissenschaft vom Abendland.' *Frankfurter Allgemeine Zeitung,* June 19, 2008.

——. 'The Social Conditions of the Arabic-(Hebrew-)Latin Translation Movements in Medieval Spain and in the Renaissance.' In *Wissen über Grenzen: Arabisches Wissen und lateinisches Mittelalter,* ed. A. Speer and L. Wegener. Miscellanea Mediaevalia 33 (Berlin, 2006), 68–86, 806.

——. 'The Soul's Faculties.' In *The Cambridge History of Medieval Philosophy,* ed. R. Pasnau (Cambridge, 2010), 1:305–319.

Hasse, D. N., and A. Bertolacci, eds. *The Arabic, Latin and Hebrew Reception of Avicenna's Metaphysics* (Berlin, 2012).

Hasse, D. N., and A. Büttner. 'Notes on Anonymous Twelfth-Century Translations of Philosophical Texts from Arabic into Latin on the Iberian Peninsula.' In *The Arabic, Latin and Hebrew Reception of Avicenna's Physics and Cosmology,* ed. D. N. Hasse and A. Bertolacci (Berlin, forthcoming).

Hasselhoff, G. K. 'The Reception of Maimonides in the Latin World: The Evidence of the Latin Translations in the 13th–15th Century.' *Materia Giudaica* 6 (2001): 258–280.

Hayoun, M.-R. 'Le commentaire de Moïse de Narbonne (1300–1362) sur le Ḥayy ibn Yaqẓān d'Ibn Ṭufayl (mort en 1185).' *Archives d'histoire doctrinale et littéraire du moyen âge* 55 (1988): 23–98.

Hayoun, M.-R., and A. de Libera. *Averroès et l'averroïsme.* Que sais-je? 2631 (Paris, 1991).

Heiberg, J. L., and E. Wiedemann. 'Ibn al Haiṭams Schrift über parabolische Hohlspiegel.' *Bibliotheca mathematica* 3rd ser. 10 (1909–1910): 201–237.

Heilen, S. *"Hadriani genitura": Die astrologischen Fragmente des Antigonos von Nikaia.* 2 vols. (Berlin, 2015).

———. 'Lorenzo Bonincontris Schlußprophezeiung in *De rebus naturalibus et divinis.*' In *Zukunftsvoraussagen in der Renaissance,* ed. K. Bergdolt and W. Ludwig (Wiesbaden, 2005), 309–328.

———. 'Some Metrical Fragments from Nechepsos and Petosiris.' In *La poésie astrologique dans l'Antiquité. Actes du colloque organisé les 7 et 8 décembre 2007,* ed. J.-H. Abry, I. Böhm, and W. Hübner (Paris, 2011), 23–93.

Heischkel, E. 'Die Geschichte der Medizingeschichtschreibung.' In *Einführung in die Medizinhistorik,* ed. W. Artelt (Stuttgart, 1949), 202–237.

Heller, J. L. See Meyer, F. G.

Hellmann, G. 'Aus der Blütezeit der Astrometeorologie: J. Stöfflers Prognose für das Jahr 1524.' In *Beiträge zur Geschichte der Meteorologie* (Berlin, 1914–1922), 1.1:3–102.

———. *Versuch einer Geschichte der Wettervorhersage im 16. Jahrhundert* (Berlin, 1924).

Helm, J. 'Die Galenrezeption in Philipp Melanchthons *De anima* (1540/1552).' *Medizinhistorisches Journal* 31 (1996): 298–321.

Herczeg, Á. 'Johannes Manardus: Hofarzt in Ungarn und Ferrara im Zeitalter der Renaissance.' *Janus* 33 (1929): 52–78, 85–130.

Hermelink, H. 'Tabulae Jahen.' *Archive for History of Exact Sciences* 2 (1964): 108–112.

Heyd, W. *Geschichte des Levantehandels im Mittelalter* (Stuttgart, 1879).

Hilty, G. *Aly Aben Ragel: el libro conplido en los iudizios de las estrellas* (Madrid, 1954).

Hirai, H. 'The New Astral Medicine.' In *A Companion to Astrology in the Renaissance,* ed. B. Dooley (Leiden, 2014), 267–286.

Hirschberg, J. 'Über das älteste arabische Lehrbuch der Augenheilkunde.' *Sitzungsberichte der Königlich Preussischen Akademie der Wissenschaften* 49 (1903): 1080–1094.

Hissette, R., ed. *Averrois Commentum medium super libro Peri Hermeneias Aristotelis* (Louvain, 1996).

——, ed. *Averrois Commentum medium super libro Praedicamentorum Aristotelis* (Louvain, 2010).

——. 'Des éditions d'Aristote-Averroès par Lorenzo Canozi (1472–75) et Andrea Torresano (1483).' *Gutenberg Jahrbuch* 87 (2012): 105–122.

——. 'Guillaume de Luna—Jacob Anatoli—Jacob Mantinus: À propos du commentaire moyen d'Averroès sur le *De interpretatione*.' *Bulletin de Philosophie Médiévale* 32 (1990): 142–158.

——. 'Jacob Mantino tributaire de Nicoleto Vernia? Le cas du commentaire moyen d'Averroès sur les *Prédicaments* de Urb. lat. 221.' *Miscellanea Bibliothecae Apostolicae Vaticanae* 13 (2006): 265–286.

——. 'Jacob Mantino utilisateur de Nicoleto Vernia?' *Bulletin de Philosophie Médiévale* 47 (2005): 157–161.

——. 'Le *corpus averroicum* des manuscrits vaticans Urbinates latins 220 et 221 et Nicoleto Vernia.' *Miscellanea Bibliothecae Apostolicae Vaticanae* 3 (=Studi e Testi 333), (1989): 257–356.

Hofmeier, T. See Bachmann, M.

Hoppe, H. A. *Drogenkunde: Band 1, Angiospermen* (Berlin, [8]1975).

Hübner, W. 'Astrologie in der Renaissance.' In *Zukunftsvoraussagen in der Renaissance,* ed. K. Bergdolt and W. Ludwig (Wiesbaden, 2005), 241–279.

Hudry, F. See d'Alverny, M.-T.

Hugonnard-Roche, H. 'The Influence of Arabic Astronomy in the Medieval West.' In *Encyclopedia of the History of Arabic Science,* ed. R. Rashed. 3 vols. (London, 1996), 1:284–305.

Humphreys, R. S. *Islamic History: A Framework for Inquiry* (Princeton, [2]1991).

Hyman, A. 'Aristotle's Theory of the Intellect and Its Interpretation by Averroes.' In *Studies in Aristotle,* ed. D. J. O'Meara (Washington, DC, 1981), 161–191.

Hyman, A., and J. J. Walsh, eds. *Philosophy in the Middle Ages: The Christian, Islamic, and Jewish Traditions* (Indianapolis, 1997).

Idel, M. 'Jewish Mystical Thought in the Florence of Lorenzo il Magnifico.' In *La cultura ebraica all'epoca di Lorenzo il Magnifico,* ed. D. Liscia Bemporad and I. Zatelli (Florence, 1998), 17–42.

Ijsewijn, J. 'Die humanistische Biographie.' In *Biographie und Autobiographie in der Renaissance,* ed. A. Buck. Wolfenbütteler Abhandlungen zur Renaissanceforschung 4 (Wiesbaden, 1983), 1–19.

Illgen, H. O. 'Die abendländischen Rhazes-Kommentatoren des XIV. bis XVII. Jahrhunderts.' In *Muḥammad ibn Zakarīyāʾ al-Rāzī (d. 313/925): Texts and Studies,* ed. F. Sezgin. Islamic Medicine 25. 2 vols. (Frankfurt am Main, 1996), 2:113–120.

Imbach, R. 'L'averroïsme latin du XIIIe siècle.' In *Gli studi di filosofia medievale fra otto e nove cento, contributo a un bilancio storiografico,* ed. R. Imbach and A. Maierù (Rome, 1991), 191–208.

———. 'Lulle face aux Averroïstes parisiens.' In *Raymond Lulle et le pays d'Oc* (Toulouse, 1987), 261–282.

*Index Londinensis to Illustrations of Flowering Plants, Ferns and Fern Allies.* 7 vols. (Oxford, 1921–1941).

Ivry, A. L. 'Averroes' Middle and Long Commentaries on the *De anima.*' *Arabic Sciences and Philosophy* 5 (1995): 75–92.

———. 'Averroes and the West: The First Encounter/Nonencounter.' In *A Straight Path: Studies in Medieval Philosophy and Culture,* ed. R. Link-Salinger (Washington, DC, 1988), 142–158.

———. 'Remnants of Jewish Averroism in the Renaissance.' In *Jewish Thought in the Sixteenth Century,* ed. B. D. Cooperman (Cambridge, MA, 1983), 243–265.

Jacquart, D. 'The Influence of Arabic Medicine in the Medieval West.' In *Encyclopedia of the History of Arabic Science,* ed. R. Rashed. 3 vols. (London, 1996), 3:963–984.

———. 'La réception du *Canon* d'Avicenne: comparaison entre Montpellier et Paris aux XIIIe et XIVe siècles.' In *Histoire de l'école médicale de Montpellier.* Actes du 110e Congrès national des Sociétés savantes (Montpellier, 1985), 2:69–77.

———. *La science médicale occidentale entre deux renaissances (XIIe s.–XVe s.)* (Aldershot, 1997), esp. art. 1, 'À l'aube de la renaissance médicale des XIe–XIIe siècles: l' "Isagoge Johannitii" et son traducteur'; art. 8, 'Note sur la traduction latine du *Kitāb al-Manṣūrī* de Rhazès';

art. 10, 'La coexistence du grec et de l'arabe dans le vocabulaire médical du latin médiéval: l'effort linguistique de Simon de Gênes'; and art. 11, 'Arabisants du Moyen Age et de la Renaissance: Jérôme Ramusio († 1486), correcteur de Gérard de Crémone († 1187).'

——. 'Remarques préliminaires à une étude comparée des traductions médicales de Gérard de Crémone.' In *Traduction et traducteurs au Moyen Âge,* ed. G. Contamine (Paris, 1989), 109–118.

——. See Burnett, C.

——. See Wickersheimer, E.

Jacquart, D., and F. Micheau. *La médecine arabe et l'occident médiéval* (Paris, 1990).

Jacquart, D., and C. Thomasset. *Sexuality and Medicine in the Middle Ages* (Cambridge, 1988).

Jacquart, D., and G. Troupeau, eds. *Yūḥannā ibn Māsawayh (Jean Mesue): Le Livre des axiomes médicaux (Aphorismi)* (Geneva, 1980).

Jahier, H., and A. Noureddine, eds. *Avicenne: Poème de la médecine* (Paris, 1956).

Jardine, N. 'Problems of Knowledge and Action: Epistemology of the Sciences.' In *The Cambridge History of Renaissance Philosophy,* ed. C. B. Schmitt et al. (Cambridge, 1988), 685–711.

Jones, J. R. 'The Arabic and Persian Studies of Giovan Battista Raimondi (c.1536–1614)' (M.Phil. diss., University of London, Warburg Institute, 1981).

——. 'Piracy, War, and the Acquisition of Arabic Manuscripts in Renaissance Europe.' *Manuscripts of the Middle East* 2 (1987): 96–109.

——. 'Thomas Erpenius (1584–1624) on the Value of the Arabic Language, Translated from the Latin.' *Manuscripts of the Middle East* 1 (1986): 15–25.

Jordan, M. 'The Fortune of Constantine's *Pantegni.*' In *Constantine the African and 'Alī ibn al-'Abbās al-Maǧūsī: The* Pantegni *and Related Texts,* ed. C. Burnett and D. Jacquart (Leiden, 1994), 286–302.

Kabelitz, L., and K. Reif. 'Anthranoide in Sennesdrogen. Ein analytischer Beitrag zur Risikobewertung.' *Deutsche Apotheker Zeitung* 134 (1994): 5085–5088.

Karrer, K. *Johannes Posthius (1537–1597): Verzeichnis der Briefe und Werke mit Regesten und Posthius-Bibliographie* (Wiesbaden, 1993).

Kaufmann, D. 'Jacob Mantino: Une page de l'histoire de la Renaissance.' *Revue des études juives* 26 (1893): 30–60, 207–229.

Keil, G., and R. Peitz. '*Decem quaestiones de medicorum statu:* Beobachtungen zum Fakultätenstreit und zum mittelalterlichen Unterrichtsplan Ingolstadts.' In *Der Humanismus und die oberen Fakultäten,* ed. G. Keil et al. (Weinheim, 1987), 215–238.

Kennedy, E. S. 'Al-Battānī's Astrological History of the Prophet and the Early Caliphate.' *Suhayl* 9 (2009–2010): 13–148.

——. 'Late Medieval Planetary Theory.' *Isis* 57 (1966): 365–378 (repr. *Studies in the Islamic Exact Sciences* [Beirut, 1983], 84–97).

——. 'The World Year Concept in Islamic Astrology.' In *Studies in the Islamic Exact Sciences,* ed. E. S. Kennedy (Beirut, 1983), 351–371.

Kennedy, E. S., and D. Pingree. *The Astrological History of Māshāʾallāh* (Cambridge, MA, 1971).

Kenney, E. J. *The Classical Text: Aspects of Editing in the Age of the Printed Book* (Berkeley, 1974).

Kessel, B. A. See Butterworth, C. E.

Kessler, E. *Die Philosophie der Renaissance: das 15. Jahrhundert* (Munich, 2008).

——. 'The Intellective Soul.' In *The Cambridge History of Renaissance Philosophy,* ed. C. B. Schmitt et al. (Cambridge, 1988), 485–534.

——. 'Introducing Aristotle to the Sixteenth Century: the Lefèvre Enterprise.' In *Philosophy in the Sixteenth and Seventeenth Centuries: Conversations with Aristotle,* ed. C. Blackwell and S. Kusakawa (Aldershot, 1999), 1–21.

——. 'Nicoletto Vernia oder die Rettung eines Averroisten.' In *Averroismus im Mittelalter und in der Renaissance,* ed. F. Niewöhner and L. Sturlese (Zurich, 1994), 269–290.

——. 'Petrarca—Salutati—Bruni.' In *Contemporary Philosophy: A New Survey,* ed. G. Fløistad, vol. 6/1 (Dordrecht, 1990), 547–562.

——. 'Von der Psychologie zur Methodenlehre: Die Entwicklung des methodischen Wahrheitsbegriffes in der Renaissancepsychologie.' *Zeitschrift für philosophische Forschung* 41 (1987): 548–570.

———. See Schmitt, C. B.

Kieszkowski, B. 'Les rapports entre Elie de Medigo et Pic de la Mirandole (d'après le ms. lat. 6508 de la Bibliothèque Nationale).' *Rinascimento* 4 (1964): 41–91.

Kischlat, H. *Studien zur Verbreitung von Übersetzungen arabischer philosophischer Werke in Westeuropa 1150–1400: Das Zeugnis der Bibliotheken* (Münster, 2000).

Klein-Franke, F. *Die klassische Antike in der Tradition des Islam* (Darmstadt, 1980).

Kleinhans, A. *Historia studii linguae arabicae et collegii missionum ordinis fratrum minorum* (Florence, 1930).

Klempt, A. *Die Säkularisierung der universalhistorischen Auffassung: Zum Wandel des Geschichtsdenkens im 16. und 17. Jahrhundert* (Göttingen, 1960).

Klimaschewski-Bock, I. *Die 'Distinctio Sexta' des Antidotarium Mesuë in der Druckfassung Venedig 1561 (Sirupe und Robub): Übersetzung, Kommentar und Nachdruck der Textfassung von 1561.* Quellen und Studien zur Geschichte der Pharmazie 40 (Stuttgart, 1987).

Kluxen, W. 'Literaturgeschichtliches zum lateinischen Moses Maimonides.' *Recherches de Théologie ancienne et médiévale* 21 (1954): 23–50.

———. 'Rabbi Moyses (Maimonides): Liber de uno deo benedicto.' In *Judentum im Mittelalter,* ed. P. Wilpert. Miscellanea Mediaevalia 4 (Berlin, 1966), 167–182.

Knappich, W. *Geschichte der Astrologie* (Frankfurt am Main, 1967).

Kottek, S. S. 'Jacob Mantino, a 16th Cent. Jewish Physician and Scholar Related to Bologna.' In *Thirty-First International Congress of the History of Medicine,* ed. R. A. Bemabeo (Bologna, 1990), 179–185.

Kraemer, J. L. *Humanism in the Renaissance of Islam: The Cultural Revival during the Buyid Age* (Leiden, 1986).

Krafft, F. 'Fuchs (Fuchsius, Füchsel), Leonhart.' In *Biographisches Lexikon der Ludwig-Maximilians-Universität München,* ed. L. Boehm (Berlin, 1998), 135–142.

———. 'Tertius interveniens: Johannes Keplers Bemühungen um eine Reform der Astrologie.' In *Die okkulten Wissenschaften in der Renaissance,* ed. A. Buck (Wiesbaden, 1992), 197–225.

Krause, C. *Euricius Cordus: Eine biographische Skizze aus der Reformationszeit* (Hanau, 1863).

Kraye, J., ed. *The Cambridge Companion to Renaissance Humanism* (Cambridge, 1996).

——. 'The Immortality of the Soul in the Renaissance: Between Natural Philosophy and Theology.' *Signatures* 1 (2000): ch. 2, 1–24.

——. 'Philologists and Philosophers.' In *The Cambridge Companion to Renaissance Humanism,* ed. J. Kraye (Cambridge, 1996), 142–160.

——. 'The Philosophy of the Italian Renaissance.' In *Renaissance and Seventeenth-Century Rationalism,* ed. G. H. R. Parkinson. Routledge History of Philosophy 4 (London, 1993), 16–69.

——. 'Pietro Pomponazzi (1462–1525).' In *Philosophen der Renaissance: Eine Einführung,* ed. P. R. Blum (Darmstadt, 1999), 87–103.

——. 'The Pseudo-Aristotelian *Theology* in Sixteenth- and Seventeenth-Century Europe.' In *Pseudo-Aristotle in the Middle Ages,* ed. J. Kraye et al. (London, 1986), 265–286.

——. 'The Role of Medieval Philosophy in Renaissance Thought: The Evidence of Early Printed Books.' In *Bilan et perspectives des études médiévales (1993–1998),* ed. J. Hamesse (Turnhout, 2004), 695–714.

——. See Schmitt, C. B.

Kristeller, P. O. *Die Philosophie des Marsilio Ficino* (Frankfurt am Main, 1972).

——. 'Humanism.' In *The Cambridge History of Renaissance Philosophy,* ed. C. B. Schmitt et al. (Cambridge, 1988), 113–137.

——. 'Humanism and Scholasticism in the Italian Renaissance.' In *Studies in Renaissance Thought and Letters* (Rome, 1956), 553–583.

——. 'The Myth of Renaissance Atheism and the French Tradition of Free Thought.' *Journal of the History of Philosophy* 6 (1968): 233–243.

——. 'Paduan Averroism and Alexandrism in the Light of Recent Studies.' In *Renaissance Thought II* (New York, 1965), 111–118.

——. 'Petrarch's "Averroists": A Note on the History of Aristotelianism in Venice, Padua, and Bologna.' *Bibliothèque d'Humanisme et Renaissance* 14 (1952): 59–65.

——. 'Two Unpublished Questions on the Soul of Pietro Pomponazzi.'
   *Medievalia et Humanistica* 9 (1955): 76–101.

——. See Cassirer, E.

Krüger, K. H. *Die Universalchroniken.* 2 vols. (Turnhout, 1976/1985).

Krümmel, A. *Das "Supplementum Chronicarum" des Augustinermönches
   Jacobus Philippus Foresti von Bergamo: Eine der ältesten Bilderchroniken
   und ihre Wirkungsgeschichte* (Herzberg, 1992).

Kügelgen, A. von. *Averroes und die arabische Moderne: Ansätze zu einer
   Neubegründung des Rationalismus im Islam* (Leiden, 1994).

Kühlmann, W. et al. *Humanistische Lyrik des 16. Jahrhunderts: lateinisch und
   deutsch* (Frankfurt am Main, 1997).

Kuhn, H. C. 'Die Verwandlung der Zerstörung der Zerstörung:
   Bemerkungen zu Augustinus Niphus' Kommentar zur *Destructio
   destructionum* des Averroes.' In *Averroismus im Mittelalter und in der
   Renaissance,* ed. F. Niewöhner and L. Sturlese (Zurich, 1994),
   291–308.

——. *Venetischer Aristotelismus im Ende der aristotelischen Welt: Aspekte der Welt
   und des Denkens des Cesare Cremonini (1550–1631)* (Frankfurt am
   Main, 1996).

Kuhn, T. S. *The Structure of Scientific Revolutions,* 3rd ed. (Chicago, 1996).

Kuhne, R. 'El *Sirr ṣinā'at al-ṭibb* de Abū Bakr Muḥammad b. Zakariyyā'
   al-Rāzī.' *Al-Qantara* 3 (1982): 347–414.

Kühner, R., and C. Stegmann. *Ausführliche Grammatik der lateinischen
   Sprache.* 3 vols. (Darmstadt, ⁴1962).

Kukkonen, T. *Ibn Ṭufayl: Living the Life of Reason* (London, 2014).

Kuksewicz, Z. *De Siger de Brabant à Jacques de Plaisance: La théorie de l'intellect
   chez les Averroïstes latins des XIIIe et XIVe siècles* (Wrocław, 1968).

——. 'L'évolution de l'averroïsme latin au XIVe siècle.' In *Knowledge and the
   Sciences in Medieval Philosophy: Proceedings of the Eighth International
   Congress of Philosophy,* ed. M. Asztalos et al. 3 vols. (Helsinki, 1990),
   3:97–102.

——. 'Paul de Venise et sa théorie de l'âme.' In *Aristotelismo Veneto e scienza
   moderna,* ed. L. Olivieri. 2 vols. (Padua, 1983), 1:297–324.

——. 'Some Remarks on Erfurt Averroism.' *Studia mediewistyczne* 32 (1997): 93–122.

Kunitzsch, P. *The Arabs and the Stars: Texts and Traditions on the Fixed Stars, and Their Influence in Medieval Europe* (Northampton, 1989).

——. 'Late Traces of Arabic Influence in European Astronomy (17th–18th centuries).' In *Astronomy as a Model for the Sciences in Early Modern Times,* ed. M. Folkerts and A. Kuhne (Augsburg, 2006), 97–102.

——. 'Review of F. Sezgin, *Geschichte des arabischen Schrifttums. Bd. 7: Astrologie—Meteorologie und Verwandtes, bis ca. 430 H.*' *Zeitschrift der Deutschen Morgenländischen Gesellschaft* 132 (1982): 174–179.

Langermann, Y. T. 'Another Andalusian Revolt? Ibn Rushd's critique of al-Kindī's pharmacological computus.' In *The Enterprise of Science in Islam: New Perspectives,* ed. J. P. Hogendijk and A. I. Sabra (Cambridge, MA, 2003), 351–372.

——. *Ibn al-Haytham's On the Configuration of the World* (New York, 1990).

Langhade, J., and M. Grignaschi, eds. *Al-Fārābī: Deux ouvrages inédits sur la réthorique* (Beirut, 1971).

Lasinio, F. 'Studii Sopra Averroe.' *Annuario della Società Italiana per gli Studi Orientali* 1 (1872): 125–159.

*L'Averroismo in Italia. Convegno internazionale.* Atti dei convegni Lincei 40 (Rome, 1979).

Leclerc, L. *Histoire de la médecine arabe* (Paris, 1876).

Lehmann, P. 'Der Schriftstellerkatalog des Arnold Gheylhoven von Rotterdam.' In *Erforschung des Mittelalters, Ausgewählte Abhandlungen und Aufsätze.* 5 vols. (Stuttgart, 1959–1962), 4: 216–236.

——. 'Literaturgeschichte im Mittelalter.' In *Erforschung des Mittelalters: Ausgewählte Abhandlungen und Aufsätze.* 5 vols. (Stuttgart, 1959–1962), 1:82–113.

——, ed. *Mittelalterliche Bibliothekskataloge Deutschlands und der Schweiz,* vol. 2 (Bistum Mainz, Erfurt) (Munich, 1928).

——. 'Staindel-Funde.' In *Erforschung des Mittelalters, Ausgewählte Abhandlungen und Aufsätze.* 5 vols. (Stuttgart, 1959–1962), 4:237–256.

Leibowitz, J. O., and S. Marcus, eds. *Moses Maimonides on the Causes of Symptoms* (Berkeley, 1974).

Leinsle, U. G. *Das Ding und die Methode: Methodische Konstitution und Gegenstand der frühen protestantischen Metaphysik.* 2 vols. (Augsburg, 1985).

Lemay, R. J. *Abū Maʿšar al-Balḫī (Albumasar): Kitāb al-mudḫal al-kabīr ilā ʿilm aḥkām an-nuǧūm: Liber introductorii maioris ad scientiam judiciorum astrorum.* 9 vols. (Naples, 1995–1996).

——. 'Origin and Success of the Kitāb Thamara of Abū Jaʿfar Aḥmad ibn Yūsuf ibn Ibrāhīm from the Tenth to the Seventeenth Century in the World of Islam and the Latin West.' In *Proceedings of the First International Symposium for the History of Arabic Science, April 5–12, 1976* (Aleppo, 1978), 2:91-107.

——. 'The Teaching of Astronomy at Medieval Universities, Principally at Paris in the 14th Century.' *Manuscripta* 20 (1976): 197-217.

Leppin, V. *Antichrist und Jüngster Tag: Das Profil apokalyptischer Flugschriftenpublizistik im deutschen Luthertum 1548–1618* (Gütersloh, 1999).

Lerch, A. *Scientia astrologiae: Der Diskurs über die Wissenschaftlichkeit der Astrologie und die lateinischen Lehrbücher 1470–1610* (Leipzig, 2015).

Lhotsky, A. *Die Wiener Artistenfakultät, 1365–1497* (Graz, 1965).

Licata, G. B., ed. *L'averroismo in etá moderna (1400–1700)* (Macerata, 2013).

——. *La via della ragione: Elia del Medigo e l'averroismo di Spinoza* (Macerata, 2013).

Lichtenberg, G. C. *Sudelbücher.* In *Schriften und Briefe,* ed. W. Promies. 6 vols. (Frankfurt am Main, ³1994), vols. 1–2.

Lieberknecht, S. *Die* Canones *des Pseudo-Mesue: Eine mittelalterliche Purgantien-Lehre.* Quellen und Studien zur Geschichte der Pharmazie 71 (Stuttgart, 1995).

Liess, L. *Geschichte der medizinischen Fakultät in Ingolstadt von 1472 bis 1600* (Munich, 1984).

Lindberg, D. C. *A Catalogue of Medieval and Renaissance Optical Manuscripts* (Toronto, 1975).

——. 'Conceptions of the Scientific Revolution from Bacon to Butterfield: A Preliminary Sketch.' In *Reappraisals of the Scientific Revolution,* ed. D. C. Lindberg and R. S. Westman (Cambridge, 1990), 1-26.

——. *Theories of Vision from Al-Kindi to Kepler* (Chicago, 1976).

Lockwood, D. P. *Ugo Benzi: Medieval Philosopher and Physician, 1376–1439* (Chicago, 1951).

Lohr, C. H. *Latin Aristotle Commentaries. 1.1. Medieval Authors A–L* (Florence, 2013).

——. *Latin Aristotle Commentaries. 1.2. Medieval Authors M–Z* (Florence, 2010).

——. *Latin Aristotle Commentaries. 2. Renaissance Authors* (Florence, 1988).

——. 'Logica Algazelis: Introduction and Critical Text.' *Traditio* 21 (1965): 223–290.

——. 'Metaphysics and Natural Philosophy as Sciences: the Catholic and the Protestant Views in the Sixteenth and Seventeenth Centuries.' In *Philosophy in the Sixteenth and Seventeenth Centuries: Conversations with Aristotle,* ed. C. Blackwell and S. Kusakawa (Aldershot, 1999), 280–295.

——. *Raimundus Lullus' Compendium logicae Algazelis: Quellen, Lehre und Stellung in der Geschichte der Logik* (Freiburg im Breisgau, 1967).

——. 'Renaissance Latin Translations of the Greek Commentaries on Aristotle.' In *Humanism and Early Modern Philosophy,* ed. J. Kraye and M. W. F. Stone (London, 2000), 24–40.

——. 'The Sixteenth-Century Transformation of the Aristotelian Natural Philosophy.' In *Aristotelismus und Renaissance: In memoriam Charles B. Schmitt,* ed. E. Kessler et al. Wolfenbütteler Forschungen 40 (Wiesbaden, 1988), 89–99.

Lorch, R. 'The Astronomy of Jābir ibn Aflaḥ.' *Centaurus* 19 (1975): 85–107.

——. *Thābit ibn Qurra: On the Sector-Figure and Related Texts* (Frankfurt am Main, 2001).

Lorenz, S. 'Libri ordinarie legendi: Eine Skizze zum Lehrplan der mitteleuropäischen Artistenfakultät um die Wende vom 14. zum 15. Jahrhundert.' In *Argumente und Zeugnisse,* ed. W. Hogrebe. Studia Philosophica et Historica 5 (Frankfurt am Main, 1985), 204–258.

——. *Studium Generale Erfordense: Zum Erfurter Schulleben im 13. und 14. Jahrhundert* (Stuttgart, 1989).

Lowry, M. J. C. 'Two Great Venetian Libraries in the Age of Aldus Manutius.' *Bulletin of the John Rylands Library* 57 (1974–1975): 128–166.

Lucchetta, F. 'L'Averroismo Padovano.' *Annali della facoltà di lingue e letterature straniere di cà foscari* 20 (1981): 73–86.

——. 'Girolamo Ramusio, profilo biografico.' *Quaderni per la storia dell'Università di Padova* 15 (1982): 1–60.

——. *Il medico e filosofo bellunese Andrea Alpago († 1522), traduttore di Avicenna* (Padua, 1964).

——. 'La prima presenza di Averroè in ambito veneto.' *Studia Islamica* 46 (1977): 133–146.

——. 'Recenti studi sull'Averroismo padovano.' *L'Averroismo in Italia. Convegno internazionale.* Atti dei convegni Lincei 40 (Rome, 1979), 91–120.

Lucentini, P., and V. Perrone Compagni. *I testi e i codici di Ermete nel Medioevo* (Florence, 2001).

Ludolphy, I. 'Luther und die Astrologie.' In *"Astrologi Hallucinati": Stars and the End of the World in Luther's Time,* ed. P. Zambelli (Berlin, 1986), 101–107.

MacLean, G., ed. *Re-Orienting the Renaissance: Cultural Exchanges with the East* (Houndmills, 2005).

——. *The Rise of Oriental Travel: English Visitors to the Ottoman Empire, 1580–1720* (Basingstoke, 2004).

Maclean, I. 'Heterodoxy in Natural Philosophy: Pietro Pomponazzi, Guglielmo Gratarolo, Girolamo Cardano.' In *Heterodoxy in Early Modern Science and Religion,* ed. I. Maclean and J. H. Brooke (Oxford, 2005), 1–29.

Mägdefrau, K. *Geschichte der Botanik: Leben und Leistung großer Forscher* (Stuttgart, ²1992).

Mahlmann-Bauer, B. 'Die Bulle "contra astrologiam iudiciariam" von Sixtus V., das astrologische Schrifttum protestantischer Autoren und die Astrologiekritik der Jesuiten: Thesen über einen vermuteten Zusammenhang.' In *Zukunftsvoraussagen in der Renaissance,* ed. K. Bergdolt and W. Ludwig (Wiesbaden, 2005), 143–222.

Mahoney, E. P. 'Albert the Great and the *Studio Patavino* in the Late Fifteenth and Early Sixteenth Centuries.' In *Albertus Magnus and the Sciences,* ed. J. Weisheipl (Toronto, 1980), 537–563.

——. 'Giovanni Pico della Mirandola and Elia del Medigo, Nicoletto Vernia and Agostino Nifo.' In *Giovanni Pico della Mirandola. Atti del convegno internazionale di studi nel cinquecentesimo anniversario della morte (Mirandola, 4–8 ottobre 1994)*, ed. G. C. Garfagnini (Florence, 1997), 127–156.

——. 'Neoplatonism, the Greek Commentators, and Renaissance Aristotelianism.' In *Neaplatonism and Christian Thought*, ed. D. J. O'Meara (New York, 1982), 169–177, 264–282.

——. 'Themes and Problems in the Psychology of John of Jandun.' In *Studies in Medieval Philosophy*, ed. J. F. Wippel (Washington, DC, 1987), 278–288.

——. *Two Aristotelians of the Italian Renaissance: Nicoletto Vernia and Agostino Nifo* (Aldershot, 2000), esp. art. 1, 'Philosophy and Science in Nicoletto Vernia and Agostino Nifo'; art. 3, 'Nicoletto Vernia on the Soul and Immortality'; art. 4, 'Nicoletto Vernia's Annotations on John of Jandun's *De anima*'; art. 7, 'Agostino Nifo's Early Views on Immortality'; and art. 9, 'Antonio Trombetta and Agostino Nifo on Averroes and Intelligible Species.'

Maier, A. 'Wilhelm von Alnwicks Bologneser Quaestionen gegen den Averroismus (1323).' *Gregorianum* 30 (1949): 265–308 (repr. *Ausgehendes Mittelalter: gesammelte Aufsätze zur Geistesgeschichte des 14. Jahrhunderts*. 3 vols. (Rome, 1964–1977), 1:1–40).

Maierù, A. 'Influenze arabe e discussioni sulla natura della logica presso i latini fra XIII e XIV secolo.' In *La diffusione delle science islamiche nel Medio Evo europeo*, ed. B. Scarcia Amoretti (Rome, 1987), 243–267.

Malagola, C., ed. *Statuti delle Università e dei collegi dello studio bolognese* (Bologna, 1888).

Mancha, J. L. 'La version Alfonsi del Fī hay'at al-ʿālam (De configuratione mundi) de Ibn al-Hayṭam (Oxford, Canon. misc. 45, ff. 1r–56r).' In *"Ochava espera" y "Astrofísica": textos y estudios sobre las fuentes árabes de la astronomía de Alfonso X*, ed. M. Comes, H. Mielgo, and J. Samsó (Barcelona, 1990), 133–197.

*Manoscritti e stampe venete dell'aristotelismo e averroismo (secoli X–XVI): Catalogo di mostra presso la Biblioteca Nazionale Marciana* (Venice, 1958).

Mansi, G. D. *Sacrorum conciliorum nova et amplissima collectio*. 32 vols. (Venice, 1758–1798, repr. Paris, 1901–1927).

Marcus, S. See Leibowitz, J. O.

Markowski, M. 'Die mathematischen und Naturwissenschaften an der Krakauer Universität im XV. Jahrhundert.' *Mediaevalia Philosophica Polonorum* 18 (1973): 121–130.

Marmura, M. E. 'Some Remarks on Averroës's Statements on the Soul.' In *Averroës and the Enlightenment,* ed. M. Wahba and M. Abousenna (New York, 1996), 279–291.

———. See Avicenna.

Martin, A. *Averroès: Grand Commentaire de la* Mètaphysique *d'Aristote (Tafsīr mā baʿd aṭ-ṭabīʿat): livre lam-lambda traduit de l'arabe et annoté* (Paris, 1984).

Martin, C. 'Humanism and the Assessment of Averroes in the Renaissance.' In *Renaissance Averroism and Its Aftermath: Arabic Philosophy in Early Modern Europe,* ed. A. Akasoy and G. Giglioni (Dordrecht, 2013), 65–79.

———. 'Providence and Seventeenth-Century Attacks on Averroes.' In *Averroes' Natural Philosophy and Its Reception in the Latin West,* ed. P. J. J. M. Bakker (Leuven, 2015), 193–212.

———. *Renaissance Meteorology: Pomponazzi to Descartes* (Baltimore, 2011).

———. 'Rethinking Renaissance Averroism.' *Intellectual History Review* 17, no. 1 (2007): 3–28.

———. *Subverting Aristotle: Religion, History, and Philosophy in Early Modern Science* (Baltimore, 2014).

Matsen, H. S. *Alessandro Achillini (1463–1512) and His Doctrine of "Universals" and "Transcendentals": A Study in Renaissance Ockhamism* (Lewisburg, 1974).

Mavroudi, M. *A Byzantine Book on Dream Interpretation: The Oneirocriticon of Achmet and Its Arabic Sources* (Leiden, 2002).

Mazzocco, A., ed. *Interpretations of Renaissance Humanism.* 3 vols. (Leiden, 2006).

McLaughlin, M. 'Biography and Autobiography in the Italian Renaissance.' In *Mapping Lives: The Uses of Biography,* ed. P. France and W. St. Clair (Oxford, 2002), 37–65.

McVaugh, M. R. *Arnaldi de Villanova Opera Medica Omnia: II Aphorismi de Gradibus* (Granada, 1975).

——. 'Historical Awareness in Medieval Surgical Treaties (12th–14th Centuries).' In *Geschichte der Medizingeschichtsschreibung: Historiographie unter dem Diktat literarischer Gattungen von der Antike bis zur Aufklärung,* ed. T. Rütten (Remscheid, 2009), 171–199.

McVaugh, M. R., and L. Ferre. *The* Tabula antidotarii *of Armengaud Blaise and Its Hebrew Translation.* Transactions of the American Philosophical Society 90, 6 (Philadelphia, 2000).

Mendelsohn, A. See Burnett, C.

Menéndez Pidal, G. 'Cómo trabajaron las escuelas Alfonsíes.' *Nueva revista de filología hispánica* 5 (1951): 363–380.

Mentgen, G. *Astrologie und Öffentlichkeit im Mittelalter* (Stuttgart, 2005).

Meserve, M. *Empires of Islam in Renaissance Historical Thought* (Cambridge, MA, 2008).

Meyer, E. H. F. *Geschichte der Botanik.* 4 vols. (Königsberg, 1854–1857, repr. Amsterdam, 1965).

Meyer, F. G., E. E. Trueblood, and J. L. Heller. *The Great Herbal of Leonhart Fuchs: de historia stirpium commentarii insignes, 1542.* 2 vols. (Stanford, 1999).

Micheau, F. "Alī ibn al-ʿAbbās al-Maǧūsī et son milieu.' In *Constantine the African and ʾAlī ibn al-ʿAbbās al-Maǧūsī: The* Pantegni *and Related Texts,* ed. C. Burnett and D. Jacquart (Leiden, 1994), 1–15.

——. See Jacquart, D.

Mikkeli, H. *An Aristotelian Response to Renaissance Humanism: Jacopo Zabarella on the Nature of Arts and Sciences* (Helsinki, 1992).

Millás Vallicrosa, J. M. 'La traducción castellana del *Tratado de agricultura* de Ibn Wāfid.' *al-Andalus* 8 (1943): 281–332.

——. *Las traducciones orientales en los manuscritos de la Biblioteca Catedral de Toledo* (Madrid, 1942).

Minio-Paluello, L. *Opuscula: The Latin Aristotle* (Amsterdam, 1972).

Monarca, G. *Agostino Nifo: vita ed opere, traccia per una riscoperta* (Scauri, 1975).

Monfasani, J. 'Aristotelians, Platonists, and the Missing Ockhamists: Philosophical Liberty in Pre-Reformation Italy.' *Renaissance Quarterly* 46 (1993): 247–276.

Moraux, P. *Alexandre d'Aphrodise: Exégète de la Noétique d'Aristote* (Liège, 1942).

Morelon, R. *Thābit ibn Qurra: Œuvres d'astronomie* (Paris, 1987).

Morton, A. G. *History of Botanical Science: An Account of the Development of Botany from Ancient Times to the Present Day* (London, 1981).

Moureau, S. 'Some Considerations concerning the Alchemy of the *De anima in arte alchemiae* of Pseudo-Avicenna.' *Ambix* 56, no. 1 (2009): 49–56.

Moyer, A. 'Renaissance Representations of Islamic Science: Bernardino Baldi and His *Lives of Mathematicians*.' *Science in Context* 12 (1999): 469–484.

Muckle, J. T. 'Isaac Israeli: *Liber de definicionibus*.' *Archives d'histoire doctrinale et littéraire du moyen âge* 12–13 (1937–1938): 299–340.

Mugnai Carrara, D. *La biblioteca di Nicolò Leoniceno: Tra Aristotele e Galeno: cultura e libri di un medico umanista* (Florence, 1991).

Müller, M. *Johann Albrecht von Widmanstetter 1506–1557: Sein Leben und Wirken* (Bamberg, 1908).

Müller-Jahncke, W.-D. *Astrologisch-magische Theorie und Praxis in der Heilkunde der frühen Neuzeit* (Stuttgart, 1985).

Munk, S. 'Ibn-Roschd.' In *Dictionnaire des sciences philosophiques*, vol. 3 (Paris, 1847), 157–172.

Muñoz Sendino, J. *La Escala de Mahoma: Traducción del árabe al castellano, latín y francés, ordenada por Alfonso X el Sabio; edición, introducción y notas* (Madrid, 1949).

Nallino, C. A. *Al-Battānī sive Albatenii: Opus Astronomicum. Ad fidem codicis escurialensis arabice editum, latine versum, adnotationibus instructum* (Milan, 1903, repr. Hildesheim, 1977).

Nante, A., C. Cavalli, and P. Gios, eds. *Pietro Barozzi: un vescovo del Rinascimento* (Padua, 2012).

Nardi, B. *Saggi sull'Aristotelismo Padovano del secolo XIV al XVI* (Florence, 1958).

——. *Sigieri di Brabante nel pensiero del Rinascimento italiano* (Rome, 1945).

——. *Studi su Pietro Pomponazzi* (Florence, 1965).

Narducci, E. 'Vita di Pitagora scritta da Bernardino Baldi, tratta dall'Autografo ed annotata.' *Bullettino di Bibliografia e di Storia delle Scienze Matematiche e Fisiche* 20 (1887): 197–308.

———. 'Vite inedite di Matematici Italiani scritte da Bernardino Baldi.' *Bullettino di Bibliografia e di Storia delle Scienze Matematiche e Fisiche* 19 (1886): 335–406, 437–489, 521–640.

Nauck, E. T. 'Der Ingolstädter medizinische Lehrplan aus der Mitte des 16. Jahrhunderts.' *Sudhoffs Archiv* 40 (1956): 1–15.

———. *Zur Geschichte des medizinischen Lehrplans und Unterrichts der Universität Freiburg i. Br.* (Freiburg im Breisgau, 1952).

Nelson, J. K. 'Filippino Lippi e i contesti della pittura a Firenze e Roma (1488–1504).' In *Filippino Lippi,* ed. P. Zambrano and J. K. Nelson (Milan, 2004), 367–611.

Nenci, E. *Bernardino Baldi (1553–1617): Studioso rinascimentale: poesia, storia, linguistica, meccanica, architettura* (Milan, 2005).

Newman, W. R. 'Arabo-Latin Forgeries: The Case of the *Summa perfectionis* (with the Text of Jābir ibn Ḥayyān's *Liber regni*).' In *The 'Arabick' Interest of the Natural Philosophers in Seventeenth-Century England,* ed. G. A. Russell (Leiden, 1994), 278–296.

———. *The* Summa Perfectionis *of Pseudo-Geber: A Critical Edition, Translation and Study* (Leiden, 1991).

Newton, F. 'Constantine the African and Monte Cassino: New Elements and the Text of the *Isagoge.*' In *Constantine the African and 'Alī ibn al-'Abbās al-Maǧūsī: The* Pantegni *and Related Texts,* ed. C. Burnett and D. Jacquart (Leiden, 1994), 16–47.

Nickl, P. See Schönberger, R.

Niewöhner, F. *Veritas sive Varietas: Lessings Toleranzparabel und das Buch von den drei Betrügern* (Heidelberg, 1988).

Niewöhner, F., and L. Sturlese, eds. *Averroismus im Mittelalter und in der Renaissance* (Zurich, 1994).

North, J. D. 'Astrology and the Fortunes of Churches.' In *Stars, Minds, and Fate: Essays in Ancient and Medieval Cosmology* (London, 1989), 59–89.

———. *Chaucer's Universe* (Oxford, 1988).

———. *The Fontana History of Astronomy and Cosmology* (London, 1994).

——. *God's Clockmaker: Richard of Wallingford and the Invention of Time* (London, 2005).

——. 'The *Quadrivium.*' In *A History of the University in Europe,* ed. W. Rüegg, vol. 2 (Cambridge, 1992), 337–359.

Norton, F. J. *Italian Printers: 1501–1520* (London, 1958).

Noureddine, A. See Jahier, H.

Novati, F. ed. *Epistolario di Coluccio Salutati.* Fonti per la storia d'Italia: Epistolari, secoli XIV–XV (Rome, 1896).

Nuovo, A. *Alessandro Paganino (1509–1538)* (Padua, 1990).

——. 'Il Corano arabo ritrovato.' *La Bibliofilía* 89 (1987): 237–271 (repr. in Nuovo, *Alessandro Paganino,* 107–131).

Nutton, V. 'The Changing Language of Medicine, 1450–1550.' In *Vocabulary of Teaching and Research between Middle Ages and Renaissance,* ed. O. Weijers. Etudes sur le vocabulaire intellectuel du moyen age 8 (Turnhout, 1995), 184–198.

——. 'Greek Science in the Sixteenth-Century Renaissance.' In *Renaissance and Revolution: Humanists, Scholars, Craftsmen and Natural Philosophers in Early Modern Europe,* ed. J. V. Field et al. (Cambridge, 1993), 15–28.

——. 'Hellenism Postponed: Some Aspects of Renaissance Medicine, 1490–1530.' *Sudhoffs Archiv* 81 (1997): 158–170.

——. 'Humanist Surgery.' In *The Medical Renaissance of the Sixteenth Century,* ed. A. Wear et al. (Cambridge, 1985), 75–99.

——. 'John Caius and the Linacre Tradition.' *Medical History* 23, no. 4 (1979): 373–391.

——. 'Mattioli and the Art of the Commentary.' In *La complessa Scienza dei Semplici: Atti delle Celebrazioni per il V Centenario della Nascita di Pietro Andrea Mattioli,* ed. D. Fausti (Siena, 2004), 133–147.

——. 'The Rise of Medical Humanism: Ferrara, 1464–1555.' *Renaissance Studies* 11 (1997): 2–19.

——. 'Wittenberg Anatomy.' In *Medicine and the Reformation,* ed. O. P. Grell and A. Cunningham (London, 1993), 11–32.

Oestmann, G. *Heinrich Rantzau und die Astrologie: ein Beitrag zur Kulturgeschichte des 16. Jahrhunderts* (Braunschweig, 2004).

Offenberg, A. K. 'Untersuchungen zum hebräischen Buchdruck in Neapel um 1490.' In *Buch und Text im 15. Jahrhundert: Arbeitsgespräch in der Herzog August Bibliothek Wolfenbüttel vom 1.–3. März 1978,* ed. L. Hellinga and H. Härtel (Hamburg, 1981), 129–141.

Öhlschlegel, C. *Studien zu Lorenz Fries und seinem 'Spiegel der Arznei'* (Tübingen, 1985).

Oliva, C. 'Note sull'insegnamento di Pietro Pomponazzi.' *Giornale critico della filosofia italiana* 7 (1926): 83–103, 179–190, 254–275.

O'Malley, C. D. *Andreas Vesalius of Brussels 1514–1564* (Berkeley, 1964).

Opitz, K. *Ar-Razi (Razes), Über die Pocken und die Masern (ca. 900 n. Chr.): aus dem Arabischen übersetzt* (Leipzig, 1911).

Orthmann, E. 'Astrologie und Propaganda: Iranische Weltzyklusmodelle im Dienst der Fāṭimiden.' *Die Welt des Orients* 36 (2006): 131–142.

Ottosson, P.-G. *Scholastic Medicine and Philosophy: A Study of Commentaries on Galen's Tegni (ca. 1300–1450)* (Naples, 1984).

Pachnicke, C. See Brinkhus, G.

Pagallo, G. F. 'L'*animus averroisticus* di Nicoletto Vernia e il vescovo Pietro Barozzi: alcuni ritocchi al quadro d'insieme (1487–1499).' In *Pietro Barozzi: un vescovo del Rinascimento,* ed. A. Nante, C. Cavalli, and P. Gios (Padua, 2012), 125–149.

——. 'Sull' autore (Nicoletto Vernia?) di un'anonima e inedita *quaestio* sull' anima del secolo XV.' In *La filosofia della natura nel medioevo: atti del terzo Congresso Internazionale di Filosofia Medioevale* (Milan, 1966), 670–682.

Paganini, G. 'Umanesimo e "Denkstil": Su una storia recente dell'etica umanistica italiana.' *Giornale critico della filosofia italiana,* 7 ser. vol. 7 (2011): 661–670.

Pagel, W. 'The History of Mineral Terminology.' *History of Science* 12 (1974): 70–76 (review of D. Goltz, *Studien zur Geschichte der Mineralnamen* [Wiesbaden, 1972]).

——. 'Medical Humanism: A Historical Necessity in the Era of the Renaissance.' In *Essays on the Life and Work of Thomas Linacre c. 1460–1524,* ed. F. Maddison, M. Pelling, and C. Webster (Oxford, 1977), 375–386.

Palmer, R. 'Medical Botany in Northern Italy in the Renaissance.' *Journal of the Royal Society of Medicine* 78 (1985): 149–157.

——. 'Nicolò Massa, His Family and His Fortune.' *Medical History* 25 (1981): 385–410.

——. 'Pharmacy in the Republic of Venice in the Sixteenth Century.' In *The Medical Renaissance of the Sixteenth Century,* ed. A. Wear (Cambridge, 1985), 100–117.

——. 'Texts and Documents: Physicians and Surgeons in Sixteenth-Century Venice.' *Medical History* 23 (1979): 451–460.

Park, K. 'Albert's Influence on Late Medieval Psychology.' In *Albertus Magnus and the Sciences,* ed. J. Weisheipl (Toronto, 1980), 510–535.

——. 'The Imagination in Renaissance Philosophy' (M.Phil. diss., University of London, Warburg Institute, 1974).

——. 'The Organic Soul.' In *The Cambridge History of Renaissance Philosophy,* ed. C. B. Schmitt et al. (Cambridge, 1988), 464–484.

——. 'Picos *De imaginatione* in der Geschichte der Philosophie.' In G. F. Pico della Mirandola, *Über die Vorstellung/De imaginatione, lateinisch-deutsche Ausgabe,* ed. E. Kessler, C. B. Schmitt, and K. Park (Munich, 1984), 16–43.

Paschini, P. *Domenico Grimani: cardinale di S. Marco († 1523)* (Rome, 1943).

Pattin, A. 'Un grand commentateur d'Aristote: Agostino Nifo.' In *Historia Philosophiae Medii Aevi: Studien zur Geschichte der Philosophie des Mittelalters,* ed. B. Mojsisch and O. Pluta. 2 vols. (Amsterdam, 1991), 2:787–803.

Pedersen, F. S. *The Toledan Tables: A Review of the Manuscripts and the Textual Versions with an Edition.* 4 vols. (Copenhagen, 2002).

Peitz, R. See Keil, G.

Perfetti, S. *Aristotle's Zoology and Its Renaissance Commentators (1521–1601)* (Leuven, 2000).

Perrone Compagni, V. See Lucentini, P.

Pescasio, L. *Cardinale Ercole Gonzaga: presidente del Concilio di Trento (1505–1563)* (Suzzara, Mantua, 1999).

Peters, C. 'Johannan b. Serapion.' *Le Muséon: Revue d'études orientales* 55 (1942): 139–142.

Petersen, P. *Geschichte der aristotelischen Philosophie im protestantischen Deutschland* (Leipzig, 1921).

Petoletti, M. 'Les Recueils *De viris illustribus* en Italie (XIVe–XVe siècles).' In *Exempla docent. Les exemples des philosophes de l'Antiquité à la Renaissance,* ed. T. Ricklin and M. Gadebusch Bondio (Paris, 2006), 335–353.

Pfeiffer, R. *History of Classical Scholarship: From the Beginnings to the End of the Hellenistic Age* (Oxford, 1968).

——. *History of Classical Scholarship: From 1300 to 1850* (Oxford, 1976).

Pianetti, E. ' "Fra" Jacopo Filippo Foresti e la sua opera nel quadro della cultura Bergamasca.' *Bergomum* 33 (1939): 100–109, 147–174.

Piemontese, A. M. 'Venezia e la diffusione dell'alfabeto arabo nell'Italia del Cinquecento.' *Quaderni di studi arabi* 5–6 (1987–1988): 641–653.

Pierro, F. 'Abramo di Meir de Balmes (1460–1523): medico, filosofo e grammatico Ebreo della scuola Napoletana.' *Atti del Congresso Nazionale di Storia della Medicina* 19 (1965): 360–379.

Pine, M. L. *Pietro Pomponazzi: Radical Philosopher of the Renaissance* (Padua, 1986).

Pingree, D. E. 'The Diffusion of Arabic Magical Texts in Western Europe.' In *La diffusione delle scienze islamiche nel Medio Evo europeo,* ed. B. Scarcia Amoretti (Rome, 1987), 57–102.

——. *From Astral Omens to Astrology: from Babylon to Bīkāner* (Rome, 1997).

——. 'Hellenophilia versus the History of Science.' *Isis* 83 (1992): 554–563.

——. 'Historical Horoscopes.' *Journal of the American Oriental Society* 82 (1962): 487–502.

——. 'The *Liber Universus* of ʿUmar Ibn al-Farrukhān aṭ-Ṭabarī.' *Journal for the History of Arabic Science* 1 (1977): 8–12.

——. 'Māshāʾallāh: Some Sasanian and Syriac Sources.' In *Essays on Islamic Philosophy and Science,* ed. G. F. Hourani (New York, 1975), 5–14.

——. 'Māshāʾallāh's (?) Arabic Translation of Dorotheus.' In *La Science des Cieux: Sages, mages, astrologues,* ed. R. Gyselen. Res Orientales 12 (Bures-sur-Yvette, 1999), 191–209.

——. See Kennedy, E. S.

Pinto, O. 'Una rarissima opera araba stampata a Roma nel 1585.' *Studi Bibliografici* (Florence, 1967), 47–51.

Piron, S. 'Olivi et les averroïstes.' *Freiburger Zeitschrift für Philosophie und Theologie* 53 (2006): 251–309.

Polzer, J. 'The "Triumph of Thomas" Panel in Santa Caterina, Pisa: Meaning and Date.' *Mitteilungen des Kunsthistorischen Institutes in Florenz* 37 (1993): 29–70.

Poppi, A. *Introduzione all'aristotelismo padovano* (Padua, ²1991).

——. 'L'Averroismo nella filosofia francescana.' In *L'Averroismo in Italia. Convegno internazionale*. Atti dei convegni Lincei 40 (Rome, 1979), 175–220 (repr. in Poppi, *La filosofia nello studio francescano*, 219–270).

——. *La filosofia nello studio francescano del santo a Padova* (Padua, 1989).

——. *Saggi sul pensiero inedito di Pietro Pomponazzi* (Padua, 1970).

——. See Pomponazzi, P.

Pormann, P. E. 'The Dispute between the Philarabic and Philhellenic Physicians and the Forgotten Heritage of Arabic Medicine.' In *Islamic Medical and Scientific Tradition: Critical Concepts in Islamic Studies,* ed. P. E. Pormann. 3 vols. (London, 2011), 2:283–316.

——. 'Yūḥannā ibn Sarābiyūn: Further Studies into the Transmission of His Works.' *Arabic Sciences and Philosophy* 14 (2004): 233–262.

Porter, R. 'The Scientific Revolution: A Spoke in the Wheel?' In *Revolution in History,* ed. R. Porter and M. Teich (Cambridge, 1986), 290–316.

Pozzi, G. 'Appunti sul *Corollarium* del Barbaro.' In *Tra latino e volgare: per Carlo Dionisotti,* ed. G. Bernardoni Trezzini et al. 2 vols. (Padua, 1974), 2:619–640.

Prelog, J. 'Die Handschriften und Drucke von Walter Burleys Liber de vita et moribus philosophorum.' *Codices manuscripti* 9 (1983): 1–18.

Primavesi, O. 'Aristotle, *Metaphysics* A. A New Critical Edition with Introduction.' In *Aristotle's Metaphysics Alpha. Symposium Aristotelicum,* ed. C. Steel (Oxford, 2012), 385–516.

Procter, E. S. *Alfonso X of Castile: Patron of Literature and Learning* (Oxford, 1961).

Pruckner, H. *Studien zu den astrologischen Schriften des Heinrich von Langenstein* (Leipzig, 1933).

Puig Montada, J. *Averroes: Epítome de física* (Madrid, 1987).

——. 'Continuidad medieval en el Renacimiento: El caso de Elia del Medigo.' *La Ciudad de Dios: Revista Augustiniana* 206 (1993): 47–64.

——. 'Eliahu del Medigo, traductor del epítome de Averroes Acerca del Alma.' *Ciudad de Dios: Revista agustiniana* 219 (2006): 713–729.

——. 'Materials on Averroes's Circle.' *Journal of Near Eastern Studies* 5 (1992): 241–260.

——. 'On the Chronology of Elia del Medigo's Physical Writings.' In *Jewish Studies at the Turn of the Twentieth Century,* ed. J. Targarona Borrás and A. Sáenz-Badillos, vol. 2 (Leiden, 1999), 54–56.

Putallaz, F.-X. 'Censorship.' In *The Cambridge History of Medieval Philosophy,* ed. R. Pasnau and C. van Dyke (Cambridge, 2010), 99–113.

Quaritch, B. *Arabic Science and Medicine: A Collection of Manuscripts and Early Printed Books Illustrating the Spread and Influence of Arabic Learning in the Middle Ages and the Renaissance.* Bernard Quaritch Ltd. Catalogue 1186 (London, 1993).

Ragep, F. J. 'Copernicus and His Islamic Predecessors: Some Historical Remarks.' *History of Science* 45 (2007): 65–81.

——. 'Tusi and Copernicus: The Earth's Motion in Context.' *Science in Context* 14, nos. 1–2 (2001): 145–163.

Ragnisco, P. 'Documenti inediti e rari intorno alla vita ed agli scritti di Nicoletto Vernia e di Elia del Medigo.' *Atti e memorie della R. Accademia di scienze, lettere ed arti in Padova* N.S. 7 (1890/1891), 275–302.

Rahman, F. 'Ibn Sīnā.' In *A History of Muslim Philosophy,* ed. M. M. Sharif. 2 vols. (Wiesbaden, 1963), 1:480–506.

——. *Prophecy in Islam: Philosophy and Orthodoxy* (Chicago, 1958).

Raimondi, E. *Codro e l'umanesimo a Bologna* (Bologna, 1950).

Ramminger, J. 'A Commentary? Ermolao Barbaro's Supplement to Dioscorides.' In *On Renaissance Commentaries,* ed. M. Pade (Hildesheim, 2005), 65–85.

——. 'Zur Entstehungsgeschichte des Dioskurides von Ermolao Barbaro (1453–1493).' *Neulateinisches Jahrbuch* 1 (1999): 189–204.

Randall, J. H. Jr. *The Making of the Modern Mind: A Survey of the Intellectual Background of the Present Age* (Cambridge, 1940).

——. *The School of Padua and the Emergence of Modern Science* (Padua, 1961).

——. See Cassirer, E.

Rashed, R., ed. *Encyclopedia of the History of Arabic Science.* 3 vols. (London, 1996).

——, ed. *Oeuvres philosophiques et scientifiques d'al-Kindī, vol. 1: L'optique et la Catoptrique* (Leiden, 1997).

——, ed. *Thābit ibn Qurra. Science and Philosophy in Ninth-Century Baghdad* (Berlin, 2009).

Rauchenberger, D. *Johannes Leo der Afrikaner: Seine Beschreibung des Raumes zwischen Nil und Niger nach dem Urtext.* Orientalia biblica et christiana 13 (Wiesbaden, 1999).

Reeds. K. M. *Botany in Medieval and Renaissance Universities* (New York, 1991).

——. 'Renaissance Humanism and Botany.' *Annals of Science* 33 (1976): 519–542.

Reeve, M. D. 'Classical Scholarship.' In *Renaissance Humanism,* ed. J. Kraye (Cambridge, 1996), 20–46.

Reif, K. See Kabelitz, L.

Reinhardt, L. *Kulturgeschichte der Nutzpflanzen.* 2 vols. (Munich, 1911).

Renan, E. *Averroès et l'Averroïsme: essai historique* (Paris, ³1867, repr. Frankfurt am Main, 1985).

Rice, E. F. 'Humanist Aristotelianism in France: Jacques Lefèvre d'Étaples and His Circle.' In *Humanism in France at the End of the Middle Ages and in the Early Renaissance,* ed. A. H. T. Levi (New York, 1970), 132–149.

Ricklin, T., and M. Gadebusch Bondio, eds. *Exempla docent: Les exemples des philosophes de l'Antiquité à la Renaissance* (Paris, 2006).

Rico, F. 'Aristoteles Hispanus: En torno a Gil de Zamora, Petrarca y Juan de Mena.' *Italia medioevale e umanistica* 10 (1967): 143–164.

Riddle, J. M. 'Dioscorides.' In *Catalogus translationum et commentariorum: Medieval and Renaissance Latin Translations and Commentaries,* ed. F. E. Cranz and P. O. Kristeller (Washington, DC, 1960–), 4:1–143.

——. 'The Introduction and Use of Eastern Drugs in the Early Middle Ages.' *Sudhoffs Archiv* 49 (1965): 185–198.

——. See Wilcox, J.

Rigo, C. 'Zur Rezeption des Moses Maimonides im Werk des Albertus Magnus.' In *Albertus Magnus: Zum Gedenken nach 800 Jahren: Neue Zugänge, Aspekte und Perspektiven,* ed. W. Senner et al. (Berlin, 2001), 29–66.

Risse, W. 'Averroismo e Alessandrinismo nella logica del Rinascimento.' *Filosofia* 15 (1964): 15–30.

Rizzo, S. *Il lessico filologico degli umanisti* (Rome, 1973).

Rodríguez, D. C. *Juan de Segovia y el problema islámico* (Madrid, 1952).

Roling, B. *Aristotelische Naturphilosophie und christliche Kabbalah im Werk des Paulus Ritius* (Tübingen, 2007).

——. 'Mediatoris fungi munere: Synkretismus im Werk des Paolo Ricci.' In *Christliche Kabbala,* ed. W. Schmidt-Biggemann (Ostfildern, 2003), 77–100.

——. 'Prinzip, Intellekt und Allegorese im Werk des christlichen Kabbalisten Paolo Ricci (gest. 1541).' In *An der Schwelle zur Moderne: Juden in der Renaissance,* ed. G. Veltri and A. Winkelmann (Leiden, 2003), 155–187.

Rose, P. L. *The Italian Renaissance of Mathematics: Studies on Humanists and Mathematicians from Petrarch to Galileo* (Geneva, 1975).

Rosenfeld, B. A., and A. P. Youschkevitch. 'Geometry.' In *Encyclopedia of the History of Arabic Science,* ed. R. Rashed. 3 vols. (London, 1996), 2:447–494.

Rosner, F. *The Medical Aphorisms of Moses Maimonides: Maimonides' Medical Writings* (Haifa, 1989).

Ross, B. 'Giovanni Colonna, Historian at Avignon.' *Speculum* 45 (1970): 533–563.

Roth, C. *The Jews in the Renaissance* (Philadelphia, 1959).

Roth, U. 'Die astronomisch-astrologische "Weltgeschichte" des Nikolaus von Kues im Codex Cusanus 212. Einleitung und Edition.' *Mitteilungen und Forschungsbeiträge der Cusanus-Gesellschaft* 27 (2001): 1–29.

Ruderman, D. B. 'The Italian Renaissance and Jewish Thought.' In *Renaissance Humanism: Foundations, Forms and Legacy,* ed. A. Rabil. 3 vols. (Philadelphia, 1988), 1:382–433.

Rummel, E. *The Humanist-Scholastic Debate in the Renaissance and Reformation* (Cambridge, MA, 1995).

Ruska, J. *Al-Rāzī's Buch* Geheimnis der Geheimnisse. Quellen und Studien zur Geschichte der Naturwissenschaften und der Medizin 6 (Berlin, 1937).

——. 'Pseudepigraphe Rasis-Schriften.' *Osiris* 7 (1939): 31–93.

——. *Übersetzung und Bearbeitungen von al-Rāzī's Buch* Geheimnis der Geheimnisse. Quellen und Studien zur Geschichte der Naturwissenschaften und der Medizin 4 (Berlin, 1935).

Russell, G. A., ed. *The 'Arabick' Interest of the Natural Philosophers in Seventeenth-Century England* (Leiden, 1994).

Rutkin, H. D. 'Astrology.' In *The Cambridge History of Science. Volume 3: Early Modern Science,* ed. K. Park and L. Daston (Cambridge, 2006), 541–561.

——. 'Astrology, Natural Philosophy and the History of Science, c. 1250–1700: Studies Toward an Interpretation of Giovanni Pico della Mirandola's *Disputationes adversus astrologiam divinatricem*' (PhD diss., Indiana University, 2002).

——. 'The Use and Abuse of Ptolemy's *Tetrabiblos* in Renaissance and Early Modern Europe: Two Case Studies (Giovanni Pico della Mirandola and Filippo Fantoni).' In *Ptolemy in Perspective: Use and Criticism of His Work from Antiquity to the Nineteenth Century,* ed. A. Jones (Dordrecht, 2010), 135–149.

Rütten, T., ed. *Geschichte der Medizingeschichtsschreibung: Historiographie unter dem Diktat literarischer Gattungen von der Antike bis zur Aufklärung* (Remscheid, 2009).

Sabra, A. I. 'The Andalusian Revolt Against Ptolemaic Astronomy: Averroes and al-Biṭrūjī.' In *Transformation and Tradition in the Sciences: Essays in Honor of I. Bernard Cohen,* ed. E. Mendelsohn (Cambridge, 1984), 133–153.

——. 'The Authorship of the *Liber de crepusculis,* an Eleventh-Century Work on Atmospheric Refraction.' *Isis* 58 (1967): 77–85.

——. *The Optics of Ibn Al-Haytham: Books I–III, On Direct Vision.* 2 vols. (London, 1989).

Said, E. W. *Orientalism: Western Conceptions of the Orient. With a New Afterword* (London, 1995).

Salatowsky, S. *De Anima: Die Rezeption der aristotelischen Psychologie im 16. und 17. Jahrhundert* (Amsterdam, 2006).

Saliba, G. *A History of Arabic Astronomy: Planetary Theories during the Golden Age of Islam* (New York, 1994).

——. *Islamic Science and the Making of the European Renaissance* (Cambridge, MA, 2007).

Salman, D. 'Algazel et les Latins.' *Archives d'histoire doctrinale et littéraire du moyen âge* 11 (1935/1936): 103–127.

——. 'Le "Liber exercitationis ad viam felicitatis" d'Alfarabi.' *Recherches de Théologie Ancienne et Mediévale* 12 (1940): 33–48.

Samsó, J. See Vernet, J.

Sarton, G. *The Appreciation of Ancient and Medieval Science during the Renaissance (1450–1600)* (Philadelphia, 1955).

——. 'Arabic Science and Learning in the Fifteenth Century, Their Decadence and Fall.' In *Homenaje a Millás Vallicrosa.* 2 vols. (Barcelona, 1956), 2:303–324.

——. 'The Scientific Literature transmitted through the Incunabula (An Analysis and Discussion Illustrated with Sixty Facsimiles).' *Osiris* 5 (1938): 41–245.

Savage-Smith, E. 'Medicine.' In *Encyclopedia of the History of Arabic Science,* ed. R. Rashed. 3 vols. (London, 1996), 3:903–962.

Scarabel, A. 'Une "Édition critique" latine du *Mudḫal* d'al-Qabīsī à Venise à la veille de la Renaissance.' *Quaderni di studi arabi* 14 (1996): 5–18.

Schacht, J. 'Ibn al-Nafīs, Servetus and Colombo.' *Andalus* 22 (1957): 317–336.

Scheible, H. 'Aristoteles und die Wittenberger Universitätsreform.' In *Humanismus und Wittenberger Reform,* ed. M. Beyer and G. Wartenberg (Leipzig, 1996), 123–144.

Schipperges, H. 'Diaetetische Leitlinien in der Heilkunde des Manardus.' In *Atti del convegno internazionale per la celebrazione del V. centenario della nascità di Giovanni Manardo, 1462–1536* (Ferrara, 1963), 252–257.

——. *Die Assimilation der arabischen Medizin durch das lateinische Mittelalter.* Sudhoffs Archiv Beihefte 3 (Wiesbaden, 1964).

——. *Ideologie und Historiographie des Arabismus.* Sudhoffs Archiv Beihefte 1 (Wiesbaden, 1961).

Schmidt, E. A. Ullmann, M. *Aristoteles in Fes: zum Wert der arabischen Überlieferung der Nikomachischen Ethik für die Kritik des griechischen Textes* (Heidelberg, 2012).

Schmitt, C. B. *The Aristotelian Tradition and Renaissance Universities* (London, 1984), esp. art. 1, 'Aristotelianism in the Veneto and the Origins of Modern Science: Some Considerations on the Problem of Continuity'; art. 8, 'Renaissance Averroism Studied through the Venetian Editions of Aristotle-Averroes (with Particular Reference to the Giunta Edition of 1550–1552); and art. 12, 'Thomas Linacre and Italy.'

——. *Aristotle and the Renaissance* (Cambridge, MA, 1983).

——. *John Case and Aristotelianism in Renaissance England* (Kingston, 1983).

——. *Reappraisals in Renaissance Thought* (London, 1989).

——. 'The Rise of the Philosophical Textbook.' In *The Cambridge History of Renaissance Philosophy,* ed. C. B. Schmitt et al. (Cambridge, 1988), 792–804.

——. *Studies in Renaissance Philosophy and Science* (London, 1981), esp. art. 4, 'Reappraisals in Renaissance Science'; and art. 5, 'Philosophy and Science in Sixteenth-Century Universities: Some Preliminary Comments.'

——. 'Theophrastus.' In *Catalogus translationum et commentariorum: Mediaeval and Renaissance Latin Translations and Commentaries,* ed. P. O. Kristeller and F. E. Cranz (Washington, 1960–), 2:239–322.

——. See Copenhaver, B. P.

——. See Cranz, F. E.

Schmitt, C. B. and D. Knox. *Pseudo-Aristoteles Latinus. A Guide to Latin Works Falsely Attributed to Aristotle before 1500* (London, 1985).

Schmitt, C. B., Q. Skinner, E. Kessler, and J. Kraye, eds. *The Cambridge History of Renaissance Philosophy* (Cambridge, 1988).

Schmitz, R. 'Der Anteil des Renaissance-Humanismus an der Entwicklung von Arzneibüchern und Pharmakopöen.' In *Das Verhältnis der Humanisten zum Buch,* ed. F. Krafft and D. Wuttke (Boppard, 1977), 227–243.

——. *Geschichte der Pharmazie. Band I: von den Anfängen bis zum Ausgang des Mittelalters* (Eschborn, 1998).

———. 'Pharmazie und angewandte Naturwissenschaften in ihrer Beziehung zum Renaissance-Humanismus.' In *Humanismusforschung seit 1945: Ein Bericht aus interdisziplinärer Sicht,* ed. A. Buck (Boppard, 1975), 185–191.

Schneider, J. H. J. *Al-Fārābī: De scientiis secundum versionem Dominici Gundisalvi* (Freiburg im Breisgau, 2006).

Schnell, B. 'Arzt und Literat: Zum Anteil der Ärzte am spätmittelalterlichen Literaturbetrieb.' *Sudhoffs Archiv* 75 (1991): 44–57.

Schönberger, R., and P. Nickl. *Über die Ewigkeit der Welt: Texte von Bonaventura, Thomas von Aquin und Boethius von Dacien* (Frankfurt am Main, 2000).

Schramm, M. *Ibn Al-Haythams Weg zur Physik* (Wiesbaden, 1963).

Schroeder, F. M., and R. B. Todd. *Two Greek Aristotelian Commentators on the Intellect: The* De intellectu *Attributed to Alexander of Aphrodisias and Themistius' Paraphrase of Aristotle,* De anima *3.4–8* (Toronto, 1990).

Schullian, D. M., and L. Belloni, eds. *Giovanni Tortelli: On Medicine and Physicians. Gian Giacomo Bartolotti: On the Antiquity of Medicine. Two Histories of Medicine of the XVth Century* (Milan, 1954).

Schülting, S., S. L. Müller, and R. Hertel, eds. *Early Modern Encounters with the Islamic East: Performing Cultures* (Farnham, 2012).

Schulze, W. *Reich und Türkengefahr im späten 16. Jahrhundert: Studien zu den politischen und gesellschaftlichen Auswirkungen einer äußeren Bedrohung* (Munich, 1978).

Schum, W. *Beschreibendes Verzeichniss der Amplonianischen Handschriften-Sammlung zu Erfurt* (Berlin, 1887).

Secret, F. 'Guillaume Postel et les Etudes Arabes à la Renaissance.' *Arabica* 9 (1962): 21–36.

———. *Les Kabbalistes chrétiens de la Renaissance* (Paris, 1964).

Sensi, M. 'Niccolò Tignosi da Foligno: L'opera e il pensiero.' *Annali della Facultà di lettere e Filosofia, Università degli studi di Perugia* 9 (1971–1972): 359–495.

Sezgin, F. *Geschichte des arabischen Schrifttums.* 15 vols. (Leiden, 1967–2010).

Shank, M. 'Setting up Copernicus? Astronomy and Natural Philosophy in Giambattista Capuano da Manfredonia's *Expositio* on the *Sphere.*' *Early Science and Medicine* 14 (2009): 290–315.

Shields, C. 'Some Recent Approaches to Aristotle's *De Anima*.' In Aristotle, *De Anima Books II and III (with Passages from Book I)*, trans. D. W. Hamlyn (Oxford, 1993), 157–281.

Siraisi, N. *Avicenna in Renaissance Italy: the Canon and Medical Teaching in Italian Universities after 1500* (Princeton, 1987).

——. *History, Medicine, and the Traditions of Renaissance Learning* (Ann Arbor, 2007).

——. *Medieval and Early Renaissance Medicine: An Introduction to Knowledge and Practice* (Chicago, 1990).

——. *Taddeo Alderotti and His Pupils: Two Generations of Italian Medical Learning* (Princeton, 1981).

Sirat, C. *A History of Jewish Philosophy in the Middle Ages* (Cambridge, 1985).

——. 'Les traducteurs juifs à la cour des rois de Sicile et de Naples.' In *Traduction et traducteurs au Moyen Âge,* ed. G. Contamine (Paris, 1989), 169–191.

Sirat, C., and M. Geoffroy. *L'Original arabe du grand commentaire d'Averroès au De anima d'Aristote. Prémisse de l'édition* (Paris, 2005).

Skinner, Q. See Schmitt, C. B.

Smithuis, R. 'Abraham ibn Ezra's Astrological Works in Hebrew and Latin: New Discoveries and Exhaustive Listing.' *Aleph* 6 (2006): 239–338.

Smoller, L. A. *History, Prophecy, and the Stars: The Christian Astrology of Pierre d'Ailly, 1350–1420* (Princeton, 1994).

Spina, G., and F. Guido. 'Il commento di Giovanni Manardo al *De simpliciis* di Mesue.' In *Atti del convegno internazionale per la celebrazione del V. centenario della nascità di Giovanni Manardo, 1462–1536* (Ferrara, 1963), 270–272.

Spruit, L. *Agostino Nifo: De intellectu* (Leiden, 2011).

——. *Species intelligibilis: From Perception to Knowledge. Volume 1: Classical Roots and Medieval Discussions* (Leiden, 1994).

——. *Species intelligibilis: From Perception to Knowledge. Volume 2: Renaissance Controversies, Later Scholasticism, and the Elimination of Intelligible Species in Modern Philosophy* (Leiden, 1995).

Stannard, J. 'Dioscorides and Renaissance Materia Medica.' *Analecta Medico-Historica* 1 (1968): 1–21.

Steel, C. 'Siger of Brabant versus Thomas Aquinas on the Possibility of Knowing the Separate Substances.' In *Nach der Verurteilung von 1277: Philosophie und Theologie an der Universität von Paris im letzten Viertel des 13. Jahrhunderts,* ed. J. A. Aertsen, K. Emery Jr., and A. Speer. Miscellanea Mediaevalia 28 (Berlin, 2001), 211–231.

Steel, C., and G. Guldentops. 'An Unknown Treatise of Averroes against the Avicennians on the First Cause. Edition and Translation.' *Recherches de Théologie et Philosophie médiévales* 64 (1997): 86–135.

Stegemann, V. *Beiträge zur Geschichte der Astrologie I: Der griechische Astrologe Dorotheos von Sidon und der arabische Astrologe Abu 'l-Ḥasan 'Ali ibn abi 'r-Riǧāl, genannt Albohazen* (Heidelberg, 1935).

———. *Dorotheos von Sidon und das sogenannte Introductorium des Sahl ibn Bišr* (Prague, 1942).

Steinschneider M. *Die arabischen Übersetzungen aus dem Griechischen* (Graz, 1960) (reprint of articles published 1889–1896).

———. *Die europäischen Übersetzungen aus dem Arabischen bis Mitte des 17. Jahrhunderts.* In *Sitzungsberichte der Kaiserlichen Akademie der Wissenschaften in Wien, philosophisch-historische Klasse* 149, no. 4 (1904), and 151, no. 1 (1905) (repr. Graz, 1956), 1–84 and 1–108.

———. *Die hebraeischen Übersetzungen des Mittelalters und die Juden als Dolmetscher* (Berlin, 1893, repr. Graz, 1956).

———. 'Die Metaphysik des Aristoteles in jüdischen Bearbeitungen.' In *Jubelschrift zum Neunzigsten Geburtstag des Dr. L. Zunz* (Berlin, 1884), 1–35.

———. 'Die toxikologischen Schriften der Araber bis Ende des XII. Jahrhunderts.' *Virchows Archiv* 52 (1871): 340–375, 467–503.

———. *Vite di matematici arabi tratte da un'opera inedita di Bernardino Baldi* (Rome, 1874) (separate print of an article in *Bulletino di bibliografia e di storia delle scienze matematiche e fisiche* 5 [1872], 427–534).

———. 'Zur Oculistik des 'Isa Ben Ali (9. Jahrh.) und des sogenannten Canamusali.' *Janus* 11 (1906): 399–408.

Stelcel, I. *Codex diplomaticus universitatis studii generalis Cracoviensis: pars tertia ab anno 1471 usque ad annum 1506* (Kraków, 1880).

Stern, S. M. See Altmann, A.

Stopka, K. 'Les programmes d'études dans les universités médiévales et dans l'université de Cracovie en particulier.' In *L'université et la ville au moyen âge et d'autres questions du passé universitaire* (Kraków, 1993), 99–114.

Straberger-Schneider, J. Der *'Liber aggregatus in medicinis simplicibus'* des *Pseudo-Serapion aus der Mitte des 13. Jahrhunderts: Mit einer deutschen Teilübersetzung nach der Druckfassung von 1531* (Marburg, 2000 [microfiche edition]).

Strauss, H. A. Strauss-Kloebe, S. *Die Astrologie des Johannes Kepler: eine Auswahl aus seinen Schriften* (Fellbach, ²1981).

Stübler, E. *Geschichte der medizinischen Fakultät der Universität Heidelberg: 1386–1925* (Heidelberg, 1926).

———. *Leonhart Fuchs: Leben und Werk.* Münchner Beiträge zur Geschichte und Literatur der Naturwissenschaften und Medizin 13/14 (Munich, 1928).

Sturlese, L. See Niewöhner, F.

Sudhoff, W. 'Die Lehre von den Hirnventrikeln in textlicher und graphischer Tradition des Altertums und des Mittelalters,' *Archiv für Geschichte der Medizin* 7 (1914), 149–205.

Swerdlow, N. 'The Derivation and First Draft of Copernicus' Planetary Theory.' *Proceedings of the American Philosophical Society* 97 (1973): 423–512.

Sylla, E. 'Astronomy at Cracow University in the late Fifteenth Century: Albert of Brudzewo and John of Glogów.' Forthcoming in the Proceedings of the Eighteenth Colloquium of the SIEPM in Lodz, September 2011.

Talkenberger, H. *Sintflut: Prophetie und Zeitgeschehen in Texten und Holzschnitten astrologischer Flugschriften 1488–1528* (Tübingen, 1990).

Tamani, G. 'I libri ebraici del cardinal Domenico Grimani.' *Annali di Ca' Foscari* 34, no. 3 (1995): 5–52.

———. 'I libri ebraici di Pico della Mirandola.' In *Giovanni Pico della Mirandola: Atti del convegno internazionale di studi nel cinquecentesimo anniversario della morte (Mirandola, 4–8 ottobre 1994),* ed. G. C. Garfagnini (Florence, 1997), 491–530.

———. 'I traduttori ebrei e l'aristotelismo averroizzante a Padova nella prima metà del Cinquecento.' In *Aristotle and the Aristotelian Tradition.*

*Innovative Contexts for Cultural Tourism,* ed. E. De Bellis (Soveria Mannelli, 2008), 437–452.

——. *Il Canon medicinae di Avicenna nella tradizione ebraica: le miniature del manoscritto 2197 della Biblioteca Universitaria di Bologna* (Padua, 1988).

——. 'Le traduzioni ebraico-latine di Abraham De Balmes.' In *Biblische und Judaistische Studien: Festschrift für Paolo Sacchi,* ed. A. Vivain (Frankfurt am Main, 1990), 613–635.

——. 'Le versione ebraiche del *Canone della medicina* di Avicenna.' In *Autori classici in lingue del vicino e medio oriente,* ed. G. Fiaccadori (Rome, 2001), 301–314.

——. 'Traduzioni ebraico-latine di opere filosofiche e scientifiche.' In *L'hébreu au temps de la Renaissance: Ouvrage collectif recueilli et édité,* ed. I. Zinguer (Leiden, 1992), 105–114.

——. 'Una traduzione ebraico-latina delle *Questioni sulla Fisica* di Averroè.' *Italia. Studi e ricerche sulla cultura e sulla letteratura degli ebrei d'Italia* 13–15 (2001): 91–101.

Tarán, L., and D. Gutas. *Aristotle: Poetics: Editio Maior of the Greek Text with Historical Introductions and Philological Commentaries* (Leiden, 2012).

Targioni Tozzetti, A. 'Sulla coltivazione della Sena (Cassia obovata) nelle Maremme Toscane.' *Atti dell'I.E.R. Accademia economico-agraria dei Georgofili di Firenze* 25 (1847): 17–26.

Taylor, R. C. ' "The Future Life" and Averroës's Long Commentary on the *De Anima* of Aristotle.' In *Averroës and the Enlightenment,* ed. M. Wahba and M. Abousenna (New York, 1996), 263–277.

——. 'Intelligibles in Act in Averroes.' In *Averroès et les averroïsmes juif et latin: Actes du colloque tenu à Paris, 16–18 juin 2005,* ed. J.-B. Brenet (Turnhout, 2007), 111–140.

——. 'Personal Immortality in Averroes' Mature Philosophical Psychology.' *Documenti e studi sulla tradizione filosofica medievale* 9 (1998): 87–110.

——. See Averroes.

Temkin, O. *Galenism: Rise and Decline of a Medical Philosophy* (Ithaca, 1973).

Tester, J. *A History of Western Astrology* (Woodbridge, 1987).

Tezmen-Siegel, J. *Die Darstellungen der septem artes liberales in der Bildenden Kunst als Rezeption der Lehrplangeschichte* (Munich, 1985).

Theuerkauf, G. 'Soziale Bedingungen humanistischer Weltchronistik: Systemtheoretische Skizzen zur Chronik Nauclerus.' In *Landesgeschichte und Geistesgeschichte: Festschrift für Otto Herding zum 65. Geburtstag,* ed. K. Elm (Stuttgart, 1977), 317–340.

Thomasset, C. See Jacquart, D.

Thorndike, L. *A History of Magic and Experimental Science.* 8 vols. (New York, 1923–1958).

——. 'John of Seville.' *Speculum* 34 (1959): 20–38.

——. 'Latin Manuscripts of Works by Rasis at the Bibliothèque Nationale, Paris.' *Bulletin of the History of Medicine* 32 (1958): 54–67.

——. 'The Latin Translations of Astrological Works by Messahala.' *Osiris* 12 (1956): 49–72.

——. *Science and Thought in the Fifteenth Century: Studies in the History of Medicine and Surgery, Natural and Mathematical Science, Philosophy and Politics* (New York, 1929).

——. *The Sphere of Sacrobosco and Its Commentators* (Chicago, 1949).

——. *University Records and Life in the Middle Ages* (New York, 1944).

Tinto, A. *La Tipografia medicea orientale* (Lucca, 1987).

Toomer, G. J. *Eastern Wisedome and Learning: the Study of Arabic in Seventeenth-Century England* (Oxford, 1996).

——. 'A Survey of the Toledan Tables.' *Osiris* 15 (1968): 5–174.

Tournoy-Thoen, L. 'Nicolaus Clenardus.' In *Leuven, Stedelijk Museum, 550 Jaar Universiteit Leuven* (Lembeke, 1976), 204–205.

Toynbee, P. 'Dante's Obligations to Alfraganus in the *Vita nuova* and *Convivio.' Romania: revue trimestrielle* 24 (1895): 413–432.

Troupeau, G. 'Du syriaque au latin par l'intermédiaire de l'arabe: Le *Kunnāš* de Yūḥannā ibn Sarābiyūn.' *Arabic Sciences and Philosophy* 4 (1994): 267–278.

——. See Jacquart, D.

Trueblood, E. E. See Meyer, F. G.

Tyler, V. E. et al., eds. *Pharmacognosy* (Philadelphia, ⁹1988).

Ullmann, M. *Die Medizin im Islam.* Handbuch der Orientalistik, Abt. 1, Erg.-Bd. 6, Abschnitt 1 (Leiden, 1970).

———. *Die Natur- und Geheimwissenschaften im Islam.* Handbuch der Orientalistik, Abt. 1, Erg.-Bd. 6, Abschnitt 2 (Leiden, 1972).

———. *Islamic Medicine* (Edinburgh, 1978).

———. 'Yūḥannā ibn Sarābiyūn: Untersuchungen zur Überlieferungsgeschichte seiner Werke.' *Medizinhistorisches Journal* 5 (1970): 278–296.

Walsh, J. J. See Hyman, A.

Vadet, J.-C. 'Les aphorismes latins d'Almansor, essai d'interprétation.' *Annales Islamologiques* 5 (1963): 31–130.

Van Steenberghen, F. *Introduction à l'étude de la philosophie médiévale* (Louvain, 1974).

———. *La philosophie au XIIIe siècle* (Louvain, ²1991).

———. *Maître Siger de Brabant* (Louvain, 1978).

Vanden Broecke, S. *The Limits of Influence: Pico, Louvain, and the Crisis of Renaissance Astrology* (Leiden, 2003).

Vasoli, C. 'Le débat sur l'astrologie à Florence: Ficin, Pic de la Mirandole, Savonarole.' In *Divination et controverse religieuse en France au XVIe siècle.* Cahiers V. L. Saulnier 4 (Paris, 1987), 19–33.

———. *Studi sulla cultura del Rinascimento* (Manduria, 1968).

Veit, R. 'Andrea Alpago und Schah Ismāʿīl: Der Begründer der Safavidendynastie im Zeugnis eines Venezianer Gesandtschaftsarztes, Händlers und Informanten.' In *Das Charisma: Funktionen und symbolische Repräsentationen,* ed. P. Rychterová et al. (Berlin, 2008), 457–465.

———. *Das Buch der Fieber des Isaac Israeli und seine Bedeutung im lateinischen Westen: Ein Beitrag zur Rezeption arabischer Wissenschaft im Abendland* (Wiesbaden, 2003).

———. 'Der Arzt Andrea Alpago und sein medizinisches Umfeld im mamlukischen Syrien.' In *Wissen über Grenzen: Arabisches Wissen und lateinisches Mittelalter,* ed. A. Speer and L. Wegener. Miscellanea Mediaevalia 33 (Berlin, 2006), 305–316.

Vercellin, G. ed. *Il Canone di Avicenna: fra Europa e Oriente nel primo Cinquecento: l'Interpretatio Arabicorum nominum di Andrea Alpago* (Torino, 1991).

Vernet, J. *Die spanisch-arabische Kultur in Orient und Okzident* (Zurich, 1984).

Vernet, J., and J. Samsó. 'The Development of Arabic Science in Andalusia.' In *Encyclopedia of the History of Arabic Science,* ed. R. Rashed. 3 vols. (London, 1996), 1:243–275.

Villaverde Amieva, J. C. 'Review of L. F. Aguirre de Cárcer, *Ibn Wāfid: Kitāb al-adwiya al-mufrada,* 2 vols. (Madrid, 1995).' *Aljamía* 9 (1997): 111–118.

Vitkus, D. J. 'Early Modern Orientalism: Representations of Islam in Sixteenth- and Seventeenth-Century Europe.' In *Western Views of Islam in Medieval and Early Modern Europe,* ed. D. R. Blanks and M. Frassetto (New York, 1999), 207–230.

Volger, L. *Der Liber fiduciae de simplicibus medicinis des Ibn al-Jazzār* (Würzburg, 1941).

Wack, M. F. *Lovesickness in the Middle Ages: The* Viaticum *and Its Commentaries* (Philadelphia, 1990).

Warburg, A. M. 'Heidnisch-antike Weissagung in Wort und Bild zu Luthers Zeiten.' In *Gesammelte Schriften.* 2 vols. (Leipzig, 1932), 2:487–541.

Wear, A. 'Medicine in Early Modern Europe, 1500–1700.' In *The Western Medical Tradition 800 BC to AD 1800,* ed. L. I. Conrad et al. (Cambridge, 1995), 207–361.

Wegele, F. X. von. *Geschichte der Universität Würzburg.* 2 vols. (Würzburg, 1882).

Weisser, U. 'Paravicius Immortal?' *Sudhoffs Archiv* 72 (1988): 239–240.

Wels, H. *Aristotelisches Wissen und Glauben im 15. Jahrhundert: ein anonymer Kommentar zum Pariser Verurteilungsdekret von 1277 aus dem Umfeld des Johannes de Nova Domo: Studie und Text* (Amsterdam, 2004).

——. *Die Disputatio de anima rationali secundum substantiam des Nicolaus Baldelli S.J. nach dem Pariser Codex B.N. lat. 16627: eine Studie zur Ablehnung des Averroismus und Alexandrismus am Collegium Romanum zu Anfang des 17. Jahrhunderts* (Amsterdam, 2000).

Wichtl, M., ed. *Teedrogen und Phytopharmaka: Ein Handbuch für die Praxis auf wissenschaftlicher Grundlage* (Stuttgart, ³1997).

Wickersheimer, E. *Dictionnaire biographique des médecins en France au moyen âge*. 2 vols. (Paris, 1936, repr. Geneva, 1979), with *Supplément*, ed. D. Jacquart (Geneva, 1979).

———. 'Laurent Fries et la querelle de l'arabisme en médecine (1530).' *Les cahiers de Tunisie* (1955): 96–103.

———. 'Les tacuini sanitatis et leur traduction allemande par Michel Herr.' *Bibliothèque d'Humanisme et Renaissance* 12 (1950): 85–97.

Wiedemann, E. See Heiberg, J. L.

Wilcox, J., and J. M. Riddle. *Qustā ibn Lūqā's Physical Ligatures and the Recognition of the Placebo Effect: With an Edition and Translation* (Leiden, 1995).

Wilson, N. G. *From Byzantium to Italy: Greek Studies in the Italian Renaissance* (London, 1992).

Wirmer, D. *Averroes: Über den Intellekt: Auszüge aus seinen drei Kommentaren zu Aristoteles' De Anima. Arabisch—Lateinisch—Deutsch* (Freiburg im Breisgau, 2008).

Wisnovsky, R. *Avicenna's Metaphysics in Context* (Ithaca, 2003).

Wittern-Sterzel, R. 'Kontinuität und Wandel in der Medizin des 14. bis 16. Jahrhunderts am Beispiel der Anatomie.' In *Mittelalter und frühe Neuzeit: Übergänge, Umbrüche und Neuansätze,* ed. W. Haug (Tübingen, 1999), 550–571.

Woepcke, M. F. 'Notice sur quelques manuscrits arabes relatifs aux mathématiques, et récemment acquis par la bibliothèque impériale.' *Journal asiatique* 5 ser. 19 (1862): 101–127.

Wolfson, H. A. 'Revised Plan for the Publication of a *Corpus Commentariorum Averrois in Aristotelem.*' *Speculum* 38 (1963): 88–104.

———. 'The Twice-Revealed Averroes.' *Speculum* 36 (1961): 373–392.

Wright, W. *A Grammar of the Arabic Language*. 2 vols. (Cambridge, [3]1962).

Yamamoto, K. See Burnett, C.

Yano, M. See Burnett, C.

Youschkevitch, A. P. See Rosenfeld, B. A.

Zahlten, J. 'Disputation mit Averroes oder Unterwerfung des "Kommentators": Zu seinem Bild in der Malerei des Mittelalters und der

Renaissance.' In *Wissen über Grenzen. Arabisches Wissen und lateinisches Mittelalter,* ed. A. Speer and L. Wegener. Miscellanea Mediaevalia 33 (Berlin, 2006), 717–744.

Zambelli, P., ed. *'Astrologi hallucinati': Stars and the End of the World in Luther's Time* (Berlin, 1986).

——. 'Eine Gustav-Hellmann-Renaissance? Untersuchungen und Kompilationen zur Debatte über die Konjunktion von 1524 und das Ende der Welt auf deutschem Sprachgebiet.' *Jahrbuch des italienisch-deutschen historischen Instituts in Trient* 18 (1992): 413–455.

——. 'Fine del mondo o inizio della propaganda?' In *Scienze, credenze occulte, livelli di cultura,* ed. P. Zambelli (Florence, 1982), 291–368.

——. 'Giovanni Mainardi e la polemica sull'astrologia.' In *L'opera e il pensiero di Giovanni Pico Della Mirandola nella storia dell'umanesimo.* 2 vols. (Florence, 1965), 2:205–279 (repr. in Zambelli, *L'ambigua natura della magia,* ch. 4).

——. 'I problemi metodologici del necromante Agostino Nifo.' *Medioevo* 1 (1975): 129–171.

——. 'Introduction: Astrologers' Theory of History.' In *'Astrologi Hallucinati': Stars and the End of the World in Luther's Time,* ed. P. Zambelli (Berlin, 1986), 1–28.

——. *L'ambigua natura della magia: filosofi, streghe, riti nel Rinascimento* (Milan, 1991).

——. *L'apprendista stregone: astrologia, cabala e arte lulliana in Pico della Mirandola e seguaci* (Venice, 1995).

——. 'Many Ends for the World.' In *'Astrologi Hallucinati': Stars and the End of the World in Luther's Time,* ed. P. Zambelli (Berlin, 1986), 239–263.

——. *The* Speculum astronomiae *and Its Enigma: Astrology, Theology and Science in Albertus Magnus and His Contemporaries* (Dordrecht, 1992).

Zedler, B. H., ed. *Averroes'* Destructio destructionum philosophiae Algazelis *in the Latin Version of Calo Calonymos* (Milwaukee, 1961).

Zhiri, O. *L'Afrique au miroir de l'Europe: Fortunes de Jean Léon l'Africain à la Renaissance* (Geneva, 1991).

Zimmermann, A. 'Albertus Magnus und der lateinische Averroismus.' In *Albertus Magnus, Doctor universalis: 1280/1980,* ed. G. Meyer (Mainz, 1980), 465–493.

Zonta, M. 'Aristotle's *Physics* in Late Medieval Jewish Philosophy (14th–15th Century) and a Newly Identified Commentary by Yehudah Messer Leon.' *Micrologus* 9 (2001): 203–217.

——. 'The Autumn of Medieval Jewish Philosophy: Latin Scholasticism in Late 15th-Century Hebrew Philosophical Literature.' In *"Herbst des Mittelalters"? Fragen zur Bewertung des 14. und 15. Jahrhunderts,* ed. J. A. Aertsen and M. Pickavé. Miscellanea Mediaevalia 31 (Berlin, 2004), 474–492.

——. *Il* Commento medio *di Averroè alla* Metafisica *di Aristotele nella tradizione ebraica. Edizione delle versioni ebraiche medievali di Zeraḥyah Ḥen e di Qalonymos ben Qalonymos con introduzione storica e filologica.* 3 vols. (Pavia, 2011).

——. *La filosofia antica nel Medioevo ebraico: Le traduzioni ebraiche medievali dei testi filosofici antichi* (Brescia, 1996).

——. 'Osservazioni sulla traduzione ebraica del Commento grande di Averroè al De anima di Aristotele.' *Annali di Ca' Foscari* 33, no. 3 (1994): 15–28.

## MANUSCRIPTS

Cesena, Biblioteca Malatestiana, MS D.XXV.4: Averroes, *Colliget,* trans. Bonacosa.

Dijon, Bibliothèque municipale, 449: Omar Tiberiadis, *Iudicia,* trans. Salio of Padua.

Florence, Bibliotheca Medicea Laurentiana, MS Plut. 36, cod. 35 (Ms 184), ff. 31r–69v: Leo Africanus, *De viris quibusdam illustribus apud arabes.*

Florence, Biblioteca nazionale centrale, MS con. soppr. J.II.10: Omar Tiberiadis, *De nativitatibus.*

Genua, Biblioteca Universitaria, MS A.IX.29, ff. 79v–116r: Ibn Ṭufayl, *Ḥayy ibn Yaqẓān,* Latin, trans. anonymous.

Granada, Colección del Sacro Monte, MS nr. I: Averroes, *Kitāb al-Kullīyāt.*

Leiden, Universiteitsbibliotheek, MS Or. 2073: Averroes, middle commentaries on the Organon.

Milan, Biblioteca Ambrosiana, MS G. 290: Averroes, *Liber modorum rationis,* trans. Abraham de Balmes.

Nuremberg, Stadtbibliothek, MS Cent. VI 2: Avicenna, *Canon*.

Paris, Bibliothèque nationale, MS arabe 2897: Avicenna, *Qānūn,* with interlinear Latin trans. by Girolamo Ramusio.

Paris, Bibliothèque nationale, MS hébr. 886 (Or. 112): Averroes, Comm. mag. Metaph. Hebrew.

Paris, Bibliothèque nationale, MS lat. 6504: Averroes, Comm. mag. Metaph. Latin.

Paris, Bibliothèque nationale, MS lat. 15453: Averroes, Comm. mag. Metaph. Latin.

Passau, Staatliche Bibliothek, MS addition in Inc. 65 (Trithemius), ff. 142–179r: Johann Staindel, *Suppletio virorum illustrium*.

Vatican City, Biblioteca Apostolica Vaticana, MS Vat. lat. 3897, ff. 32–65: Avempace, *Epistola expeditionis,* trans. Abraham de Balmes.

Vatican City, Biblioteca Apostolica Vaticana, MS Vat. lat. 4548: Averroes, Comm. med. Phys. trans. Abraham de Balmes.

Vatican City, Biblioteca Apostolica Vaticana, MS Vat. lat. 4549: Averroes, Comm. med. Animal. trans. Elia del Medigo; Averroes, *Tractatus de intellectu speculativo,* trans. Elia del Medigo.

Vatican City, Biblioteca Apostolica Vaticana, MS Vat. lat. 4551: Averroes, Comm. med. An. trans. anonymous.

Vatican City, Biblioteca Apostolica Vaticana, MS Vat. lat. 4554: Algazel, *Liber intentionum philosophorum* (Maqāṣid al-falāsifa) (with commentary by Moses Narboni), trans. anonymous.

Vatican City, Biblioteca Apostolica Vaticana, MS Vat. lat. 4566: Alhazen, *Liber de mundo et celo,* trans. Abraham de Balmes.

Vatican City, Biblioteca Apostolica Vaticana, MS Vat. lat. 12055: Alfarabi, *De intellectu,* trans. Abraham de Balmes.

Vatican City, Biblioteca Apostolica Vaticana, MS Vat. ottob. lat. 2016: Averroes, *De qualitate esse mundi,* trans. Calo Calonymos ben David.

Vatican City, Biblioteca Apostolica Vaticana, MS Vat. ottob. lat. 2060: Averroes, *Quaesita naturalia,* trans. Abraham de Balmes; Averroes, *Liber modorum rationis,* trans. Abraham de Balmes.

Venice, Biblioteca Marciana, MS lat. 2656 (VI, 105), ff. 156r–160v: Vernia, *Utrum anima intellectiva . . . eterna atque unica sit in omnibus hominibus*.

Venice, Biblioteca Marciana, MS lat. 2666 (L. VI, lvi): Avicenna, *De animalibus*.

Vienna, Österreichische Nationalbibliothek, MS hebr. 114, ff. 1–89v: Averroes, Comm. mag. An. post. Hebrew.

# ACKNOWLEDGMENTS

The idea for this book emerged during my years at the Warburg Institute in the mid-1990s. Since then, Charles Burnett has been a constant source of advice, for which I am deeply grateful. I am also grateful to Andreas Speer, who for several years was a colleague full of encouragement and counsel on the same Würzburg corridor. Special thanks go to Dimitri Gutas, who has stimulated this study in many ways ever since its beginnings.

A less comprehensive version of this book was submitted as a Habilitationsschrift to the Philologische Fakultät of the Albert-Ludwigs-Universität Freiburg im Breisgau in 2004. I am very grateful to the three reviewers, Paul Gerhard Schmidt, Maarten Hoenen, and Charles Burnett. I recall with gratitude the encouragement by Schmidt, who died too early to see the book in print. It is impossible to thank all those from whose advice I have benefited over the years: the many friends and colleagues in my academic stations in Göttingen, New Haven, London, Tübingen, Freiburg im Breisgau, and Würzburg, and the many friends and colleagues in Greek, Arabic, Hebrew, Latin, Medieval, and Renaissance studies, philosophers, historians, and philologists, who have responded to previous presentations of my research. I owe special thanks to the members of Glenn Most's Leibniz Kreis, who have discussed several chapters of the text with me, and to my collaborators on two research projects: the Arabic and Latin Glossary in Würzburg, especially to Katrin Fischer, and the Ptolemaeus Arabus et Latinus in Munich and Würzburg. In preparing the index, I was greatly helped by the student assistants of the Institute of Philosophy in Würzburg. I am grateful to Markus Asper, Jon Bornholdt, David Cory, Therese Cory, Silvia Di Donato, Peter Godman, David Juste, Andreas Lerch, Jörn Müller, Darrel Rutkin, and Anna-Katharina Strohschneider, who have read chapters of the book. All mistakes, of course, remain my own.

My research was supported by the Deutsche Forschungsgemeinschaft, which financed the postdoctoral fellowship at the Graduiertenkolleg Ars und Scientia im Mittelalter und der frühen Neuzeit at Tübingen University (1998–1999), by the Dr. Meyer-Struckmann-Stiftung (1999–2000), and by the VolkswagenStiftung as part of a Lichtenberg professorship grant (2005–2010). I gratefully acknowledge this support. I am particularly grateful for the support of the VolkswagenStiftung, which has given me the freedom to design my research agenda beyond disciplinary constraints.

The greatest gratitude I feel is expressed in the dedication.

# INDEX OF NAMES

# GENERAL INDEX